John T. Tanacredi • Mark L. Botton
Paul K. S. Shin • Yumiko Iwasaki
Siu Gin Cheung • Kit Yue Kwan
Jennifer H. Mattei
Editors

# International Horseshoe Crab Conservation and Research Efforts: 2007-2020

## Conservation of Horseshoe Crabs Species Globally

Springer

# Remembering Two Friends

Friendships … always impossible to identify all the required (if any) attributes that will meld into a personal relationship with anyone. Dr. Jayant Mishra, from Pondicherry University of India, was a wonderful human being. I first met him at the Dowling Conference in 2007 and then several other times (too few to be sure) in Hong Kong and Sasebo Japan. A "competition" for the best youth drawing of horseshoe crab resulted in a win for his son and the emblem on T-shirts distributed to all horseshoe crab aficionados. He had a sense of humor, and when we gathered in Hong Kong for dinner and reviewed the listings of "exotic species" to be served, we debated whether or not our children, and for me, my grandchildren, would eat what was on the menu. I did not have the opportunity to attend the China conference in 2019, so I missed the rare opportunity to meet with Jayant again, as we had connected about "the next international horseshoe crab conference" hopefully to be in India and we should talk about this over our IUCN-SSG telephone calls! He will be missed.

On May 28, 2020, Carl N. Shuster Jr., at the age of 100, passed away. His friendship goes back to my days in the National Park Service and taking courses at Rutgers under Hal Haskin in preparation of my Ph.D. I loved that Carl's signature coupled with his own developed horseshoe crab logo (Figs. 1 and 2) were characteristic of this truly gentile man, scientist, and a true field scientist, who always used the word "cordially" with his salutation on any correspondence (Fig. 3). He would always make me feel as if I made a contribution to the science and to what scientists do, contributing to the never-ending need for more scientific knowledge. He was truly insatiable in his search for "the details" regarding horseshoe crabs. His enthusiasm was infectious, and the only panacea he self-prescribed was being in a "discussion" with you! Any question, any observation, any opportunity to share his thoughts and insights about horseshoe crabs were cherished interactions. There were so many topics to discuss with Carl. I was a flight meteorologist with the US Navy "Hurricane Hunters" out of Jacksonville Naval Air Base from 1968 through 1970, and for as long as I knew him, our military experiences would always be a point for a fleeting few minutes of conversation.

Rachel Carson, in *The Edge of the Sea*, noted, "The horseshoe crab is an example of an animal that is very tolerant to temperature change. It has a wide range as a species, and its northern forms can survive being frozen into ice in New England, while its southern representatives thrive in tropical waters of Florida and southward to Yucatán. I remember wading on a June day over Bird Shoal, which lies off the town of Beaufort in North Carolina, where at low tide, acres of sand bottom are covered only by a few inches of water. Near the shore I found two sharply defined grooves in the sand; my index finger could have measured their span. Between the grooves was a faint, irregular line. Step by step, I was led out across the flat by the tracks; finally, at the temporary end of the trail, I came upon a young horseshoe crab, heading seaward." This acknowledged trail of discovery of horseshoe crabs I believe best describes Carl Schuster's legacy, inquisitive, experiential, and always some discovery, all the qualities of an excellent scientist. I am honored to have known him, and his legacy through his research and the establishment of the Carl N. Shuster Jr., Horseshoe Crab Reserve (see Figs. 4 and 5) will live with me the rest of my life.

John T. Tanacredi, Ph.D. (2021)

6/07

We go back to some of the beginnings of horsecrab conservation - you have been an effective advocate - which I greatly respect -

Carl N. Shuster

# THE AMERICAN HORSESHOE CRAB

Edited by

Carl N. Shuster, Jr.

Robert B. Barlow

H. Jane Brockmann

HARVARD UNIVERSITY PRESS

Cambridge, Massachusetts, and London, England

2003

Monday 21 Dec 2009

Dear John—

The autographed Symposium treatise arrived Friday & we got 20" of snow in our driveway over the weekend.

I should have thanked the editors for dedicating the book to Sekiguchi-san and me long before this—but the autograph and dedication are much appreciated.

My big chore right now is contributing to the Xiphosura section for the revised Handbuch der Zoologie. Cordially, Carl

10 June 2016

Dear John,

Thank you for participating in the June 4th celebration of *Limulus*. We were interested in learning about your CERCOM program.

Your gift is much appreciated; very colorful and beautifully done.

Cordially

Carl

# Carl N. Shuster, Jr.
# Horseshoe Crab Reserve

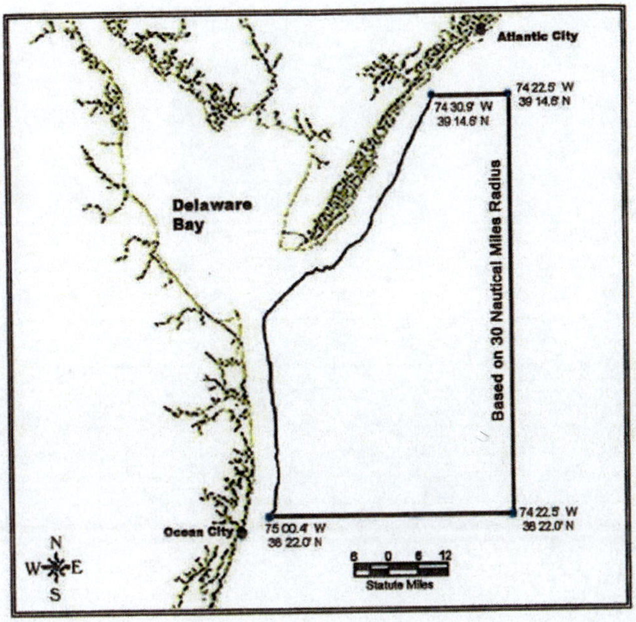


Presented to Dr. Carl N. Shuster Jr. for his contributions to horseshoe crab biology and conservation on this 28th day of March, 2001

William T. Hogarth, Ph.D.

William T. Hogarth
Acting Assistant Administrator for Fisheries

# Preface

Noble Laureate and theoretical physicist Murray Gell-Mann in his 1994 work, *The Quark and the Jaguar: Adventures in the Simple and the Complex*, identifies, as it has been emphasized by historians of science, that the essential feature of science is that its theories are falsifiable. Horseshoe crab survivability of 445 million years, resulting in their biological endurance through five mass-extinction events, has set in motion theory on how they have survived for so long. Thus unwrapping their paleo-history of resilience to share a varied list of possible causative factors contributing to mass extinctions and their survival of meteorite impacts, climate catastrophes, and volcanism challenges their being continuously classified as "threatened," of "critical concern," or "in poor condition." These anthropogenic factors, such as habitat loss or the lack of protecting their breeding and nursery habitats, or the collection for fish bait or "exotic food" consumption, all result in animals being permanently removed from the breeding population! The extinction of these important species will have negative consequences: cessation of the production of *Limulus amebocyte lysate* (LAL), which is essential in the fight to stem the surging COVID-19 pandemic, and the loss of their ecological role throughout the majority of their range.

With 2019–2020 immortalized in human history as the beginning of a global pandemic unprecedented in modern history, protecting horseshoe crabs will eclipse all ecological events affecting society for several generations. Loss of life measured in millions. Natural and cultural resources pushed to limits of their resiliency. A silent siren, horseshoe crabs' history and contribution to human health and life reveal to the world LAL's practical importance in the development of immunizations for disease treatment and prevention. Horseshoe crab blue blood has at last earned its moniker of true "bluebloods." This richness continues to affect global societal resiliency.

To take a quote from essayist Adam Bly's *Science Is Culture*, this tagline seems self-evident; science is culture. In the last four decades, science has transformed the social, political, economic, and aesthetic nature of our intellectual landscape. It has reshaped our understanding of who we are and where we come from. It has modernized our system of values and helped us determine how we regard our planet and one another. Findings in science have undeniably affected the state of the world. Knowledge of science is an overwhelming and universal agent of change. Today, science affects and influences every single person on the planet.

Today, we think scientifically about issues like health and the environment. We also think scientifically about the size of cities, the reason for poverty, the basis of mortality, the resilience of markets, and much more. Science is a methodology and a philosophy rooted in evidence, kept in check by persistent inquiry, and bounded by the constraints of a self-critical and rigorous practice. Science is a lens through which we can visualize and solve complex problems, establish international relationships, and embolden (even reignite) human endeavors. More than anything, what this lens offers us is a limitless capacity to handle all that comes our way no matter how complex or unanticipated.

Yet even this optimistic humanism may very well be jeopardized by the alarming decreases in global biodiversity. If today we are perpetuating a colossal diversity loss on the ecological landscape, we must have even greater protections for all species. This compendium is an attempt to help chronicle preservation efforts for just four species of horseshoe crabs that exist on Earth. Though this goal may appear quixotic, these sentinel organisms may very well be the silent siren even after surviving five mass-extinction events, to be lost not in past geological time, but lost in our time resultant of an unparalleled Anthropocene extinction.

This volume is unique in how proceedings are generally presented. It includes materials from an array of conference presentations, workshops, position papers, exploratory colloquia, symposia, videography documentation, photos, testimonials, translations, and media items. Most revealing, is the documentation of the comradery, collegiality, and collaborative efforts experienced by those who study, research, educate, observe, and commit themselves to further the understanding of these organisms. They have saved millions of lives due to their heteropathic role, their seminal existence in the framework of estuarine and coastal ecology, and, lastly, their contributions to conservation biology, natural resources protection, biological diversity, and ecological health, which collectively provide an unrivaled role over any single species.

One of the fascinating outcomes of the work with horseshoe crabs has been the international inclusion through the establishment of the IUCN Species Survival Commission (SSC) Horseshoe Crab Scientific Specialist Group (SSG). The group aims to protect horseshoe crabs globally through collaborative effort in conservation of their populations and habitats, and in raising public awareness of their importance in evolutionary history, marine coastal ecology, and biomedical uses. Through the effort of the group first established in 2011, the status of IUCN Red List of Threatened Species for two species of horseshoe crabs has recently been updated as "Vulnerable to extinction" (*Limulus polyphemus*) and "Endangered" (*Tachypleus tridentatus*). Various international meetings and symposia have also been organized to further promote exchanges and collaborations among scientists, conservationists, educators, and government officials in the work for the protection of the existing four horseshoe crab species. Part of this continuing effort is the result of completing most recently two international workshops (Sasebo, Japan, 2015, and

Qinzhou, China, 2019), which preceded the Covid-19 pandemic and the first International Horseshoe Crab Day (June 20, 2020), with a variety of webinars (hosted in China, Malaysia, India, and Indonesia) and a full day of local horseshoe crab awareness activities and programs in many other countries. Each of these two workshops attracted some 150 participants from various countries to present their current research findings and conservation activities, and discuss issues and future efforts in the protection of horseshoe crabs globally. The latest meeting in China also resulted in the designation of June 20 every year as International Horseshoe Crab Day, so as to promote and celebrate the conservation efforts for these species with the public in different countries. These workshops also further initiated an expansion in interest from the overall scientific community while being the primary driving force embraced by Springer Nature to publish this work. Their continued foresight and support have been monumental.

There have been disagreements filled with passion on both sides of the issue regarding conservation of horseshoe crabs, from the price of the collection of animals from their breeding sites to restaurants and YouTube food posts on the best way to consume these animals, especially the gravid females, to the harvest of thousands of animals a month from Southeast Asia, to the extraction of blood and then consumption as exotic menus, to the construction of thousands of new housing units adjacent to the coastline resulting in loss breeding ground. Not even a meteorite creating the massive Chicxulub crater in the Gulf of Mexico 65 million years ago, which ultimately resulted in the demise of the dinosaurs, could eliminate the four horseshoe crab species from the face of the Earth. In spite of a lack of a strong conservation strategy which charaterizes the risk of extinction, we need to prescribe ways to reduce this risk.

In this unprecedented time of the COVID-19 pandemic, the significance of horseshoe crabs' aid in detection of critical health impacts from other disease concerns amplifies the importance of horseshoe crabs to human health. Each year in the USA, 275,000 deaths occur due to sepsis, of which one out of three people going into hospitals die of this infection. The blue blood pharmaceutical endotoxin detecting LAL is critically important in hospitals to detect this major cause of human mortality and will be indispensable in the development of the billions of immunization products required to quell the COVID-19 pandemic. Molloy College graduates hundreds of nurses each year and is a major academic institution providing healthcare professionals to the world. All our nursing students know about LAL and understand how important a tool it is in providing the safest possible environment for those in their care.

Sir Isaac Newton coined a phrase of praise noted in 1675 that I believe is appropriate about the scientists and educators contributing to this work as it allows us to see further because we all "stand on the shoulders of giants." Carl N. Shuster, Robert Barlow, Anne Rudloe, Koichi Sekiguchi, Toshifumi Wada, Tomio Itow, and Jayant Mishra, with their research on horseshoe crabs, were inspirational to all of us who have followed.

This work is still only the beginning of the exploration and narration of the story of horseshoe crabs, and hopefully will be inspire more scientists to explore these amazing animals.

West Sayville, NY, USA — John T. Tanacredi
New York, NY, USA — Mark L. Botton
Hong Kong, China — Paul K. S. Shin
West Sayville, NY, USA — Yumiko Iwasaki
Hong Kong, China — Siu Gin Cheung
Qinzhou, People's Republic of China — Kit Yue Kwan
Fairfield, CT, USA — Jennifer H. Mattei

# Contents

**Part I Evolution Genetics & Adaptation**

**Variation in the American Horseshoe Crab: A Spiny Variant from the Southeast** ........ 3
H. Jane Brockmann, Carl N. Shuster Jr, Patrick Norby, and Dorothea Saunders

**Genetic Structure of the Tri-Spine Horseshoe Crab in the Seto Inland Sea, Japan: Is the Current Population at Kasaoka in the Eastern Area Native or Re-established?** ........ 19
Shinichi Watanabe, Jun J. Sato, Keiki Kawazu, Eri Horinouchi, Shin Nishida, Koujirou Azumakawa, Satoshi Morinobu, and Norimichi Souji

**Molecular Identification and Phylogenetic Analysis of Indian Mangrove Horseshoe Crab** ........ 29
B. Bal, B. P. Das, Alok Prasad Das, and B. Akbar John

**Population Genetics and Movement Show Metapopulation Dynamics of Mid-Atlantic Region Horseshoe Crabs** ........ 41
Eric M. Hallerman, David N. Hata, Michael S. Eackles, and Timothy L. King

**Microbiome Associated with Laboratory Bred Horseshoe Crabs: A Case Study** ........ 59
Vishal Shah, John T. Tanacredi, and Viraj Joshi

**Population Genetic Structure of Juvenile Tri-Spine Horseshoe Crabs in Hong Kong** ........ 69
Alice K. Y. Chan, Clement K. M. Tsui, K. L. Pang, Kit Yue Kwan, Richard Y. C. Kong, S. G. Cheung, and Paul K. S. Shin

**Applying Records of Extant and Extinct Horseshoe Crab Abnormalities to Xiphosurid Conservation** . . . 85
Russell D. C. Bicknell, Stephen Pates, Dorkas Kaiser, Sara Zakrzewski, and Mark L. Botton

**Part II Population Dynamics & Ecology**

**Activity Rhythm of the Tri-Spine Horseshoe Crab *Tachypleus tridentatus* in the Seto Inland Sea, Western Japan, Monitored with Acceleration Data-Loggers** . . . 107
Shinichi Watanabe, Saori Oyamada, Keisuke Mizuta, Koujirou Azumakawa, Satoshi Morinobu, and Norimichi Souji

**Sediment-Molting Relationship of Juvenile *Limulus polyphemus* (Arthropoda, Merostomata) and Predation Against them** . . . 123
Yumiko Iwasaki

**Conservation and Restoration of Estuaries and Coasts: Horseshoe Crabs as Flagship Species** . . . 135
Jennifer H. Mattei, Mark L. Botton, Satoquo Seino, Alfredo Quarto, Jaruwan (Ning) Enright, J. Vanitha, Hsu Chia Chi, Lester Tan, Joe Cheung, and John T. Tanacredi

**Preliminary Characterization of Hemocyte and Immunity of Asian Horseshoe Crabs, *Tachypleus tridentatus*, and *Carcinoscorpius rotundicauda* in Captivity** . . . 161
Xiaoyong Xie, Kit Yue Kwan, Jinxiang Zhong, Mujiao Xie, Guoling Ye, and Yuyuan Bao

**Visually Guided Behaviour of Juvenile Tri-spine Horseshoe Crab *Tachypleus tridentatus*: Implications for Designing Racks for Oyster Cultivation Compatible with Horseshoe Crab Conservation** . . . 177
H. K. Chan, H. S. Lo, S. Y. Ng, R. -F. Chen, and S. G. Cheung

**Condition of Horseshoe Crabs as Bycatch from Gillnet Fisheries in Mayangan Waters of Subang Regency, West Java, Indonesia** . . . 189
Fahresa N. Supadminingsih, Mochammad Riyanto, and Ronny Irawan Wahju

**Occurrence and Morphometric Analysis of Horseshoe Crabs Along Bhadrak Coast of Odisha, India** . . . 203
Shaik Rahimuddin, Siddhartha Pati, Sanatan Tudu, Ayaskanta Pramanik, Apratim Sai Rajesh, Nirlipta Mishra, Gobinda Chandra Biswal, Anil Chatterji, and Bisnu Prasad Dash

**Preliminiary Investigation of *Tachypleus tridentatus* (Tri-spine HSC at Sone Tidal Flat, Fukuoka Japan)** . . . 215
Osamu Hayashi

**Distribution of Horseshoe Crabs, *Carcinoscorpius rotundicauda*, and *Tachypleus gigas* in Singapore, Including Coney Island** . . . . . . . . . . . 247
Jasmin Lim, Yi Long Ng, Oliver Chang, Prashant Shingate, Boon Hui Tay, Qian Tang, Frank Rheindt, Hor Yee Tong, Karenne Tun, Byrappa Venkatesh, and Laura Yap

**An Assessment of Horseshoe Crab Breeding Habitat and Trends on Long Island, New York, 2003–2017** . . . . . . . . . . . . . . . . . . . . . . . . . . . . 261
J. Wisnewski and J. T. Tanacredi

**Part III Conservation Efforts & Managment Education**

**Identifying the Major Threats to American Horseshoe Crab Populations, with Emphasis on Delaware Bay** . . . . . . . . . . . . . . . . . . . . . . 315
Mark L. Botton, Robert E. Loveland, Daphne Munroe, David Bushek, and James F. Cooper

**The Need to Establish Horseshoe Crab Global Biorepository** . . . . . . . . . . 345
B. Akbar John, Qian Tang, and Mike Eackles

**Economics of the Limulus/Tachypleus Amebocyte Lysate (LAL/TAL) Industry Relative to the Sustainability of Horseshoe Crabs Worldwide** . . . . . . . . . . . . . . . . . . . . . . . . . . . . . . . . . . . . . . . . . . 351
Thomas J. Novitsky

**LAL/TAL and Animal-Free rFC-Based Endotoxin Tests: Their Characteristics and Impact on the Horseshoe Crab Populations in the United States and Asia** . . . . . . . . . . . . . . . . . . . . . . . . . . 369
Glenn A. Gauvry, Thomas Uhlig, and Karolina Heed

**The Power of Citizen Science: 20 Years of Horseshoe Crab Community Research Merging Conservation, Education, and Management** . . . . . . . . 391
Jennifer H. Mattei, Jo-Marie E. Kasinak, Samah Senbel, and Kirk Bartholomew

**Relative Abundance of Horseshoe Crabs in the Delaware Bay Region: A Critical Factor for Adaptive Resource Management** . . . . . . . . . . . . . . . 415
David N. Hata and Eric M. Hallerman

**Detrimental Coastline Alteration at Horseshoe Crab Spawning Grounds in Pahang, South China Sea** . . . . . . . . . . . . . . . . . . . . . . . . . . . . 435
Nurul Ashikin Mat Zauki, Shantika Maylana Sastraprawira, Mohammad Yahya Ibrahim Alkhalili, Nur Fairuz-Fozi, Siddhartha Pati, Salwa Shahimi, Chong Ju Lian, Hisham Atan Edinur, Akbar John, Ahmed Jalal Khan Chowdhury, Behara Satyanarayana, and Bryan Raveen Nelson

**Assessing the Use of Acoustic Sampling for Locating Horseshoe Crabs During Migrations Using Side Scan Sonar** ..... 453
Stephen D. Nagiewicz, Vince Capone, Peter F. Straub, Steven P. Evert, and Nate Robinson

**Asian Horseshoe Crab Conservation: Knowledge, Attitudes, and Intentions of Local College Students in Northern Beibu Gulf, China** ..... 469
Zhou Wu, Shuyan Huang, Yijian Fu, Junfu He, Wenquan Zhen, Xiaoyong Xie, Chun-Chieh Wang, Junhua Zhu, and Kit Yue Kwan

**Fishery Bycatch Jeopardizes Indian Horseshoe Crab *Tachypleus gigas* Along the Northeast Coast of Odisha, India** ..... 485
Siddhartha Pati, Sanatan Tudu, Apratim Sai Rajesh, Satyabrata Meher, Bryan Raveen Nelson, and Bisnu Prasad Dash

**Citizen Science Approach for Indonesian Horseshoe Crab Conservation** ..... 497
Yusli Wardiatno, Naila Khuril Aini, Ali Mashar, Zairion Zairion, Peter Funch, Yuni Puji Hastuti, and Untung Sangaji

**Cumulative Modifications of the Endangered Horseshoe Crab *Tachypleus tridentatus* Habitat by Small-Scale Developments at Tsuyazaki Cove in Fukuoka, Japan** ..... 511
Shinji Itaya, Toshifumi Wada, and Mari Shuuno

**Exploitation Status of Two Species of Horseshoe Crabs from Subang, West Java, Indonesia** ..... 531
Lusita Meilana, Qinhua Fang, Lukman Guam Hakim, Okoli Moses Ugochukwu, Ali Mashar, and Yusli Wardiatno

**Captive Breeding Success and Pitfalls for Horseshoe Crab, *Limulus polyphemus*, at CERCOM, Molloy College** ..... 551
K. Maurelli

**City-Wide Efforts in the Conservation of Horseshoe Crab (*Tachypleus tridentatus*) in Imari City, Saga, Japan** ..... 561
Koyo Funai

**Part IV Biology & Physiology**

**Temperature and Salinity Preferences of Adult American Horseshoe Crabs (*Limulus polyphemus*) in the Great Bay Estuary, New Hampshire U.S.A** ..... 581
Helen Cheng, Vilma Vaattovaara, Meghan Connelly, Brianna Looney, Christopher C. Chabot, and Winsor H. Watson III

**The Effects of Roundup® (a Glyphosate-Based Herbicide) on the Survival and Development of American Horseshoe Crab, *Limulus polyphemus*** . . . . . . . . . . 599
Hope VanDerwater, Karine Khoder, and Mark L. Botton

**Feeding Ecology and Dietary Preferences of *Tachypleus gigas* from East Malay Peninsula** . . . . . . . . . . 609
Akbar John, Bryan Raveen Nelson, Hassan I. Sheikh, S. Hajisamae, and Jalal Khan

**Patterns of Flatworm (*Bdelloura candida*) Infestation in a Population of Horseshoe Crabs (*Limulus polyphemus*) from Jamaica Bay, New York** . . . . . . . . . . 621
Royall McMahon Ward and Mark L. Botton

**Trophic Relationship of Sympatric Juvenile Asian Horseshoe Crabs in Beibu Gulf, Southwestern China** . . . . . . . . . . 633
Menghong Hu, Kit Yue Kwan, Zhe Xie, Siu Gin Cheung, Youji Wang, and Paul K. S. Shin

***Tachypleus gigas* and *Carcinoscorpius rotundicauda* Selection and Consumption of Tissue from Four Different Bivalve Species in Captivity** . . . . . . . . . . 651
Anis Syahira Abdul Halim, Farah Najihah Razali, Fisal Ahmad, Noraznawati Ismail, Tuan Chilek Tuan Zainazor, Ahmad Shamsuddin Ahmad, and Faridah Mohamad

**Research Progress and Prospect of Tachypleus Amebocyte Lysate in China** . . . . . . . . . . 661
Chang Liu, Ximei Liu, Youji Wang, Jie Song, Jinfeng He, Zongguang Tai, Quangang Zhu, and Menghong Hu

**Ecology of Juvenile American Horseshoe Crabs (*Limulus polyphemus*) at Plumb Beach, Jamaica Bay, New York** . . . . . . . . . . 669
Christina P. Colon, Mark L. Botton, Peter Funch, Emil Hoffgaard, Kaur Mandeep, and Kera Mansfield

**Correction to: Feeding Ecology and Dietary Preferences of *Tachypleus gigas* from East Malay Peninsula** . . . . . . . . . . C1

# About the Editors

**John T. Tanacredi,** Ph.D., is Professor of Earth and Environmental Sciences in the Biology, Chemistry, and Environmental Studies Department and Director of the Center for Environmental Research and Coastal Oceans Monitoring (CERCOM) Field Station at Molloy College, Rockville Centre, New York. He has held research associate positions at the Wildlife Conservation Society, New York Aquarium – Osborn Laboratories of Marine Science; at The American Museum of Natural History Invertebrate Zoology Department; at the Lamont Doherty Earth Observatory, Columbia University; and, has co-chaired the Conservation Committee of the Explorers Club. He has been a research ecologist for 26 years in the US National Park Service, serving as chief of the Division of Natural Resources and as one of the NPS's Coastal Natural Resource Specialist duty stationed for 9 years at the Graduate School of Oceanography, University of Rhode Island, Narragansett, RI. He was an environmental analyst of the U.S. Coast Guard, Bridge Administration, preparing NEPA environmental impact statements for bridge and highway construction in six eastern US states, and was a U.S. Navy Flight Meteorologist "Hurricane Hunter."

Dr. Tanacredi received his doctorate in environmental health engineering from Polytechnic University; M.S. degree in environmental health sciences from Hunter College, Institute of Health Sciences, CUNY; and B.S. in biological sciences from Richmond College, CUNY.

He has continued for the last 20 years at CERCOM, his Long Island horseshoe crab habitat inventory of some 115 locations, tracking horseshoe crab breeding conditions and habitat. He is one of the founding members of the IUCN-SSG (International Union for the Conservation of Nature – Scientific Specialists Group for Horseshoe Crabs) and principal coordinator of several international conferences on horseshoe crab conservation and biology.

He has published over 70 peer-reviewed scientific research publications, written and edited 7 books, and has been included and interviewed in a host of TV, radio, and social media outlets.

His co-edited the book *Conservation and Biology of Horseshoe Crabs*, Springer, 2009, which was one of the initiating factors in conducting the subsequent Asian Horseshoe Crab Conferences providing considerable support for including all four horseshoe crab species on the IUCN's "Red List." His latest book, *The Redesigned Earth: An Introduction to Ecology for Engineers as if the Earth Really Mattered* (2019), by Springer Nature, has received rave reviews. https://link.springer.com/book/10.1007/978-3-030-31237-4. Facebook https://www.facebook.com/CERCOMMCLI Twitter https://twitter.com/CERCOMMCLI, Instagram http://www.instagram.com/molloycollegecercom/, www.molloy.edu/cercom

**Mark L. Botton** is Professor of Biology in the Department of Natural Sciences at Fordham University – Lincoln Center, New York City, and co-director of the environmental science program. He received his bachelor's degree in biology from Stony Brook University, his master's degree in biology from Brooklyn College, and his Ph.D. in zoology from Rutgers University. Dr. Botton has published over 70 articles and book chapters on various aspects of horseshoe crab biology, including feeding ecology, mating behavior, the effects of pollution on developmental success, and population and conservation biology. He is the co-chairman of the Horseshoe Crab Specialist Group, IUCN Species Survival Commission.

**Paul K. S. Shin** is a retired associate professor at the City University of Hong Kong, Hong Kong SAR, China, and a fellow of the Chartered Institution of Water and Environmental Management in the UK. He is a benthic ecologist with research interest in community structure analysis, marine pollution, and coastal conservation. He has been involved in the study and conservation of horseshoe crabs for the past 15 years. In collaboration with the Ocean Park Conservation Foundation, Hong Kong, he and Dr. S.G. Cheung initiated the Juvenile Horseshoe Crab Rearing Programme to promote conservation education to young generations through hands-on experience in taking care of juvenile horseshoe crabs at schools with subsequent release of these juveniles back to the wild. Dr. Shin received the Medal of Honour from the Hong Kong SAR Government in 2015 for his contribution to conservation and public services. Currently, he serves as the co-chair (South East Asia) of the IUCN SSC Horseshoe Crab Specialist Group.

**Yumiko Iwasaki,** Ph.D., Based on the fossil specimens from the Altiplano, Bolivia, she worked on the phylogenetic studies of the Devonian trilobite group, Phacopidae, and a reconstruction of the Devonian paleobiogeography at the American Museum of Natural History, NY, with Dr. Niles Eldredge who taught her the value of trilobites as (once) living beings. After receiving a Ph.D. in geology and paleontology (Invertebrates) from the Earth and Environmental Sciences, Graduate Center, City University of New York, she was hired as a full-time faculty at Dowling College, located along the shore of Great South Bay, Long Island, NY, where her research interest extended to the behavioral studies of "living fossil," horseshoe crabs. Her research on their juveniles continued at Fordham College at Lincoln Center, NY, where she worked as a visiting scholar. A successful coordination between the Japanese host team and IUCN Horseshoe Crab SSG Steering Committee for the third International Workshop on the Science and Conservation of Horseshoe Crabs in Sasebo, Japan, in 2015 granted her to serve as a co-chair. Since then, she has been active in connecting the horseshoe crab communities around the globe. Currently, she is a research affiliate at Molloy College & CERCOM (Center for Environmental Research and Coastal Oceans Monitoring), IUCN SSC Horseshoe Crab Specialist Group member, and an advisory committee member at Japan Society for the Conservation of Horseshoe Crab.

**Siu Gin Cheung** is a marine biologist with primary interests in the physiological and behavioral ecology of marine invertebrates and their responses to human disturbances such as microplastic pollution, ocean acidification, and hypoxia. He started working on the ecology and conservation of horseshoe crabs in 2005 and is currently a steering committee member of the Horseshoe Crab Specialist Group under the IUCN Species Survival Commission (SSC). In collaboration with the Ocean Park Conservation Foundation of Hong Kong, he has been running a very successful educational outreach program for more than 10 years. This program aims to nurture secondary students in their responsibilities and commitment to marine conservation by rearing juvenile horseshoe crabs in their schools. Currently, he is working on the habitat utilization of juvenile horseshoe crabs and the conflict with oyster cultivation.

**Kit Yue Kwan** is a marine ecologist based at Beibu Gulf University, Guangxi, China and has a long-standing interest on finding workable solutions to tackle issues of biodiversity conservation in Asian regions. After completing his Bachelor of Science in Environmental Science and Management at City University of Hong Kong, he continued his Ph.D. study on biology and ecology of Asian horseshoe crabs. His findings contributed to the conservation and management of the juvenile populations in Hong Kong waters, as well as explored the use of the juvenile hemolymph constituents as a health indicator for wild populations. Dr. Kwan is now a steering committee member of the Horseshoe Crab Specialist Group under the IUCN Species Survival Commission. Recently, he has organized the fourth International Workshop on the Science and Conservation of Horseshoe Crabs in China, assisted the establishment of International Horseshoe Crab Day, and now leading the Asian Horseshoe Crab Observation Network program. His recent research work focuses on developing minimum standards for collecting/reporting juvenile population information and conceptual frameworks for responsible stock enhancement programs for Asian horseshoe crab conservation.

**Jennifer H. Mattei** is Professor of Biology at Sacred Heart University in Fairfield, Connecticut, USA, and has a wide range of interests in population ecology, restoration, and conservation, with over 25 years of experience working in coastal ecosystems. After completing her Master of Forest Science at Yale and Ph.D. in ecology and evolution at Stony Brook, Dr. Mattei was a postdoc at Rutgers University where she was part of a team of scientists to be the first to successfully restore coastal forest habitat on top of closed sections of the largest landfill in the USA on Staten Island, NY. In 1998, Dr. Mattei started Project *Limulus* (www.projectlimulus.org), a community research program within Long Island Sound involving horseshoe crab ecology. Dr. Mattei is a steering committee member of the Horseshoe Crab Specialist Group, IUCN Species Survival Commission. Her research with Connecticut citizen scientists found that overharvest and loss of habitat are the major factors causing the population's decline. Currently, her research on coastal restoration employs a whole ecosystem approach that includes installing oyster reefs, saltmarsh, coastal dune grasslands, pollinator meadow, and coastal forest habitats (www.sacredheart.edu/livingshorelines).

# An Homage to Carl Shuster's Contributions to Horseshoe Crab Biology

## Dr. Carl N. Shuster Jr.: A Brief Biography

Dr. Carl N. Shuster Jr., who passed away on May 28, 2020, at the age of 100 years, was truly a towering presence in the world of horseshoe crab biology since his early studies in the 1940s. Born in Randolph, Vermont, on November 16, 1919, Carl received his B.S. from Rutgers University in 1942. He proudly served in the United States Army Air Force during WWII, flying 27 combat missions over Europe, and earning the Distinguished Flying Cross. Carl returned to Rutgers for his M.S. degree in 1948, conducting a careful anatomical study of the digestive system of the American horseshoe crab, *Limulus polyphemus*, under the guidance of Dr. Thurlow C. Nelson. Carl continued his studies of horseshoe crabs for his Ph.D. at NYU through 1955, while holding summer research fellowships at the Woods Hole Oceanographic Institution.

In 1955, Carl became director of the nascent Marine Sciences program at the University of Delaware, where he remained until 1963. After nine years with the US Public Health Service in Rhode Island, Carl moved to the Washington, DC area working for the US EPA, the Federal Power Commission, and later as an environmental consultant.

Carl's heart never strayed far from the horseshoe crabs in the Delaware Estuary. Carl authored close to 100 publications in his long and distinguished career, and his book, *The American Horseshoe Crab* (Harvard Press, 2003), is an indispensable resource for all researchers in the field. Carl's work established the vital role that horseshoe crabs play in the Delaware Bay ecosystem. During the late 1980s and 1990s, when there were highly charged and contentious debates about the emerging horseshoe crab bait fishery, Carl was a voice of reason whose encyclopedic knowledge was respected by watermen, conservationists, and fishery scientists.

**Fig. 1** Carl holding a large female horseshoe crab, ca. 1951 (Glenn Gauvry)

Carl urged his younger colleagues to first "know your animal," and he always stressed the importance of making long-term studies of nature rather than rushing to judgment based on casual or incomplete observations. Carl took great pleasure in sharing his knowledge with students of all ages, and he was instrumental in developing the award-winning "Green Eggs and Sand" educational curriculum. Carl was a dear friend and mentor to so many of us, and he leaves behind an incredible legacy (Fig. 1).

## Reflections on Dr. Carl N. Shuster Jr., by Mark L. Botton (Fordham University) and Bob Loveland (Rutgers University)

We all went to Rutgers University at different times, conducting research at the Cape Shore Oyster Research Lab on the Cape May Peninsula of Delaware Bay. It was Carl who was the first person to begin a research program featuring the little studied creatures known locally as "King Crabs." Mark, as a student with Dr. Harold Haskin, first met Carl on the beaches of Delaware Bay in 1977 during the filming of

**Fig. 2** Carl at Imari Bay, Japan in July1994 (Mark L. Botton)

a "Life on Earth" segment for the BBC. He was quickly persuaded by Carl to begin working on horseshoe crabs for his Ph.D. thesis. Over the ensuing years, Carl, Mark, and Bob expanded their research to include the east coast of North America, and the inland seas of Asia. Together, the three of us worked on projects that took us to Virginia and New Hampshire (Fig. 2).

Carl was a tall gentleman, soft spoken and attentive to one's opinion. After completing his undergraduate studies at Rutgers in 1942, Carl flew 27 combat missions in the nose of a B-45 toward the end of WWII. How he ever tucked his 6'5" frame into the nose of a bomber mystifies us. Carl, his wife Helen, and their five sons lived on the Busch Campus of Rutgers University while he was a Ph.D. student. His children went to school in Piscataway, NJ, where Helen volunteered in the improvement of the schools. Even after his retirement, Carl and Helen often rented a cottage close to the Delaware Bay beach so that he could continue pursuing his interest in horseshoe crabs.

We will always remember Carl for his kindness, a gentleman's attitude and respect for one's opinion, and for many hours of fascinating conversations about the evolution of horseshoe crabs and the mystery of their origin in North America. Above all, Carl was a treasured friend and an inspiration to both of us and countless others.

## Celebrating the Life of Dr. Carl N. Shuster Jr., by Glenn Gauvry (Ecological Research Development Group) and Gary Kreamer (Delaware Division of Fish and Wildlife, retired)

May 28, 2020, marked the passing of a legendary man and scientist. Dr. Carl N. Shuster Jr., renowned "founding father" of horseshoe crab science, died in his Arlington, Virginia, home at the age of 100. For over 70 years, Carl devoted his extraordinary energy, sharply-focused mind, and exceptional capacity for scientific insight into studying the biology and ecology of the American horseshoe crab, *Limulus polyphemus*. Along the way, Dr. Shuster authored scores of scientific articles; presented at innumerable regional, national, and international conferences; and contributed substantially to the cause of education and conservation of this species (Fig. 3).

As sometimes happens in science, Carl got his start working with horseshoe crabs serendipitously. After growing up a farm boy in central New Jersey, he attended Rutgers University, with plans to pursue a career in the science of farm animals. After receiving his undergraduate degree in 1942, Carl's education was interrupted by World War II. Upon returning, Carl undertook graduate work at Rutgers under the guidance of oyster expert Dr. Thurlow Nelson, who one day (as

**Fig. 3** Presenting at the 1st International Workshop on the Science and Conservation of Horseshoe Crabs held at Dowling College, New York, in June 2009 (Mark L. Botton)

Carl later recounted) provided him with a "jar of gunk" and told him to "study this." That jar of gunk contained the eggs, embryos, and associated bay-bottom detritus of *Limulus,* and the rest, as the cliché goes, became "history."

Years (and decades of dedicated research on *Limulus*) later, in recognition of Dr. Shuster's outstanding contributions to the field of horseshoe crab science and conservation, the Atlantic States Marine Fisheries Commission (in 2001), as part of their coastwide management plan for horseshoe crabs, set aside hundreds of square miles in the Atlantic Ocean off the New Jersey and Delaware coast as the "Carl N. Shuster, Jr. Horseshoe Crab Reserve." In protecting a large population of overwintering adult and sub-adult horseshoe crabs from harvest, this reserve has played a huge role in helping rebuild and sustain spawning capacities of *Limulus.*

Not unlike the animal that became his life's work, Dr. Shuster could be described as an ecological generalist. His capacity for applying a broad sweep of scientific knowledge and real-life, natural-history based experience, along with a grounded, common sense approach to issues, were his trademarks, enabling him to identify meaningful pathways for research and postulate unique and logical explanations to patterns observed that – in this time of scientist specialization – it is hard to imagine anyone else achieving. One of his favorite mantras to young people, on first encountering horseshoe crabs, was: "if you really want to know this animal, you've got to get down on the beach, eyeball to eyeball with it."

At a time when most experts, after devoting a long and dedicated career to their special calling, moved on to a well-earned time of quiet retirement, Dr. Shuster, at the age of 80, merely redirected his unique intellect and substantial energies to a different horseshoe crab cause: education. In 2000, as awareness of the significance of horseshoe crabs to shorebird ecology and human uses emerged, an educational initiative was launched to bring this story, and the management controversy surrounding it, into the school curriculum in the Delaware Estuary area. That project, later dubbed "Green Eggs & Sand," featured development of a comprehensive curriculum on horseshoe crabs that would be distributed at full-weekend workshops for educators up and down the Atlantic coast. Over two decades, more than 1000 educators from 25 US states and 4 foreign nations attended these workshops, resulting in the curriculum reaching more than 100,000 students coastwide.

A key ingredient in the success of Green Eggs & Sand was the recruitment and involvement of scientists in sharing their knowledge and passion for horseshoe crabs with teachers at these workshops, and Dr. Shuster was absolutely the catalyst for that. Over the seminal years of Green Eggs & Sand, throughout the 1980s, Dr. Shuster hit the workshop trail to present at 25 workshops in 6 states. At every venue, he had some illuminating and relevant information and perspectives on the horseshoe crab situation to present. Carl's "talks" were always unique, well-articulated, and thought-provoking, but it was on the beach with horseshoe crab in hand that Carl was especially adept at working his magic. Field sessions at these workshops featured Carl "holding court" on the beach with live Limuli in hand, guiding teachers in observation and interpretation of finer details of the unique structure-to-function anatomy and ecology of the animal. After long days of seminars and in-the-field activities, Carl also made himself available and approachable to teachers and fellow

**Fig. 4** Carl discussing horseshoe crabs during a Green Eggs and Sand Workshop in 2011 (Mike Oates)

scientists during meals and late-night workshop "down-times," answering questions they had, swapping stories of various research adventures and endeavors, and otherwise making the experience more memorable for all engaged.

In addition to pitching in full-tilt at workshops, Dr. Shuster was integrally involved in development of the curriculum at various levels. This included: supplying his horseshoe crab model and other resources to the lessons, ground-truthing text for scientific accuracy, and contributing to key video components. In recognition of the key role Carl played as scientific advisor to the project, and for his part in the various awards and honors it received over the years, the Green Eggs & Sand curriculum was specifically dedicated to Dr. Shuster in 2003 (Fig. 4).

## Memories of Carl, by Mike Oates (Filmmaker and Documentarian)

Drinking Mike's Hard Limeade and watching Jurassic Park, designing and building a video benthic sled, picking blue crabs at horseshoe crab festivals, dusting off live crabs half-submerged in his basement shower stall, feeding horseshoe crabs leftover cocktail shrimp, advising Sir David Attenborough on a BBC shoot, teaching 4-year-olds on the beach at Fortescue, New Jersey, listening to watermen, touring the 8th Army Air Force Museum in GA, Green Eggs and Sand Workshop field trips, and enjoying his "good fellowship."

## Personal Reflections on Dr. Carl N. Shuster Jr., by Gary Kreamer (Delaware Division of Fish and Wildlife, retired)

I first met Dr. Shuster in 2000, when launching a new horseshoe crab education initiative entitled "Green Eggs & Sand." Then at the age of 80, Dr. Shuster became our lead scientist for the project, contributing material, and ground-truthing content, for the curriculum, presenting at teacher workshops, and otherwise advising, supporting, and inspiring all educators on our team. Over its seminal years, Dr. Shuster hit the workshop trail to present at 25 Green Eggs & Sand workshops in 6 states. At every venue, he had illuminating and relevant information and perspectives on horseshoe crabs to present.

My personal collection of Dr. Shuster's memories from those times runs deep and wide as well, spanning nearly two decades and thousands of miles of I-95 travels to workshop venues up and down the Atlantic coast, with many memorable Carl stories entertaining me along the way. Like all great teachers, he challenged me to be a better observer, look at things in new ways, ask good questions, and trust my brain, gut, and common sense in seeking the best answers. In these and other ways, he has made me a better horseshoe crab scientist and educator. His influence is evident, and his teaching carries on through every presentation I make, school group I take to see the spawning spectacle, and teacher workshops I do.

## Remembrances, by Dr. Dick Weber (University of Delaware, retired)

Dr. Shuster has been a special resource for all of us who have investigated aspects of horseshoe crab biology. I have benefitted greatly from Carl's knowledge, enthusiasm, insights, and friendly encouragement. Through Carl, I have met other people with similar qualities, who share a common interest in horseshoe crabs. This has been stimulating, and has broadened my knowledge.

## An Appreciation, by Dr. David Smith (U.S. Geological Survey)

Certainly, Carl deserves great admiration and respect for all that he has contributed to our knowledge and understanding of horseshoe crab biology and conservation. However, Carl – like the great leaders – also inspires deep affection because he gives so freely and selflessly of himself as a colleague, a teacher, and a mentor. To appreciate Carl's gifts to us, imagine if he had pursued a different subject of study instead of *Limulus*. How much poorer would our understanding be? How many of us would have found their current path in his absence?

## Remembering Carl Shuster, by Dr. Ruth Carmichael (Dauphin Island Sea Lab)

My best personal memory of Carl is that he was always (and has always remained) inclusive about his research. He was so incredibly warm to me when I started studying horseshoe crabs as a graduate student with no knowledge whatsoever of these amazing beasts. Carl was completely open and accessible and willing to share everything he knew and to encourage new lines of study. His warm and welcoming approach to scientific inquiry is rare these days with so much competition for funding and recognition. I have always admired his utter lack of concern for those issues and focus on what is meaningful, what will move us ahead in our studies and understanding. Through the years, Carl has sent me innumerable manuscripts, notes, pictures, and documents just because he thought I would be interested. I have read everything he has ever sent to me because he is right; it has all been useful and interesting.

I particularly enjoyed the items he designed and penned himself. I use one of his bivalve dissection cutouts every year as part of the curriculum in my invertebrate zoology class. I copy the two-sided paper images and give them with instructions to the students. His hand-drawn diagrams rival anything you can find today in a textbook, and the 20-something students still get a huge kick out of the craft of cutting it out and putting together the paper clam. It's too cool for words. Carl is a real treasure!

## Personal Reflections, by Glenn Gauvry (Ecological Research and Development Group)

My nonprofit organization, ERDG, was in its infancy in 1996. I stopped for lunch at the local convenience store in the small Delaware Bay coastal town of Broadkill Beach, where I lived at the time, and was told by the owner that an older man had stopped in and purchased a few of our Just flip 'em!™ T-shirts and left a small pewter horseshoe crab pin along with his card. As I fiddled with the pin, I looked at the card, curious as to who had left this gift and had taken interest in our work. I was astonished to read the name Carl N. Shuster, Ph.D., the world's authority on horseshoe crabs. After much procrastination, I found the courage to call this remarkable man, who has dedicated a lifetime to unraveling the mysteries of the horseshoe crab species. To my relief, I was greeted by a warm and inviting voice, a man secure with himself, without the need to impress. We talked for quite some time about horseshoe crabs, and our call concluded with Dr. Shuster inviting me on a road trip to introduce me to others whom he felt could be helpful in the advancement of our organization's conservation efforts. That summer we took that road trip, and as we drove, we talked mostly about horseshoe crabs and how our lives had led us to the paths we found ourselves following. The miles and days passed all too quickly, and Dr. Shuster's relaxed and unpretentious manner of sharing information was like lounging in a comfortable chair with a good book. We have been friends ever since.

## Remembering Carl Shuster, by H. Jane Brockmann (University of Florida, retired)

I first met Carl in 1991 when I was studying the mating behavior of horseshoe crabs at Cape Henlopen, Delaware. He was on his way to a horseshoe crab workshop in Cape May, NJ, to give a public lecture. He traveled in an enormous station wagon filled with boxes of specimens, reprints, brochures, drawings, and articles about horseshoe crabs. He also had quite a store of memorabilia – horseshoe crab pins, bookmarks, coasters, mugs, and the like, which he readily gave out to horseshoe crab enthusiasts (including me). This was typical Carl, eager to share his knowledge and fascination for horseshoe crabs with colleagues, students, and the public.

In the mid-1990s, Carl led an effort to put together a book on horseshoe crabs. He wanted a reference that students, colleagues, and the public would enjoy – but at the same time it needed to be authoritative and comprehensive. Carl's encyclopedic knowledge of the literature, his broad understanding of the animal, and his tireless enthusiasm allowed him to push the project forward. While I tried to help, he did most of the organizing, editing, and indexing and helped with the writing of many chapters. He also shared many of his beautiful drawings with the authors. In the making for more than 10 years, *The American Horseshoe Crab* (Harvard University Press, 2003) was the pinnacle of Carl's lifework (Figs. 5, 6, 7, and 8).

**Fig. 5** Carl posing next to a large statue of a Red Knot at the DuPont Nature Center, Delaware, in May 2007 (Glenn Gauvry)

**Fig. 6** Carl and his wife Helen at a celebration of Carl's 97th birthday in June 2016 (Mark L. Botton)

**Fig. 7** Carl and Helen with their five sons and family at Carl's 97th birthday in June 2016 (Mark L. Botton)

**Fig. 8** The front cover of The American Horseshoe Crab, published by Harvard University Press in 2003

In 2005, while on a trip to the Georgia coast, Mike Oates and Carl observed the exceptionally spiny juvenile horseshoe crabs that are characteristic of this region. Carl remembered he had once read a report about a spiny variant of *Limulus* and he managed to find his notes (see his notes, Fig. 9) from a nearly 200 year-old publication. Was this a different species or simply a spiny variant of *Limulus polyphemus*? Since I had been studying the horseshoe crabs from this population for several years and was regularly visiting Georgia and North Florida, we agreed that I would continue to gather data on this question. This is the subject of Carl's final publication on horseshoe crabs and our contribution to the present book (Chapter 1 in this volume) (Fig. 10).

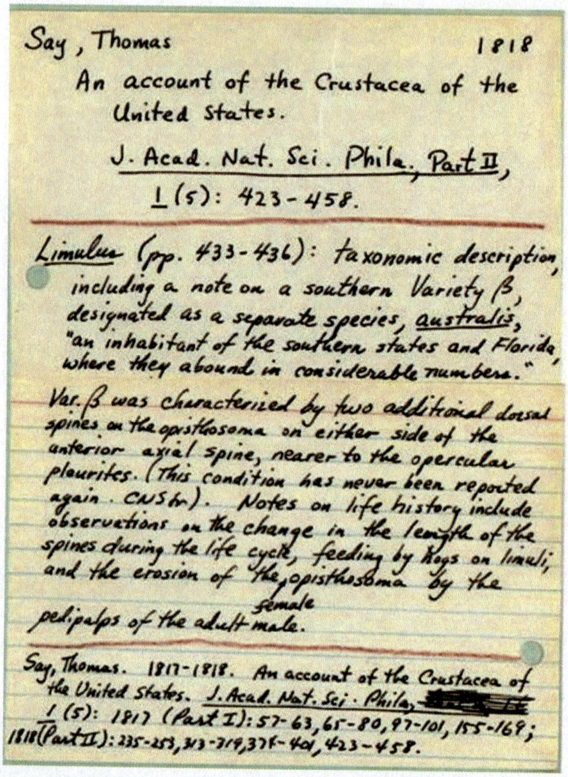
Say, Thomas 1818

An account of the Crustacea of the United States.

J. Acad. Nat. Sci. Phila, Part II, 1 (5): 423-458.

Limulus (pp. 433-436): taxonomic description, including a note on a southern Variety β, designated as a separate species, australis, "an inhabitant of the southern states and Florida, where they abound in considerable numbers."

Var. β was characterized by two additional dorsal spines on the opisthosoma on either side of the anterior axial spine, nearer to the opercular pleurites. (This condition has never been reported again. CNSJr). Notes on life history include observations on the change in the length of the spines during the life cycle, feeding by hogs on limuli; and the erosion of the female opisthosoma by the pedipalps of the adult male.

Say, Thomas. 1817-1818. An account of the Crustacea of the United States. J. Acad. Nat. Sci. Phila, [redacted] 1 (5): 1817 (Part I): 57-63, 65-80, 97-101, 155-169; 1818 (Part II): 235-253, 313-319, 374-401, 423-458.

**Fig. 9** Carl's handwritten notes from the Shuster Collection (see ERDG website horseshoe crab.org) on the presence of a spiny variant in the Georgia population of horseshoe crabs

## Dr. Carl N. Shuster Jr.: A Selected Bibliography

M.L. Botton, C.N. Shuster Jr, K. Sekiguchi, H. Sugita, Amplexus and mating behavior in the Japanese horseshoe crab, *Tachypleus tridentatus*, Zool. Sci. **13**, 151–159 (1996)

K. Sekiguchi, C.N. Shuster Jr (2009) Limits on the global distribution of horseshoe crabs (Limulacea): Lessons learned from two lifetimes of observations: Asia and America, in *Biology and Conservation of Horseshoe Crabs*, ed by J.T. Tanacredi, M.L. Botton, D.R. Smith (Springer, New York, 2009), pp 5–24

C.N. Shuster Jr, On morphometric and serological relationships within the Limulidae, with particular reference to *Limulus polyphemus* (L.), Ph.D. Dissertation, Rutgers University, 1955

C.N. Shuster Jr, A pictorial review of the natural history and ecology of the horseshoe crab *Limulus polyphemus*, with reference to other Limulidae, in *Physiology and Biology of Horseshoe Crabs*, ed. by J Bonaventura, C Bonaventura, S Tesh (Alan R Liss, New York, 1982), pp. 1–52

**Fig. 10** Carl with a very large horseshoe crab from Tybee Island, Georgia in 2005. Horseshoe crabs from the Southeast Atlantic region (Georgia and South Carolina) are typically the largest found along the entire coastline, an observation first made by Carl in his Ph.D. dissertation (H. Jane Brockmann)

C.N. Shuster Jr, The Delaware Bay area, U.S.A.: A unique habitat of the American horseshoe crab, *Limulus polyphemus*, in *Changing Global Perspectives on Horseshoe Crab Biology, Conservation and Management*, ed by R.H. Carmichael, M.L. Botton, P.K.S. Shin, S.G. Cheung (Springer International, Switzerland, 2015), pp. 15–39

C.N. Shuster Jr, M.L. Botton, A contribution to the population biology of horseshoe crabs, *Limulus polyphemus* (L.), in Delaware Bay. Estuaries **8**, 363–372 (1985)

C.N. Shuster Jr, R.B. Barlow, H.J. Brockmann (eds.), *The American Horseshoe Crab* (Harvard Press, Cambridge, 2003)

New York, NY, USA

Mark L. Botton

# Part I
# Evolution Genetics & Adaptation

# Variation in the American Horseshoe Crab: A Spiny Variant from the Southeast

**H. Jane Brockmann, Carl N. Shuster Jr, Patrick Norby, and Dorothea Saunders**

## 1 Introduction

American naturalist Thomas Say (1787–1834) traveled to the coastal islands of Georgia, USA (Ossabaw, Wassaw, St. Simons, Cumberland), and Spanish Florida (Amelia, Fernandina, Fort George) during the winter and spring of 1817–1818 to collect invertebrate specimens (Weiss and Ziegler 1931; Porter 1983; Stroud 1992). Being familiar with the horseshoe crabs of Delaware Bay near his home in Philadelphia, Say knew that the southern horseshoe crabs were different: they had an extra pair of spines along the anterior margin of the opisthosoma (Say 1818) (Fig. 1a). Say referred to this population as "Var. β" of *Limulus polyphemus*, and he proposed that it be considered a new species, *Limulus australis*. He described Var. β as "an inhabitant of the southern states and Florida, where they abound in considerable numbers." De Kay (1844) cited Say (1818) but suggested that Var. β was a variety of *L. polyphemus* from "Carolina and Florida." Shuster (2015) rediscovered this "long-spined variant" on a juvenile specimen collected near Skidaway Island, Georgia (Fig. 1a, b). Here we discuss the range and frequency of Var. β from the population of horseshoe crabs in which Say made his initial observations. We then evaluate the taxonomic status of *L. australis* using morphological, behavioral, and genetic data.

H. J. Brockmann (✉) · P. Norby
Department of Biology, University of Florida, Gainesville, FL, USA
e-mail: hjb@ufl.edu

C. N. Shuster Jr
Virginia Institute of Marine Science, William & Mary, Gloucester Point, VA, USA

D. Saunders
Marine Education Center & Aquarium, University of Georgia, Skidaway Island, GA, USA



J. T. Tanacredi et al. (eds.), *International Horseshoe Crab Conservation and Research Efforts: 2007-2020*, https://doi.org/10.1007/978-3-030-82315-3_1

**Fig. 1** (**a**). Photograph of two molted juvenile exoskeletons, left from Delaware Bay (i.e., from the mid-Atlantic population of *L. polyphemus*) and right Var. β from Savannah, Georgia (i.e., from the Southeast Atlantic population). In addition to the "extra" spines on the Georgia specimen, the dorsal spines of Var. β are longer than those typical of juvenile *L. polyphemus* (i.e., juveniles from the mid-Atlantic population). (**b**). Molted exoskeletons of juvenile *L. polyphemus* from Wassaw Sound, Georgia. The one on the right shows the "extra" spines on the opercular pleurite typical of Var. β, and the one on the left does not have opercular pleurite spines. (**c**). Adults and late-stage juveniles (this molted exoskeleton from Savannah has a prosoma width of 25 cm) have much smaller spines and smaller opercular pleurite spines than younger juveniles

American horseshoe crabs are subdivided into eight genetically distinct populations along the Atlantic and Gulf coasts of North America (King et al. 2005, 2015). These populations differ in body size and shape (Shuster 1979; Riska 1981; Botton and Loveland 1992; Brockmann and Smith 2009; Faurby et al. 2011; Smith and Brockmann 2014), color (Sokoloff 1978), physiology and development (Mayer 1914; Shuster 1954, 2015; Vasquez et al. 2017), behavior (Rudloe 1985; Penn and Brockmann 1994; Anderson et al. 2017), sperm quantity and quality (Sasson and Brockmann 2016), and genetics (Selander et al. 1970; Saunders et al. 1986; Avise 1992; Pierce et al. 2000; King et al. 2005, 2015), but none of the differences are thought to elevate any of the populations to the level of a species. Is Var. β a cryptic species of *L. polyphemus* from the coastal islands of Georgia and Florida as Say (1818) thought, or is Var. β an example of phenotypic variation within one species as De Kay (1844) suspected?

In their range-wide analyses of genetic variation in *Limulus polyphemus*, King et al. (2005, 2015) defined a Southeast Atlantic population of horseshoe crabs based on tissue samples from South Carolina (Bulls Bay and Beaufort) and Georgia (Savannah and Sapelo Island). They demonstrated that this population is genetically differentiated from the mid-Atlantic population to the north and the Florida East (Indian River Lagoon) population to the south. This Southeast Atlantic population is where Var. β was observed. This population is also characterized by having the largest individuals anywhere within the species' range (Shuster 1979; Riska 1981; Botton and Loveland 1992; Brockmann and Smith 2009; Smith and Brockmann 2014). They contrast sharply with the horseshoe crabs found just to the south in the Indian River Lagoon (Florida East population), which are among the smallest found along the Atlantic coast (Shuster 1979; Riska 1981; Smith and Brockmann 2014). In addition to tissue samples from Sapelo Island (Southeast Atlantic population) and Indian River Lagoon (Florida East population), Norby (2017) and Norby et al. (in prep) collected tissue samples from sites in Georgia (GA) and Florida (FL), including Cumberland Island, GA; St. Augustine, FL; New Smyrna Beach, FL; and Canaveral National Seashore, FL. Norby (2017) demonstrated that the Southeast Atlantic and Florida East populations are genetically distinct but overlap and interbreed producing viable hybrid offspring in North Florida between New Smyrna Beach and Cumberland Island. We examine here the distribution, frequency, and genetics of horseshoe crabs from these Georgia and Florida sites and evaluate the taxonomic status of Var. β.

## 2 Methods

**A Note on Taxonomy** Say's 1817–1818 expedition was sponsored by the newly established Philadelphia Academy of Natural Sciences (Stroud 1992), and therefore, his specimens should have been deposited in the Academy's collections. However, "Most of Say's material went with him to New Hope and was lost" (personal communication from Paul Callomon, Collections Manager, Academy of Natural Sciences, Philadelphia). [In 1826, Say moved from Philadelphia to New Hope, Indiana, to join a Utopian community founded by Robert Owen.] Say sent the type specimen of *Limulus australis* to W.E. Leach at the British Museum and labeled it as from "Florida," but it is "not now traceable" (Gunther 1904–1912). We have examined two juvenile specimens from the invertebrate collection of the American Museum of Natural History, New York, that were collected in October 1972 on St. Catherine's Island, Georgia: one (juvenile) has the distinct Var. β spines that Say described and one does not. It seems that both *L. polyphemus* and Var. β (*L. australis*?) occur within the same population of horseshoe crabs from the coastal region of Georgia and Florida (Fig. 1b).

**The Var. β Spines** American horseshoe crabs from all populations have three dorsal axial spines along the midline of the prosoma (ocellar, mid-cardiac, and post-cardiac prosomal spines) and three spines along the midline of the opisthosoma (anterior, medial, and posterior opisthosomal spines) (Shuster 1955; Fig. 2). In addition, there are spines medial to each of the lateral eyes (ophthalmic spines) and on each post-ophthalmic ridge at the posterior margin of the prosoma (post-ophthalmic

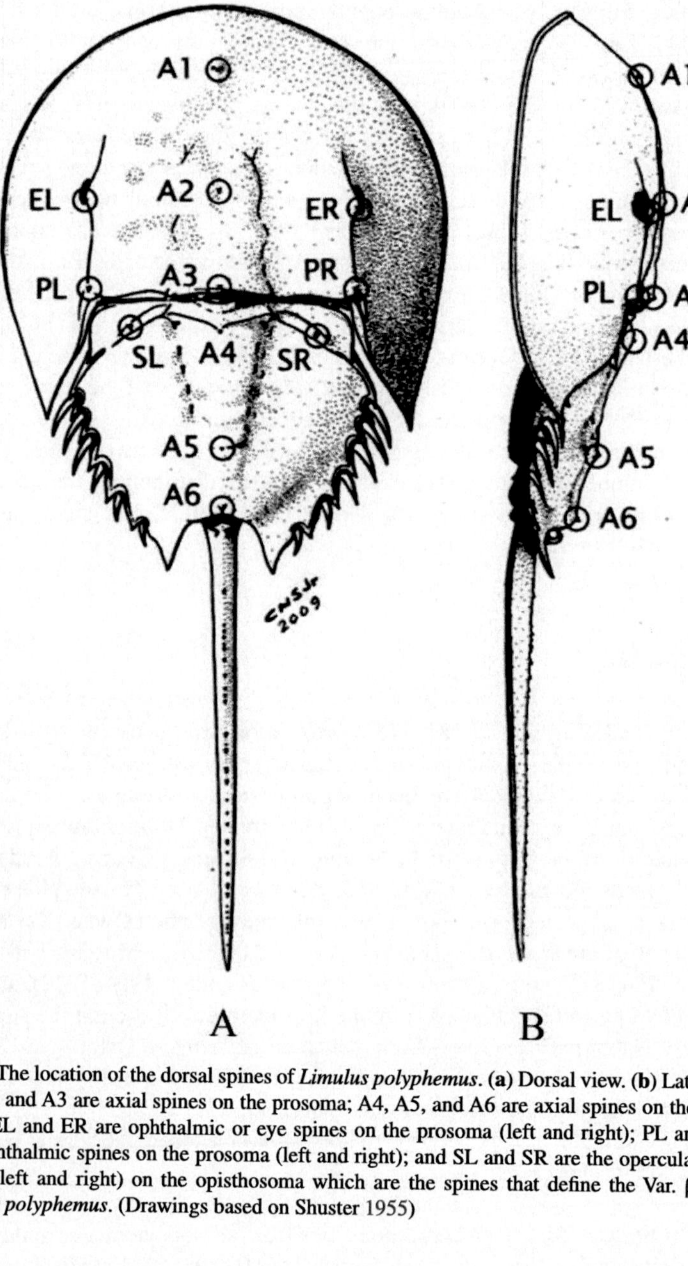


**Fig. 2** The location of the dorsal spines of *Limulus polyphemus*. (**a**) Dorsal view. (**b**) Lateral view. A1, A2, and A3 are axial spines on the prosoma; A4, A5, and A6 are axial spines on the opisthosoma; EL and ER are ophthalmic or eye spines on the prosoma (left and right); PL and PR are post-ophthalmic spines on the prosoma (left and right); and SL and SR are the opercular pleurite spines (left and right) on the opisthosoma which are the spines that define the Var. β form of *Limulus polyphemus*. (Drawings based on Shuster 1955)

spines) (Fig. 2). In the Var. β form, there is an additional pair of dorsal spines on the opercular pleurite (Patten 1912), i.e., the anterior margin of the opisthosoma between the axial and post-ophthalmic spines (Fig. 2). In all populations, juvenile horseshoe crabs have more pronounced spines than adults (Fig. 1c; Say 1818; Barthel 1974). Spines begin to develop during the late embryonic stages and increase in size with each molt such that the early juvenile stages have the longest and sharpest spines relative to their body size (Shuster 1955, 1982; Fig. 1a). The opercular pleurite spines are still clearly visible on some adult individuals although they are relatively much smaller than on juveniles (pers. obs.). Molted exoskeletons of juvenile horseshoe crabs from all populations reveal considerable structural variation of the opercular pleurites even among juveniles of the same instar (pers. obs.): there may be a thickening of the cuticle, a ridge, blunt nodules, or spines present. The spines of Var. β are pointed, sharp protrusions on the opercular pleurites (Fig. 2). Occasionally, an individual had a spine on only one opercular pleurite rather than on both; these individuals were recorded as Var. β. Throughout, we refer to the opercular pleurite spines of Var. β as "extra" spines.

**Data Collection** We collected data on the presence or absence of opercular pleurite spines, i.e., "extra" spines, from sites in GA and FL. When possible, we also recorded the size of each specimen, i.e., prosomal width at the widest point measured ventrally and/or interocular distance measured between the two ophthalmic spines (these two measures of size are strongly correlated). We collected data from live, nesting horseshoe crabs whenever possible, but live overturned individuals were also included in our data sets, as were dead animals particularly at sites where few live individuals were observed. At some GA sites, we also collected juveniles, measured their size, and determined whether they had "extra" spines or not.

The data were collected opportunistically whenever we visited coastal Georgia or the Atlantic coast of North Florida (GPS coordinates given in Table 1). At each site, we made every effort to record an unbiased sample, i.e., all animals present on the beach. Due to the opportunistic nature of our data collection, sample sizes vary for different measures at the various sites:

1. Wassaw Sound, GA. Size and spine data were collected from live adults at Dead Man's Key or Hammock Island on 4 May 2012 and Cabbage Island on 4 May 2012, 27 April 2013, 3 May 2014, 2 May 2015, and 23 April 2016. These data were collected in association with field trips for Green Eggs and Sand teacher workshops. Since we did not want to disturb nesting animals, we collected only interocular distance as a measure of size. We also collected molted exoskeletons opportunistically from Wassaw Island during the fall in 2008–2011.
2. Ossabaw Sound, GA. Size and spine data were collected from live adults at Raccoon Key on 3 May 2014 and also in association with a Green Eggs and Sand field trip. Molted exoskeletons were collected at this site opportunistically during the fall 2008–2011.
3. Tybee Island, GA. Size and spine data were collected from live adults on 3 May 2012.
4. Little St. Simons Island, GA. Size and spine data were collected from live animals on 30 April 2017 in association with a Green Eggs and Sand field trip.

**Table 1** Frequency of Var. β (adult individuals with "extra" spines) at different sites along the Atlantic coast

| Location (listed north to south) | State | GPS – north | GPS – west | % Females with "extra" spines (N) | % Males with "extra" spines (N) | Combined male and female % with "extra" spines (N) |
|---|---|---|---|---|---|---|
| Wassaw Sound Islands, Raccoon Key, and Tybee Island | GA | 31.94554 | −80.93 | 57.1 (35) | 53.7 (95) | 54.6 (130)[a] |
| Sapelo and Little St. Simons Islands | GA | 31.38068 | −81.28 | 56.1 (66) | 52.1 (165) | 53.2 (235*)[a] |
| Cumberland Island | GA | 30.72303 | −81.46 | 46.6 (73) | 40 (60) | 41.1 (146*)[b] |
| St. Augustine | FL | 29.8998 | −81.29 | 25 (24) | 22.2 (27) | 23.5 (51)[c] |
| New Smyrna Beach | FL | 29.02162 | −80.92 | 7.2 (125) | 4.8 (124) | 6.0 (249)[d] |
| Mosquito Lagoon | FL | 28.85772 | −80.78 | 4.7 (106) | 4.5 (112) | 4.6 (218)[d] |
| Northern Indian River Lagoon | FL | 28.69618 | −80.77 | 0 (46) | 1.9 (52) | 1.0 (98) [e] |

(N) refers to the sample size. * indicates that additional data were taken from dead carapaces to determine the presence of "extra" spines, but it was not possible to determine sex on these specimens. [abcde] letters indicate sites that are significantly different in the frequency of extra spines; see text for $\chi^2$ values

5. Sapelo Island, GA. Size and spine data and tissue samples for genetic analyses were collected by Daniel Sasson, Sheri Johnson, Matthew D. Smith, and Jane Brockmann from live animals along Sapelo Island beaches: Cabretta on 22 May 2009 and Nanny Goat on 23–24 May 2009 and 13–16 May 2010; and data from dead animals were collected at the south end of the island on 21 April 2012. The tissue samples were part of the population genetics study by Patrick Norby (2017).
6. Cumberland Island, GA. Size and spine data were collected from Cumberland Island National Seashore beaches on 12–16 April 2014 from a mix of alive and dead individuals by Jane Brockmann and Devon Pharo. They also collected tissue samples for genetic analysis from live individuals, which were included in the population genetics study by Patrick Norby (2017).
7. St. Augustine, FL, at Anastasia State Park (north end of Conch Island on Salt Run). Tissue samples and size and spine data were collected from live animals on 16 and 30 September and 1 October 2016, 11 March 2017, and 28–29 March 2018 by Patrick Norby for use in his study on population genetics (Norby 2017; Norby et al. in prep.).
8. New Smyrna Beach, FL (shoreline and island beaches in Indian River North). Size and spine data were collected from a mix of alive and dead animals on 28 April 2014, 2–3 and 17–19 April 2015, 3–4 May 2015, 9–11 March 2016, 27–28 March 2017, and 24–27 April 2017 as part of the Marine Discovery

Center (MDC) Horseshoe Crab citizen scientist breeding survey and tagging program. Tissue samples were also collected for the population genetics study by Patrick Norby (2017).

9. Mosquito Lagoon, FL, Canaveral National Seashore (at Parking lot 5). Tissue samples were collected from live individuals, and spine and size data were collected from a mix of alive and dead animals on 19 April 2015, 23 February 2016, 10 March 2016, 28–29 March 2017, and 25–27 April 2017 also as part of the MDC program and population genetics studies.
10. Indian River, Merritt Island National Wildlife Refuge. These specimens were collected by trawling in Northern Indian River (north of Titusville) and brought to the Whitney Marine Laboratory where they were held in running seawater tanks for 2 weeks. We collected tissue samples (used in the population genetics study) and spine and size data on 2 April 2013 and 8 April 2014 for use in this study.

**Assortative Pairing** We recorded the presence or absence of "extra" spines on the male and female of attached pairs found nesting on or approaching breeding beaches. Even when we were unable to record size data, we nonetheless kept track of whether a female with "extra" spines was paired with a male that also had "extra" spines. These data were collected opportunistically.

**Genetic Analysis** We used data previously collected and analyzed by Norby (2017; Norby et al., in prep.) to test for an association between the presence or absence of "extra spines" and population identity. Norby (2017) and Norby et al. (in prep) analyzed genetic variation at 10 microsatellite loci in 219 individuals from 6 of the sites listed above (Sapelo, Cumberland, St. Augustine, New Smyrna Beach, Mosquito Lagoon, and Indian River). Using the clustering program STRUCTURE, they calculated the proportion of each individual's genotype that was derived from either of the two populations in this region: Southeast Atlantic or Florida East (as defined by King et al. 2005, 2015). Similar to other population assignment methods, this admixture value provides a metric for determining to which population each individual belongs. For the present study, we compared the admixture values of individuals with "extra" spines to those without. For each site, we calculated the median percent admixture derived from the Southeast Atlantic population in individuals with "extra" spines versus those without. Unfortunately, although we had tissue samples from Sapelo Island, the association between the genetic samples and the presence or absence of "extra" spines was lost.

**Statistical Analyses** Frequency, size, and genetic data in this study were often found to be bimodal or not normally distributed. For this reason, we use nonparametric tests throughout.

## 3 Results

**Frequency of Var. β** The frequency of Var. β declines north to south along the Atlantic coast from northern GA to central FL (Table 1). In northern and central GA (Wassaw to Sapelo) 52–57% of adults had “extra” spines, but in southern GA (Cumberland Island), only 40–47% had “extra” spines. Farther south, in Florida, the frequency of those with “extra” spines declined further to 22–25% at St. Augustine, to 5–7% at New Smyrna Beach and Mosquito Lagoon, and to 1% in Indian River Lagoon. The northern and central Georgia sites did not differ in the frequency of “extra” spines, but they had a significantly higher frequency than horseshoe crabs from Cumberland Island ($\chi^2 = 6.0$, d.f. = 2, $P = 0.05$) which was significantly higher than those from St. Augustine ($\chi^2 = 4.3$, d.f. = 1, $P = 0.04$). Horseshoe crabs from St. Augustine had a significantly higher frequency of “extra” spines than those from New Smyrna Beach ($\chi^2 = 23.3$, d.f. =2, $p < 0.001$), and the Mosquito Lagoon site did not differ from New Smyrna Beach.

At the Wassaw Sound and Ossabaw Sound (GA) sites where we had data from both adults and the molted exoskeletons of juveniles, the frequency of adults with “extra” spines (Table 1) did not differ from the frequency of juveniles with “extra” spines (45%, $N = 101$; $\chi^2 = 1.75$, d.f. =1, $P = 0.185$). In the juvenile sample, we found that the frequency of single spines (rather than the usual pair of spines) was 7%.

At the three Georgia sites, adult males and females were equally likely to have “extra” spines (females 52.3%, males 50.3%; $\chi^2 = 0.11$, d.f. = 1, $P = 0.743$; Table 1). Similarly at the three Florida sites, adult males and females were equally likely to have “extra” spines (7.8% females, 6.6% males; $\chi^2 = 0.14$, d.f. = 1, $p = 0.7$). At most sites, there was no difference in adult size between those with (Var. β) and without “extra” spines, but at New Smyrna Beach, females with “extra” spines were significantly larger than females without (Table 2).

**Assortative Pairing** At both the GA and FL sites, pairing was random with respect to the presence or absence of “extra” spines (Table 3).

**Genetic Analysis** The presence of “extra” spines was strongly associated with individuals whose genotypes were predominately from the Southeast Atlantic population: those with “extra” spines had a median admixture of 82% alleles from the Southeast Atlantic population (18% from the Florida East population), whereas those without “extra” spines had a median admixture of 10% from the Southeast Atlantic population (90% from the Florida East population; Table 4 and Fig. 3). At sites where the genotype was strongly Southeast Atlantic (Cumberland, St. Augustine), there was no association between “extra” spines and genotype. However, at sites where individuals were mostly of the Florida East genotype (New Smyrna Beach, Mosquito Lagoon), “extra” spines were strongly associated with those individuals that had a larger proportion of Southeast Atlantic alleles (Table 4).

**Table 2** Median sizes of female and male horseshoe crabs with and without "extra" spines at different sites along the Atlantic coast. Bold indicates a significant difference

| Location | Female IO with "extra" spines (N) | Female IO without "extra" spines (N) | Male IO with "extra" spines (N) | Male IO without "extra" spines (N) | Female PW with "extra" spines (N) | Female PW without "extra" spines (N) | Male PW with "extra" spines (N) | Male PW without spines (N) |
|---|---|---|---|---|---|---|---|---|
| Wassaw Sound, Raccoon Key, and Tybee Island | 19.0 (20) | 18.1 (15) | 13.2 (51) | 13.3 (44) | -- | -- | -- | -- |
| Sapelo and Little St. Simons Islands | 18.1 (31) | 18.6 (25) | 13.5 (78) | 13.5 (71) | 29.5 (31) | 30.4 (25) | 23.0 (78) | 23.3 (71) |
| Cumberland Island | 18.0 (33) | 17.5 (40) | 13.4 (24) | 13.3 (36) | 29.3 (33) | 28.7 (40) | 23.5 (24) | 22.9 (36) |
| St. Augustine | 16.7 (7) | 14.3 (21) | 12.9 (6) | 12.9 (16) | 26.8 (7) | 23.2 (21) | 22.1 (6) | 21.9 (16) |
| New Smyrna Beach | **17.9** (9)[a] | **12.0** (116)[a] | 9.9 (6)[d] | 8.8 (118)[d] | **28.0** (9)[b] | **19.8** (116)[b] | 17.5 (6)[c] | 15.0 (118)[c] |
| Mosquito Lagoon | 11.3 (4) | 12.0 (50) | 8.9 (4) | 8.5 (56) | 18.3 (4) | 19.4 (50) | 15.1(4) | 14.5 (56) |
| Northern Indian River Lagoon | N/A | 11.6 (46) | 7.5 (1) | 9.0 (52) | N/A | 18.7 (46) | 13.9 (1) | 14.9 (52) |

(N) refers to the sample size. N/A means that there were no animals in this category. -- refers to missing data. Sizes are measured as PW (prosomal width) at the widest point and IO the distance between the lateral eyes (from the left eye spine to the right eye spine). The size data are not normally distributed; [a]Mann-Whitney U Statistic = 207.500, T = 881.500 n(small) = 9 n(big) = 116 ($P$ = 0.003); [b]Mann-Whitney U Statistic = 186.000, T = 903.000 n(small) = 9 n(big) = 116 ($P$ = 0.001); [c]Mann-Whitney U Statistic = 192.000, T = 537.000 n(small) = 6 n(big) = 118 ($P$ = 0.060); [d]Mann-Whitney U Statistic = 217.000, T = 512.000 n(small) = 6 n(big) = 118 ($P$ = 0.112)

# 4 Discussion

Our results are consistent with the hypothesis that Var. β is a polymorphism within the Southeast Atlantic population of *L. polyphemus* and not a distinct species. Our genetic data revealed that those with "extra" spines were not a distinct genotype but that "extra" spines were strongly associated with the Southeast Atlantic population. At the Cumberland Island site, where all individuals were of Southeast Atlantic ancestry (defined as having alleles similar to those from Sapelo Island, GA), about half the individuals also had "extra" spines, i.e., they were Var. β (Table 1). At this site, there were no genetic, sex, or size differences between those with and those without "extra" spines (Tables 2, 3, and 4). However, at New Smyrna Beach where

**Table 3** Assortative pairing between females and males (natural attached pairs observed at breeding beaches) with and without "extra" spines

| Location | % Females with "extra" spines (N) that are paired with males with "extra" spines | % Females without "extra" spines (N) that are paired with males without "extra" spines | Fisher exact test *P*= |
|---|---|---|---|
| GA sites (Wassaw, Sapelo, Cumberland) | 62% (42) | 54% (37) | 0.180 |
| FL sites (St. Augustine and New Smyrna Beach) | 27% (11) | 91% (79) | 0.102 |

(*N*) refers to the sample size upon which the percentage is based

**Table 4** The association between population (Southeast Atlantic or Florida East) and the presence of "extra" spines at different locations along the Atlantic coast

| Location | Median percent Southeast Atlantic alleles (CI) | Median percent Southeast Atlantic alleles in individuals with "extra" spines (CI) | Median percent Southeast Atlantic alleles in individuals without "extra" spines (CI) | Mann-Whitney U Test, *P*=[1] |
|---|---|---|---|---|
| Sapelo Island, GA | 96% (94, 98%)[a] *N* = 29 | -- | -- | -- |
| Cumberland Island, GA | 96% (89, 97%) [a] *N* = 31 | 96% (88, 97%) *N* = 12 | 94% (94, 98%) *N* = 19 | 0.53[1] |
| St. Augustine, FL | 86% (18, 95%) [b] *N* = 51 | 93% (71, 97%) *N* = 12 | 85 (9, 93%) *N* = 39 | 0.08[1] |
| New Smyrna Beach, FL | 7% (3, 26%) [c] *N* = 78 | 14.4% (7, 76%) *N* = 17 | 6.3% (3, 16%) *N* = 61 | **0.023**[1] |
| Mosquito Lagoon, FL | 4% (2, 10%) [c] *N* = 30 | 0 | 3.9% (2, 10%) *N* = 30 | |
| Total (all sites combined) | | 82% (24, 97%) *N* = 41 | 10% (4, 87%) N = 149 | **0.001**[2] |

*CI* confidence interval, -- missing data. [abc]Significant differences between sites in the % Southeast Atlantic alleles, Kruskal-Wallis one-way analysis of variance on Ranks. H = 88, d.f. = 3, $P < 0.001$ with pairwise multiple comparisons between sites using Dunn's method
[1]Differences in the median percent Southeast Atlantic alleles for those with and without "extra" spines, Mann-Whitney U Test. [2]For the combined data, Mann-Whitney U Statistic = 1780.00, T = 5190.00 n(small) = 41, n(big) = 149, $P < 0.001$. [12]Significant differences in bold

there was admixture of alleles from the Southeast Atlantic and Florida East populations, "extra" spines were more likely to be found on individuals of largely Southeast Atlantic ancestry (but a few individuals of largely Florida East ancestry did have "extra" spines; Table 1 and Fig. 3). Although juveniles have longer and sharper spines than adults (Barthel 1974), the frequency of juveniles with "extra" spines did not differ from adults. Observations (HJB unpub. data) of juvenile molted

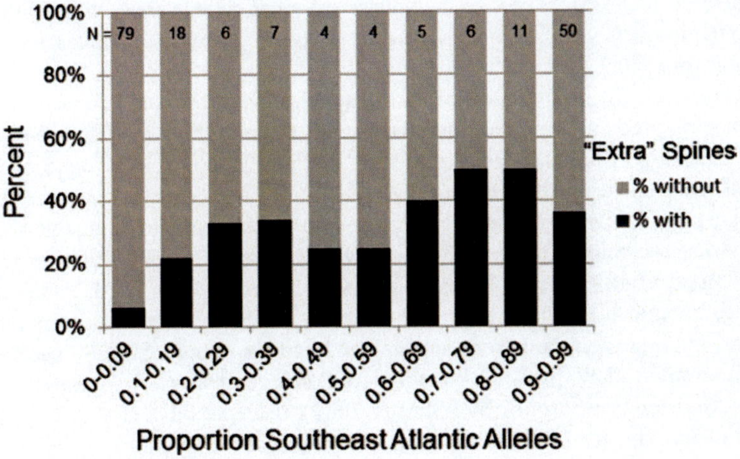


**Fig. 3** The association between genotype and "extra" spines. These data are from two North Florida sites with significant admixture of alleles from the Southeast Atlantic and Florida East populations; New Smyrna Beach, FL; and St. Augustine. N refers to the number of samples analyzed for each grouping of Southeast Atlantic alleles

exoskeletons from Delaware Bay (Primehook Beach) resulted in no individuals with "extra" spines ($N = 100$), and of adults (Slaughter Beach), only one individual had "extra" spines ($N = 145$). Similarly, observations of adults ($N = 136$) and juvenile molts ($N = 153$) from Seahorse Key on the west coast of Florida revealed no individuals with "extra" spines (HJB unpub. data). Taken together, our results show that the "extra" spine trait is a polymorphism found in about half the individuals from the Southeast Atlantic population.

Both at sites where the majority of individuals were of Southeast Atlantic ancestry and at sites where most individuals were of Florida East ancestry, pairing occurred at random with respect to the presence or absence of "extra" spines (Table 3). This is consistent with the view that Var. β is a polymorphism among individuals from the Southeast Atlantic population and that there is no evidence of incipient speciation.

The function of the dorsal spines on horseshoe crabs is unknown, but five hypotheses have been proposed to explain the presence of spines on fossil chelicerates, trilobites, and other arthropods (Park and Choi 2010; Fortey 2014). Those most likely to be relevant to the "extra" spines on horseshoe crabs include (a) protection and (b) locomotor or hydrodynamic advantage. (c) Strongly spinose trilobites may have derived an advantage from camouflage by attracting epibionts or against some backgrounds (Fisher 1979), but juvenile horseshoe crabs possess a layer of slime that prevents epibionts from settling on their exoskeletons (Harrington et al. 2008), and they usually remain buried (Eldredge 1970). (d) It is unlikely that the "extra" spines on horseshoe crabs serve a sensory function since there is no distal perforation or other opening on these spines that might suggest chemoreception (Hayes 1985), and the spines do not move, which means that mechanoreception is unlikely

(Kaplan et al. 1976). (e) Since the spines are most conspicuous on juveniles and since the frequency of spines does not differ between males and females, there is no support for sexual selection as an explanation (Knell and Fortey 2005).

Spines have evolved on a wide range of species where they are known to function in reducing predation by gape-limited predators (Huntingford and Coyle 2007; Fine et al. 2014), so it is reasonable to assume that the "extra" spines found on Var. β horseshoe crabs might function to deter predators. The "extra" spines are longest relative to body size on the smaller juveniles (less than about 9 cm), making it more likely that the spines function to deter juvenile predation. Like trilobites and the early xiphosurans (Beecher 1898; Fisher 1977; Park and Choi 2010; Fortey 2014), young juvenile horseshoe crabs fold ventrally when disturbed with their dorsal spines, making them both very spikey and larger in diameter. Many species that might be deterred by spines are known to feed on juvenile horseshoe crabs including crabs (Shuster 1982; Botton and Shuster 2003), fish (de Sylva et al. 1962), sharks (Botton and Shuster 2003), herons and ibis (personal observation), alligators (Reid and Bonde 1990), and sea turtles (Keinath 2003). Gulls (Botton 1993), hogs (Say 1818), and raccoons (pers. obs.) also feed on horseshoe crabs in shallow water near shore. However, we are not aware of any predators that occur in the Southeast Atlantic region that are not found elsewhere along the coast, and therefore, predation seems a weak explanation for the increased spinosity of the Southeast Atlantic horseshoe crabs.

However, spines may serve to protect juveniles in other ways, such as by affecting their locomotor or hydrodynamic properties or their ability to burrow under the substrate (Eldredge 1970; Fisher 1975, 1977; Park and Choi 2010). Small juveniles are easily swept along or overturned by currents and waves, and dorsal spines may prevent skidding on the prosoma when they try to right themselves (Barthel 1974). Dorsal spines may also prevent back-slip in the sediment while foraging or "bulldozing" forward just under the mud or sand surface, creating their characteristic trails (Barthel 1974). Juveniles spend most of their lives buried in mud or sand (Eldredge 1970), and extra dorsal spines may act to hold the substrate in place so that they remain buried when faced with waves, currents, or other disturbances (pers. comm. Kevin Laurie), thus making them less detectable by predators. Consistent with this hypothesis is the fact that traits that increase crypticity, which are found in a wide range of species, are often polymorphic (Bond 2007), like the "extra" spines. For the locomotor, hydrodynamic, or burying function to be a viable hypothesis, however, there must be some physical differences between the Southeast Atlantic region and elsewhere along the Atlantic coast that would be likely to affect juvenile horseshoe crabs. One difference is the extreme tidal range of the semidiurnal tides and the much faster tidal currents that occur in the South Atlantic Bight compared with mid-Atlantic and Florida coastal regions (Blanton et al. 2004; Lynch et al. 2004).

Although hydrodynamic advantage seems a viable hypothesis, it is not clear why the trait is polymorphic within the Southeast Atlantic population. In general, trait variability can be maintained in populations by a wide array of evolutionary and developmental processes (Bond 2007; Franks and Oxford 2017). For example, in some marine invertebrates, a spiny or hardened defensive morph arises through phenotypic plasticity, induced during development by the presence of predator cues (Lively 1986; Harvell 1990; Sepulveda et al. 2012) or other threats such as wave-related abrasion (Bayer et al. 1997). Frequency-dependent selection acts on the developmental process to maintain the two morphs together in the same population (Gross 1996; Brockmann and Taborsky 2008). We have no information on the developmental plasticity or heritability of "extra" spines or the selective pressures that may favor this polymorphic trait in the Southeast Atlantic population.

We conclude that "Var. β," which Thomas Say thought was a new species *Limulus australis*, is not a separate or cryptic species but a polymorphic or variable character found in the Southeast Atlantic population of *Limulus polyphemus*. The "extra" dorsal spines that are characteristic of this variant are very rarely found in other populations along the Atlantic coast and may arise through developmental variation, mutation, or admixture with Southeast Atlantic horseshoe crabs. Phenotypic variation (polymorphism) within populations of horseshoe crabs is known for other characters including eye shape (Weiner and Chamberlain 1994), behavior (Wasserman 1983; Rudloe 1985; Brockmann and Penn 1992; Penn and Brockmann 1995; Johnson and Brockmann 2012; Anderson et al. 2017), development (Patten 1896), life history patterns (Sasson et al. in prep.), and proteins (Selander et al. 1970). The selective pressures that favor "extra" spines and the Var. β morph in the population are not known, but the extreme tidal inundation and currents found in the South Atlantic Bight are a likely factor because of their suspected effect on juvenile locomotion, righting, and burying behavior.

One particularly interesting aspect of this study is that we have documented stable and consistent genetic and phenotypic (size, "extra" spines) differences between adjacent populations of horseshoe crabs (Tables 1, 2, and 4). These differences occur despite the fact that individuals from the two populations are clearly interbreeding (Table 3), i.e., trait differences are being maintained in the face of considerable gene flow between populations. This means that there is strong local adaptation to maintain the population differences (Tigano and Friesen 2016). Although we do not understand the selective pressures that are acting to maintain these trait differences, our results demonstrate that there is strong adaptation to local environmental conditions in the two populations. Genetic structuring and phenotypic differences between horseshoe crab populations suggest that management efforts should be targeted within regions at a local scale (King et al. 2005, 2015; Smith et al. 2016).

## Literature Cited

R.L. Anderson, W.H. Watson, C.C. Chabot, Local tidal regime dictates plasticity of expression of locomotor activity rhythms of American horseshoe crabs, *Limulus polyphemus*. Mar. Biol. **164**, 63–75 (2017)

J.C. Avise, Molecular population structure and the biogeographic history of a regional fauna: A case history with lessons for conservation biology. Oikos **63**, 62–76 (1992)

K.W. Barthel, *Limulus*: A living fossil. Horseshoe crabs aid interpretation of an upper Jurassic environment (Solnhofen). Naturwissenschaften **61**, 428–433 (1974)

M.M. Bayer, C.D. Todd, J.E. Hoyle, J.F.B. Wilson, Wave-related abrasion induces formation of extended spines in a marine bryozoan. Proc. R. Soc. Lond. B **264**, 1605–1611 (1997)

C.E. Beecher, The origin and significance of spines: A study in evolution. Am. J. Sci. **6**, 1–20 (1898)

B.O. Blanton, F.E. Werner, H.E. Seim, R.A. Luettich Jr., D.R. Lynch, K.W. Smith, G. Voulgaris, F.M. Bingham, F. Way, Barotropic tides in the South Atlantic Bight. J. Geophys. Res. **109**, C12024 (2004) (17 pp)

A. Bond, The evolution of color polymorphism: Crypticity, searching images, and apostatic selection. Annu. Rev. Ecol. Evol. Syst. **38**, 489–514 (2007)

M.L. Botton, Predation by herring gulls and great black-backed gulls on horseshoe crabs. Wilson Bulletin **105**, 518–521 (1993)

M.L. Botton, R.E. Loveland, Body size, morphological constraints, and mated pair formation in four populations of horseshoe crabs (*Limulus polyphemus*) along a geographic cline. Mar. Biol. **112**, 409–415 (1992)

M.L. Botton, C.N. Shuster Jr., Horseshoe crabs in a food web: Who eats whom? in *The American horseshoe crab*, ed. by C. N. Shuster Jr., R. B. Barlow, H. J. Brockmann, (Harvard University Press, Cambridge, 2003), pp. 133–153

H.J. Brockmann, D. Penn, Male mating tactics in the horseshoe crab, *Limulus polyphemus*. Anim. Behav. **44**, 653–665 (1992)

H.J. Brockmann, M. Taborsky, Alternative reproductive tactics and the evolution of alternative allocation phenotypes, in *Alternative Reproductive Tactics. An Integrative Approach*, ed. by R. F. Oliveira, M. Taborsky, H. J. Brockmann, (Cambridge University Press, Cambridge, 2008), pp. 25–51

H.J. Brockmann, M.D. Smith, Reproductive competition and sexual selection in horseshoe crabs, in *Biology and Conservation of Horseshoe Crabs*, ed. by J. T. Tanacredi, M. Botton, D. R. Smith, (Springer, New York, 2009), pp. 199–221

J.E. De Kay, *Zoology of New-York or the New-York Fauna* (Part VI. Crustacea, Carroll and Cook, Printers to the Assembly, Albany, NY, 1844)

D.P. de Sylva, F.A. Kalber, C.N. Shuster Jr., *Fishes and Ecological Conditions in the Shore Zone of the Delaware River Estuary with Notes on Other Species Collected in Deeper Water* (University of Delaware Marine Laboratories, Lewes, Delaware, 1962)

N. Eldredge, Observations on burrowing behavior in *Limulus polyphemus* (Chelicerata, Merostomata), with implications on the functional anatomy of trilobites. Am. Mus. Novitates **2436**, 1–17 (1970)

S. Faurby, K.K. Nielsen, S. Bussarawit, I. Intanai, N. van Cong, C. Pertoldi, P. Funch, Intraspecific shape variation in horseshoe crabs: The importance of sexual and natural selection for local adaptation. J. Exp. Mar. Biol. Ecol. **407**, 131–138 (2011)

M.L. Fine, S. Lahiri, A.D.H. Sullivan, M. Mayo, S.H. Newton, E.N. Sismour, Reduction of the pectoral spine and girdle in domesticated channel catfish is likely caused by changes in selection pressure. Evolution **68**, 2102–2107 (2014)

D.C. Fisher, Swimming and burrowing in *Limulus* and *Mesolimulus*. Fossils Strata **4**, 281–290 (1975)

D.C. Fisher, Functional significance of spines in the Pennsylvanian horseshoe crab *Euproops danae*. Paleobiology **3**, 175–195 (1977)

D.C. Fisher, Evidence for subaerial activity of *Euproops danae* (Merostomata, Xiphosurida), in *Mazon Creek Fossils*, ed. by M. H. Nitecki, (Academic Press, New York, 1979), pp. 379–447

R. Fortey, The palaeoecology of trilobites. J. Zool. **292**, 250–259 (2014)

D.W. Franks, G.S. Oxford, The co-evolution of anti-predator polymorphisms in sympatric populations. Biol. J. Linn. Soc. **122**, 729–737 (2017)

M.R. Gross, Alternative reproductive strategies and tactics: Diversity within sexes. Tr. Ecol. Evol. **11**, 92–97 (1996)

C.L.G. Gunther, *The History of the Collections Contained in the Natural History Departments* (Trustees of the British Museum, London, 1904–1912)

J.M. Harrington, M. Leippe, P.B. Armstrong, Epithelial immunity in a marine invertebrate: A cytolytic activity from a cuticular secretion of the American horseshoe crab, *Limulus polyphemus*. Mar. Biol. **153**, 1165–1171 (2008)

C.D. Harvell, The ecology and evolution of inducible defenses. Q. Rev. Biol. **65**, 323–340 (1990)

W.F. Hayes, Chemoreceptor sensillum structure in *Limulus*. J. Morphol. **119**, 121–142 (1985)

F. Huntingford, S. Coyle, Antipredator defences in sticklebacks: Trade-offs, risk sensitivity, and behavioral syndromes, in *Biology of the Three-Spined Stickleback*, ed. by S. Ostlund-Nilsson, I. Mayer, F. Huntingford, (CRC Press, New York, 2007), pp. 127–156

S.L. Johnson, H.J. Brockmann, Alternative reproductive tactics in female horseshoe crabs. Behav. Ecol. **23**, 999–1008 (2012)

E. Kaplan, R.B. Barlow Jr., S.C. Chamberlain, D.J. Stelzner, Mechanoreceptors on the dorsal carapace of *Limulus*. Brain Res. **109**, 615–622 (1976)

J.A. Keinath, Predaton of horseshoe crabs by loggerhead turtles, in , ed. by C. N. Shuster Jr., R. B. Barlow, H. J. Brockmann, (The American horseshoe crab. Harvard University Press, Cambridge, 2003), pp. 152–153

T.L. King, M.S. Eackles, A.W. Aunins, H.J. Brockmann, E.M. Hallerman, B.B. Brown, Conservation genetics of the American horseshoe crab (*Limulus polyphemus*): Allelic diversity, zones of genetic discontinuity, and regional differentiation, in *Changing Global Perspectives on Biology, Conservation and Management of Horseshoe Crabs*, ed. by R. H. Carmichael, M. L. Botton, P. K. S. Shin, S. G. Cheung, (Springer, New York, 2015), pp. 65–96

T.L. King, M.S. Eackles, A.P. Spidle, H.J. Brockmann, Regional differentiation and sex-biased dispersal among populations of the horseshoe crab *Limulus polyphemus*. Trans. Am. Fish. Soc. **134**, 441–465 (2005)

R.J. Knell, R.A. Fortey, Trilobite spines and beetle horns: Sexual selection in the Palaeozoic. Biol. Lett. **1**, 196–199 (2005)

C.M. Lively, Predator-induced shell dimorphism in the acorn barnacle *Chthamalus anisopoma*. Evolution **40**, 232–242 (1986)

D. Lynch, K. Smith, B. Blanton, F. Werner, R. Luettich, Forecasting the coastal ocean: Resolution, tide and operational data in the South Atlantic Bight. J. Atmos. Oceanic Tech. **21**, 1074–1085 (2004)

G. Mayer, The effects of temperature upon tropical marine animals. Pap Tortugas Lab, Carnegie Inst Publ **183**, 1–14 (1914)

P.L. Norby, G. Paulay, H.J. BrockmannFine-scale Analysis of a Genetic Discontinuity in the American Horseshoe Crab (*Limulus polyphemus*) (in prep)

P.L. Norby Genetic and phenotypic analyses of the American horseshoe crab (*Limulus polyphemus*) across the northeast Florida phylogenetic break. MS thesis, University of Florida (2017)

T.-Y. Park, D.K. Choi, Two middle Cambrian diceratocephalid trilobites, *Cyclolorenzella convexa* and *Diceratocephalus cornutus*, from Korea: Development and functional morphology. Lethaia **43**, 73–87 (2010)

W. Patten, Variations in the development of *Limulus polyphemus*. J. Morphol. **12**, 17–148 (1896)

W. Patten, *The Evolution of the Vertebrates and Their Kin* (P Blakiston's Son & Co, Philadelphia, 1912)

D. Penn, H.J. Brockmann, Nest-site selection in the horseshoe crab, *Limulus polyphemus*. Biol. Bull. **187**, 373–384 (1994)

D. Penn, H.J. Brockmann, Age-biased stranding and righting in horseshoe crabs, *Limulus polyphemus*. Anim. Behav. **49**, 1531–1539 (1995)

J.C. Pierce, G. Tan, P.M. Gaffney, Delaware Bay and Chesapeake Bay populations of the horseshoe crab *Limulus polyphemus* are genetically distinct. Estuaries **23**, 690–698 (2000)

C.M. Porter, Following Bartram's "Track": Titian Ramsay Peale's Florida journey. Fl. Hist. Q. **61**, 431–444 (1983)

J.P. Reid, R.K. Bonde, *Alligator mississippiensis* (American alligator) diet. Herp. Rev. **21**, 59 (1990)
B. Riska, Morphological variation in the horseshoe crab *Limulus polyphemus*. Evolution **35**, 647–658 (1981)
A. Rudloe, Variation in the expression of lunar and tidal behavioral rhythms in the horseshoe crab, *Limulus polyphemus*. Bull. Mar. Sci. **36**, 388–395 (1985)
D.A. Sasson, H.J. Brockmann, Geographic variation in sperm and ejaculate quantity and quality of horseshoe crabs. Behav. Ecol. Sociobiol. **70**, 1715–1725 (2016)
D.A. Sasson, S.L. Johnson, M.D. Smith, H.J. Brockmann. Seasonal variation in horseshoe crab spawning (*Limulus polyphemus*) (in prep)
N.C. Saunders, L.G. Kessler, J.C. Avise, Genetic variation and geographic differentiation in mitochondrial DNA of the horseshoe crab, *Limulus polyphemus*. Genetics **112**, 613–627 (1986)
T. Say, An account of the crustaceans of the United States. J Acad Nat Sci Philadelphia **1**, 423–458 (1818)
R.K. Selander, S.Y. Yang, R.C. Lewontin, W.E. Johnson, Genetic variation in the horseshoe crab (*Limulus polyphemus*), a phylogenetic "relic". Evolution **24**, 402–414 (1970)
R.D. Sepulveda, C.G. Jara, C.S. Gallardo, Morphological analysis of two sympatric ecotypes and prdator-induced phenotypic plasticity in *Acanthina monodon* (Gastropoda: Muricidae). J. Moll. Stud. **78**, 173–178 (2012)
C.N. Shuster Jr., A horseshoe "crab" grows up. Ward's Nat Sci Bull **28**, 1–6 (1954)
C.N. Shuster Jr., *On morphometric and serological relationships within the Limulidae, with particular reference to Limulus polyphemus (L)*. PhD thesis (New York University, 1955)
C.N. Shuster Jr., Distribution of the American horseshoe "crab", *Limulus polyphemus* (L), in *Biomedical Applications of the Horseshoe Crab (Limulidae)*, ed. by E. Cohen, (Alan R Liss, Inc., New York, 1979), pp. 3–26
C.N. Shuster Jr., A pictorial review of the natural history and ecology of the horseshoe crab *Limulus polyphemus*, with reference to other Limulidae, in *Physiology and Biology of Horseshoe Crabs: Studies on Normal and Environmentally Stressed Animals*, ed. by J. Bonaventura, C. Bonaventura, S. Tesh, (Alan R Liss Inc, New York, 1982), pp. 1–52
C.N. Shuster Jr., The Delaware Bay area, USA: A unique habitat of the American horseshoe crab, *Limulus polyphemus*, in *Changing Global Perspectives on Biology, Conservation and Management of Horseshoe Crabs*, ed. by R. H. Carmichael, M. L. Botton, P. K. S. Shin, S. G. Cheung, (Springer, New York, 2015), pp. 15–39
D.R. Smith, H.J. Brockmann, M.A. Beekey, T.L. King, M.J. Millard, J. Zaldívar-Rae, Conservation status of the American horseshoe crab (*Limulus polyphemus*): A regional assessment. Rev. Fish. Biol. Fisheries **26**, 1–41 (2016)
M.D. Smith, H.J. Brockmann, The evolution and maintenance of sexual size dimorphism in horseshoe crabs: An evaluation of six functional hypotheses. Anim. Behav. **96**, 127–139 (2014)
A. Sokoloff, Observations on populations of the horseshoe crab *limulus* (=Xiphosura) *polyphemus*. Res. Popul. Ecol. **19**, 222–236 (1978)
P.T. Stroud, *Thomas Say, New World Naturalist* (University of Pennsylvania Press, Philadelphia, 1992)
A. Tigano, V.L. Friesen, Genomics of local adaptation with gene flow. Mol. Ecol. **25**, 2144–2164 (2016)
M.C. Vasquez, H.J. Brockmann, D. Julian, Between-population differences in multi-stressor tolerance during embryo development in the American horseshoe crab, *Limulus polyphemus*. Estuar. Coasts **40**, 1487–1501 (2017)
G.S. Wasserman, Daily and seasonal variations in *limulus* visual behavior. J. Interdiscipl. Cycle Res. **14**, 173–194 (1983)
H.B. Weiss, G.M. Ziegler, *Thomas Say, Early American Naturalist* (Charles C Thomas Publisher, Springfield, 1931)
W.W. Weiner, S.C. Chamberlain, The visual fields of American horseshoe crabs - 2 different eye shapes in *Limulus polyphemus*. Vis. Neurosci. **11**, 333–346 (1994)

# Genetic Structure of the Tri-Spine Horseshoe Crab in the Seto Inland Sea, Japan: Is the Current Population at Kasaoka in the Eastern Area Native or Re-established?

**Shinichi Watanabe, Jun J. Sato, Keiki Kawazu, Eri Horinouchi, Shin Nishida, Koujirou Azumakawa, Satoshi Morinobu, and Norimichi Souji**

## 1 Introduction

The tri-spine horseshoe crab *Tachypleus tridentatus* is distributed from coastal Southeast Asia to western Japan (Sekiguchi and Shuster Jr 2009). The Japanese populations represent the northernmost range of this species. The currently recognized spawning sites in Japan are intermittently observed along the coastal region of northern Kyushu and in the Seto Inland Sea (Nishida et al. 2015; Laurie et al. 2019). During the past half century, the population size of each habitat had decreased rapidly and drastically, due to loss of tidal flats and spawning beaches and the deterioration of coastal waters largely caused by human activity (Sekiguchi 1988; Itow 1993; Laurie et al. 2019). Most habitats in the Seto Inland Sea were destroyed during the period of rapid economic growth after World War II. Until the 1970s, Kasaoka City in Okayama Prefecture (see Fig. 1) had one of the largest populations of *T. tridentatus* in Japan. One of the spawning sites at Kasaoka was designated as a natural reserve in 1928 to protect horseshoe crabs. However, most of the bay area

S. Watanabe (✉) · K. Kawazu · E. Horinouchi
Department of Marine Bio-Science, Faculty of Life Science and Biotechnology, Fukuyama University, Hiroshima Prefecture, Hiroshima, Japan

J. J. Sato
Department of Biotechnology, Faculty of Life Science and Biotechnology, Fukuyama University, Hiroshima Prefecture, Hiroshima, Japan

S. Nishida
Biology, Science Education, Faculty of Education, University of Miyazaki, Miyazaki Prefecture, Miyazaki, Japan

K. Azumakawa · S. Morinobu · N. Souji
Kasaoka City Horseshoe Crab Museum, Okayama Prefecture, Okayama, Japan



J. T. Tanacredi et al. (eds.), *International Horseshoe Crab Conservation and Research Efforts: 2007-2020*, https://doi.org/10.1007/978-3-030-82315-3_2

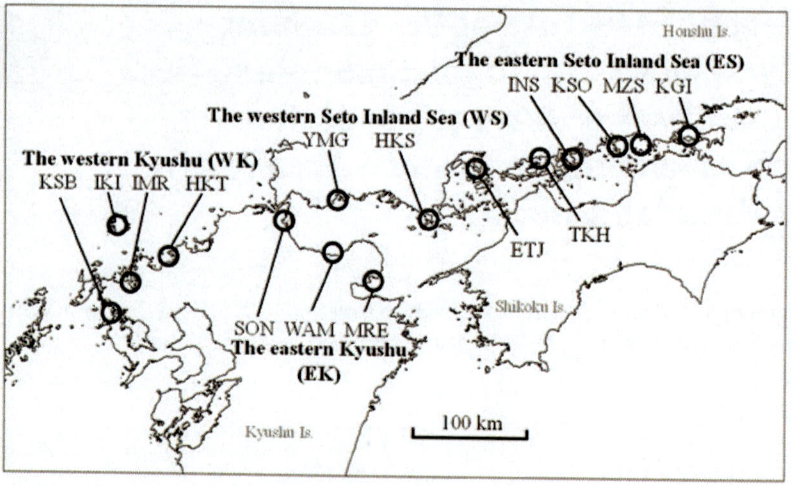


**Fig. 1** Locations of 13 sampling sites of *Tachypleus tridentatus* divided into four geographic groups in western Japan. KGI Kugui, MZS Mizushima, KSO Kasaoka, INS Innoshima Is, ES in the eastern Seto Inland Sea, HKS Hirao Bay and Kaminoseki YMG Yamaguchi Bay, WS in the western Seto Inland Sea, SON Sone tidal flat, WAM Wama tidal flat, MRE Morie Bay, EK in the eastern Kyushu, HKT Hakata Bay, IMR Imari Bay, IKI Iki Is, KSB Kujukushima-Sasebo Bay, WK in the western Kyushu, TKH locations of two populations between ES and EW, ETJ Takehara and Etajima studied in Ohtsuka et al. (2017)

(1800 hectares) was drained by the large-scale land reclamation from 1969 (Sekiguchi 1988), which was turned into a narrow waterway later called the Konoshima Channel. After the completion of land reclamation, almost all coastal areas suitable for horseshoe crab adults to spawn and juveniles to grow disappeared. Furthermore, water pollution had become worse after the land reclamation (Itow 1993), probably due to a loss of the self-purification function of tidal flat (Souji 2015). Tsuchiya and Asano (1993) reported annual changes of the numbers of egg clusters and juveniles of horseshoe crabs found at the Konoshima Channel after land reclamation. Their study showed that after the completion of land reclamation, the number of egg clusters had significantly decreased from 2232 clusters in 1979 to 60 clusters in 1982. The numbers of juvenile horseshoe crabs in tidal flats also had decreased drastically; no juvenile was found from 1980 to 1989. The breeding population of native horseshoe crabs at Kasaoka disappeared in the early 1990s.

In the meantime, Kasaoka City and some associations with local volunteers had worked to conserve and recover the horseshoe crab population. In 1975, Kasaoka City established Horseshoe Conservation Center (later developed into Kasaoka City Horseshoe Crab Museum) which started studies on the recovery of population together with local volunteers (Tsuchiya 2009; Souji 2015). One of the associations "Kabutogani wo Aisurukai" meaning a community of people who love horseshoe crabs had conducted a reintroduction program which released a total of 254 sub-adult and adult horseshoe crabs to the Konoshima Channel at four times between

1995 and 1998 (Souji 1997; Kasaoka City Board of Education 2001). All individuals reintroduced were caught by local fishery near Yamaguchi Bay (the western Seto Inland Sea in Fig. 1). For several years after the reintroduction program, spawning eggs of horseshoe crabs had been recognized again in the channel. Although the number of egg clusters found was 60 in 1996, it decreased to two in 1997 and remained at a low level from 1998 to 2004 (Souji and Morinobu 2003a). Then, no spawning activity was observed from 2004 to 2010 (Souji and Morinobu 2004, 2005, 2006, 2007, 2008, 2009).

For another reintroduction program, Kasaoka City Horseshoe Crab Museum developed techniques to rear thousands of juveniles that were to be released to tidal flats in the Channel (Souji and Morinobu 2000, 2002, 2003b). Juvenile horseshoe crabs were raised in captivity from fertilized eggs obtained from adult pairs, most of which were caught in the sea off Kasaoka. Exceptionally, one or two adult pairs caught in Iki (the western Kyushu in Fig. 1) were used for breeding in each year between 2004 and 2014, while enough adult pairs were not obtained in the sea off Kasaoka. Over a thousand of juveniles raised to 2–7 instar stages were released to the tidal flats every year since 2000 (Souji and Morinobu 2000, 2002, 2003b; Souji 2015). A decade after the juvenile releasing program, 2–21 egg clusters had been recognized again from the Channel in each year between 2009 and 2015 (Souji 2015). During the same period, number of subadult and adult horseshoe crabs caught as fisheries bycatch in the sea off Kasaoka had increased: 5–41 individuals in each year (Souji 2015). It is likely that the juvenile releasing program had contributed to the recovery of the local population of *T. tridentatus* in Kasaoka area.

According to the IUCN/SSC (2013) guidelines, reintroduction of species aims to re-establish a viable population within its indigenous range while assuring conservation benefit. However, it also carries risks from ecological, social, and economic perspectives. Thus, reintroduction should have a feasible design depending on the situation of focal species. For one of the essential points, founder selection should be aimed to provide adequate genetic diversity. Populations within species may sometimes be sufficiently differentiated in terms of adaptive characteristics or genetic composition, requiring separate and specialized management (Frankham et al. 2010). If founders from distantly related populations are mixed, there may be genetic incompatibilities (IUCN/SSC 2013). When reintroducing horseshoe crabs at Kasaoka, the genetic composition of the source populations was not considered; breeding individuals were translocated from other populations. To preserve the local genetic diversity, introduced individuals should not be from other genetically different populations.

Molecular genetic analysis is a powerful tool for investigating genetic structure, diversity, and differentiation in individuals and groups. Nishida et al. (2015) described the genetic structure of *T. tridentatus* populations in Japan using mitochondrial (mt) DNA AT-rich region and microsatellite DNA markers. The AT-rich region (control region) is a highly variable, noncoding region that is useful for phylogeographic and population genetic surveys of *T. tridentatus* (Yang et al. 2007; Nishida and Koike 2009; Weng et al. 2013; Nishida et al. 2015). Nishida et al. (2015) identified nine haplotypes (AT1 to AT9: GenBank accession Nos. AB353283 to AB353289, LC002283 and LC002284) in Japanese populations based on the sequence of the AT-rich region. All haplotypes were closely connected by mostly a single substitution. The dominant haplotype (AT1) was observed in all localities examined in Nishida et al. (2015). The eastern group from the western Seto Inland Sea (WS), the eastern Kyushu (EK), and Hakata (HKT) in the western Kyushu (WK) (Fig. 1) was monomorphic for AT1, resulting in a low haplotype diversity. On the other hand, the western group from WK excluding HKT (Fig. 1) was comprised of 4–5 haplotypes, and consequently higher haplotype diversity than in the eastern group. Nishida et al. (2015) suggested that the far eastern habitat in the Seto Inland Sea (the eastern Seto Inland Sea: ES in Fig. 1) exhibited differentiations, but they could not perform an analysis in detail because population size was very small, and it was difficult to collect enough native individuals for analysis. In this study, we examined genetic structure of *T. tridentatus* in the Seto Inland Sea including the eastern population. From the results, we discuss the origin of the current population in the eastern area (at Kasaoka in particular). Furthermore, we would like to offer our opinion to improve the conservation of the horseshoe crab population in the eastern Seto Inland Sea.

## 2 Materials and Methods

We analyzed mtDNA of *Tachypleus tridentatus* from 13 localities in four geographic groups in the western Japan: (1) the eastern and (2) western Seto Inland Sea (ES and WS), and off the coast of the (3) eastern and (4) western Kyushu (EK and WK) (Fig. 1). We obtained blood samples of 5 subadult and 46 adult horseshoe crabs from five localities in ES and one locality (Kaminoseki) in WS (Fig. 1). The animals were caught in fixed fishing nets or trawl fishing as bycatch for local fisheries. Blood sampling was carried out according to the procedures in Nishida and Koike (2009). Total genomic DNA was extracted from the blood samples using the conventional phenol–chloroform method (Sambrook and Russell 2001). In addition to the above samples, we also used 111 samples analyzed in Nishida et al. (2015), all of which were identified as a haplotype AT1. Almost all the samples were extracted from blood of individuals of various ages living in WS, EK, and WK (see Table 6.1 in Nishida et al. 2015).

All PCR reactions were performed with the KOD-Plus-NEO DNA Polymerase Kit (Toyobo, Osaka, Japan) in an automated thermal cycler (Life Touch thermal cycler, Bioer Technology, Hangzhou, China). The PCR reaction mixture consisted of KOD-Plus-NEO buffer, 1.5 mM $MgSO_4$, 0.2 mM of each dNTP, 0.2 μM of forward and reverse primers, 1.0 U of KOD-Plus-Neo DNA polymerase, and 0.1–0.2 μg of template total genomic DNA to a total volume of 50 μL. We used a primer set: Hb-12S and Hb-trna designed by Yang et al. (2007), which amplified a total fragment of 686 base pairs (bp) including the complete AT-rich region (369 bp) with a partial sequence of the flanking region of 12S rRNA (183 bp) and tRNA genes (134 bp). The PCR condition was as follows: a 2-min denaturation period at 94 °C,

35 cycles of denaturation at 98 °C for 15 s, annealing at 50 °C for 30 s, extension at 68 °C for 45 s, and a 10-min extension period at 68 °C. Direct sequencing was carried out using a BigDye Terminator v3.1 Cycle Sequencing Kit (Applied Biosystems™), according to the manufacturer's protocol. The nucleotide sequences were determined from both ends by direct sequencing of the PCR products with an ABI 3130 Genetic Analyzer (Applied Biosystems™). Alignment of the sequence data was performed manually using MEGA X (Kumar et al. 2018).

Genetic variations in the Seto Inland Sea were expressed as gene (haplotype) diversity (*h*, Nei 1987), which was estimated from the frequency distribution of haplotypes in each of three geographic groups within the Seto Inland Sea (ES, WS, and EK). The diversity of WK was not estimated because we used only samples consisted of a haplotype AT1 excluding other AT series normally found in WK. The analysis of molecular variance (AMOVA, Excoffier et al. 1992) was performed to differentiate molecular variance among the geographic groups. To evaluate genetic differentiation, pairwise $\Phi_{PT}$ values and *Nei's* genetic distance between the geographic groups were calculated by GenAlEx 6.5 (Peakall and Smouse 2012). The $\Phi_{PT}$ value, an $F_{ST}$ analog, suppresses within-population variance and simply calculates population differentiation based on the genotypic variance. The probability values estimated by 1000 permutations were used to determine whether the partitioning of variance components was significant.

## 3 Results

We identified only one haplotype AT1 in all of 162 samples in the sequence of the AT-rich region. However, we found variable sites with nucleotide length polymorphisms by single base repeat numbers in the flanking region of 12S rRNA located in the upstream of the AT-rich region, which could subdivide AT1 into five haplotypes (ST1 to 5, Table 1). These haplotypes newly obtained in this study were deposited in the DDBJ/Genbank/ENA international DNA database with accession numbers LC523832 to LC523836.

**Table 1** Variable sites of haplotypes: ST1 to 5 with nucleotide length polymorphisms by single base repeat numbers in the flanking region of 12S rRNA located in the upstream of the AT-rich region of mtDNA in the Japanese populations of *Tachypleus tridentatus*

| Nucleotide position | | | | |
|---|---|---|---|---|
| Haplotype | −103 | −102 | −101 | −40 |
| ST01 | A | A | A | – |
| ST02 | • | • | – | • |
| ST03 | • | – | – | • |
| ST04 | – | – | – | • |
| ST05 | – | – | – | A |

Nucleotide positions are coordinated to the AT-rich region of Nishida and Koike (2009). A dot indicates the same nucleotide as in the ST01. Dashes indicate gaps in the alignment

Among the five haplotypes, ST3 was the most dominant haplotype in all four geographic groups (68% in all samples, Table 1). The frequency of ST3 in ES was lowest (45%) whereas those in other regions were higher: 85% in WS, 71% in EK, and 80% in WK. Instead, the frequency of ST4 is highest (37%) in ES, compared with the other regions (15% in WS, 18% in EK, and 20% in WK). ST3 and 4 were found in all four regions; whereas ST2 was found in ES and EK, ST1 and 5 were found only in ES. The haplotype diversity (*h*) was highest in ES (0.6624), followed by EK (0.4630) and WS (0.2585).

From the results of AMOVA evaluated genetic differentiation among the groups, significant $\Phi_{PT}$ values were detected between ES and WS (0.168, $P < 0.001$), and between ES and EK (0.133, $P < 0.05$); while it was not significant between WS and EK (0.012, $P = 0.223$). A similar pattern of differentiation among the groups was observed using *Nei's* genetic distance; those were larger between ES and WS (0.135), and between ES and EK (0.100), than between WS and EK (0.002).

## 4 Discussion

According to the previous studies that examined the AT-rich region of *T. tridentatus*, the populations in the Seto Inland Sea may have been strongly influenced by the founder effect, thereby monomorphic for AT1 (Nishida and Koike 2009; Nishida et al. 2015). In this study, we also identified only AT1 in the AT-rich region of the populations in the Seto Inland Sea. However, we found variable sites in the flanking region of 12S rRNA, which could subdivide AT1 into the novel five haplotypes (ST1 to 5). Although it was difficult to identify the genealogy of obtained haplotypes with such nucleotide length polymorphisms only, we presume that the most dominant haplotype ST3 in all four geographic groups was ancestral and evolved to other four haplotypes. The other haplotypes ST2 and 4 were found in two and four geographic groups, respectively, whereas ST1 and 5 were found only in the eastern Seto Inland Sea (ES); resulting in the high genetic diversity in ES. The results of AMOVA and genetic distances between the geographic groups also indicated the genetic differentiation between the populations of the eastern and western Seto Inland Sea (WS and ES). Our results thus suggest that the horseshoe crab population in the eastern Seto Inland Sea may have a unique genetic constitution reflecting independent evolutionary history from other geographic groups. It should, however, be noted that our preliminary study provides only a small fraction of the mtDNA, leaving a concern that we did not fully capture the genetic diversity. Nishida et al. (2015) used nuclear compound microsatellite markers as well as mtDNA markers and detected genetic subdivisions in Japanese populations of *T. tridentatus* other than the eastern Seto Inland Sea. We are planning to apply such a genome-wide biparentally genetic markers to re-evaluate the genetic structure of the populations in the Seto Inland Sea and to try to detect the genetic difference of the eastern genetic group (Table 2).

**Table 2** Distribution of the haplotypes: ST1 to 5 for Japanese populations of *Tachypleus tridentatus* in 13 localities divided into four geographic groups

| Local population | Haplotype | | | | | |
|---|---|---|---|---|---|---|
| | ST1 | ST2 | ST3 | ST4 | ST5 | Total |
| Eastern Seto Inland Sea (ES): | 1 | 5 | 22 | 18 | 3 | 49 |
| Kugui (KGI) | | | | | 1 | 1 |
| Mizushima (MZS) | | | 2 | | | 2 |
| Kasaoka (KSO) | 1 | 5 | 19 | 18 | 2 | 45 |
| Innoshima is. (INS) | | | 1 | | | 1 |
| Western Seto Inland Sea (WS): | | | 29 | 5 | | 34 |
| Hirao-Kaminoseki (HKS) | | | 21 | 5 | | 26 |
| Yamaguchi (YMG) | | | 8 | | | 8 |
| Eastern Kyushu (EK): | | 3 | 20 | 5 | | 28 |
| Sone (SON) | | 2 | 9 | 4 | | 15 |
| Wama (WAM) | | 1 | 6 | | | 7 |
| Morie (MRE) | | | 5 | 1 | | 6 |
| Western Kyushu (WK): | | | 41 | 10 | | 51 |
| Hakata (HKT) | | | 8 | 2 | | 10 |
| Imari (IMR) | | | 14 | 4 | | 18 |
| IKI is. (IKI) | | | 13 | | | 13 |
| Kujukushima-Sasebo (KSB) | | | 6 | 4 | | 10 |
| Total | 1 | 8 | 112 | 38 | 3 | 162 |

In the introduction described above, we explained the history of horseshoe crab population in ES (mainly Kasaoka). The breeding population disappeared in the early 1990s; it has gradually been recovered in the last decade presumably due to the reintroduction program of juveniles released to tidal flats. At the early stage of the program, adult and subadult individuals were translocated from WS: Yamaguchi Prefecture (Kasaoka City Board of Education 2001). In addition, some adult pairs from WK were used for breeding juvenile. If these individuals were the founders for current population in ES, the haplotypes found in ES were also common in the source populations. Contrary to this assumption, the genetic constitution in ES was different from the other regions and the unique haplotypes (ST1 and 5) in ES were not found in other populations. We suspect two possible reasons why the unique haplotypes were observed only in ES.

First, the native population of horseshoe crabs with endemic genetic structure may have existed in ES; it has been recovered independently from the reintroduction program. Although eggs and juveniles disappeared at Kasaoka in the 1980s (Tsuchiya and Asano 1993), subadult and adult individuals had still been caught in the sea off Kasaoka: 21–76 individuals in each year from 1984 to 1992 (Sekiguchi 1993). It is, thus, likely that the native endemic population might have survived in the sea off Kasaoka, but the native individuals could have not found suitable sites for spawning during the period. For the past two decades, the water quality in the Konoshima Channel had gradually been improved with the spread of sewerage in

Kasaoka City; various marine animals including horseshoe crabs came back to the Channel (Souji 2015). There is a possibility that the survivors were able to find suitable spawning sites in the Channel during this decade.

Second, the unique haplotypes in ES may have been originated from the founders translocated from other populations. It is probable that the unique haplotypes were rare and recently became extinct in the source populations. To reveal the origin of unique haplotypes found in ES, further research is required to examine genetic structure of populations in the whole Seto Inland Sea with more samples. In particular, we are interested in local populations on the border of eastern and western Seto Inland Sea: e.g., Takehara and Etajima in Hiroshima Prefecture (Fig. 1). These local populations also seem to have recently recovered and are genetically uniform with a haplotype AT1 (Ohtsuka et al. 2017). Examinations using the genetic polymorphisms (ST1 to 5) newly found in this study would contribute to revealing the history of this species in the Seto Inland Sea. It could, however, be argued that the former assumption is more plausible; the current population at Kasaoka includes native and endemic genotypes. Because the local populations in Hiroshima, where reintroduction program had not been conducted, also seem to have recovered during this decade (Ohtsuka et al. 2017). It also indicates that marine environments in the Seto Inland Sea have recently been improved, thereby the horseshoe crab populations have recovered in the eastern Seto Inland Sea.

In either case, currently the unique haplotypes of *T. tridentatus* may exist only in the eastern Seto Inland Sea. Thus, it is critically important to preserve the genetic diversity of this species in this area. In this study, genetic differentiation was found between the eastern and western populations in the Seto Inland Sea. Therefore, we should not translocate individuals from genetically different populations, even from Yamaguchi Prefecture to Kasaoka, and vice versa. In the current status of the horseshoe crab population at Kasaoka, it is likely that the breeding population was established and has been maintained with a small number. At the current situation, to increase the number of breeding individuals, we suggest that it is essential to preserve and improve habitats of spawning beaches and tidal flats where juveniles could grow up rather than increasing the effort for the reintroduction. For these purposes, we have conducted monitoring surveys for counting egg clusters in spawning beaches (Souji 2015) and for estimations of mortality and growth rates of juveniles released to the tidal flats (Watanabe et al. submitted). Our results in the present and future studies will provide useful information for considering a reintroduction program in restoring population of horseshoe crabs.

**Acknowledgments** We thank S. Oyamada, K. Mizuta, and other members in Laboratory of Marine Animal Behavior, Fukuyama University, M. Mizukami in the Research Institute of Marine Bioresources in Fukuyama University, H. Kohama and M. Takashima in the Kaminoseki Nature Conservation Association, for their support and collection of samples. We also thank Y. Iwasaki for valuable comments on this manuscript. This work was partially supported by the Japan Society for the Promotion of Science (JSPS) research grants (23710286), Research Center for Green Science at Fukuyama University, and the Private University Research Branding Project from Ministry of Education, Culture, Sports, and Technology of Japan.

## References


L.G. Excoffier, P.E. Smouse, J.M. Quattro, Analysis of molecular variance inferred from metric distances among DNA haplotypes: Application to human mitochondrial DNA restriction data. Genetics **131**, 479–491 (1992)

R. Frankham, J.D. Ballou, D.A. Briscoe, *Introduction to Conservation Genetics*, 2nd edn. (Cambridge University Press, Cambridge, 2010)

T. Itow, Crisis in the Seto Inland Sea: The decimation of the horseshoe crab. EMECS Newsl **3**, 10–11 (1993)

IUCN/SSC, *Guidelines for reintroductions and other conservation translocations ver. 1.0* (IUCN Species Survival Commission, Gland, Switzerland, 2013)

Kasaoka City Board of Education, *Conservation Guideline for the Natural Monument: The Breeding Site of Horseshoe Crabs* (Kasaoka City Board of Education, Okayama, 2001) (in Japanese)

S. Kumar, G. Stecher, M. Li, et al., MEGA X: Molecular evolutionary genetics analysis across computing platforms. Mol. Biol. Evol. **35**, 1547–1549 (2018)

K. Laurie, C.-P. Chen, S.G. Cheung et al., *Tachypleus tridentatus*. The IUCN Red List of Threatened Species 2019: e.T21309A133299524 (2019)

M. Nei, *Molecular Evolutionary Genetics* (Columbia University Press, New York, 1987)

S. Nishida, H. Koike, Genetic structure of Japanese populations of *Tachypleus tridentatus* by mtDNA AT-rich region sequence analysis, in *Biology and Conservation of Horseshoe Crabs*, ed. by J. Tanacredi, M. Botton, D. Smith, (Springer, Heidelberg, 2009), pp. 183–196

S. Nishida, K. Kuroyanagai, H. Koike, Genetic features of *Tachypleus tridentatus* in Japan and an alien (non-native) population founded at Ise-Mikawa Bay, Chubu Region, central Japan, in *Changing Global Perspectives on Horseshoe Crab, Biology, Conservation and Management*, ed. by R. H. Carmichael, M. L. Botton, P. K. S. Shin, S. G. Cheung, (Springer, Heidelberg, 2015), pp. 129–146

S. Ohtsuka, N. Nishihara, R. Hirayama, et al., Occurrence of endangered tri-spine horseshoe crab in Etajima and Takehara Cities of Hiroshima Prefecture, western Japan. JPN J. Benthol. **72**, 16–26 (2017) (in Japanese with English abstract)

R. Peakall, P.E. Smouse, GenAlEx 6.5: Genetic analysis in Excel. Population genetic software for teaching and research – An update. Bioinformatics **28**, 2537–2539 (2012)

J. Sambrook, D.W. Russell, *Molecular Cloning: A Laboratory Manual*, 3rd edn. (Cold Spring Harbor Laboratory Press, ColdSpring Harbor, New York, 2001)

K. Sekiguchi, *Biology of Horseshoe Crabs* (Science House Co., Tokyo, 1988)

K. Sekiguchi, Japanese horseshoe crabs since 1989, in *Present Situation of Horseshoe Crabs in Japan*, ed. by K. Sekiguchi, 2nd edn., (Research club to protect horseshoe crab in Japan, Okayama, 1993), pp. 179–223. (in Japanese)

K. Sekiguchi, C.N. Shuster Jr., Limits on global distribution of horseshoe crabs (Limulacea): Lessons learned from two lifetimes of observations: Asia and America, in *Biology and Conservation of Horseshoe Crabs*, ed. by J. T. Tanacredi, M. L. Botton, D. R. Smith, (Springer, New York, 2009), pp. 5–24

N. Souji, Follow-up survey in 1996 after the released program of horseshoe crabs, in *A Report of Research on Horseshoe Crab*, vol. 6, (1997), pp. 15–21. (in Japanese)

N. Souji, *Mystery of Horseshoe Crabs: Why they Have Lived for 200 Million Years Without Change in the Form* (Seibundo Shinkosha Publishing, Tokyo, 2015) (in Japanese)

N. Souji, S. Morinobu, Follow-up survey in 1999 after the released program of juvenile horseshoe crabs, in *A Report of Research on Horseshoe Crab*, vol. 7, (2000), pp. 50–60. (in Japanese)

N. Souji, S. Morinobu, Results of survey on juvenile horseshoe crabs in 2000 and 2001, in *A Report of Research on Horseshoe Crab*, vol. 8, (2002), pp. 14–19. (in Japanese)

N. Souji, S. Morinobu, Situation of spawning of horseshoe crabs in 2002, in *A Report of Research on Horseshoe Crab*, vol. 9, (2003a), pp. 7–9. (in Japanese)

N. Souji, S. Morinobu, Results of survey on juvenile horseshoe crabs in 2002, in *A Report of Research on Horseshoe Crab*, vol. 9, (2003b), pp. 10–13. (in Japanese)
N. Souji, S. Morinobu, Situation of spawning of horseshoe crabs in 2003, in *A Report of Research on Horseshoe Crab*, vol. 10, (2004), pp. 5–6. (in Japanese)
N. Souji, S. Morinobu, Situation of spawning of horseshoe crabs in 2004, in *A Report of Research on Horseshoe Crab*, vol. 11, (2005), pp. 6–7. (in Japanese)
N. Souji, S. Morinobu, Situation of spawning of horseshoe crabs in 2005, in *A Report of Research on Horseshoe Crab*, vol. 12, (2006), pp. 8–9. (in Japanese)
N. Souji, S. Morinobu, Situation of spawning of horseshoe crabs in 2006, in *A Report of Research on Horseshoe Crab*, vol. 13, (2007), pp. 8–9. (in Japanese)
N. Souji, S. Morinobu, Situation of spawning of horseshoe crabs in 2007, in *A Report of Research on Horseshoe Crab*, vol. 14, (2008), pp. 9–10. (in Japanese)
N. Souji, S. Morinobu, Situation of spawning of horseshoe crabs in 2008, in *A Report of Research on Horseshoe Crab*, vol. 15, (2009), pp. 12–13. (in Japanese)
K. Tsuchiya, The history of horseshoe crab research and conservation in Japan, in *Biology and Conservation of Horseshoe Crabs*, ed. by J. T. Tanacredi, M. L. Botton, D. R. Smith, (Springer, New York, 2009), pp. 559–570
K. Tsuchiya, M. Asano, Transition and present situation of spawning sites and habitats of horseshoe crabs in Kasaoka Bay, in *Present situation of horseshoe crabs in Japan*, ed. by K. Sekiguchi, (2nd, Research club to protect horseshoe crab in Japan, Okayama, 1993), pp. 45–55. (in Japanese)
S. Watanabe, H. Yoshikawa, K. Azumakawa, et al., Population structure and growth of juvenile tri-spine horseshoe crabs released in the Konoshima tidal flat, Kasaoka, Okayama, Japan. JPN J. Conserv. Ecol. (submitted) (in Japanese with English abstract)
Z.H. Weng, Y.J. Xie, Z.Q. Xiao, et al., Microsatellite and mitochondrial DNA analysis of the genetic structure of Chinese horseshoe crab (*Tachypleus tridentatus*) in southeast China coast. Afr. J. Biotechnol. **12**, 2088–2099 (2013)
M.-C. Yang, C.A. Chen, H.-L. Hsieh, et al., Population subdivision of the tri-spine horseshoe crab, *Tachypleus tridentatus*, in Taiwan Strait. Zool. Sci. **24**, 219–224 (2007)

# Molecular Identification and Phylogenetic Analysis of Indian Mangrove Horseshoe Crab

**B. Bal, B. P. Das, Alok Prasad Das, and B. Akbar John**

## 1 Introduction

Horseshoe crabs are frequently taken as major illustration for invertebrates that stay alive for a lengthy period of time with no major differences in their structure; it is well-known as 'living fossils' a 'phylogenetic relicts' (John et al. 2016; Shuster and Barlow 2004; Selander et al. 1970). Horseshoe crabs have been used for exploration on visualization, cellular phagocytosis, neurobiology, and in environmental science (Das et al. 2015; Shuster et al. 2004). In the remnant evidence primeval, these species are previously renowned since the Ordovician period (Van et al. 2010) and recent structures which are identical of current species shows throughout Upper Jurassic (Avise et al. 1994; Briggs et al. 2005; Sekiguchi and Sugita 1980). It seems like horseshoe crabs preserve a stagnant morphology or moderately unaffected approximately for the ancient times 450 million years and now are at the focus of competing resource uses (Mitsugu et al. 2003; Fisher 1984).The development of self-determining range optima is forced by hereditary connection between the physiological and phylogenetic limitations (Sekiguchi and Sugita 1980; Fairbairn et al. 2007). Due to the maternal inheritance, primitive evolutionary speed in contrast to the nuclear DNA, and be short of intermolecular genetic recombination, mitochondrial DNA (mt DNAs) have been broadly used for learning of inhabitants structure

B. Bal · A. P. Das (✉)
Department of Chemical & Polymer Engineering, Tripura University, Agartala, Tripura, India

B. P. Das
Department of Bioscience & Biotechnology, Fakir Mohan University, Balasore, Odisha, India

B. Akbar John
INOCEM Research Station (IRS), Kulliyyah of Science, International Islamic, University Malaysia (IIUM), Kuantan, Pahang, Malaysia



J. T. Tanacredi et al. (eds.), *International Horseshoe Crab Conservation and Research Efforts: 2007-2020*, https://doi.org/10.1007/978-3-030-82315-3_3

and phylogenetic relationships at diverse taxonomic point (Avise 2000). The mtDNA, which is characteristically a single-rounded duplex molecule, vary in the size in about 15–17 kb and programming genes are of around 15 proteins, 22 tRNAs, and 2 rRNAs (Wolstenholme 1992; Boore et al. 1999). Horseshoe crabs have rearrangements of genes and principal mitochondrial genome sequences which are helpful in renovation of arthropod phylogeny. Complete mitochondrial genome orders have been resolute for around 10 crustaceans, as well as 6 decapods. A variety of mitochondrial DNA sequence reorganization have originated in the hermit crab *Paguruslongi carpus*as described by (Hickerson and Cunningham 2000), the crayfish *Cherax destructor* (Miller et al. 2004), and the swimming crab *Portunustrituber culatus* according to (Yamauchi et al. 2003). A vibrational methodology, Fourier-transform infrared (FTIR) spectroscopy, facilitates the differentiation of the elements present in the sample (Wu et al. 2015). Nucleic acids, phospholipids, proteins, and many other organic molecules have been analysed by this technique. In order to resolve the argument concerning beginning, descent, and different biological compound present in Indian horseshoe crab cell lysate, we investigated the phylogenetic relations between the recent species with new molecular hereditary facts (Faurbyb et al. 2012; Kumar et al. 2016) and FTIR analysis of amebocyte lysate. In this investigation, we informed the partial mtDNA sequences of the horseshoe crabs. Theses sequences were compared with each other and used for the analysis of evolutionary relation amongst them (Scheme 1).

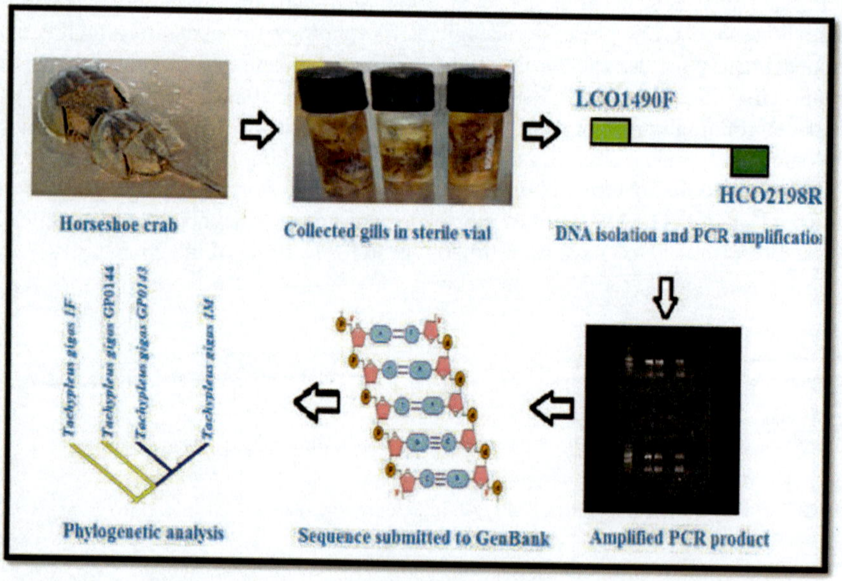


**Scheme 1** Molecular identification of Mangrove horseshoe crab

## 2 Materials and Methods

### 2.1 *Collection of Horseshoe Crabs*

Five Horseshoe crabs were collected from diverse geological region, including Chandipur coast (21.44 N and 87.02 E), Paradip coast (20.16 N and 86 40 E), Gopalpur coast (19.27 N and 84.92 E) (Fig. 1) of Bay of Bengal. Collection of these specimens were carried by fishermen in their innate costal territory and then acknowledged for their gender, measured, and weighed. On reaching laboratory, horseshoe crabs were first checked for any injuries. The next primary step was to scrub of carapace of horseshoe crabs to remove any unwanted growth that occurred due to various parasites. This helps in maintaining clean water in aquacultures and spreading of infection from one crab to another. All these procedures were carried out with utmost care so as not to hurt any horseshoe crab during the process. The horseshoe crab gills (1 × 1 cm) were cut through a sterilized scissor and collected in a sterile vial which was instantly kept at −80 °C until the extraction of DNA.

### 2.2 *Isolation of Horseshoe Crab DNA from the Gills*

After a short inspection of an peripheral morphology for the recognition of the specimen, previously frozen samples stored at −80 °C were homogenized in 2 ml of pre-chilled lysis buffer composed of (0.03 M Tris, 0.01 M EDTA, 0.25 M sucrose at pH 7.6) protocol designated by (Ling et al.; 2011). With little modification, the homogenate was moved to a chilled 2 ml tube and centrifuged at 10,000 rpm for

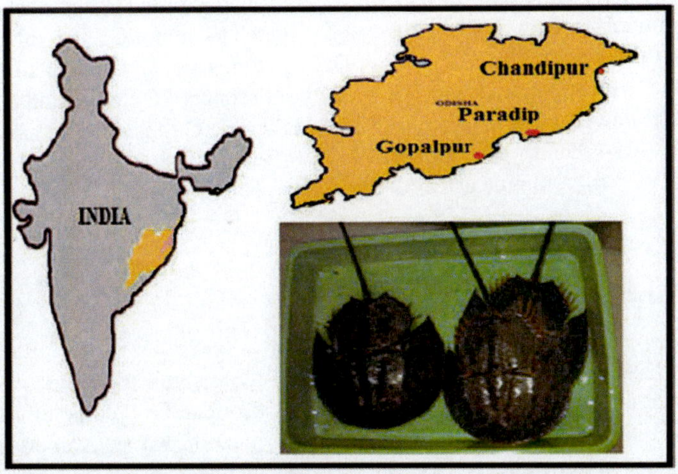


**Fig. 1** Nesting sites of Horseshoe shoe crab

5 minute at 4 °C direct to pellet down the cellular fragments. The consequential supernatant was again recovered through centrifugation at 12,000 rpm for 15 minute at 4 °C, thereby pelleting down the mitochondria. These pallets were utilized to isolate the mtDNA by phenol-chloroform DNA extraction method followed by 5 μl of protianase k was added and directed to the incubation at 50 °C for 4 hours. 10 μl of lysozyme was added and again incubated for an hour at 37 °C. Equal volume of phenol chloroform was added into the samples after an hour of incubation. It was properly mixed for 5 minutes manually after which it was allowed to be stable for 10–15 minutes at room temperature. This step was followed by centrifugation of the sample at 12000 rpm for 5 minute at 4 °C. After centrifugation samples were separated in two different layers, the upper aqueous layer containing the DNA was transferred into another fresh 2 ml centrifuged tube. 750 μl of prechilled 70% ethanol and 50 μl of 3 M sodium acetate was added for the precipitation and stored at −20 °C. DNA pellet was obtained after centrifuging the sample at 12000 rpm at 4 °C for 12 minute. The supernatant was discarded and 1 ml of 70% chilled ethanol was added to the pellet. The resulting solution was further centrifuged 12,000 rpm for 10 minute at 4 °C and the supernatant was discarded. The pellets containing mtDNA was dissolved in 40 μl elution buffer and stored at −20 °C for long-term preservation.

## 2.3 *PCR Amplification of Isolated DNA*

A single fragment of the mt DNA, cytochrome c oxidase I (COI) subunit were amplified using a primer set LCO1490F (5'-GGTGAACAAATCATAAAGA TATTGG-3′) and HCO2198R (5′- TAAACTTCAGGGTGACCAAAAAATCA-3) (Folmer et al. 1994). Amplification of these fragment were carried out by thermal cycler (Applied Biosystem). 25 ml PCR master mix composed of 2.5 ml of 10× buffer (1.5 mmol/L Mgcl2), 1 ml of every primer (10 mmol/L), 0.5 ml of dNTP (10 mmol/L each), 0.5 μl (1 U) of Taq DNA polymerase, DNA template, to make the volume with Millipore water followed by subsequent PCR conditions; 5 minute at 95 °C followed by denaturation (32 cycles at 95 °C (45 second), annealing at 53 °C for (45 second), extension at 72 °C (45 second) and a concluding primer extension step for 5 minute at 72 °C (Hoh 2005; Kumar et al. 2002).

## 2.4 *Purification of PCR Products*

Desired DNA fragments were excised through sterile cutter from the agarose gel. The weight of the excised DNA fragments was taken and 3 volume of QG buffer was added into the gel as prescribed by the purification Kit (QIAamp® genomic DNA kits), Germany. The excised gel slice was incubated at 50 °C for 10 minute.

Regular vortexing was carried out to ensure proper mixing. The colour of the gel turned yellow after its complete dissolution. 1 volume of isopropanol was added and mixed properly followed by centrifugation in the provided purification column at 10000 rpm for 1 minute at room temperature. Flow through was leftover and the column was positioned into another fresh tube. 750 μl of 10 Mm Tris-Cl was mixed and centrifuged for 2 minute at 12000 rpm. The DNA was eluted by the addition of 100 μl of elution buffer and incubated at room temperature for 10 minutes and then the sample was stored at −20 °C.

### *2.5 Molecular Identification and Phylogenetic Analysis*

Mitochondrial partial DNA sequences were obtained through Sanger Sequencing by Eurofin Scientific, Bangalore, India. The sequences were evaluated and taxonomical categorization was accomplished using Basic Local Alignment Search Tool (BLAST) analysis in opposition to genetic information acquired from National Center for Biotechnology Information (NCBI) (Ghosh et al. 2015). Ten neighbouring matches were aligned with the CLC sequence viewer. Consequential results were utilized to build the phylogenetic tree by using neighbourhood joining method (Kumar et al. 2016).

## 3 Result and Discussion

### *3.1 DNA Isolation and PCR Amplification*

Amplified PCR products were examined on 1.2% agarose gel, staining with Ethidium bromide with a 1Kb DNA marker under Ultraviolet transillumination (Fig. 2). Resulting mtDNA of Horseshoe crabs were found 700bpb and size analysed by utilizing Gel-Pro Analyzer 415 after carrying out agarose gel electrophoresis.

### *3.2 Molecular Characterization and Phylogenetic Analysis*

Molecular classification was done for all isolated mitochondrial DNA sequence of M3, M4, TG1F, TG1M, and TG2M collected horseshoe crab species and it was identified as *Tachypleus gigas* (Table 1). The sequences were put forward to GenBank submission and the acquired accession numbers were KU880543, KU880544, KT380891, KT380892, and KT380893. M3 partial mt DNA sequence was examined for its evolutionary relationship and demonstrated that it belong to

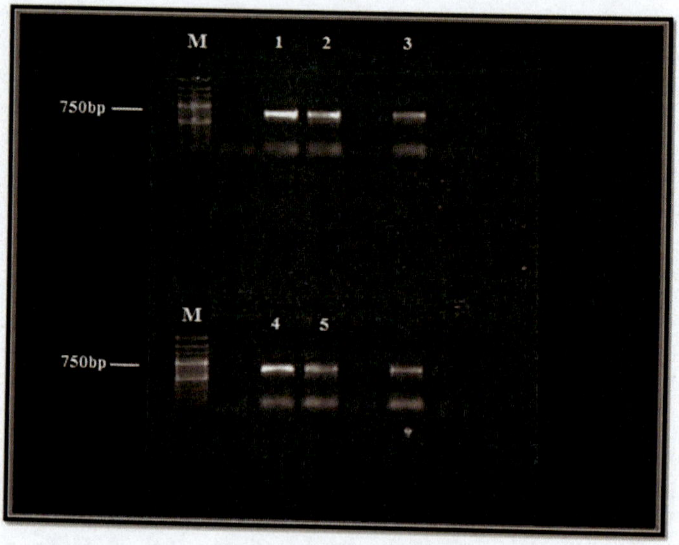


**Fig. 2** DNA bands observed under UV light after agarose gel electrophoresis

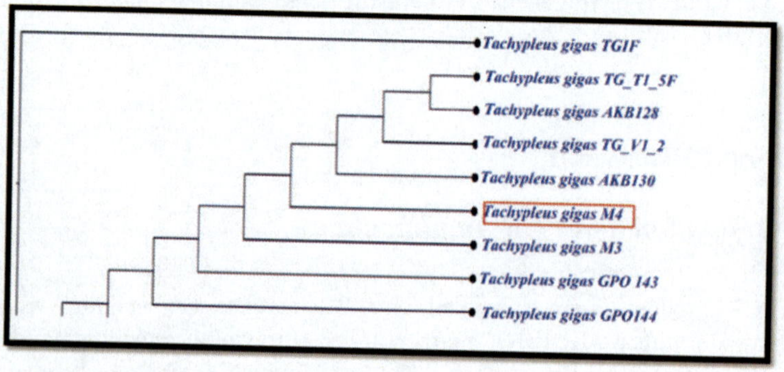


**Fig. 2** (continued)

*Tachypleus gigas*. *Tachypleus gigas* GPO 143 and *Tachypleus gigas* having accession no FJ860267.1 having 99% sequence similarity are its nearest phylogenetic species. Other closest families' were *Tachypleus gigas* AKB125, *Tachypleus gigas* TG1F, *Tachypleus gigas* TG2M, and *Tachypleus gigas* TG1M. M4 partial mtDNA sequence was studied for its evolutionary connection and confirmed that it comes under *Tachypleus gigas*. *Tachypleus gigas* GPO 144 and *Tachypleus gigas* having accession no FJ860267.1 having 99% sequence similarity are its nearest phylogenetic species. Other closest families are *Tachypleus gigas* AKB125, *Tachypleus gigas* TG1F, *Tachypleus gigas* TG2M, and *Tachypleus gigas* TG1M.

The phylogenetic study of the TG1F sequence revealed that its closest phylogentic relative *Tachypleus gigas* isolates GPO143 and *Tachypleus gigas* AKB130

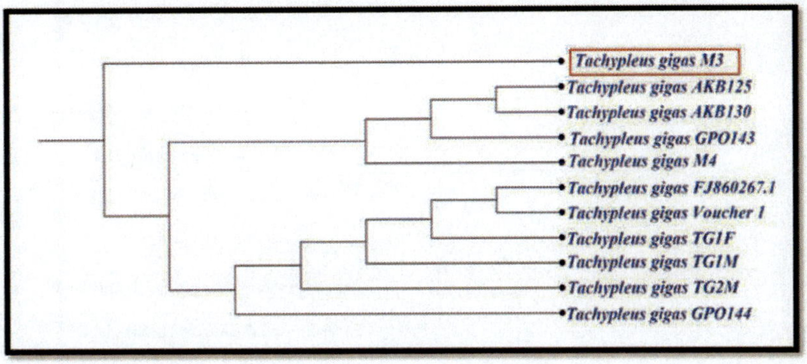


**Fig. 2** (continued)

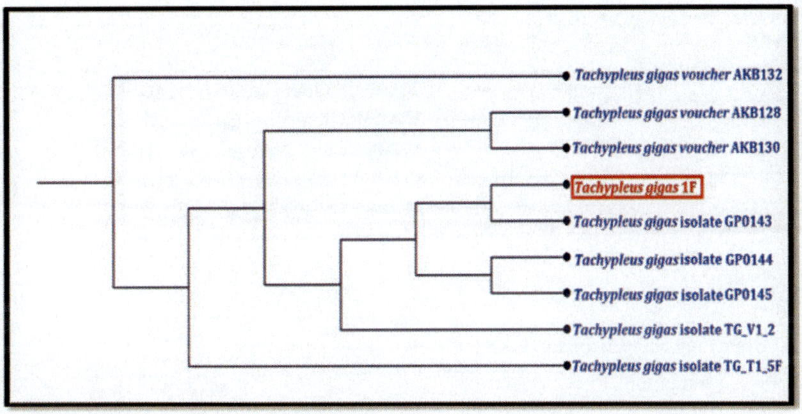


**Fig. 2** (continued)

mitochondrion, partial genome having a similarity index at 99%. Its further neighbouring is *Tachypleus gigas* AKB132, *Tachypleus gigas* AKB128, *Tachypleus gigas* GPO144. TG1M belongs to the genus *Tachypleus gigas.* Its closest sequence with *Tachypleus gigas* TG_VI_2 and *Tachypleus gigas* GPO 143 having 99% sequence similarity. The closest matches are *Tachypleus gigas GPO145, Tachypleus gigas GPO144, Tachypleus gigas AKB130, and Tachypleus gigas AKB128.* The phylogenetic analysis of the TG2M sequence revealed that its closest phylogentic relative *Tachypleus gigas* isolates GPO145 and *Tachypleus gigas* AKB132 mitochondrion, partial genome contains a similarity at 99%. Other closest phylogenetic related species are *Tachypleus gigas* TG_T1_5F, *Tachypleus gigas* TG_V1_2, *Tachypleus gigas* GPO144, and *Tachypleus gigas AKB128* (Fig. 3). Different results are obtained by using COI primers in terms of genomic variation but agreed that the most evolutionary differences are amongst the habitat rather than amongst the mangrove

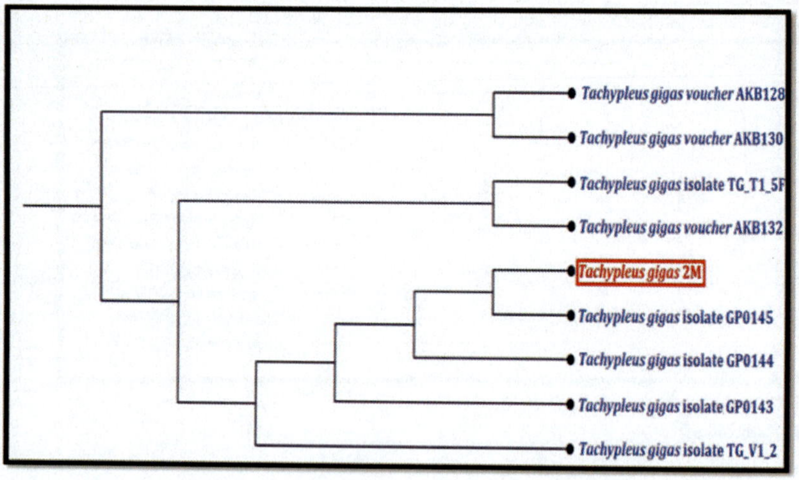


**Fig. 2** (continued)

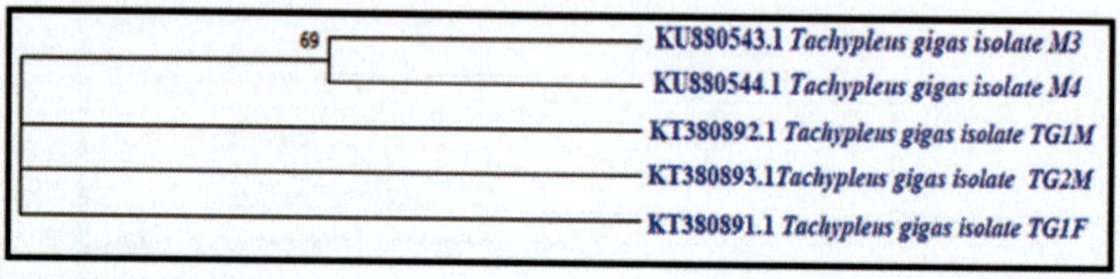


**Fig. 2** (continued)

**Table 1** Homology comparison of Horseshoe crabs mitochondrial DNA

| Sl. no | Code | Accession no | Similarity | % of similarity | Max score | Query coverage |
|---|---|---|---|---|---|---|
| 1 | TG1F | KT380891 | *Tachypleus gigas*GPO145 | 99% | 1140 | 99% |
| 2 | TG1M | KT380892 | *Tachypleus gigas GP0144* | 99% | 1133 | 99% |
| 3 | TG2M | KT380893 | *Tachypleus gigas GP0144* | 99% | 1158 | 99% |
| 4 | M3 | KU880543 | *Tachypleus gigas*GPO145 | 99% | 1201 | 100% |
| 5 | M4 | KU880543 | *Tachypleus gigas*GPO145 | 99% | 1127 | 93% |

horseshoe crab species. Earlier work on molecular phylogenetics and revision of horseshoe crab phylogeny has clearly demonstrated the monophyletic origin and distant genetic relatedness of *Limulus polyphemus* with its Asian conspecifics (John et al. 2016; Yunus et al. 2011). Compared to the genomic information found in this

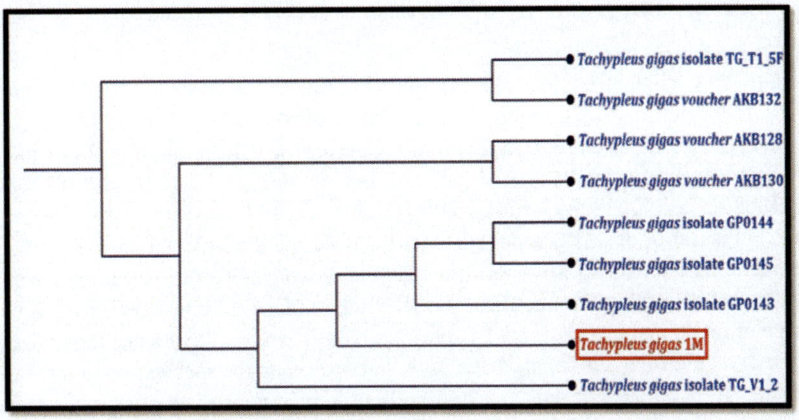


**Fig. 3** Phylogeetic tree constructed by neighbour joining method: (**a**) Phylogenetic tree of M3 (**b**) phylogenetic tree of M4 (**c**) phylogenetic tree of TG1F (**d**) phylogenetic tree of TG1M (**e**) phylogenetic tree of TG2M (**f**) Evolutionary analysis of Indian Mangrove horseshoe crab

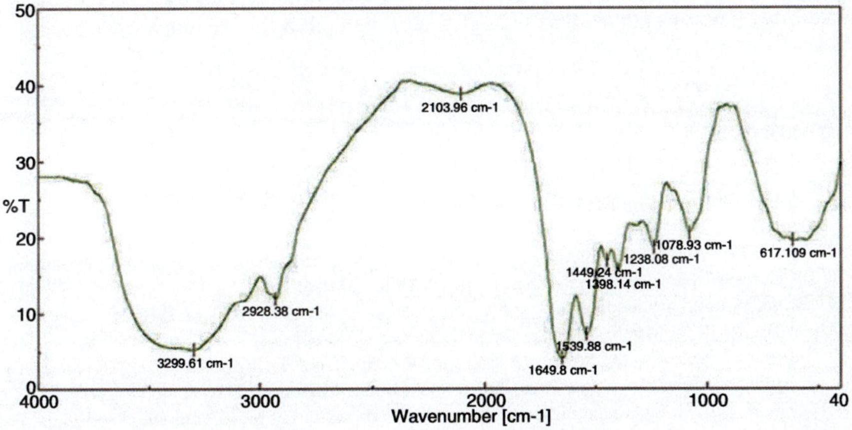


**Fig. 4** FTIR spectra of horseshoe crab lysate

investigation, the terrestrial signals are also apparent by cladding *T. gigas* collected from geologically nearer inhabitants such as Malaysia. As horseshoe crab grows, it roams from its natural beach to deeper into the sea water, and returns to the beach to spawn. The results suggest horseshoe crabs do not migrate so far along the coastline, therefore reassuring inbreeding amongst the individuals from neighbouring beaches, which reduced the whole chromosomal variation with having 99% sequence similarity amongst the populations and at the same time improved the probabilities of inbreeding (Fig. 4).

## 4 Conclusion

In summary, current exploration exposed that the entire collected five specimens were identified as *Tachypleus gigas*. The achieved horseshoe crab mt DNA sequences were submitted in National Centre for Biotechnology Information (NCBI) GenBank having the accession number KU880543, KU880544, KT380891, KT380892, and KT380893. Earlier it was explained by (Avise et al.) that the horseshoe crabs morphostatis was not sufficient to describe the little variation in molecular stage. The universal DNA primers, HCO2198and LCO1490, amplified around 700-bp length of the mitochondrial cytochrome oxidase subunit I gene from a wide range of metazoan invertebrates. At present, by utilizing these two primers, the evolutionary relations amongst these five horseshoe crab species were revealed. Molecular techniques have been extensively used in genetic diversity studies, inhabitant's configuration prediction, as well as their phylogeographic analysis. The sequences were analysed and taxonomical classification was accomplished using BLAST algorithm tools. Different molecular methods of DNA isolation, sequencing were worked to recognize the specimen, which confirms to be a competent implement of recognition. Collectively with the molecular processes, it will permit for deep investigation of huge number of samples. These proposed methodology exhibits numerous advantages such as a short detection time, exquisite sensitivity.

## References


J.C. Avise, *Phylogeography: The History and Formation of Species* (Harvard University Press, Cambridge, 2000)

J.C. Avise, W.S. Nelson, H. Sugita, A speciational history of living fossils – Molecular evolutionary patterns in horseshoe crabs. Evolution 48, 1986–2001 (1994)

J.L. Boore, Animal mitochondrial genomes. Nucleic Acids Res. **27**, 1767–1780 (1999)

D.E.G. Briggs, R.A. Moore, J.W. Shultz, G. Schweigert, Mineralization of softpart anatomy and invading microbes in the horseshoe crab Mesolimulus from the Upper Jurassic Lagerstatte of Nusplingen, Germany. Home Proc. Royal Soc. London B Biolog. **272**, 627–632 (2005)

A.P. Das, B. Bal, P.S. Mahapatra, Horseshoe crabs in modern day biotechnological applications. Changing global perspectives on biology, in *Conservation and Management of Horseshoe Crabs*, vol. 26, (Springer, 2015), pp. 463–474

D.J. Fairbairn, W.U. Blanckenhorn, T. Székely, *Sex, Size, and Gender Roles: Evolutionary Studies of Sexual Size Dimorphism* (Oxford University Press, Oxford, 2007)

O.S. Faurbyb, S. Bussarawitc, F. Funchd, Molecular phylogeny of extant horseshoe crabs (Xiphosura, Limulidae) indicates Paleogene diversification of Asian species. Mol. Phylogenet. Evol. **62**(1), 21–26 (2012)

D.C. Fisher, The Xiphosurida: Archetypes or bradytely? in *Living Fossils*, ed. by N. Eldredge, S. M. Stanley, (Springer, 1984), pp. 106–212

O. Folmer, M. Black, W. Hoeh, R. Lutz, R. Vrijenhoek, DNA primers for amplification of mitochondrial cytochrome c oxidase subunit I from diverse metazoan invertebrates. Mol. Mar. Biol. Biotechnol. **3**(5), 294–299 (1994)

S. Ghosh, S. Mohanty, S. Nayak, L.B. Sukla, A.P. Das, Molecular identification of indigenous manganese solubilising bacterial biodiversity from manganese mining deposits. J. Basic Microbiol. (2015). https://doi.org/10.1002/jobm.201500477

M.J. Hickerson, C.W. Cunningham, Dramatic mitochondrial gene rearrangements in the hermit crab Paguruslongicarpus (Crustacea, nomura). Mol. Biol. Evol. **17**, 639–644 (2000)

B.P. Hoh, *Isolation, Characterization and Genetic Linkage Mapping of Microsatellite Markers in Mystus Nemurus*. (Ph.D. thesis, Universiti Putra Malaysia, Malaysia 2005)

B.A. John, H.I. Sheikh, K.C.A. Jalal, K. Zaleha, B.Y. Kamaruzzaman, Revised phylogeny of extant xiphosurans (horseshoe crabs), in *DNA Barcoding in Marine Perspectives*, (Springer, 2016), pp. 113–130

S.V. Kumar, S.G. Tan, S.C. Quah, K. Yusoff, Isolation of microsatellite markers in mungbean, Vign radiate. Mol. Ecol. Notes **2**, 96–98 (2002)

M.S. Kumar, S. Ghosh, S. Nayak, A.P. Das, Recent advances in biosensor based diagnosis of urinary tract infection. Biosens. Bioelectron. **80**, 497–510 (2016)

L.P. Ling, A.B. Adibah, G.U. Tan, A. Christianus, Q.J. Faridah, Isolation by the 5′ anchored PCR technique and characterization of eighteen microsatellite loci in horseshoe crab (*Tachypleus gigas*). J. Genet. **90**, 101–104 (2011)

A.D. Miller, T.T.T. Nguyen, B.C. Burridge, C.M. Austin, Complete mitochondrial DNA sequence of the Australian freshwater crayfish, Cherax destructor (Crustacea: Decapoda: Parastacidae): A novel gene order revealed. Gene **331**, 65–72 (2004)

M. Mitsugu, M.U. Yamauchia, M. Miyab Nishidaa, Complete mitochondrial DNA sequence of the swimming crab, Portunustrituberculatus (Crustacea: Decapoda: Brachyura). Gene **311**, 129–135 (2003)

K. Sekiguchi, H. Sugita, Systematics and hybridization in the 4 living species of horseshoe crabs. Evolution **34**, 712–718 (1980)

R.K. Selander, S.Y. Yang, R.C. Lewontin, W.E. Johnson, Genetic variation in the horseshoe crab (Limulus polyphemus), a phylogenetic "relic". Evolution **24**(2), 402–414 (1970)

C.N. Shuster, R.B. Barlow, H.J. Brockmann, *The American Horseshoe Crab* (Harvard University Press, Cambridge, MA, 2004)

P. Van Roy, P.J. Orr, J.P. Botting, L.A. Muir, J. Vinther, B. Lefebvre, K. Hariri, Briggs, DEG, Ordovician faunas of Burgess Shale type. Nature **465**, 215–218 (2010)

D.R. Wolstenholme, Animal mitochondrial DNA: Structure and evolution. Int. Rev. Cytol. **141**, 173–216 (1992)

B.B. Wu, Y.P. Gong, H. Wu, Y.Y. Chen, F.F. Chen, L.T. Jin, B.R. Cheng, Hu, B. Xiong, Fourier transform infrared spectroscopy for the distinction of MCF-7 cells treated with different concentrations of 5-fluorouracil. J. Transl. Med. **13**, 108 (2015)

M.M. Yamauchi, M. Miya, M. Nishida, Complete mitochondrial DNA sequence of the swimming crab, *Portunustrituberculatus* (Crustacea: Decapoda: Brachyura). Gene **311**, 129–135 (2003)

K. Yunus, A. John, Z. Kasim, A.J. Khan Chowdhury, Molecular phylogeny of horseshoe crab. Asian J. Biotechnol. **3**(3), 302–309 (2011)

# Population Genetics and Movement Show Metapopulation Dynamics of Mid-Atlantic Region Horseshoe Crabs

**Eric M. Hallerman, David N. Hata, Michael S. Eackles, and Timothy L. King**

## 1 Introduction

Horseshoe crabs *Limulus polyphemus* comprise an economically and ecologically important marine resource in the mid-Atlantic region of the United States (Walls et al. 2002). Horseshoe crab populations are targeted by biomedical and bait fisheries, and its large spawning assemblages attract ecotourism along the shores of Delaware Bay (Manion et al. 2000). Ecologically, horseshoe crabs provide a vital link in near-shore marine food webs. Additionally, horseshoe crab eggs provide a major food item for migratory shorebirds, including the threatened red knot *Calidrus canutus rufa* (Botton et al. 1994; Botton and Harrington 2003; USFWS 2003; Karpanty et al. 2007; USFWS 2014). Effective ecosystem management for the Delaware Bay calls for maintaining a sufficient population of horseshoe crabs to sustain both itself and shorebirds consuming its eggs (McGowan et al. 2011a, 2011b). Against the background of declining horseshoe crab numbers, the Atlantic States Marine Fisheries Commission (ASMFC 1998) adopted the Interstate Fishery Management Plans for Horseshoe Crab with seven subsequent addenda. In 2000, Addendum I established state-by-state harvest quotas for crabs harvested for bait. In 2004, Addendum III 2004 reduced harvest quotas, implemented seasonal bait harvest closures, and revised monitoring. In 2006, Addendum IV reduced quotas in New Jersey and Delaware and added restrictions on harvest of horseshoe crab

E. M. Hallerman (✉) · D. N. Hata
Department of Fish and Wildlife Conservation, Virginia Polytechnic Institute and State University, Blacksburg, VA, USA
e-mail: ehallerm@vt.edu

M. S. Eackles · T. L. King
U.S. Geological Survey, Leetown Science Center-Aquatic Ecology Branch, Kearneysville, WV, USA



J. T. Tanacredi et al. (eds.), *International Horseshoe Crab Conservation and Research Efforts: 2007-2020*, https://doi.org/10.1007/978-3-030-82315-3_4

populations in Maryland and Virginia. The addendum allowed a male-only bait fishery in New Jersey and Delaware; Delaware subsequently allowed such a fishery, but New Jersey imposed a moratorium on harvest of horseshoe crabs in 2008. In 2013, the Horseshoe Crab Management Board adopted use of the Adaptive Resource Management (ARM) Framework (ASMFC 2009) to set horseshoe crab harvest levels for the Delaware Bay area. The ARM Framework was established through Addendum VII, incorporating both shorebird and horseshoe crab abundance levels to set optimized harvest levels for horseshoe crabs of Delaware Bay origin. For the 2016, 2017, and 2018 fishing seasons, harvest in the Delaware Bay area was limited to 500,000 male horseshoe crabs and no female horseshoe crabs. This total harvest is allocated among the four states that harvest horseshoe crabs from the Delaware Bay crab population, New Jersey, Delaware, Maryland, and Virginia. Ongoing monitoring tracks the relative abundance of horseshoe crabs in the Delaware Bay and nearby coastal areas (Hata and Hallerman 2018). An overarching conclusion of recent technical assessments is that management of horseshoe crabs should be both regional and embayment-specific (HSC-TC 2008). The fisheries for coastal horseshoe crab stocks raise the question of the degree to which exploited stocks represent populations that are demographically linked to the Delaware Bay population that is the particular focus of management interest.

## *1.1 Population Genetics*

The fishery management plan for horseshoe crabs (ASMFC 1998) identified stock identification as a principal research need for regional management of the species. Population genetic analysis offers the basis for defining biologically based management units (MUs) and provides an evolutionary framework for developing and evaluating conservation priorities (Ryder 1986; Moritz 1994). Application of a growing range of molecular genetic markers and data analytics has led to greater understanding of population genetic structure in *L. polyphemus*. Allozyme variation among four collections of crabs. (Selander et al. 1970) suggested that Atlantic Ocean and Gulf of Mexico populations were genetically differentiated. Mitochondrial DNA (mtDNA) restriction site polymorphism (Saunders et al. 1986) showed a major genetic discontinuity distinguishing northern from southern populations, with the phylogeographic break occurring around Cape Canaveral, Florida. Sequence variation at the mtDNA *COI* region (Pierce et al. 2000) showed little evidence of gene flow between Delaware and Chesapeake Bay populations, although lack of geographic variation at nuclear randomly amplified polymorphic DNA (RAPD) markers implied that gene flow may be sex-biased. Variation at 13 microsatellite DNA markers among 1684 horseshoe crabs from 33 spawning assemblages from northern Maine to the Yucatan Peninsula of Mexico (King et al. 2005, 2015) showed shallow, but statistically significant, differentiation within regions and highly significant differences among regions, suggesting a series of genetic discontinuities across the species' range; the authors suggested that these regional groupings may warrant recognition as MUs. Among the proposed MUs was a mid-Atlantic assemblage of populations that spawn on suitable beaches from coastal southern New Jersey to Delaware Bay to coastal Maryland and Virginia as far south as the southern tip of Assateague Island. While some substructure was apparent, population genetic data supported the hypothesis that migration tends to homogenize allele frequencies along this section of the Atlantic coast. The data also supported the interpretation that males tend to migrate farther than females, a hypothesis that would be supported by direct observation of movement and participation in spawning. Application of Bayesian coalescent-based analysis of these data to infer historic demography (Faurby et al. 2010) suggested that mid-Atlantic populations experienced a demographic bottleneck within the past 150 years, likely caused by overharvest.

## 1.2 *Movement*

While population genetic studies capture the effects of genetically effective migration over the long term, mark-recapture studies quantify movement in real time, with the possibility that the movement observed may result in genetically effective migration. Movement studies conducted in the mid-Atlantic region showed that while most crabs do not move great distances, a small proportion does. Tagging efforts targeting spawning Delaware Bay horseshoe crabs include projects by the U.S. Geological Survey, Delaware Department of Natural Resources, Maryland Department of Natural Resources, and a range of non-governmental organizations. Biomedical companies (BioWhittaker-Cambrex-Lonza; c.f., Swan 2005) and Virginia Tech University sometimes mark crabs caught in offshore trawls. Having marked horseshoe crabs at six sites, Shuster Jr (1985) found that individuals marked in both the Chesapeake Bay and Delaware Bay were collected at Tom's Cove on Assateague Island in Virginia. Among crabs marked and released in the Delaware Bay, Swan (2005) found that 75% traveled 0–20 km, 21% 20–50 km, 2.6% 50–100 km, and 1.4% over 100 km. For crabs marked at Tom's Cove, Virginia, Grogan (2004) reported a mean distance between mark and recapture sites of 68 km, with a maximum of 494 km. Results from these studies suggest that Delaware Bay crabs disperse as far as Chincoteague and Assateague islands and vice versa. The ASMFC Horseshoe Crab Technical Committee (2008) recommended increased mark-recapture effort to determine how many Delaware Bay crabs are harvested in other states. While other projects had resulted in already-tagged horseshoe crabs in the area south of the Delaware Bay, there were no ongoing studies to mark and recapture them there. Against this background, we focused on the spawning assemblage in Tom's Cove, a Virginia bay toward the southern end of the distribution of "Delaware Bay" crabs inferred by Shuster Jr (1985) and King et al. (2005, 2015). Marking crabs there would support observation of their subsequent movement. By referring to a recapture reports database maintained by the U.S. Fish and Wildlife Service (USFWS), individual movements among regional populations would be demonstrated.

### 1.3 Objectives

Against this background, we conducted population genetics and movement studies to contribute to characterization of the Mid-Atlantic assemblage of horseshoe crab populations. We focused upon: (1) characterizing the sources of origin for horseshoe crabs taken in seven regional fisheries and (2) characterizing movement of crabs marked in a key regional spawning assemblage, and subsequently recaptured across the region. We discuss our findings in the context of implications for regional horseshoe crab management.

## 2 Methods

### 2.1 Population Genetics

To characterize the likely spawning assemblage of origin for horseshoe crabs taken in seven key regional fisheries, we used a Bayesian assignment algorithm to assign each fished individual's multilocus microsatellite DNA genotype to a population or assemblage of populations. Data for reference spawning populations (King et al. 2005, 2015) regards variation at 13 microsatellite loci among 1229 horseshoe crabs sampled from 28 spawning assemblages across the range of the species (Table 1 and Fig. 1). Possible populations or population assemblages of origin were based on the results of Faurby et al. (2010) and included Hog Bay, Maine; Gulf of Maine, Northeast (Cape Cod to northern New Jersey), Outer Delaware Bay (coastal regions from southern New Jersey to Shackleford Banks, NC), Delaware Bay, Inner Chesapeake Bay, Southeast (South Carolina and Georgia), Florida – Atlantic coast, Florida – Gulf coast, and Yucatan Peninsula, Mexico.

New microsatellite data were collected for fished populations. The Long Island, New York sample came from a survey conducted along western Long Island on October 22, 2005 (Fig. 1). The New Jersey coastal sample came from a survey conducted north of the Delaware Bay on September 23, 2005. The Ocean City, Maryland collections came from a trawl fishery for the biomedical industry in 2005 and 2006. The Chincoteague Island, Virginia sample came from a dredge fishery for the biomedical industry. The Assateague Island, VA collection came from a hand-collection conducted in Tom's Cove in 2006. The North Carolina sample came from a USFWS trawl conducted in 2007. Molecular genetic analysis of all these samples followed the methods used by King et al. (2015).

We used the Bayesian assignment algorithm implemented in STRUCTURE ver. 2.3.1 (Pritchard et al. 2000) to assign the multilocus genotypes for individual crabs collected in these five fisheries to their most likely source; as appropriate, individuals could be partially assigned to multiple source populations. The first 100,000 iterations of implementation of the algorithm were discarded as burn-in, followed by 200,000 recorded iterations; 10 independent runs were conducted. This replication allowed us to estimate confidence in assignment probabilities.

**Table 1** Location and sample size for 25 spawning assemblages of American horseshoe crabs *Limulus polyphemus* genotyped at 13 microsatellite DNA loci (King et al. 2005, 2015). Samples from seven fisheries were genotyped for this study

| | Collection abbreviation and site | Sample size |
|---|---|---|
| | *Spawning assemblages*: | |
| 1 | MEH, Hog Bay, Franklin, Maine | 47 |
| 2 | MET, Thomas Point Beach, Maine | 45 |
| 3 | MEM, Middle Bay, Brunswick, Maine | 48 |
| 4 | NHS, Chadman's Landing, Squamscott River, New Hampshire | 48 |
| 5 | MAP, Pleasant Bay, Massachusetts | 48 |
| 6 | RIN, Green Island, Narragansett Bay, Rhode Island | 48 |
| 7 | CTH, Housatonic River, Milford Point, Connecticut | 48 |
| 8 | NYP, Great Peconic Bay, Long Island, New York | 48 |
| 9 | NJF, Fortescue Beach, New Jersey | 48 |
| 10 | NJR, Reeds Beach, New Jersey | 48 |
| 11 | NJH, Highs Beach, New Jersey | 49 |
| 12 | DEK, Kitts Hummock Beach, Delaware | 36 |
| 13 | DEB, Big Stone Beach, Delaware | 31 |
| 14 | DEF, Fowler Beach, Delaware | 47 |
| 15 | MDT, Turkey Point, Chesapeake Bay, Maryland | 30 |
| 16 | MDF, Flag Pond State Park, Chesapeake Bay, Maryland | 29 |
| 17 | VKI, Kiptopeake State Park, Chesapeake Bay, Virginia | 48 |
| 18 | NCS, Shackleford Banks, North Carolina | 55 |
| 19 | SBU, Bulls Bay, South Carolina | 53 |
| 20 | SBE, Beaufort, South Carolina | 48 |
| 21 | GSA, Savannah, Georgia | 48 |
| 22 | FLM, Mosquito Lagoon, Indian River, Florida (Atlantic coast) | 46 |
| 23 | FLC, Seahorse Key, Cedar Keys NWR, Florida (Gulf coast) | 46 |
| 24 | FLJ, St. Joseph Bay, Florida (Gulf coast) | 23 |
| 25 | MXY, Ria Lagartos and San Felipe, Yucatan, Mexico | 20 |
| | *Fisheries* | |
| 26 | NYL, Offshore trawl, New York | 46 |
| 27 | NJC, Trawl offshore Cape May, New Jersey | 48 |
| 28 | MD5, Ocean City, Maryland – 2005 | 48 |
| 29 | MD6, Ocean City, Maryland – 2006 | 48 |
| 30 | VCH, Chincoteague Island, Virginia dredge | 46 |
| 31 | VAI, Tom's Cove, Assateague Island, Virginia hand-harvest | 48 |
| 32 | FWS, U.S. Fish and Wildlife Service, 2007 trawl | 48 |

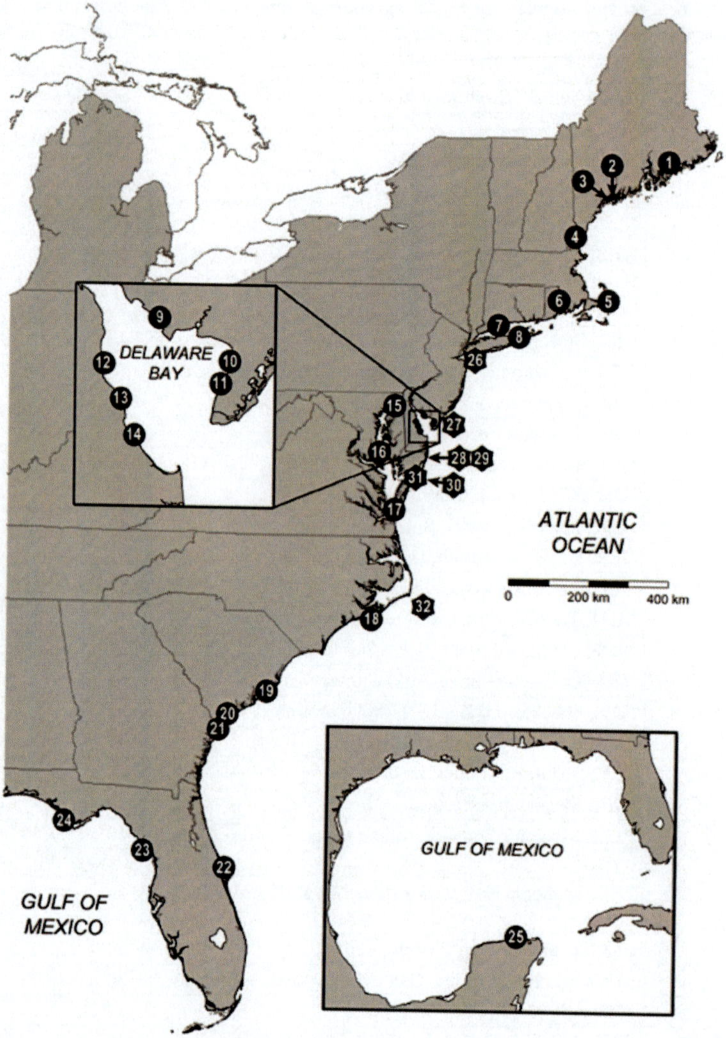


**Fig. 1** Collection sites for *Limulus polyphemus* along the Atlantic and Gulf coasts of the United States. Collection names and numbers are provided in Table 1. The site for collection from Mexico's Yucatan Peninsula is depicted in the inset. Spawning collections are depicted with circles, fisheries with stars

## 2.2 *Movement*

Horseshoe crabs were collected in Tom's Cove in the Chincoteague Island National Wildlife Refuge – Assateague Island National Seashore in Virginia. Tom's Cove is a shallow, protected embayment approximately 1 km$^2$ in area with excellent spawning habitat, especially along its southern shore (Fig. 2). We made hand collections of spawners in shallow water during spawning runs at the new and full moons in May 2008, that is, before the opening of the commercial fishery on June 8. Numbers on any existing tags were recorded. Button tags (Jersey Cape Diagnostics, Cape May, NJ) were attached to unmarked individuals by drilling a small hole in the posterior of the prosoma with a hand-held drill and inserting the tag, which anchored itself. Tags instructed collectors to report sightings to the USFWS in Annapolis, Maryland. We collected data on sex and measured prosomal width of all individuals.

We directly observed multiple uses of the Tom's Cove spawning beach by individual crabs both within and between spawning runs during 2008. Recaptures also were recorded after the fishery opened on June 8. Commercial bait-fishers harvested crabs for 10 nights between June 8 and 20. Following the recommendation of Grogan (2004), we offered commercial fishers a small reward ($1 per tag) for tags returned with information on the date and place of capture, along with information on total numbers of male and female crabs captured.

The USFWS office in Annapolis, MD, maintains a database of horseshoe crab tags deployed and subsequently observed. For existing tags that we observed on

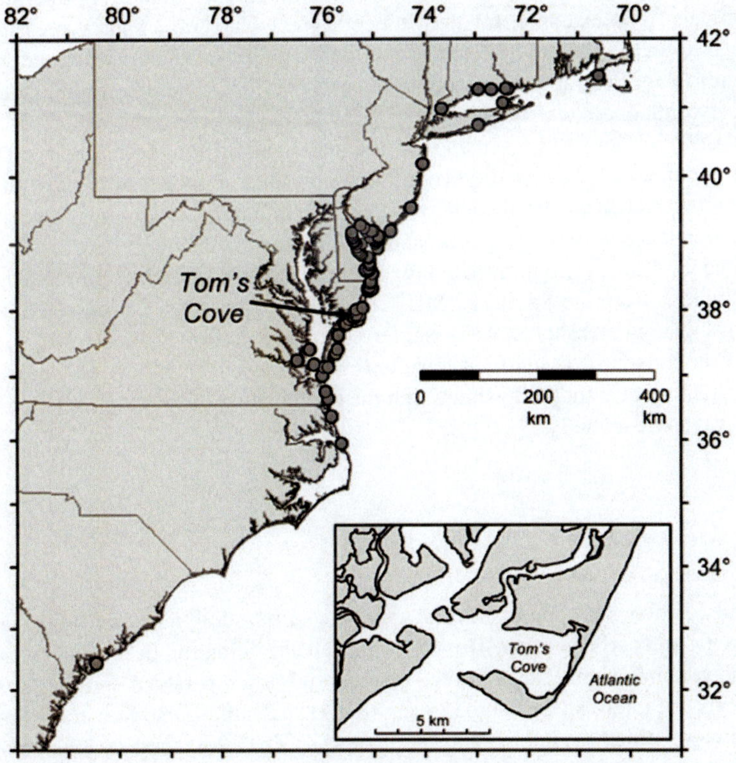


**Fig. 2** Resight locations for horseshoe crabs marked and released in Tom's Cove, Virginia, 2008

crabs spawning or harvested in Tom's Cove, the USFWS provided location, date, and other data associated with the original tagging, which allowed us to directly infer the origins of spawners observed in Tom's Cove. The USFWS subsequently reported to us observations of crabs tagged in Tom's Cove through 2017 following their post-tagging dispersal. They also provided updated, unreported recapture data for crabs tagged previously by coauthor David Hata during the annual horseshoe crab trawl survey (Hata and Hallerman 2018), by Walls (2001), or Grogan (2004).

# 3 Results

## 3.1 *Population Genetics*

The original data on multilocus microsatellite DNA genotypes of spawning horseshoe crabs are reported in King et al. (2005, 2015). In our analysis, assignment test-based results show the estimated contributions of regional spawning assemblages to the seven harvests (Table 2). Because many independent runs of the assignment algorithm were made, confidence in the respective point estimates could be estimated; the example, for the Chincoteague Island trawl fishery presented in Table 3 shows the breadth of the 95% confidence intervals. Hence, while the spawning assemblages contributing to a particular fishery could be determined with reasonable confidence, the magnitudes of the relative contributions could not. Such uncertainty typifies results of using the assignment algorithm for modestly differentiated populations (Waples and Gaggiotti 2006). Overall, our results show that the respective fisheries captured approximately 90% of landed individuals from spawning assemblages within the mid-Atlantic MU. Three of the offshore trawl harvests included modest contributions from the Gulf of Maine MU, and all four had modest contributions from the Southeast MU. The inshore Tom's Cove hand-catch fishery upon spawning horseshoe crabs was inferred to draw entirely from the mid-Atlantic MU. Comparison of assignment results for the trawl fisheries off of Ocean City, Maryland, in 2005 and 2006 shows that the composition of the catch varies temporally as well as spatially.

## 3.2 *Movement*

Horseshoe crabs spawned on several evenings surrounding new and full moons in late spring and early summer. There was a small spawning run in Tom's Cove at the full moon in late April (Chris Cuono, commercial fisher, personal communication). Our mark-recapture study handled over 18,000 spawning horseshoe crabs through the four spawning runs in May and June (Table 4). Over 2000 crabs were marked on the May new moon run. A "nor'easter" storm during the second week of May disturbed and chilled coastal waters, leading to but a small run at the May full moon.

**Table 2** Inferred proportions of management unit (MU) of origin for horseshoe crabs taken in seven fisheries within the Mid-Atlantic region

| | Fishery | | | | | | |
|---|---|---|---|---|---|---|---|
| Candidate source | Long Island, NY | New Jersey coast | Ocean City, MD 2005 | Ocean City, MD 2006 | Chinco-teague Island, VA | Assa-teague Island, VA | USFWS trawl, NC |
| Gulf of Maine MU | 0.071 | 0.027 | 0.0222 | 0.0930 | 0.085 | 0.000 | 0.000 |
| Hog Bay | 0.000 | 0.0000 | 0.000 | 0.000 | 0.000 | 0.000 | 0.000 |
| Gulf of Maine | 0.071 | 0.027 | 0.022 | 0.093 | 0.085 | 0.000 | 0.000 |
| Mid-Atlantic MU | 0.881 | 0.909 | 0.889 | 0.894 | 0.918 | 1.000 | 0.978 |
| Northeast | 0.225 | 0.289 | 0.244 | 0.234 | 0.340 | 0.657 | 0.211 |
| Delaware Bay | 0.441 | 0.381 | 0.467 | 0.426 | 0.339 | 0.061 | 0.282 |
| Inner Chesap. Bay | 0.000 | 0.000 | 0.067 | 0.116 | 0.000 | 0.000 | 0.098 |
| Mid-coastal | 0.214 | 0.249 | 0.111 | 0.128 | 0.239 | 0.281 | 0.387 |
| Southeast MU | 0.048 | 0.065 | 0.089 | 0.021 | 0.000 | 0.000 | 0.022 |
| Florida Atlantic MU | 0.000 | 0.000 | 0.000 | 0.000 | 0.000 | 0.000 | 0.000 |
| Florida Gulf MU | 0.000 | 0.000 | 0.000 | 0.000 | 0.000 | 0.000 | 0.000 |
| Yucatan, Mexico MU | 0.000 | 0.000 | 0.000 | 0.000 | 0.000 | 0.000 | 0.000 |

**Table 3** Inferred proportion of management unit (MU) of origin for horseshoe crabs in the Chincoteague Island trawl fishery, showing uncertainties associated with point estimates

| Candidate Source | Estimate | (95% CI) |
|---|---|---|
| Gulf of Maine MU | | |
| Hog Bay | 0.0000 | (0.000–0.000) |
| Gulf of Maine | 0.0775 | (0.000–0.196) |
| Mid-Atlantic MU | | |
| Northeast | 0.3187 | (0.006–0.638) |
| Delaware Bay | 0.2786 | (0.002–0.638) |
| Inner Chesapeake Bay | 0.0000 | (0.000–0.134) |
| Mid-coastal | 0.3251 | (0.001–0.503) |
| Southeast MU | 0.0003 | (0.000–0.127) |
| Florida Atlantic MU | 0.0000 | (0.000–0.030) |
| Florida Gulf MU | 0.0000 | (0.000–0.000) |
| Yucatan, Mexico MU | 0.0000 | (0.000–0.000) |

**Table 4** Nightly captures of spawning horseshoe crabs

| Date | Total number | Females | Males | Females resighted[a] | Males resighted[a] | Comments |
|---|---|---|---|---|---|---|
| *Mark phase* | | | | | | |
| May 4 | 831 | 389 | 442 | 0 | 0 | |
| May 5 | 360 | 164 | 196 | 13 | 15 | New moon |
| May 6 | 952 | 388 | 564 | 41 | 41 | |
| May 19 | 8 | 4 | 4 | 0 | 0 | |
| May 20 | 6 | 3 | 3 | 0 | 1 | Full moon |
| May 21 | 473 | 157 | 316 | 2 | 6 | |
| May 22 | 50 | 24 | 26 | 4[b] | 1 | |
| May 23 | 102 | 44 | 58 | 0 | 2[c] | |
| June 2 | 2082 | 681 | 1401 | 24[d] | 46[e] | |
| June 3 | 1656 | 511 | 1145 | 41[g] | 110[i] | New moon |
| June 4[f] | 0 | 0 | 0 | 0 | 0 | |
| June 5 | *2054* | *393* | *1661* | 52[h] | *321*[j] | |
| Subtotals | 8574 | 2758 | 5816 | 177 | 543 | |
| *Recapture phase* | | | | | | |
| June 8 | 2500[k] | 318 | 2182 | 42 | 513 | |
| June 9 | 2500[k] | 526 | 1974 | 36 | 311 | |
| June 10 | 912 | 174 | 738 | 6 | 26 | |
| June 12 | 233 | 54 | 179 | 0 | 2 | |
| June 13 | 121 | 46 | 75 | 1 | 4 | |
| June 15 | 15 | 7 | 8 | 0 | 0 | |
| June 17 | 501 | 183 | 318 | 1 | 4 | Full moon |
| June 18 | 831 | 290 | 541 | 4 | 20 | |
| June 19 | 707 | 255 | 452 | 3 | 14 | |
| June 20 | *1176* | *376* | *800* | *6* | *26* | |
| Subtotals | 9496 | 2229 | 7267 | 99 | 920 | |
| **Totals** | 18,070 | 4987 | 13,083 | 276 | 1463 | |

[a]Marked individuals resighted on subsequent evenings; eight of these individuals had previously been marked; the bulk were marked in this study
[b]One resighted female had spawned during the May new moon
[c]One resighted male had spawned during the May new moon
[d]Eleven resighted females had spawned during the May new moon, and five during the May full moon
[e]Thirty resighted males had spawned during the May new moon, and 18 during the May full moon
[f]Stormy, no fishing
[g]Five resighted females had spawned during the May new moon, and five during the May full moon
[h]Four resighted females had spawned during the May new moon, and four during the May full moon
[i]Twelve resighted males had spawned during the May new moon, and nine during the May full moon
[j]Twenty resighted males had spawned during the May new moon, and 12 during the May full moon
[k]Maximum permitted catch

Over 5000 crabs were tagged in the large run that came at the June new moon. Overall, we marked 8574 horseshoe crabs across these three runs. The sex ratio among marked crabs was 2.11 males per female. During the marking period, we observed few crabs that had already been marked in earlier studies. Six crabs that were observed and released during the marking period had been marked in Delaware Bay in 2003. Two crabs had been captured, bled by a biomedical company, marked, and released off of Ocean City, Maryland in October 2007.

Many marked crabs were recaptured in the local bait fishery, which included the end of the June new-moon run and the subsequent full-moon run. The five cooperating fishers in Tom's Cove hand-captured 9496 crabs over 10 nights, of which 1019 (10.7%) we had tagged. Sex ratios were skewed for both landings and marked crabs; there were 7267 males, of which 920 (12.7%) had been marked; and 2229 females, of which 99 (4.4%) had been marked. The sex ratio in the harvest was skewed, 3.26 males per female, more highly skewed than in the marking period earlier in the spawning season. The 2008 harvest included 31 crabs that had been marked earlier by Walls (2001) or Grogan (2004).

Mark-recapture efforts covered two new and two full moons with their associated spawning peaks; if the spawning assemblage was one well-mixed population, one would expect the number of resighted individuals to increase more or less smoothly within and across spawning periods. However, while numbers of crabs resighted tended to increase *within* spawning periods, they did not increase smoothly *across* spawning periods (Table 3). This observation suggests that while individuals spawning multiple times within a spawning period were relatively common, individuals returning to spawn in multiple spawning periods were relatively uncommon. Males tended to be recaptured more frequently than females, which is consistent with the view that during spawning periods, males hold just offshore waiting for prospective mates, and hence would be subject to a higher likelihood of recapture. In contrast, females spawned and left the beach area.

After spawning, crabs marked in Tom's Cove dispersed, and reports of recapture of marked individuals yielded information on that dispersal. In total, 655 tags were reported 679 times through 2017: 503 males, 163 females, and 13 of unknown sex. Within 30 days of tagging, 295 tags were resighted. Researchers and beach combers sighted 20 dead horseshoe crabs and six live crabs. Biomedical companies observed and released 10 live crabs, while commercial and sport fishers released eight alive and retained 251 crabs for bait. Although most resightings were within or near Tom's Cove, two males were reported about 52 km north at the northern end of Assateague Island near Ocean City, MD 11 and 29 days after tagging, one male was observed 108 km away on a spawning beach in Delaware Bay 30 days after tagging, and one male was reported by a horseshoe crab spawning surveyor only 3 days after tagging, 281 km north at Belmar, NJ.

Within 6 months, 50 marked crabs were caught by the biomedical fishery, 40 in Tom's Cove and 10 near Ocean City, MD, a distance of approximately 60 kilometers. The annual Virginia Tech trawl survey in September 2008 captured two marked crabs off of Assateague Island and eight off of Ocean City, MD. The USFWS database had records of 48 marked crabs reported by beachcombers, mostly found dead. Forty-three were found nearby, in Tom's Cove, Chincoteague Island, Assateague Island, or Nelsonia, VA. To the northeast, one was observed at Ocean City Beach, MD, and three in Delaware Bay (one each at Slaughter Beach, Fowlers Beach, and Prime Hook Beach, DE). Three tagged crabs were reported by sport fishermen, two in Tom's Cove, and one on Assateague Island near Berlin, MD. One tagged crab was observed by a shorebird surveyor in Tom's Cove. Five crabs marked during spawning at Tom's Cove (four males and one female) were observed by spawning surveyors on Skimmer Island, Isle of Wight Bay, near Ocean City, MD, from 26 to 41 days later. One questionable recovery of a dead female was reported 574 km away in Massachusetts 110 days after tagging. To the south, one female was reported from Cobb Island, VA, near the southern end of the Eastern Shore, 65 days after tagging. Of those at-large for 30 days to 6 months, males on average traveled 13.5 km, while females traveled 17.2 km. We note that 17 of the horseshoe crabs caught by biomedical companies near Tom's Cove 0–6 months after tagging were reported as re-released at Cape Charles, at the entrance to Chesapeake Bay.

After 6 months, beach combers, researchers, and unknown reporters observed 135 live and 50 dead horseshoe crabs, 15 loose tags, and two of unknown status. Biomedical companies released seven live individuals, and fishers found three tags and one dead crab, released 25 live horseshoe crabs, and kept 21 individuals. Of the horseshoe crabs at-large for more than 6 months, males traveled an average of 104 km, whereas females traveled 72 km. Seven males and five females were reported alive along the Atlantic coast of the Eastern Shore south of Tom's Cove, in addition to five dead males, one dead female and two loose tags. Four males and two females were reported alive from within Chesapeake Bay, five dead individuals and one tag were reported from the Virginia and North Carolina coasts south of Chesapeake Bay, and one live male was reported from St. Helena Sound, SC, 767 km away and 7.9 years after tagging. North of Delaware Bay, dead crabs were reported from New Jersey (2 individuals) and Long Island (2 crabs). Four live males and two live females were reported from Connecticut. Of those live resightings more than 6 months after tagging, males traveled on average 106 km, while females averaged 83 km. Excluding the male resighted in South Carolina, males averaged 102 km.

Live individuals resighted during the May–June spawning period in subsequent years after tagging were reported from known or likely spawning beaches at Tom's Cove (10 males, 5 females, 2 unknown), Isle of Wight and Assawoman Bays, MD (9 males, 3 females), Indian River and Rehoboth Bays, DE (6 males), and Delaware Bay (86 males, 10 females), and possible spawning beaches in Connecticut (3 males, 2 females). This indicates that the spawning population extends north, but the absence of resightings associated with spawning to the south may simply reflect fewer opportunities for encounters.

# 4 Discussion

## *4.1 Population Genetics*

Eggs deposited by horseshoe crabs spawning in Delaware Bay provide an important feeding resource for migrating red knots (Botton et al. 1994; Botton and Harrington 2003; USFWS 2003; Karpanty et al. 2007; USFWS 2014). However, "Delaware

Bay" horseshoe crabs are not comprised only of assemblages spawning in the Bay itself. Rather, Delaware Bay crabs are part of a larger metapopulation of spawning assemblages ranging from southern coastal New Jersey to coastal Virginia (King et al. 2005, 2015). As distance from the Delaware Bay increases, other sources also contribute to the crabs spawning or harvested at a site.

Our results show that regional fisheries draw approximately 90% of the crabs harvested from spawning assemblages within the mid-Atlantic MU proposed by King et al. (2015), with some contributions from adjacent MUs to the north and south (Table 2). This finding supports the recognition of the mid-Atlantic populations as an MU, that is, a unit that is demographically independent from other such assemblages; the criterion for demographic independence is often taken at fewer than 10% of individuals being migrants from other units (Hastings 1993).

Metapopulation dynamics, characterized by migration among local demes, would tend to buffer the demographics of exploited horseshoe crab populations. Demographic linkage to coastal horseshoe crab populations effectively increases the number of crabs in the "Delaware Bay" population so critical to recovery of red knot. With effective monitoring of relative abundance of adult horseshoe crabs in the Delaware Bay region, limited harvest can go forward while protecting red knots and other species dependent upon horseshoe crabs. The Adaptive Resource Management Framework utilized by the ASMFC recognizes that Delaware Bay crabs are part of a larger metapopulation and applied population genetic estimates of the contributions from regional fisheries within recommended harvest limits.

## 4.2 *Movement*

Mark-recapture studies at various locations across their range have provided insights into the basic biology of horseshoe crabs (Rudloe 1980; Shuster Jr 1985; Brockman 1990; Grogan 2004; Swan 2005; Moore and Perrin 2007). There have been mark-recapture studies focusing on the regional biomedical fisheries for horseshoe crabs (Walls 2001; Grogan 2004) and some crabs are marked during the annual Virginia Tech trawl survey. We build upon this marking effort. The few crabs captured in Tom's Cove in this study that were already tagged had been marked in the Delaware Bay and off of Ocean City, MD. Although 80,000 horseshoe crabs had been marked along the Atlantic coast when we embarked upon marking (Sheila Eyler, USFWS, personal communication), they represented only a small proportion of the adult stock, which was estimated to number in the millions within Delaware Bay alone (Smith et al. 2006). Hence, the number of already-marked individuals observed in this study under-represents the true proportion of in-migrants. We note that the biomedical industry handles tens of thousands of crabs each year. At one time, these operations were required to mark all crabs handled, but now mark only a small proportion. Given management interest in estimating numbers and delineating the spatial extent of the Delaware Bay MU that they exploit, biomedical companies could be required to mark all crabs handled.

Grogan (2004) marked 621 crabs spawning in Tom's Cove in 2004. Fishers there caught "hundreds" of marked crabs in the 2–3 years after that marking effort (Chris Cuono, fisher, personal communication), but did not report the recaptures to the USFWS. That 31 crabs marked by Grogan (2004) in Tom's Cove subsequently were recaptured there suggests considerable spawning site fidelity. Walls (2001) and Grogan (2004) reported movement of 12,500 horseshoe crabs marked after capture in biomedical fisheries and released at Chincoteague, VA, or Ocean City, MD, between 1999 and 2004. Grogan (2004) reported the recaptures of those marked crabs through 2004, showing a tendency for migration to the northeast. Horseshoe crabs tagged and released from Chincoteague, Virginia, after being bled by Cambrex in the years 1999–2002 were resighted from Corolla, NC, to Long Island, NY, with most resightings from just south of Chincoteague, VA,to Cape May, NJ (Grogan 2004). Horseshoe crabs tagged and released from Ocean City, Maryland, after being bled by Cambrex in the years 1999–2002 and 2004 were resighted from Virginia Beach, VA to the central New Jersey coast, with most resightings from just south of Chincoteague, VA, to Cape May, NJ (Grogan 2004). Our analysis of subsequent tag returns showed that these crabs were resighted 53 additional times, once in Virginia, twice in Maryland, nine times in Delaware, 22 New Jersey, 17 in New York, and twice in Rhode Island. By 2010, resightings included 30 in Virginia, 6 in Maryland, 30 in Delaware, 28 in New Jersey, 29 in New York, and two in RI. Of these 165 resightings, 70 were in Virginia, 6 in Maryland, 30 in Delaware, 28 in New Jersey, 29 in New York, and 2 in Rhode Island; none were to the south. Individuals tagged in the Delaware Bay area during the annual horseshoe crab trawl survey have been resighted from Connecticut to Virginia Beach, VA, and individuals tagged along Long Island have been resighted from Atlantic City, NJ, to Rhode Island Sound (Hata and Hallerman, unpublished data). In this study, 94% of the 679 resighting reports were between Wallop's Island adjacent to Tom's Cove and Delaware Bay, 4% were south of Wallop's Island, and 2% were north of Delaware Bay.

## 4.3 Management Perspectives

Horseshoe crab managers must understand the degree to which exploited coastal populations are demographically connected to the Delaware Bay spawning population that is critical to the sustainability of the fishery and the threatened *rufa* red knot (ASMFC 2009; McGowan et al. 2011a, 2011b). Previous work (King et al. 2005, 2015) has shown that while there is some degree of subpopulation structure, that mid-Atlantic populations are genetically distinct from assemblages to the northeast and south. Our analyses of landings in five selected regional fisheries shows that the take is approximately 90% comprised of crabs from the mid-Atlantic MU. Fishery managers and law enforcement officers would want to be able to identify the sources of origin of landed horseshoe crabs that arouse suspicion of having been transferred from a protected area or regulated fishery; however, we were not able to identify the subpopulations of origin for crabs in the mid-Atlantic MU with precision, certainly not to the standards expected for prosecution of alleged crime.

Observations of migrating crabs participating in spawning at sites, not where they were marked, suggests genetically effective migration. We observed small numbers of crabs migrating into Tom's Cove from Delaware Bay and coastal waters. Crabs marked in Tom's Cove were observed spawning near Ocean City, Maryland, Indian River Bay, Delaware, Delaware Bay, and possibly Long Island Sound. Mark-recapture studies of crabs handled by the biomedical industry (Walls 2001; Grogan 2004) support the hypothesis that assemblages of spawners across the region are linked by dispersal of individual crabs. Crabs marked in Virginia and Maryland were observed to tend to disperse to the northeast, although none were observed to have passed Cape Cod, an area of discontinuity noted by King et al. (2005, 2015) in population genetics studies.

The inferences reached from earlier population genetic and movement studies informed the spatial boundaries of the Adaptive Resource Management approach adopted by the Atlantic States Marine Fisheries Commission (2009). Our inferences of harvest composition and movement of crabs tagged in Tom's Cove contribute to our understanding of demographic processes ongoing within the Delaware Bay metapopulation and support the case for management of regional spawning assemblages as one MU.

**Acknowledgements** This project was supported by a grant from the National Marine Fisheries Service. Special thanks go to Rick Robins (Chesapeake Bay Packing) and Dave Smith (U.S.G.S. – Leetown Science Center) for their assistance with the conception and design of the movement study. The Assateague Island National Sea Shore, Chincoteague National Wildlife Refuge, and Virginia Marine Resources Commission granted scientific collection permits for this study. We gratefully acknowledge the skill and assistance of horseshoe crab fishers Chris, Amy and Amber Cuono, Randy Birch, Mike Gibson, and Ray Winbrow, Jr.; and of volunteers Jeanne Covert, Ester and Simon Hallerman, Koan Heindel, Pridasang Kadkhuntod, Caitlin Kelliher, Doug King, Pawapol Kongchum, Teeraphan Laomettachit, Steve Nash, Mike Oates, Lindsey Pierce, Audrey and Christy Ramsey, Rick Robins, Samara Silva, and Evoy Zaniboni Filho. Sheila Eyler and Joshua Newhard (U.S. Fish and Wildlife Service) provided data on recaptures of horseshoe crabs and Sheila Eyler assisted with quality assessment of the dataset. Lindsey Pierce and Tamar Hallerman assisted with data entry. Rick Robins and Sheila Eyler provided useful comments on the movement portion of the manuscript. EH's participation in this study was supported in part by the U.S. Department of Agriculture National Institute of Food and Agriculture. Authors EH and ME mourn the untimely passing of authors Tim King and Dave Hata, whose efforts were critical to the realization of this study.

## References

ASMFC (Atlantic States Marine Fisheries Commission), Interstate Fishery Management Plan for Horseshoe Crab. Fishery Management Report No. 32, ASMFC, Arlington, VA. http://www.asmfc.org/species/horseshoe-crab (1998)

ASMFC (Atlantic States Marine Fisheries Commission), A framework for adaptive management of horseshoe crab harvest in the Delaware Bay constrained by red knot conservation. Stock Assessment Report No. 09-02 (Supplement B). ASMFC, Arlington, VA. http://www.asmfc.org/species/horseshoe-crab (2009)

M.L. Botton, R.E. Loveland, T.R. Jacobsen, Site selection by migratory shorebirds in Delaware Bay as its relationship to beach characteristics and abundance of horseshoe crab (*Limulus polyphemus*) eggs. Auk **111**, 605–616 (1994)

M.L. Botton, B.A. Harrington, Synchronies in migration: Shorebirds, horseshoe crabs, and Delaware Bay, in *The American Horseshoe Crab*, ed. by C. N. Shuster Jr., R. B. Barlow, H. J. Brockmann, (Harvard University Press, Cambridge, MA, 2003), pp. 5–32

H.J. Brockman, Mating behavior of horseshoe crabs (*Limulus polyphemus*). Behaviour **114**, 206–220 (1990)

S. Faurby, T.L. King, M. Obst, E.M. Hallerman, C. Pertoldi, P. Funch, Population dynamics of American horseshoe crabs – Historic climatic events and recent demographic pressures. Mol. Ecol. **19**, 3088–3100 (2010)

W.N. Grogan, *A Mid-Atlantic Study of the Movement Patterns and Population Distribution of the American Horseshoe Crab (Limulus Polyphemus)*. (MS Thesis, Virginia Polytechnic Institute and State University, Blacksburg, VA, 2004)

A. Hastings, Complex interactions between dispersal and dynamics – Lessons from coupled logistic equations. Ecology **74**, 1362–1372 (1993)

D.N. Hata, E.M. Hallerman, *Preliminary Results of the 2017 Horseshoe Crab Trawl Survey: Report to the Atlantic States Marine Fisheries Commission Horseshoe Crab and Delaware Bay Ecology Technical Committees* (ASMFC, Arlington, VA, 2018). 29 p

HSC-TC (Horseshoe Crab Technical Committee), Draft Horseshoe Crab Technical Committee Report, Baltimore, MD, January 17, 2008 (2008)

S.M. Karpanty, J.D. Fraser, J. Berkson, L.J. Niles, A. Dey, E.P. Smith, Horseshoe crab eggs determine red knot distribution in Delaware Bay. J Wildl. Mgmt. **70**, 1704–1710 (2007)

T.L. King, M.S. Eackles, A.W. Aunins, H.J. Brockmann, E. Hallerman, B.B. Brown, Conservation genetics of the horseshoe crab (*Limulus polyphemus*): Allelic diversity, zones of genetic discontinuity, and regional differentiation, in *Changing Global Perspectives on Biology, Conservation and Management of Horseshoe Crabs*, ed. by R. Carmichael, M. L. Botton, P. Shin, S. G. Cheung, (Springer, Berlin, 2015), pp. 65–96

T.L. King, M.S. Eackles, A.P. Spidle, H.J. Brockmann, Regional differentiation and sex-biased dispersal among populations of the horseshoe crab, *Limulus polyphemus*. Trans. Am. Fish Soc. **134**, 441–465 (2005)

M.M. Manion, R.A. West, R.E. Unsworth, Economic assessment of the Atlantic coast horseshoe crab fishery. Division of Economics, U.S. Fish and Wildlife Service, Arlington, VA. 71 p (2000)

C.P. McGowan, J.E. Hines, J.D. Nichols, J.E. Lyons, D.R. Smith, K.S. Kalasz, L.J. Niles, A.D. Dey, N.A. Clark, P.W. Atkinson, C.D. Minton, Demographic consequences of migratory stopover: Linking red knot survival to horseshoe crab spawning abundance. Ecosphere **2**(6), 1–22 (2011a)

C.P. McGowan, D.R. Smith, J.A. Sweka, J. Martin, J.D. Nichols, R. Wong, J.E. Lyons, L.J. Niles, K. Kalasz, J. Brust, M. Klopfer, Multispecies modeling for adaptive management of horseshoe crabs and red knots in the Delaware Bay. Nat. Resour. Model. **24**(1), 117–156 (2011b)

S. Moore, S. Perrin, Seasonal movement and resource-use patterns of resident horseshoe crab (*Limulus polyphemus*) populations in a Maine, USA estuary. Estuar. Coasts **30**, 1016–1026 (2007)

C. Moritz, Defining 'evolutionary significant units' for conservation. Trends Ecol. Evol. **9**, 373–375 (1994)

J.P. Pierce, G. Tan, P.M. Gaffney, Delaware Bay and Chesapeake Bay populations of the horseshoe crab (*Limulus polyphemus*) are genetically distinct. Estuaries **23**, 690–698 (2000)

J.K. Pritchard, M. Stephens, P. Donnelly, Inference of population structure using multilocus genotype data. Genetics **155**, 945–959 (2000)

A. Rudloe, The breeding behavior and patterns of movement of horseshoe crabs, *Limulus polyphemus*, in the vicinity of breeding beaches in Apalachee Bay, Florida. Estuaries **3**, 177–187 (1980)

O. Ryder, Species conservation and systematics: The dilemma of subspecies. Trends Ecol. Evol. **1**, 9–10 (1986)

N.C. Saunders, L.G. Kessler, J.C. Avise, Genetic variation and geographic differentiation in mitochondrial DNA of the horseshoe crab, *Limulus polyphemus*. Genetics **112**, 613–627 (1986)

R.K. Selander, S.Y. Yang, R.C. Lewontin, W.E. Johnson, Genetic variation in the horseshoe crab (*Limulus polyphemus*), a phylogenetic "relic". Evolution **24**, 402–414 (1970)

C.N. Shuster Jr., Introductory remarks on the distribution and abundance of the horseshoe crab, *Limulus polyphemus*, spawning in the Chesapeake Bay area, in *The Chesapeake: Prologue to the Future. Proceedings of the Chesapeake Bay Symposium, National Marine Educators Conference*, (1985), pp. 34–38

D.R. Smith, M.J. Millard, S. Eyler, Abundance of adult horseshoe crabs (*Limulus polyphemus*) in Delaware Bay estimated from a bay-wide mark-recapture study. Fish. Bull. **104**, 456–464 (2006)

B.L. Swan, Migrations of adult horseshoe crabs, *Limulus polyphemus*, in the Middle Atlantic Bight: A 17-year tagging study. Estuaries **28**, 28–40 (2005)

USFWS (U.S. Fish and Wildlife Service), Delaware Bay shorebird – horseshoe crab assessment report and peer review. Migratory Bird Publication R9-03/02. (USFWS, Arlington, VA, 2003)

USFWS (U.S. Fish and Wildlife Service), Endangered and threatened wildlife and plants: Threatened species status for the *rufa* red knot. Fed. Regist. **79**(238), 73706–73748 (2014)

E.A. Walls, *A Horseshoe Crab Demographic Study*. (MS Thesis, Virginia Polytechnic Institute and State University, Blacksburg, VA, 2001)

E.A. Walls, J. Berkson, S.A. Smith, The horseshoe crab, *Limulus polyphemus*: 200 million years of existence, 100 years of study. Rev. Fish. Sci. **10**, 39–73 (2002)

R.S. Waples, O. Gaggiotti, What is a population? An empirical evaluation of some genetic methods for identifying the number of gene pools and their degree of connectivity. Mol. Ecol. **15**, 1419–1439 (2006)

# Microbiome Associated with Laboratory Bred Horseshoe Crabs: A Case Study

Vishal Shah, John T. Tanacredi, and Viraj Joshi

## 1 Introduction

Horseshoe crabs (*Limulus polyphemus*) are widely regarded as the exemplars of bradytelic evolution (Xuhua 2000). They are prototypical 'living fossils' of biology textbooks, palaeontological treatises, and popular natural history accounts, with fossils dating back 445 million years (Rudkin et al. 2008). The population of *Limulus* has seen a sharp reduction in recent times over much of its natural habitat in the Atlantic Ocean (Tanacredi et al. 2009). All four species of horseshoe crabs have been subject to considerable conservation research, as well as levels of sustainability over the last three decades (Tanacredi 2002; Tanacredi et al. 2009). The primary reason is understood to be over-harvest of the organism for commercial applications (Smith et al. 2016). As the global community continues to learn more about physiology, genetics, and health of organisms, in an effort to conserve the organisms, it is equally important to understand the microorganisms living in and on the horseshoe crabs. Interactions between multicellular organisms and their associated microbes are fundamentally important aspects from the development of an organism to systems ecology (McFall-Ngai et al. 2013). The health and well-being of the organisms depend on these interactions. No information is available in the literature on the microbial assemblage associated with *Limulus.* Limited

V. Shah (✉)
College of the Sciences and Mathematics, West Chester University, West Chester, PA, USA
e-mail: vshah@ccp.edu

J. T. Tanacredi
Centre for Environmental Research and Coastal Oceans Monitoring, Molloy College, Rockville Centre, New York, NY, USA

V. Joshi
Unionville High School, Kennett Square, PA, USA



J. T. Tanacredi et al. (eds.), *International Horseshoe Crab Conservation and Research Efforts: 2007-2020*, https://doi.org/10.1007/978-3-030-82315-3_5

information available in the literature on the microbiome present in the aquatic animals suggests that the relationship between the symbiotic bacteria and the environment in aquatic animals is more complex than in terrestrial animals (Rungrassamee et al. 2013; Zhang et al. 2016). Moreover, it is known that aquatic animal harbours more specific microbial flora than the microorganisms found in the water and that the microbiome is site-specific (Zhang et al. 2016). With increasing temperature due to global climate change, acidification of the oceans, and increasing pollution, microbial assemblages associated with *Limulus* could be impacted. These could very well have a large impact on the health and community structure of the organism. While our understanding of microbiome environmental changes in marine environment is meagre, Liu et al. (2010) and Witt et al. (2011) have revealed how microbial assemblages in open waters and the Great Barrier Reef are affected by ocean acidification. It would not be far-fetched to state that microbiome associated with the aquatic animals would also be sensitive to environmental changes. In this case study, we report the composition of the microbiome associated with a male and a female *Limulus* that were laboratory breed.

## 2 Methods

**Sample Collection** A male and female horseshoe crab was selected for sampling, and the same crabs were sampled both times in April and October months of 2016. Horseshoe crabs were estimated to be approximately 15 years old (Fig. 1) and had been maintained for captive spawning events for approximately 6–8 months at the CERCOM facility of Molloy College, NY. Sexually mature adults were maintained in troughs at constant saltwater temperature of 14 °C and salinity of 26 ppt. The subject pair had not produced a spawning event in approximately 7 months prior to the collection of microbial swab sample. The crabs were brought on a sterile (alcohol washed) dissecting table and using swab samples collected for each horseshoe crab from mouth, eye, anal, gills, and heart regions. For mouth samples, swabs were collected twice – a topical sample and a sample approximately 2 inches into mouth opening. During sample collection, it was not possible to avoid contamination between the two samples, and hence, we are not distinguishing the two samples in this study. They are labelled Mouth 1 and Mouth 2 sample for each round of study. Overall, the sampling frequency ($n$) for each site was 4, except for the mouth where $n$ was 8. From the tank, 50 mL of water sample was also collected in a sterile tube at both time points.

**Microbial DNA Extraction and Sequencing** DNA was extracted from the swabs and water using PowerSoil™ DNA isolation kits (MO BIO Laboratories Inc., Carlsbad, CA). DNA amplification of the 16S rRNA was performed at Molecular Research Laboratory (Shallowater, TX, USA). The 16S rRNA gene V4 variable region was amplified using the PCR primers 515f (GTGCCAGCMGCCGCGGTAA)/806r (GGACTACHVGGGTWTCTAAT)

**Fig. 1** Female (left) and male (right) horseshoe crabs used to collect sample in this study

(Caporaso et al. 2012). PCR amplifications were conducted using a single step 30 cycle PCR using the HotStarTaq Plus Master Mix Kit (Qiagen, USA) under the following conditions, 94 °C for 3 minutes, followed by 28 cycles of 94 °C for 30 seconds, 53 °C for 40 seconds, and 72 °C for 1 minute, after which a final elongation step at 72 °C for 5 minutes was performed. Negative PCR controls were included in all amplification reactions (Santiago-Rodriguez et al. 2015). After amplification, PCR products were checked in 2% agarose gel to determine the success of amplification and the relative intensity of the bands. Multiple PCR products were pooled together in equal proportions based on their molecular weight and DNA concentrations. Pooled samples were purified using Agencourt AMPure beads (Agencourt Bioscience Corporation, MA, USA). The pooled and purified PCR products were used to prepare the DNA library following Illumina MiSeq DNA library preparation protocol using the MiSeq reagent kit V3 (2X300 bp) for paired-end reads on a MiSeq following the manufacturer's guidelines. In summary, sequences were joined and depleted of barcodes, and then sequences <150 bp were removed.

**Taxonomical Analyses of Bacterial Communities** FASTQ files corresponding to the horseshoe crab's samples were joined using the QIIME pipeline using join_paired_ends.py (Caporaso et al. 2010) and depleted of barcodes, and then sequences <150 bp were removed. Sequences were denoised, OTUs generated, and chimeras removed. Operational taxonomic units (OTUs) were defined by clustering at 3% divergence (97% similarity). De novo operational bacterial and operational taxonomic units (OTUs) were selected using pick_de_novo_otus.py workflow. 16S rRNA taxonomy was defined by ≥97% similarity to reference sequences. The phylogenetic composition of the microcommunities present in the samples was characterized using summarize_taxa_through_plots.py up to the genus level. In order to assign presumptive bacterial species, BLASTn of the 16S rRNA gene sequences was performed against the NCBI 16S rRNA gene database downloaded using CLC Genomics Workbench 8.0 (CLC bio USA, Cambridge, MA, USA). Rarefaction curves and diversity metrics were calculated in QIIME (Caporaso et al. 2010). Tree cluster analysis was carried out using Statistica (Release 10.0) software, selecting Ward's method as the amalgamation rule and the linkage distance measured as Euclidean units.

## 3 Results and Discussion

Clustering analysis was performed using the bacterial diversity at the genus level. When the analysis was performed with the microbiome in the water samples alongside those on horseshoe crabs, the water samples were not part of any clusters and were outliers. To allow ease of following the results, we did not include water samples in further analysis. Clustering analysis results for microbiome on horseshoe crabs indicate that samples collected from different sites cluster differently, demonstrating clearly the difference in bacterial biota in different sites (Fig. 2). For the most part, bacterial biota seems to be the same for male (M) and female (F) sites as they clustered together. However, for the majority of sites, the microbiome from the same sites at different sampling times did not cluster together, indicating a shift in microbiome as a function of time (Fig. 2). PCoA plots support the clustering results with bacterial biota from round 1 and round 2 distinctly clustering (Fig. 3). The results clearly indicate that there is a sampling time dependency and a change in microbiome of the horseshoe crabs observed across all sites. While the exact reasons for the change in microbiome in the horseshoe crabs are not clear, we hypothesize that the saltwater well, which provides all water for each aquaculture trough, may have profound influence along with the possible changes in the horseshoe crab metabolism as the breeding season approaches. It is important to note that residence time of water varies during a breeding event, after a feeding cycle, or if both events are occurring concurrently.

Multiple rarefaction curves using the Shannon index (Fig. 4) showed that plateau of diversity was achieved at around 5902 sequences per sample. This value would be considered as the minimum sampling depth to capture diversity. Our rarefaction was performed at a minimum of 42,000 sequences per sample, which therefore is a realistic panorama of each sample's diversity. The curves clearly indicate that female horseshoe crabs have more microbial diversity than male crabs. The anal region of the female harbours the highest microbial diversity (particularly when sampled in October). The second mouth samples had the least bacterial diversity.

Analysis of the 16s rRNA gene revealed that members of 54 phyla were detected across all samples, with *Proteobacteria* being the most abundant phyla present across all the sites (Table 1 lists the complete microbiome at the phyla level). *Acidobacteria*, *Bacteroidetes*, *Cyanobacteria*, and Nitrospirae were revealed as the

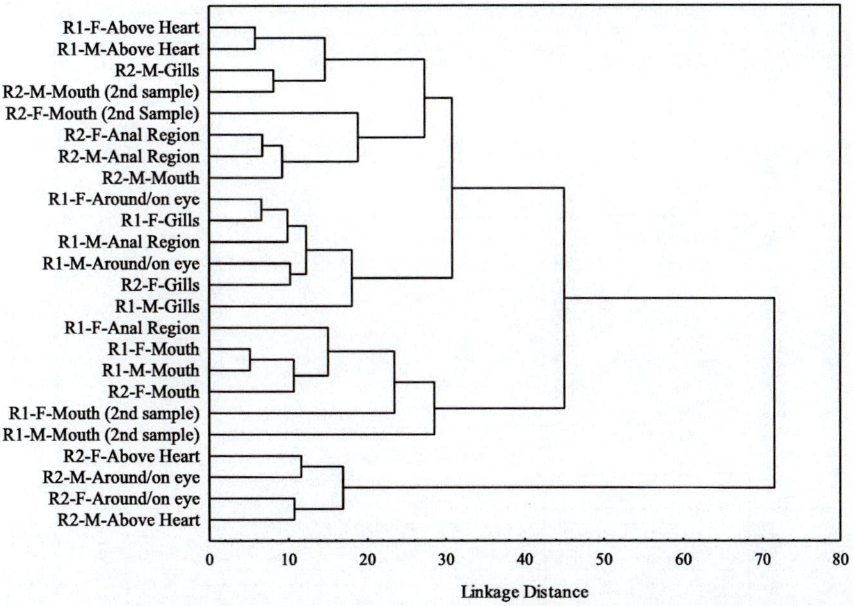


**Fig. 2** A hierarchical clustering dendrogram showing the differences between the microbiome at various sites of horseshoe crabs. (R1, samples collected in April; R2, samples collected in October; M, male; F, female)

other phyla in the top five most prevalent ones in horseshoe crabs (Table 1). Unclassified bacteria were around 1% in all the samples. There was no correlation between the microbiome found in the water sample and those on the horseshoe crabs.

*Proteobacteria* being the most abundant phyla provides us with a further understanding of why the blue blood of the horseshoe crab contains a sophisticated immune system capable of detecting trace amounts of lipopolysaccharide (LPS) molecules (Muta and Iwanaga 1996). All organisms in the *Proteobacteria* phyla are gram-negative in nature, having LPS layer in the outer membrane. Also, many of the organisms in the phyla are parasitic in nature. Kaiser and Benner (2008) reported that the marine environment is dominated by gram-negative organisms. Evolutionarily, it would be advantageous for horseshoe crabs to develop a system to protect itself from an invasion by abundant organisms present on itself and in its surroundings. Modern horseshoe crabs seem to have developed a system where if any of the external flora is released into the blue blood, hemocytes circulating in the hemolymph detect the trace amounts of lipopolysaccharide molecules of gram-negative bacteria and respond quickly to coagulate the organisms. The organisms are then killed by antimicrobial molecules released by various cells in the blood (Levin 1988; Muta and Iwanaga 1996). Not surprisingly, the evolutionary war between horseshoe crabs and *Proteobacteria* continues even today, with many of the organisms in the microbiome capable of causing disease in a healthy crab (Nolan et al. 2009).

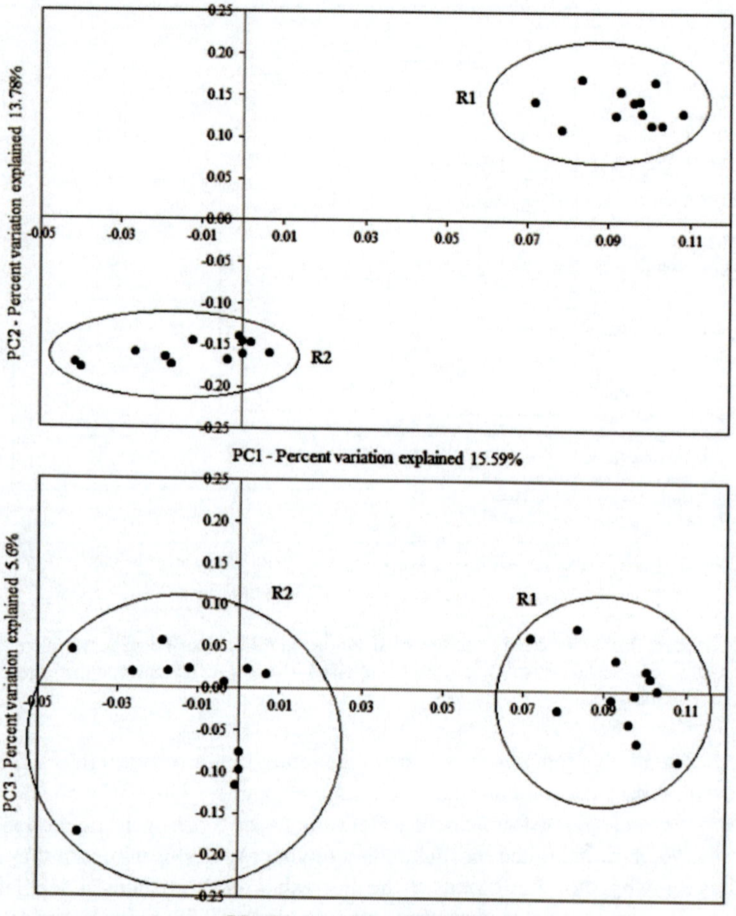


**Fig. 3** Unweighted PCoA analysis of the microbiome of each site of horseshoe crabs for samples collected in April (R1) and October (R2)

Analysis of the microbiome at the genus level (Fig. 5) suggests that genera involved in nitrogen fixation (*Cyanobacteria*), ammonia oxidation (*Nitrosomonas*), and nitrite oxidization (*Nitrospira*) are present as normal flora. Presence of *Nitrosomonas* and *Nitrospira* in all the samples analysed suggests that indeed the organisms could be having a synergistic relationship with the horseshoe crab, providing an inorganic source of nitrogen. It is important to note that in the water samples analysed during both sampling periods, all three microorganisms were absent (Table S2). Lee and Childress (1994) have reported similar synergy between marine invertebrates and microbial symbionts for assimilation of inorganic nitrogen. Horseshoe crabs are shelf animals and are present at the edge of the continental shelf, at depths of 290 m or more during non-spawning seasons (Botton and Ropes

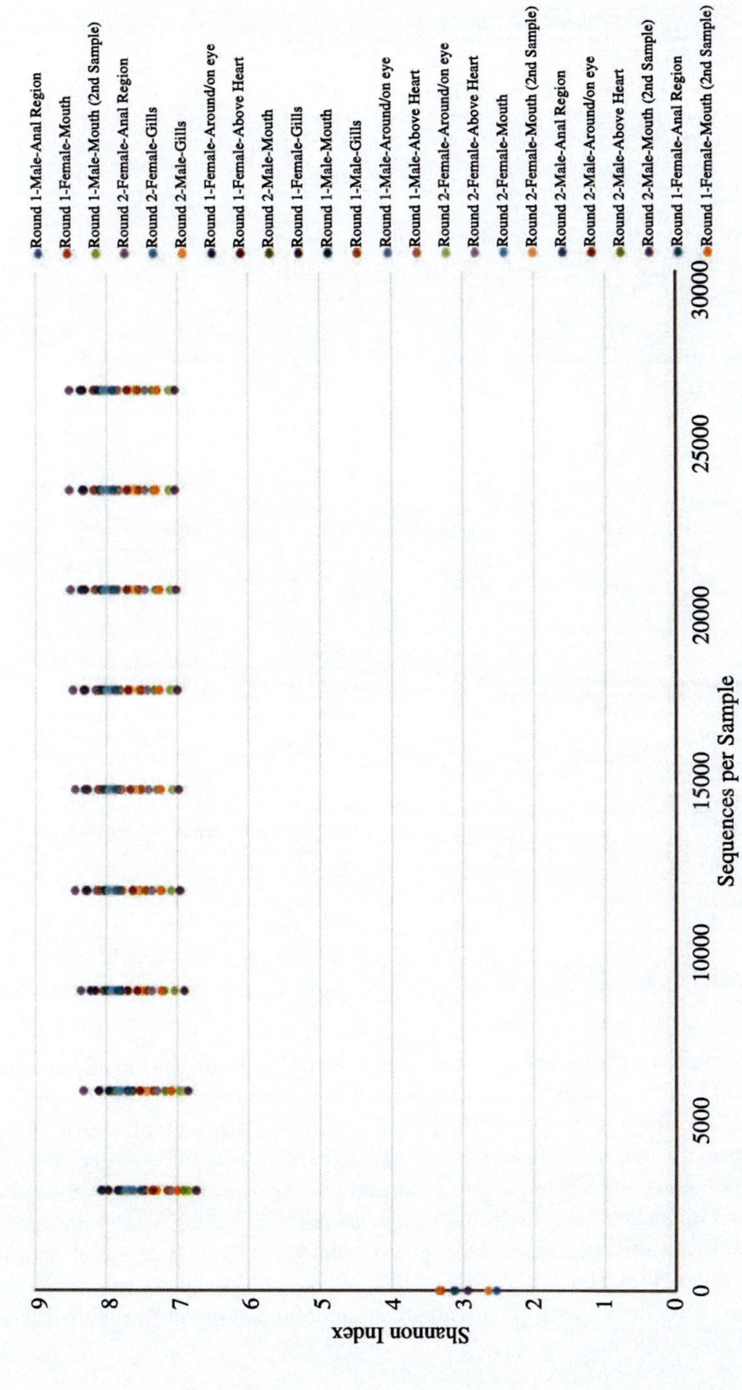


**Fig. 4** Microbial richness inferred using Shannon index for each site of horseshoe crabs. (R1, samples collected in April; R2, samples collected in October)

**Table 1** Top five most abundant microbial phyla across all sites of horseshoe crabs

| | *Acidobacteria* | *Bacteroidetes* | *Cyanobacteria* | *Nitrospirae* | *Proteobacteria* |
|---|---|---|---|---|---|
| R1-F-above heart | 2.5% | 2.1% | 0.5% | 14.6% | 71.9% |
| R1-F-anal region | 1.4% | 5.4% | 2.0% | 6.0% | 67.0% |
| R1-F-around/on eye | 2.0% | 6.7% | 0.6% | 5.5% | 74.9% |
| R1-F-gills | 3.6% | 2.9% | 0.1% | 5.8% | 75.0% |
| R1-F-mouth | 1.4% | 8.2% | 0.1% | 2.2% | 76.5% |
| R1-F-mouth (second sample) | 0.4% | 8.6% | 0.1% | 0.5% | 65.4% |
| R1-M-above heart | 1.0% | 3.7% | 0.6% | 14.6% | 69.5% |
| R1-M-anal region | 2.8% | 4.1% | 0.2% | 6.8% | 70.5% |
| R1-M-around/on eye | 2.0% | 5.0% | 0.5% | 6.8% | 77.5% |
| R1-M-gills | 2.5% | 4.5% | 0.1% | 5.2% | 77.5% |
| R1-M-mouth | 4.0% | 5.3% | 0.3% | 2.3% | 77.3% |
| R1-M-mouth (second sample) | 0.7% | 4.9% | 0.1% | 4.2% | 60.2% |
| R2-F-above heart | 1.9% | 4.1% | 17.6% | 10.7% | 55.3% |
| R2-F-anal region | 5.2% | 11.4% | 4.9% | 8.5% | 57.2% |
| R2-F-around/on eye | 1.3% | 7.7% | 28.8% | 5.0% | 48.4% |
| R2-F-gills | 4.5% | 4.3% | 1.4% | 11.5% | 67.5% |
| R2-F-mouth | 3.9% | 9.6% | 1.6% | 3.7% | 68.3% |
| R2-F-mouth (second sample) | 2.6% | 11.7% | 0.8% | 3.4% | 74.6% |
| R2-M-above heart | 1.8% | 5.8% | 24.4% | 4.3% | 56.8% |
| R2-M-anal region | 5.8% | 11.2% | 9.1% | 9.8% | 55.5% |
| R2-M-around/on eye | 3.0% | 12.3% | 12.7% | 4.6% | 59.9% |
| R2-M-gills | 5.9% | 5.2% | 1.3% | 15.6% | 60.7% |
| R2-M-mouth | 2.5% | 11.3% | 4.6% | 6.7% | 62.7% |
| R2-M-mouth (second sample) | 3.4% | 4.2% | 1.4% | 13.5% | 68.6% |

1987). In fact, horseshoe crabs have been observed at depths of 1097 m (Botton and Ropes 1987). At such depths, the presence of organic nitrogen is low, requiring them to utilize inorganic nitrogen from its surroundings. This symbiotic relationship between bacteria and horseshoe crabs could also explain how the crabs could have survived periods of mass extinction when organic nitrogen could be less in the natural habitat. Organisms from the genus *Roseovarius* are also significantly present in all sites. Interestingly, organisms belonging to this genus have been isolated from deep-sea water sites (Lai et al. 2011, Jung et al. 2012). Organisms from the *Marinicella* and *Marinobacter* genus are other predominant organisms in the horseshoe crab microbiome.

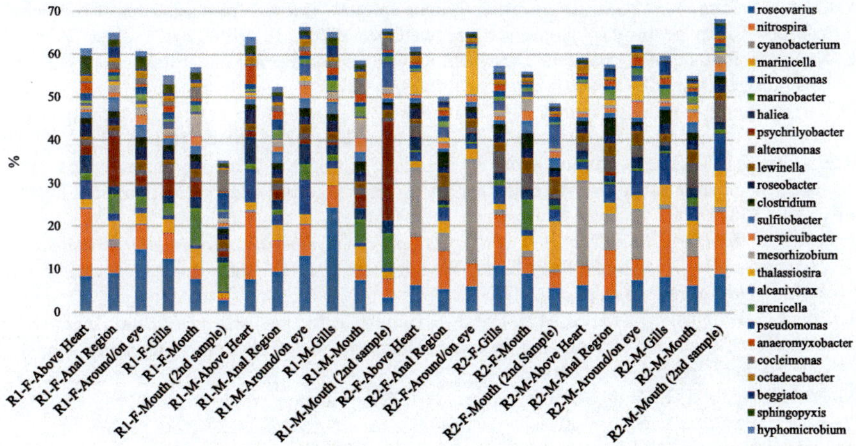


**Fig. 5** Bacterial phylogenetic composition for each site of horseshoe crabs at the genus level. (R1, samples collected in April; R2, samples collected in October; M, male; F, female)

In summary, our results suggest that the microbiome associated with horseshoe crabs could have evolutionary significance. Considering the horseshoe crabs to have originated hundreds of million years back, many of the microorganisms present could be unique and provide detailed insight on the primitive life present on the Earth during that historic period. We also understand that the microbiome has a dynamic nature and can be influenced by a large variety of factors (Cheng et al. 2015). Further studies are warranted, comparing the microbiome of horseshoe crabs raised in captivity and those found in the wild. Also, we need to monitor how the microbiome changes as the crabs undergo developmental changes. Only then can we establish a baseline microbiome of horseshoe crabs and better understand the relationship between the host and associated microbiome.

**Acknowledgement** VS thanks the Foundation for Ecological Research in the Northeast for funding the proposal. JT thanks Molloy College for funding research at CERCOM.

## References

M.L. Botton, J.W. Ropes, Populations of Horseshoe Crabs, *Limulus Polyphemus,* on the northwestern Atlantic continental shelf. Fish. Bull. **85**(4), 805–812 (1987)

J.G. Caporaso, J. Kuczynski, J. Stombaugh, K. Bittinger, F.D. Bushman, E.K. Costello, et al., QIIME allows analysis of high-throughput community sequencing data. Nat. Methods **7**(5), 335–336 (2010)

J.G. Caporaso, C.L. Lauber, W.A. Walters, et al., Ultra-high-throughput microbial community analysis on the Illumina HiSeq and MiSeq platforms. ISME J. **6**(8), 1621–1624 (2012)

Y. Cheng, S. Fox, D. Pemberton, C. Hogg, A.T. Papenfuss, K. Belov, The Tasmanian devil microbiome—implications for conservation and management. Microbiome **3**(1), 76 (2015)

Y.T. Jung, S. Park, J.H. Yoon, Roseovarius litoreus sp. nov., isolated from seawater of southern coast of Korean peninsula. Antonie Van Leeuwenhoek **102**(1), 141–148 (2012)

K. Kaiser, R. Benner, Major bacterial contribution to the ocean reservoir of detrital organic carbon and nitrogen. Limnol. Oceanogr. **53**(1), 99 (2008)

Q. Lai, H. Zhong, J. Wang, J. Yuan, F. Sun, et al., Roseovarius indicus sp. nov., isolated from deep-sea water of the Indian Ocean. Int. J. Syst. Evol. Microbiol. **61**(9), 2040–2044 (2011)

R.W. Lee, J.J. Childress, Assimilation of inorganic nitrogen by marine invertebrates and their chemoautotrophic and methanotrophic symbionts. Appl. Environ. Microbiol. **60**(6), 1852–1858 (1994)

J. Levin, in *The Horseshoe Crab: A Model for Gram-Negative Sepsis in Marine Organisms and Humans*, ed. by J. Levin, H. R. Buller, J. W. Ten Cate, S. J. H. VanDeventer, A. Sturk, (Alan R. Liss, New York, 1988), pp. 3–15

J. Liu, M.G. Weinbauer, C. Maier, M.H. Dai, J.-P. Gattuso, Effect of ocean acidification on microbial diversity and on microbe-driven biogeochemistry and ecosystem functioning. Aquat. Microb. Ecol. **61**, 291–305 (2010)

M. McFall-Ngai, M.G. Hadfield, T.C.G. Bosch, H.V. Carey, T. Domazet-Loso, et al., Animals in a bacterial world, a new imperative for the life sciences. PNAS **110**, 3229–3236 (2013)

T. Muta, S. Iwanaga, Clotting and immune defense in Limulidae, in *Invertebrate Immunology*, (Springer, Berlin Heidelberg, 1996), pp. 154–189

M. Nolan, S.A. Smith, D.R. Smith, Clinical evaluation, common diseases, and veterinary care of the horseshoe crab, Limulus polyphemus, in *Biology and Conservation of Horseshoe Crabs*, ed. by J. T. Tanacredi, M. L. Botton, (Springer Science, New York, 2009), pp. 479–499

D.M. Rudkin, G.A. Young, G.S. Nowlan, The oldest horseshoe crab: A new Xiphosurid from late Ordovician Konservat-Lagerstatten deposits, Manitoba, Canada. Paléo **51**, 1–9 (2008)

W. Rungrassamee, A. Klanchui, S. Chaiyapechara, S. Maibunkaew, S. Tangphatsornruang, P. Jiravanichipaisal, et al., Bacterial population in intestines of black tiger shrimp (*Penaeus monodon*) under different growth stages. PLoS One **8**, e60802 (2013)

T.M. Santiago-Rodriguez, G. Fornaciari, S. Luciani, S.E. Dowd, G.A. Toranzos, I. Marota, et al., Gut microbiome of an 11th century A.D. Pre-Columbian Andean Mummy. Plos One **10**(9), e0138135 (2015). https://doi.org/10.1371/journal.pone.0138135

D.R. Smith, M.A. Beekey, H.J. Brockmann, T.L. King, M.J. Millard & J.A. Zaldívar-Rae 2016. Limulus polyphemus. The IUCN Red List of Threatened Species 2016: https://doi.org/10.2305/IUCN.UK.2016-1.RLTS.T11987A80159830.en. Downloaded on 02 March 2017

J.T. Tanacredi, Horseshoe Crabs imperiled? in *Limulus in the Limelight*, (Springer US, 2002), p. 240

J. T. Tanacredi, M. L. Botton, D. R. Smith (eds.), *Biology and Conservation of Horseshoe Crabs* (Springer, New York, 2009)

V. Witt, C. Wild, K.R. Anthony, G. Diaz-Pulido, S. Uthicke, Effects of ocean acidification on microbial community composition of, and oxygen fluxes through, biofilms from Great Barrier Reef. Environ. Microbiol. **13**, 2976–2989 (2011)

X. Xuhua, Phylogenetic relationship among horseshoe crab species: Effect of substitution models on phylogenetic analyses. Syst. Biol. **49**, 87–100 (2000)

M. Zhang, Y. Sun, L. Chen, C. Cai, F. Qiao, Z. Du, E. Li, Symbiotic bacteria in gills and guts of Chinese Mitten Crab (*Eriocheir sinensis*) differ from the free-living bacteria in water. PLoS One **11**, e0148135 (2016)

# Population Genetic Structure of Juvenile Tri-Spine Horseshoe Crabs in Hong Kong

**Alice K. Y. Chan, Clement K. M. Tsui, K. L. Pang, Kit Yue Kwan, Richard Y. C. Kong, S. G. Cheung, and Paul K. S. Shin**

## 1 Introduction

Of the three Asian species of horseshoe crabs, the populations of tri-spine horseshoe crab *Tachypleus tridentatus* have declined substantially, particularly in southern mainland China, Taiwan, and Hong Kong, due to the destruction and degradation of natural habitats, as well as overharvesting for biomedical applications and use as local food (Weng et al. 2012b; Hsieh and Chen 2015; Kwan et al. 2016; Liao et al. 2019). The recent assessment update has placed *T. tridentatus* in the Endangered (EN) category on the IUCN Red List (Laurie et al. 2019). Previous studies suggested that both juvenile and adult horseshoe crabs tend not to travel long distance for survival, growth, and reproduction (Pierce et al. 2000, Yang et al. 2007; Kwan

A. K. Y. Chan · K. Y. Kwan
Department of Chemistry (formerly Biology and Chemistry), City University of Hong Kong, Hong Kong SAR, China

C. K. M. Tsui
Department of Pathology and Laboratory Medicine, University of British Columbia, Vancouver, BC, Canada

K. L. Pang
Institute of Marine Biology and Center of Excellence for the Oceans, National Taiwan Ocean University, Keelung City, Taiwan

R. Y. C. Kong · S. G. Cheung · P. K. S. Shin (✉)
Department of Chemistry (formerly Biology and Chemistry), City University of Hong Kong, Hong Kong SAR, China

State Key Laboratory in Marine Pollution, City University of Hong Kong, Hong Kong SAR, China

 
J. T. Tanacredi et al. (eds.), *International Horseshoe Crab Conservation and Research Efforts: 2007-2020*, https://doi.org/10.1007/978-3-030-82315-3_6

et al. 2015). Hence, large-scale coastal construction and reclamation activities near their habitats could drive local horseshoe crab populations to extinction.

To revert the decline of horseshoe crab populations, ex-situ conservation measures of artificial breeding through the spawning of adult *T. tridentatus* and the culture of their eggs, larvae and juveniles in the laboratory have been conducted in Hong Kong and other places in Asia (Chen et al. 2010; Hu et al. 2011; Kwan et al. 2014; Chen et al. 2016). In Hong Kong, after a rearing period of 10 months, the cultured juveniles at their sixth instar stage were released to their natural shore habitat, as part of the effort to increase the natural population (Kwan et al. 2016). Genetic variability and structure among local populations should be considered when introducing these cultured juveniles to the environment (Mamuris et al. 2001; Laikre et al. 2010). Understanding the genetic variation of the species would therefore be useful to complement existing conservation measures, especially in developing new management initiatives for *T. tridentatus* populations. Previous studies on the genetic structure of the American horseshoe crab *Limulus polyphemus* populations along the east coast of the United States reported a minimal contemporary gene flow, such as between north and south of Florida (Saunders et al. 1986), as well as between Delaware Bay and Chesapeake Bay (Pierce et al. 2000). However, the information on the genetic structure of *T. tridentatus* populations is limited. Sugawara et al. (1988) reported that there are no genetic differences detected among *T. tridentatus* populations of Hakata Bay and Kasaoka Bay in Japan and Xiamen in China, suggesting that the genetic relationship of populations from Japanese and Chinese waters is close. By using amplified fragment length polymorphism, Xu et al. (2011) drew a similar conclusion that the genetic diversity of *T. tridentatus* populations from three southeastern coastal locations in southern China, including Pingtan, Hong Kong, and Beihai, was minimal. However, the recent application of sequence analysis of mitochondrial (mt) DNA, including mt cytochrome c oxidase subunit I (COI) fragment and a mtDNA AT-rich region, on *T. tridentatus* populations demonstrated significant genetic population subdivisions between Kinmen Island and Tiexianwei on Magong Island in Taiwan Strait (Yang et al. 2007), between Ninghai and Danzhou along the Chinese coast (Weng et al. 2012a), as well as between eastern and western shores around Kyushu, Japan (Nishida and Koike 2009). In order to determine population genetic structure on a smaller spatial scale, we aimed to determine the genetic variability and phylogeographic structure of juvenile *T. tridentatus* in Hong Kong using the AT-rich region of the mtDNA gene. Such genetic information, supplemented with previous ecological data, can be utilized in developing better conservation management strategies.

## 2 Materials and Methods

### 2.1 *Field Collection of Juvenile Horseshoe Crabs*

A total of 33 juvenile *T. tridentatus* (> seventh instar stage with prosomal width ≥ 35 mm) were collected in the summer (June–September) of 2012 from six horseshoe crab spawning/nursery shores in Hong Kong, including Ha Pak Nai and Pak Nai in Deep Bay as well as San Tau, Sham Wat, Yi O, and Shui Hau along the coastline of Lantau Island (Fig. 1). Juveniles were sampled in this study because of the very low density of adult *T. tridentatus* remaining in Hong Kong waters (Li 2008; Kwan et al. 2016). The relatively low sample numbers in this study were also due to a lack of sufficient larger juveniles found, which were required for subsequent genetic analysis. All the sampling locations are the most important local nursery habitats and support relatively higher densities of juvenile *T. tridentatus* in a variety of growth stages (Kwan et al. 2016). All of these shores are largely protected from strong wave action, providing a total shore area of 2.1–11.4 ha for the growth of dwarf mangroves, seagrasses, and many intertidal organisms, including fiddler crabs, oysters, and gobiid fish. Figure 1 also illustrates the hydrographic conditions

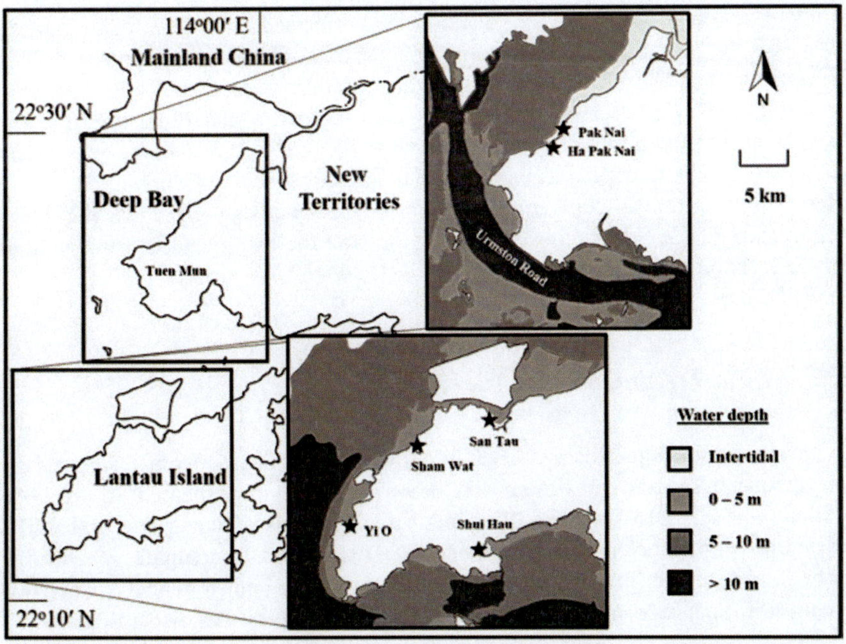


**Fig. 1** Hydrographic conditions at Deep Bay and Lantau Island, Hong Kong. Star symbols represent the sampling locations for juvenile *T. tridentatus*

of western waters in Hong Kong near the juvenile sampling locations. Lantau Island and Tuen Mun in Hong Kong are separated by Urmston Road, a relatively deep (11–33 m) and wide (1.0–3.2 km) local waterway (Marine Department HKSAR 2011).

## 2.2 DNA Extraction and Analysis

The total genomic DNA sample was extracted from juvenile hemolymph and muscle tissue, using DNeasy Blood and Tissue Kit (Qiagen, California, USA). To avoid any significant tissue damage caused by the bleeding process on smaller-sized juveniles, only individuals larger than seventh instar stage were collected. The juveniles were immediately released back to their corresponding shores after hemolymph extraction and DNA isolation. Samples were subsequently suspended in 50 μl diethylpyrocarbonate water, and the nucleic acid concentration was quantified using NanoDrop spectrophotometer (Thermo Scientific, Massachusetts, USA). The mtDNA AT-rich region was amplified using Hb-12S Sequence (5′-GTCTAACCGCGGTAGCTGGCAC-3′) and Hb-tRNA Sequence (5′-GAGCCCAATAGCTTA-AATTAGCTTA-3′), which were designed for the mitochondrial genome of *L. polyphemus* (Lavrov et al. 2000; Yang et al. 2007). Polymerase chain reaction (PCR) reaction was performed using the SapphireAMP® fast PCR master mix (25 μl master mix, 0.2 μM forward and reverse primer, 100 ng template). The PCR reaction was set with an initial denaturation of 94 °C for 1 min, and followed by 30 cycles of 98 °C for 5 s, 55 °C for 5 s, and 72 °C for 10 s. The amplicon size and purity of each sample were checked by 1% agarose gel electrophoresis. The PCR amplicons were purified and sequenced by Genomics BioSci & Tech Co., Ltd. (Taipei, Taiwan) with the same primers using ABI PRISM BigDye Terminator Cycle Sequencing Kit V3.1 following the manufacturer's instructions (Life Technologies, California, USA).

## 2.3 Data Treatment and Analysis

Sequences were aligned using Geneious 6 (http://www.geneious.com/). We used the neighbor-joining and maximum likelihood methods implemented in MEGA6 (Tamura et al. 2013) to infer the genetic relationships of horseshoe crabs. The Kimura two-parameter model (Kimura 1980) was used to compute genetic distances, and the robustness of branches was assessed by bootstrap analysis of 1000 replicates. Sequence similarity among individual samples was also calculated by MEGA6. A haplotype network analysis was constructed using the Median-Joining method in NETWORK 4.6.1.0 (Fluxus-Technology). The sequences were deposited in Genbank (Benson et al. 2013).

## 3 Results

The PCR product from primer Hb-12S and Hb-tRNA was approximately 650 bp after removing the ambiguous flanking regions. Eight variable nucleotide positions were identified among the 33 juveniles collected from different nursery shores, resulting in eight haplotypes (GenBank accession no: KY014236-KY014243, Figs. 2, 3, and Table 1). Genetic diversity of juvenile populations among the six sampling locations was high, and six of the haplotypes were found to be unique to specific juvenile populations, such as haplotype 2 which was only recorded at Ha Pak Nai, haplotype 4 at Pak Nai, haplotype 5–7 at Shui Hau, and haplotype 8 at San Tau (Fig. 3 and Table 1). The most dominant haplotype (i.e., haplotype 1) was detected in juveniles from Ha Pak Nai and Pak Nai in Deep Bay and Yi O on the west coast of Lantau Island (Fig. 3 and Table 1). Juvenile populations from San Tau, Sham Wat, and Yi O on Lantau Island were only composed of one haplotype, while the individuals from other shores made up by at least two different haplotypes (Fig. 2 and Table 1). The genetic variability among populations was small as all sequence similarities were above 99% (Appendix 1).

The phylogenetic relationship of juvenile horseshoe crabs in Hong Kong using the Median-Joining method revealed six groups of juveniles and two isolated individuals (PN3 and SH8) (Fig. 4). This result was largely congruent to the haplotype network pattern of individuals (Fig. 3 and Table 1). *T. tridentatus* populations in Hong Kong can thus be divided into three genetic units: Deep Bay (Pak Nai and Ha

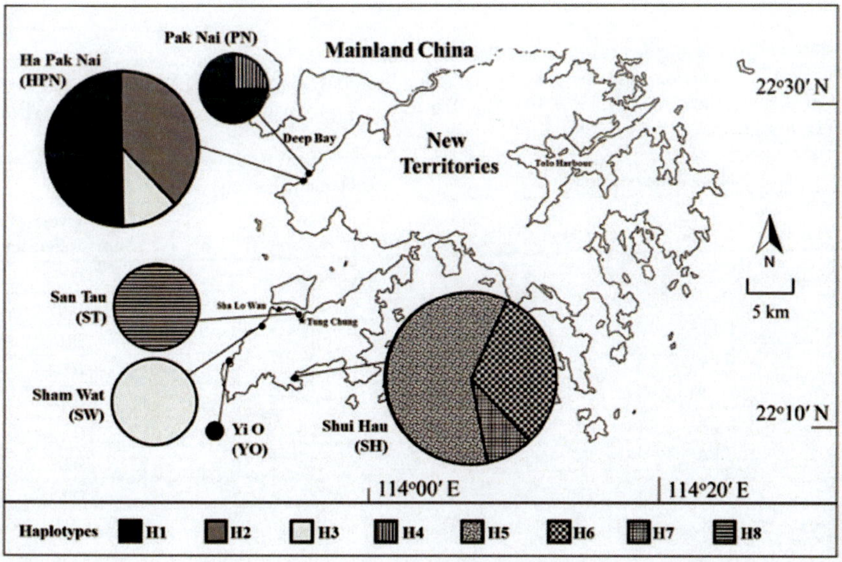


**Fig. 2** Geographic distribution and viability of mtDNA haplotypes for juvenile *T. tridentatus* in Hong Kong. Star symbols represent other breeding sites of juvenile *T. tridentatus* in Hong Kong. The size of the circle is proportional to the sample size

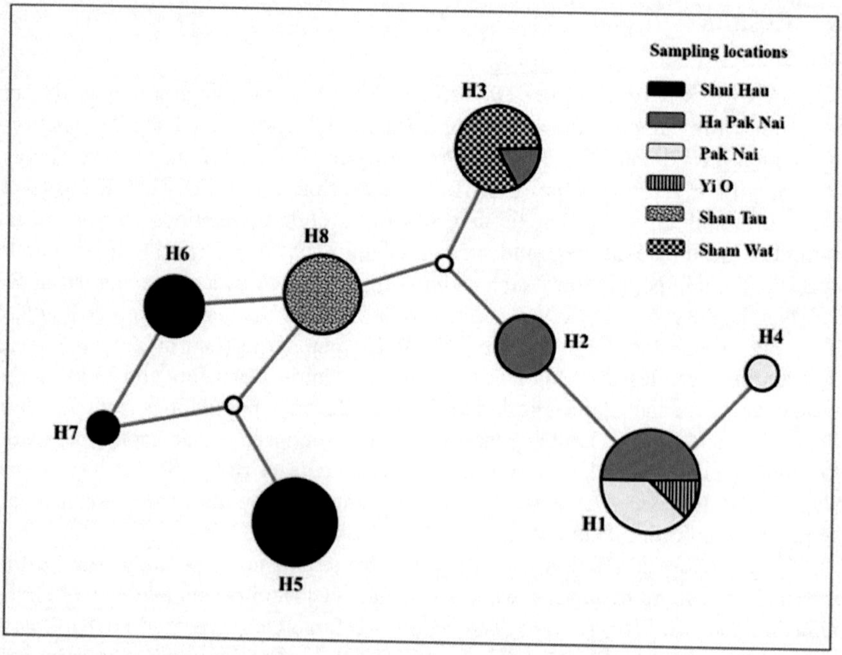


**Fig. 3** Haplotypes of horseshoe crab calculated with median joining algorithms using fragment of mitochondrial gene sequence. The size of the circle is proportional to the haplotype frequency. Each line represents a mutational step

**Table 1** The nucleotide positions and distribution of eight haplotypes in mtDNA AT-rich region of juvenile *T. tridentatus* in Hong Kong. Sampling locations: Ha Pak Nai (HPN; $n = 8$), Pak Nai (PN; $n = 4$), Shui Hau (SH; $n = 10$), San Tau (ST; $n = 5$), Sham Wat (SW; $n = 5$), and Yi O (YO; $n = 1$)

| | Nucleotide positions | | | | | | | | Distribution of haplotypes | | | | | | | |
|---|---|---|---|---|---|---|---|---|---|---|---|---|---|---|---|---|
| Haplotypes | 6 | 56 | 242 | 313 | 332 | 403 | 449 | 456 | HPN | PN | SH | ST | SW | YO | Sum | Overall frequency |
| 1 | C | – | T | T | C | A | T | T | 4 | 3 | | | | 1 | 8 | 0.24 |
| 2 | C | – | T | C | C | A | T | T | 3 | | | | | | 3 | 0.09 |
| 3 | C | – | C | C | C | A | T | C | 1 | | | | 5 | | 6 | 0.18 |
| 4 | C | – | T | T | C | G | T | T | | 1 | | | | | 1 | 0.03 |
| 5 | T | A | T | C | C | G | C | C | | | 6 | | | | 6 | 0.18 |
| 6 | C | A | T | C | T | A | T | C | | | 3 | | | | 3 | 0.09 |
| 7 | C | A | T | C | T | G | T | C | | | 1 | | | | 1 | 0.03 |
| 8 | C | A | T | C | C | A | T | C | | | | 5 | | | 5 | 0.15 |

Pak Nai), northwestern (San Tau and Sham Wat), and southern (Shui Hau) coasts of Lantau Island. Since there was only one sample collected from Yi O in this study, more individuals from Yi O should be collected and re-examined in order to confirm

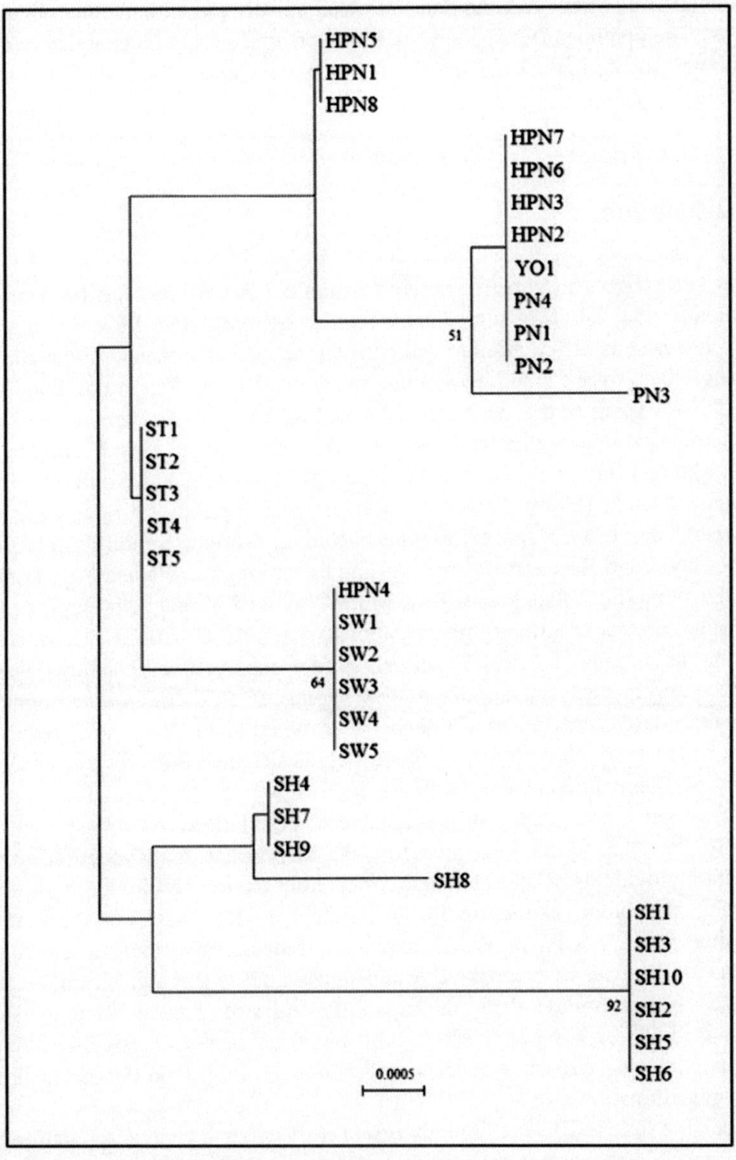


**Fig. 4** The phylogenetic relationship of horseshoe crabs in Hong Kong using the Neighbor-Joining method. The analysis involved 33 sequences of 651 nucleotides in the dataset. The abbreviation of sampling locations is described in Fig. 2

their genetic pattern. The combination of haplotypes for individuals from Pak Nai and Ha Pak Nai in Deep Bay were relatively similar, in which haplotype 1 was the major genetic constituent, whereas the phylogenetic relationship between individuals from Sham Wat and San Tau was relatively closer than other populations (Fig. 4). The juveniles from Shui Hau, however, appeared to be genetically distinct from other populations as haplotypes 5–7 were only detected in Shui Hau population (Figs. 3 and 4).

## 4 Discussion

To our knowledge, no genetic study on juvenile *T. tridentatus* has been reported. Population subdivisions of adult *T. tridentatus* were reported in Taiwan Strait, where significant genetic differentiation was detected between Kinmen and Tiexianwei on Magong Island over a small geographic range of 150 km (Yang et al. 2007). Yang et al. (2007) explained that the restricted gene flow between Kinmen and Tiexianwei may be due to the geographically isolated Tiexianwei population within the semi-closed Magong Bay on the southwestern part of Magong Island. Similar findings in Nishida and Koike (2009) showed that *T. tridentatus* populations in Japanese waters are separated into two genetic groups, including western region (Kujukushima-Sasebo Bay, Imari Bay, and Kafuri Bay) and eastern region (Hakata Bay, Tsuyazaki tidal flat, Sone tidal flat, Wama Bay, Morie Bay, and Yamaguchi Bay), in which geographic distances among these locations were only 40–70 km. There were no apparent geographic barriers observed, except the relatively shallow water at 20–30 m depth. The analysis of mtDNA sequences of *T. tridentatus* populations along the coast of mainland China between Ninghai in Zhejiang Province and Beihai in Guangxi Zhuang Autonomous Region detected only several genetically distinct subpopulations (Weng et al. 2012b, 2013). Weng et al. (2012a) explained that the genetic homogeneity of horseshoe crab populations in Chinese waters was possibly attributed to the frequent artificial transport and reintroduction of horseshoe crab individuals to other habitats for economic purposes, including food consumption and blood extraction for biomedical applications. For the American horseshoe crab *L. polyphemus*, their haplotype frequency distribution between populations in the upper Chesapeake Bay and Delaware Bay was significantly different because of their limited dispersal capability and strong natal homing behavior (Pierce et al. 2000). King et al. (2015) also found a series of genetic discontinuities for *L. polyphemus* across the species ranges, due to the Florida Peninsula as a barrier to gene flow.

Albeit the sample size in this study is relatively small, a genetic separation based on the mtDNA AT-rich region of juvenile *T. tridentatus* populations could be observed on a small geographic scale in Hong Kong. Although the geographic distances among these three genetic units were small (about 22–26 km), the formation of three genetically distinct populations implies limited genetic exchange among populations, at least for female populations, as the mitochondrial genome is typically maternally inherited (Pierce et al. 2000). One of the possible explanations is that horseshoe crabs, or at least females, in Deep Bay, as well as on northwestern and southern coasts of Lantau Island are resident and largely inhabit the same estuary or bay. This explanation is further supported by the tracking investigations using telemetry technologies on year-round movements of *L. polyphemus* (James-Pirri 2010; Schaller et al. 2010). Previous capture-recapture studies on adult *L. polyphemus* have demonstrated that these animals can have a restricted movement range close to their original tagging location and tend to repeatedly return to the same beach to spawn in the same or subsequent seasons (Rudloe 1980, James-Pirri et al. 2005). Preliminary adult tagging experiments in Hong Kong also suggested that adult *T. tridentatus* were commonly recorded on the same or nearby shore within a 5 km radius from their points of release (Kwan et al. 2016). During high tides in spring and summer (April–September), eggs are deposited near high tide mark at a depth of 10–15 cm, and spend approximately 30 days to hatch into trilobite larvae (Chiu and Morton 2003). The life-history of horseshoe crabs is lengthy, requiring about 16 to 18 molts and feeding on the intertidal areas for 10–15 years prior to returning back to shallow waters (Hu et al. 2015).

The Shui Hau population located on the southern side of Lantau Island also appears to be geographically isolated. Limited gene flow among the three populations in Hong Kong may be related to the relatively deeper water mass (10–33 m) between Tuen Mun and Lantau Island (Fig. 1). While the effect of water depth on movement of adult *T. tridentatus* is virtually unknown, it is generally believed that adult horseshoe crabs migrate offshore in deep water (>20 m) in winter and move to shallower water during late spring for spawning (Sekiguchi 1988). A recent study using acoustic telemetry also revealed that there was no significant difference in the seasonal residency of adult *T. tridentatus* in a coastal embayment in Japan (Wada et al. 2016).

The low haplotype diversity observed in San Tau and Sham Wat on the northwestern coast of Lantau Island is likely attributed to a population bottleneck. Recent territory-wide population surveys in 2012 and 2014 revealed low juvenile *T. tridentatus* density at San Tau (0.16–1.41 individuals/100 $m^2$) and Sham Wat (0.08–0.31 individuals/100 $m^2$) (Kwan et al. 2016). The survey results also suggested that juveniles in Hong Kong are vulnerable to local extirpation due to the small and discrete juvenile populations, as well as relatively few recruits (Kwan et al. 2016). Although surveys on the adult *T. tridentatus* population in Hong Kong have yet to be conducted, it is generally believed that adult densities are very low, according to scattered observations of spawning pairs and interviews with local fishermen. According to the genetic data of *T. tridentatus* in this study, albeit preliminary, it seems that there are at least three local groups of juvenile horseshoe crab populations, and these groups should be considered as distinct management and recovery units for conservation purposes. For *T. tridentatus* populations in Deep Bay, the relocation of individuals from the adjacent bays or estuaries in Deep Bay, supplemented by artificial breeding and restocking program, appears to be a promising option for future restoration efforts. However, the less diverse haplotype pattern of San Tau and Sham Wat populations may imply the low effective population size of females at these two shores and that they are highly vulnerable to local extinction. As suggested from juvenile horseshoe crab population survey results in Kwan et al. (2016), at least two marine protected areas at Deep Bay and the northwestern coast of Lantau Island (Tung Chung) should be established to protect the remaining *T. tridentatus* populations in Hong Kong. The present findings on the genetic structure of juvenile horseshoe crab populations provide further evidence for the need to designate such marine protected areas as soon as possible.

**Acknowledgments** Assistance from Angie NY Wong and Vicky WK Chu on sample collection is much appreciated. This research was supported by the Ocean Park Conservation Foundation, Hong Kong (OPCFHK).

**Appendix 1. Genetic sequence similarities among different tri-spine horseshoe crab individuals in Hong Kong. Sampling locations: Ha Pak Nai (HPN), Pak Nai (PN), Shui Hau (SH), San Tau (ST), Sham Wat (SW), and Yi O (YO)**

| | YO1 | HPN8 | HPN1 | HPN2 | HPN3 | HPN4 | HPN5 | HPN6 | HPN7 | PN1 | PN2 | PN3 | PN4 | SH1 | SH3 | SH9 |
|---|---|---|---|---|---|---|---|---|---|---|---|---|---|---|---|---|
| HPN8 | 99.8 | | | | | | | | | | | | | | | |
| HPN1 | 99.8 | 100 | | | | | | | | | | | | | | |
| HPN2 | 100 | 99.8 | 99.8 | | | | | | | | | | | | | |
| HPN3 | 100 | 99.8 | 99.8 | 100 | | | | | | | | | | | | |
| HPN4 | 99.5 | 99.7 | 99.7 | 99.5 | 99.5 | | | | | | | | | | | |
| HPN5 | 99.8 | 100 | 100 | 99.8 | 99.8 | 99.7 | | | | | | | | | | |
| HPN6 | 100 | 99.8 | 99.8 | 100 | 100 | 99.5 | 99.8 | | | | | | | | | |
| HPN7 | 100 | 99.8 | 99.8 | 100 | 100 | 99.5 | 99.8 | 100 | | | | | | | | |
| PN1 | 100 | 99.8 | 99.8 | 100 | 100 | 99.5 | 99.8 | 100 | 100 | | | | | | | |
| PN2 | 100 | 99.8 | 99.8 | 100 | 100 | 99.5 | 99.8 | 100 | 100 | 100 | | | | | | |
| PN3 | 99.8 | 99.7 | 99.7 | 99.8 | 99.8 | 99.4 | 99.7 | 99.8 | 99.8 | 100 | 100 | | | | | |
| PN4 | 100 | 99.8 | 99.8 | 100 | 100 | 99.5 | 99.8 | 100 | 100 | 100 | 100 | 100 | | | | |
| SH1 | 99.1 | 99.2 | 99.2 | 99.1 | 99.1 | 99.2 | 99.2 | 99.1 | 99.1 | 99 | 99 | 99 | 99 | | | |
| SH3 | 99.1 | 99.2 | 99.2 | 99.1 | 99.1 | 99.2 | 99.2 | 99.1 | 99.1 | 99 | 99 | 99 | 99 | 100 | | |
| SH9 | 99.4 | 99.5 | 99.5 | 99.4 | 99.4 | 99.5 | 99.5 | 99.4 | 99.4 | 99 | 99 | 99 | 99 | 99 | 99 | |
| SH10 | 99.1 | 99.2 | 99.2 | 99.1 | 99.1 | 99.2 | 99.2 | 99.1 | 99.1 | 99 | 99 | 99 | 99 | 100 | 100 | 99 |
| SH2 | 99.1 | 99.2 | 99.2 | 99.1 | 99.1 | 99.2 | 99.2 | 99.1 | 99.1 | 99 | 99 | 99 | 99 | 100 | 100 | 99 |

| | YO1 | HPN8 | HPN1 | HPN2 | HPN3 | HPN4 | HPN5 | HPN6 | HPN7 | PN1 | PN2 | PN3 | PN4 | SH1 | SH3 | SH9 |
|---|---|---|---|---|---|---|---|---|---|---|---|---|---|---|---|---|
| SH4 | 99.4 | 99.5 | 99.5 | 99.4 | 99.4 | 99.5 | 99.5 | 99.4 | 99.4 | 99 | 99 | 99 | 99 | 99 | 99 | 100 |
| SH5 | 99.1 | 99.2 | 99.2 | 99.1 | 99.1 | 99.2 | 99.2 | 99.1 | 99.1 | 99 | 99 | 99 | 99 | 100 | 100 | 99 |
| SH6 | 99.1 | 99.2 | 99.2 | 99.1 | 99.1 | 99.2 | 99.2 | 99.1 | 99.1 | 99 | 99 | 99 | 99 | 100 | 100 | 99 |
| SH7 | 99.4 | 99.5 | 99.5 | 99.4 | 99.4 | 99.5 | 99.5 | 99.4 | 99.4 | 99 | 99 | 99 | 99 | 99 | 99 | 100 |
| SH8 | 99.2 | 99.4 | 99.4 | 99.2 | 99.2 | 99.4 | 99.4 | 99.2 | 99.2 | 99 | 99 | 99 | 99 | 100 | 100 | 100 |
| ST1 | 99.5 | 99.7 | 99.7 | 99.5 | 99.5 | 99.7 | 99.7 | 99.5 | 99.5 | 100 | 100 | 99 | 100 | 100 | 100 | 100 |
| ST2 | 99.5 | 99.7 | 99.7 | 99.5 | 99.5 | 99.7 | 99.7 | 99.5 | 99.5 | 100 | 100 | 99 | 100 | 100 | 100 | 100 |
| ST3 | 99.5 | 99.7 | 99.7 | 99.5 | 99.5 | 99.7 | 99.7 | 99.5 | 99.5 | 100 | 100 | 99 | 100 | 100 | 100 | 100 |
| ST4 | 99.5 | 99.7 | 99.7 | 99.5 | 99.5 | 99.7 | 99.7 | 99.5 | 99.5 | 100 | 100 | 99 | 100 | 100 | 100 | 100 |
| ST5 | 99.5 | 99.7 | 99.7 | 99.5 | 99.5 | 99.7 | 99.7 | 99.5 | 99.5 | 100 | 100 | 99 | 100 | 100 | 100 | 100 |
| SW1 | 99.5 | 99.7 | 99.7 | 99.5 | 99.5 | 100 | 99.7 | 99.5 | 99.5 | 100 | 100 | 99 | 100 | 99 | 99 | 100 |
| SW2 | 99.5 | 99.7 | 99.7 | 99.5 | 99.5 | 100 | 99.7 | 99.5 | 99.5 | 100 | 100 | 99 | 100 | 99 | 99 | 100 |
| SW3 | 99.5 | 99.7 | 99.7 | 99.5 | 99.5 | 100 | 99.7 | 99.5 | 99.5 | 100 | 100 | 99 | 100 | 99 | 99 | 100 |
| SW4 | 99.5 | 99.7 | 99.7 | 99.5 | 99.5 | 100 | 99.7 | 99.5 | 99.5 | 100 | 100 | 99 | 100 | 99 | 99 | 100 |
| SW5 | 99.5 | 99.7 | 99.7 | 99.5 | 99.5 | 100 | 99.7 | 99.5 | 99.5 | 100 | 100 | 99 | 100 | 99 | 99 | 100 |

| | SH10 | SH2 | SH4 | SH5 | SH6 | SH7 | SH8 | ST1 | ST2 | ST3 | ST4 | ST5 | SW1 | SW2 | SW3 | SW4 |
|---|---|---|---|---|---|---|---|---|---|---|---|---|---|---|---|---|
| HPN8 | | | | | | | | | | | | | | | | |
| HPN1 | | | | | | | | | | | | | | | | |
| HPN2 | | | | | | | | | | | | | | | | |
| HPN3 | | | | | | | | | | | | | | | | |
| HPN4 | | | | | | | | | | | | | | | | |
| HPN5 | | | | | | | | | | | | | | | | |
| HPN6 | | | | | | | | | | | | | | | | |
| HPN7 | | | | | | | | | | | | | | | | |
| PN1 | | | | | | | | | | | | | | | | |
| PN2 | | | | | | | | | | | | | | | | |
| PN3 | | | | | | | | | | | | | | | | |
| PN4 | | | | | | | | | | | | | | | | |
| SH1 | | | | | | | | | | | | | | | | |
| SH3 | | | | | | | | | | | | | | | | |
| SH9 | | | | | | | | | | | | | | | | |
| SH10 | | | | | | | | | | | | | | | | |
| SH2 | 100 | | | | | | | | | | | | | | | |
| SH4 | 99.4 | 99 | | | | | | | | | | | | | | |
| SH5 | 100 | 100 | 99 | | | | | | | | | | | | | |
| SH6 | 100 | 100 | 99 | 100 | | | | | | | | | | | | |
| SH7 | 99.4 | 99 | 100 | 99 | 99 | | | | | | | | | | | |
| SH8 | 99.5 | 100 | 100 | 100 | 100 | 100 | | | | | | | | | | |
| ST1 | 99.5 | 100 | 100 | 100 | 100 | 100 | 100 | | | | | | | | | |
| ST2 | 99.5 | 100 | 100 | 100 | 100 | 100 | 100 | 100 | | | | | | | | |
| ST3 | 99.5 | 100 | 100 | 100 | 100 | 100 | 100 | 100 | 100 | | | | | | | |
| ST4 | 99.5 | 100 | 100 | 100 | 100 | 100 | 100 | 100 | 100 | 100 | | | | | | |
| ST5 | 99.5 | 100 | 100 | 100 | 100 | 100 | 100 | 100 | 100 | 100 | 100 | | | | | |
| SW1 | 99.2 | 99 | 100 | 99 | 99 | 100 | 99 | 100 | 100 | 100 | 100 | 100 | | | | |
| SW2 | 99.2 | 99 | 100 | 99 | 99 | 100 | 99 | 100 | 100 | 100 | 100 | 100 | 100 | | | |
| SW3 | 99.2 | 99 | 100 | 99 | 99 | 100 | 99 | 100 | 100 | 100 | 100 | 100 | 100 | 100 | | |
| SW4 | 99.2 | 99 | 100 | 99 | 99 | 100 | 99 | 100 | 100 | 100 | 100 | 100 | 100 | 100 | 100 | |
| SW5 | 99.2 | 99 | 100 | 99 | 99 | 100 | 99 | 100 | 100 | 100 | 100 | 100 | 100 | 100 | 100 | 100 |

## References

D.A. Benson, M. Cavanaugh, K. Clark, I. Karsch-Mizrachi, D.J. Lipman, J. Ostell, E.W. Sayers, GenBank. Nucleic Acids Res. **41**, D36–D42 (2013)

C.P. Chen, R.F. Chen, P.H. Chen, H.J. Liu, H.L. Hsieh, Intermediate culture of juvenile horseshoe crab (*Tachypleus tridentatus*) mixed with juvenile spotted babylon (*Babylonia areolata*) for restocking horseshoe crab populations. Aquacul. Aquar. Conserv. Legis. **9**, 623–633 (2016)

Y. Chen, C.W. Lau, S.G. Cheung, C.H. Ke, P.K.S. Shin, Enhanced growth of juvenile *Tachypleus tridentatus* (Chelicerata: Xiphosura) in the laboratory: A step towards population restocking for conservation of the species. Aquat. Biol. **11**, 37–46 (2010)

H.M.C. Chiu, B. Morton, The sediment and hydrographic characteristics of three horseshoe crab nursery beaches in Hong Kong. J. Ocean Univ. Qingdao. **2**, 35–43 (2003)

H.L. Hsieh, C.P. Chen, Current status of *Tachypleus tridentatus* in Taiwan for Red List Assessment, in *Changing Global Perspectives on Horseshoe Crab Biology, Conservation and Management*, ed. by R. H. Carmichael, M. L. Botton, P. K. S. Shin, S. G. Cheung, (Springer, Switzerland, 2015), pp. 383–396

M. Hu, B.K.Y. Kwan, Y. Wang, S.G. Cheung, P.K.S. Shin, Population structure and growth of juvenile horseshoe crabs *Tachypleus tridentatus* and *Carcinoscorpius rotundicauda* (Xiphosura) in southern China, in *Changing Global Perspectives on Horseshoe Crab Biology, Conservation and Management*, ed. by R. H. Carmichael, M. L. Botton, P. K. S. Shin, S. G. Cheung, (Springer, Switzerland, 2015), pp. 167–180

M. Hu, Y. Wang, S.G. Cheung, P.K.S. Shin, Comparison of different frozen natural foods on survival and growth of juvenile Chinese horseshoe crab *Tachypleus tridentatus* (Leach, 1819): Implications on laboratory culture. Aquacul. Res. **4**, 567–573 (2011)

M.J. James-Pirri, Seasonal movement of the American horseshoe crab *Limulus polyphemus* in a semi-enclosed bay on Cape Cod, Massachusetts (USA) as determined by acoustic telemetry. Curr. Zool. **56**, 575–586 (2010)

M.J. James-Pirri, K. Tuxbury, S. Marino, Spawning densities, egg densities, size structure, and movement patterns of spawning horseshoe crabs, *Limulus polyphemus*, within four coastal embayments on Cape Cod, Massachusetts. Estuaries **28**, 296–313 (2005)

M. Kimura, A simple method for estimating evolutionary rates of base substitutions through comparative studies of nucleotide sequences. J. Mol. Evol. **16**, 111–120 (1980)

T.L. King, M.S. Eackles, A.W. Aunins, H.J. Brockmann, E. Hallerman, B.L. Brown, Conservation genetics of the American horseshoe crab (*Limulus polyphemus*): Allelic diversity, zones of genetic discontinuity, and regional differentiation, in *Changing Global Perspectives on Horseshoe Crab Biology, Conservation and Management*, ed. by R. H. Carmichael, M. L. Botton, P. K. S. Shin, S. G. Cheung, (Springer, Switzerland, 2015), pp. 65–96

B.K.Y. Kwan, A.K.Y. Chan, S.G. Cheung, P.K.S. Shin, Hemolymph quality as indicator of health status in juvenile Chinese horseshoe crab *Tachypleus tridentatus* (Xiphosura) under laboratory culture. J. Exp. Mar. Biol. Ecol. **457**, 135–142 (2014)

B.K.Y. Kwan, H.Y. Hsieh, S.G. Cheung, P.K.S. Shin, Present population and habitat status of potentially threatened Asian horseshoe crabs *Tachypleus tridentatus* and *Carcinoscorpius rotundicauda* in Hong Kong: A proposal for marine protected areas. Biodivers. Conserv. **25**, 673–692 (2016)

B.K.Y. Kwan, P.K.S. Shin, S.G. Cheung, Preliminary home range study of juvenile Chinese horseshoe crab *Tachypleus tridentatus* (Xiphosura) using passive tracking methods, in *Changing Global Perspectives on Horseshoe Crab Biology, Conservation and Management*, ed. by R. H. Carmichael, M. L. Botton, P. K. S. Shin, S. G. Cheung, (Springer, Switzerland, 2015), pp. 149–166

L. Laikre, M.K. Schwartz, R.S. Waples, N. Ryman, GeM Working Group, Compromising genetic diversity in the wild: Unmonitored large-scale release of plants and animals. Trends Ecol. Evol. **25**, 520–529 (2010)

K. Laurie, C.-P. Chen, S.G. Cheung, V. Do, H. Hsieh, A. John, F. Mohamad, S. Seino, S. Nishida, P. Shin, M. Yang, Tachypleus tridentatus. The IUCN Red List of Threatened Species 2019: e.T21309A133299524. https://doi.org/10.2305/IUCN.UK.2019-1.RLTS.T21309A133299524.en (2019)

D.V. Lavro, J.L. Boore, W.M. Brown, The complete mitochondrial DNA sequence of the horseshoe crab *Limulus polyphemus*. Mol. Biol. Evol. **17**, 813–824 (2000)

H.Y. Li, *The Conservation of Horseshoe Crabs in Hong Kong*. (MPhil thesis, City University of Hong Kong, 277 pp, 2008)

Y. Liao, H.-L. Hsieh, S. Xu, Q. Zhing, J. Lei, M. Liang, H. Fang, L. Xu, W. Lin, X. Xiao, C.-P. Chen, S.G. Cheung, B.K.Y. Kwan, Wisdom of crowds reveals decline of Asian horseshoe crabs in Beibu Gulf, China. Oryx **53**, 222–229 (2019)

Z. Mamuris, A.I. Sfougaris, C. Stamatis, Genetic structure of Greek brown hare (*Lepus europaeus*) populations as revealed by mtDBNA RFLP-PCR analysis: Implications for conserving genetic diversity. Biol. Conserv. **101**, 187–196 (2001)

Marine Department, The Government of the Hong Kong Special Administrative Region (HKSAR), *Charts for Local Vessels: Hong Kong Waters* (Government Logistics Department, Hong Kong, 2011)

S. Nishida, H. Koike, Genetic structure of Japanese populations of *Tachypleus tridentatus* by mtDNA AT-rich region sequence analysis, in *Biology and Conservation of Horseshoe Crabs*, ed. by J. T. Tanacredi, M. L. Botton, D. R. Smith, (Springer, New York, 2009), pp. 183–196

J.C. Pierce, G. Tan, P.M. Gaffiney, Delaware Bay and Chesapeake Bay populations of the horseshoe crab *Limulus polyphemus* are genetically distinct. Esturaries **23**, 690–698 (2000)

A. Rudloe, The breeding behavior and patterns of movement of horseshoe crabs, *Limulus polyphemus*, in the vicinity of breeding beaches in Apalachee Bay, Florida. Estuaries **3**, 177–183 (1980)

N.C. Saunders, L.G. Kessler, J.C. Avise, Genetic variation and geographic different in mitochondrial DNA of the horseshoe crab, *Limulus polyphemus*. Genetics **112**, 613–627 (1986)

S.Y. Schaller, C.C. Chabot, W.H. Watson III, Seasonal movements of American horseshoe crabs *Limulus polyphemus* in the Great Bay Estuary, New Hampshire (USA). Curr Zool **56**, 587–598 (2010)

K. Sekiguchi, *Biology of Horseshoe Crabs* (Science House Co. Ltd., Tokyo, 1988), 428 pp

K. Sugawara, H. Yonekawa, Y. Tagashima, K. Sekiguchi, Mitochondrial DNA polymorphisms, in *Biology of Horseshoe Crabs*, ed. by K. Sekiguchi, (Science House Co. Ltd, Tokyo, 1988), pp. 375–382

K. Tamura, G. Stecher, D. Peterson, A. Filipski, S. Kumar, MEGA6: Molecular evolutionary genetics analysis version 6.0. Mol. Biol. Evol. **30**, 2725–2729 (2013)

T. Wada, T. Mitsushio, S. Inoue, H. Koike, R. Kawabe, Movement patterns and residency of the critically endangered horseshoe crab *Tachypleus tridentatus* in a semi-enclosed bay determined using acoustic telemetry. PLoS One **11**, e0147429 (2016)

Z. Weng, Z. Xiao, Y. Xie, Z. Wang, J. Gui, Genetic difference of Chinese horseshoe crab (*Tachypleus tridentatus*) in southeast coast of China based on mitochondrial COI gene analysis. Acta Oceanol. Sin. **31**, 132–137 (2012a)

Z.H. Weng, Y.J. Xie, Z.Q. Xiao, L.M. Huang, J. Li, S.H. Wang, Y.Z. Zhang, Distribution and resource of Chinese horseshoe crab (*Tachypleus tridentatus*) in Fujian and other coast water of China. Chinese J. Zool. **47**, 40–48 (2012b) (in Chinese with English abstract)

Z.H. Weng, Y.J. Xie, Z.Q. Xiao, Z.Y. Wang, J.F. Gui, Microsatellite and mitochondrial DNA analysis of the genetic structure of Chinese horseshoe crab (*Tachypleus tridentatus*) in Southeast China coast. Afr. J. Biotechnol. **12**, 2088–2099 (2013)

Q.A. Xu, F. Chen, P.K.S. Shin, S.G. Cheung, Y. Chen, C.H. Ke, AFLP analysis of genetic variation among three natural populations of horseshoe crab *Tachypleus tridentatus* along Chinese coast. Chin. J. Oceanol. Limnol. **29**, 284–289 (2011)

M.C. Yang, C.A. Chen, H.L. Hsieh, C.P. Chen, Population subdivision of the tri-spine horseshoe crab, *Tachypleus tridentatus,* in Taiwan Strait. Zool. Sci. **24**, 219–224 (2007)

# Applying Records of Extant and Extinct Horseshoe Crab Abnormalities to Xiphosurid Conservation

**Russell D. C. Bicknell, Stephen Pates, Dorkas Kaiser, Sara Zakrzewski, and Mark L. Botton**

## 1 Introduction

Horseshoe crabs (Xiphosurida) have been extensively studied by both biologists and palaeontologists (Bicknell and Pates 2020). As such, the anatomical (Owen 1872; Lankester 1881; Patten 1894; Shultz 2001; Bicknell et al. 2018a, 2018c; Bicknell et al. 2019b; Bicknell and Pates 2019a), biochemical (Kaplan et al. 1977), biomechanical (Bicknell et al. 2018b; Razak and Kassim 2018), developmental (Sekiguchi et al. 1988; Farley 2010; Kaiser and Schoppe 2018), ecological (Sokoloff 1978; Shuster Jr. 1982; Dietl et al. 2000; Shuster Jr. and Sekiguchi 2009; Botton et al. 2010; Akbar John et al. 2012; Fairuz-Fozi et al. 2018), genetic (Sokoloff 1978; Obst et al. 2012), morphometric (Riska 1981; Vijayakumar et al. 2000; Srijaya et al. 2010; Bicknell 2019; Bicknell and Pates 2019b; Bicknell et al. 2019d, 2019e), physiological (Redmond et al. 1982), and palaeontological (Woodward 1879; Fisher 1984; Selden and Siveter 1987; Hauschke and Wilde 1991; Bicknell et al. 2018d, 2019a, 2019c) aspects of horseshoe crabs are well documented. Despite this extensive research program, at least one facet of horseshoe crabs

R. D. C. Bicknell (✉)
Palaeoscience Research Centre, School of Environmental and Rural Science, University of New England, Armidale, NSW, Australia

S. Pates
Department of Zoology, University of Cambridge, Downing Street, Cambridge, UK

D. Kaiser
Katala Foundation Inc., Puerto Princesa City, Palawan, Philippines

Western Philippines University, San Juan, Palawan, Philippines

S. Zakrzewski · M. L. Botton
Department of Natural Sciences and Environmental Science Program, Fordham University, New York, NY, USA



J. T. Tanacredi et al. (eds.), *International Horseshoe Crab Conservation and Research Efforts: 2007-2020*, https://doi.org/10.1007/978-3-030-82315-3_7

anatomy—abnormalities—has been underexplored. Currently, van der Meer Mohr (1935), Shuster Jr. (1982), Jell (1989), Bicknell et al. (2018d), and Bicknell and Pates (2019a) are the main documents considering abnormal xiphosurids. Prior to the recent research, the majority of documented abnormalities were recorded for the telson (Shuster Jr. 1982; Botton and Loveland 1989; Shuster Jr. and Sekiguchi 2004), with rarer examples for the cephalothorax (=prosoma) and thoracetron (=opisthosoma) (Jell 1989; Brockmann 1990; Schaller et al. 2004, 2005). Revised interest in documenting abnormal specimens, especially through using museum collections, has produced a larger summary of horseshoe crabs with malformations (Bicknell et al. 2018d; Bicknell and Pates 2019a). This has extended to palaeontological examples, highlighting that injuries and malformations observed today have a fossil record extending back to at least the Carboniferous (Bicknell et al. 2018d). Five abnormal fossil specimens have been depicted to-date: a *Mesolimulus walchi* (Desmarest, 1822) with a cephalothoracic injury, another *M. walchi* specimen with a cephalothoracic-thoracetronic injury, two *Euproops danae* (Meek and Worthen, 1865) specimens with cephalothoracic injuries, and a *E. danae* specimen with putative cephalothoracic scrapes (Bicknell et al. 2018d). In addition, microborings have been reported on xiphosurids from the Jurassic aged Kcynia Formation, Poland (Błażejowski et al. 2019). Xiphosurid abnormalities have also been used to understand the development of trilobitic abnormalities (Jell 1989; Bicknell et al. 2018d; Bicknell and Pates 2019a). Further documentation of abnormal xiphosurid features may therefore increase our knowledge regarding how extinct arthropod groups recovered from injuries. Here, we present new abnormal extant specimens of *Limulus polyphemus* (Linnaeus, 1758) *and Tachypleus tridentatus* (Leach, 1819) explore their recovery from abnormalities and also consider species-specific abnormality patterns. Further, we also present the first example of a completely recovered horseshoe crab telson in a juvenile *T. tridentatus* specimen and three as-of-yet un- documented abnormal fossil specimens—two *M. walchi* and one *E. danae*. With these data, our chapter explores how studying horseshoe crab abnormalities can aid xiphosurid conservation efforts.

## 1.1 Abnormalities: Causes and Forms

Following Owen (1985), Babcock (1993), Bicknell et al. (2018d), and Bicknell and Pates (2019a), abnormities are described as malformed or absent exoskeletal sections, as well as additional or unusual growths. There are three main likely causes of abnormalities: injuries, teratologies, and pathologies (Bicknell et al. 2018d), although these cannot always be distinguished (Owen 1985; Bicknell et al. 2018d; Pates and Bicknell 2019).

Horseshoe crabs injuries are defined as mechanical damage induced by predators, anthropic tampering, environmental factors, or self-inflicted during moulting, mating, or burrowing complications. Injuries are generally 'V'-, 'W'-, 'U'-, or 'L'-shaped embayments in the xiphosurid exoskeleton and usually located on a dorsal region. Here, we extend this definition to include holes in the dorsal exoskeleton. Appendage injuries (documented herein) truncate the appendage, often at the end of a segment. Scar formation (cicatrisation) often occurs at the injury site, with partial regeneration of injuries during subsequent moults. Teratologies for horseshoe crabs are related to developmental and genetic aberrations and produce abnormal spines, exoskeletal sections, or appendages. Finally, pathologies for horseshoe crabs are atrophied tissue from disease or abnormal exoskeletal growth around parasitic intrusions.

## 2 Methods

Specimens documented here were collated from four different sources. *Tachypleus tridentatus* specimens were documented and photographed by D. Kaiser in 2001 at BM Beach, Palawan, Philippines, with a Canon EOS 500 N (Kaiser 2002). Further examples were photographed by Kevin Laurie at Ha Pak Nai, Deep Bay, Hong Kong, with a Canon EOS 70D. Adult *Limulus polyphemus* were examined, documented, and photographed by M. L. Botton and S. Zakrzewski at Plumb Beach, Jamaica Bay, New York, between June 4th and 28th, 2018. Abnormalities impacting the lateral compound eyes, cephalothorax, thoracetron, telson, and cephalothoracic appendages were recorded. Using criteria established by Duffy et al. (2006), the condition of the dorsal carapace was scored between 1 (glossy, smooth surface), 2 (intermediate), and 3 (darkened, pitted, eroded). Additional juvenile moults were collected at this location in July and August 2018. Live specimens with pronounced abnormalities were photographed with either a Fujifilm Finepix (M. L. Botton) or a Nikon 4100 DSLR (R. D. C. Bicknell) and moults were photographed with a Nikon 4100 DSLR (R. D. C. Bicknell). Distributions of appendage, cephalothoracic, thoracetronic, and telson abnormalities documented in Kaiser (2002) and in the adult Jamaica Bay 2018 population were compiled and plotted. Fossil specimens were identified by reviewing *Mesolimulus walchi* and *Euproops danae* specimens in the Carnegie Museum of Natural History (CM) collection and Yale Peabody Museum Division of Invertebrate Paleontology (YPM IP) collection. Abnormal specimens were photographed with a Nikon 4100 DSLR (CM) or Canon D60 DSLR (YPM IP). Images from the YMP IP were stacked using Helicon Focus 6 (Helicon Soft Limited) software.

## 3 Results

A selection of notable *Limulus polyphemus* abnormalities are presented to expand knowledge on the topic: four cephalothoracic, one thoracetronic, two telson abnormalities, and examples of abnormal and damaged appendages. The most extreme cephalothoracic abnormality (Fig. 1a) is a 'U'-shaped embayment in the left side of the cephalothorax that extends to the cardiac lobe. This embayment is recovered half-way to the cephalothoracic border. Two examples of holes in the cephalothorax are noted (Fig. 1b, c). The adult male had a 36 mm wide, ovate hole on the right side of the cephalothorax that removed the right lateral compound eye and shows evidence of recovery (Fig. 1b). The second example is a juvenile specimen with a large (13 mm) ovate hole on the left-central cephalothoraxcic region that removed the left lateral compound eye (Fig. 1c). The hole extends proximal close to the cardiac lobe and shows signs of recovery. The last cephalothoracic abnormality is noted on a juvenile moult having two small (<1 mm) sized 'U'-shaped indentations on the anterior left cephalothoraxcic region, adjacent to a 2.5 mm long section of buckled dorsal exoskeleton (Fig. 2d). The documented thoracetronic abnormality is present on a juvenile moult with a 'U'-shaped indentation on the left posterior thoracetronic region that removed moveable spines and shows signs of cicatrisation (Fig. 2c). The two documented examples of telson abnormalities are on juvenile moults (Fig. 2a, b). The first specimen has a right deflected telson 12 mm along the telson length that offsets the telson by 35° relative to the sagittal line (Fig. 2a). The second specimen has a stunted, 7 mm long telson (Fig. 2b). Two specimens with appendage abnormalities were adult male specimens. The first has a stunted claw on the left pedipalp and a broken right pedipalp, at least five truncated walking legs and a truncated right pushing leg that is partly cicatrised (Fig. 1d). The second specimen has fully chelate male pedipalps, comparable to normal walking legs (Fig. 1e).

Notable *Tachypleus tridentatus* abnormalities depicted here consist of two dorsal exoskeletal and three telson abnormalities, and a failed moult. The most extreme example (Fig. 3a) displays a small 'V'-shaped indentation on the right cephalothorax, as well as removal of the telson and left half of the thoracetron. The other cephalothoracic abnormality was a specimen with a 'U'-shaped indentation on the left cephalothoracic region and a 'W'-shaped indentation on the right cephalothoraxcic region (Fig. 3b). The third specimen appeared to have failed ecdysis and has the older (smaller) exuvium located within the newer (larger) exoskeleton (Fig. 3c). The fourth specimen displays a telson that is curved to the right side. The deflection begins 18 mm from the telson-thoracetron embayment with the distal telson tip orientated at ~90° relative to the sagital line (Fig. 3d). The fifth specimen shows a dorsally deflected telson and with last fifth of the telson kinked vertically (Fig. 3e). The final specimen documented here shows the first evidence for complete telson recovery during a moulting event. Prior to moulting (Fig. 4a), the juvenile had a

**Fig. 1** (continued) extended to the cardiac lobe. (**b**) Ovate hole in right cephalothoracic region, removing right lateral compound eye. Injury is recovering (white arrow). (**c**) Juvenile with ovate hole on left side of the cephalothorax, removing left lateral compound eye. Hole shows signs of recovery (white arrow). (**d**) Male specimen with two damaged male pedipalp claws (black arrows), five damaged walking legs (white arrows), and one damaged pushing leg (grey arrow). (**e**): Male specimen with symmetrical claws on the pedipalps (white arrows). Scale bars: (**a, d, e**) = 50 mm; (**b**) = 40; (**c**) = 10 mm. (Image credit: (**a, b, d, e**) Mark Botton; (**c**) Russell Bicknell. (**a, b, d, e**) were not accessioned due to limits on collecting live specimens. (**c**) Housed at Fordham University)

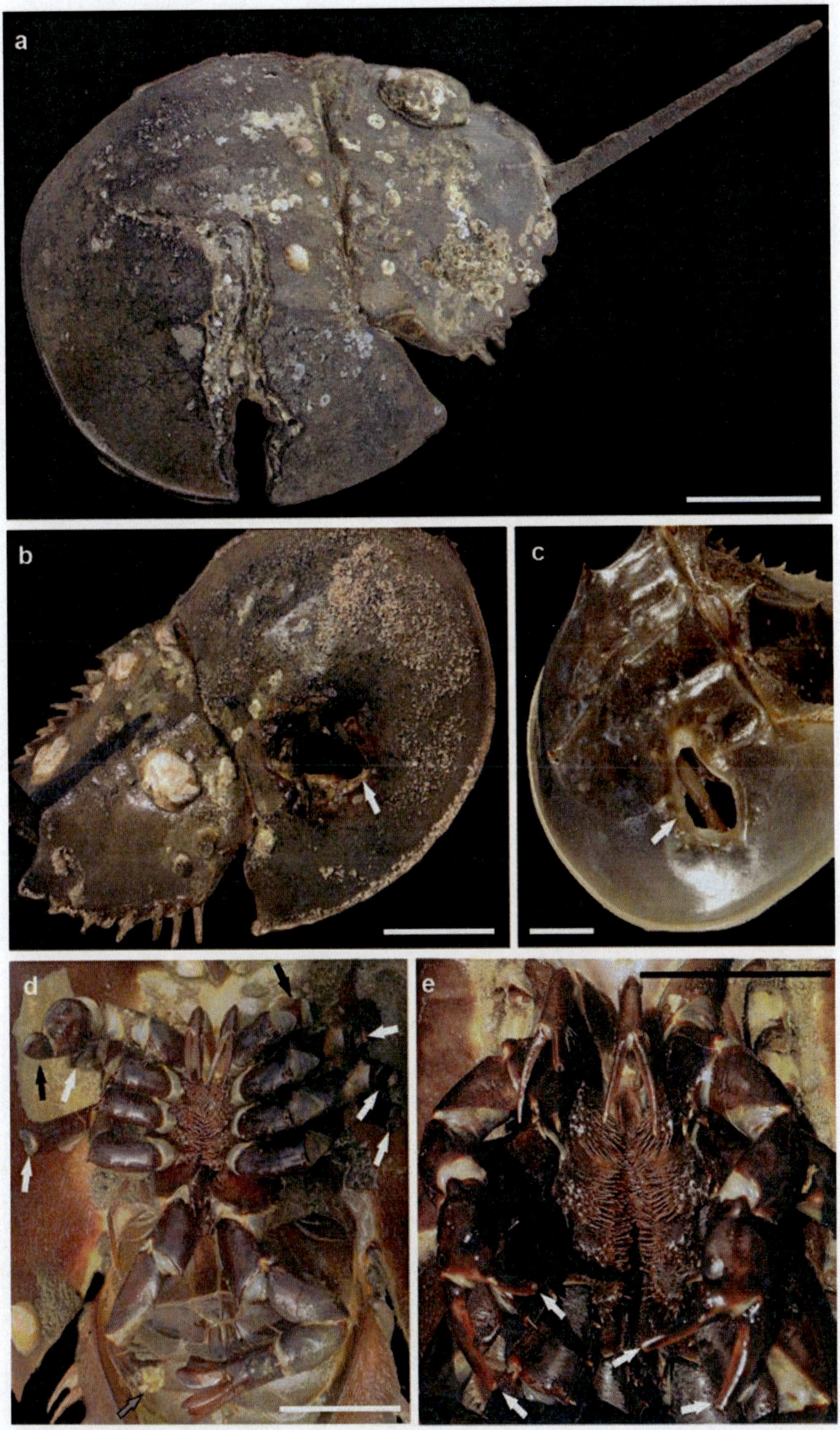


**Fig. 1** *Limulus polyphemus* from Jamaica Bay with cephalothoracic and appendage injuries from the 2018 field season. (**a**) Partly recovered, large, 'U'-shaped cephalothoracic injury that originally

**Fig. 2** *Limulus polyphemus* juvenile moults from Jamaica Bay with pronounced abnormalities. (**a**) Right-kinked telson, offsetting telson spine by 35°. (**b**) Specimen with stunted telson.

stunted, 2 mm long, telson. This same specimen was collected after two moulting events and the telson had recovered entirely (27 mm), retaining a small kink at the terminus (Fig. 4b).

## 3.1 *Injury Patterns*

The abnormality datasets show different patterns for the two species (Fig. 5). Within the *Tachypleus tridentatus* specimens ($n = 374$, of which 266 were injured) documented in Kaiser (2002) were exclusively juvenile, the most common abnormalities were thoracetronic (50%) and telson abnormalities (20%). The adult *Limulus polyphemus* (48 females, 235 males, total of 283 individuals) from the 2018 Jamaica Bay field season showed highest incidences of cephalothoracic (35%) and telson abnormalities (27.5%). Furthermore, 52% of *L. polyphemus* specimens had eye abnormalities (decay and/or epibiont overgrowth). In both datasets, select specimens showed bilaterally expressed abnormalities (e.g., Fig. 3b).

In more detail, the mean carapace condition of Jamaica Bay *Limulus polyphemus* was 2.5 on a scale of 1–3. Only 6 (3.8%) of the adults were new recruits (category 1), compared with 58 (36.8%) intermediate (category 2) and 95 (59.7%) older (category 3) age classes. Injuries to eyes (decay and/or epibiont coverage), dorsal exoskeleton, telson, and appendages each became more frequent as the animal's presumed age increased (Fig. 6). Claspers and the fourth walking leg are the most frequently damaged cephalothoracic appendages. Both show a striking sexual disparity: 8.1% of males (19 males) had damaged pedipalps (this morphology is not observed in female xiphosurids) and 9.7% (23) of males had a damaged fourth walking leg, compared to 2.0% (1) of females. A total of 3.5% (10 individuals) have a damaged first walking leg; 4.9% (14 individuals) have a damaged second walking leg; 6.0% (17 individuals) have a damaged third walking leg. There was no notable difference between the injury frequency in males and females for these appendages. However, this might reflect the abundance of male individuals when compared to females. In rare cases, such as the specimen depicted in Fig. 1d, a single animal shows multiple damaged appendages.

## 3.2 *Abnormal Fossil Specimens*

Three dimensionally (3D) preserved fossil horseshoe crabs present an excellent opportunity to identify abnormalities. Two examples of cephalothoracic abnormalities and one telson abnormality are documented here. A *Mesolimulus walchi*

**Fig. 2** (continued) **(c)** 'U'-shaped injury to left posterior thoracetron with removal of movable spines (white arrow). **(d)** 'U'-shaped indentations on left anterior cephalothorax (white arrows) and cephalothoracic buckling (grey arrow). Scale bars: **(a–c)** = 10 mm; **(d)** = 5 mm. (Image credit: Russell Bicknell. Specimens housed at Fordham University)

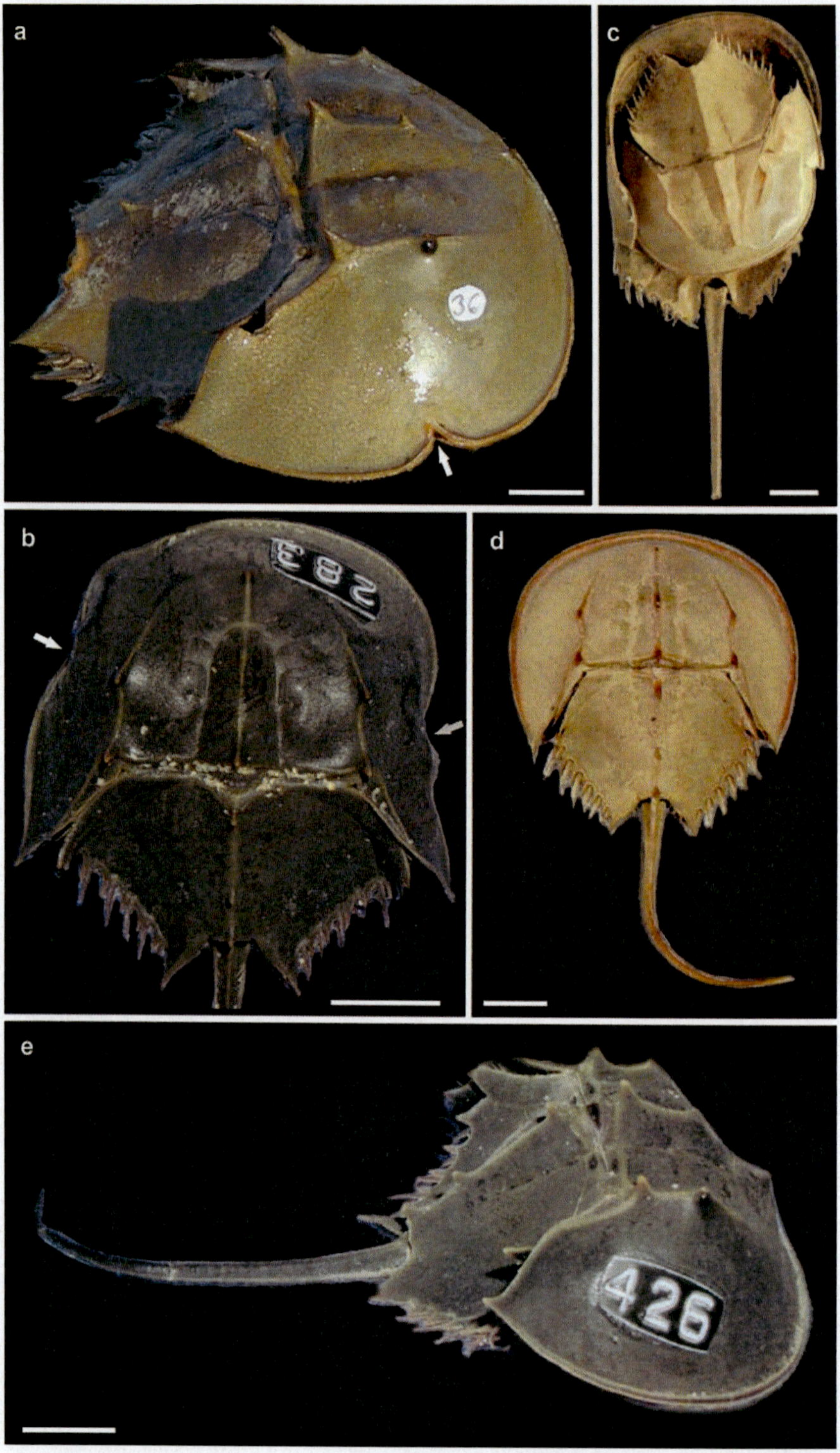


**Fig. 3** Abnormal *Tachypleus tridentatus* specimens. (**a**) 'V'-shaped indentation on the right side of the cephalothorax (white arrow), removal of telson and posterior left thoracetronic region.

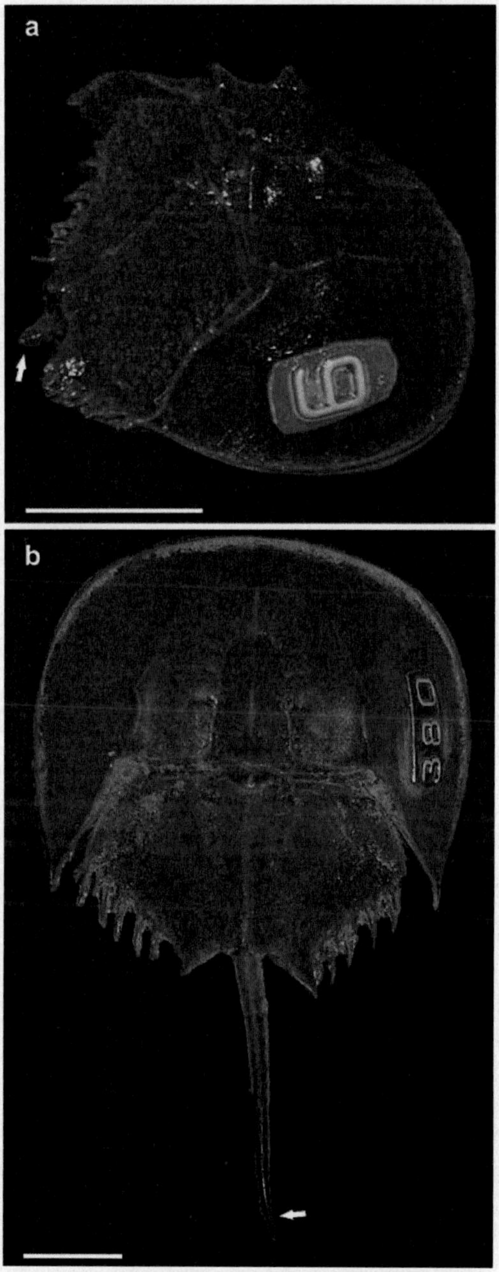


**Fig. 4** *Tachypleus tridentatus* specimen from BM Beach, Palawan, Philippines, showing signs of complete telson recovery. (**a**) Stunted telson (white arrow). (**b**) After two moults showing complete recovery of telson with a small kink at terminus (white arrow). Photos converted to grey-scale. Scale bars: (**a, b**) =10 mm. (Image credit: Dorkas Kaiser. Specimens were not accessioned into a collection)

**Fig. 3** (continued) (**b**) 'U'-shaped indentation on left side of the cephalothorax (white arrow) and 'W'-shaped indentation on right side of the cephalothorax (grey arrow). (**c**) Failed moulting event, preserving old exuvium inside younger one. (**d**) Strongly right-curved telson. (**e**) Telson point deflected vertically. (**a, b, e**) from BM Beach, Palawan, Philippines. (**c, d**) from Ha Pak Nai, Deep Bay, Hong Kong. Scale bars: (**a**) = 20 mm; (**b–e**) =10 mm. (Image credit: (**a, b, e**) Dorkas Kaiser; (**c, d**) Kevin Laurie). Specimens were not accessioned into a collection

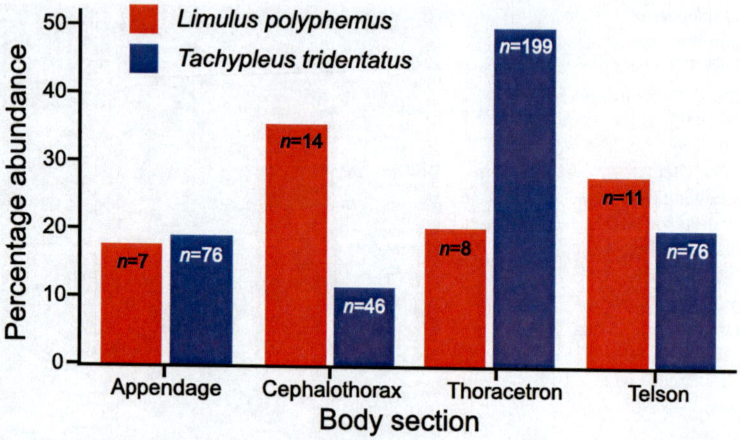


**Fig. 5** Plot of abnormality distributions on *Limulus polyphemus* and *Tachypleus tridentatus* samples. *Tachypleus tridentatus* abnormalities are more commonly located on the thoracetron. *Limulus polyphemus* abnormalities are more commonly located on the cephalothorax. Note that all *L. polyphemus* specimens were adults, *T. tridentatus* while specimens were juvenile. Specimens with injuries on two sections of the exoskeleton were counted as multiple observations

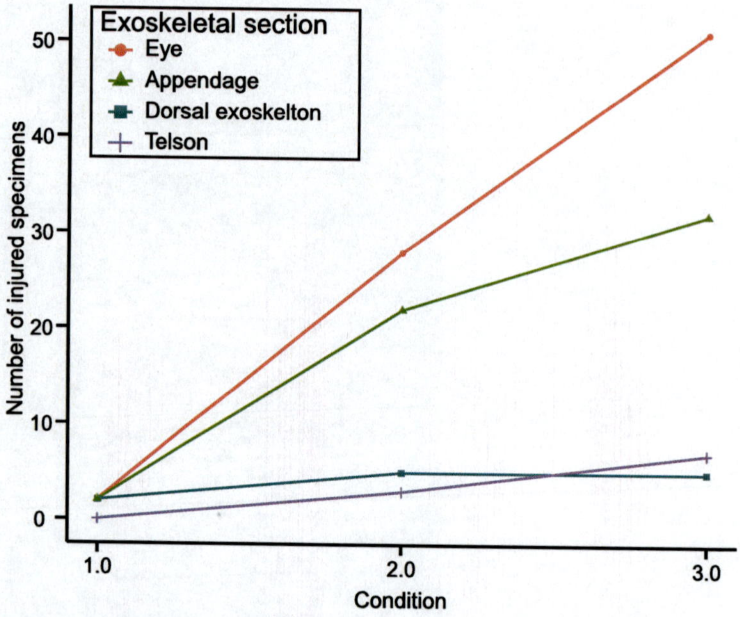


**Fig. 6** Plot of adult *Limulus polyphemus* exoskeletal condition and abnormalities as a function of presumed "age" from a sample of 159 injured individuals at Jamaica Bay. Abnormalities increase as the species have poorer exoskeletons, indicative of an older age. Specimens with injuries on two sections of the exoskeleton were counted as multiple observations

specimen preserved in 3D from the Solnhofen Limestone (Jurassic, late Kimmeridgian-early Tithonian; CM 28512, Fig. 7a, b) has a 'V'-shaped indentation on the left side of the cephalothorax. The indentation extends 3 mm into the cephalothorax and does not exceed the cephalothoracic doublure. The indentation has slightly raised relief and is potentially cicatrised. A *Euproops danae* specimen (YPM IP 50578, Fig. 7c, d) from the siderite nodules of the Mazon Creek *Konservat Lagerstätte,* Carbondale Formation (Pennsylvanian, Moscovian) (Fisher 1979; Anderson 1994; Clements et al. 2018) has a small 'V'-shaped indentation on the right genal spine. The indentation extends 1 mm into the genal spine and lacks any obvious cicatrisation. A further *M. walchi* specimen (CM 28496) has a curved telson (Fig. 8a, b). The telson begins to curve 27 mm from the thoracetron-telson embayment and deflects to the right of specimen's sagittal axis. This produces an offset of ~12° relative to sagittal plane.

## 4 Discussion

The cephalothoracic and thoracetronic abnormalities considered here are examples of injuries. The 'V'- and 'W'-shaped embayments, as well as the cephalothoracic buckling (Fig. 2d; Jell 1989), reflect mechanical damage. The large holes in cephalothoracic sections cannot easily be ascribed to atrophial or developmental issues as they are substantial (13–36 mm wide), show evidence of cicatrisation, and recovery (Fig. 1). These were likely produced by predatory attacks (see Walls et al. 2002) or damage caused by mechanized fishing gear. 'U'-shaped injuries (Fig. 2c) record either tearing of softer exoskeleton after moulting, or an earlier injury that has undergone substantial regeneration and recovery (Bicknell and Pates 2019a). Figure 3c is a unique example of a moulting event that failed to completely remove the previous exuvium. This specimen likely died soon after moulting, as it would have been unable to move or feed.

Abnormal cephalothoracic appendages are common among adult horseshoe crabs and include missing, partially broken, or misshapen legs (Fig. 1d, e). The pedipalp claw is frequently damaged or missing in adult males likely due to the mechanical stress associated with amplexus (e.g., Brockmann 1990). However, appendage damage may have also been caused by poor handling by anglers after bycatch. Males with missing or damaged claspers are unable to form a firm attachment to a female, but can nonetheless reproduce as satellite males: unattached males that release sperm in the vicinity of a mated pair (Duffy et al. 2006). The behavioural changes due to loss or damage of the other cephalothoracic appendages merits further investigation. Indeed, we would predict an adverse effect on feeding as walking legs chelae assist in moving food to the mouth (Botton et al. 2003). Further, if the coxal sections of walking legs are absent or damaged, the ability for an individual to masticate prey would be highly reduced (Bicknell et al. 2018b, 2018c).

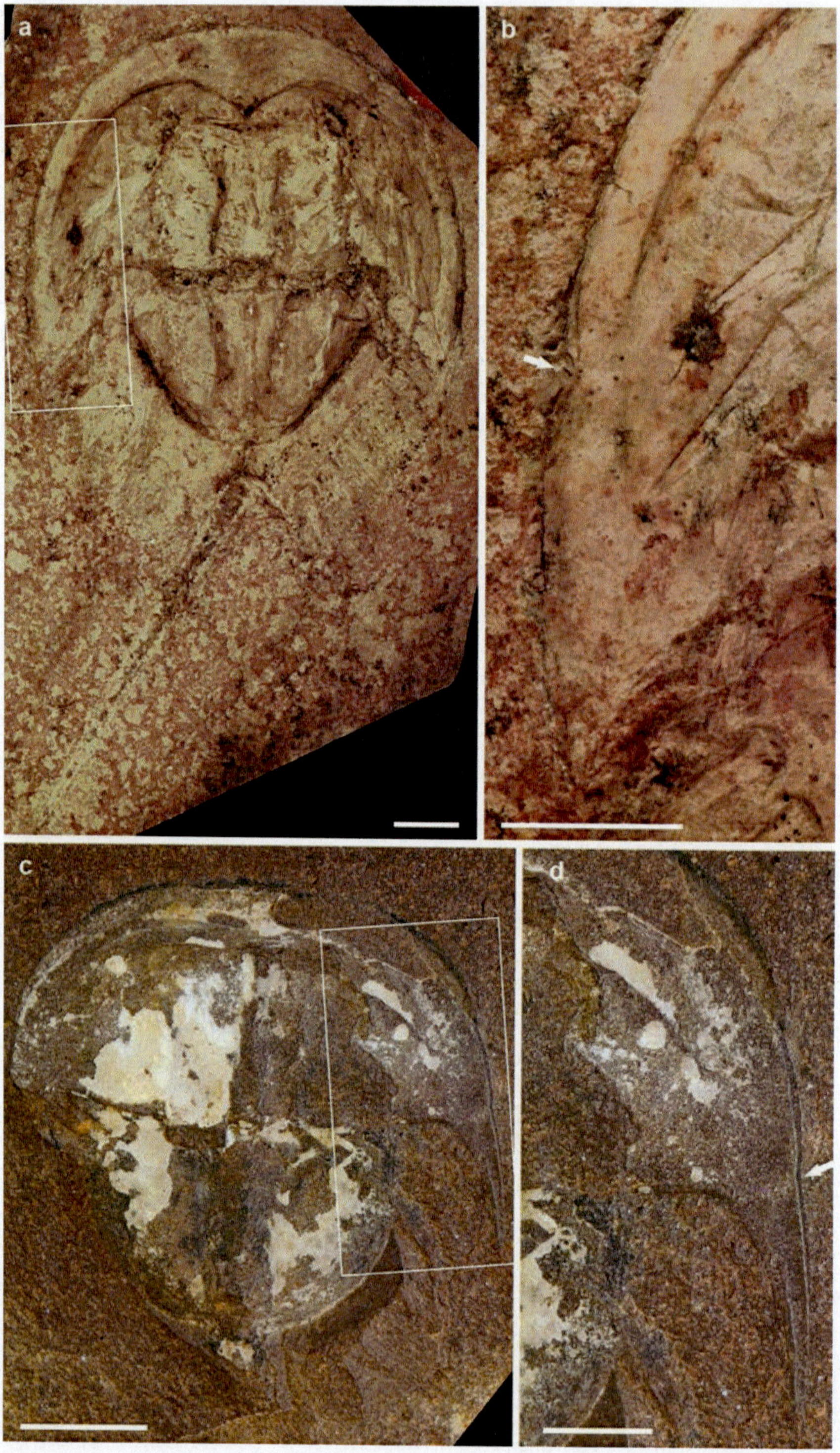

**Fig. 7** Fossil xiphosurids with cephalothoracic abnormalities considered injuries. **(a, b)** *Mesolimulus walchi* with a 'V'-shaped indentation on left side of the cephalothorax (CM 28512).

Telson abnormalities are very morphologically variable, with known examples of stunted (Botton and Loveland 1989; Figs. 2b, 4a), kinked (Fig. 2a), curved (Bicknell and Pates 2019a; Fig. 3d), and bifurcated (Osburn 1911; Gudger 1935; Shuster Jr. 1982; Bicknell et al. 2018d) morphologies. Despite the variable and striking nature of these abnormalities, the means of developing them is unclear, especially as laboratory studies have not yet explored how these structures are produced. However, moulting complications and propagation of abnormalities during subsequent moulting events are a suggested mechanism (Gudger 1935; Bicknell et al. 2018d). Either during, or post-ecdysis, when the exoskeleton is softer and easier to damage (Brandt 2002; Bicknell et al. 2018d), telsons may incur defects, resulting in injuries and deformation. One likely scenario would involve a recently moulted horseshoe crab that is overturned by wave action and subsequently used the telson to right itself (Vosatka 1970). During this event, the telson spine may get caught in the substrate and due to the pliable nature of newly moulted exoskeleton

**Fig. 8** *Mesolimulus walchi* specimen with deflected telson (CM 28496). (**a**) Complete specimen. (**b**) Close up of telson. Black arrow indicates curve onset. Photographs converted to grey scale. Scale bars: (**a**) = 20 mm; (**b**) = 5 mm. (Image credit: Russell Bicknell)

**Fig. 7** (continued) (**a**) Complete specimen. (**b**) Close up on injury (white arrow). (**c, d**) *Euproops danae* with a 'V'-shaped indentation on right genal spine (YPM IP 50578). (**c**) Complete specimen. (**d**) Close up on injury (white arrow). Scale bars: (**a, b**) =10 mm; (**c**) = 5 mm; (**d**) = 2.5 mm. (Image credit: Russell Bicknell) become curved upward or laterally (Figs. 2a, 3d, e). Direct observations of live species are needed to confirm that telsons can produce the observed range of injury morphologies.

## 4.1 Patterns of Abnormalities on Extant Species

Previous research has suggested that telson injuries were potentially the most common abnormalities observed on extant horseshoe crabs (Gudger 1935; Shuster Jr. 1982; Bicknell et al. 2018d) and that the slender, elongate morphology of the telson allowed for easy identification of such abnormalities. However, subsequent work suggested that an abundance of telson abnormalities could instead record species-specific patterns (Bicknell and Pates 2019a). Data from Bicknell and Pates (2019a) suggest that *Limulus polyphemus* specimens commonly have telson injuries; an observation aligning with statements in Shuster Jr. (1982) and Botton and Loveland (1989). In our sample of adults from Jamaica Bay, cephalothoracic injuries were more common than telson abnormalities. *Tachypleus tridentatus* specimens assessed in Bicknell and Pates (2019a) suggested no preference for abnormality location. This is contradicted by the Kaiser (2002) data, in which thoracetronic abnormalities were the most commonly noted. The difference in the results of these two studies likely reflects the low sample size of *T. tridentatus* in Bicknell and Pates (2019a) ($n = 3$). Regardless, the patterns documented here show an abnormality locality bias. The bias may reflect environments inhabited by the different species, the comparison of adult with more juvenile populations, potential predators, or a species-specific pre-disposition for select abnormalities. Further examination of spawning populations, ideally of similar ages, may highlight more patterns in injury locality and bias.

## 4.2 Fossil Xiphosurids

Examples of abnormal extinct xiphosurids are rare. However, the few known specimens represent important evidence to support trace fossil records of predation on extinct xiphosurids (Diedrich 2011; Oukassou et al. 2018) and the history of abnormalities in horseshoe crabs. Presently, eight examples of injuries have now been documented and seven are 'V'- or 'W'-shaped injuries on the dorsal exoskeleton (Bicknell et al. 2018d). Such shapes contrast 'U'-shaped injuries noted on select extant specimens (Fig. 2c, d; Bicknell et al. 2018d, fig. 2). Smaller 'U'-shaped indentations are less likely to be preserved in the fossil record as exoskeletal recovery over moulting events decreases angularity of indentations making them harder to confidently differentiate from distortion during preservation (Bicknell et al. 2018d). No examples of cephalothoracic injuries on fossil horseshoe crabs extend to the cardiac lobe, contrasting rare extant examples (Fig. 1a). Fossil xiphosurids may therefore have been less resilient to larger injuries, a facet related to their smaller size relative to extant forms.

The first example of an extinct xiphosurid with an abnormal telson is reported (CM 28496). This fossil specimen confirms that extinct xiphosurids can record injuries on all dorsal exoskeletal sections. The curved *Mesolimulus walchi* telson is comparable to a curved telson specimen figured in Bicknell and Pates (2019a, fig. 7F, G), highlighting evidence for curved telson abnormalities extends to at least the Jurassic. Taphonomy might explain this abnormal telson; however, given a direct modern analogue, a biological explanation is more plausible. Furthermore, a taphonomic cause predicts that morphologies would be observed more often. Botton and Loveland (1989) documented wild adult *Limulus polyphemus* specimens along Delaware Bay and noted that individuals with abnormal telson spines were more commonly stranded, likely because they were ineffective at righting themselves (Vosatka 1970). Inverted individuals were also more susceptible to predation by Great Black-backed Gulls (*Larus marinus* Linnaeus, 1758) and Herring Gulls (*Larus argentatus* Pontoppidan, 1763) (Botton and Loveland 1989). Assuming that the telson in fossil forms functioned comparably taxon to extant taxa, CM 28496 would potentially have been less fit than other co-occurring *M. walchi* individuals.

Xiphosurid body fossils are rare, even in deposits that permit soft-bodied, cuticular preservation (Anderson 1994; Babcock and Merriam 2000; Rudkin and Young 2009; Bicknell and Pates 2020). As such, unless xiphosurids are common in a given deposit, the likelihood of identifying abnormal fossil specimens is low. In those limited deposits where xiphosurids are abundantly preserved, more abnormal fossil specimens can be uncovered. The Solnhofen Limestone has presented a fortunate opportunity to identify these rare features as extensive quarrying of the lithographic limestone has resulted in large collections of *M. walchi* in multiple institutions (Arratia and Tischlinger 2010). Large collections of *Euproops danae* from the Mazon Creek *Konservat Lagerstätte* in the YPM IP (>500 specimens) (Haug and Rötzer 2018), or the Field Museum, Chicago (>500 specimens) (Fisher 1977) are other avenues worth exploring for abnormalities. Finally, xiphosurans from the Lower and Upper Fezouata formations, Morocco (Lower Ordovician), could be explored as hundreds of specimens are known from these deposits (Van Roy et al. 2010), despite the uncertain taxonomic affinities of these forms.

## 4.3 Conservation Implications

Horseshoe crabs now face serious threats to their populations. Habitat alteration, such as land reclamation, harvesting for blood extraction, food (horseshoe crab eggs are considered a delicacy in Southeast Asia), damage from fishermen, and other human interactions have placed these organisms in serious danger (Botton 2001; Hsieh and Chen 2009; Akbar John et al. 2011, 2018; Nelson et al. 2015; Kwan et al. 2016; Fairuz-Fozi et al. 2018). We propose that documentation of carapace condition and abnormalities in populations will allow researchers to identify those spawning areas that require conservation attention. To do this, baseline values of abnormal, ideally injured, specimens, as well as carapace condition, need to be gathered over field seasons at key spawning sites. These values can then be used to determine whether a population is maintaining a consistent abundance of injured specimens. Populations displaying increases in injuries, or constantly high injury frequencies, represent areas that require conservation attention. In addition, a high number of embryonic abnormalities is a potential indicator of poor water quality. These data should also be collected and assessed alongside information on adult specimens (Botton and Itow 2009). This approach, in conjunction with larger conservation efforts, will contribute to preventing the extinction of the animals that have already survived five mass extinctions.

**Acknowledgements** This research was supported by funding from an Australian Postgraduate Award (to R.D.C.B.), a University of New England Postdoctoral Fellowship (to R.D.C.B.), a Charles Schuchert and Carl O. Dunbar Grants-in-Aid award (to R.D.C.B.), Studienstiftung des deutschen Volkes (to D. K.), a Dean's Summer Undergraduate Research Grant (to S.Z.), an Oxford-St Catherine's Brade-Natural Motion Scholarship (to S.P), and an Alexander Agassiz Postdoctoral Fellowship (to S.P). We thank Susan Butts and Jessica Utrup for use of the YMP IP camera and the help with the collection. We thank Albert Kollar for help with the CM collection. We thank Kevin Laurie for photos of abnormal specimens. Finally, we thank the two anonymous referees for their comments that helped improve the manuscript.

## References

B. Akbar John, K.C.A. Jalal, K. Zaleha, P. Armstrong, B.Y. Kmaruzzaman, Effects of blood extraction on the mortality of Malaysian horseshoe crabs (*Tachypleus gigas*). Mar. Freshw. Behav. Physiol. **44**(5), 321–327 (2011). https://doi.org/10.1080/10236244.2011.642505

B. Akbar John, B.Y. Kamaruzzaman, K.C.A. Jalal, K. Zaleha, Feeding ecology and food preferences of *Carcinoscorpius rotundicauda* collected from the Pahang nesting grounds. Sains Malaysiana **41**(7), 855–861 (2012)

B. Akbar John, B.R. Nelson, H.I. Sheikh, S.G. Cheung, Y. Wardiatno, B.P. Dash, K. Tsuchiya, Y. Iwasaki, S. Pati, A review on fisheries and conservation status of Asian horseshoe crabs. Biodivers. Conserv. **27**(14), 3573–3598 (2018)

L.I. Anderson, Xiphosurans from the Westphalian D of the Radstock Basin, Somerset Coalfield, the South Wales Coalfield and Mazon Creek, Illinois. Proc. Geol. Assoc. **105**(4), 265–275 (1994). https://doi.org/10.1016/s0016-7878(08)80179-4

G. Arratia, H. Tischlinger, The first record of Late Jurassic crossognathiform fishes from Europe and their phylogenetic importance for teleostean phylogeny. Fossil Record **13**(2), 317–341 (2010). https://doi.org/10.1002/mmng.201000005

L.E. Babcock, Trilobite malformations and the fossil record of behavioral asymmetry. J. Paleontol. **67**(2), 217–229 (1993)

L.E. Babcock, D.F. Merriam, Horseshoe crabs (Arthropoda: Xiphosurida) from the Pennsylvanian of Kansas and elsewhere. Trans. Kans. Acad. Sci. **103**(1), 76–94 (2000). https://doi.org/10.2307/3627941

R.D.C. Bicknell, L. Lustri, T. Brougham, Revision of "*Bellinurus*" *carteri* (Chelicerata: Xiphosura) from the Late Devonian of Pennsylvania, USA. Comptes. Rendus. Palevol. **18**(8), 967–976 (2019a)

R.D.C. Bicknell, Xiphosurid from the Upper Permian of Tasmania confirms Palaeozoic origin of Austrolimulidae. Palaeontol. Electron. **22**(3), 1–13 (2019). https://doi.org/10.26879/1005

R.D.C. Bicknell, L. Amati, J. Ortega Hernández, New insights into the evolution of lateral compound eyes in Palaeozoic horseshoe crabs. Zool. J. Linnean Soc. **187**(4), 1061–1077 (2019b). https://doi.org/10.1093/zoolinnean/zlz065

R.D.C. Bicknell, T. Brougham, S. Charbonnier, F. Sautereau, T. Hitij, N.E. Campione, On the appendicular anatomy of the xiphosurid *Tachypleus syriacus* and the evolution of fossil horseshoe crab appendages. Sci. Nat. **106**(7), 38 (2019c). https://doi.org/10.1007/s00114-019-1629-6

R.D.C. Bicknell, A.J. Klinkhamer, R.J. Flavel, S. Wroe, J.R. Paterson, A 3D anatomical atlas of appendage musculature in the chelicerate arthropod *Limulus polyphemus*. PLoS ONE **13**(2), e0191400 (2018a). https://doi.org/10.1371/journal.pone.0191400

R.D.C. Bicknell, J.A. Ledogar, S. Wroe, B.C. Gutzler, W.H. Watson III, J.R. Paterson, Computational biomechanical analyses demonstrate similar shell-crushing abilities in modern and ancient arthropods. Proc. R. Soc. B Biol. Sci. **285**(1889), 20181935 (2018b). https://doi.org/10.1098/rspb.2018.1935

R.D.C. Bicknell, J.R. Paterson, J.-B. Caron, C.B. Skovsted, The gnathobasic spine microstructure of Recent and Silurian chelicerates and the Cambrian artiopodan *Sidneyia*: Functional and evolutionary implications. Arthropod Struct. Dev. **47**(1), 12–24 (2018c). https://doi.org/10.1016/j.asd.2017.12.001

R.D.C. Bicknell, S. Pates, Abnormal extant xiphosurids in the Yale Peabody Museum Invertebrate Zoology collection. Bull. Peabody Mus. Nat. Hist. **60**(1), 41–53 (2019a). https://doi.org/10.3374/014.060.0102

R.D.C. Bicknell, S. Pates, Xiphosurid from the Tournaisian (Carboniferous) of Scotland confirms deep origin of Limuloidea. Sci. Rep. **9**(1), 17102 (2019b)

R.D.C. Bicknell, S. Pates, Pictorial atlas of fossil and extant horseshoe crabs, with focus on Xiphosurida. Front. Earth Sci. (2020). https://doi.org/10.3389/feart.2020.00098

R.D.C. Bicknell, S. Pates, M.L. Botton, Abnormal xiphosurids, with possible application to Cambrian trilobites. Palaeontol. Electron. **21**(2), 1–17 (2018d). https://doi.org/10.26879/866

R.D.C. Bicknell, S. Pates, M.L. Botton, *Euproops danae* (Belinuridae) cluster confirms deep origin of gregarious behaviour in xiphosurids. Arthropoda Selecta **28**(4), 549–555 (2019d)

R.D.C. Bicknell, J. Žalohar, P. Miklavc, B. Celarc, M. Križnar, T. Hitij, A new limulid genus from the Strelovec Formation (Middle Triassic, Anisian) of northern Slovenia. Geol. Mag. **156**(12), 2017–2030 (2019e). https://doi.org/10.1017/S0016756819000323

B. Błażejowski, P. Gieszcz, A.P. Shinn, R.M. Feldmann, E. Durska, Environment deterioration and related fungal infection of Upper Jurassic horseshoe crabs with remarks on their exceptional preservation. Palaeogeogr. Palaeoclimatol. Palaeoecol. **516**, 336–341 (2019)

M.L. Botton, The conservation of horseshoe crabs: What can we learn from the Japanese experience? in *Limulus in the Limelight*, ed. by J. T. Tanacredi, (Springer, New York, 2001), pp. 41–51. https://doi.org/10.1007/0-306-47590-1_4

M.L. Botton, T. Itow, The effects of water quality on horseshoe crab embryos and larvae, in *Biology and Conservation of Horseshoe Crabs*, ed. by J. T. Tanacredi, M. L. Botton, D. R. Smith, (Springer, New York, 2009), pp. 439–454

M.L. Botton, R.E. Loveland, Reproductive risk: High mortality associated with spawning by horseshoe crabs (*Limulus polyphemus*) in Delaware Bay, USA. Mar. Biol. **101**(2), 143–151 (1989). https://doi.org/10.1007/BF00391453

M.L. Botton, C.N. Shuster Jr., J.A. Keinath, Horseshoe crabs in a food web: Who eats whom, in *The American Horseshoe Crab*, ed. by C. N. Shuster Jr., R. B. Barlow, H. J. Brockmann, (Harvard University Press, Cambridge, 2003), pp. 133–153

M.L. Botton, R.A. Tankersley, R.E. Loveland, Developmental ecology of the American horseshoe crab *Limulus polyphemus*. Curr. Zool. **56**(5), 550–562 (2010)

D.S. Brandt, Ecydsial efficiency and evolutionary efficacy among marine arthropods: Implications for trilobite survivorship. Alcheringa **26**(3), 399–421 (2002). https://doi.org/10.1080/03115510208619264

H.J. Brockmann, Mating behavior of horseshoe crabs, *Limulus polyphemus*. Behaviour **114**(1), 206–220 (1990). https://doi.org/10.1163/156853990x00121

T. Clements, M. Purnell, S. Gabbott, The Mazon Creek Lagerstätte: A diverse late Paleozoic ecosystem entombed within siderite concretions. J. Geol. Soc., 1–11 (2018). https://doi.org/10.1144/jgs2018-088

A.-G. Desmarest, Les crustacés proprement dits, in *Histoire naturelle des crustacés fossiles, sous les rapports zoologiques et geologiques*, ed. by A. Brongniart, A.-G. Desmarest, (F.-G. Levrault, Paris, 1822), pp. 67–142

C.G. Diedrich, Middle Triassic horseshoe crab reproduction areas on intertidal flats of Europe with evidence of predation by archosaurs. Biol. J. Linn. Soc. **103**(1), 76–105 (2011). https://doi.org/10.1111/j.1095-8312.2011.01635.x

J. Dietl, C. Nascimento, R. Alexander, Influence of ambient flow around the horseshoe crab *Limulus polyphemus* on the distribution and orientation of selected epizoans. Estuaries **23**(4), 509–520 (2000). https://doi.org/10.2307/1353142

E.E. Duffy, D.J. Penn, M.L. Botton, H.J. Brockmann, R.E. Loveland, Eye and clasper damage influence male mating tactics in the horseshoe crab, *Limulus polyphemus*. J. Ethol. **24**(1), 67–74 (2006). https://doi.org/10.1007/s10164-005-0163-5

N. Fairuz-Fozi, B. Satyanarayana, N.A.M. Zauki, A.M. Muslim, M.-L. Husain, S. Ibrahim, B.R. Nelson, *Carcinoscorpius rotundicauda* (Latreille, 1802) population status and spawning behaviour at Pendas coast, Peninsular Malaysia. Glob. Ecol. Conserv. **15**, e00422 (2018). https://doi.org/10.1016/j.gecco.2018.e00422

R.D. Farley, Book gill development in embryos and first and second instars of the horseshoe crab *Limulus polyphemus* L. (Chelicerata, Xiphosura). Arthropod Struct. Dev. **39**(5), 369–381 (2010). https://doi.org/10.1016/j.asd.2010.04.001

D.C. Fisher, Functional significance of spines in the Pennsylvanian horseshoe crab *Euproops danae*. Paleobiology **3**(2), 175–195 (1977). https://doi.org/10.1017/s009483730000525x

D.C. Fisher, Evidence for subaerial activity of *Euproops danae* (Merostomata, Xiphosurida), in *Mazon Creek Fossils*, ed. by M. H. Nitecki, (Elsevier, New York, 1979), pp. 379–447. https://doi.org/10.1016/b978-0-12-519650-5.50022-9

D.C. Fisher, The Xiphosurida: Archetypes of bradytely? in *Living Fossils*, ed. by N. Eldredge, S. M. Stanley, (Springer, New York, 1984), pp. 196–213. https://doi.org/10.1007/978-1-4613-8271-3_23

E.W. Gudger, Horseshoe crabs with forked tails. Bull. N. Y. Zool. Soc. **38**, 170–173 (1935)

C. Haug, M.A.I.N. Rötzer, The ontogeny of the 300 million year old xiphosuran *Euproops danae* (Euchelicerata) and implications for resolving the *Euproops* species complex. Dev. Genes Evol. **228**(1), 63–74 (2018). https://doi.org/10.1007/s00427-018-0604-0

N. Hauschke, V. Wilde, Zur Verbreitung und Ökologie mesozoischer Limuliden. Neues Jahrb. Geol. Palaontol. Abh. **183**(1–3), 391–411 (1991)

H.-L. Hsieh, C.-P. Chen, Conservation program for the Asian horseshoe crab *Tachypleus tridentatus* in Taiwan: Characterizing the microhabitat of nursery grounds and restoring spawning grounds, in *Biology and Conservation of Horseshoe Crabs*, ed. by J. T. Tanacredi, M. L. Botton, D. R. Smith, (Springer, New York, 2009), pp. 417–438. https://doi.org/10.1007/978-0-387-89959-6_26

P.A. Jell, Some aberrant exoskeletons from fossil and living arthropods. Mem. Queensl. Mus. **27**(2), 491–498 (1989)

D. Kaiser, Zur Ökologie und Morphologie von *Tachypleus tridentatus* (Leach, 1819) (Chelicerata: Merostomata) im Gezeitenwatt der Ostküste von Palawan (Philippinen). Diploma Thesis, University of Rostock, (2002)

D. Kaiser, S. Schoppe, Postembryonic development of *Tachypleus tridentatus* (Merostomata: Xiphosura) in a nursery habitat in the Philippines. J. Threat. Taxa **10**(15), 12916–12932 (2018). https://doi.org/10.11609/jott.4125.10.15.12916-12932

R. Kaplan, S.S.L. Li, J.M. Kehoe, Molecular characterization of limulin, a sialic acid binding lectin from the hemolymph of the horseshoe crab, *Limulus polyphemus*. Biochemistry **16**(19), 4297–4303 (1977). https://doi.org/10.1021/bi00638a026

B.K.Y. Kwan, H.-L. Hsieh, S.G. Cheung, P.K.S. Shin, Present population and habitat status of potentially threatened Asian horseshoe crabs *Tachypleus tridentatus* and *Carcinoscorpius rotundicauda* in Hong Kong: A proposal for marine protected areas. Biodivers. Conserv. **25**(4), 673–692 (2016). https://doi.org/10.1007/s10531-016-1084-z

E.R. Lankester, *Limulus* an Arachnid. Q. J. Microsc. Sci. **23**, 504–649 (1881)

W.E. Leach, Entomostraca, in *Dictionaire des Science Naturelles*, ed. by F. Levrault, vol. 14, (Levrault and Schoell, Paris, 1819), pp. 524–543

C. Linnaeus, *Systema naturæ per Regna Tria naturæ, Secundum Classes, Ordines, Genera, Species, Cum Characteribus, Differentiis, Synonymis, Locis*, 10th edn. (Laurentius Salvius, Holmia, 1758). https://doi.org/10.5962/bhl.title.559

F.B. Meek, A.H. Worthen, Notice of some new types of organic remains, from the Coal Measures of Illinois. Proc. Acad. Natl. Sci. Phila. **17**(1), 41–48 (1865)

B.R. Nelson, B. Satyanarayana, J.M.H. Zhong, F. Shaharom, M. Sukumaran, A. Chatterji, Episodic human activities and seasonal impacts on the *Tachypleus gigas* (Müller, 1785) population at Tanjung Selangor in Peninsular Malaysia. Estuar. Coast. Shelf Sci. **164**, 313–323 (2015). https://doi.org/10.1016/j.ecss.2015.08.003

M. Obst, S. Faurby, S. Bussarawit, P. Funch, Molecular phylogeny of extant horseshoe crabs (Xiphosura, Limulidae) indicates Paleogene diversification of Asian species. Mol. Phylogenet. Evol. **62**(1), 21–26 (2012). https://doi.org/10.1016/j.ympev.2011.08.025

R.C. Osburn, An unusual horseshoe crab. Bull. N. Y. Zool. Soc. **48**, 808 (1911)

M. Oukassou, A. Lagnaoui, A. Charrière, H. Saber, W.B. Haouz, G.D. Gierliński, H. Klein, H. Ibouh, New evidence of xiphosurids from the Middle Jurassic of Morocco: Palaeoenvironmental, palaeoecological and palaeobiogeographical implications. Palaeogeogr. Palaeoclimatol. Palaeoecol. **516**, 268–283 (2018). https://doi.org/10.1016/j.palaeo.2018.11.034

A.W. Owen, Trilobite abnormalities. Trans. R. Soc. Edinb. Earth Sci. **76**(2–3), 255–272 (1985). https://doi.org/10.1017/s0263593300010488

R. Owen, On the anatomy of the American King-crab (*Limulus polyphemus*, Latr.). Trans. Linn. Soc. Lond. **28**(3), 459–506 (1872). https://doi.org/10.5962/bhl.title.11275

S. Pates, R.D.C. Bicknell, Elongated thoracic spines as potential predatory deterrents in olenelline trilobites from the lower Cambrian of Nevada. Palaeogeogr. Palaeoclimatol. Palaeoecol. **519**(2019), 295–306 (2019)

W. Patten, On the morphology and physiology of the brain and sense organs of *Limulus*. Q. J. Microsc. Sci. **35**, 1–96 (1894)

E.L. Pontoppidan, *Den danske atlas eller Konge-Riget Dannemark*, vol 1 (A. H, Godiche, Kiøbenhavn, 1763)

M.R.M. Razak, Z. Kassim, Feeding mechanisms of adult tropical horseshoe crab, *Tachypleus gigas* toward feeds' conditions. ASM Sci. J. **11**(2), 76–85 (2018)

J.R. Redmond, D.D. Jorgensen, G.B. Bourne, Circulatory physiology of *Limulus*. Prog. Clin. Biol. Res. **81**, 133–146 (1982)

B. Riska, Morphological variation in the horseshoe crab *Limulus polyphemus*. Evolution **35**(4), 647–658 (1981). https://doi.org/10.2307/2408237

D.M. Rudkin, G.A. Young, Horseshoe crabs–an ancient ancestry revealed, in *Biology and Conservation of Horseshoe Crabs*, ed. by J. T. Tanacredi, M. L. Botton, D. R. Smith, (Springer, New York, 2009), pp. 25–44. https://doi.org/10.1007/978-0-387-89959-6_2

S.Y. Schaller, P. Thayer, S. Hanson, Maine horseshoe crab (*Limulus polyphemus*) spawning surveys, 2003, in *Report to Maine Department of Marine Resources*, (Bar Mills Ecological, Buxton, ME, 2004), pp. 1–26

S.Y. Schaller, P. Thayer, S. Hanson, S. LaTulippe, E. Solet, Maine horseshoe crab (*Limulus polyphemus*) spawning surveys, 2004, in *Report to Maine Department of Marine Resources*, (Bar Mills Ecological, Buxton, ME, 2005), pp. 1–35

K. Sekiguchi, H. Seshimo, H. Sugita, Post-embryonic development of the horseshoe crab. Biol. Bull. **174**(3), 337–345 (1988). https://doi.org/10.2307/1541959

P.A. Selden, D.J. Siveter, The origin of the limuloids. Lethaia **20**(4), 383–392 (1987). https://doi.org/10.1111/j.1502-3931.1987.tb02060.x

J.W. Shultz, Gross muscular anatomy of *Limulus polyphemus* (Xiphosura, Chelicerata) and its bearing on evolution in the Arachnida. J. Arachnol. **29**(3), 283–303 (2001). https://doi.org/10.1636/0161-8202(2001)029[0283:gmaolp]2.0.co;2

C.N. Shuster Jr., K. Sekiguchi, Basic habitat requirements of the extant species of horseshoe crabs (Limulacea), in *Biology and Conservation of Horseshoe Crabs*, ed. by J. T. Tanacredi, M. L. Botton, D. R. Smith, (Springer, New York, 2009), pp. 115–129. https://doi.org/10.1007/978-0-387-89959-6_7

C.N. Shuster Jr., A pictorial review of the natural history and ecology of the horseshoe crab *Limulus polyphemus*, with reference to other Limulidae. Prog. Clin. Biol. Res. **81**, 1–52 (1982)

C.N. Shuster Jr., K. Sekiguchi, Growing up takes about ten years and eighteen stages, in *The American Horseshoe Crab*, ed. by C. N. Shuster Jr., R. B. Barlow, H. J. Brockmann, (2004), pp. 103–132

A. Sokoloff, Observations on populations of the horseshoe crab *Limulus* (= *Xiphosura*) *polyphemus*. Res. Popul. Ecol. **19**(2), 222–236 (1978). https://doi.org/10.1007/BF02518829

T.C. Srijaya, P.J. Pradeep, S. Mithun, A. Hassan, F. Shaharom, A. Chatterji, A new record on the morphometric variations in the populations of horseshoe crab (*Carcinoscorpius rotundicauda* Latreille) obtained from two different ecological habitats of Peninsular Malaysia. Our Nat. **8**(1), 204–211 (2010). https://doi.org/10.3126/on.v8i1.4329

J.C. van der Meer Mohr, Sur quelques malformations chez la limule, *Tachypleus gigas*. Misc. Zool. Sumatrana **87**, 1–3 (1935)

P. Van Roy, P.J. Orr, J.P. Botting, L.A. Muir, J. Vinther, B. Lefebvre, K. El Hariri, D.E.G. Briggs, Ordovician faunas of Burgess Shale type. Nature **465**(7295), 215–218 (2010). https://doi.org/10.1038/nature09038

R. Vijayakumar, S. Das, A. Chatterji, A.H. Parulekar, Morphometric characteristics in the horseshoe crab *Tachypleus gigas* (Arthropoda: Merostomata). Indian J. Mar. Sci. **29**, 333–335 (2000)

E.D. Vosatka, Observations on the swimming, righting, and burrowing movements of young horseshoe crabs, *Limulus polyphemus*. Ohio J. Sci. **70**(5), 276–283 (1970)

E.A. Walls, J. Berkson, S.A. Smith, The horseshoe crab, *Limulus polyphemus*: 200 million years of existence, 100 years of study. Rev. Fish. Sci. **10**(1), 39–73 (2002). https://doi.org/10.1080/20026491051677

H. Woodward, Contributions to the knowledge of fossil Crustacea. Q. J. Geol. Soc. **35**(1), 549–556 (1879). https://doi.org/10.1144/gsl.jgs.1879.035.01-04.37

# Part II
# Population Dynamics & Ecology

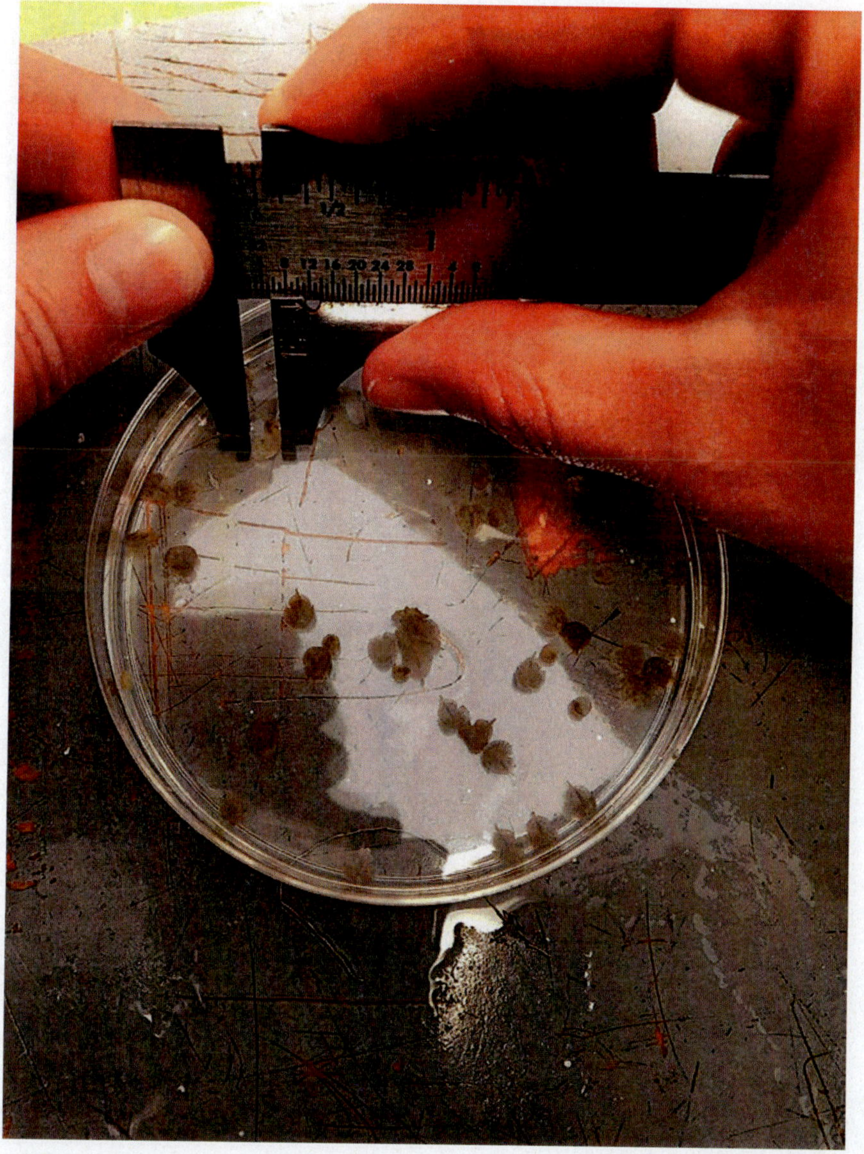

# Activity Rhythm of the Tri-Spine Horseshoe Crab *Tachypleus tridentatus* in the Seto Inland Sea, Western Japan, Monitored with Acceleration Data-Loggers

**Shinichi Watanabe, Saori Oyamada, Keisuke Mizuta, Koujirou Azumakawa, Satoshi Morinobu, and Norimichi Souji**

## 1 Introduction

All lives evolved on our planet are exposed to periods of day and night. Thus, endogenous *circadian* rhythms (24-h) can be found everywhere in bacteria, plants, and animals (Foster and Kreitzman 2004). Intertidal zone is the coastal region which alternately submerged and then exposed by the sea twice each lunar day (24.8-h). Organisms living there often show a tidal rhythm (12.4-h) in their behavior and physiology (for reviews, see Palmer 1976; Naylor 2010). Activity rhythms of the intertidal animals have been most studied in crustaceans. Fiddler crabs have clear, endogenous, *circatidal* rhythms that free-run in constant conditions (the rhythms persist in constant conditions with periods close to 12.4-hours) (Bennett et al. 1957; Barnwell 1966). The green shore crab, *Carcinus maenas* also exhibits circatidal locomotive activity rhythms, with peaks at the time of high tide (Naylor 1958). For another model species, activity rhythm of the American horseshoe crab, *Limulus polyphemus* has recently been well-documented. Horseshoe crabs (the class Merostomata) are an ancient group of marine arthropods which has had little change in external morphology since the Cretaceous (Botton et al. 1996). The four extant species are assigned to three genera in two families (Sekiguchi and Shuster 2009). *Limulus polyphemus*, only an extant species in the family Limulidae, migrates into the intertidal zone along the eastern coast of North America in the late spring to early summer for breeding. Their breeding activity is synchronized to high tides.

S. Watanabe (✉) · S. Oyamada · K. Mizuta
Department of Marine Bio-Science, Faculty of Life Science and Biotechnology, Fukuyama University, Hiroshima Prefecture, Hiroshima, Japan

K. Azumakawa · S. Morinobu · N. Souji
The Kasaoka Horseshoe Crab Museum, Okayama Prefecture, Okayama, Japan

 
J. T. Tanacredi et al. (eds.), *International Horseshoe Crab Conservation and Research Efforts: 2007-2020*, https://doi.org/10.1007/978-3-030-82315-3_8

Adult pairs appear 1–2 h before high tide and return to deep water about 2 h after high tide (Barlow et al. 1986; Penn and Broc kmann 1994).

Activity rhythms of adult *Limulus polyphemus* had been monitored in laboratory works exposed to artificial light/dark (24.0-h) and tidal (12.4-h) cycles (Chabot et al. 2004, 2007, 2008). Most individuals clearly synchronized their activity to the imposed artificial tidal cycles, some of which showed clear evidence of entrainment after the artificial tides were terminated. These results demonstrate that the endogenous tidal (12.4-h) clock that influences locomotion in *Limulus* that can be entrained by imposed artificial tides. However, many individuals also exhibit a daily pattern of activity, either to be more active during the day or night (Chabot et al. 2004, 2007). In the recent works, these activity patterns could result from light influencing the expression of one of the two coupled circalunidian clocks (24.8 h) that are 180° out of phase with each other (Chabot and Watson 2010, 2014).

In comparison to *Limulus*, activity rhythms of other species in Merostomata are poorly studied. The three Indo-Pacific species of horseshoe crabs (*Tachypleus tridentatus*, *Tachypleus gigas*, and *Carcinoscorpius rotundicauda*) are distantly related to *L. polyphemus* (Sekiguchi and Shuster 2009). The divergence between the Asiatic and American lineages probably occurred during the Jurassic period. The tri-spine horseshoe crab, *Tachypleus tridentatus* is the largest species in its body size with the widest distribution ranging from coastal Southeast Asia to western Japan. The behaviors separated by at least 100 million years of isolation between the two groups of horseshoe crabs (Botton et al. 1996).

Living organisms have adapted to the temporal structure of their environment in different ways with the overall periodicity of the environment. The biotic factors, such as food supply, competition for food, predator pressure, vary on a daily or lunar basis (Erkert 1982). Biological rhythms of various intertidal animals have been studied mainly in the physiological and biochemical aspects clarifying the mechanisms (e.g. Naylor 2010; Chabot and Watson 2014). However, their ecological importance and evolutionary significance are poorly discussed. The main purpose of our study is to provide information for the activity rhythms of *T. tridentatus* and to compare the results with those of *Limulus* to understand the evolution of biological clocks in Merostomata. Our interests are not only for examining proximate factors influencing the biological clocks of *T. tridentatus* but also for examining evolutionary ecological functions of the activity rhythms expressed by this species in its environments.

In this study, activity of *T. tridentatus* was monitored with an animal-borne data-logger. The use of animal-borne data-loggers (a.k.a. bio-logging) to study the physiology and behavior of various species has increased over the past decades. As a result, we are beginning to gain new insights into the behavior of these animals in their natural habitats (for reviews, see Cooke 2008; Ropert-Coudert et al. 2009; Rutz and Hays 2009; Bograd et al. 2010). Accelerometer tags are particularly useful to measure physical movements of animals. Their use in animal behavior studies has increased considerably during the recent decades (e.g., Yoda et al. 2001; Wilson et al. 2006; Watanabe et al. 2005, 2012; Moreau et al. 2009). In the preset study, we employed an accelerometer to monitor activity rhythm of *T. tridentatus* in unrestrained conditions. Thus, we expect that activity of horseshoe crabs in our study to be expressed more naturally, and therefore providing more reliable information to understand the ecological functions of its activity rhythm.

# 2 Materials and Methods

## *2.1 Experimental Conditions*

Fifteen adult male (mean ± s.d. =1164 ± 113 g, range = 880–1300 g) and 11 female (2473 ± 439 g, 1660–3100 g) *T. tridentatus* were used for our experiments. The animals were caught using fixed fishing nets or trawling as bycatch around Kasaoka Bay (34.46°N, 133.54°E) in the Seto Inland Sea, western Japan, and brought to the Kasaoka Horseshoe Crab Museum, Okayama prefecture, Japan (http://www.city.kasaoka.okayama.jp/site/kabutogani/).

Activity of *T. tridentatus* was investigated in two experimental conditions. The animals were exposed to photoperiodic (light/dark) cycles in laboratory experiments and to both photoperiod and tidal cycles in field manipulated experiments. These experiments were conducted during the spawning season of this species (July and August; Shuster and Sekiguchi 2009) from 2012 to 2014.

In the laboratory experiments, 10 individual horseshoe crabs (5 males and 5 females) were placed in a recirculating aquarium (300 cm × 100 cm, 80 cm deep). The bottom was covered with 20 cm thick layer of fine-grained sand where animals dug deep enough to cover themselves. The water temperature was controlled between 22 and 26 °C. Lighting was provided by four 100-W fluorescent bulbs suspended above the aquarium, which were switched on at 08:00 and off at 18:00 every day.

In the field manipulated experiments exposed to natural photoperiod and tidal cycles, 19 individuals of horseshoe crabs (11 males and 8 females) were placed in an experimental pond (12 m × 18 m) in an exterior area of the museum. The experimental pond is connected to Kasaoka Bay by a canal. The water level changes between 0.5 and 3.0 m depending on ebbing and rising tides synchronizing with natural tidal cycles in Kasaoka Bay (SioMieYell, http://sio.mieyell.jp/). The pond floor was deposited with ca. 1.0 m thick layer of fine-grained sand and mud where the animals feed freely and burrow for resting and spawning. One to four individuals were placed in the pond during the same recording period. Water temperature changes seasonally between 18 °C and 33 °C with a small daily change ±1.6 °C. *Tachypleus tridentatus* can be active in water temperature higher than 18 °C (Nishii 1975; Lee and Morton 2009; Watanabe et al. unpublished data). Change of water depth in the experimental pond was recorded by a water level logger (HOBO U20–001-02-Ti, Onset Corporation, Pocasset, MA) set on the bottom of the pond. Squid fillets and polychaetes were fed once every 3 days during the day in both experimental conditions.

## 2.2 Activity Recording

The locomotion activity of a horseshoe crab was recorded using an acceleration data-logger (HOBO Pendant G data-logger, 58 × 33 × 23 mm, 18 g, Onset Corporation, Pocasset, MA). The device is considerably adequate in size for adult horseshoe crabs (approx. 0.6–2.0% of body weight of the animals in our experiments) to monitor their activities. The device records acceleration and inclination through measurement of an analogue signal in each of its three axes (Moreau et al. 2009). These signals are converted to gravity units ranging from −3 g to +3 g (1 g = 9.8 m $s^{-2}$). The logger's memory of 64 kB permits to record 65,400 data points at user-determined intervals of 1 s to 18 h. In our experiments, each devise was set to record along only longitudinal axis (Fig. 1a) at 1-min interval to record up to 45 days.

A data-logger fixed on meshed PVC sheet (4 cm × 6 cm) with two plastic cable ties was attached using two-component epoxy glue (Quick 5, Konishi Co., Ltd) on the dorsal carapace of each animal (Fig. 1a). Before glue curing, the animals were kept in a plastic container with ca. 10 cm depth water at least for 30 min before released to the pond.

## 2.3 Data Analysis

The acceleration loggers were retrieved from recaptured animals or when they became naturally unglued and floated to the water surface. Data were readout through USB data transfer to PC by HOBO ware Pro (Onset Corporation, Pocasset, MA) and saved in ASCII format for further processing. Activity data were analyzed using customized macro programs in IGOR Pro ver. 6 (Wave Metrics, Inc., Lake Oswego, OR).

Acceleration signals are reflected by both accelerations related to changes in the movements of animals, that is, dynamic accelerations, and gravitational acceleration resulting from changes in the body posture (Yoda et al. 2001; Watanabe et al. 2005, 2012). The resolution of acceleration sensor: 0.025 g (0.245 m $s^{-2}$) is sufficient to determine whether an animal was being active or inactive. Amounts of change between consecutive data points (*delta XG*) were calculated; the value 0.1 g (0.98 m $s^{-2}$) was used as a criterion for active or inactive because the distribution of amounts of changes was bimodal with higher values (>0.1 g) corresponding to walking or digging on floor of the pond (Fig. 1b). If the *delta XG* in *p* and *p-1* was >0.1 g, the animal was determined active at the point (*p*). Number of active points was summed into 10-min bins for actograms double-plotted (refer to be described in Chabot et al. 2004, 2007, 2008, for more details) to improve visualization of the patterns for the rhythmicity of activity (see the results in Figs. 2, 3, and 4; upper left panels).

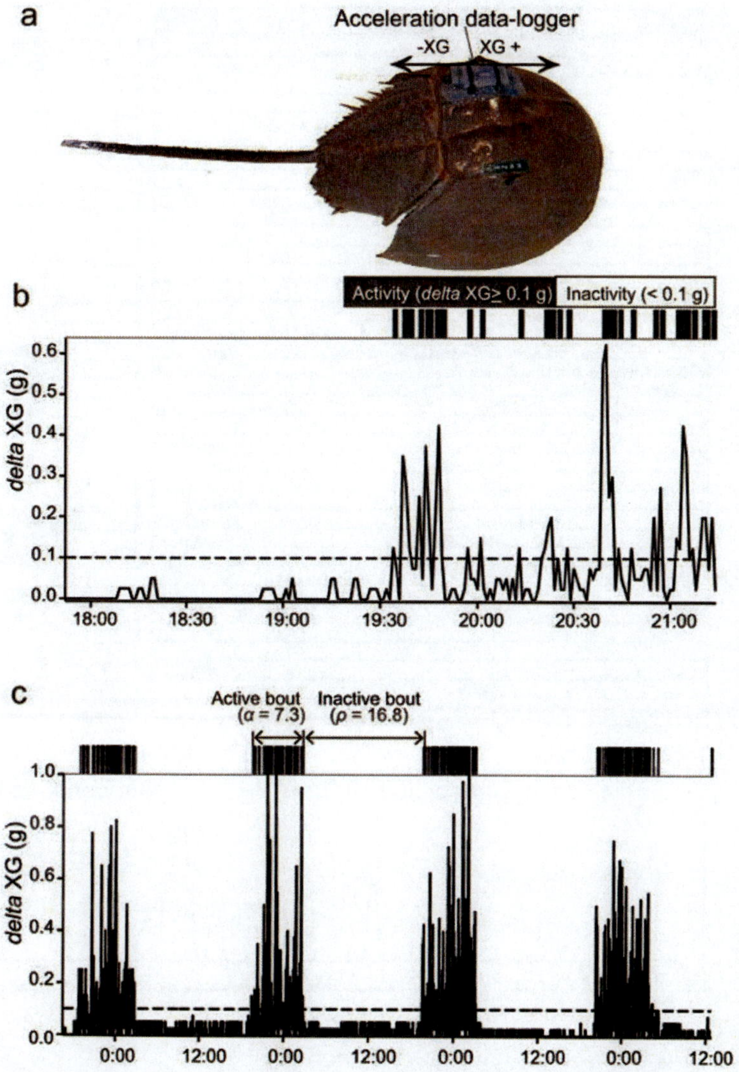


**Fig. 1** Examples of activity data of tri-spine horseshoe crab, *Tachypleus tridentatus*. **(a)** An acceleration logger glued on the dorsal carapace of a female horseshoe crab to record its acceleration along the longitudinal axis (XG) at 1-min interval. **(b)** An extended record for 3.5 h in **(c)** approx. 3.5-days excerpt. Amount of change between consecutive data points (*delta* XG) was calculated. The dashed lines in **(b)** and **(c)** indicate the criterion with higher values (*delta* XG >0.1 g) corresponding to be active. Lengths of active bouts ($\alpha$) and inactive bouts ($\rho$) are measured **(c)**

To determine a preference for activity during the daytime (06:00–18:00, *diurnality*) or night-time (18:00–06:00, *nocturnality*), cumulated number of active points during each of daytime and night-time was calculated (see the results in Figs. 2, 3,

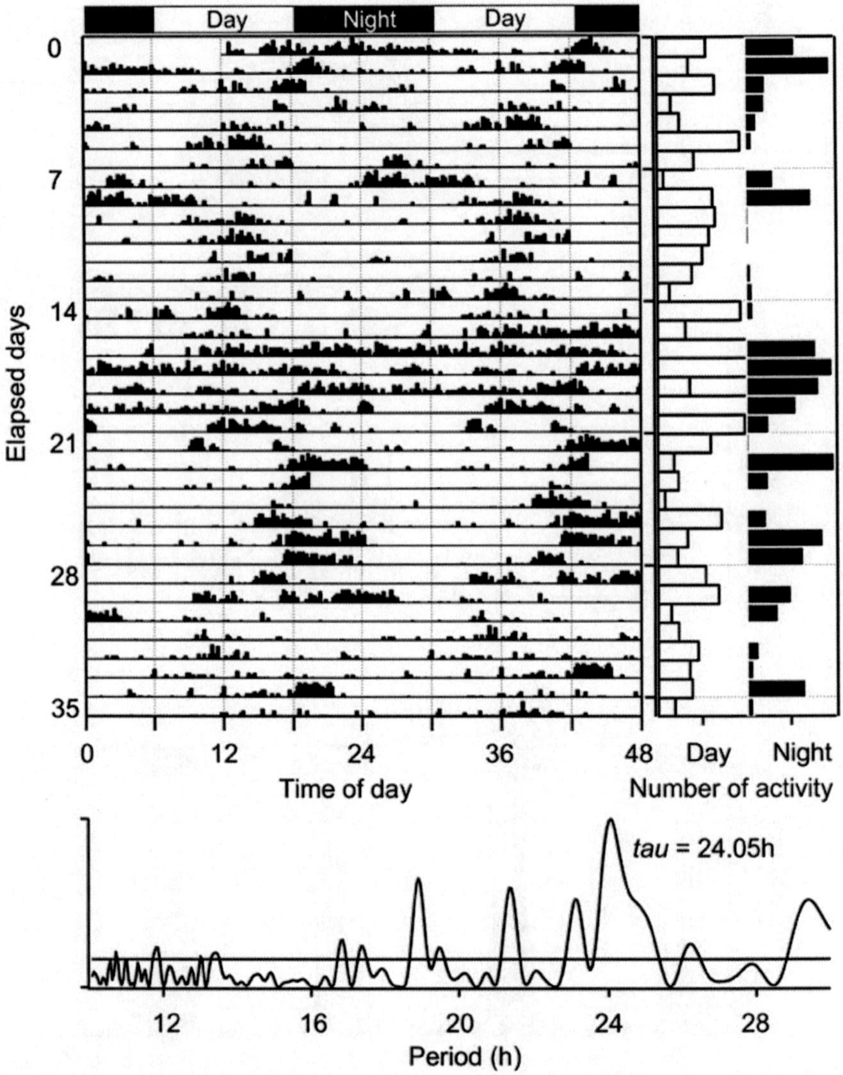


**Fig. 2** An example of *circadian* rhythm expressed by *Tachypleus tridentatus* exposed to only light/dark cycles in the laboratory experiment. Upper left panel: actograms double-plotted to improve visualization of the patterns. Black/white bars at the top indicate daytime (06:00–18:00) and night-time (18:00–06:00). Upper right panel: cumulated number of activity points during each of daytime and night-time. Lower panel: Lomb–Scargle periodogram analyses of the actogram; vertical scale is relative Q (*p*). Largest peak value above horizontal line of significance ($P < 0.01$) indicated by numerical value (*tau*)

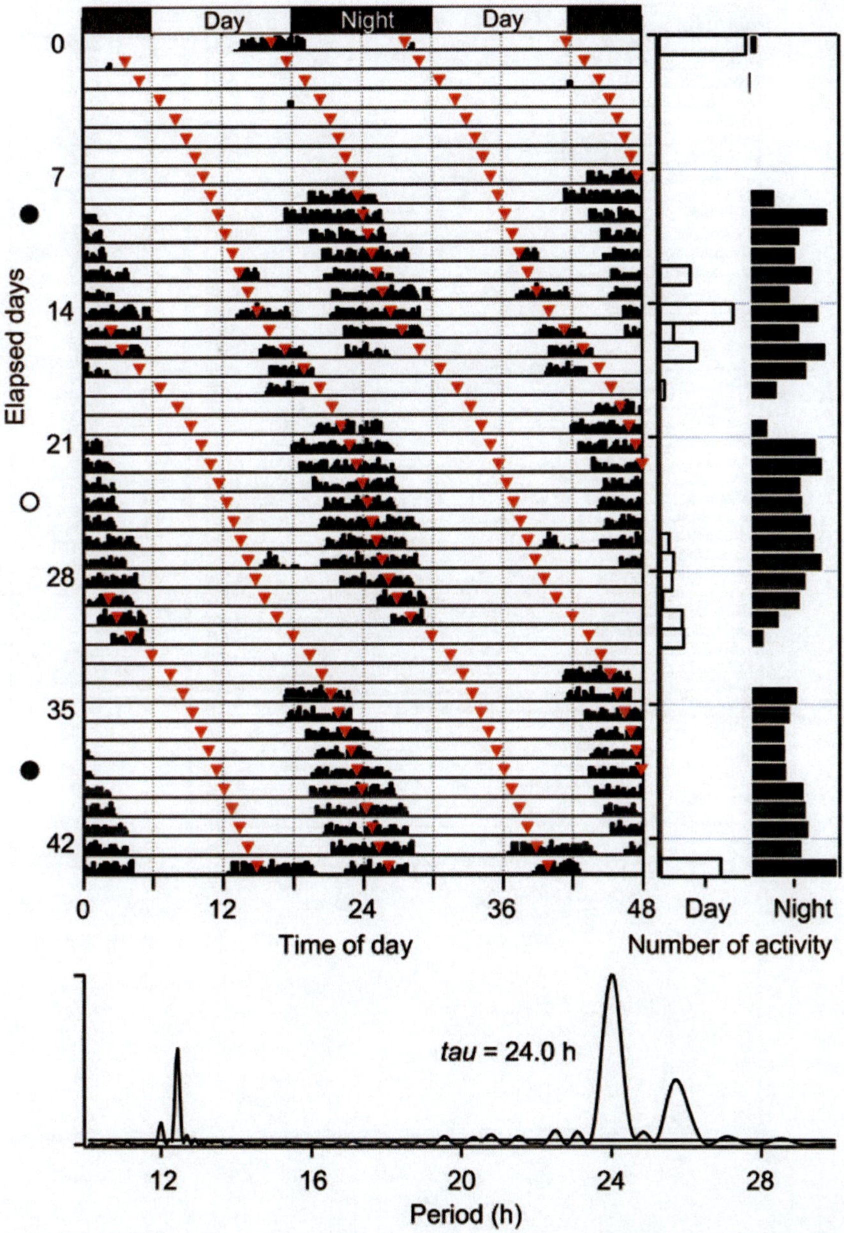


**Fig. 3** An example of *circadian* rhythm expressed by *Tachypleus tridentatus* exposed to natural photoperiod and tidal cycles in the field experiment. Upper left panel: actograms double-plotted to improve visualization of the patterns. Black/white bars at the top indicate daytime (06:00–18:00) and night-time (18:00–06:00). Timing of high tides is indicated by triangles in the figure. Black/white circles on the left indicate new moon and full moon. Upper right panel: cumulated number of activity points during each of daytime and night-time. Lower panel: Lomb–Scargle periodogram analyses of the actogram; vertical scale is relative Q ($p$). Largest peak value above horizontal line of significance ($P < 0.01$) indicated by numerical value (*tau*)

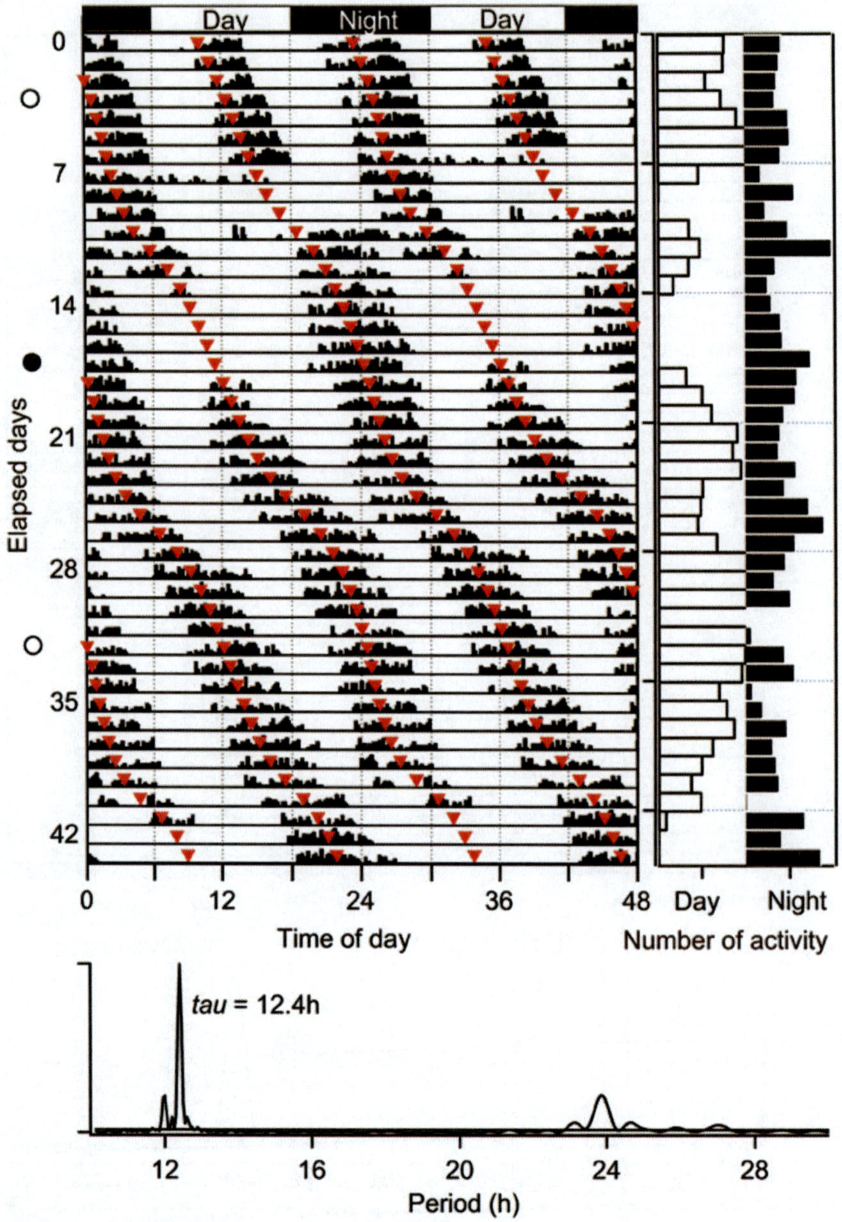


**Fig. 4** An example of *circatidal* rhythm expressed by *Tachypleus tridentatus* exposed to photoperiod and tidal cycles in the field experiment. See legend in Fig. 3 for more detail

and 4; upper right panels). To determine statistical significance ($P < 0.05$) between the means during the daytime and night-time, Wilcoxon-signed rank test, a non-parametric statistical hypothesis test to compare two dependent samples, was used (Zar 2010). Percentage of active points for night-time (nocturnal activity rate) in each experimental trial was also calculated as an index of degree of nocturnality.

Activity of horseshoe crabs represented periodicity with active bouts after inactivity bouts at least 3 hours (Fig. 1c). The time lengths (h) of active bouts ($\alpha$) and inactive bouts ($\rho$) were measured (Fig. 1c). To determine statistical significance ($P < 0.05$) for the means of $\alpha$ and $\rho$ between the laboratory and field experimental conditions, Mann-Whitney *U* test, a non-parametric statistical hypothesis test to compare two independent samples, was used (Zar 2010).

In addition to calculate duration of each activity and inactivity bout, rhythmicity of activity throughout each experimental trial was examined by Lomb–Scargle periodogram analysis (Lomb 1976) as described in Chabot et al. (2008). Cumulated number of active points was summed every hour, and the maximal value of primary component of rhythmicity (*tau*, $P < 0.01$) was determined by the periodogram (see the results in Figs. 2, 3, and 4; lower panels). Rhythmicity of activity was determined as the *circadian* with *tau* nearly equal to 24-h or the *circatidal* with *tau* nearly equal to 12.4-h (see the results in Figs. 3 and 4; upper left panels as typical examples) for each trial. Visualization by actograms double-plotted and periodogram analysis were performed using custom-build macro programs written in IGOR Pro.

## 3 Results

A total of 487 and 569 days of activity data was obtained from 14 and 28 experimental trials in the laboratory and field experiments, respectively. In the laboratory experiments exposed to LD cycles, the animals exhibited rhythmicity in their activity but with large variations among trials with active bouts ($\alpha$ = 9.8 ± 8.7 h, *mean* ± *s.d.*) and inactive bouts ($\rho$ = 20.3 ± 31.7 h).

In the field experiments exposed to LD and tidal cycles, the animals exhibited clear rhythmicity of their activity with active bouts ($\alpha$ = 8.3 ± 12.7 h) and inactive bouts ($\rho$ = 12.1 ± 3.8 h). Means of $\alpha$ and $\rho$ were not significantly different between the laboratory and field experimental conditions ($\alpha$: $U = 221$, $P = 0.5131$, $\rho$: $U = 194$, $P = 0.9574$).

Lomb–Scargle periodogram analyses indicated that significant activity rhythms in the 24-h range (23.1–24.1 h) were found in all laboratory experimental trials exposed to only LD cycles. However, the timing of onset and offset of active bouts were not synchronized to LD cycles (see an example in Fig. 2: upper left panel). Preference for activity during each of daytime or night-time in a day (i.e., diurnality or nocturnality) was different among the individuals and trials. Out of a total of 14 trials, the animals exhibited diurnality in eight trials (nocturnal activity rate: 8.2%–39.2%, $P < 0.05$) and nocturnality in one trial (80.5%, $P < 0.01$); in the remaining five trials. The activity levels were not statistically different between daytime and night-time (37.2%–56.5%, $P > 0.05$).

Twelve of 19 animals exposed to both LD and tidal cycles expressed significant circadian rhythms in the 24-h range (23.8–24.4 h, $P < 0.01$) in 16 of 28 (57%) trials (see Fig. 3: lower panel as a typical example). Eight animals expressed significant circatidal rhythms in the 12.4-h range (12.3–12.5 h, $P < 0.01$) in 12 (43%) trials (see Fig. 4: lower panel as a typical example). Three animals expressed both circadian and circatidal rhythms in different trials. Eighteen of 19 animals expressed nocturnality in 26 (93%) trials (nocturnal activity rate: 66.2%–99.0%, $P < 0.05$), while no animal expressed diurnality. In the remaining two trials, the activity levels were not statistically different between daytime and night-time ($P > 0.05$).

In the actograms double-plotted for the animals with 24-h rhythms (as shown in Fig. 3: upper left panel), activity bouts were more likely to be shown around high tides during the night. These animals generally began to move about 2.5 h before high tides during the night. On the other hand, active bouts were shown around high tides during both day and night in those for animals with 12.4-h rhythms (as shown in Fig. 4: upper left panel).

# 4 Discussion

## 4.1 *Activity Rhythm of Tachypleus tridentatus*

Here we present the first report of the activity rhythm in the tri-spine horseshoe crab, *T. tridentatus*. The individuals exposed to both light/dark (LD) and tidal cycles exhibited 24-h (i.e., circadian) rhythms and nocturnality in more than half (57%) experimental trials. Their activity rhythms synchronized to high tide during the night. In the remaining experiments (43%), the animals exhibited 12.4-h (i.e., circatidal) rhythms synchronized to high tide during both day and night.

Activity rhythm of the American horseshoe crab, *L. polyphemus,* has been well-studied in laboratory works (e.g., Chabot et al. 2004, 2007, 2008; Chabot and Watson 2010, 2014). It is likely that there are some similarities and differences in the activity rhythms and their possible external synchronizing cues (*Zeitgebers*) between these species. In the experiments of *Limulus* exposed to LD and artificial tidal cycles, most adult *Limulus* exhibited clear circatidal (12.4-h) rhythms. Their activity synchronizes to high tide during both light and dark periods (Chabot et al. 2007, 2008). After a period with water-level fluctuation, their activity rhythms entrain in constant water level, which indicate that *Limulus* possess an endogenous circatidal (12.4-h) clock that drive its locomotion activity (Chabot et al. 2008).

In our study, *T. tridentatus* exposed to LD cycles in the constant water level exhibited circadian rhythms out of synchronization with LD cycles, that is, their activity was free-running with 24-h cycles. No animals expressed 12.4-h rhythms of their activity in the constant water level. It is, thus, more likely that *T. tridentatus* possess an endogenous circadian (24-h) clock, but LD cycles do not act solely as a

*Zeitgeber* in its activity. In contrast, it was reported that LD cycles can solely affect locomotion activity of *Limulus*. Many individuals in the present study expressed daily rhythms of activity when exposed to LD cycles in constant water level. Nearly half exhibited a preference for diurnal movements, while the others exhibited no preference. However, in *Limulus*, these daily patterns are much less apparent when tidal cycles are present (Chabot et al. 2008). Thus, the authors suggested that the tidal cues override the influence of LD cycles, which produce circatidal rhythms of their activity of *Limulus*.

In this study, we did not evaluate the solo effect of tidal cycles that *T. tridentatus* exposed to tidal cycles under LL or DD conditions. Thus, the effects of LD and tidal cycles cannot be separated. However, we presume that tidal cues combined with LD cycles regulate activity rhythms in *T. tridentatus*. It is similar to that suggested in *Limulus* (described as above, Chabot et al. 2008). It is, however, likely that the effects of LD cycles differ between the two species. LD cycles combined with tidal cues give stronger negative effects in locomotion activity of *T. tridentatus* compared with *Limulus*. In the experiments on *Limulus* exposed to artificial tidal cycles under constant light conditions (LL), the animals exhibit circatidal rhythms synchronizing with tidal cycles.

Contrary to this assumption, several individuals of *T. tridentatus* expressed 12.4-h rhythms or both 24-h and 12.4-h in different trials. In the studies on *Limulus*, most individuals exhibited circatidal rhythms, but many also exhibit a daily pattern of activity, either to be more active during the day or night (Chabot et al. 2004, 2007). The variations in activity rhythms in *Limulus* had been presumed as resulted from congenital traits such as individual and geographical differences, or from problems of monitoring techniques and experimental conditions. In their more recent works, it is hypothesized that a combination of daily and tidal activity patterns could result from light influencing the expression of one of the two coupled circalunidian clocks (24.8 h), resulting in one activity bout being diminished or suppressed relative to the other (Chabot and Watson 2010, 2014).

In our results, some individuals expressed both circadian and circatidal rhythms in different trials. All animals in this study were obtained within the narrow geographic range (approximately 20 km × 30 km) in the eastern part of Seto Inland Sea. In addition, genetic variation of *T. tridentatus* in the Seto Inland Sea is considerably much lower than those of other populations (Nishida and Koike 2009; Nishida et al. 2015). Thus, it is unlikely that different activity patterns are resulted from such congenital traits. Moreover, our recent works suggest that tidal level fluctuations due to lunar cycles influence whether *T. tridentatus* express circadian or circatidal rhythms of its activity (Watanabe et al. unpublished data) that support the two coupled circalunidian clocks hypothesis supposed to be in *Limulus*. Additional experiments are needed to identify entraining agents of activity rhythms of *T. tridentatus* and to understand the mechanisms of its biological clock.

## 4.2 *Ecological Functions of the Activity Rhythm in the Seto Inland Sea*

Our second purpose of this study is to understand the evolutionary ecological functions of the activity rhythms of *T. tridentatus*. Locomotion activity of *Limulus* had been monitored with experimental activity boxes and running wheels in restrained conditions in Chabot et al. (2004, 2007, 2008). Range of water level is limited with small capacity of water tanks (normally depth less than 50 cm) in the laboratory experiments while sea surface level fluctuates largely depending on lunar cycles in natural environments. In this study, we employed an accelerometer to monitor activity rhythm of *T. tridentatus* in unrestrained conditions. Thus, we expect that activity of horseshoe crabs in our study to be expressed more naturally, providing reliable information to understand the ecological functions of its activity rhythm.

Interspecific differences of chronotypes, that is, diurnality or nocturnality, have been investigated in various taxonomic groups to understand the ecological niches in its ecosystems. Our results suggest that *T. tridentatus* is primarily nocturnal, while most *Limulus* is active both during the day and night. Both diurnality and nocturnality are found among taxonomically similar species (Kronfeld-Schor et al. 2001), or even in a same species (Yokota and Oishi 1991). The choice of temporal ecological niche may be important to determine either diurnality or nocturnality. It may depend on complex interactions with coexisting species of their prey, predators, and competitors. Both *T. tridentatus* and *Limulus* probably adapted to similar environments and ecological niches in these coastal ecosystems. However, it is likely that local differences of their main prey and predators produce different chronotypes. *Limulus* is opportunistic foragers that can take advantage of a wide range of locally available prey (Botton and Shuster 2003), including bivalves, polychaetes, crustaceans, and gastropods (Botton 2009). It is likely that *T. tridentatus* is also an opportunistic forager, but its diet in the Seto Inland Sea is more highly consisted of benthic polychaetes (Souji et al. 2005, 2007, 2008). Although species identification of polychaetes in the diet samples was insufficient, most polychaete species in the diet are presumably nocturnal such as other large polychaete *Nereis virens* (Last and Olive 1999, 2004). If so, *T. tridentatus* may become more active feeding on nocturnal polychaetes crawling out of tidal flat sediments during the high tides at night. A similar example was reported in shorebirds *Charadrius alexandrines* as a predator for polychaetes. The shorebirds increase nocturnal feeding rates corresponding to nocturnal activity in polychaetes (Kuwae 2007). Thus, there is the possibility that *T. tridentatus* in the Seto Inland Sea evolved to synchronize their activity rhythms with those of their main prey.

For another possible reason, there is little quantitative information about adult mortality due to predation on horseshoe crabs. However, it is more likely that potential predators are less for *T. tridentatus* than for *Limulus*. It was reported that *Limulus* is eaten by sharks, large crustaceans, sea birds, and sea turtles (Botton and Shuster 2003). There is little information that large sharks and sea turtles migrate to the Seto Inland Sea (Shigeta 2008; Ishihara et al. 2014). They seem to be transient visitors from the Pacific Ocean. Therefore, it is unlikely that *T. tridentatus* tends to be nocturnality due to avoiding from their potential predators.

Horseshoe crabs should be considered as keystone species in the coastal ecosystem, that is, an ecologically important species in the environment (reviewed by Botton 2009). Thus, identifying its activity is important to understand the coastal ecosystem. In addition, we believe that our bio-logging techniques will be a powerful tool for monitoring biological rhythms of keystone species in coastal ecosystems.

**Acknowledgments** This work was supported by the Japan Society for the Promotion of Science (JSPS) research grants (19255001 and 23710286), Research Center for Green Science at Fukuyama University, the program "Bio-Logging Science of the University of Tokyo (UTBLS)", and the Private University Research Branding Project from Ministry of Education, Culture, Sports, and Technology of Japan.

## References

R.B. Barlow, M.K. Powers, H. Howard, L. Kass, Migration of *Limulus* for mating: Relation to lunar phase, tide height, and sunlight. Biol. Bull. **171**, 310–329 (1986)

F.H. Barnwell, Daily and tidal patterns of activity in individual fiddler crab (Genus *Uca*) from the Woods Hole region. Biol. Bull. **130**, 1–17 (1966)

M.F. Bennett, J. Shriner, R.A. Brown, Persistent tidal cycles of spontaneous motor activity in the fiddler crab, *Uca pugnax*. Biol. Bull. **112**, 267–275 (1957)

S.J. Bograd, B.A. Block, D.P. Costa, B.J. Godley, Biologging technologies: New tools for conservation. Introduction. Endanger. Species Res. **10**, 1–7 (2010)

M.L. Botton, The ecological importance of horseshoe crabs in estuarine and coastal communities: A review and speculative summary, in *Biology and Conservation of Horseshoe Crabs*, ed. by J. T. Tanacredi, M. L. Botton, D. R. Smith, (Springer, New York, 2009), pp. 45–63

M.L. Botton, C.N. Shuster Jr., Horseshoe crabs in a food web: Who eats whom? in *The American Horseshoe Crab*, ed. by C. N. Shuster Jr., R. B. Barlow, H. J. Brockmann, (Harvard University Press, Cambridge, 2003), pp. 133–153

M.L. Botton, C.N. Shuster Jr., K. Sekiguchi, H. Sugita, Amplexus and mating behavior in the Japanese horseshoe crab, *Tachypleus tridentatus*. Zool. Sci. **13**, 151–159 (1996)

S.J. Cooke, Biotelemetry and biologging in endangered species research and animal conservation: Relevance to regional, national, and IUCN Red List threat assessments. Endanger. Species Res. **4**, 165–185 (2008)

C.C. Chabot, S.H. Betournay, N.R. Braley, W.H. Watson III, Endogenous rhythms of locomotion in the American horseshoe crab, *Limulus polyphemus*. J. Exp. Mar. Biol. Ecol. **345**, 79–89 (2007)

C.C. Chabot, J. Kent, W.H. Watson III, Circatidal and circadian rhythms of locomotion in *Limulus polyphemus*. Biol. Bull. **207**, 72–75 (2004)

C.C. Chabot, S.J. Skinner, W.H. Watson III, Rhythms of locomotion expressed by *Limulus polyphemus*, the American horseshoe crab: I. synchronization by artificial tides. Biol. Bull. **215**, 34–45 (2008)

C.C. Chabot, W.H. Watson III, Circatidal rhythms of locomotion in the American horseshoe crab *Limulus polyphemus*: Underlying mechanisms and cues that influence them. Curr Zool **56**, 499–517 (2010)

C.C. Chabot, W.H. Watson III, Biological rhythms in Intertidal animals, in *Annual, Lunar and Tidal Clocks Patterns and Mechanisms of Nature's Enigmatic Rhythms*, ed. by H. Numata, B. Helm, (Springer, New York, 2014), pp. 41–64

H.G. Erkert, Ecological aspects of bat activity rhythms, in *Ecology of Bats*, ed. by T. H. Kunz, (Plenum Press, New York, 1982), pp. 201–242
R.G. Foster, L. Kreitzman, *Rhythms of Life: The Biological Clocks that Control the Daily Lives of Every Living Thing* (Profile Books, London, 2004)
T. Ishihara, N. Kamezaki, Y. Matsuzawa, A. Ishizaki, Assessing the status of Japanese coastal fisheries and sea turtle bycatch. Wildlife Human Soc **2**, 23–35 (2014) (in Japanese and English abstract)
N. Kronfeld-Schor, T. Dayan, R. Elvert, A. Haim, N. Zisapel, G. Heldmaier, On the use of the time axis for ecological separation: Diel rhythms as an evolutionary constraint. Am. Nat. **158**, 451–457 (2001)
T. Kuwae, Diurnal and nocturnal feeding rate in Kentish plovers *Charadrius alexandrinus* on an intertidal flat as recorded by telescopic video systems. Mar. Biol. **151**, 663–673 (2007)
K.S. Last, P.J.W. Olive, Photoperiodic control of growth and segment proliferation by Nereis (*Neanthes*) virens in relation to state of maturity and season. Mar. Biol. **134**, 191–199 (1999)
K.S. Last, P.J.W. Olive, Interaction between photoperiod and an endogenous seasonal factor in influencing the diel locomotor activity of the benthic polychaete *Nereis virens* Sars. Biol. Bull. **206**, 103–112 (2004)
C.N. Lee, B. Morton, Emergence behavior of juvenile *Tachypleus tridentatus* under simulated tidal conditions in the laboratory and at two different sediment temperatures, in *Biology and Conservation of Horseshoe Crabs*, ed. by J. T. Tanacredi, M. L. Botton, D. R. Smith, (Springer, New York, 2009), pp. 275–283
N.R. Lomb, Least-squares frequency analysis of unequally spaced data. Astrop. Spa Sci. **39**, 447–462 (1976)
M. Moreau, S. Siebert, A. Buerkert, E. Schlecht, Use of a tri-axial accelerometer for automated recording and classification of goats' grazing behaviour. Appl. Anim. Behav. Sci. **119**, 158–170 (2009)
E. Naylor, Tidal and diurnal rhythms of locomotory activity in *Carcinus maenas* (L.). J. Exp. Biol. **35**, 602–610 (1958)
E. Naylor, *Chronobiology of Marine Organisms* (Cambridge University Press, Cambridge, 2010)
S. Nishida, H. Koike, Genetic structure of Japanese populations of *Tachypleus tridentatus* by mtDNA AT-rich region sequence analysis, in *Biology and Conservation of Horseshoe Crabs*, ed. by J. T. Tanacredi, M. L. Botton, D. R. Smith, (Springer, New York, 2009), pp. 183–196
S. Nishida, K. Kuroyanagai, H. Koike, Genetic features of *Tachypleus tridentatus* in Japan and an alien (non-native) population founded at Ise-Mikawa Bay, Chubu Region, central Japan, in *Changing Global Perspectives on Horseshoe Crab, Biology, Conservation and Management*, ed. by R. H. Carmichael, M. L. Botton, P. K. S. Shin, S. G. Cheung, (Springer, New York, 2015), pp. 129–146
H. Nishii, *Kabutogani-Jiten (Encyclopedia of the Horseshoe Crabs)* (Private publication, Kasaoka, 1975) (in Japanese)
J.D. Palmer, *An Introduction to Biological Rhythms* (Academic Press, New York, 1976)
D. Penn, H.J. Brockmann, Nest-site selection in the horseshoe crab, *Limulus polyphemus*. Biol. Bull. **187**, 373–384 (1994)
Y. Ropert-Coudert, M. Beaulieu, N. Hanuise, A. Kato, Diving into the world of biologging. Endanger. Species Res. **10**, 21–27 (2009)
C. Rutz, G.C. Hays, New frontiers in biologging science. Biol. Lett. **5**, 289–292 (2009)
K. Sekiguchi, C.N. Shuster Jr., Limits on the global distribution of horseshoe crabs (Limulacea): Lessons learned from two lifetimes of observations: Asia and America, in *Biology and Conservation of Horseshoe Crabs*, ed. by J. T. Tanacredi, M. L. Botton, D. R. Smith, (Springer, New York, 2009), pp. 5–24
T. Shigeta, Recent ecological problems of the fishes in the Seto Inland Sea, Japan. Nippon Suisan Gakkaishi **74**, 868–872 (2008) (in Japanese)

C.N. Shuster Jr., K. Sekiguchi, Basic habitat requirements of the extant species of horseshoe crabs (Limulacea), in *Biology and Conservation of Horseshoe Crabs*, ed. by J. T. Tanacredi, M. L. Botton, D. R. Smith, (Springer, New York, 2009), pp. 115–129

N. Souji, S. Morinobu, A. Sawamoto, Fecal analysis of the horseshoe crab, in *A Report of Research on Horseshoe Crab*, vol. 11, (2005), pp. 8–12. (in Japanese)

N. Souji, S. Morinobu, A. Sawamoto, Fecal analysis of the horseshoe crab, in *A Report of Research on Horseshoe Crab*, vol. 13, (2007), pp. 10–15. (in Japanese)

N. Souji, S. Morinobu, A. Sawamoto, Fecal analysis of the horseshoe crab, in *A Report of Research on Horseshoe Crab*, vol. 14, (2008), pp. 10–15. (in Japanese)

K. Yoda, Y. Naito, K. Sato, A. Takahashi, J. Nishikawa, Y. Ropert-Coudert, M. Kurita, Y. Le Maho, A new technique for monitoring the behaviour of free-ranging Adélie penguins. J. Exp. Biol. **204**, 685–690 (2001)

T. Yokota, T. Oishi, Seasonal change in the locomotor activity rhythm of the medaka, *Oryzias latipes*. Int. J. Biometeorol. **36**, 39–44 (1991)

S. Watanabe, M. Izawa, A. Kato, Y. Ropert-Coudert, Y. Naito, A new technique for monitoring the detailed behaviour of terrestrial animals: A case study with domestic cat. Appl. Anim. Behav. Sci. **94**, 117–131 (2005)

S. Watanabe, K. Sato, P.J. Ponganis, Activity time budget during foraging trip of emperor penguins. PLoS One **7**, e50357 (2012)

R.P. Wilson, C.R. White, F. Quintana, L.G. Halsey, N. Liebsch, G.R. Martin, P.J. Butler, Moving towards acceleration for estimates of activity-specific metabolic rate in free-living animals: The case of the cormorant. J. Anim. Ecol. **75**, 1081–1090 (2006)

J.H. Zar, *Biostatistical Analysis*, 5th edn. (Prentice Hall/Pearson, NJ, 2010)

# Sediment-Molting Relationship of Juvenile *Limulus polyphemus* (Arthropoda, Merostomata) and Predation Against them

**Yumiko Iwasaki**

## 1 Introduction

Most studies on the molting of the American horseshoe crabs (*Limulus polyphemus*) are focused on molt-inducing hormones or ecdysones (Herman 1972; Jegla 1982), mechanism of molting (Shuster Jr. and Sekiguchi 2003), or growth and development (ibid; Shuster Jr. 1982; Botton et al. 2003a, b). A few briefly mentioned molting behavior of juveniles (Shuster Jr. 1982; Shuster Jr. and Sekiguchi 2003), but no particular literature has been found on this topic.

Rustán et al. (2011) reported peculiar trilobite molts that exhibit an infaunal molting pattern. They hypothesized from their observation on these molts from northwestern Argentina that trilobite species, *Paciphacops* spp., burrowed themselves into the soft substrate prior to molting in order to avoid predators like fish and ammonoids that were progressively evolving in the Silurian-Devonian Sea, some 420 million years ago. At first, the authors suggested that this behavior implicated a phylogenetic significance within the phacopid lineage which includes *Paciphacops*. However, an observation of the same molting pattern in another phacopid species, *Viaphacops* from the Middle Devonian of Bolivia, some 30 million years later in a different oceanic basin (Rustán et al. 2012a) now questions the validity of its phylogenetic characterization. In addition, some species of calmoniids, a sister group of the phacopids, also from the Middle Devonian of Bolivia, are found to possess the same pattern of molts (Rustán et al. 2012b). The pattern of infaunal molting therefore implies a paleoecological evolutionary adaptation across different lineages of trilobites.

Y. Iwasaki (✉)
Center for Environmental Research and Coastal Oceans Monitoring, CERCOM, Molloy College, West Sayville, NY, USA

 
J. T. Tanacredi et al. (eds.), *International Horseshoe Crab Conservation and Research Efforts: 2007-2020*, https://doi.org/10.1007/978-3-030-82315-3_9

This study investigated the molting behavior of juvenile horseshoe crabs. This group of animals is evolutionarily the closest living relative of trilobites, and is therefore used to model trilobite behaviors. Rustán et al. (2011) suggested increased predation as a cause for infaunal behavior in trilobites when they were most vulnerable during molting. Although Lee and Morton (2009) proposed that juvenile horseshoe crabs may burrow in sediments to avoid predation, not much is known about predation against juvenile horseshoe crabs (Botton et al. 2003b). Because sediment type will affect the growth and survival of juvenile horseshoe crabs (Hong et al. 2009), sediment grain size was also be analyzed to see their preference to burrow in this study.

# 2 Materials and Methods

## 2.1 *Experimental Setup*

Three 10-gallon (approx.. 37.85 L) experimental tanks were set up at two locations; two at Center for Estuarine, Environmental and Coastal Oceans Monitoring (CEECOM), Dowling College (now as Center for Environmental Research and Coastal Oceans Monitoring, CERCOM, Molloy College), West Sayville, New York, and one at Fordham University at Lincoln Center, New York, USA. Each tank started with twelve 2nd and 3rd instars that were artificially hatched at CEECOM in 2012. Sediment in the size of 1 mm, 0.5 mm, and 0.25 mm was provided for them to burrow in (Table 1). Instant Ocean® was used to make saltwater of 30% in

**Table 1** Details of tank setup at two locations. Two tanks were set up at CEECOM and one at Fordham University

| | Tank A | Tank B | Tank C |
|---|---|---|---|
| Location | CEECOM | CEECOM | Fordham |
| Saltwater | Natural Seawater | Natural Seawater | Artificial |
| # of HSCs (initial) | 12 | 12 | 12 |
| Growth Stage | $2^{nd}$ and $3^{rd}$ instars | $2^{nd}$ and $3^{rd}$ instars | $2^{nd}$ and $3^{rd}$ instars |
| AverageSalinity (‰) | 26 | 26 | 30 |
| Average Water Temperature (°C) | 25.63 | 24.55 | 25.25 |
| Average DO (mg/L) | 7.47 | 8.35 | 7.97 |
| Water Level | Full tank | Full tank | 5 cm above substrate |
| Sediment Sizes (mm) | 1, 0.5, 0.25 and less | 1, 0.5, 0.25 and less | 1, 0.5, 0.25 and less |
| Sediment Depth in Tank (cm) | 5 | 5 | 5 |

CEECOM tanks had natural seawater while Fordham tank was with artificial water. The substrate condition was kept the same, however, for the CEECOM tanks, the available sand was of coral. For Fordham tank, sand from Jamaica Bay, NY, was sieved accordingly for the use

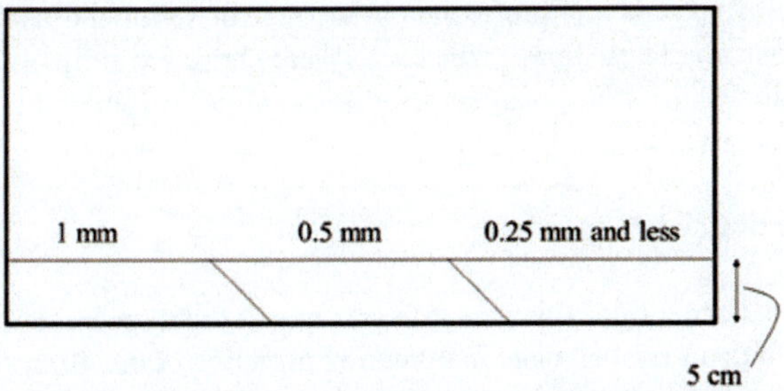


**Fig. 1** Sediment layout. Three size classes of sediment was laid on the bottom of the tank, in the depth of up to 5 cm for the horseshoe crab juveniles to burrow in. This is to investigate which size of sediment they prefer to burrow in

salinity in the tank at Fordham University, which is the same salinity level of natural seawater used at CEECOM, and also an ideal level according to Sekiguchi et al. (1988). This is also consistent with the seawater measured at Plumb Beach, New York, USA, where juvenile HSCs can be found. Sediment of three size classes was laid on the bottom of the tanks, up to 5 cm in depth (Fig. 1). Then the saltwater was added to cover 5 cm above the substrate, barely enough to provide a comfortable habitat for the subjects while allowing observation to be made easily. An aerator was installed to ensure a high dissolved oxygen level, but no water filter was used to avoid any agitation in the water which might cause the juvenile HSCs to burrow. In order to give an optimal condition for molting, 12 h/12 h light-dark cycle was applied (Schreibman and Zarnoch 2009). The water temperature was brought up above 22 °C (ranging from 22 °C to 27 °C, average: 25.25 °C), using the heat produced by the incandescent light bulb used for lighting. A tank heater was also used occasionally. The HSCs were fed with frozen brine shrimp outside of the tank to prevent the increase of ammonia level from the leftover food. They were initially fed with 1 cube of brine shrimp (approximately 3 g) every other day, and the tank water was changed 80% to 95% during that time. Later, during the horseshoe crab spawning season in 2013, several juveniles (4th–6th instars) were captured at Plumb Beach for a further experiment.

## 2.2 *Experimental Procedures*

### Preference in Substrate Texture

Sediment type will affect the growth and survival of the juvenile horseshoe crabs (Hong et al. 2009) and here sediment provided was sieved into 1 mm, 0.5 mm, and 0.25 mm and then systematically laid on the bottom of the tank in three sections

(Fig. 1) to see if there is any preference in the size of sediment juvenile horseshoe crabs would burrow. These three size classes were chosen to simulate a natural environment for horseshoe crab juveniles (Botton et al. 1988; John et al. 2012).

#### Introduction of Predators

The primary experiment of this research was to test the hypothesis on the effect of predation for burrowing behavior in juvenile horseshoe crabs. Botton et al. (2003b) identified juvenile blue crabs, spider crabs, hermit crabs, and fiddler crabs to be predators of juvenile horseshoe crabs. One male juvenile stone crab (*Menippe mercenaria*) caught in Great South Bay, Long Island, New York, and kept at CEECOM, two adult female blue crabs (*Callinectes sapidus*) purchased from market, and one juvenile female blue crab (about 7 cm in carapace width) and several hermit crabs (*Pagurus longicarpus*) collected outside CEECOM were introduced one at a time in each tank. No water filter was installed in the horseshoe crab tanks with predators in them, so as to let horseshoe crabs receive chemosensory signals from the predators. Because the adult blue crabs were too large to move around in the tank, their tank water was applied to see any chemosensory reaction of juvenile horseshoe crabs. Other predators were put in a black mesh cage (10 cm × 10 cm × 7 cm) and dropped into the horseshoe crab tanks. These animals were kept in a different tank and fed with fresh clams, flounders, and tilapia. The latter two were frozen.

## 3 Results

### 3.1 *Preference in Substrate Texture*

#### Second and Third Instars

The juvenile horseshoe crabs seemed to have no preference of sediment size varying in 1 mm, 0.5 mm and 0.25 mm. They burrowed randomly, showing no particular pattern. After 2 weeks, the substrate was homogenized. Dissolved oxygen was maintained at 6.0–7.0 mg/L while it was about 2.0–3.0 mg/L in the upper 2 cm of the substrate, at the depth where second and third instars nested. These numbers were lower when the water temperature was higher. The second and third instars were too small to alter the profile of the substrate.

#### Fourth to Sixth Instars

Larger juveniles also did not show any preference in sediment size to burrow. However, they tended to bulldoze the substrate flat.

## 3.2 *Introduction of Predators*

#### Juvenile Stone Crab

The presence of a male stone crab did not affect the behavior of juvenile horseshoe crabs. The stone crab acted very timidly whenever a horseshoe crab approached its cage.

#### Adult Female Blue Crabs

Next, two adult female blue crabs bought at a fish market were introduced to the horseshoe crab tank. Because of their size, they were not placed in the mesh cage; assuming they would release certain chemicals to signal their presence to their prey. The water in which they stayed was taken out with a pipette and was slowly squirted over each horseshoe crab. There was no reaction from any of the horseshoe crabs.

#### Juvenile Female Blue Crab

One juvenile female blue crab was captured outside of CEECOM. It was placed in a mesh cage and dropped into the horseshoe crab tank. Horseshoe crabs exhibited no change in their behavior, but occasionally bumped themselves into the cage.

#### Hermit Crabs

Hermit crabs do eat juvenile horseshoe crabs (Botton 2009). Several hermit crabs (the aperture of the shells they were in varied around 2 cm) were caught outside of CEECOM and three of them were introduced into one of the horseshoe crab tanks at CEECOM where three horseshoe crabs were in their second to third instar stage. After only a couple of days together, horseshoe crab juveniles were all consumed by unfed hermit crabs (Lundstrom, pers. comm.). In the meantime, the juvenile horseshoe crabs at Fordham University went through a few ecdysial processes and became too large for hermit crabs (same size as CEEECOM hermit crabs) to prey. The hermit crabs were put in the mesh cage and dropped into the horseshoe crab tanks. No reaction was observed on the horseshoe crab behavior.

# 4 Discussion

## 4.1 Frequency of Molting and Death

A few observations were made on the molting of juvenile horseshoe crabs. First, they molt on the substrate, not within. Second, most death occurred during molting. And third, they have synchronized ecdysis. After a month at Fordham laboratory, horseshoe crabs started molting almost simultaneously one after another during the second week of March. The second such event was in the second week of May. According to Haug et al. (2013), the advantage of synchronized molting is to deter predation pressure. But in most crustaceans, together with fossil arthropods, such as trilobites, synchronized ecydisis is related to reproductive cycle (Raviv et al. 2008; Speyer and Brett 1985). It is also during the molting process when mortality rate increases (Faizul et al. 2011). The ecdysial rate increased when the water temperature was at or around 23 °C. This is considerably lower than the optimal temperature for the larval development which is 30 °C (Laughlin 1983).

## 4.2 Chemosensory Reaction

Animals, especially benthic marine animals like horseshoe crabs, rely on chemical cues to collect information about their environment (Webster and Weissburg 2001). There are many studies involving chemical signals produced by horseshoe crabs that attract their predators (Ferrari and Targett 2003), or to attract their mates (Hassler and Brockmann 2001; Schwab and Brockmann 2007; Saunders et al. 2010), but a few studies mentioned chemical cues released from predators and the subjects were often crustaceans (Hay 2009, 2011). In this experiment, all the predators must have released chemical signals that are different from horseshoe crabs' own, but none of the horseshoe crab reacted to foreign species. In case of hermit crabs, they were put directly into the horseshoe crab tank without being in a cage, with three third instar horseshoe crabs. After only a couple of days (Lundstrom, pers. comm.), all the horseshoe crabs were consumed. This may mean that horseshoe crab juveniles either detected no chemical signals from the hermit crabs or their chemosensory reaction was not developed in these horseshoe crabs. It was not directly observed, but one would guess that horseshoe crabs may not have even burrowed to hide from the hermit crabs. Further studies for larger juvenile groups that tend to burrow in sediment are necessary to see if there is a tendency among them to hide from predators.

### 4.3 *Visual Acuity*

According to Schwab and Brockmann (2007) and Saunders et al. (2010), mature male horseshoe crabs, together with chemical signals, use visual cues to locate spawning females. However, although late embryos can perceive lights with their lateral eyes and seemingly, object recognition will start when ommatidia develop at the end of the larval trilobite stage (Harzsch et al. 2006). Their horizontal visual resolution is also less than that of adults while their vertical visual resolution is same as adult because of the same number of vertical column of ommatidia (Meadors et al. 2001). It may have been that the horseshoe crabs were not able to see the predators in the mesh cage, and that the cage was just an immovable obstacle. A yellow square object (about 4 cm × 3 cm × 1 cm) was dropped in to one of the horseshoe crab tanks, but it did not cause any stir among juvenile horseshoe crabs at any growth stage. Eventually, they knocked it down and buried in sand.

### 4.4 *Role of Temperature*

According to Chen et al. (2010), the rate of growth in juvenile *Tachypleus tridentatus* was increased by three-fold when the water temperature was raised. For this particular species, 28 °C was the optimal water temperature. In the tank at Fordham University, juvenile horseshoe crabs burrowed more when the water temperature was below 20 °C and above 33 °C, and were most active at 23–26 °C. When colder water was pipetted over an individual, it reacted to the temperature difference and burrowed in sediment; the colder water was 20 °C and the tank water was 25 °C. This indicates that juvenile horseshoe crabs are sensitive to the temperature change in water. The difference in water temperature also influences the level of dissolved oxygen (DO). However, despite the importance of high DO for the juveniles in natural environment (between 8 and 14 mg/L, Morton and Lee 2011), the activities in the laboratory samples were not affected by the low DO (6.0–7.0 mg/L). On an anecdotal note, when the tank heater broke, they gathered around the heated coils and springs, some even tried to "taste" them by grabbing them with their legs and bringing to the gnathobase. It is here observed that they may be thermal sensitive. This needs a further investigation.

## 5 Conclusion

A few observations were made on the molting of juvenile horseshoe crabs: (1) that they molt on the substrate, not within; (2) that most death occurs during molting; and (3) that they have synchronized ecdysis. According to Haug et al. (2013), the advantage of synchronized molting is to deter predation pressure, but also for reproductive cycle (Speyer and Brett 1985; Raviv et al. 2008). Faizul et al. (2011) reported that it is during the molting process when mortality rate increased, and this study, although much smaller in scale, resulted in the same conclusion. The molting rate increased when the water temperature was at or around 23 °C. This is considerably lower than the optimal temperature for the larval development which is 30 °C (Laughlin 1983). The horseshoe crabs molted only on the substrate.

In this study, juvenile horseshoe crabs, either laboratory-reared or wild, did not react to the chemical signals from their potential predators or visual cues from foreign objects in their habitat. When the temperature changed drastically, as and when water change in the tank took place, or when they were returned from a feeding dish to the tank, they quickly burrowed in sediment. It seems that they only reacted to any physical disturbance in the water or sediment. It was also found that they were easily attracted to foreign objects such as thermometer or square object, often checking them with their limbs, a common behavior shared by Xiphosura (Ax 2000). The survivability of juveniles once they hatched increases to 97.5% after settling on substrate and their high-mortality rate was often the result of failed ecdysis (Botton et al. 2003a, b). In this study, some died during molting, but in nature, their behavior of not hiding and coming toward objects to feel them with their legs could also contribute to their survival. The experiment must be designed better to test these points in the future study.

## 6 Future Study

### *6.1 Sediment Size*

In this study, sediment size varied slightly and the range was narrow, from 1 mm to 0.25 mm. Although this range is favorable for juvenile horseshoe crabs (see John et al. 2012), they showed no preference in sediment size. This may be the artifact that the size range was too narrow. In the future study, the range must be wider, for example, 2 mm, 0.5 mm, and 0.125 mm, or even finer. The Silurian and Devonian fossils from Bolivia and Argentina that exhibit a possible infaunal molting pattern have the matrix of silty sandstone to sandy siltstone with grain size ranging from coarse to medium silt (Iwasaki, unpub. data). Although this does not mean that trilobite species lived in such fine sediment particles, but determining the preference of sediment size for juvenile horseshoe crabs is as important as finding whether they molt in the sediment.

### 6.2 Role of Temperature for Burrowing Activities in Juvenile Horseshoe Crabs

The second improvement to be made is the testing of temperature range for juvenile horseshoe crab burrowing activities. As it was observed here that juvenile horseshoe crabs can be sensitive to the change in water temperature and that they tend to burrow more often when the water temperature is below 20 °C and above 33 °C. Between these temperatures, they were very active and ecdysis happened to occur in this range. This means that they might molt less frequently when the extreme temperature would be set, but it could solicit them to molt within the sediment. Together with sediment size, water temperature can be simulated in future experiments.

**Acknowledgments** The author would like to thank J. T. Tanacredi (Molloy College/CERCOM) and M. L. Botton (Fordham University) for their logistic help, S. Portilla and A. Lundstrom (then Dowling College/CEECOM) for their laboratory work. Thanks extend to the reviewers of this manuscript. The study was funded by the Foundation for Ecological Research in the Northeast (FERN), LI, NY.

## References

P. Ax, Xiphosura – Arachnida, in *Multicellular Animals, The Phylogenetic System of the Metazoa*, ed. by P. Ax, vol. II. 420p, (Springer, Berlin, Heidelberg, 2000), pp. 89–98

M.L. Botton, The ecological importance of horseshoe crabs in estuarine and coastal communities: A review and speculative summary, in *Biology and Conservation of Horseshoe Crabs*, ed. by J. T. Tanacredi, M. L. Botton, D. R. Smith, (Springer, Dordrecht, Heidelberg, London and New York, 2009), pp. 45–63

M.L. Botton, R.E. Loveland, T.R. Jacobsen, Beach erosion and geochemical factors: Influence on spawning success of horseshoe crabs (*Limulus polyphemus*) in Delaware Bay. Mar. Biol. **99**, 325–332 (1988)

M.L. Botton, R.E. Loveland, A. Tiwari, Distribution, abundance, and survivorship of young-of-the-year in a commercially exploited population of horseshoe crabs *Limulus polyphemus*. Marine Ecol. Prog. Ser. **265**, 175–184 (2003a)

M.L. Botton, C.N. Shuster Jr., J.A. Keinath, Horseshoe crabs in a food web: Who eats whom? in *The American Horseshoe Crab*, ed. by C. N. Shuster Jr., R. B. Barlow, H. J. Brockman, (Harvard University Press, Cambridge, MA/London, UK, 2003b), pp. 133–153

Y. Chen, C.W. Lau, S.G. Cheung, Enhanced growth of juvenile *Tachypleus tridentatus* (Chelicerata: Xiphosura) in the laboratory: A step towards population restocking for conservation of the species. Aquat. Biol. **11**(1), 37–46 (2010)

M.I.M. Faizul, A. Christianus, S.M.N. Amni, S. Shafeeqa, T. Rishzuan, Evaluation of growth performance of *Tachypleus gigas* (Muller, 1785) under two different culture system. J. Fish. Aquat. Sci. **6**(7), 788–794 (2011)

K.M. Ferrari, N.M. Targett, Chemical attractants in horseshoe crab, Limulus polyphemus, eggs: The potential for an artificial bait. J. Chem. Ecol. **29**(2), 477–496 (2003)

S. Harzsch, K. Vilpoux, D.C. Blackburn, D. Platshetzki, N.L. Brown, R. Melzer, K.E. Kempler, B.A. Battelle, Evolution of arthropod visual systems: Development of the eyes and central visual pathways in the horseshoe crab Limulus polyphemus Linnaeus, 1758 (Chelicerata, Xiphosura). Dev. Dyn. **235**, 2641–2655 (2006)

C. Hassler, H.J. Brockmann, Evidence for use of chemical cues by male horseshoe crabs when locating nesting females (*Limulus polyphemus*). J. Chem. Ecol. **27**(11), 2319–2335 (2001)

J.T. Haug, J.-B. Caron, C. Haug, Demecology in the Cambrian – Synchronized molting in arthropods from the Burgess Shale. BioMed Central Biol. **11**(64), 1–10 (2013)

M.E. Hay, Marine chemical ecology: Chemical signals and cues structure marine populations, communities, and ecosystems. Ann. Rev. Marine Sci. **1**, 193–212 (2009)

M.E. Hay, Crustaceans as powerful models in aquatic chemical ecology, in *Chemical Communication in Crustaceans*, ed. by M. Breithaupt, M. Thiel, (Springer, New York, 2011), pp. 41–62

W.M.S. Herman, Molt initiation in response to phytoecdysones and low doses of animal ecdysones in the chelicerate arthropod, *Limulus polyphemus*. Gen. Comp. Endocrinol. **18**, 301–305 (1972)

S. Hong, S.H.X. Zhang, Y. Zhao, X. Yongzhuang, Y. Zhang, H. Xu, Effect of sediment type on growth and survival of juvenile horseshoe crabs (*Tachypleus tridentatus*), in *Biology and Conservation of Horseshoe Crabs*, ed. by J. T. Tanacredi, M. L. Botton, D. R. Smith, (Springer, Dordrecht/Heidelberg/London/New York, 2009), pp. 535–540

T.C. Jegla, A review of the molting physiology of the trilobite larva of Limulus, in *Physiology and Biology of Horseshoe Crabs: Studies on Normal and Environmentally Stressed Animals*, ed. by J. Bonaventura, C. Bonaventura, S. Tesh, (Alan R. Liss, Inc., New York, 1982), pp. 83–101

B.A. John, S. Sihin, B.Y. Kamaruzzaman, K.C.A. Jalal, Sediment profiling on the nesting grounds of horseshoe crabs at East Peninsular Malaysia. Int. J. Biol. **4**, 159–165 (2012)

R. Laughlin, The effects of temperature and salinity on larval growth of the horseshoe crab *Limulus polyphemus*. Biol. Bull. **164**(1), 93–103 (1983)

C.N. Lee, B. Morton, Emergence behavior of juvenile *Tachypleus tridentatus* under simulated tidal conditions in the laboratory and at two different sediment temperatures, in *Biology and Conservation of Horseshoe Crabs*, ed. by J. T. Tanacredi, M. L. Botton, D. R. Smith, (Springer, Dordrecht/Heidelberg/London/New York, 2009), pp. 275–283

S. Meadors, C. McGuiness, F.A. Dodge, R.B. Barlow, Growth, visual field, and resolution in the juvenile *Limulus* lateral eye. Biol. Bull. **201**, 272–274 (2001)

B. Morton, C.N. Lee, Spatial and temporal distributions of juvenile horseshoe crabs (Arthropoda: Chelicerata) approaching extirpation along the northwestern shoreline of the New Territories of Hong Kong SAR, China. J. Nat. Hist. **45**(3–4), 227–251 (2011)

S. Raviv, S. Parnes, A. Sagi, Coordination of reproduction and molt in decapods, in *Reproductive Biology of Crustaceans*, ed. by E. Mente, (Science Publishers, Enfield, NH, 2008), pp. 365–390

J.J. Rustán, D. Balseiro, Y. Iwasaki, R.D. Foglia, Infaunal molting in *Viaphacops orurensis* (Bonarelli 1921) and its evolutionary implications. The 5th Conference on Trilobites and Their Relatives, 1st–4th July 2012, Prague, Czech Republic, Abstracts, p. 51. Abstract (2012a)

J.J. Rustán, R.D. Foglia, D. Balseiro, Y. Iwasaki, Infaunal molting in calmoniids trilobites: The polyphyletic origin of a defensive character in the context of the Middle Paleozoic Marine Revolution. The Palaeontological Association, 56th Annual Meeting (16th–18th December 2012), University College Dublin, Programme and Abstracts, pp. 85–86. Abstract (2012b)

J.J. Rustán, D. Balseiro, B. Weisfeld, R.D. Foglia, N.E. Vaccari, Infaunal molting in Trilobita and escalatory responses against predation. Geology **39**(5), 495–498 (2011)

K.M. Saunders, H.J. Brockmann, W.H. Watson III, S.H. Jury, Male horseshoe crabs *Limulus polyphemus* use multiple sensory cues to locate mates. Curr. Zool. **56**(5), 485–498 (2010)

M.P. Schreibman, C.B. Zarnoch, Aquaculture methods and early growth of juvenile horseshoe crabs (*Limulus polyphemus*), in *Biology and Conservation of Horseshoe Crabs*, ed. by J. T. Tanacredi, M. L. Botton, D. R. Smith, (Springer, Dordrecht/Heidelberg/London/New York, 2009), pp. 501–511

R.L. Schwab, H.J. Brockmann, The role of visual and chemical cues in the mating decisions of satellite male horseshoe crabs, *Limulus polyphemus*. Anim. Behav. **74**(4), 837–846 (2007)

K. Sekiguchi, H. Seshimo, H. Sugita, Post-embryonic development of the horseshoe crab. Biol. Bull. **174**, 337–345 (1988)

C.N. Shuster Jr., A pictorial review of the natural history and ecology of the horseshoe crab *Limulus polyphemus*, with reference to other Limulidae, in *Physiology and Biology of Horseshoe Crabs: Studies on Normal and Environmentally Stressed Animals*, ed. by J. Bonaventura, C. Bonaventura, S. Tesh, (Alan R. Liss, Inc., New York, 1982), pp. 1–52

C.N. Shuster Jr., K. Sekiguchi, Growing up takes about ten years and eighteen stages, in *The American Horseshoe Crab*, ed. by C. N. Shuster Jr., R. B. Barlow, H. J. Brockman, (Harvard University Press, Cambridge, MA/London, UK, 2003), pp. 103–132

S.E. Speyer, C.E. Brett, Clustered trilobite assemblages in the middle Devonian Hamilton Group. Lethaia **18**, 85–103 (1985)

D.R. Webster, M.J. Weissburg, Chemosensory guidance cues in a turbulent chemical odor plume. Limunol. Oceanograp. **46**(5), 1034–1047 (2001)

# Conservation and Restoration of Estuaries and Coasts: Horseshoe Crabs as Flagship Species

**Jennifer H. Mattei, Mark L. Botton, Satoquo Seino, Alfredo Quarto, Jaruwan (Ning) Enright, J. Vanitha, Hsu Chia Chi, Lester Tan, Joe Cheung, and John T. Tanacredi**

## 1 Introduction

Many coastal and estuarine habitats: sandy and cobble beaches, saltmarshes, seagrasses, mangroves, and mudflats have been reduced and degraded around the world requiring adaptive management, conservation, and restoration (Zedler 2017). The

J. H. Mattei (✉)
Sacred Heart University, Department of Biology, Fairfield, CT, USA
e-mail: matteij@sacredheart.edu

M. L. Botton
Fordham University, Department of Natural Sciences, New York, NY, USA
e-mail: botton@fordham.edu

S. Seino
Graduate School of Engineering, Kyushu University, Fukuoka, Japan
e-mail: seino@civil.kyushu-u.ac.jp

A. Quarto · J. (. Enright
Mangrove Action Project, Seattle, WA, USA
e-mail: mangroveap@olympus.net

J. Vanitha · H. C. Chi · L. Tan
Nature Society (Singapore), Singapore, Singapore
e-mail: lestertan@nss.org.sg

J. Cheung
Ocean Park Conservation Foundation Hong Kong, Hong Kong, SAR China

The Nature Conservancy Hong Kong Foundation Ltd, Hong Kong, SAR China

J. T. Tanacredi
Department of Earth and Environmental Studies, Molloy College,
Rockville Centre, NY, USA
e-mail: jtanacredi@molloy.edu



J. T. Tanacredi et al. (eds.), *International Horseshoe Crab Conservation and Research Efforts: 2007-2020*, https://doi.org/10.1007/978-3-030-82315-3_10

level of ecosystem services provided by these habitats as fish and shellfish nurseries, coastal protection, erosion control, carbon sequestration, water purification, and recreation/tourism are often undervalued and when measured accurately prove more valuable than terrestrial habitats (Barbier 2017; Barbier et al. 2011). With expanding coastal megacities, increasing levels of pollution and shoreline armoring, estuaries need "flagship species" to be recognized that are valued culturally, economically, and ecologically in order to impress upon people the importance of these coastal habitats. One taxonomic group that is important in all of these areas and should take the role of the flagship species for estuaries and coasts is the horseshoe crab.

The concept of a flagship species is often associated with charismatic terrestrial vertebrates (typically mammals), such as pandas, tigers, elephants, lions, or polar bears. Flagship species belong to a larger category of "focal species," which "… for ecological or social reasons, are believed to be valuable for the understanding, management and conservation of natural environments" (Zacharias and Roff 2001, p. 59). Some flagship species have the dual function of "umbrella species," ones whose protections will ensure conservation of other co-occurring species (Kalinkat et al. 2017). By virtue of their size, appearance, perceived threatened status, and familiarity with the public, flagship species are important vehicles in fundraising campaigns and education about the importance of preserving natural resources. By focusing public awareness on the conservation of flagship species, the status of other species which share the habitat may be improved. For instance, large megafauna (e.g., tigers, manatees, dolphins) may be more effective as flagship species for habitats dominated by mangrove trees than the myriad small resident invertebrates and fishes (Thompson and Rog 2019). The challenge in finding suitable flagship or umbrella species in aquatic habitats is that many species of ecological importance are taxonomically very distant from humans (i.e. not mammals), and may be small and/or cryptic (Kalinkat et al. 2017). Shokri et al. (2009) proposed using sygnathid fishes (pipefish and seahorses) as flagship species for the conservation of seagrass habitats. The main appeal of these species is their aesthetics and their use in traditional Chinese medicine but support for the conservation of seagrass has not improved with worldwide losses of seagrass beds estimated to be 1.5% per year (Gómez 2018).

There are four extant species of horseshoe crab, three in Asian coastal habitats: (1) tri-spine horseshoe crab *Tachypleus tridentatus*, (2) coastal horseshoe crab *T. gigas*, and (3) mangrove horseshoe crab, *Carcinoscorpius rotundicauda*. The fourth species, the American horseshoe crab *Limulus polyphemus*, inhabits the coasts of the eastern United States and parts of the Gulf of Mexico. At first glance, horseshoe crabs would appear to be unlikely candidates to enlist as a flagship species for the conservation of coastal and estuarine habitats, as they do not fit the typical profile as a physically attractive animal with features (such as large eyes or fur) that are common to most other flagship species. However, as identified by numerous authors (e.g. Shuster et al. 2005; Chen et al. 2009; Tanacredi et al. 2009; Carmichael et al. 2015, this volume) horseshoe crabs have immense ecological, evolutionary, economic, and cultural significance which serve as a counterweight to this disadvantage. Horseshoe crabs are "living fossils" with remarkable persistence over 445 million years of geological changes (Rudkin and Young 2009). Their evolutionary history of *baúplan* stasis, association with deep time, surviving the past five mass extinctions has fascinated many people from paleontologists to children visiting natural history museums. Horseshoe crab blood products have improved the health of billions of people. Their blood contains a material known as LAL (*Limulus* amoebocyte lysate) that is the worldwide industry standard to detect gram-negative bacterial endotoxins in pharmaceutical products (Krisfalusi-Gannon et al. 2018). The use of LAL for vaccine testing has not only prevented billions of people from becoming ill but has also protected our domesticated companions and agricultural livestock (Beekey and Mattei 2015). LAL and TAL (*Tachypleus* derived amoebocyte lysate) will be used to help test COVID-19 vaccines before being distributed to the world's populations (Gorman 2020). This demonstrates, once again, how important coastal and estuarine habitats are that allow horseshoe crabs to survive and reproduce for the benefit of all people to remain healthy during the current and future pandemics. There has never been a better time to draw peoples' attention to the plight of horseshoe crab habitats. Additional economic benefits include use as bait in commercial fisheries for eels, whelks, and octopus (Smith et al. 2016), and horseshoe crabs are consumed by people in many countries in Southeast Asia (John et al. 2018).

Ecologically, when horseshoe crabs are abundant, they are a dominant species responsible for increased biodiversity/biomass in coastal spawning habitats (Mattei et al. this volume). Various aspects of this food web have been investigated by researchers in North America (Walls et al. 2002; Botton and Shuster 2003; Carmichael et al. 2004; Odell et al. 2005; Botton 2009; Niles et al. 2009) and Asia (Debnath et al. 1989; Kwan et al. 2015; Fan et al. 2017). *Limulus* eggs are vital food for migratory shorebirds, fish, and numerous invertebrate species as observed in the Delaware Bay estuary, USA. Dense aggregations of horseshoe crabs can be significant predators and bioturbators of estuarine habitats (Botton 1984; Kraeuter and Fegley 1994). Predation on adult horseshoe crabs is seldom reported (Botton 2009), with two notable exceptions. *Limulus polyphemus* forms an important part of the diet of federally endangered loggerhead (*Caretta caretta*) and Kemp's Ridley (*Lepidochelys kempii*) sea turtles in the middle Atlantic region of the United States (Keinath et al. 1987; Seney and Musick 2007). *Limulus* that are overturned during spawning activity often fall prey to large seagulls (Botton and Loveland 1989). As large numbers of horseshoe crabs die onshore during spawning events, their carcasses are scavenged by everything from bald eagles to carrion beetles, flies, and worms which in turn expand the food web.

All species of horseshoe crab spend their lives in coastal and estuarine habitats. Spawning generally takes place on sandy and cobble estuarine beaches for the tri-spine horseshoe crab *T. tridentatus*, coastal horseshoe crab *T. gigas*, and American horseshoe crab *L. polyphemus*, and mangrove forests for the mangrove horseshoe crab, *C. rotundicauda*. However, *Limulus* has also been observed, across its North American range, to spawn in saltmarshes, eel grass beds, and mudflats (pers. obs. J. Mattei, M. Botton, C. Chabot, J. Brockman, D. Sasson). In addition, horseshoe crabs have a relatively long juvenile growth and development period that spans

8–10 years. Although juvenile habitat requirements are not well studied, juvenile horseshoe crabs of all four species have been observed foraging in mudflats (Hu et al. 2015), flat sandy areas (at least through 4 years of age), and *Limulus* juveniles have been observed in saltmarsh fens (Mattei, unpublished data; Colón et al. this volume). Thus, throughout their lifetime of over 20 years, horseshoe crab individuals require between two and four different habitat types within coastal and estuarine ecosystems.

Chen et al. (2009) first considered the tri-spine horseshoe crab as a flagship species for marine conservation in Taiwan, and in this chapter, we expand this to include all four species of horseshoe crab as flagship species for the conservation of estuaries and coasts. The three Asian horseshoe crab species are declining in all areas of Asia where studies have been conducted (John et al. 2018; Laurie et al. 2019), and the American horseshoe crab populations are declining in most regions of North America (Smith et al. 2016). There are many factors linked to their decline, including overharvesting for bait, human consumption, biomedical uses, pollution, and shoreline hardening/reclamation (Seino et al. 2003; Smith et al. 2016; John et al. 2018). However, the overarching factor of decline of horseshoe crab populations is the loss of coastal and estuarine habitats.

The workshop on which this chapter is based was held in September 2016 as part of the IUCN World Conservation Congress in Honolulu, Hawaii. The title of this workshop, "Conservation and Restoration of Estuarine, Beach and Mangrove Habitats: Horseshoe Crabs as a Flagship Species," reflects our belief that the broader goals of conservation and restoration of coastal and estuarine habitats can be facilitated by having horseshoe crabs as flagship species due to their dependence on these habitats for their continued existence.

## 2 Sandy/Cobble Beaches

Beaches form along shorelines where waves deposit sediments, and the beach morphology is shaped by ocean waves during shoaling, breaking, and swash. All beaches consist of sediment ranging in size from sand up to cobbles, and along some shorelines, mixed with larger rock and boulders. In mid-latitudes, beaches are composed of eroded grains of siliceous or quartz sand grains. Tropical beaches are generally composed of carbonate sediment or shell fragments (Short 2012).

### 2.1 *Ecosystem Services of Beaches*

Beaches and associated dunes provide protection of the upland from storms by forcing waves to shoal and break before reaching human developments. Beaches are important foraging areas for shorebirds, crabs, and finfish due to the benthic animals and microalgae living on or within the sand (King et al. 2018).

## 2.2 What Are the Causes of Loss/Decline of Beaches?

Changes in sea level associated with global climate change threatens the continued persistence of sandy estuarine beaches which are the primary spawning habitat for three of the world's four horseshoe crab species, *Tachypleus tridentatus*, *T. gigas*, and *Limulus polyphemus*. To protect coastal property and infrastructure, shorelines are frequently armored, an alteration that greatly decreases their ecological value relative to the beaches or marshes they replace (e.g., Dugan et al. 2008). Hardening the shoreline makes it unsuitable as spawning habitat for horseshoe crabs (Botton 2001; Seino 2001; Seino and Uda 2007; Jackson and Nordstrom 2015; Loveland and Botton 2015). Less intrusive methods, such as beach nourishment and living shorelines, are an alternative approach for coastal protection that may cause less damage to biota. The outcome of some of these projects in the United States and Japan may be useful in the design of future shoreline restoration efforts.

**Case Study: Shoreline Armoring, the Japanese Experience**

Over a half of the Japanese coast (except island areas) were artificialized in the early 1990s for protection of erosion and reclamation (Environmental Agency, Government of Japan 1994). National seashore protection policies were oriented to harmonize with the ecosystem along with amendments of the Seacoast Act in 1999. *Tachypleus tridentatus* was designated as the target species for conservation of estuaries and sandy beaches by national governmental technical guidelines (Seino et al. 2004). However, erosion control of sandy beaches was executed with solid concrete structures. Moreover, the 2011 Great East Japan Earthquake and tsunami spurred the Japanese government to build bigger and wider concrete sea walls. In northeast Japan, efforts to fend off future disasters are focusing on a nearly 402 km chain of cement sea walls, at places nearly five stories high (~15 m, Fig. 1). Opponents of this US$6.8 billion plan argue that the massive concrete barriers will damage the marine ecology and aesthetics of the area, hinder vital fisheries and the protective value to residents is not known; residents are encouraged to relocate to higher ground if possible.

Japan's southern inland sea has experienced urbanization, development, and land reclamation for agriculture. Port construction and waterway dredging have caused deformity and loss of geomorphological continuity of the sea bottom. Every river has dams, weirs, and armored shorelines. Thus, sediment transport downstream has been halted (Seino et al. 2000).

*Tachypleus tridenta*tus, the tri-spine horseshoe crab, is now an endangered species in Japan with a declining population due to the loss of spawning habitats (i.e., coarse sandy beaches) since the 1990s. Sea level rise and coastal erosion are increasing the threat of horseshoe crab extinction in Japan. The ecological restoration of estuaries is needed. The possibility of beach nourishment is being discussed. Small-scale trials of beach nourishment projects have been executed during the 1980s as

Fig. 1 Some views of coastal Japan. (a) typical artificialized beach along the coast of Japan; (b) giant sea wall construction after the Great East Japanese Earthquake of 2011

mitigation of reclamation. Continuous extensions of roads and highways have reclaimed many sandy beaches. Often, local inhabitants would request their beaches to be restored through governmental nourishment projects (Ishikawa et al. 2013). Dredged sand from harbors and dam reservoirs may be used but also may have contaminants and/or have high silt components. Very fine-grained sediments may not be the best beach composition for spawning horseshoe crabs. Restoration of sand transport within the watershed and conservation of the sea bottom are essential, but the current social system of water resource and disaster reduction should be modified with nature-based solutions. In recent years, Japanese academics and environmental policies have become to be interested in ECO-DRR (Ecosystem-based Disaster Risk Reduction) and Green Infrastructure. In part, these strategies involve re-thinking the proper placement of revetments to preserve the continuity between the land and sea (Fig. 2). Sandy beaches for horseshoe crab will benefit if we avoid coastal overexploitation.

Ecological restoration must be attempted to allow this species to survive in Japan. Mass mortality of *T. tridentatus* was observed at Sone tidal flat in northern

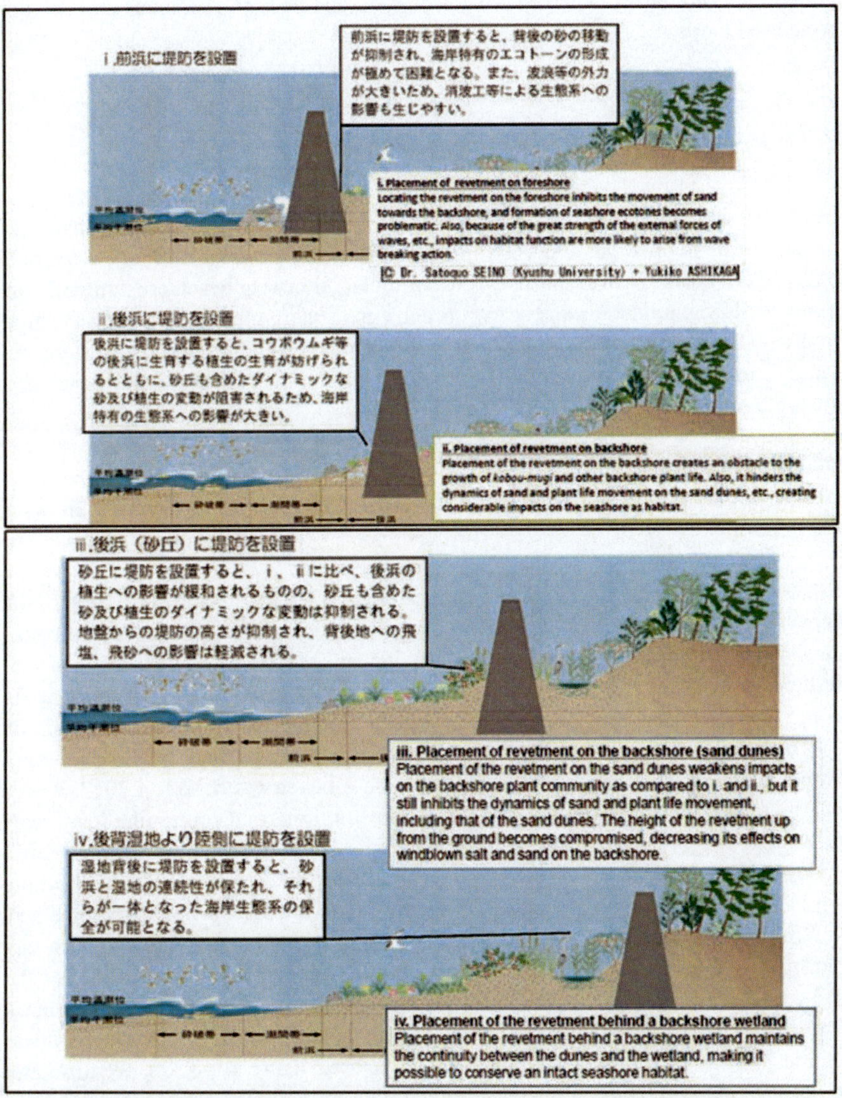


**Fig. 2** Strategies for the placement of revetments along the coast of Japan is part of the national guidelines of disaster recovery construction

Kyushu. This population lives at the northern limit in Asia. In 2016, western Japan experienced the hottest summer on record with 1 day in August reaching 41 °C. Global climate change research and studies of local human impacts that affect horseshoe crab biology and ecology are being revealed by international cooperation. In this decade, floods and landslide disasters caused by intense rainstorms had previously been recorded only once a half century or rarely (e.g., every 350 years) are occurring more frequently. Almost all horseshoe habitats in western Japan are impacted by climate changes. In many areas, tidal mudflats that used to be horseshoe crab nurseries have come to be regarded as high-risk areas and become reclaimed lands.

## 2.3 *Restoration of Beaches in General*

Sea level rise is coupled with the overall problem of global climate change, and hence to greenhouse gas emissions (IPCC 2018). In the final analysis, we would collectively need to lower our use of fossil fuels, otherwise beach erosion will continue. Building bulkheads and sea walls ends up causing more erosion and attempts to divide the land from the sea which in turn damages and reduces ecosystem services. Large-scale techniques like beach nourishment may slow or delay beach erosion, but are expensive (Thuy et al. 2018).

### Beach Nourishment and Horseshoe Crabs: A Case History from New York City, USA

Sea level rise associated with global climate change is a world-wide threat to coastal ecosystems. The materials used in coastal restoration make a difference to the species that use the area. In particular, we provide an example of how the grain size of sediments used during beach nourishment activities may be more important for horseshoe crab spawning and egg development than the volume of sand brought in.

This study took place on Plumb Beach, Jamaica Bay, New York, where we compared spawning abundance and egg density on a beach nourished in 2012, with a nearby reference site beginning in 2011 (pre-restoration) and continuing for 4 years post-restoration (Botton et al. 2018). The lengths of the nourishment site and reference site were similar, 0.9 and 0.85 km, respectively. Shoreline counts of spawning adult crabs took place on 12 dates each spring, that were selected to bracket the new and full moons, which are the anticipated peaks of spawning activity. To assess egg densities, we collected a series of 30 core samples every 2–3 weeks during the spawning season in 2012 (pre-nourishment) and 2013–2015 (post-nourishment). Cores were 3.8 cm in diameter and we sampled to a depth of 20 cm. The hardness of the surface mid-beach sediment was measured using a pocket penetrometer (Forestry Suppliers Inc., model 77,114). Sediment samples from each site were sieved in the laboratory and standard grain-size statistics were calculated (see Botton et al. 1988 for details of all methodologies).

The newly restored beach area sustained very low horseshoe crab density compared to the reference beach, even 4 years after the restoration. There was only a slight increase in the density of spawning females on the restored beach before and after beach nourishment, but spawning densities were consistently higher at the reference site throughout the study. Similarly, beach nourishment in Fall 2012 was not followed by a detectable increase in horseshoe crab egg deposition in the first three post-nourishment seasons, 2013–2015. The nourishment site had significantly fewer horseshoe crab eggs, embryos, and trilobite larvae both before and after the restoration project. These differences in spawning and egg densities between beaches have persisted through 2019 (M. Botton and C. Colón, unpublished data).

We believe that differences in sediment texture between sites may explain the differential use of the restored and reference sites. Hard-packed fine sand of relatively uniform texture was found at the restored beach. In contrast, the reference site has a less compacted, coarser, and more heterogeneous mix of particle sizes. The very dense sand at the restored beach has small pore spaces for oxygen and water exchange, leading to occasional anoxic conditions that are unfavorable for horseshoe crab egg development. These results are consistent with the study by Botton et al. (1988) which showed that horseshoe crabs tend to avoid depositing their eggs on beaches that have high levels of hydrogen sulfide, which is an indicator of anoxic conditions.

In summary, this case history showed that beach restoration at Plumb Beach did not result in significant benefits to horseshoe crabs as assessed by spawning and egg densities. Subtle differences in sediment texture over relatively short distances can be detected by horseshoe crabs and may underlie their selection of specific nesting sites. This study illustrates the importance of selecting the proper type of sediment for beach nourishment projects to create suitable conditions for spawning horseshoe crabs.

## 3 Saltmarshes, Seagrasses

Saltmarshes and seagrasses are highly productive biodiverse ecosystems, consisting of a variety of plant species that provide important interconnected coastal habitats. Tidal saltmarsh communities are found in gently sloping areas above mudflats between mean high water neap tides and mean high water spring tides. Seagrasses (i.e., submerged aquatic vegetation or SAV) are found further offshore and remain submerged to depths of 1–3 m. Saltmarshes and seagrasses occur worldwide, particularly in middle to high latitudes, in sheltered locations; however, diverse seagrass species are also found throughout the tropics (Short et al. 2007; Mcowen et al. 2017).

## 3.1 Ecosystem Services of Marshes and Seagrasses

Saltmarshes and seagrasses are extremely valuable ecosystems, contributing to local and global economies. Services include fish/shellfish production, building materials and fodder for livestock, ecotourism, wildlife habitat, carbon sequestration/storage, and protection of coastal areas from natural disasters, such as erosion and storm surges (Short et al. 2007; Mcowen et al. 2017).

## 3.2 What Are the Causes of Loss/Decline in General?

Even though the estimates of ecosystem services of global saltmarsh and seagrass habitats range in trillions of dollars, they are declining around the world with estimates at 50% of their global historical coverage (Duarte et al. 2008; Crooks et al. 2011). These coastal habitats are subjected to high levels of disturbance that vary both in frequency and intensity including hurricanes, runoff, sedimentation, sea level rise, as well as coastal development and other human uses including fishing/harvesting, aquaculture, marine trades, and access for recreational activities. Long-term disturbance, historical loss and alteration of habitats, and urbanization have increased the need for coastal habitat restoration (Greipsson 2011; Davis and Kidd 2012).

## 3.3 Restoration

### Case Study: Nature-Based Restoration in Stratford, Connecticut, USA

Restoration of highly disturbed coastal habitats involves activities that include remediation (e.g., removal of contaminants), rehabilitation (e.g., dune construction and prescribed burning), and revegetation (e.g., planting of native beach grasses and upland woody species). Unfortunately, confidence levels are often low regarding the target goals of many restoration projects. This is due to the uncertainty regarding natural variation in temporal and spatial conditions with respect to hydrology, weather, growth and reproduction patterns of plants and animals, effects of natural predators and invasive species, and changes in the surrounding landscape due to human activity (Thom 2000). The development of successful restoration projects requires an understanding of the state of the system, the prognosis for further development, the measures needed to correct problems, and a long-term commitment via monitoring and adaptive management (Mattei 2019).

Living shorelines are nature-based solutions to man-made problems along our highly populated coasts. Coastal development, pollution, and uncontrolled harvest of seafood have changed the conformation of our coastlines. One would need to go back to the late 1700s to find descriptions of what our natural coastal habitats looked like because we have changed every habitat in our coastal ecosystem (Ermgassen et al. 2012). For example, we overharvested the oysters and stripped out nearly all of the oyster beds and reefs. We have filled in and taken out the majority of our saltmarshes and vegetated dunes, many replaced with human developments. These coastal habitats were the natural buffers of wave energy and coastal storms. In the past, wave energy would dissipate well before it hit the shoreline and upland.

Living shorelines can be used for capping heavy metals, slowing coastal erosion, increasing sediment deposition, and increasing wildlife habitat structure and function. Living shorelines in New England, USA, consist of temperate coastal habitats including oyster reef, *Spartina* saltmarsh grasses and upland beach grass, shrubs, and trees. The model used at Stratford Point, Connecticut, employs reef balls (www.reefball.org) operating as artificial oyster reefs, to slow the incoming waves, protect newly restored marsh grasses, and provide fish habitat (Fig. 3). Oysters and blue mussels are slowly colonizing the reef. The sediments that have accumulated 2 years after installation (~30 cm) are now deep enough for horseshoe crab spawning to be successful in the area (Fig. 4). The numbers of spawning horseshoe crabs have increased in the area over the past 4 years and biodiversity has dramatically increased (Mattei 2019).

**Fig. 3** Stratford Point living shoreline, August 2019, Connecticut, USA (photo credit: J. Mattei)

**Fig. 4** Spawning horseshoe crabs found in accumulated sediments around reef (left). Shellfish colonization of reef (right) (photos by J. Mattei)

# 4 Mangrove Forest Ecosystem

Mangrove forests are tidal tropical and subtropical coastal wetlands found in estuaries, deltas, lagoons, or surrounding islands found in the Indo West Pacific (58 species) and the Atlantic East Pacific (12 species) along the coastal areas of Mexico, Central America, and parts of the Caribbean. Mangrove trees are uniquely adapted for growth in salt water.

## *4.1 Relevance of Mangrove Restoration for Conservation of Horseshoe Crabs*

Mangrove forests are important for healthy coastal ecosystems in the tropics and sub-tropics. These forest wetlands support an immense array of marine and coastal life, serving as marine nurseries, nesting, and feeding grounds for migratory shore birds, one of the last remaining refuges for Bengal tigers in the Sundarbans of the Bay of Bengal, and a wide variety of other mammals dependent on a healthy mangroves and related habitats including manatees, proboscis monkeys, lemurs, a myriad of insects, amphibians, reptiles, sea turtles, and mangrove horseshoe crabs.

Mangroves also support the health and productivity of coral reefs and sea grass beds by filtering out sediment and absorbing excess nutrients from river water. In addition, mangroves play an important, life-supporting role for countless coastal communities who depend on mangroves for life and livelihoods. Mangroves are recognized for their important role in mitigating climate change, sequestering up to five times more carbon than other forest ecosystems, storing that carbon in their peat soils for hundreds, if not thousands of years (Howard et al. 2017). Mangroves are also living buffers against the forces of storms and waves that can otherwise erode and devastate coastlines (del Valle et al. 2020).

Horseshoe crabs have been around for over 400 million years, evolving in the shallow seas of the Paleozoic Era (540–248 million years ago). Yet these resilient creatures that survived the five previous mass extinction events may not survive the present one because of loss of habitat. For one species of horseshoe crab, that loss intertwines its fate with that of the mangrove forests of Southeast Asia, as industrial shrimp aquaculture, tourism, urban, and agricultural expansion all threaten this habitat. Healthy, intact mangrove forests play an ancient and complex role for *Carcinoscorpius rotundicauda*, popularly called the "mangrove horseshoe crab." These horseshoe crabs are found only in Southeast Asia. The fate of *C. rotundicauda* is tied closely to the decline of mangroves, because this species spawns only along the creek banks in areas of mangroves (Vestbo et al. 2018). The other two species of Asian horseshoe crabs, *Tachypleus gigas* and *T. tridentatus* use sandy beaches for spawning, but the loss of mangroves also impact them indirectly due to altered hydrodynamics, changes in nutrient cycling, erosion of beaches, and altered food webs.

As loss of (mangrove) habitat has been and still is a major driver of horseshoe crab extinction globally, effective conservation of existing primary forests and restoration of degraded forests is essential.

## What Are the Causes of Mangrove Decline and Loss?

It is estimated that less than 15 million hectares remain worldwide—less than half their original area (Strong and Minnemeyer 2015). Although the overall losses of mangroves are declining, this average masks regions and countries where mangroves are being converted at 5–8% per annum. In Southeast Asia, this is mainly driven by shrimp aquaculture, rice and oil palm production (Richards and Friess 2016), and also timber and fuel wood extraction, charcoal production, urban expansion, pollution, coastal road construction, and other industrial and infrastructure developments.

## Community-Based Ecological Mangrove Restoration: An Effective Approach to Mangrove Restoration

Non-Governmental Organization (NGO) Mangrove Action Project (MAP) (Wodehouse 2020) and others (Lewis 2009; Primavera et al. 2012) have argued that the long-term conservation of primary mangrove forests should be the primary objective, followed by effective restoration of degraded or destroyed mangroves. While efforts to conserve mangroves have increased with the growing awareness of their value, many restoration efforts still fail completely or fail to meet their objectives. Specifically, traditional mangrove restoration efforts (line planting of mid-zone mangrove species, such as *Rhizophora* sp. on all sites and at all elevations relative to sea level) often fail or fail to reach their objectives. Restoration efforts often fail to resolve the underlying problems or stressors that caused their loss in the first place, such as over-exploitation, local poverty, unusually high salinity or disturbed hydrology and fail to understand why a site is not regenerating naturally. Planting projects fail due to poor site selection, poor site-species matching, and a poor understanding of mangrove biology and ecology (Wodehouse 2020). Furthermore, if successful, traditional planting establishes monoculture plantations that lack biodiversity and full mangrove ecosystem function.

Community-Based Ecological Mangrove Restoration (CBEMR) is an alternative, holistic approach to mangrove restoration that tries to mitigate underlying mangrove stressors and works with nature to facilitate natural regeneration if the site is not propagule-limited. This might be done by restoring the hydrology and improving the drainage or adjusting a site's topography so that mangroves may regenerate naturally, resulting in more biodiverse ecosystem (Fig. 5). Key to the success of this process is to conduct a stakeholder analysis at the outside and involve the local communities and other stakeholders with power and interest in the project from the outset.

**Fig. 5** Photo sequence of a successful Community-Based Ecological Mangrove Restoration project that the Mangrove Action Project helped initiate in El Salvador 10 years ago resulting in good mangrove recovery after the blocked hydrology was restored by volunteers from the resident communities. (Association Manglar/Eco-Viva, El Salvador)

# 5 Intertidal Flats: Important for Juvenile Horseshoe Crab Growth and Development

Found worldwide, intertidal flats are a common feature of many estuarine environments. These habitats occur in relatively protected areas, allowing for the accumulation of fine sands and mud. Mudflats are areas of land that are flooded at high tide and are formed by a buildup of fine sediment carried in by tides and rivers.

## *5.1 Importance of Mudflats*

Despite the rather barren appearance of many sand flats and mudflats, they often harbor abundant benthic infauna that are well-adapted to survive frequent sediment disturbances caused by waves, storms, and bioturbation (Zajac and Whitlatch 2003). At low tide, flats are important feeding and resting habitats for birds (Burger et al. 1977; Fonseca et al. 2017), and at high tide when the flats are submerged, the resident fauna can be preyed upon by rays and other types of fishes (Cross and Curran 2000). Indeed, some intertidal flats have such abundant benthic resources that they are crucial foraging sites for shorebirds in estuaries throughout the world, such as the Bay of Fundy, Canada (Hicklin 1987), Wadden Sea, Netherlands (Kraan et al. 2009), Banc d'Arguin, Mauritania (Lourenço et al. 2015), and Copper River Delta, Alaska (Gill Jr and Handel 1990). Tidal flats provide a wealth of additional ecosystem services, including nutrient cycling, carbon sequestration, wave attenuation, sediment storage, and protection of the shoreline against erosion (Miththapala 2013). These systems are threatened by sea level rise, invasive species, shoreline development, sand extraction, and river diversions that affect the hydrological regime and sediment budget (Miththapala 2013; Itaya et al. this volume).

Tidal flats are a critical link in the life history of horseshoe crabs. During the spring spawning season and well into the summer, adult *Limulus* often aggregate on tidal flats during low tides, where they feed on mollusks and other invertebrates (Botton 1984). Young-of-the-year *Limulus*, as well as juveniles up to about 3 years of age, are commonly found on tidal flats during the summer months (Botton et al. 2003; Colón et al. this volume). Tidal flats are also critical habitats for juvenile Asian horseshoe crab species (Hu et al. 2015; Cartwright-Taylor et al. 2011; Kwan et al. 2016). New hatchlings of *Tachypleus tridentatus* disperse from spawned sandy beaches to neighboring mud tidal flats to settle there and grow (Maeda et al. 2000). Tidal flows are essential to support hatchling ecology. Consequently, modifications of coastal land forms by development can adversely change critical processes of the early life history of horseshoe crabs.

### 5.2 *Restoration and Conservation of Mudflats*

The management of tidal wetlands including mudflats has a long history in North America that is summarized by Tiner (2013) but very little information exists on mudflat restoration. The World Bank has called for the worldwide conservation and management of coastal wetlands for "ecosystem mitigation" to help combat sea level rise through carbon sequestration (Crooks et al. 2011) by our estuaries and coastal habitats.

## 6 Horseshoe Crabs Outreach: Connect People with Their Coastal Environments

### 6.1 *Beaches/Marshes/Seagrasses*

The members of the International Union for Conservation of Nature (IUCN) Horseshoe Crab Specialist Group invited researchers and community members from around the world to celebrate the value of horseshoe crabs in our coastal ecosystems and to designate them as "flagship species" for the conservation of estuarine and coastal ecosystems. Many community-based activities have recently been launched and promoted due to the first International Horseshoe Crab Day on June 20, 2020 (https://www.youtube.com/watch?v=bDjqXfEWy4Q&list=PLdrZ71BTGCwwZpDoS1nBNeiu6GM95eiUY). Over 50 videos have been combined and posted from 11 different countries.

Project *Limulus* (www.projectlimulus.org) is a model community research program whose participants learn the ecological importance of horseshoe crabs to the Long Island Sound ecosystem in the United States through hands-on activities such as tagging and conducting spawning surveys. In 2012, a total of 18 nonprofit organizations encompassing more than 400 volunteers participated in 234 horseshoe crab spawning surveys across 30 Connecticut beaches. Volunteers tagged and released approximately 9000 horseshoe crabs (Mattei et al. 2015). Parts of this program have been replicated in Malaysia, Hong Kong, and India (Faridah et al. 2015) and community members have been encouraged to tag and report recaptured horseshoe crabs as well as help with hatching of eggs and juvenile release programs.

### 6.2 *Mangroves*

Community-Based Ecological Mangrove Restoration (CBEMR) involves a more methodological ecosystem approach than the usual monoculture restoration efforts, incorporating natural mangrove dispersal and ecological recovery. The key is in the restoration of the hydrology and topography where needed of the area being considered for restoration, and then working with nature itself to help facilitate regeneration of the area's naturally occurring mangrove species. Adequate monitoring and evaluation follow this for 3 the 5 years at each site to assess progress and take corrective action to ensure success,

CBEMR is based on principles of community engagement and empowerment, recognizing that sustainable restoration requires the active participation of the affected local communities. The importance of local community involvement in mangrove conservation and restoration cannot be overstated, as it is these local communities who reside on-site and have most to gain from a healthy, living mangrove buffer, including improved livelihoods from increased wild fisheries and protection from storms and wave surges, These communities also possess important local knowledge of their community base and surroundings and are more able to monitor and assess the status of their mangrove areas on an ongoing basis.

If the mangrove horseshoe crab is to survive this present sixth extinction, surely effective conservation and restoration of its spawning grounds in the mangroves must occur. Hand planting mangrove seedlings in mud flats, seagrass beds, or other inappropriate coastal zones will not accomplish the objectives of long-term, biodiverse habitat restoration. An ecosystem-based restoration and management approach such as represented by CBEMR is urgently needed.

## Horseshoe Crabs, Mangroves, and Citizen Science in Singapore

Coastal development has been occurring in Singapore since agricultural times, but the rate at which development has been happening in the last few decades is rapid. Mangrove forests are being cut down for new buildings and ports. A stretch of mangroves adjacent to Sungei Buloh Wetlands Reserve known as Mandai Mangroves (72.8 ha park) is an important habitat for migratory birds and horseshoe crabs. Mangrove horseshoe crabs gather at these mudflats in high densities. It is imperative to preserve what is left in order to prevent the extinction of the mangrove horseshoe crab, *Carcinoscorpius rotundicauda* and preserve other fish and wildlife habitat. In addition to the importance of the mangrove habitats for horseshoe crabs, Mandai mudflats and mangroves have cultural and historical significance. Conserving the habitat is crucial for future generations. Mangrove forests can also prevent coastal erosion and faster recovery from storm events (i.e., resiliency).

Singapore is a small-island city-state of 582 $km^2$ with about 90% of our shoreline covered by mangroves when it was founded in 1819. Over the years, increase in population and the need for housing, infrastructure, industrial, shrimp farms, other land use development and extensive land reclamation have resulted in the loss of 70–80% of these mangroves. Even now with increasing demand for water for our five million population, the damming of many of our rivers to create water catchment/containment areas has resulted in diminished sedimentation for our vulnerable remaining mangroves.

The past 10 years that our Horseshoe Crab Rescue & Research (HSC R&R) has been monitoring the *C. rotundicauda* population at the Mandai mudflats, we are witness to the gradual degradation of the mudflats and dying back of the adjacent mangrove habitat. Although we still have a sizable population of mangrove horseshoe crabs at the mudflats, we are concerned that without the protective nursery afforded by the roots of the adjacent healthy mangrove habitat, the survival of horseshoe crabs, other fish, and intertidal creature hatchlings will decline. HSC R&R volunteers help release horseshoe crabs that become entangled in ghost fishing nets (Fig. 6) With the possible loss of most, if not all, of the more fragile species in this ecosystem, the impact on the overall biodiversity of this habitat may not only effect the sustainability of the horseshoe crab population, which acts as a flagship species here, but also have adverse consequences on this area as a feeding and foraging area for the shorebirds of our Sungei Buloh Wetland Reserve, located a few kilometers to the west on the same shoreline, and also as an important pit stop for migratory birds on the East Asian Australasian Flyway (EAAF).

On a group of islets off main island Singapore, the authorities have built a surrounding sea wall to create a land fill. Along the shoreline of this sea-walled landfill a mono-culture of mangrove has been planted. As this instant mangrove is meant as a biological barrier to help contain and monitor possible landfill seepage, it does not qualify as mangrove restoration.

However, on another larger islet off main island Singapore, a collaboration, involving National Parks, Singapore, the Geography Department of the National University of Singapore and Nature Society (Singapore) (NSS), known as the Restore Urban Mangroves (RUM) project has just been initiated (Friess 2017). This is a hydrology-based method to guide mangrove restoration of an area of degraded shoreline and NSS volunteers are involved in the replanting effort. We look forward to sharing the results of this project in the near future.

The hallmark of this cycle of climate change is the significant contribution of mankind's excesses to its accelerated and pronounced nature. Flora and fauna that are slow or unable to adapt to the effects of global warming will perish. It is

**Fig. 6** Rescuing mangrove horseshoe crabs entangled in discarded fishing nets in Singapore (Nature Society Singapore)

appropriate, opportune, and timely for enlightened governments to provide committed leadership together with the multitude of nature, environmental, and biodiversity conservation NGOs with their many passionate volunteers to employ nature-based solutions to, hopefully, stem the tide of rising sea levels and species loss.

In 2016, the Police Coast Guard erected a fence surrounding the north-western coast of Singapore to keep illegal immigrants out. This has indirectly contributed to the decline of abandoned nets and traps left by fishermen as the mudflats and mangroves are behind the fence. During the same period, an invasive mussel, *Mytella strigata,* hitched a ride into Singapore and established a strong colony throughout the north-western coastline. It has blanketed the mudflats and encroached onto the horseshoe crabs, impairing movements and latching onto its book gills. The population of the mussels has since declined and stabilized within several parts of Singapore. Late in 2018, The National Parks Board (NParks) announced that Mandai Mangrove and Mudflat will be conserved as a Nature Park.

### *6.3 Mudflats: Horseshoe Crabs and Marine Conservation in Hong Kong*

At Ocean Park Conservation Foundation, Hong Kong (OPCFHK), we use juvenile horseshoe crab that are artificially bred in the laboratory at City University of Hong Kong, and then raised in local secondary schools to educate children about the importance of coastal habitats (see Kwan et al. 2017 for details of this program). The children come out to release their horseshoe crab "babies" and see the degraded mudflats (Fig. 7). The children wanted their "babies" to survive so they had incentive and helped pick up oyster farm debris, garbage, and plastics. The children experienced how they could now make a difference in the condition of horseshoe crab habitat and positively affect their survival. Children teach their parents and share to the public of what they experienced and become ambassadors for preserving coastal habitats to save the horseshoe crabs. More attention has been raised among the community by these children which attracted more stakeholders to join us restoring the habitat in recent years.

## 7 Summary and Conclusions

The four species of extant horseshoe crabs are reliant on an interdependent set of nearshore estuarine and marine habitats, which are under increasing stress from shoreline development and sea level rise. For *Tachypleus tridentatus*, *T. gigas*, and *Limulus polyphemus*, sandy beaches are critical for reproduction and intertidal flats

**Fig. 7** At Ocean Park Conservation Foundation, Hong Kong children released laboratory reared horseshoe crabs

and salt marshes for juvenile nursery habitat. Beach nourishment is preferable to hardening the shoreline, but to have maximum benefit, these shore protection projects should consider the texture of the sediment and not merely the quantity of sand added to a beach. Experiences from Japan show that if revetments are needed, they must be properly placed to preserve the continuity between the land and sea. The fate of *Carcinoscorpius rotundicauda* is tightly coupled to the mangrove ecosystems in Southeast Asia. Efforts to restore mangrove forests may be well-intentioned, but if best practices are not observed, the results may be disappointing. Lastly, we are encouraged by the enthusiastic public participation in the United States, Singapore, Hong Kong, and elsewhere which have embraced the concept of horseshoe crabs as a flagship species for coastal conservation.

**Acknowledgments** This chapter is based on the Knowledge Café roundtable discussion co-sponsored by the IUCN Horseshoe Crab Specialist Group, Mangrove Action Project, and Nature Society (Singapore), which was held on September 3, 2016 in Honolulu, Hawaii, during the IUCN World Conservation Congress. We thank all of the participants at the roundtable for their input. We thank Chang-Po Chen and Hwey-Lian Hsieh for initiating the conversation about using horseshoe crabs as a flagship species for coastal conservation, and D. Wodehouse for his comments on an earlier draft of this manuscript.

## References


E.B. Barbier, S.D. Hacker, C. Kennedy, et al., The value of estuarine and coastal ecosystem services. Ecol. Monogr. **81**, 169–183 (2011)

E.B. Barbier, Marine ecosystem services. Curr. Biol. **27**(11), 507–510 (2017)

M.A. Beekey, J.H. Mattei, The mismanagement of Limulus polyphemus in Long Island Sound, U.S.A.: What are the characteristics of a population in Decline? In R. Carmichael, et al., Changing global perspectives on biology, conservation and management of horseshoe crabs. Springer, New York, NY. pp. 433–461 (2015)

M.L. Botton, The importance of predation by horseshoe crabs, *Limulus polyphemus*, to an intertidal sand flat community. J. Mar. Res. **42**, 139–161 (1984)

M.L. Botton, The conservation of horseshoe crabs: What can we learn from the Japanese experience? in *Limulus in the Limelight*, ed. by J. T. Tanacredi, (Kluwer Academic/Plenum Publ, New York, 2001), pp. 41–51

M.L. Botton, The ecological importance of horseshoe crabs in estuarine and coastal communities: A review and speculative summary, in *Biology and Conservation of Horseshoe Crabs*, ed. by J. T. Tanacredi, M. L. Botton, D. R. Smith, (Springer, New York, 2009), pp. 45–64

M.L. Botton, C.P. Colón, J. Rowden, et al., Effects of a beach nourishment project in Jamaica Bay, New York on horseshoe crab (*Limulus polyphemus*) spawning activity and egg deposition. Estuar. Coasts **41**, 974–987 (2018)

M.L. Botton, R.E. Loveland, Reproductive risk: High mortality associated with spawning by horseshoe crabs (*Limulus polyphemus*) in Delaware Bay, USA. Mar. Biol. **101**, 143–151 (1989)

M.L. Botton, R.E. Loveland, T.R. Jacobsen, Beach erosion and geochemical factors: Influence on spawning success of horseshoe crabs (*Limulus polyphemus*) in Delaware Bay. Mar. Biol. **99**, 325–332 (1988)

M.L. Botton, R.E. Loveland, A. Tiwari, Recruitment and survival of young-of-the-year in a commercially exploited population of horseshoe crabs (*Limulus polyphemus*). Mar. Ecol. Prog. Ser. **265**, 175–184 (2003)

M.L. Botton, C.N. Shuster, Horseshoe crabs in a food web: Who eats whom? in *The American Horseshoe Crab*, ed. by C. N. Shuster, R. B. Barlow, H. J. Brockmann, (Harvard Press, Cambridge, 2003), pp. 133–153

J. Burger, M.A. Howe, D.C. Hahn, J. Chase, Effects of tide cycles on habitat selection and habitat partitioning by migrant shorebirds. Auk **94**, 743–758 (1977)

R.H. Carmichael, D. Rutecki, B. Annett, et al., Position of horseshoe crabs in estuarine food webs: N and C stable isotopic study of foraging ranges and diet composition. J. Exp. Mar. Biol. Ecol. **299**, 231–253 (2004)

R. H. Carmichael, M. L. Botton, P. K. S. Shin, S. G. Cheung (eds.), *Changing Global Perspectives on Horseshoe Crab Biology, Conservation and Management* (Springer, Switzerland, 2015)

L. Cartwright-Taylor, Y.V. Bing, H.C. Chi, L.S. Tee, Distribution and abundance of horseshoe crabs *Tachypleus gigas* and *Carcinoscorpius rotundicauda* around the main island of Singapore. Aquat. Biol. **13**, 127–136 (2011)

C.-P. Chen, H.-L. Hsieh, H.-Y. Chen, et al., The conservation network of horseshoe crab *Tachypleus tridentatus* in Taiwan, in *Biology and Conservation of Horseshoe Crabs*, ed. by J. T. Tanacredi, M. L. Botton, D. R. Smith, (Springer, New York, 2009), pp. 543–557

C.P. Colón, M.L. Botton, P. Funch, et al., Ecology of Juvenile American Horseshoe Crabs (*Limulus polyphemus*) at Plumb Beach, Jamaica Bay, New York. In: (this volume)

S. Crooks, D. Herr, J. Tamelander, et al., Mitigating climate change through restoration and management of coastal wetlands and near-shore marine ecosystems: Challenges and opportunities. Environment department papers; no. 121. Marine ecosystem series. World Bank; Washington, DC (2011)

R.E. Cross, M.C. Curran, Effects of feeding pit formation by rays on an intertidal meiobenthic community. Estuar. Coast. Shelf Sci. **51**, 293–298 (2000)

J. Davis, I.M. Kidd, Identifying major stressors: The essential precursor to restoring cultural ecosystem services in a degraded estuary. Estuar. Coasts **35**, 1007–1017 (2012)
R. Debnath, S.K. Nag, A. Choudhury, et al., Feeding habit and digestive physiology of the Indian horseshoe crab, *Tachypleus gigas* (Muller). Indian J. Physiol. Allied Sci. **43**, 44–49 (1989)
A. del Valle, M. Eriksson, O. Ishizawa, et al., Mangroves protect coastal economic activity from hurricanes. Proc. Natl. Acad. Sci. **117**(1), 265–270 (2020)
C.M. Duarte, W.C. Dennison, R.J.W. Orth, T.J.B. Carruthers, The charisma of coastal ecosystems: Addressing the imbalance. Estuar. Coasts **31**(2), 233–238 (2008)
J.E. Dugan, D.M. Hubbard, I.F. Rodil, et al., Ecological effects of coastal armoring on sandy beaches. Mar. Ecol. **29**(suppl. 1), 160–170 (2008)
Environmental Agency, Government of Japan. Seacoast Survey Report in The 4th National Survey on the Natural Environment (in Japanese) http://www.biodic.go.jp/reports2/4th/kaigan/4_kaigan.pdf (1994)
P.S.E.Z. Ermgassen, M.D. Spalding, B. Blake, et al., Historical ecology with real numbers: Past and present extent and biomass of an imperiled estuarine habitat. Proc. R. Soc. B **279**, 3393–3400 (2012)
L.-F. Fan, C.-P. Chen, M.C. Yang, et al., Ontogenetic changes in dietary carbon sources and trophic position of two co-occurring horseshoe crab species in southwestern China. Aquat. Biol. **26**, 15–26 (2017)
M. Faridah, N. Ismail, B. Ahmad, et al., The population size and movement of coastal horseshoe crab, *Tachypleus gigas* (Muller) on the east coast of Peninsular Malaysia, in *Changing Global Perspectives on Horseshoe Crab Biology, Conservation and Management*, ed. by R. H. Carmichael, M. L. Botton, P. K. S. Shin, S. G. Cheung, (Springer, New York, 2015), pp. 213–228
J. Fonseca, B. Baso, D. Serrano, J.G. Navedo, Effects of tidal cycles on shorebird distribution and foraging behaviour in a coastal tropical wetland: Insights for carrying capacity assessment. Estuar. Coast. Shelf Sci. **198**, 279–287 (2017)
D.A. Friess, Mangrove rehabilitation along urban coastlines: A Singapore case study. Reg. Stud. Mar. Sci. **16**, 279–289 (2017)
R.E. Gill Jr., C.M. Handel, The importance of subarctic intertidal habitats to shorebirds: A study of the central Yukon-Kuskokwim Delta, Alaska. Condor **92**, 709–725 (1990)
C. Gómez, Seagrass Meadows: The Marine Powerhouses. IUCN news releases: https://iucnrle.org/blog/seagrass-meadows-the-marine-powerhouses/ (2018)
R. Gorman, Atlantic horseshoe crabs and endotoxin testing: Perspectives on alternatives, sustainable methods, and the 3Rs (replacement, reduction, and refinement). Front. Mar. Sci. 7, 1–11 (2020)
S. Greipsson, *Restoration Ecology* (Jones & Bartlett Learning, Sudbury, MA, 2011)
P.W. Hicklin, The migration of shorebirds in the Bay of Fundy. Wilson Bull. **99**, 540–570 (1987)
J. Howard, A.E. Sutton-Grier, D. Herr, et al., Clarifying the role of coastal and marine systems in climate mitigation. Front. Ecol. Environ. **15**(1), 42–50 (2017)
M. Hu, B.K.Y. Kwan, Y. Wang, S.G. Cheung, et al., Population structure and growth of juvenile horseshoe crabs *Tachypleus tridentatus* and *Carcinoscorpius rotundi*cauda (Xiphosura) in Southern China, in *Changing Global Perspectives on Horseshoe Crab Biology, Conservation and Management*, ed. by R. H. Carmichael, M. L. Botton, P. K. S. Shin, S. G. Cheung, (Springer, Switzerland, 2015), pp. 167–180
IPCC, Global warming of 1.5°C. intergovernmental panel on climate change, 616 p. https://www.ipcc.ch/sr15/download/#full. Accessed 2 Aug 2020 2018
M. Ishikawa, S. Seino, H. Tomida, Improvement of a horseshoe crab spawning site, in *Proceeding of the Global Congress on ICM: Lessons Learned to Address New Challenges*, vol. 1, (2013), pp. 569–579
S. Itaya, T. Wada, M. Shuuno, Cumulative modifications of the endangered horseshoe crab *Tachypleus tridentatus* habitat by small scale developments at Tsuyazaki Cove in Fukuoka, Japan. In: (this volume)

N.L. Jackson, K.F. Nordstrom, Strategies to conserve and enhance sandy barrier habitat for horseshoe crabs (*Limulus polyphemus*) on developed shorelines in Delaware Bay, United States, in *Biology and Conservation of Horseshoe Crabs*, ed. by J. T. Tanacredi, M. L. Botton, D. R. Smith, (Springer, New York, 2015), pp. 399–416

B.A. John, B.R. Nelson, H.I. Sheikh, et al., A review on fisheries and conservation of Asian horseshoe crabs. Biodivers. Conserv. **27**, 3573–3598 (2018)

G. Kalinkat, J.S. Cabral, W. Darwall, et al., Flagship umbrella species needed for the conservation of overlooked aquatic biodiversity. Conserv. Biol. **31**, 481–485 (2017)

J.A. Keinath, J.A. Musick, R.A. Byles, Aspects of the biology of Virginia's sea turtles: 1979–1986. Va. J. Sci. **38**, 329–336 (1987)

P.G. King, C. Nelsen, J.E. Dugan, et al., Valuing beach ecosystems in an age of retreat. Shore Beach **86**, 1–15 (2018)

C. Kraan, J.A. van Gils, B. Spaans, et al., Landscape-scale experiment demonstrates that Wadden Sea intertidal flats are used to capacity by molluscivore migrant shorebirds. J. Anim. Ecol. **78**, 1259–1268 (2009)

J.N. Kraeuter, S.R. Fegley, Vertical disturbance of sediments by horseshoe crabs (Limulus Polyphemus) during their spawning season. Estuaries **17**, 288–294 (1994)

J. Krisfalusi-Gannon, W. Ali, K. Dellinger, et al., The role of horseshoe crabs in the biomedical industry and recent trends impacting species sustainability. Front Mar Sci **5**, 185 (2018). https://doi.org/10.3389/fmars.2018.00185

B.K.Y. Kwan, S.G. Cheung, P.K.S. Shin, A dual stable isotope study for diet composition of juvenile Chinese horseshoe crab *Tachypleus tridentatus* (Xiphosura) on a seagrass-covered intertidal mudflat. Mar. Biol. **162**, 1137–1143 (2015)

B.K.Y. Kwan, J.H.Y. Cheung, A.C.K. Law, et al., Conservation education program for threatened Asian horseshoe crabs: A step towards reducing community apathy to environmental conservation. J. Nat. Conserv. **35**, 53–65 (2017)

B.K.Y. Kwan, H.-L. Hsieh, S.G. Cheung, P.K.S. Shin, Present population and habitat status of potentially threatened Asian horseshoe crabs *Tachypleus tridentatus* and *Carcinoscorpius rotundicauda* in Hong Kong: A proposal for marine protected areas. Biodivers. Conserv. **25**, 673–692 (2016)

K. Laurie, C.-P. Chen, S.G. Cheung, et al., *Tachypleus tridentatus* (errata version published in 2019). The IUCN Red List of Threatened Species 2019: e.T21309A149768986. https://doi.org/10.2305/IUCN.UK.2019-1.RLTS.T21309A149768986.en (2019)

R.R. Lewis, Methods and criteria for successful mangrove forest restoration, in *Coastal Wetlands: An Integrated Ecosystem Approach*, ed. by G. M. E. Perillo, E. Wolanski, D. R. Cahoon, M. M. Brinson, (Elsevier, Oxford, UK, 2009), pp. 787–800

P.M. Lourenço, T. Catry, T. Piersma, J.P. Granadeiro, Comparative feeding ecology of shorebirds wintering at Banc d'Arguin, Mauritania. Estuar. Coasts **39**, 855–865 (2015)

R.E. Loveland, M.L. Botton, Sea level rise in Delaware Bay: Adaptations of spawning horseshoe crabs (*Limulus polyphemus*) to the glacial past, and the rapidly changing shoreline of the bay, in *Changing Global Perspectives on Horseshoe Crab Biology, Conservation and Management*, ed. by R. H. Carmichael, M. L. Botton, P. K. S. Shin, S. G. Cheung, (Springer, Switzerland, 2015), pp. 41–63

K. Maeda, S. Seino, S. Nishihara, A. Hino, Ecology of hatched larvae of the horseshoe Crab *Tachypleus tridentatus* (Leach) in relation to the physical environment. Benthos Res. **55**, 15–24 (2000)

J.H. Mattei, M. Botton, M. Beekey, C. Colón, Horseshoe crab research in urban estuaries: Challenges and opportunities, in *Changing Global Perspectives on Horseshoe Crab Biology, Conservation and Management*, ed. by R. Carmichael, M. L. Botton, S. G. Cheung, (Springer, Switzerland, 2015), pp. 537–555

J.H. Mattei, Audubon CT In-Lieu Fee Fund Final Report. Regulatory Division US Army Corps of Engineers, 696 Virginia Road, Concord, MA (2019)

J.H. Mattei, J.M. Kasinak, S. Senbel, K. Bartholomew, The power of citizen science: 20 years of horseshoe crab community research merging conservation, education and management. In: (This Volume)

C.J. Mcowen, L.V. Weatherdon, J.W. Van Bochove, et al., A global map of saltmarshes. Biodivers Data J. (5), e11764 (2017)

S. Miththapala, Tidal Flats. Coastal Ecosystem Series (Volume 5). Columbo, Sri Lanka, IUCN. iii + 48 pp (2013)

L.J. Niles, J. Bart, H.P. Sitters, et al., Effects of horseshoe crab harvest in Delaware Bay on Red Knots: Are harvest restrictions working? Bioscience **59**, 153–164 (2009)

J. Odell, M.E. Mather, R.M. Muth, A biosocial approach for analyzing environmental conflicts: A case study of horseshoe crab allocation. Bioscience **55**, 735–748 (2005)

J.H. Primavera, J.P. Savaris, B.E. Bajoyo, et al., Manual for Community-Based Mangrove Rehabilitation, 1st ed, London, Zoological Society of London, Mangrove Manual Series. ZSL, London, UK (2012)

D.R. Richards, D.A. Friess, Rates and drivers of mangrove deforestation in Southeast Asia, 2000–2012. Proc. Natl. Acad. Sci. **113**, 344–349 (2016)

D.M. Rudkin, G.A. Young, Horseshoe crabs – An ancient ancestry revealed, in *Biology and Conservation of Horseshoe Crabs*, ed. by J. T. Tanacredi, M. L. Botton, D. R. Smith, (Springer, New York, 2009), pp. 25–44

S. Seino, Flows related to morphology and ecology of horseshoe crab. Nagare **20**(5), 365–374 (2001). The Japanese Society of Fluid Mechanics (in Japanese) Nagare. https://archive.iii.kyushuu.ac.jp/public/kRhQAAYIakAwHMBhMFzLinrUjm6wKW1l1zyYysTEUFu

S. Seino, T. Uda, Y. Tsuchiya, et al., Field observation of geomorphological features of the spawning site and dispersion of hatchlings of the horseshoe crab *Tachypleus tridentatus*-towards mitigation planning for the rare species- Ecol. Civil. Eng. **3**(1), 7–19 (2000). https://archive.iii.kyushu-u.ac.jp/public/pR88AAvILs-AclgBMVNzrgTrh9C6z1q1crXwGZDTs4vi

S. Seino, T. Uda, Y. Tsuchiya, K. Tsuchiya, Conservation history of horseshoe crab *Tachypleus tridentatus* and its spawning ground, a designated natural monument in Kasaoka Bay in Okayama Prefecture. Asian Pacific Coasts **18**, 551–556 (2003)

S. Seino, T. Uda, S. Sato, et al., Threatened species habitat conservation methodology based on ecological engineering in nature-oriented coastal affairs. Annual J. Coastal Eng. JSCE **51**(2), 1341–1345. (in Japanese) DOI (2004). https://doi.org/10.2208/proce1989.51.1341

S. Seino, T. Uda, Conservation planning for endangered species of horseshoe crab in terms of geomorphology of habitats and life history. Proc. 4th Int. Conf Asian Pacific Coasts **2007**, 2075–2086 (2007). https://archive.iii.kyushu-u.ac.jp/public/CR18AANIQo-AuEABkgxzJhPrx2-6cme1NuzqB9FIK642

E.E. Seney, J.A. Musick, Historical diet analysis of loggerhead sea turtles (*Caretta caretta*) in Virginia. Copeia **2007**(2), 478–489 (2007)

M.R. Shokri, W. Gladstone, J. Jelbart, The effectiveness of seahorses and pipefish (Pisces: Syngnathidae) as a flagship group to evaluate the conservation value of estuarine seagrass beds. Aquat Conserv **19**, 588–595 (2009)

F. Short, T. Carruthers, W. Dennison, M. Waycott, Global seagrass distribution and diversity: A bioregional model. J. Exp. Mar. Biol. Ecol. **350**, 3–20 (2007)

A.D. Short, Coastal processes and beaches. Nat. Educ. Knowled. **3**(10), 15 (2012)

C. N. Shuster, R. B. Barlow, H. J. Brockmann (eds.), *The American Horseshoe Crab* (Harvard Press, Cambridge, 2005)

D.R. Smith, H.J. Brockmann, M.A. Beekey, et al., Conservation status of the American horseshoe crab, (*Limulus polyphemus*): A regional assessment. Rev. Fish Biol. Fish. **27**, 135–175 (2016)

A. Strong, A. Minnemeyer, Satellite data reveals state of world's mangrove forests. World Resources Institute. https://www.wri.org/blog/2015/02/satellite-data-reveals-state-world-s-mangrove-forests (2015)

J. T. Tanacredi, M. L. Botton, D. R. Smith (eds.), *Biology and Conservation of Horseshoe Crabs* (Springer, New York, 2009)

R. Thom, Adaptive management of coastal ecosystem restoration projects. Ecological Engineering, **15**, 365–372 (2000)

M.T.T. Thuy, T. Nagasawa, H. Tanaka, N.T. Viet, Sandy beach restoration using beach nourishment method: A case study of Nha Trang Beach, Vietnam, in *Tropical Coastal and Estuarine Dynamics. Journal of Coastal Research*, Special Issue No. 81, ed. by R. Almar, L. P. Almeida, N. Trung Viet, M. Sall, (2018), pp. 57–66

B.S. Thompson, S.M. Rog, Beyond ecosystem services: Using charismatic megafauna as flagship species for mangrove forest conservation. Environ Sci Policy **102**, 9–17 (2019)

R. Tiner, Tidal Wetland Conservation and Management, in *Tidal Wetlands Primer: An Introduction to Their Ecology, Natural History, Status, and Conservation*, (University of Massachusetts Press, Amherst, MA, 2013)

S. Vestbo, M. Obst, F.J.Q. Fernandez, et al., Present and potential future distributions of Asian horseshoe crabs determine areas for conservation. Front. Mar. Sci. **5**(164), 1–16 (2018)

E.A. Walls, J. Berkson, S.A. Smith, The horseshoe crab, *Limulus polyphemus*: 200 million years of existence, 100 years of study. Rev. Fish. Sci. **10**, 39–73 (2002)

D.C.J. Wodehouse, Towards successful community mangrove management and rehabilitation. Thesis for the Degree of Doctor of Philosophy, MSc, B. Comm. (Hons.), Tech Cert. School of Natural Sciences, Bangor University College of Environmental Sciences and Engineering, Bangor, Gwynedd. LL57 2UW (2020)

M.A. Zacharias, J.C. Roff, Use of focal species in marine conservation and management: A review and critique. Aquat. Conserv. **11**, 59–76 (2001)

R.N. Zajac, R.B. Whitlatch, Community and population-level responses to disturbance in a sandflat community. J. Exp. Mar. Biol. Ecol. **294**, 101–125 (2003)

J.B. Zedler, What's new in adaptive management and restoration of coasts and estuaries? Estuar. Coasts **40**, 1–21 (2017)

# Preliminary Characterization of Hemocyte and Immunity of Asian Horseshoe Crabs, *Tachypleus tridentatus*, and *Carcinoscorpius rotundicauda* in Captivity

**Xiaoyong Xie, Kit Yue Kwan, Jinxiang Zhong, Mujiao Xie, Guoling Ye, and Yuyuan Bao**

## 1 Introduction

Horseshoe crabs belong to the Phylum Arthropoda, Class Merostomata, Order Xiphosura, and Family Limulidae. There are only four extant species of horseshoe crab: the Atlantic horseshoe crab *Limulus polyphemus* found on the eastern coast of North America and Gulf of Mexico; the tri-spine horseshoe crab *Tachypleus tridentatus*, the coastal horseshoe crab *T. gigas*, and the mangrove horseshoe crab *Carcinoscorpius rotundicauda* occur in the Indo-Pacific region, ranging from southern Japan to the Bay of Bengal (Smith et al. 2017; John et al. 2018). Horseshoe crabs are among the few animals to be characterized as "living fossils," as their

X. Xie · M. Xie · G. Ye
Key Laboratory of South China Sea Fishery Resources Exploitation and Utilization, Ministry of Agriculture, Guangdong Provincial Key Laboratory of Fishery Ecology and Environment, South China Sea Fisheries Research Institute, Chinese Academy of Fishery, Guangzhou, China

K. Y. Kwan (✉)
Guangxi Key Laboratory of Beibu Gulf Marine Biodiversity Conservation, College of Marine Sciences, Beibu Gulf University, Qinzhou, China
e-mail: kityuekwan@bbgu.edu.cn

J. Zhong
Guangdong Marine Development and Planning Research Centre, Guangzhou, China

Y. Bao
Key Laboratory of South China Sea Fishery Resources Exploitation and Utilization, Ministry of Agriculture, Guangdong Provincial Key Laboratory of Fishery Ecology and Environment, South China Sea Fisheries Research Institute, Chinese Academy of Fishery Sciences, Guangzhou, China

Guangdong Marine Development and Planning Research Centre, Guangzhou, China



J. T. Tanacredi et al. (eds.), *International Horseshoe Crab Conservation and Research Efforts: 2007-2020*, https://doi.org/10.1007/978-3-030-82315-3_11

oldest fossils can be dated to the Ordovician period, 445 million years ago (Rudkin and Young 2009). Horseshoe crabs are important benthic predators in coastal and estuarine ecosystems (Botton 2009), and as the potential indicator species (Chen et al. 2015) or sentinel species (Kwan et al. 2018) to reflect the general health of the ecosystems.

The biomedical use of horseshoe crabs was discovered by Bang (1956), and further discussed in Levin and Bang (1964), which described the hemocytes present in horseshoe crab hemolymph and their endotoxin-induced coagulation by releasing an enzyme coagulation cascade. This discovery has led to the development of *Limulus* amoebocyte lysate (LAL), a widely used reagent in the detection of bacterial endotoxin in medical diagnosis and in drug assays. Horseshoe crab hemolymph was then demonstrated to contain a variety of active substances such as horseshoe crab factors, antibacterial and antiviral compounds that have great potential for medical development (Iwanaga 2002). The use of LAL reagent as an official test for pyrogen was approved by the Food and Drug Administration, USA, in 1987, and led to the substantial development in LAL industry to meet the growing market demands. The number of *L. polyphemus* individuals harvested for LAL production increased from approximately 32,000 individuals in 2005 to recently circa 60,000 individuals per annum, with an estimated biomedical mortality at 15% (Gauvry 2015). The Asian pharmaceutical market is growing rapidly, and in China, *Tachypleus tridentatus* individuals used for TAL production was estimated to reach 1.5 million per annum to fulfill the demand for testing more than 300 injectable pharmaceuticals listed in Chinese Pharmacopoeia (Li et al. 2018; Yan et al. 2018). Due to the different bleeding practices in China and possibly other Asian places, *T. tridentatus* bleeding can cause 100% mortality, and the individuals may be then diverted to the food market. Consequently, overfishing for bleeding has been identified as one of the main attributes to the substantial decline of Asian horseshoe crab populations (John et al. 2018; Liao et al. 2019; Zhu et al. 2020). *Tachypleus tridentatus* is listed as Endangered in 2019 on the IUCN Red List (Laurie et al. 2019).

The decreasing *T. tridentatus* population number along the Chinese coast has resulted in the refusal of TAL firms to return the bled horseshoe crabs to sea under the absence of local regulatory policy. The pond culture practice has been proposed as the alternative to facilitate sustainable bleeding of Asian horseshoe crabs, in which the wild-caught subadults and adults are kept in outdoor ponds in coastal area for generally more than a year. The individuals will be transported to bleed in nearby facilities and returned to the ponds repeatedly within a fixed time interval (Xie X. 2018, pers. comm.). Concerns are raised about whether hemolymph quality of captive horseshoe crabs can be maintained under prolonged culture, in which the issue is primary to TAL production and quality assurance. Previous studies have demonstrated that plasma protein level, hemocyte viability, and their morphological state can be sensitive to environmental conditions, such as the presence of captivity stress, heavy metals, and excessive nutrients (Kwan et al. 2014, 2015, 2018). James-Pirri et al. (2012) compared hemolymph constituents of wild-caught, captive control, and captive bled adult *L. polyphemus* under current or best management practices. The study found that captive bled horseshoe crabs had lower plasma protein concentration compared to other treatment groups, while the wild individuals had higher concentrations in hemolymph glucose, potassium, and creatinine. However, little is known about the effect of prolonged captivity or bleeding on hemolymph quality of Asian species.

In this study, hemocyte morphology and immune-related enzymatic activities in both serum and digestive tracts of adult *T. tridentatus* and *C. rotundicauda* under prolonged pond culture was examined. Alkaline phosphatase (AKP) and glutathione peroxidase (GSH-Px) activities were selected to reflect the general health status of captive horseshoe crabs. AKP plays important roles in nutrient metabolite transport and ion secretion (Lallès 2010; Cheng et al. 2017), which is sensitive to changes in nutritional conditions. GSH-Px activity indicates the organism's antioxidant capacity of which inflammation, hormone level, and other developmental differentiation can affect the enzymatic activity (Sugino et al. 1998; Rogers et al. 2000; Wei and Lee 2002). This preliminary study attempts to provide baseline information on hemocyte morphology and immunity of Asian horseshoe crabs kept in prolonged captivity for bleeding. *Carcinoscorpius rotundicauda* has not been utilized to serve current bleeding practices due to its smaller body size (prosomal width, PW: 11–13 cm, Fauziyah et al. 2019), but the species is still vulnerable to bleeding and exploitation when other Asian horseshoe crab populations are diminishing. The findings may be useful in developing a health monitoring system in the sustainable bleeding process and other relevant immunological studies.

## 2 Materials and Methods

### 2.1 Collection and Culture of Horseshoe Crabs

The pond culture of horseshoe crabs was initiated on November 5, 2016 on the western part of Yangjiang, a prefecture-level city in Guangdong province, China. A total of 1062 individuals of adult *T. tridentatus* (average PW: 23.5 cm; wet weight, WW: 311 g) and *C. rotundicauda* (average PW: 12.1 cm; WW: 213 g) were kept in a circa 1300 $m^2$ outdoor pond with 2.5 m depth. Individuals from both species were obtained from coastal waters in Zhanjiang, Guangdong province. The environmental conditions were maintained as follows: water temperature 16.5–31.6°, salinity 16–30 ppt, and pH 7.2–8.4. The salinity level is higher in winter compared to summer. At the bottom of the pond, a 0.3 m sediment layer (25–55% sand particles) was provided to enhance the survival of horseshoe crabs (Hong et al. 2009). The sediment grain size is relatively larger along the edges of the pond, whereas the middle of the pond contains finer particles. A waterwheel aerator was set up to increase dissolved oxygen and lower ammonia-nitrogen levels in the pond. Half of the total volume of pond water was renewed weekly during the rising/falling tides. The horseshoe crabs were fed with frozen mussel meat and small fish twice a week.

Two individuals of *T. tridentatus* and three individuals of *C. rotundicauda* were collected from the pond after one year of culture on November 29–30, 2017.

The horseshoe crabs were transported back to the laboratory and held in a 1000-L aquarium tank for three days before the bleeding to simulate the standard bleeding procedure in TAL-producing facilities. The water temperature was kept at 29–30 °C, salinity 16–30 ppt and dissolved oxygen >6 mg/L. The sediment (25–55% sand) at the bottom of the tank was disinfected by ultraviolet-C light before the experiment. The horseshoe crabs were provided clam meat daily at 16:00, and one-third of the water volume was renewed prior to the feeding.

## 2.2 *Hemolymph Sampling and Preparation*

Needles and syringes were pre-disinfected by ultraviolet-C light during the needle/syringe production, tweezers were autoclaved, and the coverslips for the blood smear were cleaned and disinfected using 75% ethanol. Approximately 5 mL of hemolymph was extracted from each individual using a 22-gauge needle from the hinge between prosoma and opisthosoma of horseshoe crabs. The hinge part was wiped with 75% ethanol followed by distilled water before and after the bleeding. The sampled hemolymph was stored on ice to reduce spontaneous cellular aggregation. Half portion of the hemolymph was used for blood smear preparation, whereas another half was centrifuged at 4 °C and 500 rpm for 3 min. The supernatant after centrifugation was used for enzymatic activity measurements. At the same time, the digestive tract was dissected and separated into esophagus, stomach, pylorus, midgut, and hindgut. The enzymatic activity of these tissues was examined immediately after the collection.

To prepare a blood smear, in the sterile room, a drop (circa 50 μL) of the collected hemolymph was placed on a slide, and a thin glass cover was added on top to form a thin hemolymph film. The blood smear was allowed to be air-dried before the staining. The Wright–Giemsa Stain Solution (Servicebio Co., Ltd., Wuhan, China) was applied following the experimental procedures described in the product protocol (http://www.servicebio.cn/html/all/cp/bl/rs/1757.html). The Wright–Giemsa stain is widely used in hematological studies to facilitate the differentiation of cell types, which is essential for accurate interpretation of their morphological characteristics and detection of abnormal cellular components for disease diagnosis (Dunning and Safo 2011). The hemolymph samples were allowed to stain for 5 min, washed with distilled water, air-dried, sealed with glycerin, and photographed using an optical microscope (AxioScope A1, ZEISS, Jena, Germany).

## 2.3 *Determination of Enzymatic Activity*

The AKP and GSH-Px activities in hemolymph and different tissues of digestive tract (i.e., esophagus, stomach, pylorus, midgut, and hindgut) of the two horseshoe crab species was determined using the AKP and GSH-Px assay kits developed from the Nanjing Jiancheng Bioengineering Institute, Wuhan, China (http://www.njjcbio.com/). The AKP activity was determined using the phenyldiphenyl phosphate method, in which a unit of enzymatic activity was defined as the amount required for either 100 ml of serum (U/100 mL) or 1 gram of tissue protein (U/g protein) to produce 1 mg of phenol following application to the substrate at 37 °C for 15 min. Double-distilled water and phenol was used as the blank and standard, respectively, for the determination. For GSH-Px activity, it was quantified based on the colorimetric method that a unit of enzymatic activity was defined as the amount required to decrease the concentration of reduced glutathione in the reaction system by 1 μmol/L at 37 °C after 5-min reaction per 0.1 ml of serum (U/0.1 ml) or per milligram of protein (U/mg protein), and after the effect of non-enzymatic reaction was subtracted. Blank and standard solutions were provided together with the testing kit. The protein content from each assay group was determined using the Coomassie Brilliant Blue method (Bradford 1976).

### 2.4 *Statistical Analysis*

Data were first checked for normality and homogeneity of variance using Shapiro–Wilk and Levene's tests, respectively. The differences in serum AKP and GSH-Px levels between species were examined by unpaired sample t-test, whereas their AKP and GSH-Px activity level differences among different digestive tract tissues and between species were compared using two-way analysis of variance (ANOVA). All the above analyses were performed using SPSS software version 22.0 (IBM, New York, USA). A significance level of $p < 0.05$ was considered in all statistical procedures.

## 3 Results

### 3.1 *Morphological Characteristics of Hemocytes*

Granulocytes, non-granulocytes, and macrophages were observed in hemolymph from both *T. tridentatus* and *C. rotundicauda* (Fig. 1). The majority of hemocytes were round or oval in shape. Granulocytes accounted for the vast majority of hemocytes, in which the number of basophil granulocytes (i.e., basophils) was considerably higher than eosinophil granulocytes (i.e., eosinophils). It is noted that the cytoplasmic granules of basophils showed different levels of stain, including darker blue and lighter blue stain. The darker-blue basophils contained a central nucleus and higher number of granules in the cell, while the lighter-blue basophils consisted of a less evident nucleus and fewer granules. Light-blue basophiles were the major cell types among the granulocytes. A similar situation was observed in eosinophils,

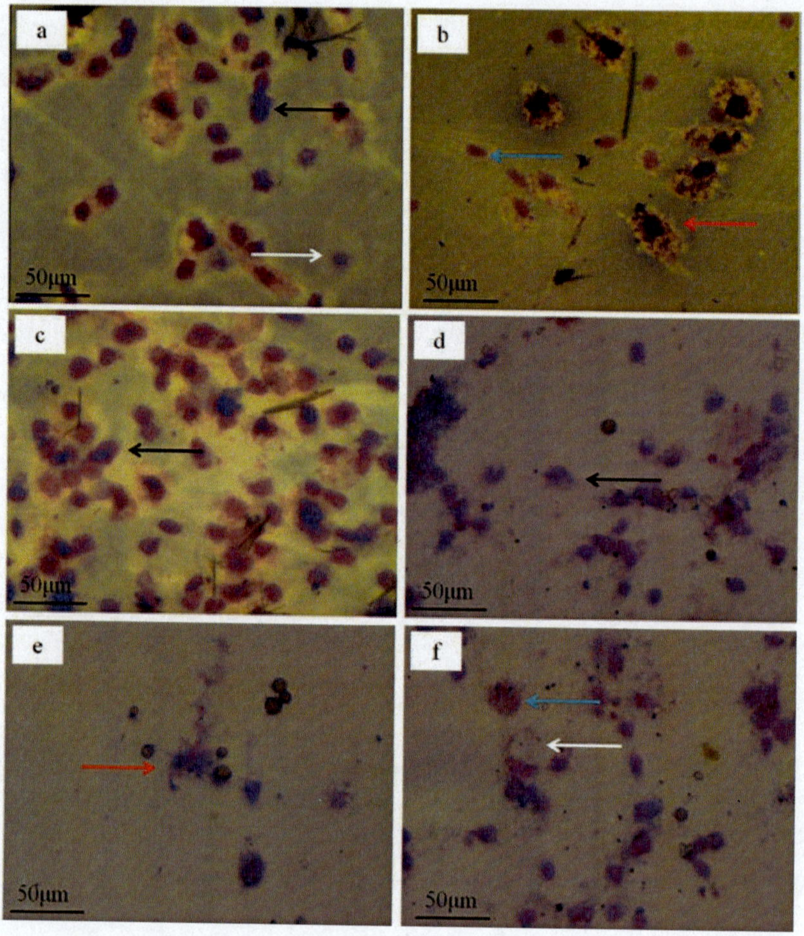


**Fig. 1** The morphology of hemocytes from *T. tridentatus* (**a–c**) and *C. rotundicauda* (**d–f**) under a light microscope. Three types of hemocytes were indicated by arrows in different colors: granulocytes (basophils: black, eosinophils: blue), non-granulocytes (white) and macrophages (red). Basophiles were stained in purple–red or light-red, whereas basophils were stained in dark- or light-blue

which included purple–red cells with evident intracellular granules and less evident nucleus, and light-red cells with a lower intracellular granule number and the nucleus located on one edge of the cytoplasm.

The number of non-granulocytes and macrophages were very few, and mainly surrounded by the basophiles (Fig. 1). The non-granulocyte had similar size to granulocyte with a central, clearly defined nucleus. The macrophages were larger in size with the nucleus located at the upper middle portion of the cell. A mixture of blue and red-stained intracellular granules was clearly observed within the macrophages (Fig. 1). No apparent inter-specific differences in staining or morphology of the hemocyte types were observed.

## 3.2 *Enzymatic Activity Determination*

The horseshoe crab AKP activity level in serum ranged 0.14–0.15 U/mL (Fig. 2), and that in different digestive tract, tissues ranged 10.89–19.37 U/g protein (Fig. 3). The AKP level was varied among individuals and tissues. The highest mean AKP activity level was recorded in the *T. tridentatus* hindgut tissue (16.13 U/g protein), whereas the lowest mean was observed in the *C. rotundicauda* midgut tissue (12.51 U/g protein). While there was no significant difference in AKP level among different digestive tract tissues and between species (species: $F = 0.411, p = 0.910$; tissue: $F = -0.089, p = 0.985$; species × tissue: $F = 0.809, p = 0.539$), a significantly higher in serum AKP level was observed in *T. tridentatus* compared to that of *C. rotundicauda* ($t = 4.871, p = 0.017$).

For the GSH-Px activity, the serum level was in the range of 29.01–36.88 U/mL (Fig. 2), whereas the digestive tract level was in the range of 22.21–49.13 U/mg protein (Fig. 4). Similar to the AKP activity, the highest and lowest mean GSH-Px level was found in the hindgut (41.40 U/mg protein) and midgut tissues (27.31 U/

**Fig. 2** Mean activity levels (± standard deviation) of alkaline phosphatase (AKP) and glutathione peroxidase (GSH-Px) in serum of *T. tridentatus* and *C. rotundicauda*. Significant differences in serum AKP and GSH-Px activity levels between species are represented by different lowercase letters

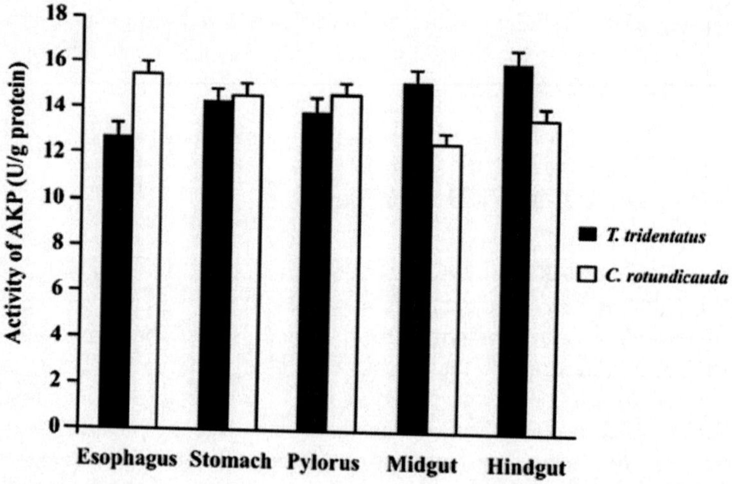


**Fig. 3** Activity levels of alkaline phosphatase (AKP, mean ± standard deviation) in different parts of digestive tract from *T. tridentatus* and *C. rotundicauda*. No significant differences in AKP levels were detected between the two horseshoe crab species and among different digestive tract tissues

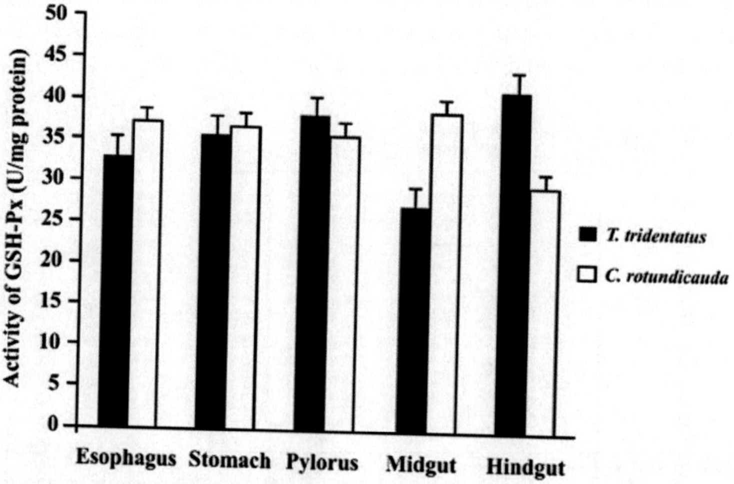


**Fig. 4** Activity levels of glutathione peroxidase (GSH-Px, mean ± standard deviation) in different parts of digestive tract from *T. tridentatus* and *C. rotundicauda*. No significant differences in GSH-Px levels were detected between the two horseshoe crab species and among different digestive tract tissues

mg protein) of *T. tridentatus*, respectively. The AKP level difference in serum between species was insignificant ($t = -1.500$, $p = 0.231$). No significant differences in GSH-Px activity levels were observed between species and among digestive tract tissues (species: $F = 0.035$, $p = 0.854$; tissue: $F = 0.188$, $p = 0.941$; species × tissue: $F = 1.527$, $p = 0.245$).

# 4 Discussion

## *4.1 Classification of Horseshoe Crab Hemocytes*

Owing to the very limited studies on their circulatory system and cellular components, at present, there is no standardized classification scheme for horseshoe crab hemocytes. Armstrong (1979) examined the hemocytes in a freshly excised gill leaflet and suggests that, *L. polyphemus* hemolymph contained solely a single cell type, the granular amebocyte (also named as granulocyte) that showing three morphological states in vitro, including the contracted, flattened granular and flattened degranulated forms. Suhr-Jessen et al. (1989) found that, in addition to granulocytes, there was another distinct cell, the plasmatocyte, which comprised 1–3% of the hemocytes in *L. polyphemus* hemolymph. Compared to granulocytes, the plasmatocytes had an euchromatic nucleus, flattened and well-developed rough endoplasmic reticulum, a greater number of free ribosomes and mitochondria, but fewer, if any, large secretory granules in the cell (Suhr-Jessen et al. 1989). Coates et al. (2012) followed the hemocyte classification described in Armstrong (1979), and modified the morphological classification into granular-spherical, granular-flat, and dendritic-like forms.

For Asian species, Toh et al. (1991) examined the hemocytes in freshly prepared hemolymph from *T. tridentatus* using an electron microscopy and classified the hemocytes into two types, including granular and non-granular hemocytes, according to the number of granules present in the cells. The population density ratio of granular to non-granular hemocytes was 100: 1, in which the ratio was consistent to that reported in Suhr-Jessen et al. (1989). Toh et al. (1991) identified two types of granules in the granulocytes, L-granules and D-granules, with differences in their maximal diameter, cross-sectioned area, identified contents, electron density, and matrix characteristics. Kwan et al. (2014, 2015, 2018) followed the granulocyte classification scheme by Coates et al. (2012) and ignored the very low occurrence of non-granulocytes present in juvenile *T. tridentatus* hemolymph. Wu et al. (2015) classified the hemocytes in both *T. tridentatus* and *C. rotundicauda* species using light-scanning electron microscopy and particle size analyzer based on their cell size, morphology, cytoplasmic granule size and quantity, and nuclear-cytoplasmic ratio. Three types of hemocytes were identified, the significantly larger-sized granular cells with abundant granules, the semi-granular cells with lesser granules, and the significantly smaller-sized hyaline cells containing few granules but a significantly greater nuclear-cytoplasmic ratio. Their morphology was indistinguishable between species and sexes, but a significantly higher hemocyte count in females of both species. The granular cell was the major cell type in both sexes of the two species. Wu et al. (2019) re-examined the results using a transmission electron microscopy and found the consistently three cell types as described in Wu et al. (2015). However, the quantification of hemocytes by flow cytometry demonstrated that the hyaline cell was dominant in numbers that might be caused by the handling process.

In this study, the hemocytes of *T. tridentatus* and *C. rotundicauda* in hemolymph were stained by the Wright–Giemsa method and examined under a light microscopy. Consistent with most relevant reports in horseshoe crabs, the granulocyte was the major cell type for captive horseshoe crabs, in which their intracellular granules were in two different staining colors, the basophil granulocyte in blue and the eosinophil granulocyte in purple–red. Similar to the previous studies, we differentiated the granulocyte from non-granulocyte according to the number of cytoplasmic granules present in the hemocytes. The third identified hemocyte type was macrophage with a mixture of blue and red-stained intracellular granules. While the classification scheme and naming of horseshoe crab hemocytes varied among studies, the three cell types mentioned in Wu et al. (2015, 2019), according to our justifications, can be considered as the granulocytes (or amebocytes) under different morphological states (Coates et al. 2012; Kwan et al. 2014). It is also important to note that the hemocyte morphology can be heavily affected by the degranulation process. The degranulation in horseshoe crab hemocytes can be induced from increased temperature (Coates et al. 2012), captivity stress (Kwan et al. 2014), deteriorated water quality (Kwan et al. 2015, 2018), and anticoagulant formulation for bleeding (Sheikh et al. 2021).

Hemocytes play a primary role in the immune defenses of varying aquatic invertebrates (Buchmann 2014; Huang and Ren 2020). A spherical-granular state appears to be the optimum, most viable cell morphology in horseshoe crabs (Armstrong 1979; Hurton et al. 2005). Wu et al. (2019) also found that granular cells contained higher reactive oxygen species, phagocytosis, and non-specific esterase. Therefore, the total and differential hemocyte counts in horseshoe crabs are fundamental in assessing their physiological state and health status, which in turn, affects the resource sustainability for bleeding and the TAL/LAL production quality. Despite the fact that the total number of different hemocyte types was not quantified in this study, it was apparent that only a very low proportion of non-granulocytes and macrophages was observed surrounded by the basophil granulocytes. Based on the preliminary morphological observations of hemocytes in this study, it seems that there was no observable difference in our captive adult horseshoe crabs compared to that reported in wild-caught horseshoe crabs in the literature. The antimicrobial and clotting substances available in basophils and eosinophils of horseshoe crabs, as well as their implication to health status indication require further investigations.

## 4.2 *Characterization of Enzymatic Activity in Different Tissues*

The AKP and GSH-Px activities have been widely utilized to reflect non-specific immunological performance in aquatic invertebrates (Javahery et al. 2019; Qi et al. 2019; Amoah et al. 2020; Xu et al. 2020). AKP is an intrinsic plasma membrane enzyme available in the cell membranes of organisms (except some plants), which plays a key role in the degradation and transport of exogenous substances, such as proteins, carbohydrates, and lipids (Blasco et al. 1993). For marine invertebrates, AKP involves in the absorption of phosphorus and calcium in seawater, and also the formation of calcium phosphate and chitin (Xu et al. 2020). Dietary supplementations with immunostimulants and essential dietary elements (Javahery et al. 2019; Qi et al. 2019) as well as changes in environmental conditions (Pinoni et al. 2005; Li et al. 2008) were demonstrated to affect the AKP activities in aquatic invertebrates. For example, arginine was demonstrated to increase the AKP activities in juvenile Chinese mitten crab, *Eriocheir sinensis*. Similar observation was reported in Xu et al. (2020) that the addition of copper supplement to *T. tridentatus* diets for two weeks had significantly increased their plasma AKP activity level. Compared with the present findings, plasma AKP activity level in *T. tridentatus* under prolonged captivity (one year) was approximately four times higher than that reported in Xu et al. (2020), which is maintained in laboratory for totally two months without copper supplementation. Significantly higher plasma AKP level was also noted in *T. tridentatus* compared to *C. rotundicauda* in this study, which may be attributed to the considerable differences in their body size. However, the AKP levels in different digestive tissues were statistically similar in both Asian species.

The GSH-Px is a primary component in animal's detoxification system by catalyzing hydrogen peroxide into water and inhibiting active oxygen and hydroxyl free radical productions (Moreira et al. 2016; Sui et al. 2017). Therefore, the GSH-Px activity can be regarded as an important indicator to reflect an organism's antioxidant capacity. A variety of factors have been showed to influence the GSH-Px activity such as inflammation (Jones et al. 1997; Rogers et al. 2000), developmental differentiation Hayashibe et al. 1990; Bravard et al. 1999), and hormonal regulation (Dougall and Nick 1991; Sampath and Perez-Polo 1997; Sugino et al. 1998). Similar to the AKP level, plasma GSH-Px activity levels in *T. tridentatus* in this study were about five times greater than that described in Xu et al. (2020) under laboratory culture for two months. The difference may be resulted from culture conditions (Capparelli et al. 2019; Schvezov et al. 2019) and diets (Wang et al. 2019; Yang et al. 2019; Liu et al. 2020). For digestive tract tissues, in this study, the GSH-Px activity levels maintained in the range of 27–41 U/mg protein regardless of species and tissue types. The GSH-Px activity levels in horseshoe crabs, in general, were higher than that of intestine tissues from other smaller-sized aquatic invertebrates, such as whiteleg shrimp, *Litopenaeus vannamei* (7–8 U/mg protein, Duan et al. 2018), and juvenile Chinese mitten crab, *Eriocheir sinensis* (8–15 U/mg protein, Lu et al. 2019). While the horseshoe crab baseline data regarding the non-specific immunological parameters are very limited, the present findings may be useful in further immunological studies for developing health monitoring system and improving sustainable bleeding practices.

**Acknowledgements** This work was supported by "Special Fund for Basic Scientific Research of Central Public Research Institutes" of the South China Sea Fisheries Research Institute, Chinese Academy of Fishery (2019TS21), Guangxi BaGui Youth Scholars Programme, and Guangxi Recruitment Program of 100 Global Experts.

## References


K. Amoah, Q.C. Huang, X.H. Dong, B.P. Tan, S. Zhang, S.Y. Chi, Q.H. Yang, H.Y. Liu, Y.Z. Yang, *Paenibacillus polymyxa* improves the growth, immune and antioxidant activity, intestinal health, and disease resistance in *Litopenaeus vannamei* challenged with Vibrio parahaemolyticus. Aquaculture **518**, 734563 (2020)

P. B. Armstrong, Motility of the Limulus blood cell. J. Cell Sci. **37**(1), 169–180 (1979)

F.C. Bang, A bacterial disease of *Limulus polyphemus*. Bull. Johns Hopkins Hosp. **98**, 325–351 (1956)

J. Blasco, J. Puppo, M.C. Sarasquete, Acid and alkaline phosphatase activities in the clam *Ruditapes philippinarum*. Mar. Biol. **115**(1), 113–118 (1993)

M.L. Botton, The ecological importance of horseshoe crabs in estuarine and coastal communities: A review and speculative summary, in *Biology and Conservation of Horseshoe Crabs*, ed. by J. T. Tanacredi, M. L. Botton, D. R. Smith, (Springer, New York, 2009), pp. 45–63.

M.M. Bradford, Rapid and sensitive method for quantification of microgram quantities of protein utilizing principle of protein-dye binding. Anal. Biochem. **72**, 248–254 (1976)

A. Bravard, F. Petridis, C. Luccioni, Modulation of antioxidant enzymes p21WAF1 and p53 expression during proliferation and differentiation of human melanoma cell lines. Free Radic. Biol. Med. **26**, 1027–1033 (1999)

K. Buchmann, Evolution of innate immunity: Clues from invertebrates via fish to mammals. Front. Immunol. **5**, 459 (2014)

M.V. Capparelli, P.K. Gusso-Choueri, D.M. de Souza Abessa, J.C. McNamara, Seasonal environmental parameters influence biochemical responses of the fiddler crab *Minuca rapax* to contamination *in situ*. Comp. Biochem. Physiol. C **216**, 93–100 (2019)

C.P. Chen, H.Q. Fan, Y.Y. Liao, G.L. Qiu, H.L. Hsieh, W.Y. Lin, Horseshoe crab, the living fossil animal facing threats to survival. Science **67**(3), 60–62 [in Chinese] (2015)

L. Cheng, Y. Chen, Y.Y. Zheng, Y. Zhan, H. Zhao, J.L. Zhou, Bioaccumulation of sulfadiazine and subsequent enzymatic activities in Chinese mitten crab (*Eriocheir sinensis*). Mar. Pollut. Bull. **121**(1–2), 176–182 (2017)

C.J. Coates, E. Bradford, C.A. Krome, J. Nairn, Effect of temperature on biochemical and cellular properties of captive *Limulus polyphemus*. Aquaculture **334–337**, 30–38 (2012)

W.C. Dougall, H.S. Nick, Manganese superoxide dismutase: A hepatic acute phase protein regulated by interleukin-6 and glucocorticoids. Endocrinology **129**, 2376–2384 (1991)

Y. Duan, Y. Wang, J. Zhang, D. Xiong, Elevated temperature disrupts the mucosal structure and induces an immune response in the intestine of whiteleg shrimp *Litopenaeus vannamei* (Boone, 1931) (Decapoda: Dendrobranchiata: Penaeidae). J Crustacean Biol **38**(5), 635–640 (2018)

K. Dunning, A.O. Safo, The ultimate Wright-Giemsa stain: 60 years in the making. Biotech. Histochem. **86**(2), 69–75 (2011)

G. Gauvry, Current horseshoe crab harvesting practices cannot support global demand for TAL/LAL: The pharmaceutical and medical device industries' role in the sustainability of horseshoe crabs, in *Changing Global Perspectives on Horseshoe Crab Biology, Conservation and Management*, ed. by R. H. Carmichael, M. L. Botton, P. K. S. Shin, S. G. Cheung, (Springer, Cham, 2015), pp. 475–482

H. Hayashibe, K. Asayama, K. Dobashi, K. Kato, Prenatal development of antioxidant enzymes in rat lung, kidney, and heart: Marked increase in immunoreactive superoxide dismutases, glutathione peroxidase and catalase in the kidney. Pediatr. Res. **27**, 472–475 (1990)

S. Hong, X. Zhang, Y. Zhao, Y. Xie, Y. Zhang, H. Xu, Effect of sediment type on growth and survival of juvenile horseshoe crabs (Tachypleus tridentatus), ed. by J. T. Tanacredi, M. L. Botton, D. R. Smith, pp. 535–540 (Springer, New York, 2009)

Y. Huang, Q. Ren, Research progress in innate immunity of freshwater crustaceans. Dev. Comp. Immunol. **104**, 103569 (2020)

L.V. Hurton, J.M. Berkson, S.A. Smith, Selection of a standard culture medium for primary culture of *Limulus polyphemus* amebocytes. Vitro Cell Dev. Biol-Anim. **41**(10), 325–329 (2005)

S. Iwanaga, The molecular basis of innate immunity in the horseshoe crab. Curr. Opin. Immunol. **14**(1), 87–95 (2002)

M.J. James-Pirri, P.A. Veillette, A.S. Leschen, Selected hemolymph constituents of captive, biomedically bled, and wild caught adult female American horseshoe crabs (*Limulus polyphemus*). Mar. Freshw. Behav. Phy. **45**(4), 281–289 (2012)

S. Javahery, A. Noori, S.H. Hoseinifar, Growth performance, immune response, and digestive enzyme activity in Pacific white shrimp, *Penaeus vannamei* Boone, 1931, fed dietary microbial lysozyme. Fish Shellfish Immunol. **92**, 528–535 (2019)

B.A. John, B.R. Nelson, H.I. Sheikh, S.G. Cheung, Y. Wardiatno, B.P. Dash, K. Tsuchiya, Y. Iwasaki, S. Pati, A review on fisheries and conservation status of Asian horseshoe crabs. Biodivers. Conserv. **27**(14), 3573–3598 (2018)

P.L. Jones, D. Ping, J.M. Boss, Tumor necrosis factor alpha and interleukin-1b regulate the murine manganese superoxide dismutase gene through a complex intronic enhancer involving C/EBP-b and NF-jB. Mol. Cell. Biol. **17**, 6970–6981 (1997)

B.K.Y. Kwan, A.K.Y. Chan, S.G. Cheung, P.K.S. Shin, Responses of growth and hemolymph quality in juvenile Chinese horseshoe crab *Tachypleus tridentatus* (Xiphosura) to sublethal tributyltin and cadmium. Ecotoxicology **24**, 1880–1896 (2015)

B.K.Y. Kwan, A.K.Y. Chan, S.G. Cheung, P.K.S. Shin, Hemolymph quality as indicator of health status in juvenile Chinese horseshoe crab *Tachypleus tridentatus* (Xiphosura) under laboratory culture. J. Exp. Mar. Biol. Ecol. **457**, 135–142 (2014)

B.K.Y. Kwan, V.K.Y. Un, S.G. Cheung, P.K.S. Shin, Horseshoe crabs as potential sentinel species for coastal health: Juvenile haemolymph quality and relationship to habitat conditions. Mar. Freshw. Res. **69**(6), 894–905 (2018)

J.P. Lallès, Intestinal alkaline phosphatase: Multiple biological roles in maintenance of intestinal homeostasis and modulation by diet. Nutr. Rev. **68**, 323–332 (2010)

K. Laurie, C.P. Chen, S.G. Cheung, V. Do, H.L. Hsieh, A. John, F. Mohamad, S. Seino, S. Nishida, P. Shin, M.C. Yang, *Tachypleus tridentatus*. The IUCN Red List of Threatened Species 2019: E.T21309A133299524 (9 May 2019). https://doi.org/10.2305/IUCN.UK.2019-1.RLTS.T21309A133299524.en

J. Levin, F.B. Bang, The role of endotoxin in the extracellular coagulation of limulus blood. Bull. Johns Hopkins Hosp. **115**, 265–274 (1964)

N. Li, Y. Zhao, J. Yang, Effects of water-borne copper on digestive and metabolic enzymes of the giant freshwater prawn *Macrobrachium rosenbergii*. Arch. Environ. Contam. Toxicol. **55**(1), 86–93 (2008)

Y.H. Li, X.Y. Xie, K.Y. Kwan, Endangered status and declaration on conservation of the "living fossil" *Tachypleus tridentatus*. Wetl. Sci. **16**(5), 690–692 (2018) [in Chinese with English abstract]

Y.Y. Liao, H.L. Hsieh, S.Q. Xu, Q.P. Zhong, J. Lei, M.Z. Liang, H.Y. Fang, L.L. Xu, W.Y. Lin, X.B. Xiao, C.P. Chen, Wisdom of Crowds reveals decline of Asian horseshoe crabs in Beibu Gulf, China. Oryx **53**(2), 222–229 (2019)

J.D. Liu, W.B. Liu, D.D. Zhang, C.Y. Xu, C.Y. Zhang, X.C. Zheng, C. Chi, Dietary reduced glutathione supplementation can improve growth, antioxidant capacity, and immunity on Chinese mitten crab, *Eriocheir sinensis*. Fish Shellfish Immunol. **100**, 300–308 (2020)

J. Lu, C. Qi, S.M. Limbu, F. Han, L. Yang, X. Wang, J.G. Qin, L. Chen, Dietary mannan oligosaccharide (MOS) improves growth performance, antioxidant capacity, non-specific immunity and intestinal histology of juvenile Chinese mitten crabs (*Eriocheir sinensis*). Aquaculture **510**, 337–346 (2019)

A. Moreira, E. Figueira, A.M.V.M. Soares, R. Freitas, The effects of arsenic and seawater acidification on antioxidant and biomineralization responses in two closely related *Crassostrea* species. Sci. Total Environ. **545**, 569–581 (2016)

S.A. Pinoni, A.L. Goldemberg, A.L. Mañanes, Alkaline phosphatase activities in muscle of the euryhaline crab *Chasmagnathus granulatus*: Response to environmental salinity. J. Exp. Mar. Biol. Ecol. **326**(2), 217–226 (2005)

Fauziyah, A.I. Purwiyanto, W.A. Putri, F. Agustriani, A.Z. Mustopa, The first investigation record of threatened horseshoe crabs in the Banyuasin estuarine, South Sumatra, Indonesia. Ecol. Montenegrina 24, 17–24 (2019)

C. Qi, X. Wang, F. Han, Y. Jia, Z. Lin, C. Wang, J. Lu, L. Yang, X. Wang, E. Li, J.G. Qin, Arginine supplementation improves growth, antioxidant capacity, immunity and disease resistance of juvenile Chinese mitten crab, *Eriocheir sinensis*. Fish Shellfish Immunol. **93**, 463–473 (2019)

R.J. Rogers, S.E. Chesrown, S. Kuo, J.M. Monnier, H.S. Nick, Cytokine-inducible enhancer with promoter activity in both the rat and human manganese-superoxide dismutase genes. Biochem. J. **347**, 233–242 (2000)

D.M. Rudkin, G.A. Young, Horseshoe crabs–an ancient ancestry revealed, in *Biology and Conservation of Horseshoe Crabs*, ed. by J. T. Tanacredi, M. L. Botton, D. R. Smith, pp. 25–44 (Springer, New York, 2009)

D. Sampath, R. Perez-Polo, Regulation of antioxidant enzyme expression by NGF. Neurochem. Res. **22**, 351–362 (1997)

N. Schvezov, G.A. Lovrich, F. Tapella, M. Gowland-Sainz, M.C. Romero, Effect of the temperature of air exposure on the oxidative stress status of commercial male southern king crab *Lithodes santolla*. Fish. Res. **212**, 188–195 (2019)

H.I. Sheikh, B.A. John, S.J. Ichwan, B.Y. Kamaruzzaaman, Effect of prolonged captivity on the hemolymph profile of Tachypleus gigas using the various anticoagulant formulations. Aquacult. Rep. **20**, 100760 (2021)

D.R. Smith, H.J. Brockmann, M.A. Beekey, T.L. King, M.J. Millard, J. Zaldívar-Rae, Conservation status of the American horseshoe crab, (*Limulus polyphemus*): A regional assessment. Rev. Fish. Biol. Fisher. **27**(1), 135–175 (2017)

N. Sugino, M. Hirosawa-Takamori, L. Zhong, C.M. Telleria, K. Shiota, G. Gibori, Hormonal regulation of copper-zinc superoxide dismutase and manganese superoxide dismutase messenger ribonucleic acid in the rat corpus luteum: Induction by prolactin and placental lactogens. Biol. Reprod. **59**, 599–605 (1998)

P. Suhr-Jessen, L. Baek, P.P. Jakobsen, Microscopical, biochemical, and immunological studies of the immune defense system of the horseshoe crab, *Limulus polyphemus*. Biol. Bull. **176**(3), 290–300 (1989)

Y.M. Sui, M.H. Hu, Y.Y. Shang, F.L. Wu, X.Z. Hu, S. Dupont, Y.J. Wang, Antioxidant response of the hard-shelled mussel *Mytilus coruscus* exposed to reduced pH and oxygen concentration. Ecotoxicol. Environ. Saf. **137**, 94–102 (2017)

Y. Toh, A. Mizutani, F. Tokunaga, T. Muta, S. Iwanaga, Morphology of the granular hemocytes of the Japanese horseshoe crab *Tachypleus tridentatus* and immunocytochemical localization of clotting factors and antimicrobial substances. Cell Tissue Res. **266**, 137–147 (1991)

X. Wang, Z. Shen, C. Wang, E. Li, J.G. Qin, L. Chen, Dietary supplementation of selenium yeast enhances the antioxidant capacity and immune response of juvenile *Eriocheir sinensis* under nitrite stress. Fish Shellfish Immunol. **87**, 22–31 (2019)

Y.H. Wei, H.C. Lee, Oxidative stress, mitochondrial DNA mutation, and impairment of antioxidant enzymes in aging. Exp. Biol. Med. **227**, 671–682 (2002)

F.L. Wu, X.Z. Huang, Q.Z. Li, M.H. Hu, W.Q. Lu, Y.J. Wang, Classification and characterization of hemocytes between *Tachypleus tridentatus* and *Carcinoscorpius rotundicauda*. Acta Hydrobiol. Sin. **39**(6), 1169–1176 (2015) [in Chinese with English abstract]

F.L. Wu, Z. Xie, M.Y. Yan, Q.Z. Li, J. Song, M.H. Hu, Y.J. Wang, Classification and characterization of hemocytes from two Asian horseshoe crab species *Tachypleus tridentatus* and *Carcinoscorpius rotundicauda*. Sci. Rep. **9**(1), 7095 (2019)

Z. Xu, Y. Wang, Y. Gul, Q. Li, J. Song, M. Hu, Effects of copper supplement on the immune function and blood-chemistry in adult Chinese horseshoe crab *Tachypleus tridentatus*. Aquaculture **515**, 734576 (2020)

M.Y. Yan, W. Yu, H.X. Gu, Y.J. Wang, Q.Z. Li, Z.F. Fan, B. Sun, M.H. Hu, Current situation of Tachypleus amebocyte lysate industry in China and relevant suggestions. J. Biol. **35**(2), 88–91 (2018) [in Chinese with English abstract]

C. Yang, R. Hao, X. Du, Q. Wang, Y. Deng, R. Sun, Response to different dietary carbohydrate and protein levels of pearl oysters (*Pinctada fucata martensii*) as revealed by GC–TOF/MS-based metabolomics. Sci. Total Environ. **650**, 2614–2623 (2019)

J.H. Zhu, Z. Wu, B.B. Feng, S.S. Deng, W.Q. Zhen, Y.Y. Liao, X.Y. Xie, K.Y. Kwan, Global conservation status of *Tachypleus tridentatus* and its introspection. Biodivers. Sci. **28**(5), 621–629 (2020) [in Chinese with English abstract]

# Visually Guided Behaviour of Juvenile Tri-spine Horseshoe Crab *Tachypleus tridentatus*: Implications for Designing Racks for Oyster Cultivation Compatible with Horseshoe Crab Conservation

**H. K. Chan, H. S. Lo, S. Y. Ng, R. -F. Chen, and S. G. Cheung**

## 1 Introduction

Vision plays an important role in horseshoe crabs. In reproduction, female *Limulus polyphemus* avoid dark female-size objects to prevent building nests on shores already occupied, whereas males use vision for locating females (Powers et al. 1991). Vision in horseshoe crabs is relied on a pair of lateral compound eyes on ophthalmic ridges and a pair of simple eyes at the anterior part of the median ridge (Sekiguchi 1988). The compound eyes of horseshoe crabs consist of individual units called ommatidia. Each ommatidium is equipped with a lens cylinder that refracts light and transmits it to the sensory cells arranged in the form of a rosette. At the centre of the rosette is a specialized cell end for connecting signals of neighbouring ommatidia to improve the contours received by the crabs that is particularly useful under water with low visibility (Barlow and Powers 2003).

As juvenile horseshoe crabs grow, so does the size of their lateral eyes. A trilobite larva (Stage I) of *L. polyphemus* has 12–14 ommatidia, and that number increases to about 600, about half the number of adult ommatidia, by the 10th instar (Meadors

H. K. Chan · H. S. Lo · S. Y. Ng
Department of Chemistry, City University of Hong Kong,
Hong Kong, People's Republic of China

R. -F. Chen
Guangxi Institute of Oceanology, Beihai City, People's Republic of China

S. G. Cheung (✉)
Department of Chemistry, City University of Hong Kong,
Hong Kong, People's Republic of China

State Key Laboratory of Marine Pollution, City University of Hong Kong,
Hong Kong, People's Republic of China
e-mail: bhsgche@cityu.edu.hk

 
J. T. Tanacredi et al. (eds.), *International Horseshoe Crab Conservation and Research Efforts: 2007-2020*, https://doi.org/10.1007/978-3-030-82315-3_12

et al. 2001). Although lateral eyes in juvenile horseshoe crabs are not fully developed, they are functional and possess contrast-sensitive vision similar to that of adults, allowing them to avoid black objects (Errigo et al. 2001; Ridings et al. 2002). This avoidance behaviour is regarded as an adaptation to reduce predation risk in juvenile horseshoe crabs (Errigo et al. 2001).

Hong Kong has a long history of oyster cultivation on mudflats in Deep Bay (Bromhall 1958). One commonly used method is the traditional bottom-laying method in which gravel, concrete tiles and posts are placed on the mudflat as oyster cultch for collecting spat (Morton and Wong 1975) which takes 4–5 years to grow to marketable size on the mudflat. The cultivated oyster beds, however, overlap with the habitats where the highest density of juvenile tri-spine horseshoe crab, *Tachypleus tridentatus*, in Hong Kong is found (Kwan et al. 2016). In a field manipulation experiment using bricks to simulate oyster cultch, Kwan et al. (2018) showed that juvenile *T. tridentatus* avoided the bricks and foraged in the outer boundary of the cultch instead of within the gaps, possibly because the juveniles avoided the hummocks developed between bricks due to the change in sedimentation. This indicates that oyster culture practices tend to decrease the habitat quality for the juvenile horseshoe crabs. To reduce the conflict between oyster cultivation and conservation of horseshoe crabs, there is a need to either relocate the oyster farms to less sensitive areas or replace the traditional bottom culture method with other methods compatible with the conservation of horseshoe crabs.

Off-bottom oyster cultivation methods include rack-and-bag, cage, and floating culture techniques (Ford 1997; Lu 2015). In these methods, oysters are cultivated in cages or bags above the bottom using gears of some kind. As oysters are typically enclosed and protected, a grower has the potential to lose less oysters during extreme weather (Lu 2015). Cages and bags are prevented from fouling by being exposed to air during low tide. Munroe et al. (2017, 2020) demonstrated that the rack-and-bag method did not interfere the movement and spawning activity of the adult horseshoe crab *L. polyphemus* on the shore. The results have raised an interest in the possibility of replacing the traditional bottom-laying cultivation method by the rack-and-bag method in Hong Kong. Since juvenile horseshoe crabs forage and grow in soft sediments in near-shore habitats for 10–15 years (Sekiguchi et al. 1988), there is a need to understand if the racks affect habitat utilization of the crabs. The racks are raised on the mudflat by supporting legs and the oyster bags on the racks provide a shading effect. Whether the juvenile horseshoe crabs avoid the shading effect and the legs of the racks will determine the compatibility between the rack-and-bag cultivation method and conservation of juvenile horseshoe crabs on the mudflat.

This study included a field experiment to understand the foraging behaviour of the juvenile *Tachypleus tridentatus* in the presence of obstacles with different colours to simulate the effect of the legs of the racks, and a laboratory experiment to understand their behavioural response to covers with different degrees of shading. The results provide useful information for designing racks for oyster cultivation that are compatible with the conservation of horseshoe crabs.

## 2 Materials and Methods

### 2.1 *Effect of Obstacles on Foraging Behaviour in the Field*

To determine how the colour and size of obstacles affected the foraging pattern of juvenile *T. tridentatus*, two types of acrylic obstacles were made, a round column and a board. The column was circular with a diameter and height of 8 cm and 5 cm, respectively, whereas the board was rectangular with dimensions of 25 cm (length) × 5 cm (height) (Ridings et al. 2002). The round column simulated a pole freshly placed on the mudflat for collecting oyster spat, whereas the board represented a pole with clusters of growing oysters. Three different colours of obstacles, black, grey, and transparent, were prepared for both the columns and boards using black or grey paint. Therefore, six treatments were set up in total. No abnormal behaviour was observed in the juvenile horseshoe crabs when a column/board with paint was placed in front of them. The field study was conducted on the mudflat of Ha Pak Nai (22° 25′ N, 113° 56′ E) during low tide in summer from July to September 2018, except on rainy days. After the feeding trail of a juvenile horseshoe crab was identified, the direction of movement of the juvenile was observed for 5 minutes. An obstacle, either a board or column, was then placed in front of the juvenile at a distance of 5–15 cm (Ridings et al. 2002). The behaviour of the juvenile, whether it collided with the obstacle or avoided it was recorded. The individuals then would turn to move around the obstacles, and the distance from the center of the obstacles to the center of the juveniles in the turn was measured. The distance travelled by and displacement of each juvenile was measured 15 minutes later after the turn. The foraging trail was traced with a string and the length measured with a ruler to the nearest 0.1 cm. Each juvenile was used only once so different individuals were exposed to either boards or columns of different colors. The total number of individuals in each treatment varied between 24 and 30. To minimize the effect of handling on the juvenile horseshoe crabs that may affect their foraging behaviour, the prosomal width (1.5–6 cm) of experimental individuals was measured with a vernier caliper at the end of each observation.

### 2.2 *Preference of Degree of Shading*

The degree of shading preferred by juvenile *T. tridentatus* was investigated using a laboratory experiment. Plastic trays with a basal area of 20 cm × 40 cm were used as experimental arenas. The tray bottoms were covered with 2 cm of fine sediment, above which was 1 cm of seawater of salinity 25 ppt. Three types of acrylic plate, each with a different degree of shading, were prepared. The first type was transparent, the second type completely black and the third one checkered with black and transparent grids (each grid has dimensions of 2.5 cm × 2.5 cm) (Fig. 1). Each board, having an area of 20 cm × 20 cm, covered half of the tray. Six treatments

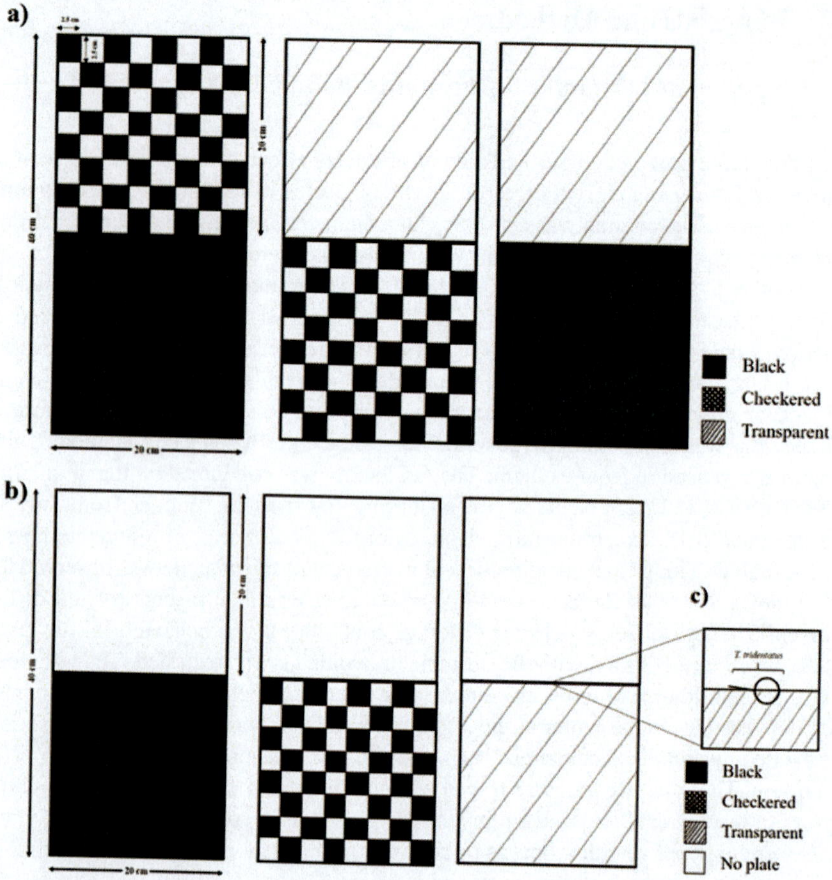


**Fig. 1** (**a**) The top view of trays with covers of different degrees of shading (black and checkered, checkered and transparent, and black and transparent). (**b**) The top view of trays with half of each tray covered by one type of board (black, checkered, and transparent) and another half uncovered. (**c**) To start each experiment, the median plane of the juvenile *T. tridentatus* was cut by two types of shadow created by the boards

were set up with three of them having two boards with different degrees of shading at 10 cm above the tray (Fig. 1a). The remaining three treatments had one board for each tray and the other half of the tray uncovered (Fig. 1b). A lamp (13 W) with white light was set above each tray to create a sharp contrast of light intensity. One 3rd instar of *T. tridentatus* (prosomal width: ~13 mm) was put in the middle of the plastic tray with the median plane at the junction of the two covers (Fig. 1c). All the juveniles were handled with great care to minimize handling stress and each juvenile was used once in the experiment to prevent bias from learning or habituation. The experiment was conducted in a windowless indoor laboratory and lasted for

6 hours from 12 noon to 6 pm. During this period, the juveniles were undisturbed and their positions recorded at the end of the experiment.

To understand the movement pattern of juvenile *T. tridentatus* under different degree of shading, another laboratory experiment was conducted with total travelling distance and displacement (i.e., net distance between the starting point and finishing point) recorded using the above set up. Each tray had half of its area shaded by a board (same as the above experiment) whereas the other half unshaded (Fig. 1b). The position of the juvenile was recorded 5 minutes after it started to move and the total distance travelled and displacement were measured to the nearest 0.1 cm using a string and ruler at the end of the experiment.

### 2.3 Statistical Analysis

Statistical analyses were conducted using Sigmaplot 12.5. Prior to the analysis, normality and homogeneity of variance of the data were tested using Shapiro–Wilk test and Levene's test, respectively. Data on the distance travelled by, and displacement of, juvenile horseshoe crabs in the presence of obstacle were analyzed by one-way ANOVA. Prior to the analysis, the data were log transformed as they did not conform to normality. The percentage of juvenile horseshoe crabs collided with the obstacle in the field was tested by Chi-square goodness of fit test. The relationship between avoidance behaviour and prosomal width was analysed using Mann–Whitney rank sum test as the data did not conform to normality even after data transformation. For the laboratory experiments, selection for covers with different degree of shading was analyzed using Chi-square goodness of fit test, whereas the distance travelled and displacement under different degree of shading was tested using student t-test.

## 3 Results

### 3.1 Effect of Obstacles on Foraging Behaviour in the Field

A higher percentage of juvenile horseshoe crabs collided with transparent obstacles (71–83%) as compared with grey or dark ones (31–46%), regardless of whether the obstacles were smaller columns or larger boards (Table 1). Such avoidance behaviour was independent of body size (prosomal width) for both columns (Black: M-W U = 1130, $p = 0.09$; Grey: M-W U = 78, $p = 0.73$; Transparent: M-W U = 65.5, $p = 0.15$) and boards (Black: M-W U = 37, $p < 0.01$; Grey: M-W U = 53.5, $p = 0.41$; Transparent: M-W U = 38, $p = 0.91$), except for boards in black that larger individuals had a higher chance of colliding with the obstacles.

**Table 1** Number and percentage of foraging juvenile horseshoe crabs colliding with the obstacles placed in front of their foraging route. The result of each treatment was analysed separately using the Chi-square goodness of fit test

| | | Sample size | No. of individuals | % of individuals | Chi-square value | Prob. level |
|---|---|---|---|---|---|---|
| Column | Black | 30 | 11 | 37 | 2.13 | 0.14 |
| | Grey | 29 | 9 | 31 | 4.17 | *<0.05* |
| | Transparent | 31 | 22 | 71 | 5.45 | *<0.05* |
| Board | Black | 26 | 12 | 46 | 0.15 | 0.70 |
| | Grey | 25 | 8 | 32 | 3.24 | 0.07 |
| | Transparent | 24 | 20 | 83 | 5.33 | *<0.05* |

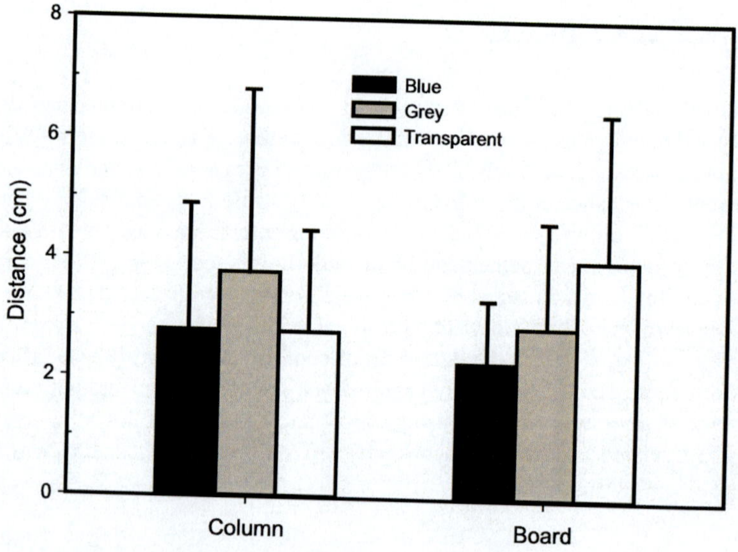


**Fig. 2** The distance (+ SD) between the obstacle and juvenile horseshoe crab when the juvenile horseshoe crab made a turn under different degrees of obstacle colour in the field. The results were not significant different ($p > 0.05$) for both the column and board

When they encountered obstacles, all the juveniles made a turn and continued to forage, irrespective of whether they collided with the obstacles. The distance between the obstacle and a juvenile horseshoe crab when a turn was made varied between 2.8 cm and 3.8 cm for the columnar obstacles and between 2.3 cm and 4.0 cm for the boards (Fig. 2). The differences in the distance among different colors were statistically indistinguishable for both the column ($F = 0.68$, df = 2, $p = 0.51$) and the board ($F = 0.87$, df = 2, $p = 0.43$). The foraging distance and displacement was not affected by the colour for both the column (distance: $F = 1.38$, df = 2, $p = 0.26$; displacement: $F = 1.74$, df = 2, $p = 0.18$) (Fig. 3a) and board (distance: $F = 1.06$, df = 2, $p = 0.35$; displacement: $F = 1.93$, df = 2, $p = 0.15$) (Fig. 3b).

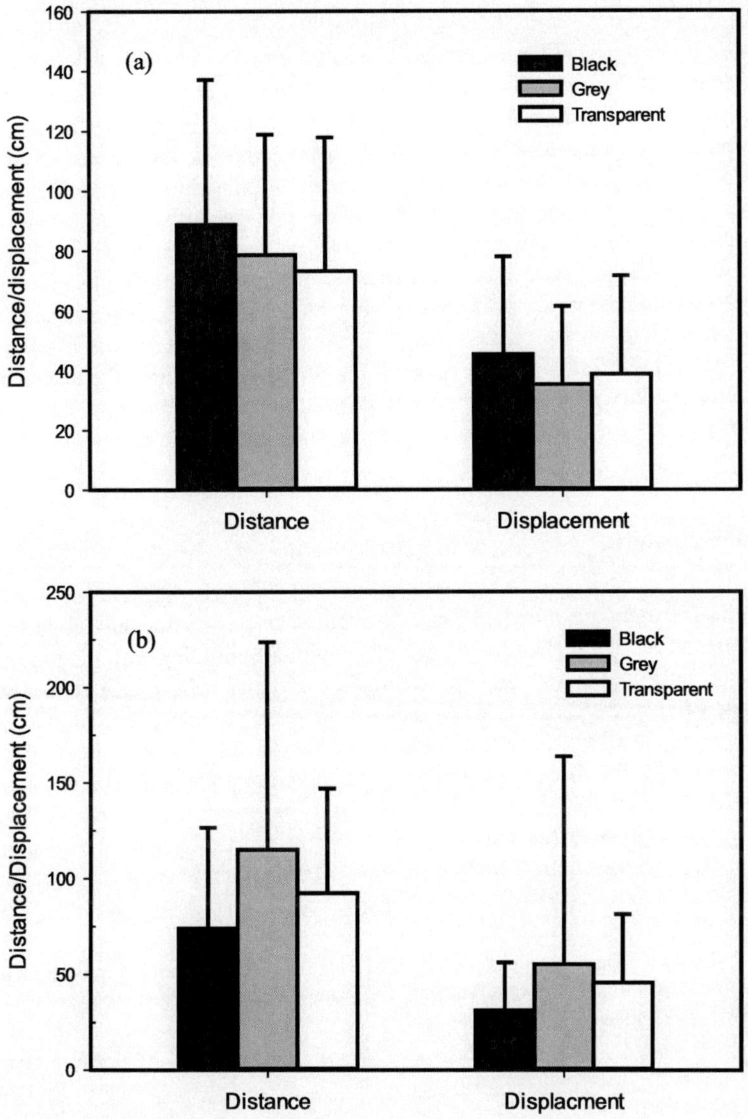


**Fig. 3** Distance travelled by and displacement (+ SD) of juvenile *T. tridentatus* when an acrylic (**a**) column or (**b**) board with different degrees of colour was placed at 5–10 cm in front of the individual. The foraging distance and displacement was not significantly affected ($p > 0.05$) by the degree of obstacle colour for both the column and board

### 3.2 *Preference of Degree of Shading*

When the juvenile horseshoe crabs in the experimental trays were allowed to choose between the area with an acrylic cover and open area after moving in the tray for 5 minutes, no preference was observed, irrespective of the degree of shading (Table 2). In another experiment in which the duration was extended to 6 hours and choices between two acrylic covers with different degrees of shading were provided in addition to choices between an acrylic cover and open area, however, no preference was observed between any two choices of acrylic cover (Table 2, $p > 0.05$). More juveniles, however, preferred staying under an acrylic cover instead of an open area, although the results were statistically significant for the black acrylic cover only (Table 2, Chi-square value = 13.2, $p < 0.001$). Neither the distance travelled nor displacement were significantly different between those juveniles choosing to stay under a cover and those under the open area (Table 3, $p > 0.05$).

## 4 Discussion

Juvenile *T. tridentatus* relied on vision to avoid obstacles on the mudflat during foraging, resulting in a higher percentage of individuals colliding with the transparent obstacles than the black or grey ones. Similar avoidance behaviour was observed in juvenile *Limulus polyphemus* when they encountered an object (Errigo et al. 2001; Ridings et al. 2002) with fewer individuals colliding with black obstacles (12%) than grey objects (25%). The higher avoidance observed in juvenile *L. polyphemus* than juvenile *T. tridentatus* as well as a better ability in discriminating black obstacles from grey ones indicate that the former probably has more effective contrast-sensitive vision. Nevertheless, both *L. polyphemus* and *T. tridentatus* show difficulty avoiding

**Table 2** Positions of the juvenile horseshoe crabs 5 minutes or 6 hours after they had been placed under covers with different degrees of shading. The result of each treatment was analysed separately using the Chi-square goodness of fit test

| | Sample size | Black board | Checkered board | Transparent board | Open area | Chi-square value | Prob. level |
|---|---|---|---|---|---|---|---|
| After 5 mins | 15 | 8 | – | – | 7 | 0.07 | 0.79 |
| | 17 | – | 11 | – | 6 | 1.47 | 0.23 |
| | 15 | – | – | 8 | 7 | 0.07 | 0.79 |
| After 6 hours | 10 | 5 | 5 | – | – | 0 | 0.99 |
| | 10 | – | 5 | 5 | – | 0 | 0.99 |
| | 10 | 4 | – | 6 | – | 0.4 | 0.53 |
| | 16 | – | – | 11 | 5 | 2.3 | 0.13 |
| | 15 | – | 9 | – | 6 | 0.6 | 0.44 |
| | 17 | 16 | – | – | 1 | 13.2 | *<0.001* |

**Table 3** Distance travelled by and displacement of juvenile horseshoe crabs 5 minutes after they had been placed under covers with different degrees of shading

| | Black board | Checkered board | Transparent board | Open area | *t* value | Prob. level |
|---|---|---|---|---|---|---|
| Mean distance travelled (cm) | 17.78 | – | – | 28.99 | −0.91 | 0.38 |
| | – | 25.92 | – | 24.57 | 0.13 | 0.90 |
| | – | – | 30.58 | 34.13 | −0.30 | 0.77 |
| Mean displacement (cm) | 6.73 | – | – | 6.89 | −0.07 | 0.95 |
| | 8.62 | – | – | 8.75 | −0.04 | 0.97 |
| | – | – | 10.26 | 10.54 | −0.08 | 0.94 |

transparent obstacles (Errigo et al. 2001; Ridings et al. 2002). Our study demonstrated that the contrast-sensitive vision was independent of body size within a range of prosomal width between 1.5 cm and 6 cm, which is equivalent to around 4th to 11th instar (Hu et al. 2015). This indicates that the compound eyes are functional in juveniles as small as 4th instar. The longest distance the juvenile horseshoe crab was from the obstacle when the juvenile horseshoe crab made a turn indicated the effectiveness of the contrast-sensitive vision. That distance was 12.0 cm in *T. tridentutus* in this study, in contrast to 18.3 cm measured for *L. polyphemus* (Errigo et al. 2001), further supporting the hypothesis that the latter has better vision than the former. Such discrepancy may also be due to the structure of the ommatidium in which more photoreceptor cells (retinular cells) are found in each ommatidium in *L. polyphemus* (~15) (Dowling 2001) than *T. tridentatus* (8) (Hong 2011).

In a field manipulation experiment, juvenile *T. tridentatus* avoided simulated oyster cultch gear on intertidal flats and foraged along the outer boundaries of farm gear instead of within the gaps of the clutch (Kwan et al. 2018). This avoidance behaviour was also demonstrated in this study and possibly an adaptation to reduce predation risk in juvenile horseshoe crabs as suggested by Errigo et al. (2001). The laboratory experiment in this study showed that the juvenile horseshoe crabs preferred staying under the cover instead of in the open air, although the results were only significant for the black cover. As dark objects themselves may appear to be a predator, the preference towards staying under a dark cover may be a kind of refuge-seeking behaviour that helps reduce predation risk of the juveniles or protecting them from hot desiccating conditions. Future research may address these possibilities.

The simulated oyster cultch gear triggered avoidance behaviour in juvenile horseshoe crabs, resulting in a reduction in the population density due to emigration (Kwan et al. 2018). This has raised a concern of the conflict between bottom oyster cultivation and conservation of horseshoe crabs. Suspended oyster farming in shallow subtidal areas has been shown to induce minor and relatively localized ecological impacts (Dumbauld et al. 2009). A variety of intertidal rack-and-bag methods have been used for a long time for oyster culture in the United States (Ford 1997), where oysters are grown in bags on short metal racks. Munroe et al. (2017) did not find any observable effect of this farming method on the spawning activity of the horseshoe crab. Although they need to occasionally navigate around or over structure in the rack-and-bag farms, resulting in a path covering 3% more distance, the effect on the horseshoe crabs was negligible (Munroe et al. 2020). No work, however, has been done on whether this cultivation method affects the foraging of juvenile horseshoe crabs. The field experiment in this study has demonstrated that juvenile *Tachypleus tridentatus* can navigate in the presence of obstacles, resulting in a negligible obstacle effect on their foraging behaviour. Therefore, the avoidance of simulated oyster clutches by juvenile horseshoe crabs demonstrated in our previous study (Kwan et al. 2018) probably was not due to the presence of obstacles on the beach, but the change in sedimentation around the cultches, resulting in the formation of hummocks. As these hummocks dry up during low tides, this area is no longer a suitable foraging ground of the juvenile horseshoe crabs which prefer areas with a film of water (Kwan et al. 2020). Using racks for collecting spat and culturing oysters can free up more space as foraging ground for juvenile horseshoe crabs without the formation of hummocks. The present laboratory study has shown that juvenile horseshoe crabs preferred staying under a dark cover. Therefore, we predict that they will not avoid staying under the racks putting up on the shore. This has raised the potential of introducing rack-and-bag methods in Hong Kong to resolve the conflict between oyster farming and conservation of horseshoe crabs. The next step will be conducting field experiments to understand the foraging behaviour and associated growth of the juvenile horseshoe crabs in the presence of the oyster rack-and-bag culture.

## References

R.B. Barlow, M.K. Powers, Seeing at night and finding mates: The role of vision, in *The American Horseshoe Crab*, ed. by C. N. Schuster, R. B. Barlow, H. J. Brockmann, (Harvard University Press, Cambridge, 2003), pp. 83–102

J.D. Bromhall, On the biology and culture of the native oyster of Deep Bay, Hong Kong, *Crassostrea* sp. Hong Kong University Fish. J. **2**, 93–107 (1958)

J.E. Dowling, *Neurons and Networks: An Introduction to Behavioral Neuroscience*, 2nd edn. (Belknap Press, 2001), p. 220

B.R. Dumbauld, J.L. Ruesink, S.S. Rumrill, The ecological role of bivalve shellfish aquaculture in the estuarine environment: A review with application to oyster and clam culture in West Coast (USA) estuaries. Aquaculture **290**, 196–223 (2009)

M. Errigo, C. McGuiness, S. Meadors, B. Mittmann, F. Dodge, R. Barlow, Visually guided behavior of juvenile horseshoe crabs. Biol. Bull. **201**, 271–272 (2001)

S.E. Ford, History and present status of molluscan shellfisheries from Barnegat Bay to Delaware Bay, in *The History, Present Condition, and Future of the Molluscan Fisheries of North and Central America and Europe*, ed. by C. L. MacKenzie, V. G. Burrell, A. Rosenfield, W. L. Hobart, vol. 1: North America, (Tech Rep. U.S. Department of Commerce, NOAA, NMFS, Seattle, WA, 1997), pp. 119–140

S.G. Hong, *Biology of Horseshoe Crabs, Tachypleus tridentatus* (Xiamen University Press, Xiamen, 2011), p. 342

M.-h. Hu, B.K.Y. Kwan, Y.-j. Wang, S.G. Cheung, P.K.S. Shin, Population structure and growth of juvenile horseshoe crabs *Tachypleus tridentatus* and *Carcinoscorpius rotundicauda* (Xiphosura) in southern China, in *Changing Global Perspectives on Horseshoe Crab*

*Biology, Conservation and Management*, ed. by R. H. Carmichael, M. L. Botton, P. K. S. Shin, S. G. Cheung, (Springer, Cham, 2015), pp. 167–180

B.K.Y. Kwan, H.K. Chan, S.G. Cheung, Habitat use of globally threatened juvenile Chinese horseshoe crab, *Tachypleus tridentatus* under the influence of simulated intertidal oyster culture structures in Hong Kong. Aquat. Conserv. Mar. Freshw. Ecosyst. **28**, 124–132 (2018)

B.K.Y. Kwan, H.-L. Hsieh, S.G. Cheung, P.K.S. Shin, Present population and habitat status of potentially threatened Asian horseshoe crabs *Tachypleus tridentatus* and *Carcinoscorpius rotundicauda* in Hong Kong: A proposal for marine protected areas. Biodivers. Conserv. **25**, 673–692 (2016)

K.Y. Kwan, W.T. Wong, P.Y. Lam, H.K. Chan, H.S. Lo, S.G. Cheung, Effects of rubbles from oyster cultivation on habitat utilization and foraging behavior of the endangered tri-spine horseshoe crab: An implication for intertidal oyster cultivation practices. J. Environ. Manage. **271**, 110925 (2020)

C. Lu, The Different Methods of Growing Oysters (2015). https://www.pangeashellfish.com/blog/the-different-methods-of-growing-oysters

S. Meadors, C. McGuiness, F.A. Dodge, R.B. Barlow, Growth, visual field, and resolution in the juvenile *Limulus* lateral eye. Biol. Bull. **201**, 272–274 (2001)

B. Morton, P.S. Wong, The Pacific oyster industry in Hong Kong. J. Hong Kong Branch of the Royal Asiatic Soc. **15**, 139–149 (1975)

D. Munroe, D. Bushek, P. Woodruff, L. Calvo, Intertidal rack-and-bag oyster farms have limited interaction with horseshoe crab activity in New Jersey, USA. Aquac. Environ. Interact. **9**, 205–211 (2017)

D.M. Munroe, T.M. Grothues, N.E. Cleary, J. Daw, S. Estrada, Oyster aquaculture does not impede spawning beach access for Atlantic horseshoe crabs *Limulus polyphemus*. Aquac. Environ. Interact. **12**, 81–90 (2020)

M.K. Powers, R.B. Barlow, L. Kass, Visual performance of horseshoe crabs day and night. Vis. Neurosci. **7**, 179–189 (1991)

C. Ridings, D. Borst, K. Smith, F. Dodge, R. Barlow, Visual behavior of juvenile *Limulus* in their natural habitat and in captivity. Biol. Bull. **203**, 224–225 (2002)

K. Sekiguchi, H. Seshimo, H. Sugita, Post-embryonic development of the horseshoe crab. Biol. Bull. **174**, 337–345 (1988)

K. Sekiguchi, *Biology of Horseshoe Crabs* (Science House Co., Ltd., Tokyo, 1988), p. 428

# Condition of Horseshoe Crabs as Bycatch from Gillnet Fisheries in Mayangan Waters of Subang Regency, West Java, Indonesia

**Fahresa N. Supadminingsih, Mochammad Riyanto, and Ronny Irawan Wahju**

## 1 Introduction

The horseshoe crab is a macrobenthic animal which is well known as a living fossil (Smith et al. 2016). Four horseshoe crabs are inhabiting the world, that is, *Limulus polyphemus* (Linnaeus, 1758) from North America (Sekiguchi and Shuster 2009; Shuster 2015), and *Tachypleus gigas* (Müller 1785), *T. tridentatus* (Leach 1819), and *Carcinoscorpius rotundicauda* (Lattreille 1802) from Asia (Sekiguchi 1988). The latter three species are known as Asian horseshoe crabs. Information on the biology, ecology, and distribution of horseshoe crabs in Indonesia has been reported from previous literature, such as the morphometric study in Pandeglang Regency, West Java and Rembang Regency, Central Java (Suparta 1992), biological reproduction, genetic and morphometric analysis, and bycatch composition in Subang Regency, West Java (Muslihah 2004; Meilana 2015; Supadminingsih et al. 2019), distribution in coastal North Java (Meilana 2015; Mashar et al. 2017), population in Jambi Province, Sumatra (Rubiyanto 2012), and genetic preservation in Northern Sumatra (Mulya 2004). Based on recent geographic surveys, horseshoe crabs have also been found in Kalimantan and Sulawesi (Meilana and Fang 2020).

In Indonesia, horseshoe crabs are protected by Ministerial Decree No. 12/Kpts-II/1987 as a genetic resource, Government Regulation No. 7/1999 as threatened species and Ministerial Decree No. P.20/MENLHK/ SETJEN /KUM.1/ 6/2018

F. N. Supadminingsih
Fisheries Science, Faculty of Agriculture, Sultan Ageng Tirtayasa University, Banten, Indonesia

M. Riyanto (✉) · R. I. Wahju
Fisheries Resources Utilization Department, Faculty of Fisheries and Marine Sciences, Bogor Agricultural University, Bogor, Indonesia
e-mail: mochammadri@apps.ipb.ac.id



J. T. Tanacredi et al. (eds.), *International Horseshoe Crab Conservation and Research Efforts: 2007-2020*, https://doi.org/10.1007/978-3-030-82315-3_13

as a protected animal. However, local people have been using horseshoe crabs as bait to catch eels (Wakefield 2012; Smith et al. 2016) and catfish (Rubiyanto 2012). They also use it as a traditional dish to enhance immunity and vitality (Meilana 2015) and treat its eggs as herbal medicine for accelerated recovery from wounds (pers. obs.). Horseshoe crabs are also facing an increasing threat of being trapped as bycatch in gillnet fishing activity (Bianchi 2002; Meilana 2015; Prasetyo et al. 2017; Supadminingsih et al. 2019).

Locally, horseshoe crabs are generally caught by fishermen as bycatch rather than the main fishing target, as reported by Meilana and Fang (2020). It is the primary factor in the decrease of horseshoe crab populations by up to 27.7%. They are usually trapped at the bottom trawl net (Badiuzzaman and Yulianto, 2014) or gillnet (Faridah et al. 2015; Fazrul et al. 2015; Prasetyo et al. 2017) which is operated to catch demersal fish. The existence of horseshoe crabs as bycatch is also observed in

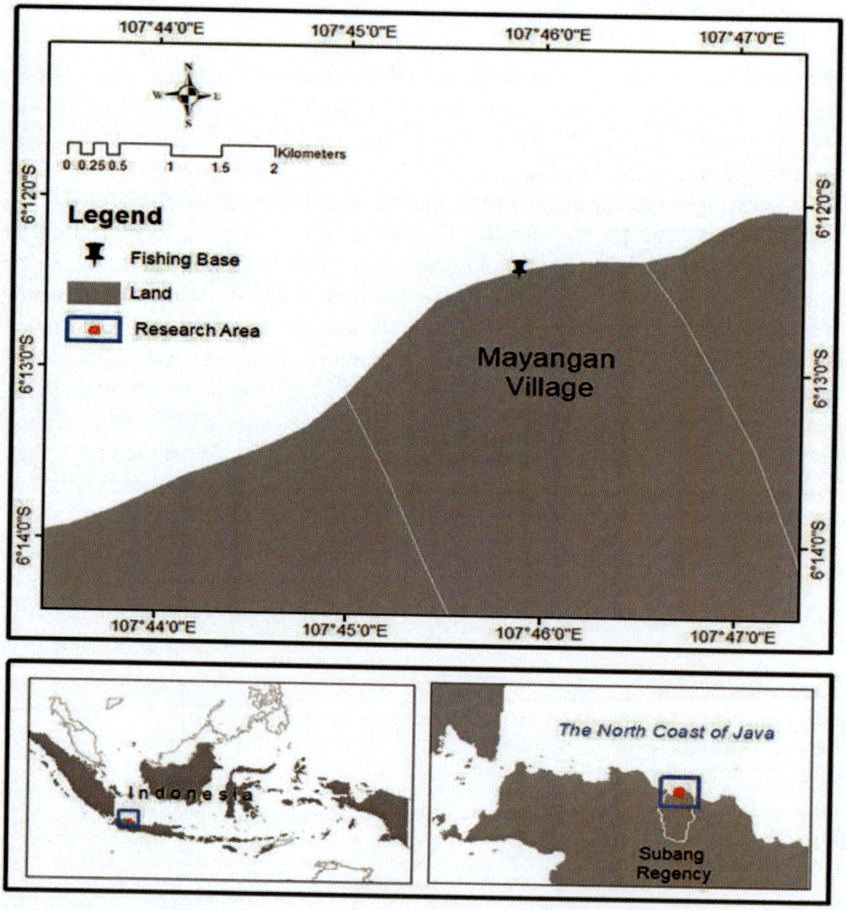


Fig. 1 Mayangan waters, Subang Regency, West Java, Indonesia

Mayangan waters, Subang Regency in West Java (06°12′08.7” S, 107°47′46.4″ E, Fig. 1). This area is known as one of the fishing areas for blue swimming crabs (*Portunus pelagicus*) as the principal fishing target (Suparta 1992). Incidentally, horseshoe crabs are also caught as bycatch during such activities. Since the local fishermen's livelihood depends mainly on the sale of blue swimming crabs, the high abundance of horseshoe crabs entangled with their fishing nets can cause damages to the fishing gear, as well as impede their fishing operations. As it is important to understand the horseshoe crab population including data on its size structure, sex ratio, and physical status (Carmichael et al. 2003), the present study attempted to obtain baseline information on the magnitude of such horseshoe crab bycatch from the local fishing communities and assess the condition of the bycatch specimens. The results of this study would indicate how horseshoe crabs may be saved from local fishing activities.

## 2 Materials and Methods

The study was conducted in June–August 2017 in Mayangan, Subang Regency, by following the fishermen's activity and counting the numbers of horseshoe crabs found in the gillnets. Mayangan village has an area of 678.37 ha with a small river flowing through a sandy shore. The shore has a flat slope and is used as a local tourist attraction, with many visitors during the weekends. The coast of Mayangan has an excellent mangrove coverage area of 290 ha (Muslihah 2004), which serves as a nursery ground for both the mud crab (*Scylla serrata*) and blue swimming crab (*P. pelagicus*), and also as a place for horseshoe crabs to spawn near the mangroves (pers. obs.). Mayangan had serious coastal erosion in 2005 and 2010, but the mangrove forest still exists (Fig. 2).

This survey was carried out by following 22 gillnet fishing operations according to the fishing ground, weather, and wave conditions. The deployment of the gillnet fishing gear usually covered 18–23 units with 414–529 m in length, 0.7 m in height,

**Fig. 2** Mayangan area with sandy and mud flats covered by mangrove forest

and circa 10 cm mesh size. The net was usually set up at intertidal area between 0.2–2 km from shore and mangroves and was deployed in 3–12 m depth for 6–12 hours. The harvest was checked the next day after the net deployment. The total catch landing was processed in the village, with blue swimming crabs, horseshoe crabs, and other bycatch species being counted and recorded based on their proportions in the total harvest. The species of horseshoe crabs trapped as bycatch were identified, sex and physical condition determined, and prosomal width (PW) measured. Each horseshoe crab was also categorized into four conditions as follows: (a) undamaged condition, where the entire body was intact and the carapace was hard, (b) molting condition, where the whole body was intact but the carapace was soft, (c) damaged or defect, where the body parts were incomplete or abnormal condition was noted, and (d) gravid condition, specifically for female horseshoe crabs where the egg grains were yellowish and visible from the genital aperture in the abdomen area. Male and female horseshoe crabs were distinguished from genital morphology and the first two pairs of legs. The present survey method, however, did not assess the potential spawning male horseshoe crabs due to the limited field time and difficulties in determining sperm maturity. All data were collected at the sampling site and entered into Microsoft Excel 2013 for analysis.

## 3 Results and Discussion

Previous research in Mayangan waters included morphometric analysis of *T. gigas* (Muslihah 2004) and study of horseshoe crab DNA (Meilana 2015), but the number of horseshoe crabs caught in gillnet fishing as bycatch has not been reported. The temperature and salinity during the 22 fishing trips of this study were 27.5–28.7 °C and 25–28 ppt. It was found that the catch of target blue swimming crabs was relatively low in comparison to the bycatch. Most of those bycatch consisted of horseshoe crabs. In total, 419 blue swimming crabs, 957 horseshoe crabs, and 702 individuals of other species, including the sea slug *Dendronotus* sp., rock snail *Murex* sp., Babylon shell *Babylonia spirata*, banana prawn *Penaeus merguiensis*, and mudskipper *Periophthalmus* sp., were captured from the gillnets. The average catch per trip for blue swimming crabs versus horseshoe crabs was 19.0 to 43.5, at a ratio of 1:2.3. Such high abundance of horseshoe crabs as bycatch (Fig. 3a) could be a threat to the fishermen as the nets were easily damaged. It also took more time to free the horseshoe crabs from the fish landing facility, and the fishermen had to work together with their neighbors to remove those trapped horseshoe crabs from their harvest (Fig. 3b).

Horseshoe crabs collected from this study were identified visually through literature to determine the species. The dominant species was *T. gigas*, comprising 870 (90.91%) in total, with the remaining 50 (5.22%) specimens of *C. rotundicauda* and 37 (3.87%) of *T. tridentatus*. In a previous study, Supadminingsih et al. (2019) reported that in addition to *T. gigas* and *C. rotundicauda*, *T. tridentatus* was present in Mayangan waters. The presence of *T. tridentatus*, however, was later confirmed

Fig. 3 (a) Horseshoe crabs entangled with the gillnet; (b) Fishermen helped by their neighbors to remove horseshoe crabs from the gillnet

by experts to avoid misidentification of *T. gigas* from morphological characteristics with smooth spines on the opisthosoma and for the juveniles as it was difficult to distinguish the sex from its genital morphology. Meilana et al. (2016) and Mashar et al. (2017) also reported that *T. gigas* was the dominant species compared with *T. tridentatus* and *C. rotundicauda* in Mayangan, Subang. All such reports indicated that Mayangan waters still support a diverse habitat for the three Asian horseshoe crab species. *T. gigas* is known as true sandy horseshoe crab or coastal horseshoe crab (Faridah et al. 2015), while *T.tridentatus* lives in sandy to muddy habitat (Mulya 2004; Shin et al. 2009; Cartwright-Taylor et al. 2011). On the other hand, *C. rotundicauda* is known as mangrove horseshoe crab and inhabits in the muddy area (Chatterji 1994). In India, *T. gigas* can co-exist with *C. rotundicauda* (Mishra 2009). The present findings were also in agreement with similar report that the three Asian horseshoe crab species were caught in the gillnet operation in Penisular Malaysia (Faridah et al. 2015).

All individuals collected were divided into size distribution with PW ranging from 6.5 to 24.4 cm. Figure 4 shows the size distribution and frequency of the horseshoe crabs in size class interval of 1.4 cm, based on PW of horseshoe crabs collected from gillnets in this study. A total of 475 individuals of *T.gigas* were recorded with the highest frequency of PW ranging from 9.5–10.9 cm, 24 *C. rotundicauda* with PW from 12.5–13.9 cm, and 11 *T. tridentatus* with PW from 14.0–15.4 cm. The distribution of the lowest frequency size was observed from 7 *T. gigas* and 1 *T.tridentatus* with PW of 6.5–7.9 cm, whereas the largest body size was observed for 1 *T. gigas* with PW of 23.0–24.4 cm.

The molting condition can be used to detect juveniles from sub-adults and to reflect the number of young juveniles which are in the active growing status (Estes Jr et al. 2015). From this study, the highest number of *T. gigas* in the class interval of 9.5–10.9 cm was categorized as juveniles, as most individuals were in the molting stage. The bycatch data also showed that *T. gigas* with PW >12.5 cm tended to have harder carapace. Although it was difficult to distinguish the specific size of juveniles

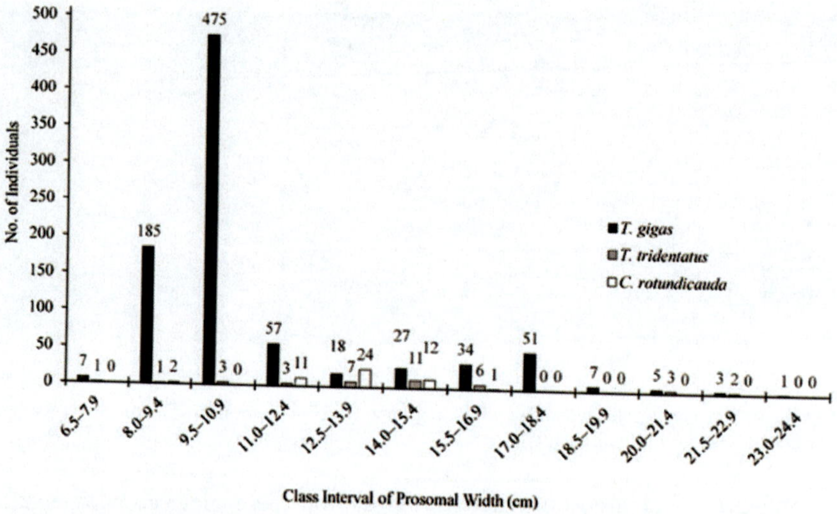


**Fig. 4** Prosomal width distribution of horseshoe crabs caught from the gillnet bycatch

from sub-adults, the smallest PW of mature *T. gigas* recorded in this study was 16.5 cm.

The body size of the horseshoe crabs can also be used to estimate their age (Cartwright-Taylor et al. 2009). According to Cartwright-Taylor et al. (2009), *C. rotundicauda* tended to be smaller than *T. gigas* and *T. tridentatus*. They reported that in Kranji, Singapore, *C. roundicauda* with a carapace width of <8 cm belonged to the juvenile stage, while that of the adult stage was >8 cm and up to 16 cm. However, this size classification may not be explicitly specific to a particular species. The observations in the present study suggested that all *C. rotundicauda* had a range of PW of 8.0–16.9 cm with a total body length of 19.0–36.5 cm. Thus, the size of *C. rotundicauda* collected from this bycatch study was primarily categorized as adults.

Manca et al. (2017) reported from Sabah, Malaysia, that adult *T. tridentatus* had PW ranging from 20.0–29.8 cm. In Japan, Wada et al. (2010) also reported the smaller size of mature *T. trdientatus* with mean PW 22.6–27.6 cm, while Harada (2009) recorded sub-adult individuals with PW of 14.2 cm, which was estimated to be at 12th instar stage and around 10 years old. In this study, it was found that only five individuals of *T. tridentatus* were categorized as adults with PW ranging from 20.0–22.9 cm, 17 individuals as sub-adults with PW ranging from 14.0–16.9 cm, and the remaining 15 individuals as juveniles with PW ranging from 6.5–13.9 cm. Hence, the present findings were comparable to that reported from literature.

In terms of body condition (Fig. 5), 360 (37.62%) horseshoe crabs collected from the bycatch had intact carapace and were undamaged, 561 (58.62%) were at molting stage, 12 (1.25%) had abnormal body form, and 24 (2.51%) were gravid females at spawning stage. The abnormal body form may be due to congenital

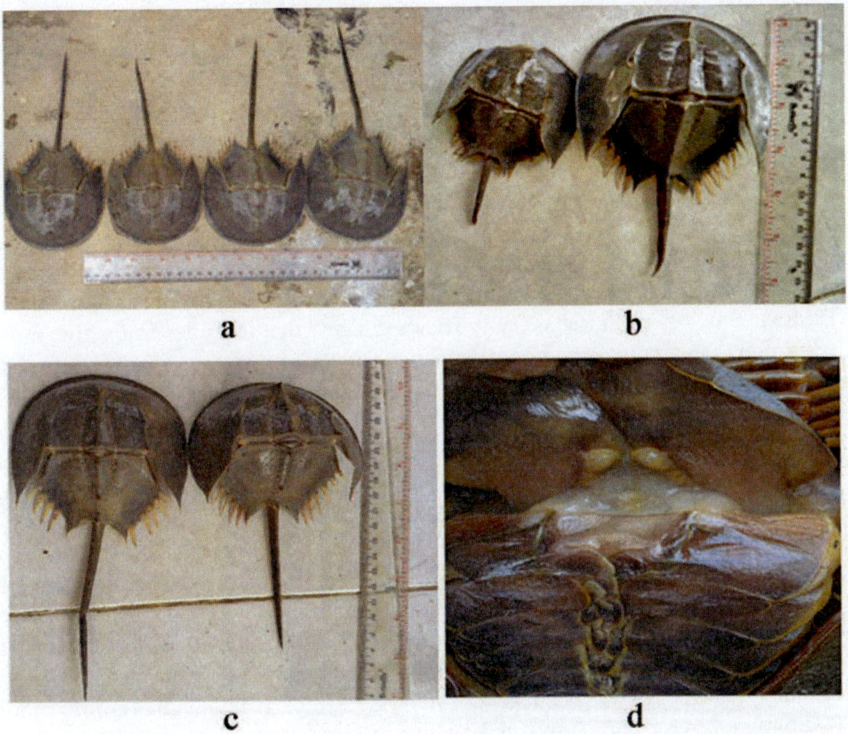


**Fig. 5** Condition of horseshoe crabs collected from gillnets: (**a**) specimens at molting stage, (**b**) specimens with abnormal, short telson, (**c**) specimens with bent telson or damaged carapace, (**d**) gravid female horseshoe crab

disabilities or damage of body parts (e.g., carapace, telson) from entanglement with the gillnet during the fishing operation.

A large number of juveniles caught in the molting condition could be related to the study period, as the entire observation was carried out from June to August, which corresponded to the dry season in Indonesia. During the dry season, Mayangan water temperature usually ranges from 28.7 to 30 °C. An increase in water temperature as well as nearshore salinity due to limited rainfall may be favorable to the growth, and hence molting, of horseshoe crabs (Watson et al. 2009; Cheng et al. 2015). Similar findings were also reported by Carmichael et al. (2003) and Estes Jr et al. (2015) for the American horseshoe crabs on the east coast of the United States. Based on personal observations, in Mayangan, a large proportion of molting horseshoe crabs was caught in summer. Meanwhile, in the rainy season with high tides, the horseshoe crabs captured were dominated by individuals with solid carapace. However, there was no specific information on the peak and low seasons of horseshoe crab abundance. A study from Muslihah (2004) in Mayangan reported that the spawning season of *T. gigas* occurs in August–October, and the highest gonad maturity index occurs in August.

As confirmed by the fishermen, horseshoe crabs are usually caught in large numbers during the dry season, most of which are small or juvenile with soft carapace. Such a soft carapace condition allows fishermen to remove horseshoe crabs from their nets quickly, although the risk of body damage is still high. It has been observed that some parts of the body of juvenile horseshoe crabs, such as the carapace and telson, are susceptible to damages during the handling and releasing of trapped individuals by the local fishing community (Fig. 6a, b). The exact number of survival rate of horseshoe crabs at the molting stage caught in gillnets could not be confirmed, but many likely died due to their weak condition, especially if they were severely damaged. This condition was also confirmed by Carmichael et al. (2003) in Pleasant Bay, Massachusetts, USA. They estimated that juveniles had a high mortality rate than adults. Based on personal observations, horseshoe crabs could survive three days on land if the fishermen did not release them on the same day during the harvest of blue swimming crabs with the bycatch. On the other hand, their damaged body parts could be bleeding. James-Pirri (2012) suggested that the bleeding may affect the physiology and activity level of the horseshoe crabs (Anderson et al. 2013). A large number of molting horseshoe crabs could also show that many juveniles were growing into sub-adults and together with many small-size juveniles being recorded, the present findings thus indicated that there has been recruitment to the local horseshoe crab population, mostly for *T. gigas*.

The presence of gravid female horseshoe crabs in Mayangan is most likely that they migrate from waters to the coast for spawning. A similar phenomenon occurred in other parts of Asia, such as in Malaysia (Robert et al. 2014), Singapore (Cartwrigth-Taylor et al. 2009), and Thailand (Fazrul et al. 2015), where summer is the peak of the spawning season for horseshoe crabs. The rising in water temperatures may provide a cue for horseshoe crabs to move to the estuary region for spawning and to continue their growth through molting. According to Smith et al. (2016), the spawning period takes place on the coast or bay protected from ocean

a. b.

**Fig. 6** (**a**) Juvenile horseshoe crabs in molting condition collected, (**b**) fisherman released the bycatch horseshoe crabs into the nearby river waves so that the eggs are safe from strong wave action. After the eggs hatch, juveniles will stay near the beach. Based on personal observations, some fishermen found horseshoe crabs eggs in the sand near the mangroves, but cannot recognize the species. The gravid females caught as bycatch are usually released similar to other entangled specimens. However, sometimes these female horseshoe crabs are retained as food, but only few people can cook them safely. There are some cases reported that people died after consuming horseshoe crabs.

Apart from the high abundance of juvenile horseshoe crabs and those at the molting stage, most specimens collected from the bycatch were females with a total of 813 individuals (85.0%), while males contributed to 144 individuals (15.0%), indicating a male-to-female sex ratio of 1:5.6. The sex ratio observed in the present survey was in agreement with previous studies (Muslihah 2004; Meilana 2015), in which females were also found to be more abundant than males in Mayangan waters. Table 1 summarizes the sex ratio data of horseshoe crab species reported from studies in Mayangan area. It was observed that the number of females is always higher than males, suggesting that one male could have been able to fertilize many female horseshoe crabs without competition. An opposite trend in this sex ratio was reported from studies in the United States, in which the number of males is higher than that of females, for example, in Florida with a male-to-female ratio of 1.4:1 (Rudloe 1980; Cohen and Brockmann 1983), Delaware Bay with a ratio of 1.5:1 to 5:1 (Shuster and Botton 1985: Smith et al. 2002), Pleasant Bay with a ratio of 9:1 (Carmichael et al. 2003), and Nauset Estuary with a ratio of 1.6:1 (James-Pirri et al. 2005). One possible reason for the observed higher male-to-female ratio in these areas could be due to the harvest of more female horseshoe crabs for biomedical and bait uses for the past 30 years (Leschen and Correia 2010; Novitsky 2015).

The sex ratio of horseshoe crabs coming onto the shore for spawning cannot be precisely determined. However, females are more abundant than males. Charmichael et al. (2003) stated that the low number of male horseshoe crabs would cause no competition in finding a female counterpart during the mating season. In this study, amplexus of horseshoe crabs was not found, but only gravid females. This could happen when mating horseshoe crabs migrated to the shore to spawn and were entangled with fishing nets, and the male horseshoe crabs were detached from their female counterparts. In Mayangan, the high abundance of female horseshoe crabs may also be caused by the absence of exploitation for large-scale trading of

**Table 1** The individual numbers of male (M) and female (F) horseshoe crab species and their sex ratio reported from Mayangan waters (– = no data)

| *T. gigas* | | | *T. tridentatus* | | M:F | *C. rotundicauda* | | | |
|---|---|---|---|---|---|---|---|---|---|
| M | F | M:F ratio | M | F | ratio | M | F | M:F ratio | Data source |
| 126 | 744 | 1:5.9 | 7 | 30 | 1:4.2 | 11 | 39 | 1:3.5 | Present study |
| 0 | 19 | 0: 9 | 0 | 17 | 0:17 | 4 | 18 | 1:4.5 | Meilana 2015 |
| 169 | 209 | 1:1.2 | – | – | – | – | – | – | Muslihah 2004 |

horseshoe crabs in the Subang Regency. In Hong Kong, Malaysia, and Thailand, female horseshoe crabs are usually targeted to serve as a desirable culinary delicacy (Shin et al. 2009; Meilana et al. 2016; John et al. 2018). This may result in fewer females observed in the field. In contrast, local villagers are less interested in horseshoe crab consumption in Subang. Moreover, some buyers from other provinces tend to obtain horseshoe crabs from where they live rather than coming to Subang for horseshoe crabs. Therefore, horseshoe crabs are regarded as non-economical catch and harmful for fishermen in Mayangan, because their entanglement as bycatch can damage the fishing nets.

Horseshoe crabs caught as bycatch in Mayangan may just serve as one case study in Indonesia or in other southeast Asian countries where similar nearshore gillnet fishing activities take place. This is a dilemma that, on the one hand, the survival of horseshoe crabs is threatened. On the other hand, fishermen are disadvantaged when they have to replace their fishing gear due to damage resulting from entanglement with horseshoe crabs and the effort in releasing the trapped horseshoe crabs. However, if this condition is left unchecked and such bycatch of horseshoe crabs continued, there will be a risk in the depletion of the local horseshoe crab population in the long term. Thus, an effort is needed to educate fishermen to save those horseshoe crabs entangled in the nets. It may also be advisable to focus the harvest area for blue swimming crabs in waters with minimal existence of horseshoe crabs or to devise new fishing gear that can prevent horseshoe crabs from being trapped during the fishing operation.

## 4 Conclusion

The fishing activity using gillnets in the Mayangan area has a high potential for horseshoe crabs to be caught as bycatch. Such bycatch can also have a negative economic impact on the local fishermen as gillnet fishing for the blue swimming crabs is the main source of their income. In particular, those horseshoe crabs at their molting stage were more vulnerable to mortality as they could be discarded as bycatch on land and suffer high mortality. During the peak season when many horseshoe crabs were trapped in the fishing nets, the fishermen were unable to release all of the individuals on the same day, leaving the horseshoe crabs entangled on the nets for several days. This could be a factor for the decline of local horseshoe crab population, even horseshoe crabs were not the fishing target as the local community is less interested in consuming or selling them. Further studies are needed to assess the economic value and survival rate of horseshoe crabs after their capture, identify the peak season of such bycatch, estimate population size through tagging and recapture, assess the stage of maturity, and investigate the spawning beaches. The stakeholders, including government conservation officials and local fisherman communities, should devise a solution to the bycatch problem since the horseshoe crab habitat and fishing ground are interlocked and must support each other. It is also necessary for fishermen to obtain basic knowledge on how to manage horseshoe crab resources in the area.

**Acknowledgments** The authors would like to thank the volunteers (Romdhon, Ilham, Algan, and Robi) and fishermen (Kardi, Jeje, Casmadi, Zaenudin, Carkim, Aang, Riki, Rustama, and Nana) for their assistance in the field, and the reviewers for their constructive comments. Thanks are also extended to Dr. Paul Shin whose advice and editing have improved the manuscript greatly.

## References

R.L. Anderson, W.H. Watson III, C.C. Chabot, Sublethal behavioral and physiological effects of the biomedical bleeding process on the American horseshoe crab. Biol. Bull. **225**, 137–151 (2013)

W.D. Badiuzzaman, T. Yulianto, Capture potential analysis of blue swimming crab at Demak waters. J. Fish. Resour. Utiliz. Manage. Technol. **3**(3), 248–256 (2014) (in Indonesian with English Abstract)

A.J. Bianchi Effects of fishing practices, gear parameters, and gear configuration on target and incidental catch in the southern flounder (*Paralichthys lethostigma*) gillnet fishery of Pamlico Sound, North Carolina. North Carolina State University, USA (2002), pp. 1–95

R.H. Charmichael, D. Rutecki, I. Valilea, Abundance and population structure of the Atlantic horseshoe crab *Limulus polyphemus* in Pleasant Bay, Cape Cod. Mar. Ecol. Prog. Ser. **246**, 255–239 (2003)

L. Cartwrigt-Taylor, J. Lee, C.C. Hsu, Population structure and breeding pattern of the mangrove horseshoe crab *Carcinoscorpius rotundicauda* in Singapore. Aquat. Biol. **8**, 61–69 (2009)

L. Cartwright-Taylor, Y.V. Bing, H.C. Chi, et al., Distribution and abundance of horseshoe crabs *Tachypleus gigas* and *Carcinoscorpius rotundicauda* around the main island of Singapore. Aquat. Biol. **13**, 127–136 (2011)

A. Chatterji, *The Horseshoe Crab – a Living Fossil* (Project Swarajya Publication, India, Orissa, 1994)

H. Cheng, C.C. Chabot, W.H. Watson III, The life history cycle of *Limulus polyphemus* in the Great Bay Estuary, New Hampshire USA, in *Changing Global Perspective on Horseshoe Crab Biology, Conservation and Management*, ed. by R. H. Carmichael, M. Botton, P. K. S. Shin, S. G. Cheung, (Springer, Cham, 2015), pp. 237–257

J.A. Cohen, H.J. Brockmann, Breeding activity and mate selection in the horseshoe crab, *Limulus polyphemus*. Bull. Mar. Sci. **33**, 274–228 (1983)

M.G. Estes, Jr, R.H. Carmichael, P.D.M. Macdonald, Molt reveals life-history patterns for juvenile American horseshoe crab in fringe habitats. In: Carmichael RH, Botton M, Shin PKS, Cheung SG (eds) Changing Global Perspective on Horseshoe Crab Biology, Conservation and Management, (Springer, Cham, 2015), pp. 255–278

M. Faridah, N. Ismail, A.B. Ahmad, et al., The population size and movement of coastal horseshoe crab, *Tachypleus gigas* (Müller) on the east coast of Peninsular Malaysia, in *Changing Global Perspectives on Horseshoe Crab Biology, Conservation and Management*, ed. by H. Charmichael Ruth, M. L. Botton, K. S. Shin Paul, C. S. Gin, (Springer, Cham, 2015), pp. 213–228

H. Fazrul, S. Hajisame, M. Ikhwanuddin, et al., Assessing impact of crab gillnet fishery to bycatch population in the lower Gulf of Thailand. Turkish J. Fish. Aquat. Sci. **15**, 761–771 (2015)

N. Harada, Growth rate for larvae of horseshoe crab (*Tachypleus tridentatus*) from a field survey. Kabutoganis **29**, 33–34 (2009)

M.J. James-Pirri, Assessment of spawning horseshoe crabs (*Limulus polyphemus)* at Cape Cod National Seashore, 2008–2009. Natural Resource Technical Report NPS/CACO/NRTR-2012/573 (2012), National Park Service, Fort Collins, Colorado, USA

M.J. James-Pirri, K. Tuxbury, S. Marino, et al., Spawning densities, egg densities, size structure, and movement patters of spawning horseshoe crabs, *Limulus polyphemus*, within four coastal embayments on Cape Cod, Massachusetts. Estuaries **28**, 296–313 (2005)

B.A. John, B.R. Nelson, H.I. Sheikh, et al., A review on fisheries and conservation status of Asian horseshoe crabs. Biodivers. Conserv. **27**, 3573–3598 (2018)

A.S. Leschen, S.J. Correia, Mortality of female horseshoe crabs (*Limulus polyphemus* ) from biomedical bleeding and handling: Implications for fisheries management. Mar. Freshw. Behav. Physiol. **43**(2), 135–147 (2010)

A. Manca, F. Mohammad, A. Ahmad, et al., Tri-spine horseshoe crab, *Tachypleus tridentatus* (L) in Sabah, Malaysia: The adult body size and population estimate. J. Asia-Pacific Biodivers. **10**, 355–361 (2017)

A. Mashar, N.A. Butet, B. Juliandi, et al., Biodiversity and distribution of horseshoe crabs in northern coast of Java and southern coast of Madura. IOP Conf. Ser.: Earth Environ. Sci. **54**, 012076 (2017)

L. Meilana, Assessment of horseshoe crabs (Xiphosura, Limulidae) morphology and genetic as basis for conservation and management in the Java Island coast. Master thesis, IPB University, Bogor, Indonesia (in Indonesian with English Summary) (2015)

L. Meilana, Y. Wardiatno, N.A. Butet, et al., Morphological character and molecular identification with COI gene marker of horseshoe crab (*Tachypleus gigas*) at coastal water of northern Java Island. Jurnal Ilmu dan Teknologi Kelautan Tropis **8**(1), 145–158 (2016) (in Indonesian with English Abstract)

L. Meilana, Q. Fang, Local knowledge-based study on the status of horseshoe crabs along the Indonesian coast. Reg. Stud. Mar. Sci. **36**, 1–9 (2020)

J.K. Mishra, Horseshoe crabs, their eco-biological status along the northeast coast of India and the necessity for ecological conservation, in *Biology and Conservation of Horseshoe Crabs*, ed. by J. T. Tanacredi, M. L. Botton, D. R. Smith, (Springer, New York, 2009), pp. 89–96

M.B. Mulya, Preservation and utilization of genetic resources of Mimi Ranti (*Carcinoscorpius rotundicauda,* L) and Mimi Bulan (*Tachypleus gigas*, M). Digital Library of Notrhern Sumatra University, Indonesia (in Indonesian) (2004)

M. Muslihah, Biologycal reproduction aspects of Mimi Bulan (*Tachypleus gigas*) in Mayangan waters, Subang Regency of West Java. Bachelor thesis, IPB University, Bogor, Indonesia (in Indonesian) (2004)

T.J. Novitsky, Biomedical implication for managing the *Limulus polyphemus* harvest along the northeas coast of the United States, in *Biology and Conservation of Horseshoe Crabs*, ed. by J. T. Tanacredi, M. L. Botton, D. R. Smith, (Springer, New York, 2015), pp. 483–500

G.D. Prasetyo, R.I. Wahju, R. Yusfiandayani, et al., Green Light Emitting Diode (LED) and its effect on sea turtle bycatch reduction of gillnet fisheries in Paloh waters. Mar. Fish. **8**(1), 87–99 (Indonesian with English Abstract) (2017)

R. Robert, A. Muhammad, P.F. Amelia-Ng, Demographics of horseshoe crab population in Kota Kinabalu, Sabah, Malaysia with emphasis on *Carcinoscorpius rotundicauda* and some aspects of its mating behaviour. Pertanika J. Trop. Agri. Sci. **37**(3), 375–388 (2014)

E. Rubiyanto, Study population of horseshoe rabs (Xiphosura) in Peninsular Kuala Tungkal, the district of Tanjung Jabung Barat, Jambi. Master thesis, IPB University, Bogor, Indonesia (in Indonesian with English Summary) (2012)

A. Rudloe, The breeding behavior and patterns of movement of horseshoe crabs *Limulus polyphemus* in the vicinity of breeding beaches in Apalachee Bay, Florida. Estuaries **3**, 177–183 (1980)

K. Sekiguchi, *Biology of Horseshoe Crabs* (Science House Co., Tokyo, 1988)

K. Sekiguchi, C.N. Shuster Jr., Limits on the global distribution of horseshoe crab (Limulidae): Lessons learned from two lifetimes of observations: Asia and America, in *Biology and Conservation of Horseshoe Crabs*, ed. by J. T. Tanacredi, M. L. Botton, D. R. Smith, (Springer, New York, 2009), pp. 5–24

P.K.S. Shin, H.Y. Li, S.G. Cheung, Horseshoe crabs in Hong Kong: Current population status and human exploitation, in *Biology and Conservation of Horseshoe Crabs*, ed. by J. T. Tanacredi, M. L. Botton, D. R. Smith, (Springer, New York, 2009), pp. 347–360

D.R. Smith, P.S. Pooler, B.L. Swan, et al., Spatial and temporal distribution of horseshoe crab (*Limulus polyphemus*) spawning in Delaware Bay: Implications for monitoring. Estuaries **38**, 115–125 (2002)

D.R. Smith, H.J. Brockmann, M.A. Beekey, et al., Conservation status of the American horseshoe crab (*Limulus polyphemus*): A regional assessment. Rev. Fish. Biol. Fisher. **27**(1), 1–4 (2016)

C.N. Shuster Jr., M.L. Botton, A contribution to the population biology of horseshoe crabs, *Limulus polyphemus* (L.), in Delaware Bay. Estuaries **8**(4), 363–372 (1985)

C.N. Shuster Jr., The Delaware Bay area, U.S.A.: a unique habitat of the American horseshoe crab, *Limulus polyphemus*, in *Changing Global Perspectives on Horseshoe Crab Biology, Conservation and Management*, ed. by R. H. Carmichael, M. L. Botton, S. PKS, S. G. Cheung, (Springer, Cham, 2015), pp. 15–40

F.N. Supadminingsih, R.I. Wahju, M. Riyanto, Composition of blue swimming crab (*Portunus pelagicus*) and horseshoe crab on the gillnet fishery in Mayangan Waters, Subang, Wes Java. AACL Bioflux **12**(1), 14–24 (2019)

I.K. Suparta, Morphometric character of horseshoe crab *Tachypleus gigas* (Müller) and *Carcinoscorpius rotundicauda* (Latreille) in coastal Pandeglang, West Java and coastal Rembang, Central Java. Bachelor thesis, Bogor Agricultural University, Indonesia (in Indonesian) (1992)

T. Wada, S. Itaya, M. Shuuno, Spawning site and annual variability of the number of reproductive visiting pairs of the horseshoe crab *Tachypleus tridentatus* along the Tsuyazaki coast in Fukuoka, Japan. Jap. J. Conserv. Ecol. **15**(2), 163–171 (2010)

K. Wakfield, *Saving the Horseshoe Crab: Designing a More Sustainable Bait for Regional Eel and Conch Fisheries* (University of Delware, USA, Sea Grant Delware, 2012)

W.H. Watson III, S.Y. Schaller, C.C. Chabot, The relationship between small-and large-scale movements of horseshoe crabs in the Great Bay estuary and *Limulus* behavior in the laboratory, in *Biology and Conservation of Horseshoe Crabs*, ed. by J. T. Tanacredi, M. L. Botton, D. R. Smith, (Springer, New York, 2009), pp. 131–147

# Occurrence and Morphometric Analysis of Horseshoe Crabs Along Bhadrak Coast of Odisha, India

**Shaik Rahimuddin, Siddhartha Pati, Sanatan Tudu, Ayaskanta Pramanik, Apratim Sai Rajesh, Nirlipta Mishra, Gobinda Chandra Biswal, Anil Chatterji, and Bisnu Prasad Dash**

## 1 Introduction

Coastal areas are characterized by high productivity and species richness (Ray 1991), and humans depend strongly on a variety of species that are living and reproducing in the coastal zone, such as fishes (Allison et al. 2009; Bell et al. 2009), sea

S. Rahimuddin (✉)
Horseshoe Crab Research Unit, Department of Biosciences and Biotechnology, Fakir Mohan University, Balasore, Odisha, India

Department of Zoology, Chandbali College, Bhubaneswar, Odisha, India

S. Pati · S. Tudu · A. S. Rajesh · G. C. Biswal
Horseshoe Crab Research Unit, Department of Biosciences and Biotechnology, Fakir Mohan University, Balasore, Odisha, India

SIAN Institute, Association for Biodiversity Conservation and Research (ABC), Balasore, Odisha, India

A. Pramanik
Horseshoe Crab Research Unit, Department of Biosciences and Biotechnology, Fakir Mohan University, Balasore, Odisha, India

A. Chatterji
SIAN Institute, Association for Biodiversity Conservation and Research (ABC), Balasore, Odisha, India

N. Mishra
Department of Zoology, Chandbali College, Bhadrak, Odisha, India

B. P. Dash (✉)
Horseshoe Crab Research Unit, Department of Biosciences and Biotechnology, Fakir Mohan University, Balasore, Odisha, India

Department of Zoology, Division of Biological Sciences, AIPH University, Bhubaneswar, Odisha, India



J. T. Tanacredi et al. (eds.), *International Horseshoe Crab Conservation and Research Efforts: 2007-2020*, https://doi.org/10.1007/978-3-030-82315-3_14

cucumbers (Purcell et al. 2013), and horseshoe crabs (Chatterji 1994). As a marine arthropod, horseshoe crab has persisted for more than 200 million years and is still enjoying its life on earth without much morphological modifications (Kin and Błażejowski 2014; Souji 2015). Four species of horseshoe crabs have been reported to exist in the world. Three species occur in the coastal waters of India, Southeast Asia, China, and Japan, namely *Carcinoscorpius rotundicauda* (Latreille 1802), *Tachypleus gigas* (Müller 1785), and *T. tridentatus* (Leach 1819). One species, *Limulus polyphemus* (Linnaeus 1758), only occurs in the coastal waters of the eastern continental shelf of North America and the Gulf of Mexico (Sekiguchi and Shuster 2009). Along the northeast coast of Bay of Bengal, India, there exists two species: the triangle-tailed horseshoe crab, *T. gigas* and the round-tailed horseshoe crab, *C. rotundicauda*. *T. gigas* lives in sandy to muddy habitats (Davidson et al. 2008; Pati et al. 2020a), whereas *C. rotundicauda* inhabits in muddy areas, commonly in brackish waters (Chatterji and Abidi 1993; Davidson et al. 2008). In some areas of India, *T. gigas* co-exists with C. *rotundicauda* (Chatterji 1999; Mishra 2009).

The global decline in horseshoe crab populations can be due to anthropogenic activities (Pati et al. 2017; John et al. 2018; Mohamed et al. 2021; Pati et al. 2022). Wild-caught horseshoe crabs are used commercially in medicine, where substances from their haemolymph (*Tachypleus* Amoebocytes Lysate) are used to test in drugs, blood products, and pharmaceutical uses (Levin et al. 2003). Many studies have provided information on different aspects of Indian horseshoe crabs particularly *T. gigas*, such as occurrence and distribution, morphometric and allometric analysis, spawning density, captive rearing, and biomedical importance (Lazarus et al. 1990; Vijayakumar et al. 2000; Sahu and Dey 2013; Chatterji and Pati 2014; Kumar et al. 2015; Biswal et al. 2016; Pati et al. 2020b; Tudu et al. 2021; Nong et al. 2021; Ayaskanta et al. 2021; Pati et al. 2021a, b). However, very limited studies on distribution and morphometric analysis of *C. rotundicauda* are documented in Odisha (Pati et al. 2015), and there is no report from Bhadrak district. This study aims to collect and record the status of both sympatric species of horseshoe crabs, *T. gigas* and *C. rotundicauda*, along the coast of Bhadrak district in Odisha. These data will hopefully help in improving the conservation and management of local horseshoe crabs and highlight their importance in maintaining ecosystem services in coastal areas of Bhadrak district.

## 2 Materials and Methods

### *2.1 Study Site*

Bhadrak district (20°43′-21°13′N and 86°6′-87°E) is located in the northeast part of Odisha state of India. Four other districts, namely Balasore, Kendrapara, Jajpur, and Keonjhar, surround the Bhadrak district, where its eastern border is bounded by the Bay of Bengal (Fig. 1). The district spreads over 2505 km$^2$, 1.61% of the total land

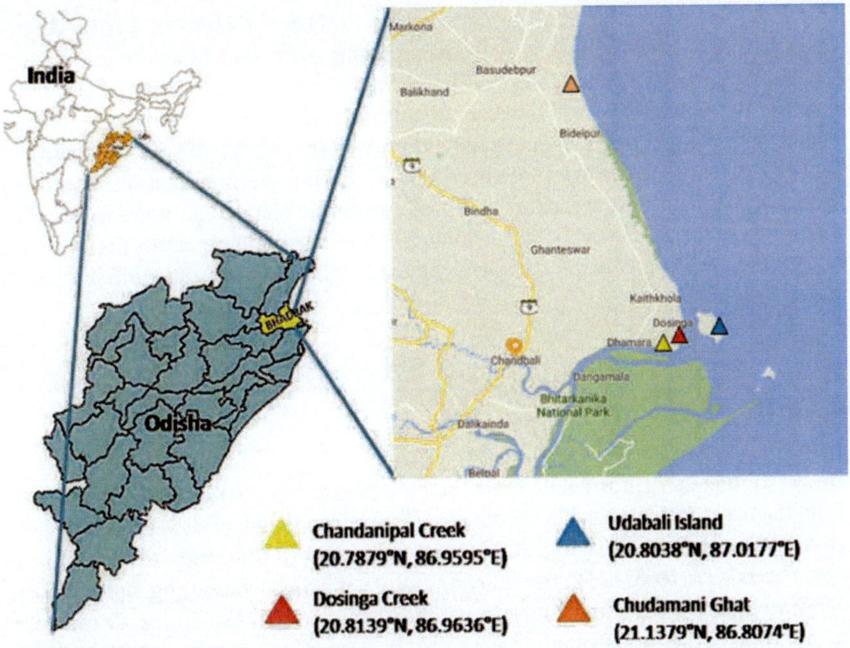


**Fig. 1** Map showing the study sites in Bhadrak district of Odisha, India

area of the state, and has 1.507 million inhabitants, 3.59% of the state's population (from 2011 Census). Most of its people live in villages and are engaged in agricultural practices as their primary occupation. Being situated close to the Bay of Bengal, the district is characterized by occasional thunderstorms, and rains and dust storms in April and May.

## 2.2 Data Collection

The present survey was carried out during February to August 2017 in four places, i.e., Chandanipal Creek (20.7879°N, 86.9595°E), Dosinga Creek (20.8139°N, 86.9636°E), Udabali or Kanika Island (20.8038°N, 87.0177°E), and Chudamani Ghat (21.1379°N, 86.8074°E) (Fig. 1). The first three places come under the Dhamara coastal region and the last is within the Basudevpur coastal region. These sites were selected based on the reported sightings and capture of horseshoe crabs by local people and fishermen. The survey was conducted by members of the author team on every fortnight based on the tidal time and new moon and full moon days of the lunar cycle. In Chandanipal and Dosinga Creeks, the samples were collected from intertidal zone towards south up to a distance of 2 km. However, the samples collected from Udabali Island covered a distance of 500 m radius. In Chudamani

Ghat, samples were collected from the one side of Gamei river bank. Live and dead horseshoe crabs were collected by the handpicking method and identified following the description in Rao and Rao (1972). The live horseshoe crabs were released into the sea after taking their body measurements. Interview with informants were conducted following established and standard procedures (Martin 1995). The male and female horseshoe crabs were identified by observing the presence or absence of monodactylus at the first pair of pedipalps, carapace shape, and gonophores. The environmental parameters of water including pH, temperature, and salinity of the sites were measured using a digital pH meter, thermometer, and refractometer, respectively.

## 2.3 Morphometric Measurements

Body parameters of horseshoe crabs including total body length (TBL), which is from the tip of the carapace to the tip of telson, telson length (TEL), prosoma width (PW), and opisthosoma width (OW) were measured in this study (Fig. 2). These parameters were recorded to the nearest centimeter using a measuring tape and were conducted on a flat surface to avoid parallax error. Data obtained for all measurements were pooled according to the species and sex of specimens, prior to calculation of their means and standard deviations. Morphometric analysis of differences in these mean body measurements of males and females in both *T. gigas* and *C. rotundicauda* was conducted using Pearson correlation and regression analysis. Individual body measurement between male and female horseshoe crabs of both species was also compared using Student's t-test. Significant level was set as $p < 0.05$.

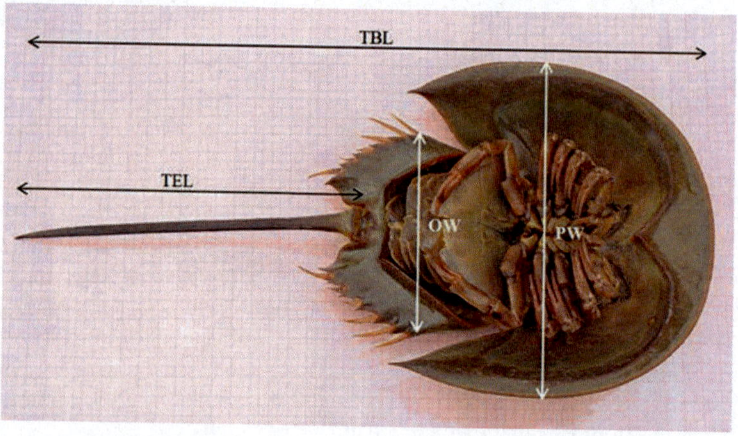


**Fig. 2** The measurement of different body parameters in horseshoe crab. TBL total body length, TEL telson length, PW prosoma width, OW opisthosoma width

## 3 Results

### 3.1 Total Samples of Horseshoe Crabs

During the study period, a total of 152 individuals of *T. gigas* and 117 individuals of *C. rotundicauda* were collected at the four sites. Of these, 21 male, 18 female individuals of *T. gigas* and 13 male, 15 female individuals of *C. rotundicauda* were recorded at Chandanipal Creek. In Dosinga Creek, 26 male, 30 female individuals of *T. gigas* and 18 male, 21 female individuals of *C. rotundicauda* were found. On Udabali Island, 11 male, 13 female individuals of *T. gigas* and 15 male, 19 female individuals of *C. rotundicauda* were noted. Similarly, at Chudamani Ghat, 16 male, 17 female individuals of *T. gigas* and 7 male, 9 female individuals of *C. rotundicauda* were encountered. Table 1 summarizes the salinity, pH, and temperature at these four survey sites with the numbers of horseshoe crabs recorded. The salinity, pH, and temperature of the four sites were approximately the same during the study period. Both *T. gigas* and *C. rotundicauda* were noted in Chandanipal and Dosinga Creeks and on Udabali Island, as the sediment comprises of sandy beaches and mudflats. However, at Chudamani Ghat the number of *C. rotundicauda* recorded

**Table 1** Environmental parameters (mean ± SD; $N = 12$) and total numbers of *T. gigas* (TG) and *C. rotundicauda* (CR) recorded at the four survey sites in Bhadrak district of Odisha, India, from February to August 2017

| | Environmental parameter | | | | | | | | |
|---|---|---|---|---|---|---|---|---|---|
| Survey site | Salinity (g/L) | pH | Temperature (°C) | Species | Sex | Live | Dead | Total | |
| Chandanipal Creek (20.7879°N, 86.9595°E) | 32.5 ± 0.7 | 8.1 ± 0.1 | 34.2 ± 0.7 | TG | M | 3 | 18 | 21 | 39 |
| | | | | | F | 4 | 14 | 18 | |
| | | | | CR | M | 2 | 11 | 13 | 28 |
| | | | | | F | 2 | 13 | 15 | |
| Dosinga Creek (20.8139°N, 86.9636°E) | 31.9 ± 0.6 | 8.1 ± 0.2 | 33.2 ± 0.5 | TG | M | 9 | 17 | 26 | 56 |
| | | | | | F | 9 | 21 | 30 | |
| | | | | CR | M | 7 | 11 | 18 | 39 |
| | | | | | F | 7 | 14 | 21 | |
| Udabali Island (20.8038°N, 87.0177°E) | 33.2 ± 0.7 | 8.2 ± 0.1 | 34.1 ± 0.7 | TG | M | 0 | 11 | 11 | 24 |
| | | | | | F | 0 | 13 | 13 | |
| | | | | CR | M | 0 | 15 | 15 | 34 |
| | | | | | F | 0 | 19 | 19 | |
| Chudamani Ghat (21.1379°N, 86.8074°E) | 32.5 ± 0.6 | 8.1 ± 0.2 | 32.4 ± 0.5 | TG | M | 5 | 11 | 16 | 33 |
| | | | | | F | 5 | 12 | 17 | |
| | | | | CR | M | 0 | 7 | 7 | 16 |
| | | | | | F | 0 | 9 | 9 | |
| Total | | | | | | 53 | 216 | 269 | 269 |

was less in comparison to that of *T. gigas*. Most of the horseshoe crabs reported were dead specimens (80.3%). On Udabali Island, no live horseshoe crabs were found, while at Chudamani Ghat only live *T. gigas* were recorded.

## 3.2 Morphometric Analysis

The morphometric data measured from 269 individuals of horseshoe crabs are summarized in Table 2, in terms of their mean values, standard deviations, and student's t-test results. Among the four body measurements, male *T. gigas* was found to have longer telson ($p < 0.05$) and wider prosoma ($p < 0.05$) than male *C. rotundicauda*, whereas there was no difference in total body length and opisthosoma width between them. There was also no statistical difference in the measured morphological parameters between female *T. gigas* and C. *rotundicauda*. For both species, females also tended to be larger than males. The mean TBL and TEL, as well as the mean PW and OW, of female *T. gigas* were significantly greater than the male counterparts at $p < 0.001$ and $p < 0.05$ level, respectively. In case of *C. rotundicauda*, the mean TEL and PW, as well as mean TBL and OW, of females were also statistically larger than males at $p < 0.001$ and $p < 0.05$ level, respectively. There was also a positive correlation ($p < 0.05$) between TBL with TEL, PW, and OW in male and female horseshoe crabs for both species. Results of regression analysis generally showed positive relationship between morphometric measurements in male and female of both horseshoe crab species, with coefficient of determination ($R^2$) >0.49. Figures 3 and 4 depict the regression plots, with $R^2 > 0.80$, between TBL and TEL, and TEL and PW, for male and female of *T. gigas* and *C. rotundicauda*.

**Table 2** The mean and standard deviation and student's t-test comparison of morphological characters of *T. gigas* and *C. rotundicauda* in Bhadrak district of Odisha, India

| | Mean ± Standard Deviation (cm) | | | |
|---|---|---|---|---|
| Body Parameter | *T. gigas* (Male) $N = 74$ | *T. gigas* (Female) $N = 78$ | *C. rotundicauda* (Male) $N = 53$ | *C. rotundicauda* (Female) $N = 64$ |
| TBL | 30.2 ± 3.3 | 36.5 ± 3.5** | 29.4 ± 2.1 | 34.6 ± 4.6# |
| TEL | 16 ± 1.4§ | 18.1 ± 1.8** | 14.8 ± 1.3 | 17.8 ± 2.1## |
| PW | 14.8 ± 1.5§ | 16.8 ± 1.7* | 13.7 ± 1.3 | 16.7 ± 1.9## |
| OW | 9.7 ± 0.9 | 11.1 ± 0.9* | 9.5 ± 1.1 | 11.4 ± 0.8# |

* and ** denote significant difference between female and male *T. gigas* at $p < 0.05$ and $p < 0.001$ respectively

# and ## denote significant difference between female than male *C. rotundicauda* at $p < 0.05$ and $p < 0.001$ respectively

§denotes significant difference between male *T. gigas* and *C. rotundicauda* at $p < 0.05$

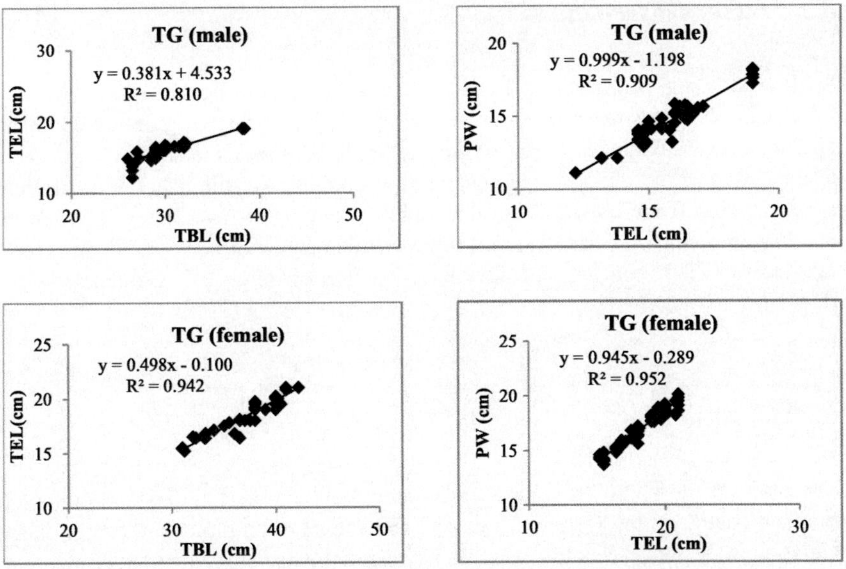


**Fig. 3** Regression analysis of TBL vs TEL and TEL vs PW in male and female *T. gigas*

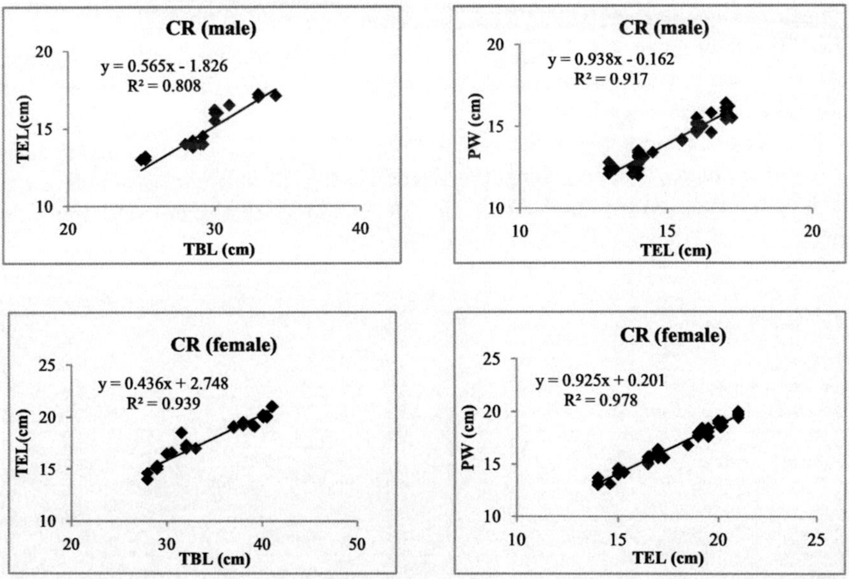


**Fig. 4** Regression analysis of TBL vs TEL and TEL vs PW in male and female *C. rotundicauda*

### 3.3 Other Observations

During sampling, physical conditions were observed in live horseshoe crabs. Some of the caught horseshoe crabs were not physically intact, showing deformed telson and/or damaged carapace (2 out of 54 numbers). Five specimens of *T. gigas* were also infested with barnacles. Through additional sightings from fishermen, different sizes (ranging from 7.4–18.5 cm of total body length) of juveniles of *C. rotundicauda* were found on Udabali Island (Fig. 5) and many larval forms of *T. gigas* (size ranges from 11–15 mm) were caught as by-catch in a fishing net at Chudamani Ghat (Fig. 6).

## 4 Discussion

The study of morphometric characters is an effective tool to understand the variation and change in the form, such as size and shape, of an organism. A significant variation in the morphometric characteristics of animals belonging to the same population has been reported by many workers (Chatterji 1994; Srijya et al. 2010; Itow et al. 2004). The variation in morphological characters can help in determining the age and sex of an individual. In this study, the average body length of female *T. gigas* was 36.5 cm and that of the male was 30.2 cm and the average body length of female *C. rotundicauda* was 34.6 cm and that of the male was 29.4 cm. The average body length of females in both *T. gigas* and *C. rotundicauda* are significantly longer than that of the males, which corroborates with the findings of other authors reported on Indian horseshoe crabs (Itow et al. 2004).

The relationship between various body parts of *T. gigas* and *C. rotundicauda* revealed an increase in body length proportion to the telson length, prosoma width, and opisthosoma width. Total body length showed a high degree of correlation with

**Fig. 5** Juvenile *C. rotundicauda*

**Fig. 6** Larvae of *T. gigas*

telson length for male (*T. gigas*, *r* = 0.90; *C. rotundicauda*, *r* = 0.89) and female (*T. gigas*, *r* = 0.97; *C. rotundicauda*, *r* = 0.96) of both species, indicating that the increase in total body length is directly proportional to the telson length which further suggested that advancement of growth is related to the growth of body dimensions. The high degree of correlation (*r* = 0.87) between carapace length with the total length in horseshoe crabs has been reported earlier. They have also showed a good correlation (*r* = 0.84) between total body length and telson length in *T. gigas* found in Balaramgadi estuary of Balasore district of Odisha. In addition, Panda and Naik (2017) have reported a good correlation between total length and telson length in *C. rotundicauda* found in Bhitarkanika mangrove area of Kendrapada district of Odisha. Both these places were around 50–60 km away from the sampling locations of this study. All the morphological parameters of this study are in good agreement with the earlier reports almost two decades ago on Indian horseshoe crabs, which may imply that there is not much genetic or environmental impact on these organisms thus far.

The coexistence of both species of horseshoe crabs along the coast of Bhadrak district of Odisha is unique, as these sites provide both sandy estuarine beaches and mudflats for spawning by *T. gigas* and *C. rotundicauda,* respectively. A similar result was reported by Sekiguchi (1988) at Digha and Junpat along the coast of West Bengal and by Chatterji (1999) at Hukitola Island of Kendrepada district of Odisha, where both *T. gigas* and *C. rotundicauda* are available. The preference for spawning on the sandy estuarine zone by *T. gigas* and mangrove mudflats by *C. rotundicauda* has also been reported by Chatterji (1994), Mishra (2009), and Itow et al. (2004).

In this study at Chudamani Ghat, the mortality of horseshoe crabs was largely due to the use of fishing nets by the local fishermen in the intertidal areas. These nets are usually fixed during low tide; the horseshoe crabs entering the intertidal zone during high tide are therefore entangled. Barnacles that encrusted on the carapace of horseshoe crabs were also observed at Chudamani Ghat. This may increase the mortality of horseshoe crabs, as such barnacle infestation is an indication that the animal was not healthy. Barnacles may impair the function of some organs on horseshoe crabs such as eyes, gills, and appendages (Botton 2009). Over 80% of horseshoe crabs found at the survey sites were dead specimens. This may be due to several anthropogenic activities, such as the use of fishing nets, motorboats, and trawlers, which are mainly related to fisheries and construction works along the coastline (Behera et al. 2015). Such high mortality could also be a concern in maintaining the local horseshoe crab population in future.

This study reports information on the occurrence and morphological characteristics of horseshoe crabs along the Bhadrak coast of Odisha. Such data may provide a baseline for the conservation of these animals, which is lacking to date. For conservation, the author team is conducting several public awareness programmes to prevent the eventual loss of horseshoe crabs. Wider surveys on horseshoe crabs and interactions with the local villagers and fishermen are essential to enhance the relationship between human society and nature

**Acknowledgements** The authors thank the Vice Chancellor, Fakir Mohan University and Principal, Chandbali College for their support to carry out this research work.

## References

E.H. Allison, A.L. Perry, M.C. Badjeck, et al., Vulnerability of national economies to the impacts of climate change on fisheries. Fish Fish. **10**, 173–196 (2009)

P. Ayaskanta, A.S. Rajesh, T. Sanatan, et al., Review on biology, distribution and conservation challenges for horseshoe crabs in India. J. Sustain. Sci. Manag. **16**, 332–346 (2021)

S. Behera, B. Tripathy, K. Sivakumar, et al., Distribution and abundance of two sympatric species of horseshoe crabs along the Odisha Coast, India, in *Changing Global Perspectives on Horseshoe Crab Biology, Conservation and Management*, ed. by R. H. Carmichael, M. L. Botton, P. K. S. Shin, S. G. Cheung, (Springer, Cham, 2015), pp. 181–191

J.D. Bell, M. Kronen, A. Vunisea, et al., Planning the use of fish for food security in the Pacific. Mar. Policy **33**, 64–76 (2009)

G.C. Biswal, B.N. Andia, S. Pati, et al., Conservation of Indian horseshoe crab, *Tachypleus gigas* through captive rearing, in *Frontiers in Life Sciences*, ed. by R. K. Behera, E. Kariali, S. K. Sahu, (Excel India Publishers, New Delhi, 2016), pp. 179–182

M. Botton, The ecological importance of horseshoe crabs in estuarine and coastal communities: A review and speculative summary, in *Biology and Conservation of Horseshoe Crabs*, ed. by T. J. Tanacredi, M. J. Botton, D. R. Smith, (Springer Science, New York, 2009), pp. 45–64

A. Chatterji, S.A.H. Abidi, The Indian horseshoe crab – A living fossil. J. Indian Ocean Stud. **1**, 43–48 (1993)

A. Chatterji, *The Horseshoe Crab – A Living Fossil* (Project Swaarajya Publication, Orissa, 1994)

A. Chatterji, New record of the sympatric distribution of two Asian species of the horseshoe crab. Curr. Sci. **77**, 746–747 (1999)

A. Chatterji, S. Pati, Allometric relationship in the adult population of the Malaysian horseshoe crab (*Tachypleus tridentatus*; Leach). Int. J. Res. **1**, 1378–1385 (2014)

G.W.H. Davidson, P.K.L. Ng, H.C. Ho, *The Singapore Red Data Book. Threatened Plants and Animals of Singapore* (Nature Society Singapore, Singapore, 2008), pp. 1–4

T. Itow, J.K. Mishra, A.T.A. Ahmed, Horseshoe crabs (King crabs) in the Bay of Bengal, South Asia. Bull. Fac. Edu. Shizuoka Univ. Nat. Sci. Ser. **54**, 13–30 (2004)

B.A. John, B.R. Nelson, H.I. Sheikh, et al., A review on fisheries and conservation status of Asian horseshoe crabs. Biodivers. Conserv. **27**, 3573–3598 (2018)

A. Kin, B. Błażejowski, The horseshoe crab of the genus Limulus: Living fossil or orstabilomorph? PLoS One **9**, e108036 (2014)

V. Kumar, S. Roy, A.K. Sahoo, et al., Horseshoe crab and its medicinal values. Int. J. Curr. Microbiol. App. Sci. **4**, 956–964 (2015)

S. Lazarus, P. Nammalwar, V.N. Pillai, et al., Occurrence of King crab, *Tachypleus gigas* (Müller), of the Northeast coast of India. In: Mathew KJ (ed) Proceedings of the First Workshop on Scientific Results of FORV Sagar Sampada, Kochi, India, 5–7 June 1989 (1990), pp 393–395

J. Levin, H.D. Hochstein, T.J. Novitsky, Clotting cells and Limulus Amebocyte Lysate: An amazing analytical tool, in *The American Horseshoe Crab*, ed. by C. N. Shuster, R. B. Barlow, H. J. Brockmann, (Harvard University Press, Cambridge, 2003), pp. 310–340

G.J. Martin, *Ethnobotany: A Conservation Manual* (Chapman and Hall, London, 1995)

J.K. Mishra, Horseshoe crabs, their ecobiological status along the northeast coast of India and the necessity for ecological conservation, in *Biology and Conservation of Horseshoe Crabs*, ed. by J. T. Tanacredi, M. L. Botton, D. R. Smith, (Springer, New York, 2009), pp. 89–96

R. Mohamed, A.P. Nur, S.I. Nor, et al., Using applied statistics for accurate size classification of the endangered Tachypleus tridentatus horseshoe crab." J. Ecol. Eng. **22**, 273–282 (2021)

W. Nong, Z. Qu, Y. Li, et al., Horseshoe crab genomes reveal the evolution of genes and microRNAs after three rounds of whole genome duplication. Commun. Biol. **4**, 83 (2021)

S. Panda, P.K. Naik, Morphometric study of horseshoe crab (*Carcinoscorpius rotundicauda*) in Odisha. IJSS J. Surg. **3**, 48–53 (2017)

S. Pati, G.C. Biswal, B.P. Dash, Availability of *Tachypleus gigas* (Müller) along the river estuaries of Balasore district, Odisha, India. Int. J. Fish. Aquat. Stud. **2**, 334–336 (2015)

S. Pati, S. Tudu, A. Rajesh, et al., Manmade activities affected the breeding ground of horseshoe crab (*Tachypleus gigas*) along Balasore coast: Call for immediate conservation. E-Planet **15**, 145–154 (2017)

S. Pati, S. Shahimi, H.A., Edinur, et al., Extraction of People's Perception Toward Horseshoe Crab Existence in Northeast Coast of India. Front. Mar. Sci. **7**, 587335 (2020a)

S. Pati, A. Chatterji, B.P. Dash, et al., Structural Characterization and Antioxidant Potential of Chitosan by γ-Irradiation from the Carapace of Horseshoe Crab. Polymers **12**, 2361 (2020b)

S. Pati, S. Shahimi, N. Debabrata, et al., Predicting Tachypleus gigas spawning distribution with climate change in Northeast Coast of India. J. Ecol. Eng. **22**, 211–220 (2021a)

S. Pati, S. Tanmay, S.I. Hassan, et. al., γ-Irradiated chitosan from Carcinoscorpius rotundicauda (Latreille, 1802) improves the shelf life of refrigerated aquatic products. Front. Mar. Sci. **8**, 664961 (2021b)

S. Pati, S. Tudu, B.R. Nelson, et al., Fishery Bycatch Jeopardizes Indian Horseshoe Crab Tachypleus gigas Along the Northeast Coast of Odisha, India, in International Horseshoe Crab Conservation and Research Efforts: 2007–2020, ed. by John Tanacredi et al. (Springer, New York, 2022)

S.W. Purcell, A. Mercier, C. Conand, et al., Sea cucumber fisheries: Global analysis of stocks, management measures and drivers of overfishing. Fish Fish. **14**, 34–59 (2013)

K.V.R. Rao, K.V.S. Rao, Studies on Indian king crabs (Arachnida, Xiphosura). Proc. Indian Nat. Sci. Acad. B. **38**, 206–211 (1972)

G.C. Ray, Coastal-zone biodiversity patterns. Bioscience **41**, 490–498 (1991)

A.C. Sahu, L. Dey, Spawning density and morphometric characteristics of the horseshoe crab *Tachypleusgigas* (Müller) on the Balasore coast of Bay of Bengal, India. Sci. Vis. **13**, 76–84 (2013)

K. Sekiguchi, Biology of horseshoe crabs, in *Biology of Horseshoe Crabs*, ed. by K. Sekiguchi, (Science House Co., Ltd., Tokyo, 1988), pp. 50–68

K. Sekiguchi, C.N. Shuster Jr., Limits on the global distribution of horseshoe crab (Limulacea): Lessons learned from two lifetimes of observations: Asia and America, in *Biology and*

*Conservation of Horseshoe Crabs*, ed. by J. T. Tanacredi, M. L. Botton, D. R. Smith, (Springer, New York, 2009), pp. 5–24
M. Souji, *The Mystery of Horseshoe Crabs: The Reason for their Unchanging Morphology over 200 Million Years* (Seibundo Shinkosha, Tokyo, 2015), p. 143
T.C. Srijaya, P.J. Pradeep, F.M. Shaharom, et al., A new record on the morphometric variations in the populations of horseshoe crab (*Carcinoscorpius rotundicauda* Latreille) obtained from two different ecological habitats of Peninsular Malaysia. Our Nature **8**, 204–211 (2010)
S. Tudu, T. Sarkar, B. Satyanarayana, et al., Is Sex Reversal Common Phenomenon in Horseshoe Crabs Too?. Ecol. Eng. Environ. Technol. **22**, 45–52 (2021)
R. Vijayakumar, S. Das, A. Chatterji, et al., Morphometric characteristics in the horseshoe crab *Tachypleus gigas* (Arthropoda: Merostomata). Indian J. Mar. Sci. **29**, 333–335 (2000)

# Preliminiary Investigation of *Tachypleus tridentatus* (Tri-spine HSC at Sone Tidal Flat, Fukuoka Japan)

**Osamu Hayashi**

## 1 Introduction

Sone Higata, or tidal flat, extends from Sone Shinden (meaning new rice paddies, indicative of land reclamation of the area), southern Ward of Kokura, Kita-Kyushu, Fukuoka, Japan (Fig. 1). It is the largest tidal flat in Kita-Kyushu, with exposed sandy-mud surface area of 517 ha during high tides. Its northern, western, and southern sides are blocked by concrete revetments, and its eastern side faces Suou-nada, which is an open sea towards Seto Inland Sea.

Since the 1990s, Sone Higata and Sone Shinden have become popular among birdwatchers who expect to see migrating birds such as sandpipers (*Limosa lapponica* and *Numenius phaeopus*) visiting the area in spring and fall, and ducks (*Tadorna tadorna* and *Anas platyrhynchos*), and seagulls (*Larus ridibundus*) that overwinter there. The author became interested in the crab species eaten by overwintering gulls and pipers (*Numenius arquata*) since 1993, and continues to observe and photograph the nature of the tidal flat, then found a juvenile horseshoe crab (*Tachypleus tridentatus*) during one of the observation trips in July 1994. There was an article in various newspapers about the "construction approval of the new Kita-Kyushu Airport by the Ministry of Land, Infrastructure, Transport and Tourism" in October, 1994. In it, a description of horseshoe crabs was found in a memorandum issued by the then environmental agency. The author was contacted by Mr. Naohiro Harada from Yamaguchi Prefecture who read the newspaper article. Mr. Harada was

Note: Translation of the main text and captions, as well as modifications made to tables were done by Yumiko Iwasaki.

O. Hayashi (✉)
Sone Tidal Flat Proud House of Horseshoe Crab, Kita-kyushu, Fukuoka, Japan



J. T. Tanacredi et al. (eds.), *International Horseshoe Crab Conservation and Research Efforts: 2007-2020*, https://doi.org/10.1007/978-3-030-82315-3_15

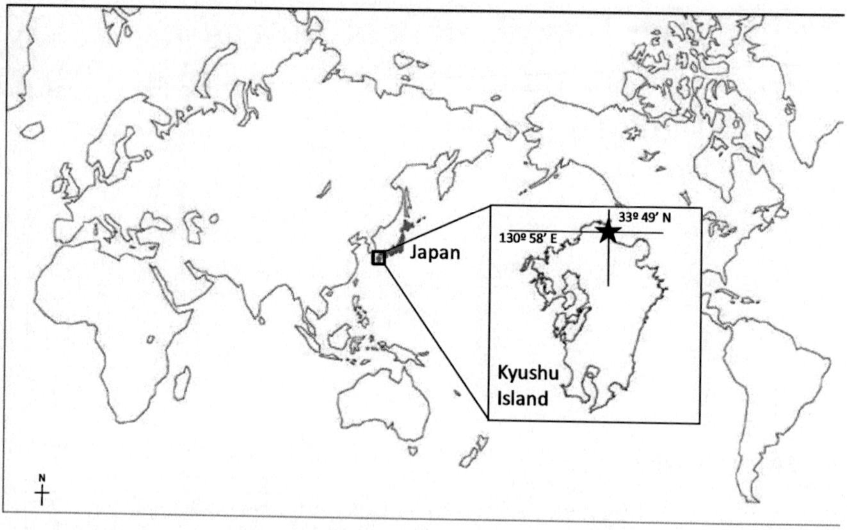


**Fig. 1** Location of Sone Higata on a global scale. Japan is shaded in gray. Sone Higata, or tidal flat, is located in the northern Kyushu Island, indicated by a star

a member of Japan Society for the Preservation of Horseshoe Crabs, or Nihon Kabutogani-wo mamoru kai.

Later, when the author took Mr. Harada to Sone Higata to accompany him with the field investigation, 1st instar horseshoe crab was dug out from the sandy substrate of the bank at the Kusami River estuary. At the southern side of the Nuki River estuary, we found about 20 of 5th to 8th instars. Upon our findings, Mr. Harada reflected that Sone Higata was one of the habitats for horseshoe crabs. Because only few reports existed about this species in Sone Higata at that time, Mr. Harada recommended to do a monitoring of this ancient species, and the author has been doing it ever since. The juvenile horseshoe crab survey has started since October 1994, and spawning survey since July 1995. The results of these surveys were taken partially for summarized reports in Japan Society's annual publication, *Kabutogani*, and *Our Natural History*, a publication from Kita-Kyushu Natural History Museum.

This report is a summary of spawning surveys and habitat conditions of juvenile horseshoe crabs at Sone Higata for the last 24 years. It is our hope that this report will help inform the current situation horseshoe crabs at Sone Higata experience today.

## 2 Spawning Survey

### 2.1 *Summary of Survey Areas*

The regular spawning survey at Sone Higata has started since the night of July 11, 1995. The target areas for the horseshoe crab survey are the estuaries of four creeks; from Chikuma River on the north, down to Ono River, Nuki River, and Kusami

**Fig. 2** Aerial view of Sone Higata with 7 survey areas for horseshoe crab spawning. An elevated concrete road was constructed from the mouth of Nuki River to the Sone fishing port in 2014, dividing Sone Higata in two. (Photo is from Google Earth)

River on the south. Later on, based on the interviews with local people who knew the area well, and on the field survey of the coastline, Yoshida Coast which runs along the northern revetment extending from the left bank of Chikuma River estuary was added in 1997, and Matsuyama Landfill at Kanda Town in 2000 (Fig. 2).

## Yoshida Coast

Yoshida Coast covers the area from the Chikuma River estuary, on the northern side of Sone Higata. Small patchy areas with sandy substrate or deposit occur along the revetment, approximately 1.0–1.5 km from the estuary where horseshoe crabs lay their eggs. In October, 1994 when a survey was conducted with Mr. Harada, there were no sandy area suitable for horseshoe crab spawning; only some drifted sea-shells were found on small sandy patches. However, it seems sand deposition began to occur on the entire tidal flat thereafter, and nowadays spawning pairs can be seen all over the tidal flat during the spawning season in July. The coast faces Mashima Island, and in between, a strong flow from Chikuma River runs towards the sea.

Drifted objects that can affect the environmental condition of the sandy areas are mostly large trash. Seaweed can also come ashore and have an impact on the beach environment, but its observed amount was little. The survey was conducted by the author from 1997 to 2001, then by Mr. Osamu Hayakawa from 2002 to 2006, then again by the author since 2006. The survey is started off by the author, followed by a few local people. Because Sone Higata is extensive, the survey sites are away from each other. Considering the total time spend for one visit to all the sites by one person, each spending time per site is minimal, and the precision of a survey could be improved.

## Ono River

Ono River is the smallest river flowing into Sone Tidal Flat. The observation point is the sandy bottom on its right bank at the estuary to 30 m south along the western revetment. Since 2002, spawning pairs and activities have been observed during daytime at the sandbar formed approximately 100 m east from the estuary, except for 2007. However, upon comparison with the pictures from 1995, the sandbar has moved northward and its elevation has lowered. After heavy rains and typhoons, there is a change in geomorphology of the area and also a deposition of trash. Birds, such as crows, often dig into the sandy substrate to extract horseshoe crab eggs. The sand deposit on the bottom under the revetment also fluctuates, and trash is carried onto it, creating unsuitable condition for horseshoe crabs to spawn.

## Nuki River

Nuki River is located in the center area along the western revetment of Sone Higata. From its mouth along the left bank extends seaward is a concrete road fixed on the seafloor. From 1995 until 2000, spawning activities were observed at three locations: (1) a small beach with sand and pebbles at the corner where the seafloor road intersects with the revetment; (2) a tiny sandy bottom under a slope which runs from the revetment down to the seafloor road; and (3) on sand deposits between scattered rubbles along the revetment on the right bank.

In the summer of 1997, the City of Kita-Kyushu artificially created a sandy beach about 100 m south from the right bank of the Nuki River estuary for a horseshoe crab spawning experiment. It was a simple construction by putting sandbags on the north and south side to block the flow of sand that was placed. The experiment lasted for only one year, and the place was left as it was. Eventually, the deposited sand was gradually carried away towards the right bank of the estuary by a northerly current along the western revetment. Since 2001, the area of sandy beach at the right bank of the estuary increased, and in 2004, the sandy beach reached 70 m from the estuary towards upriver, with a width of 6 m, providing a suitable spawning habitat for horseshoe crabs that rapidly increased their visiting frequency.

However, since the spring of 2004, deterioration of beach environment has been accelerated with rotting drifted seaweeds and trash strewn ashore. In addition, the sand supply from the experimental sandy bed had ceased, and on top of it, the recent increase of heavy storms during the rainy season had increased the volume of water flowing through the rivers, further carrying the deposited sand away from the tidal flat. Because of these phenomena, the sandy area at the right bank of the estuary had shrunk. Spawning horseshoe crabs, in search of sandy substrate, moved to the sandy area upriver under the levee on the banks, some 120–150 m from the mouth. Also, at the sand deposit on the north side of the seafloor road, 30–70 m away from the mouth, spawning horseshoe crab pairs were observed.

In October 2015, the construction for the Sone fishing port has begun. Because the seafloor road connecting to the fishing port becomes submerged during the high tides, a bridge construction was planned. In the early October, a temporary road was built along the north side of the seafloor road for construction vehicles. The road was set up on crushed stones piled up to approximately 3 m high. Because of this, a tiny sand beach at the corner where seafloor road meets the revetment had disappeared. Other constructions are going on around the Nuki River estuary, and the environment for spawning horseshoe crabs has been worsening.

## Kusami River

On the right of the Kusami River estuary is the southern revetment of Matsuyama Landfill, and on the left side, there is a guiding dike which helps the discharge of water from the river into the sea by dividing the river from the tidal flat. The Kusami River estuary is located on the south of the western revetment of Sone Higata. On the north side of the dike, a sandy beach stretches about 25 m along the revetment, and on the south side, a sand deposit occurs between scattered large rubbles. These two sandy areas are the main spawning locations at the Kusami River estuary for horseshoe crabs. Additionally, there are patches of sandy spots further up in the stream and spawning horseshoe crab pairs can be observed. Until about 2005, the deposition of sand was excellent at Kusami River, the sandy bottoms were clearly seen from the land.

In late April, 2004, an extensive area of Sone Higata south of the Nuki River estuary became covered by seaweed species, such as *Enteromorpha linza*. On the roots of this particular species, sea snails like *Cerithidea cingulata* and *Batillaria multiformis* are hitching a ride, making the roots to freely float in water column during high tides. As a result, *Enteromorpha linza* are often washed ashore. After a few days, drifted *Enteromorpha linza* begins to rot and become a massive sludge which alters the beach environment. This phenomenon is now seen every year.

Also, after the heavy rain during the rainy season or typhoons, a large quantity of trash is expected to be carried along the river into the sea, to finally arrive at the shore. Every year, a beach cleanup event is held in late May, and many volunteers participate. The local people and members of Kabutogani-wo mamoru-kai clean the beach at Sone Higata throughout the year.

However, when getting rid of drifted seaweed and garbage, large amount of sand grains attached to these unwanted objects are also eliminated from the beach, a beach cleanup became one of the reasons for the loss of sand from the beach. There are also people who take away a large volume of sand and damages to the dike at Kusami River, which is supposed to keep the sand on the beach intact, cannot prevent the sand loss into the river anymore. The result is the rapid loss of sand from the beach. Horseshoe crabs preferred to come to the southern beach at Kusami River to spawn. But its sand layer rapidly became thinner since around 2007, and their deposited egg clusters became washed away by waves. The local people who witnessed this sad incident made an effort to convince the local administration to improve the situation. In December 2011, Kokura South Ward office provided 100 $m^3$ of sand which the local citizens, administration staff, volunteers, and members from the mamoru-kai used to restore beaches. In 2012, the survey result shows that a number of spawning horseshoe crab pairs returned to the restored beaches and no more egg clusters were washed away.

After 2014, the amount of seaweeds washed ashore was reduced and since 2017, seaweeds ceased to flourish. However, after heavy rains and typhoons, trash always ends up on the beaches, and there are people who continue to steal sands. From 2016, horseshoe crab eggs and 1st instars were observed to have been washed away, indicating the recurrence of environmental destruction.

## Matsuyama Coast

Matsuyama Landfill is under the jurisdiction of another town, Kanda Town. It is located under the Matsuyama Castle ruin. The beach here was visited in 2000 during an investigation of a red tide in Sone Higata. On this beach, some round, shallow depressions, indicative of horseshoe crab spawning activities, were confirmed. This pebbly beach, stretching 70 m from the Kusami River estuary under the connection road to the Kita-Kyushu Airport was added to the survey points for spawning horseshoe crabs.

During the construction of the connection road, reddish, muddy soil was discharged onto the beach, which deteriorated the water quality along the Matsuyama Coast. Currently, the condition is back to as it was before the construction. Compared to other beaches in Sone Higata, Matsuyama Coast has poorly sorted sand with many pebbles, but many spawning horseshoe crab pairs use the beach here. Drifted items are mostly trash, with some seaweeds. Local volunteers regularly clean the beach since eight years ago to maintain its condition at its best.

## Mashima Island

Mashima island is located within Sone Higata on its northeast. It is a small island; its periphery is about 1 km. On its northeast and south sides lies a rocky coast, but about 3/2 of the island is fringed by sandy beaches. From the island's northwest, a sandbar is extending towards Ono River. Mashima is surrounded by a natural, pristine coastline which is very rare and therefore valuable within the city of Kita-Kyushu. Rare species of beach plants, such as *Lysimachia mauritiana* and *Dianthus japonicus*, are found here. However, drifted items deposited on the shores have increased these days, and an erosion of sandy beaches has begun since 2006, changing the beach environment difficult for horseshoe crabs to lay their eggs.

## 2.2 Survey Methods and Their Transition

The survey of spawning horseshoe crabs began on July 11, 1995. In the beginning, following the advice of Mr. Harada, the surveys were conducted at the estuaries of Ono River, Nuki River, and Kusami River during three high tide periods between July and early August. Each survey was done for three consecutive days. Because of the time constraint, the author carried on the investigation mostly at night to check only the spawning foams released from horseshoe crab pairs when laying eggs. In order to confirm such activities, the actual egg clusters were occasionally dug out the next day (then were put back where they were). During the survey of egg clusters in July 1997, an embryo was found developed, rotating in its shell. This meant that horseshoe crabs laid eggs earlier than July, possibly in June. The next year, the survey started in mid-June. Since 2001, the number of spawning pairs visiting Sone Higata has rapidly increased. With an anticipation of their earlier arrival than high tides, the investigation period was extended from a maximum of three days to four to five days per occasion. Sometimes, surveys took place in May and September, but these were exceptions; the basic surveys are done in June to August.

The surveys are conducted twice a day, during daytime and night-time. For the daytime investigation, it starts from the sunrise to sunset; for the night-time, from sunset to the next day's sunrise. For the night-time survey from 1995 to 2000, the trip started at the Ono River estuary half an hour before a high tide, then to the Nuki River, followed by the Kusami River estuaries. From 1997 to 2000, the Yoshida Coast was added to the investigation after Kusami River.

Due to the addition of Matsuyama Landfill and the increased number of horseshoe crab pairs to count, both schedules and a trip route were revised. Mr. Osamu Hayakawa took over the survey of Yoshida Coast from 2002 to 2005, then since 2006, the author became in charge of the area. This coastline is quite far away from the other survey areas and stretches approximately 500 m long, the investigation to and along this area is time-consuming.

To the survey sites, a car is used. The start of the survey is from Yoshida Coast, one hour before the high tide. The observation of horseshoe crab pairs and their spawning foams is done by one round trip on foot along this 500 m coastline. A record is kept when and where these were observed.

The survey area at the Ono River estuary is not extensive, so that the observation is done twice with an enough time lag in between. At the estuary of Nuki River, the survey sites are scattered. Since the site on the left bank of the river, along the north side of the seafloor road was filled in due to the temporary construction of the bridge, the investigation is currently expanded to 200 m from the western revetment along the north of the road.

Between 200 m upriver from the mouth of Nuki River, on its left bank, an observation of spawning foams has been conducted, however, no sign of horseshoe crab has been recorded for four years now. On the right bank of Nuki River, the area between the revetment and upriver, some 70 m is investigated for spawning foams. Because the survey area is small, the observation is done twice, paying an attention not to double-count the same foam. Nowadays, the area is shortened to 30 m. The experiment area of spawning horseshoe crabs set up by the city of Kita-Kyushu in 1997 was also a survey point, but all the sand was spilled out in 2002, this area naturally disintegrated into the surroundings, and it was excluded from the survey after 2003.

At the Kusami River estuary, the survey points are a 25 m stretch of beach along the western revetment, the northern side along the dike which protrudes into the tidal flat, and three sandy spots along the south of the dike towards the floodgate for a creek into Sone Shinden. Again, the observation is repeated here and any double-counting must be avoided. Since 2008, there are horseshoe crabs observed on the sandbars formed within the river, and these areas are also included in the survey. Mr. Shungo Takahashi, a chair at the Fukuoka Branch of the mamoru-kai, has been starting the investigation at Kusami River at an earlier time than the author, so that the precision and accuracy of the investigation here is higher than the other areas in Sone Higata.

Matsuyama Landfill is the last to visit for a horseshoe crab spawning survey. It is not unusual to arrive here more than one-and-half hour after a high tide. Round, shallow depressions on the sandy bottom, indicative of horseshoe crab spawning activities, can be found, but these are indirect information, and therefore are not documented; instead, observations for the pairs and spawning foams are conducted by walking the beach back and forth a few times with a sufficient amount of time in between. Here, also Mr. Takahashi has started his observation since 2013.

The author spends Sundays and Holidays for the daytime surveys. The mode of transportation is a car from one point to the other. However, unlike the survey in the dark, binoculars can be used to spot any sign of spawning horseshoe crabs at the sandbar off the estuary of Ono River and at the sandy bottom along the revetment on the right bank of the Kusami River estuary. Additionally, some members of the mamoru-kai participate in the observation, and their record is added to the survey data after a careful evaluation.

Mashima Island cannot be surveyed regularly. Between 1998 and 2000, the author visited this island and confirmed the egg clusters of horseshoe crabs. Between 2001 and 2008, Mr. Takahashi conducted an investigation once or twice a year, however, the coastline became heavily eroded and a sand accumulation has lost from the beaches and sandbars, making the survey on foot difficult. Although irregularly, Mr. Masaaki Fukuda currently surveys the island once a year by using a canoe.

The survey methods since the initial survey have been modified over the years. Visiting all the survey points to observe any horseshoe crab pairs and their spawning foams within a limited timeframe on this vast tidal flat may not give a satisfactory comprehension of their number. The volume of information on horseshoe crab spawning gathered from the local people and volunteers fluctuates each year. The number of observations per site differs as well, however, the trend in the number of spawning pairs visiting Sone Higata as a whole is well-reflected by our years of survey.

## 2.3 Results

All the data of spawning horseshoe crab pairs recorded from 1995 to 2018 were put together for an analysis in Tables 1, 2, 3, 4a, 4b, and 5; Figs. 3, 4, 5, 6, and 7.

Table 1 shows the fluctuation in number of visiting spawning horseshoe crab pairs per survey site. Figure 3 shows the overall trend of number of pairs within Sone Higata, and Fig. 4 reflects Table 1. The record of 24 years of horseshoe crab observation in Sone Higata is explained below.

From 1995 until 2000, even with the addition of Yoshida Coast in 1997, the total number of observed pairs counted less than 100 (except for 1998). Despite the fact that both the survey method and our inexperience spotting horseshoe crabs in water and spawning foam, the number of horseshoe crabs visiting Sone Higata seemed stable.

Since 2001, the number of spawning pairs in Sone Higata had increased, and a total of more than 1500 pairs were counted in 2005. The reason for the increase is yet to be investigated, but there are two possible causes. One hypothesis is that the natural environment in and around Sone Higata had improved enough for horseshoe crabs to prosper, especially for juvenile to spend years to grow. The reason for this thought is that many young sub-adults were observed among the visiting horseshoe crabs between 2001 and 2003. However, during the same period, a landfill construction was taking place offshore from Sone Higata where adult horseshoe crabs are said to be inhabiting, and therefore if the environment had improved is questionable.

Another possibility for the increased number of horseshoe crabs in Sone Higata is a migration from other areas. From the ancient times, the coast of Suounada from Moji Ward of Kita-Kyushu to Makatsu, Oita Prefecture has been known among the locals the horseshoe crab habitat (Kawahara 1993, Nishihara 1998). Based on an account from people in the fishing industry and a survey result on juvenile horseshoe crabs conducted at Shiraishi Coast in Kanda Town, there could be some areas along Suounada where horseshoe crabs visit for spawning. Among these possible areas for spawning, if there were some environmental changes that could affect horseshoe crab behavior, they could have searched a better spawning ground elsewhere, and one of them might have been Sone Higata. Unfortunately, there are no horseshoe crab survey records along Suounada, and thus a solid conclusion cannot be reached. There might be other possible explanation for the increase of spawning pairs in Sone Higata.

There was a report about the increase of spawning horseshoe crab pairs at Tatara Coast, Imari, Saga Prefecture (Sakami 2002). The increasing trend lasted for

**Table 1** Number of spawning horseshoe crab pairs at each survey site at Sone Higata, from 1995 to 2018. The data for all the tables and figures are taken from Hayashi (1995, 1997, 1998, 1999a, 1999b, 2000a, 2000b, 2001a, 2001b, 2002, 2003a, 2003b, 2003c, 2005), Hayashi et al. (2002, 2004a, 2004b, 2006, 2007), Hayashi and Shimizu (2007), Takahashi (2017), Takahasi and Hayashi (2013)

| Spawning Site | Kusami River | Nuki River | Ono River | Yoshida Coast | Matsuyama Landfill | Mashima Island | Total |
|---|---|---|---|---|---|---|---|
| 1995 | 15 | 6 | 3 | | | | 24 |
| 1996 | 34 | 9 | 2 | | | | 45 |
| 1997 | 39 | 13 | 3 | 4 | | | 59 |
| 1998 | 52 | 17 | 5 | 6 | | | 80 |
| 1999 | 72 | 24 | 7 | 13 | | | 116 |
| 2000 | 47 | 27 | 3 | 1 | | | 78 |
| 2001 | 138 | 78 | 10 | 3 | 4 | 0 | 233 |
| 2002 | 210 | 268 | 37 | 12 | 4 | 0 | 531 |
| 2003 | 322 | 219 | 32 | 240 | 31 | 68 | 912 |
| 2004 | 623 | 398 | 47 | 208 | 26 | 61 | 1363 |
| 2005 | 702 | 334 | 35 | 403 | 66 | 41 | 1581 |
| 2006 | 137 | 183 | 24 | 114 | 49 | 6 | 513 |
| 2007 | 129 | 37 | 6 | 62 | 31 | 0 | 265 |
| 2008 | 161 | 35 | 13 | 66 | 31 | 8 | 314 |
| 2009 | 205 | 169 | 13 | 54 | 32 | | 473 |
| 2010 | 231 | 69 | 20 | 152 | 35 | 4 | 511 |
| 2011 | 314 | 94 | 15 | 90 | 32 | 2 | 547 |
| 2012 | 343 | 107 | 29 | 171 | 87 | 1 | 738 |
| 2013 | 491 | 206 | 37 | 255 | 90 | | 1079 |
| 2014 | 657 | 315 | 55 | 474 | 79 | | 1580 |
| 2015 | 875 | 383 | 35 | 526 | 137 | 0 | 1956 |
| 2016 | 841 | 409 | 41 | 442 | 262 | 0 | 1995 |
| 2017 | 535 | 350 | 38 | 278 | 122 | | 1323 |
| 2018 | 367 | 147 | 27 | 144 | 76 | 2 | 763 |

**Table 2** Number of spawning horseshoe crab pairs at each survey site at Sone Higata, from 1995 to 2018. (a) Number of pairs observed during daytime. (b) Number of pairs observed during nighttime

(a)

| Spawning Site | Kusami River | Nuki River | Ono River | Yoshida Coast | Matsuyama Landfill | Mashima Island | Daytime Total |
|---|---|---|---|---|---|---|---|
| 1995 | 9 | 3 | 2 | | | | 14 |
| 1996 | 25 | 5 | 1 | | | | 31 |
| 1997 | 14 | 4 | 0 | 1 | | | 29 |
| 1998 | 6 | 2 | 0 | 0 | | | 8 |
| 1999 | 20 | 7 | 3 | 1 | | | 31 |
| 2000 | 6 | 6 | 1 | 0 | | | 13 |
| 2001 | 51 | 15 | 2 | 0 | 4 | 0 | 72 |
| 2002 | 68 | 126 | 11 | 12 | 4 | 0 | 221 |

(continued)

**Table 2** (continued)

(a)

| Spawning Site | Kusami River | Nuki River | Ono River | Yoshida Coast | Matsuyama Landfill | Mashima Island | Daytime Total |
|---|---|---|---|---|---|---|---|
| 2003 | 90 | 66 | 14 | 104 | 14 | 49 | 337 |
| 2004 | 313 | 261 | 29 | 83 | 13 | 44 | 743 |
| 2005 | 100 | 87 | 7 | 195 | 10 | 41 | 440 |
| 2006 | 46 | 99 | 9 | 29 | 27 | 5 | 215 |
| 2007 | 59 | 3 | 2 | 19 | 10 | 0 | 93 |
| 2008 | 46 | 7 | 4 | 38 | 5 | 3 | 103 |
| 2009 | 52 | 61 | 2 | 7 | 4 | | 126 |
| 2010 | 65 | 16 | 5 | 27 | 7 | 4 | 124 |
| 2011 | 129 | 20 | 3 | 21 | 8 | 2 | 183 |
| 2012 | 106 | 22 | 8 | 40 | 15 | 1 | 192 |
| 2013 | 221 | 50 | 15 | 47 | 22 | | 355 |
| 2014 | 263 | 149 | 9 | 68 | 11 | | 500 |
| 2015 | 304 | 115 | 5 | 124 | 27 | 0 | 575 |
| 2016 | 227 | 111 | 5 | 61 | 64 | 0 | 468 |
| 2017 | 142 | 110 | 9 | 20 | 25 | | 306 |
| 2018 | 144 | 58 | 12 | 60 | 26 | 2 | 302 |

(b)

| Spawning Site | Kusami River | Nuki River | Ono River | Yoshida Coast | Matsuyama Landfill | Mashima Island | Nighttime Total |
|---|---|---|---|---|---|---|---|
| 1995 | 6 | 3 | 1 | | | | 10 |
| 1996 | 9 | 4 | 1 | | | | 14 |
| 1997 | 25 | 9 | 3 | 3 | | | 40 |
| 1998 | 46 | 15 | 5 | 6 | | | 72 |
| 1999 | 52 | 17 | 4 | 12 | | | 85 |
| 2000 | 41 | 21 | 2 | 1 | | | 65 |
| 2001 | 87 | 63 | 8 | 3 | 0 | 0 | 161 |
| 2002 | 142 | 142 | 26 | 0 | 0 | 0 | 310 |
| 2003 | 232 | 153 | 18 | 136 | 17 | 19 | 575 |
| 2004 | 310 | 137 | 18 | 125 | 13 | 17 | 620 |
| 2005 | 602 | 247 | 28 | 208 | 56 | 0 | 1141 |
| 2006 | 91 | 84 | 15 | 85 | 22 | 1 | 298 |
| 2007 | 70 | 34 | 4 | 43 | 21 | 0 | 172 |
| 2008 | 115 | 28 | 9 | 28 | 26 | 5 | 211 |
| 2009 | 152 | 107 | 11 | 47 | 28 | | 345 |
| 2010 | 166 | 53 | 15 | 125 | 28 | | 387 |
| 2011 | 185 | 74 | 12 | 69 | 24 | 0 | 364 |
| 2012 | 237 | 85 | 21 | 131 | 72 | | 546 |
| 2013 | 270 | 156 | 22 | 208 | 68 | | 724 |
| 2014 | 394 | 166 | 46 | 406 | 68 | | 1080 |
| 2015 | 531 | 268 | 35 | 402 | 110 | | 1346 |
| 2016 | 614 | 298 | 36 | 381 | 198 | | 1527 |
| 2017 | 393 | 240 | 29 | 258 | 97 | | 1017 |
| 2018 | 223 | 89 | 15 | 84 | 50 | | 461 |

**Table 3** Number of spawning horseshoe crab pairs in Sone Higata on a monthly basis, from May to September. Observed period from 1995 to 2018

| | May | June | July | August | Sept. | Total |
|---|---|---|---|---|---|---|
| 1995 | | | 20 | 4 | | 24 |
| 1996 | | 1 | 22 | 22 | | 45 |
| 1997 | | 1 | 43 | 15 | 0 | 59 |
| 1998 | | 13 | 58 | 9 | | 80 |
| 1999 | | 10 | 94 | 12 | | 116 |
| 2000 | | 13 | 52 | 13 | | 78 |
| 2001 | | 18 | 193 | 22 | | 233 |
| 2002 | | 30 | 391 | 110 | | 531 |
| 2003 | | 28 | 541 | 293 | 50 | 912 |
| 2004 | | 41 | 1043 | 279 | | 1363 |
| 2005 | 2 | 316 | 1159 | 104 | 0 | 1581 |
| 2006 | 1 | 14 | 324 | 174 | | 513 |
| 2007 | | 10 | 196 | 58 | 1 | 265 |
| 2008 | | 13 | 245 | 56 | | 314 |
| 2009 | 2 | 225 | 223 | 23 | | 473 |
| 2010 | | 25 | 404 | 82 | | 511 |
| 2011 | | 8 | 295 | 243 | 1 | 547 |
| 2012 | | 9 | 601 | 128 | | 738 |
| 2013 | | 23 | 961 | 90 | | 1079 |
| 2014 | | 95 | 1208 | 277 | | 1580 |
| 2015 | | 26 | 1439 | 491 | | 1956 |
| 2016 | | 231 | 1702 | 62 | | 1995 |
| 2017 | | 406 | 854 | 63 | | 1323 |
| 2018 | 2 | 42 | 679 | 40 | | 763 |

**Table 4a** Number of pairs observed from early June to late August, between 1998 and 2018. The month for the highest number of pairs observed each year is highlighted in dark gray with white bold letters. In the months more than 1/6 of the total number of pairs observed are highlighted in light gray with bold letters. During the month of July, Sone Higata is visited by a number of spawning horseshoe crabs

| | | Early June | Late June | Early July | Late July | Early Aug. | Late Aug. | Total |
|---|---|---|---|---|---|---|---|---|
| 1998 | # pairs | 3 | 10 | 50 | 8 | 6 | 3 | 80 |
| | | 3.75% | 12.50% | **62.50%** | 10.00% | 7.50% | 3.75% | |
| 1999 | # pairs | 4 | 6 | 77 | 24 | 5 | 0 | 116 |
| | | 3.45% | 5.17% | **66.38%** | **20.70%** | 4.31% | 0% | |
| 2000 | # pairs | 8 | 18 | 27 | 18 | 7 | 0 | 78 |
| | | 10.26% | **23.08%** | **34.62%** | **23.08%** | 8.97% | 0% | |
| 2001 | # pairs | 2 | 16 | 45 | 148 | 20 | 2 | 233 |
| | | 0.86% | 6.87% | **19.31%** | **63.52%** | 8.58% | 0.86% | |
| 2002 | # pairs | 19 | 11 | 335 | 56 | 107 | 3 | 531 |
| | | 3.98% | 2.07% | **63.09%** | 10.55% | **20.15%** | 0.56% | |

(continued)

**Table 4a** (continued)

| | | Early June | Late June | Early July | Late July | Early Aug. | Late Aug. | Total |
|---|---|---|---|---|---|---|---|---|
| 2003 | # pairs | 13 | 36 | 259 | 407 | 51 | 110 | 876 |
| | | 1.48% | 4.11% | **29.57%** | **46.46%** | 5.82% | 12.56% | |
| 2004 | # pairs | 17 | 24 | 670 | 290 | 352 | 10 | 1363 |
| | | 1.25% | 1.76% | **49.16%** | **21.28%** | **25.83%** | 0.73% | |
| 2005 | # pairs | 22 | 294 | 60 | 1099 | 31 | 73 | 1579 |
| | | 1.39% | **18.62%** | 3.80% | **69.60%** | 1.96% | 4.62% | |
| 2006 | # pairs | 7 | 9 | 304 | 18 | 168 | 6 | 512 |
| | | 1.37% | 1.76% | **59.38%** | 3.52% | **32.81%** | 1.17% | |
| 2007 | # pairs | 2 | 24 | 111 | 115 | 10 | 0 | 262 |
| | | 0.76% | 9.16% | **42.37%** | **43.89%** | 3.82% | 0% | |
| 2008 | # pairs | 2 | 11 | 172 | 70 | 58 | 1 | 314 |
| | | 0.64% | 3.50% | **54.80%** | **22.29%** | **18.47%** | 0.32% | |
| 2009 | # pairs | 5 | 220 | 81 | 142 | 21 | 2 | 471 |
| | | 1.06% | **46.71%** | **17.20%** | **30.15%** | 4.46% | 0.42% | |
| 2010 | # pairs | 16 | 9 | 289 | 115 | 82 | 0 | 511 |
| | | 3.13% | 1.76% | **56.55%** | **22.50%** | 16.05% | 0% | |
| 2011 | # pairs | 1 | 7 | 125 | 123 | 263 | 16 | 535 |
| | | 0.19% | 1.31% | **23.36%** | **22.99%** | **49.16%** | 2.99% | |
| 2012 | # pairs | 4 | 5 | 327 | 274 | 125 | 2 | 737 |
| | | 0.54% | 0.68% | **44.37%** | **37.18%** | 16.96% | 0.27% | |
| 2013 | # pairs | 1 | 22 | 278 | 683 | 81 | 14 | 1079 |
| | | 0.09% | 2.04% | **25.76%** | **63.30%** | 7.51% | 1.30% | |
| 2014 | # pairs | 11 | 80 | 1028 | 189 | 268 | 0 | 1576 |
| | | 0.69% | 5.07% | **65.22%** | 11.99% | **17.00%** | 0% | |
| 2015 | # pairs | 5 | 16 | 710 | 655 | 565 | 5 | 1956 |
| | | 0.26% | 0.82% | **36.30%** | **33.49%** | **28.89%** | 0.26% | |
| 2016 | # pairs | 21 | 210 | 963 | 739 | 61 | 1 | 1995 |
| | | 1.05% | 10.53% | **48.27%** | **37.04%** | 3.06% | 0.05% | |
| 2017 | # pairs | 19 | 387 | 356 | 498 | 60 | 3 | 1323 |
| | | 1.44% | **29.25%** | **26.91%** | **37.64%** | 4.54% | 0.23% | |
| 2018 | # pairs | 16 | 32 | 630 | 48 | 32 | 3 | 761 |
| | | 2.10% | 4.20% | **82.78%** | 6.30% | 4.20% | 0.39% | |
| > the highest number observed | | | 1 | 13 | 6 | 1 | | |
| > 1/6 of the total number | | | 3 | 7 | 10 | 6 | | |
| Total | | | 4 | 20 | 16 | 7 | | |

**Table 4b** Visiting frequency patterns per tide of horseshoe crabs in Sone Higata. Pattern 1 = 1 peak; pattern 2 = 1 peak plus one small peak; pattern 3 = 1 peak plus 2 small peaks

| Patterns | | Early June | Late June | Early July | Late July | Early Aug. | Late Aug. | Peak | Years | Frequency |
|---|---|---|---|---|---|---|---|---|---|---|
| 1 | | | | ◎ | | | | 1 | 98,18 | 2 |
| 2 | ① | | | ◎ | ○ | | | 1 | 99,10,12,16 | 4 |
| | ② | | | ○ | ◎ | | | 1 | 01,03,07,13 | 4 |
| | ③ | | | ◎ | | ○ | | 1 | 02,06,14 | 3 |
| | ④ | | ○ | | ◎ | | | 1 | 05 | 1 |
| 3 | ① | | ○ | ◎ | ○ | | | 1 | 00 | 1 |
| | ② | | | ◎ | ○ | ○ | | 1 | 04,08,15 | 3 |
| | ③ | | ◎ | ○ | ○ | | | 1 | 09 | 1 |
| | ④ | | | ○ | ○ | ◎ | | 1 | 11 | 1 |
| | ⑤ | | ○ | ○ | ◎ | | | 1 | 17 | 1 |

**Table 5** Yearly change in the number of spawning horseshoe crab pairs surveyed in Sone Higata, 1995–2018. The number of observed pairs and that of surveys per year between 1995 and 2018 are shown, with the average number of pairs per survey is calculated on the right column

| year | #spawning pairs | #surveys | avg. pairs/survey |
|---|---|---|---|
| 1995 | 24 | 8 | 3 |
| 1996 | 45 | 17 | 2.6 |
| 1997 | 59 | 19 | 3.1 |
| 1998 | 80 | 23 | 3.5 |
| 1999 | 116 | 25 | 4.6 |
| 2000 | 78 | 25 | 3.1 |
| 2001 | 233 | 22 | 10.6 |
| 2002 | 531 | 41 | 12.9 |
| 2003 | 912 | 53 | 17.2 |
| 2004 | 1363 | 59 | 23.1 |
| 2005 | 1581 | 61 | 25.9 |
| 2006 | 513 | 53 | 9.7 |
| 2007 | 265 | 52 | 5.1 |
| 2008 | 314 | 48 | 6.5 |
| 2009 | 473 | 63 | 7.5 |
| 2010 | 511 | 59 | 8.7 |
| 2011 | 547 | 67 | 8.2 |
| 2012 | 738 | 45 | 16.4 |
| 2013 | 1079 | 46 | 23.5 |
| 2014 | 1580 | 56 | 28.2 |
| 2015 | 1956 | 64 | 30.6 |
| 2016 | 1995 | 70 | 28.5 |
| 2017 | 1323 | 73 | 18.1 |
| 2018 | 763 | 68 | 11.2 |

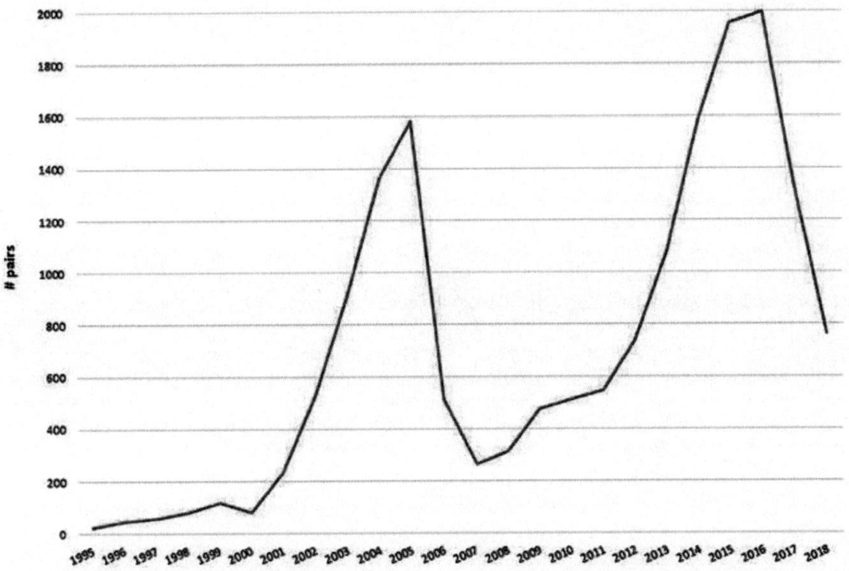


**Fig. 3** Graph showing a fluctuation in the total number of spawning pairs of horseshoe crab observed in Sone Higata. The observed period is from 1995 to 2018

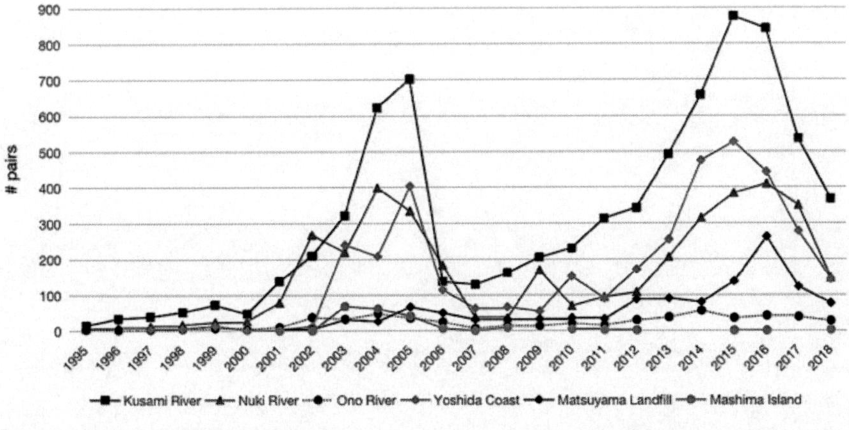


**Fig. 4** Graph showing a fluctuation in number of spawning horseshoe crab pairs observed at each site in Sone Higata

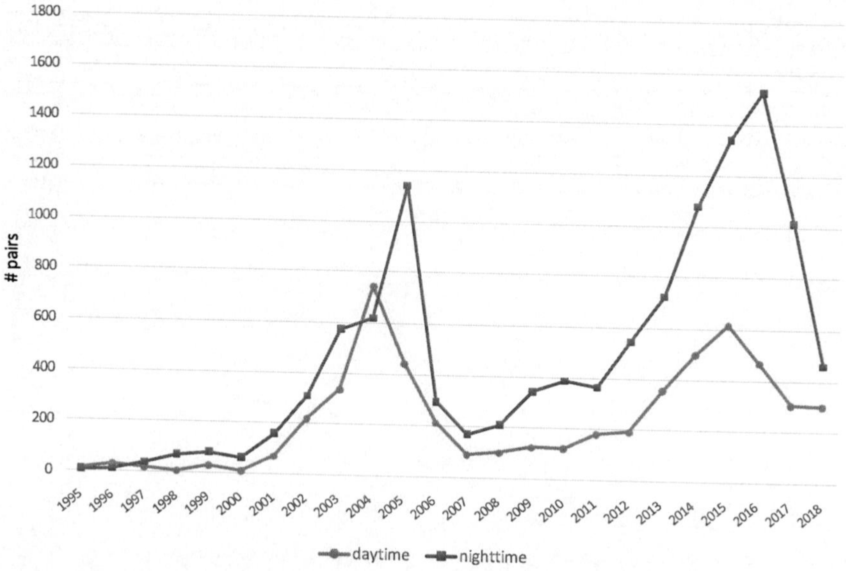


**Fig. 5** Number of spawning horseshoe crab pairs at each survey site at Sone Higata, from 1995 to 2018. (**a**) Number of pairs observed during daytime. (**b**) Number of pairs observed during nighttime

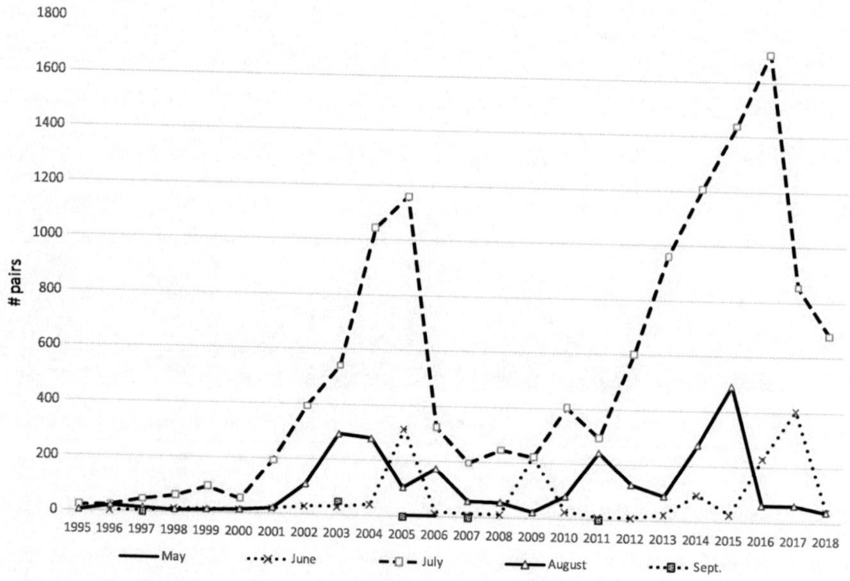


**Fig. 6** Graph showing a number of spawning horseshoe crab pairs in Sone Higata on a monthly basis, from May to September. Observed period from 1995 to 2018

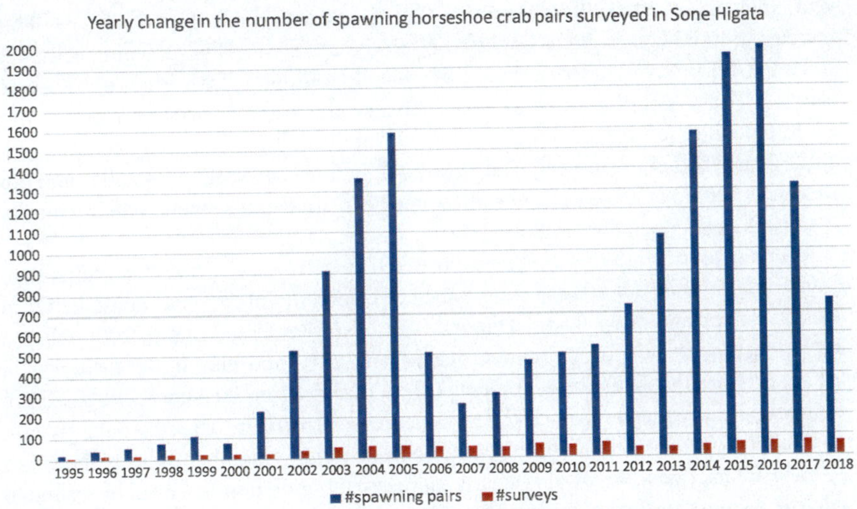


**Fig. 7** Yearly change in the number of spawning horseshoe crab pairs surveyed in Sone Higata, 1995–2018

5 years, then in the sixth year, the number suddenly decreased. The same phenomenon was expected in Sone Higata, and as a matter of fact, the number of spawning horseshoe crab pairs decreased from 1581 in 2005 to 513 pairs the next year (Table 1 and Fig. 3).

One of the reasons for the decrease could stem from the fact that a number of dead horseshoe crabs were drifted into the estuaries of Kusami River and Nuki River. Among adults, some sub-adults and larger juveniles were confirmed, and the total number was 330. More than 200 of them were female, triggering the reduction in number of pairs visiting Sone Higata.

Another reason could be the red tide which occurred in the sea surrounding Sone Higata on July 12, 2006. This is the first red tide into the tidal flat since the start of the horseshoe crab survey in 1995. It caused approximately 500 rays, such as *Dasyatis akajei* and *Aetobatus flagellum,* dead on arrival on the shores. Fish such as eels (*Anguilla japonica*, an IUCN endangered species since 2017) were found floating dead. Horseshoe crabs, on the other hand, were not found dead ashore. Juveniles in the tidal flat also were not affected by the incident. However, the red tide affected on both benthic and marine creatures to some degree. The color of red tide turned into black. In 2006, Kita-Kyushu area had no typhoons to agitate the sea, which helps changing water in and out of enclosed areas such as Sone Higata. As a result, the water in Sone Higata and the surrounding sea had rotten and emitted odor for a while, probably influencing the habitat and spawning activity of horseshoe crabs. In 2007, the observed numbers of pairs were down to 265, only 1/2 of the previous year (Table 1 and Fig. 3).

The number of pairs visiting the sandy exposures of Sone Higata increased since 2008, and between 2009 and 2011, the number was counted around 500 pairs each year, then from 2012, the number further increased (Table 1 and Fig. 3). This increasing trend is thought to be the result of the increase in the spawning activities since 2001 for 5 years. Horseshoe crabs hatch from eggs, then grow by repeated ecdysis. There is a difference in years between the sexes to become reproductive adult, but some reports say they can become adults in 10 years, then become sexually mature in their 11th year. The year 2012 is the 11th year since 2001, and our survey record up to 2016 indicates that some horseshoe crabs reached their maturity in these 5 years, then visited Sone Higata.

While a total of 1995 pairs were observed in 2016, about 500 dead horseshoe crabs, adults and sub-adults, were washed ashore mainly on the coast between Kusami River and Nuki River. This incident is similar to the one in 2005 when a rapid increase in visiting pairs took place from 2001, and then in the fifth year in 2005, a numerous dead horseshoe crabs were observed on the shores. In 2017, the number of visiting pairs was 1232. It was decreased to 763 in 2018, a similar pattern as in 2006–2007. Together with the pattern in fluctuating number of horseshoe crab pairs in Tatara Coast, Imari, this steady increase of population followed by a sudden decline is quite common among horseshoe crabs. However, along Ouki Coast in Shimono-seki, Yamaguchi Prefecture which is on the opposite side of Sone Higata, the number of horseshoe crab pairs is increasing since 2012 (Let's protect horseshoe crabs from extinction! Yamaguchi Survey report: http://www5c.biglobe.ne.jp/~h-kabu/) without any significant drop in population, and in 2018 saw the highest number of them. It might be possible that horseshoe crabs are living the shallow water off the coast of Sone Higata, but since there is no investigation into the seaward, the decline in the number of visiting pairs to Sone Higata is unknown.

Among the horseshoe crab spawning sites within Sone Higata, the Kusami River estuary is the most visited, except for the year 2002 and 2006 (Table 1 and Fig. 4). Figures 3 and 4 show the trends in the observed number of pairs at all the sites, and you can see that Kusami River is the most important point for spawning horseshoe crabs.

Nuki River had some major environmental modifications such as the construction of connection bridge to the fishing port, started in October, 2016, and the construction of bridge pier which began in May 2018. In 2002 and 2006, the estuary of Nuki River had the most pairs of horseshoe crabs (Table 1 and Fig. 3). The area is still under construction, and a special attention is paid to see any changes in the pattern of horseshoe crab sightings. Ono River was added to the survey in 1995 as the previous two. The sandy area on its right bank is very narrow, and with the reduction in the supply of sand at the sandbar continuing from the river, it has turned into a difficult spot for horseshoe crab spawning activities. At the sand bar, birds such as crows were witnessed eating horseshoe crab egg clusters.

The Yoshida Coast is the most extensive survey area within Sone Higata. Since 2003, the coast began to accumulate sand deposit, which attracted spawning pairs to visit here. The increase of sand deposition continued into 2015, but started to decrease the next year. Comparatively, the visiting pairs were reduced in number. After the heavy rain in July, 2018, a large volume of sand was washed away and the sandy area was reduced in size, which in turn, reduced the number of pairs visiting this coast. Compared to Nuki River area, this location had more pairs observed for nine times out of 24 years (Table 1 and Fig. 4).

Matsuyama Landfill is located on the southeast of the Kusami River estuary, where a number of horseshoe crab pairs have been witnessed, and because of such a location, it experiences an increase in the number of visits by horseshoe crabs. The relationships between these two locations is unclear, however, when a massive amount of trash is washed ashore at Kusami River beach, the number of pairs slightly increases on Matsuyama Landfill coast. Also, the environmental condition on this coast is stable, including the substrate full of pebbles.

Mashima Island is located at the center of the tidal flat. There is no accessible way like a bridge; no regular investigation is conducted there. In 2003, the survey was done twice and 68 pairs were confirmed, then in 2004, 61 pairs in three observations. However, the observation there is influenced by weather and schedule of people who do the survey, and there only a few people who are willing to venture into the island, the survey at this location is difficult. In addition, between 2006 and 2007, sand spill from the island became worse, making this area unsuitable for a survey.

Table 2 and Fig. 5 show the trend in number of pairs visiting Mashima Island over the years. In the year 1995 when the horseshoe crab survey has started in Sone Higata, it seems more pairs were observed during the daytime (3 observations) than night-time (5 observations), but this result may reflect the poor skills of the observers. In 1996, in order to investigate the spawning activities in details during daytime, Mr. Yuji Inoue of Wild Bird Society of Japan joined the survey. Because of this additional help, the daytime count increased. From 1998 to 2001, the survey focused on the night-time, and the daytime survey was conducted on weekends and holidays. After 2002, the survey became more frequent, thanks to the help from the members of the mamoru-kai: Ms. Masuko Aoki, Mr. Hiroyoshi Kayano, Mr. Kazuo Mamoto, Mr. Masami Yasunaga, Mr. Osamu Hayakawa, and Shungo Takahashi. In 2004, 33 daytime and 26 night-time surveys were carried out, and a total of 743 pairs (daytime) and 620 pairs (nighttime) was counted. Since then, there were more daytime surveys than night-time, however, the latter count of pairs was larger, except for 1996 and 2004 (Table 2 and Fig. 5). The reason for the larger number of pairs visiting Sone Higata at night may be the difference in the tide level between day and night; in Sone Higata, the tide level is higher at night, which provides a larger sandy bottom submerged under water for horseshoe crab spawning.

Table 3 and Fig. 6 shows monthly trend in the number of spawning horseshoe crabs. In 1995, the survey was done only in July and August, but in 1996 and 1997, due to the calendar for high tide that year, one observation was conducted in June. In July 1997, since the embryos rotating in egg shells were found among the egg clusters dug out of the sandy substrate, since 1998, a regular observation has been conducted in late June during the high tide. For 2005, 2006, 2009, and 2018, the survey took place in May. On May 5, 2005, during a juvenile survey, three adult horseshoe crabs (2 females, 1 male) were found on the south side of Mashima Island. The weather in late May of that year was sunny, and the sea temperature was higher than usual. To look for the earliest spawning activity, five spots were surveyed on May 26, and a spawning foam was seen on the right bank of the estuary of Nuki River. The egg clusters were found on the same location on the 28th at noon. In the early May of 2006 and 2009, during the juvenile survey, adults were observed coming into Sone Higata. The spawning foams and eggs were found in late May.

The investigations in September were triggered by an observation of one pair at the right bank of the Nuki River estuary on August 31, 1997. However, no sightings of spawning horseshoe crabs were confirmed. From the next year, only juvenile survey has been done in September. In late August, 2003, many spawning pairs were found and on September 2, 14 pairs were observed during the day. The survey continued into September during high tide, and 50 pairs were found. In late August, 2005, there were many spawning pairs in Sone Higata, but no spawning activities were confirmed in September. No survey in September was done in 2006; in 2007, only one pair was observed. Based on these scanty information, the survey in September didn't become a regular event.

From Table 3 and Fig. 6, you can see that except for 2009, July has the most number of horseshoe crab pairs, and for *Tachypleus tridentatus* in Japan, this month is the peak season for spawning. For 2009, the rainy season started on June 9, but from mid-June, the weather became sunny and air temperature was more than 30 °C on some days. During the high tide in late June, the survey started on June 21. It rained until the 23$^{r+d}$, but then from the 24th it became sunny and the temperature was between 28 °C and 31 °C. In late July, it rained a lot and the volume of water in rivers increased, making a hard condition for spawning activities. Therefore, the visiting number of pairs was less in July than that of June. The weather pattern also influenced the behavior of spawning horseshoe crabs.

Table 4a is a summary of the survey for 21 years from 1998 to 2018. Two high tides in a month is divided into early and late of that month, and observed number of pairs is recorded accordingly. However, one survey during high tide could continue more than one week, and it would carry over into the next month. In such a case, the date of either full moon or new moon is based on how to distinguish the month.

Since the regular spawning survey is set for the months of June to August, the record for May and September is excluded from Table 4a. For each year, a total of six high tides from June to August record 96.05–100% of horseshoe crab pairs counted, and marked 100% for 13 times during these months. This leads to a conclusion that the horseshoe crab spawning season in Sone Higata is during six high tides in June, July, and August.

In Table 4a, each month is divided into two to reflect high tides that occurred twice a month. Each year has top and bottom rows; the top shows the total number of pairs observed, and the bottom shows the ratio of the total observation. Since there were six high tides, the average of horseshoe crab count for each survey is 1/6 or 16.6%. The highest ratio is indicated by white letters with dark highlights, and the ratio above the average (i.e., 16.6%) is in bold letters with light gray highlight. Because each year, the number of visits by horseshoe crab pairs fluctuates, the ratio of the total number can show which months (or high tides) have the most visits for that year.

Based on the record, the most visits were in early July (13 times), late July (6 times), and once for late June and early August. The observations which recorded more than the average (16.6% of the total visits) were late June (3 times), early July (7 times), late July (10 times), early August (6 times), and spawning was confirmed more often in late June into early August. Especially in the month of July recorded the highest number of visits by horseshoe crabs 19 times in 21 years of observations, suggesting that this month is the peak of spawning season (Table 4a).

To further analyze the month of July, the first half of it has more than 16.6% of total visits that year 20 times while the latter half has 16 times. Also, when July had more than 50% of the total visits, it was concentrated in the first half of July and this seems to be the trend in Sone Higata.

Table 4b shows the patterns of horseshoe crab visits over the years. Combinations of largest peak and small peaks (visited pairs more than the average of 16.6%) can be found. These patterns result from the effect of the weather that year and the difference between the 2 tide levels in a month. In 2016, the high tide level was higher earlier in July which had the greatest number of pairs visited in Sone Higata that year, and the latter half of July had more than the average number (16.6%). In the next year, high tide level in late July was higher than the one in early July and had the greatest number of pairs that year. In recent years, the weather became warmer and suitable for horseshoe crab spawning activities in mid-June and on, high tides in late June and early July usually have more than the average number of visiting pairs.

The spawning period of horseshoe crabs in Sone Higata is summarized below. Spawning horseshoe crabs in the shallow water off coast start to gather together near sea area close to Mashima Island in late April to early May. Unpaired horseshoe crabs were sighted on the shore, but no pairs were observed. Although depending on the weather and other environmental conditions, usually pairs arrive in late May to early June. The peak season for their spawning is in July, but sometimes the peak lasts into early August. Except for 2003, most pairs were seen at the estuaries of Kusami River and Nuki River after August 15, then the spawning season ceases. For the spawning pairs in May, the late Prof. Koichi Sekiguchi was consulted in 2005, and it was found that there were no other records of May spawning in Japan except in Sone Higata.

Table 5 and Fig. 7 display the trend in the annual number of visiting spawning horseshoe crab pairs and those of surveys over the years. The average number of pairs was calculated by dividing the number of observed pairs by that of observations. As mentioned, there is a discrepancy between daytime and night-time visits by horseshoe crabs and by us. The calculation simply shows the overall trend within the survey period. Between 1995 and 2000, the average number of pairs was within 2.6 and 4.6; it increased to 10.6 in 2001. The increasing trend continued to 2005, with 26.8 pairs. However, it began to drop to 9.7 and 5.2 (2006 and 2007, respectively). In 2008, the total number of pairs observed was 314, an average of 6.5 per survey. In 2011, the total was 547, the average number of pairs per survey was 8.2. The result indicates that the observed pairs over the years increased by 1.74 while the average number by 1.26. Because since 2009 for 3 years, the number of surveys has increased, it may have increased the number of observed number of pairs. On average, there may have been no significant rise in number of visiting pairs between 2008 and 2011. Then in 2012, the average was 16.4 pairs visiting Sone Higata, and in 2015, it increased to 30.6 pairs. In 2016, Sone Higata had the most visiting pairs since the survey began in 1995, but its average number was 28.5, which is below the previous year. It continued to decline consecutively for 2 years; in 2017, with the average of 18.1 and in 2018 with 11.2 pairs per observation (Table 5 and Fig. 7).

It is of concern now because of the construction of the connection road extending from the mouth of Nuki River to the fishing port, the overflow from the river is blocked to the north and runs towards the southern part of Sone Higata, changing the deposition pattern of soils which may influence the condition of substrate habitat and its benthos. In addition, the rubbles laid out near the mouth of the river for the bridge may influence the course of the river flow and wash out the sand particles from the beach horseshoe crabs use for spawning. Even worse, there is a talk about an expansion of a dumping site for the excavated earth from the sea bottom for the ships, and a new construction to extend the runway for Kita-Kyushu Airport. There are so many modifications of Sone Higata, namely landfills, they could make it difficult for horseshoe crabs to come in to spawn in Sone Higata.

# 3 Juvenile Horseshoe Crab Survey

## *3.1 Changing Approach to Juvenile Horseshoe Crab Survey*

On October 16, 1994, the author accompanied Mr. Harada to Sone Higata for a horseshoe crab survey. Later, on the 23rd of the same month, 14 members of the Wild Bird Society of Japan came to join the juvenile horseshoe crab survey. Even in November, the author conducted three surveys and collected some molts. The result was published in the publication, *Our Nature*, issued from Kita-Kyushu Museum of Natural History. However, it was a preliminary report since there was no definite method for horseshoe crab inventory.

In June 1995, in order to investigate the number of juvenile horseshoe crabs, three transects on the north side of the tidal flat and four on the south were laid. A total of seven transects began at 100 m from the revetment towards the sea, with a length of 600 m and 4 m interval between each. This was conducted into 1996, but the result was unsatisfactory.

So in 1997, the purpose of the survey changed to investigate the distribution of juveniles and their morphology, instead of counting their number. Based on the data accumulated in the previous years, the tidal flat was divided into four sections to carry on our survey. From the estuary of Nuki River along the connection road, Sone Higata is divided into northern and southern areas. The entire northern area is considered Section A. The southern area, because of the frequent occurrence of juveniles, is divided into 3; Section B covers Nuki River estuary, Section C is the center area of the southern tidal flat, Section D is around Kusami River estuary.

The regular survey was conducted in the spring (late April to early June), summer (late June to early August), and fall (late August to early October), with duration of four consecutive days per survey. In 2012, the fall survey was canceled due to an injury of the author during a spawning survey. Also, the author could not have enough manpower to conduct a juvenile survey in July, 2013. Since then, the regular survey has been done for spring and fall, in addition to once a month survey from March to November.

Also, from 2002 and 2010, a winter survey (December to March) was performed. This is to find out when juveniles start moving around and to see if the tide pool temperature becomes 18 °C at which the juveniles are said to be active. However, due to weather conditions and investigator's schedule, only 10 or so surveys were performed.

After 1995, almost all the surveys are conducted by the author who realized that in order to get into Section C, Section B has to be passed. Occasionally, other people who came to the tidal flat for other reasons, such as bird watching, joined in the survey. Every day, different individual juveniles showed up on the muddy bottom of Sone Higata. The horseshoe crab juveniles are active during night-time as well, and the surveys during the daytime may not offer sufficient data to the actual number of juveniles living in the tidal flat. Additionally, there are obstacles to the investigation such as weather and climate, timing of ecdysis, etc. that current require a solution.

## 3.2 Survey Method

Upon conducting a juvenile horseshoe crab survey, survey dates were selected when the exposure of the bottom of the tidal flat is the largest at low tides, but because of the limited amount of manpower as well as the weather conditions, the survey was carried out during high tides.

The duration of each survey was 2–4 hours which was started 3 hours before the maximum low tide. The reason for this is because the juveniles become active at low tide when the sea bottom is exposed to the atmosphere, but they start dig into the substrate approximately 1 hour before the tide comes in. However, their behavior is still influenced by weather and climate (i.e., temperature); enough time is allocated for a juvenile search.

The regular spring season survey is from late April until early June, but for the juvenile survey, it is until mid-May. The regular summer survey has been halted since 2013. For the fall, except for 2012, the survey continues to be performed.

What is done for the survey is to measure the body size (width of opisthosoma, entire body length) in order to figure out their ages. The number of juveniles is also recorded, as well as their distribution within each section. The aim was to confirm the change in number of juvenile individuals and their age, in addition to the timing of their ecdysis (if they have a particular season for it), but because the influence of weather is large, detailed analyses cannot be performed at this point.

A biological investigation on juvenile horseshoe crabs is conducted once in a month from March to November on Section B where the number of juveniles is counted the most. What is done mainly is to count the number of juveniles and record their ages and see any monthly change in their occurrence. Additionally, their range of movement and their ecological relationships with other benthic species were investigated. Upon conducting a regular survey and biological investigation, weather, wind direction and speed, ambient temperature, and water temperature of the sea, as well as tide pools were recorded. As a documentation process, photographing of the environment, juveniles, and other benthic species is also done.

## 3.3 Results

Below are the results of the juvenile horseshoe crab survey conducted during spring in Sone Higata between 1999 and 2018 (Table 6). First, the number of juveniles and spawning pairs will be studied. Because it appears there is no correlation between the juveniles and pairs observed during the same season, the relationships of juveniles were made with the pairs from the previous year, suggesting that some juveniles hatched from the eggs laid by them.

Before 1999 when the number of spawning pairs began to increase, 229 juveniles were found. In 2000, the number increased to 374; there were 116 spawning pairs in the previous year. Between 1997 and 1999, the distribution of juveniles in the northern tidal flat extended into 200 m to 300 m east from the revetment, between 5 m and 300 m northward from the seafloor road, around sandbars on the west coast of Mashima Island, and around the guiding lights at the old airport. In the southern tidal flat, juveniles occur from 100 m to 300 m east from the revetment, 10 m south from the seafloor road towards Kusami River along a water course which deflects to the dike at 800 m from the seafloor road. Along the dike, many juveniles are found between 300 m and 600 m east from the revetment.

In 2000, the number of juveniles increased at around the side of the seafloor road in Section A. This area was signified with a large number of snails such as *Cerithideopsilla cingulata* living gregariously. In the southern Sone Higata, the area 100 m from the revetment was occupied by crabs (*Macrophthalmus japonicus*), from 100 m to 300 m juvenile horseshoe crabs were inhabiting, and from 300 m to 400 m, snails were dwelling in the range. In the northern tidal flat, some snail species were occupying from the revetment to 300 m into the sea, and from the seafloor road to 400 m north. In the south, Sections B and C were less populated; the horseshoe crab juveniles were somewhat more concentrated in Section D. One of the reasons for this discrepancy between the sections is the timing of ecdysis; the survey period must have matched with their molting events. When juveniles molt, they often stay and hide in one location, rather than risking exposing themselves to potential predators. This made it easy for investigators to dig in and find them for counts.

**Table 6** The result of the juvenile survey in the springtime by areas A-D in Sone Higata, 1999–2018

| **1999** | | | | | | **2000** | | | | | |
|---|---|---|---|---|---|---|---|---|---|---|---|
| Area | A | B | C | D | Total | Area | A | B | C | D | Total |
| Age | | | | | | Age | | | | | |
| 2nd instar | 5 | 7 | 5 | 10 | 27 | 2nd instar | 10 | 11 | 13 | 52 | 86 |
| 3rd instar | 0 | 0 | 0 | 0 | 0 | 3rd instar | 0 | 0 | 0 | 1 | 1 |
| 4th instar | 0 | 2 | 0 | 0 | 2 | 4th instar | 9 | 7 | 11 | 10 | 37 |
| 5th instar | 3 | 25 | 88 | 48 | 164 | 5th instar | 10 | 13 | 16 | 28 | 67 |
| 6th instar | 2 | 34 | 9 | 10 | 55 | 6th instar | 18 | 8 | 5 | 11 | 42 |
| 7th instar | 7 | 24 | 7 | 13 | 51 | 7th instar | 13 | 10 | 20 | 54 | 97 |
| 8th instar | 1 | 6 | 5 | 5 | 17 | 8th instar | 32 | 1 | 6 | 18 | 57 |
| 9th instar | 0 | 1 | 0 | 0 | 1 | 9th instar | 3 | 0 | 0 | 4 | 7 |
| 10th instar | 0 | 0 | 0 | 0 | 0 | 10th instar | 0 | 0 | 0 | 0 | 0 |
| Total | 18 | 99 | 114 | 86 | 317 | Total | 95 | 50 | 71 | 178 | 394 |
| **2005** | | | | | | **2006** | | | | | |
| Area | A | B | C | D | Total | Area | A | B | C | D | Total |
| Age | | | | | | Age | | | | | |
| 2nd instar | 0 | 0 | 0 | 4 | 4 | 2nd instar | 0 | 0 | 3 | 38 | 41 |
| 3rd instar | 0 | 1 | 2 | 2 | 5 | 3rd instar | 0 | 1 | 1 | 9 | 11 |
| 4th instar | 8 | 14 | 25 | 30 | 77 | 4th instar | 16 | 102 | 75 | 46 | 239 |
| 5th instar | 13 | 22 | 23 | 8 | 66 | 5th instar | 12 | 89 | 73 | 35 | 209 |
| 6th instar | 39 | 128 | 64 | 108 | 339 | 6th instar | 3 | 14 | 7 | 65 | 89 |
| 7th instar | 13 | 33 | 45 | 47 | 138 | 7th instar | 7 | 11 | 7 | 34 | 59 |
| 8th instar | 13 | 21 | 12 | 28 | 74 | 8th instar | 6 | 9 | 8 | 27 | 50 |
| 9th instar | 1 | 2 | 1 | 3 | 7 | 9th instar | 0 | 2 | 1 | 5 | 8 |
| 10th instar | 0 | 0 | 0 | 1 | 1 | 10th instar | 0 | 0 | 0 | 1 | 1 |
| Total | 87 | 221 | 172 | 231 | 711 | Total | 44 | 228 | 175 | 260 | 707 |
| **2012** | | | | | | **2013** | | | | | |
| Area | A | B | C | D | Total | Area | A | B | C | D | Total |
| Age | | | | | | Age | | | | | |
| 2nd instar | 0 | 0 | 0 | 0 | 0 | 2nd instar | 0 | 0 | 0 | 0 | 0 |
| 3rd instar | 0 | 25 | 7 | 0 | 32 | 3rd instar | 1 | 7 | 1 | 1 | 10 |
| 4th instar | 0 | 43 | 10 | 1 | 54 | 4th instar | 0 | 16 | 3 | 3 | 22 |
| 5th instar | 0 | 0 | 0 | 0 | 0 | 5th instar | 2 | 24 | 14 | 5 | 45 |
| 6th instar | 1 | 11 | 11 | 4 | 27 | 6th instar | 1 | 35 | 23 | 20 | 79 |
| 7th instar | 0 | 5 | 4 | 4 | 13 | 7th instar | 1 | 7 | 5 | 1 | 14 |
| 8th instar | 1 | 3 | 1 | 1 | 6 | 8th instar | 0 | 0 | 2 | 2 | 4 |
| 9th instar | 0 | 0 | 0 | 0 | 0 | 9th instar | 0 | 0 | 0 | 0 | 0 |
| 10th instar | 0 | 0 | 0 | 0 | 0 | 10th instar | 0 | 0 | 0 | 0 | 0 |
| Total | 2 | 87 | 33 | 10 | 132 | Total | 5 | 89 | 48 | 32 | 174 |

(continued)

**Table 6** (continued)

| 2017 | | | | | | 2018 | | | | | |
|---|---|---|---|---|---|---|---|---|---|---|---|
| Area | A | B | C | D | Total | Area | A | B | C | D | Total |
| Age | | | | | | Age | | | | | |
| 2nd instar | 0 | 0 | 0 | 0 | 0 | 2nd instar | 0 | 0 | 0 | 0 | 0 |
| 3rd instar | 0 | 0 | 0 | 0 | 0 | 3rd instar | 1 | 1 | 0 | 0 | 2 |
| 4th instar | 1 | 59 | 19 | 13 | 92 | 4th instar | 0 | 17 | 3 | 6 | 26 |
| 5th instar | 0 | 13 | 16 | 7 | 36 | 5th instar | 0 | 71 | 54 | 39 | 164 |
| 6th instar | 1 | 44 | 14 | 21 | 80 | 6th instar | 3 | 12 | 2 | 0 | 17 |
| 7th instar | 0 | 89 | 50 | 25 | 164 | 7th instar | 2 | 107 | 52 | 74 | 235 |
| 8th instar | 1 | 14 | 6 | 3 | 24 | 8th instar | 0 | 20 | 18 | 17 | 55 |
| 9th instar | 0 | 2 | 1 | 0 | 3 | 9th instar | 0 | 5 | 1 | 0 | 6 |
| 10th instar | 0 | 0 | 0 | 0 | 0 | 10th instar | 0 | 0 | 0 | 0 | 0 |
| Total | 3 | 221 | 106 | 69 | 399 | Total | 6 | 233 | 130 | 136 | 505 |

In 2005, the number of juveniles was 711, with 1363 spawning pairs from the previous year. In 2006, the number of juveniles was 707, while that of adult pairs was 1581. These two years recorded the highest number of horseshoe crab individuals since the survey in Sone Higata began. In the southern tidal flat, Sections B and D located at the forefront of the estuaries of Nuki River and Kusami River had the increase of pairs, followed by a sharp increase in juvenile individuals. In some areas, four to five juveniles were found within 1m$^2$.

During late April into mid-May, 2004, there was an explosion of seaweed, such as *Ulva linza*, in the south of Section B into Section D. This condition was repeated every year until 2014. A massive amount of the seaweed was drifted toward the shore and gave an impact on the beach. However, it gradually rotted away on the tidal flat and almost no change had occurred on its bottom between 2004 and 2006, and horseshoe crab juveniles were not affected, except perhaps their foraging behavior a little.

But in 2006, no juveniles were observed on the western Mashima island within the area A, and probably also along the guiding lights. They were witnessed less and less at their habitat near the levee. In the meantime, three areas in the southern tidal flat experienced almost no change in the number of juveniles. Their habitat had stayed nearly the same since 1998 and their distribution range also remained the same.

Since 2007, the number of juveniles in the southern tidal flat began to fall, and the decreasing trend became significant since 2008. In 2012, 132 juveniles were counted and the number of pairs in the previous year was 547. In 2013, 132 juveniles were observed and 738 spawning pairs were counted in the previous year. When comparing 2012 and 2000 for the juvenile number and pairs in the previous year, then number of spawning pairs had increased five-folds while the number of juveniles had decreased to one-third.

The decrease of the juveniles occurred everywhere except at area B. Areas A and D suffered the most. The bottom of area A in the northern tidal flat has always been affected by fluctuating condition of the tidal flats. Since the spring of 2004, the southern tidal flat had areas where seaweed (*Ulva linza*) grew every year, some of which worsened the condition of the tidal flat bottom from the accumulation of its rotten remains. Because of the wind and/or wave directions, area D suffered the most from this phenomenon. The rotten seaweed further influenced the tidal flat into 15–20 cm below the flat surface and benthic species such as *Macrophthalmus abbreviates* and *Glycera nicobarica* became less and less common. Southern half of area C became similar to area D, and the condition of soil in area B also became worse.

The number of juveniles increased in 2013 and this may be because the seaweed outburst was relatively low both in population and distribution in the previous year. The author suspects that one of the reasons for the decline in the juvenile count is the result of disappearance of their food, benthic species, due to the altered tidal flat environment by the increased amount of rotted seaweed.

Since 2014, the number of juveniles has been gradually increased. The explosion of seaweed in the southern tidal flat has been reduced in its expansion since the spring of 2012, then in the spring of 2015, such a burst was not observed, and this calm condition continues until today.

In 2017, 399 juveniles were observed. In the previous year, the number of spawning pairs was the most: 1995. Then in 2018, the number of juveniles was increased to 505 while that of pairs in the previous year was decreased to 1323. The area A in the northern tidal flat in 2018 was not favorable to juveniles; only three were found. On the other hand, areas B and C supported as many juveniles as those in 2005 and 2006. The seafloor morphology of area D had been changed significantly and the increase in the number of juveniles seems to be stagnant.

In 2018, there was a change in the distribution pattern in areas B and C in the southern tidal flat. The habitat of juveniles in area B had shifted southward. Until 2017, juveniles were confirmed from the mouth of Nuki River to 300 m south, but close to area C, the population density was low. However, in the spring 2018 survey, more juveniles were found close to area C and the population was reduced at the mouth of Nuki River. Additionally, near the border between areas C and D, there was almost no juveniles until 2018. This indicates that the juvenile population moved southward, restricting its distribution.

Table 7 shows the ecological survey of juveniles in area B in 2018. Within it, April 29 and September 23 are a record of regular survey. The difference between these surveys is basically the duration period; ecological survey takes approximately 2 hrs and regular survey lasts more than 3 hrs. Because of this, there is a difference in number of juveniles confirmed. The age of juveniles is discussed here based on the table.

During the winter survey on January 22, no juveniles were confirmed. In January of 2002 and 2003, even the water temperature was 14° in the tide pool, there were three 5th instars and five 6th instars were found.

**Table 7** The 2018 result of the juvenile survey in area B, Sone Higata. The juvenile surveys were conducted 9 time in the year 2018 to look at the trend in juvenile age in terms of instar stage. The number indicates the juveniles counted in each survey

| | Jan. 22 | Mar. 4 | Apr. 29 | May 14 | May 27 | June 17 | Jul. 23 | Sep. 23 | Oct. 28 |
|---|---|---|---|---|---|---|---|---|---|
| 2nd instar | 0 | 0 | 0 | 0 | 7 | 11 | 0 | 0 | 0 |
| 3rd instar | 0 | 0 | 1 | 0 | 0 | 0 | 17 | 2 | 2 |
| 4th instar | 0 | 16 | 17 | 4 | 1 | 0 | 0 | 75 | 16 |
| 5th instar | 0 | 8 | 71 | 34 | 36 | 19 | 1 | 2 | 2 |
| 6th instar | 0 | 2 | 12 | 1 | 4 | 37 | 33 | 9 | 24 |
| 7th instar | 0 | 36 | 107 | 88 | 45 | 31 | 29 | 40 | 6 |
| 8th instar | 0 | 5 | 20 | 12 | 5 | 10 | 14 | 16 | 21 |
| 9th instar | 0 | 0 | 5 | 0 | 0 | 1 | 0 | 7 | 0 |
| 10th instar | 0 | 0 | 0 | 0 | 0 | 0 | 0 | 0 | 0 |
| Total | 0 | 67 | 233 | 139 | 98 | 109 | 94 | 151 | 71 |

On March 4, 4th to 8th instars were observed, and on April 29, there was an increase in number of 5t + h instars. They may have been the 4th instars found on March 4 that molted and grew 1 instar bigger.

On May 27, 2nd instars were observed at the first time in 2018. They were hatched to 1st instars in the previous year, then wintered in the mud. They molted to become 2nd instars after winter. Usually, 2nd instars appear in late May into early June. They spend one more month to molt again to be 3rd instars. After one more month, in August to September, they become 4th instars, some of which repeat molting and become 5th instars. This is the stage of growth trend seen in Sone Higata.

The author is carefully focusing on the molting season, especially that of the 2nd instars. In the fall 2018, 90% of the 2nd year juveniles were 4th instars. In both ecological survey and regular survey in October, 2014, 90% of the 2nd year juveniles were 3rd instars. In September regular survey of 2015, 90% of the 2nd year juveniles were 3rd instars, but one month later, in the ecological survey showed that about 40% of them were 4th instars, meaning that those had molted between September and October. The September survey of 2016 indicated that the ratio of 3rd and 4th instars was 6:4, but in the next year, the ratio changed to 1:9.

There is a fluctuation in time of molting among 2nd year juveniles, and this is also true for the 3rd year juveniles. The timing may have been influenced by the weather/climate conditions, changes in the tidal flat environment, and such. Therefore, it would be necessary to consider seasonal changes and environmental conditions when analyzing the trend in juvenile molting pattern, making the study more challenging.

Based on Table 7, seemingly 3rd year juveniles that are 5th instars were thought to have molted by the end of July to be 6th instars. Furthermore, it was found that in the survey on September 27, 7th instars were found, and because almost half of them had soft shells, the molting may have occurred close to the survey date.

On the other hand, 7th instars that were found the most between March and mid-May were reduced in number since then, and there was no increase in the number of individuals between June and July, as well as in the number of 8th instars. From Table 7, molting period was not determined. However, nine molts of the 7th instars were found in juvenile survey, beach cleanup, and other activities, and it is suggested that the molting time was between early June and late July.

For the 8th instars, six individuals with soft shell were found in the survey on September 27, suggesting that the molting continues into September. It is not clear when 8th instars become 9th instars, but seven individuals of 9th instars were found on September 23.

Moreover, juveniles were usually found in September. For example, 12 juveniles were counted in September, 2017. There were areas in both northern and southern tidal flats where 9th instars were observed, but that was until 2005. There were also water routes to the offshore appearing in August and September, but there is no more confirmation of them today. The disappearance of water routes might have been caused by the change in the sea bottom condition.

The number of juveniles declined in 2016, but was recovered the next year. In these days, it is in the increasing trend. However, for the juveniles in 2018, their distribution had been restricted and the number was declining as a result.

## 4 Conclusions

Over 20 years have passed since the start of horseshoe crabs in Sone Higata. The author thought that specialists would conduct more systematic research in Sone Higata and only collected sufficient enough data to reflect the reality of the place, and therefore apologizes for the lack of more thorough analytical investigation in this chapter.

In 2018, the number of spawning pairs has been declining. The rate of decrease was lower than that of 2006–2007 which had the lowest number of pairs found, and the average number of pairs is maintained over 10%.

As for the number of pairs visiting Sone Higata, the observed spawning pairs is only reflecting the actual number. The author attempted to count the actual number in 2004 by marking the individual horseshoe crabs that came into Sone Higata. A combination of numbers and letters were marked by crayons under their compound eyes. The marked individuals divided by the number of pairs counted was calculated as 0.62, meaning 62% of actual visitors were recorded, at least that year in Sone Higata.

Using this coefficient number, the observed pairs in 2018 was 763 would make the actual number of pairs could be 472. Whether this number is small or not for Sone Tidal Flat, it is clear that because of the worsening condition of sand supply for the beaches at Yoshida Coast, Nuki River mouth, and Kusami River mouth, spawning for horseshoe crabs may have been more and more difficult.

The number of juveniles fluctuates year to year. From 2008, it was in a declining trend, then in 2011, it started to increase. However, because of the record-breaking heavy rain in northern Kyushu in 2012 which dumped massive amount of water, driftwood, trash into Sone Higata, damaging its environmental conditions. From the fall of 2016, the number of juveniles had started to recover again and it was increasing from 2017 into the spring of 2018, then crashed in the fall, despite the fact that most of the time, the population in the fall exceeds the spring. In October, 2015, the connecting road to the Sone Fishing Port was set up, dividing Sone Higata into northern and southern parts. It's been three years since then, and the decrease in the number of juveniles is a major concern now.

In the northern tidal flat, juveniles were scattered all over; in the south, their distribution became smaller and smaller.

Horseshoe crab is an important species which hasn't changed its morphology over 250 million years, and therefore is called a living fossil. Since the adults lay eggs on the beach, juveniles grow up in tidal flats, and the young and sub-adults live in the shallow sea, their habitat requires these three different environments, and their mere presence reflects the condition of the environment as well as the ecology along the coast, they are said to be an index species. Horseshoe crab is a rare species which is listed up on the Red List of the Japanese Ministry of Environment as endangered. The fishing industry is active in Suounada, and the author believes the conservation of the environment there is necessary. This report should serve as the guide for people who are to protect and conserve the nature in Sone Higata and its nearby areas.

**Acknowledgments** The author would like to thank to all the people who helped the investigation of horseshoe crabs and Sone Higata. Inoue, Y., Okamura, T., and Hayakawa, O. assisted the surveys of the spawning horseshoe crabs. Aoki, M., Kayano, K., Fukuda, M, and Yasunaga, M. supplied the results of spawning pair counts. Mizoguchi, A. helped with the PC and created tables. Lastly, co-researcher of Sone Higata, Takahashi, S. assisted with the computer input and other tasks as well. Additionally, the author thanks Iwasaki, Y. translated the Japanese manuscript into English. The author is grateful for everyone's participation in putting this chapter together.

## References

H. Kawahara, Habitat condition of juvenile horseshoe crabs in Kitsuki Bay. Current Status of Japanese Horseshoe Crabs (supplemental edition): 87–104. Japan Society of the Protection of Horseshoe Crabs (Nihon Kabutogani-wo mamoru-kai). (in Japanese) (1993)

R. Sakami, Plans for the current situation of horseshoe crabs in Imari Bay, based on the number of spawning pairs. Kabutogani **22**, 43–44 (2002). Bulletin of the Nihon Kabutogani-wo mamoru-kai. (in Japanese)

S. Takahashi, O. Hayashi, Investigation of horseshoe crab spawning in Sone Higata. Kabutogani **33**, 20–23 (2013). Bulletin of the Nihon Kabutogani-wo mamoru-kai. (in Japanese)

S. Takahashi, Report on the mass mortality of horseshoe crabs in Sone Higata in 2016. Kabutogani **37**, 16–23 (2017). Bulletin of the Nihon Kabutogani-wo mamoru-kai. (in Japanese)

S. Nishihara, Horseshoe crab survey in Morie Bay. Kabutogani **18**, 15–21 (1998). Bulletin of the Nihon Kabutogani-wo mamoru-kai. (in Japanese)

O. Hayashi, Habitat condition of horseshoe crabs in Sone Higata (result of the fall survey, 1994). Our Nat. History **52**, 4–8 (1995). Bulletin of the Fellowship of the Kita-Kyushu Museum of Natural History. (in Japanese)

O. Hayashi, Horseshoe crab survey in Sone Higata. Kabutogani **17**, 32–34 (1997). Bulletin of the Nihon Kabutogani-wo mamoru-kai. (in Japanese)

O. Hayashi, Horseshoe crab survey in Sone Higata. Kabutogani **18**, 22–26 (1998). Bulletin of the Nihon Kabutogani-wo mamoru-kai. (in Japanese)

O. Hayashi, Horseshoe crab survey in Sone Higata. Kabutogani **19**, 28–35 (1999a). Bulletin of the Nihon Kabutogani-wo mamoru-kai. (in Japanese)

O. Hayashi, The 1st instars of horseshoe crabs (*Tachypleus tridentatus*). Our Natural History, 70. Cover page with explanation. Bulletin of the Fellowship of the Kita-Kyushu Museum of Natural History. (in Japanese) (1999b)

O. Hayashi, Report of the horseshoe crab survey in Sone Higata in 1999. Kabutogani **20**, 25–29 (2000a). Bulletin of the Nihon Kabutogani-wo mamoru-kai. (in Japanese)

O. Hayashi, Current status of horseshoe crabs in Sone Higata in 1999. Our Natural History **74**, 5–11 (2000b). Bulletin of the Fellowship of the Kita-Kyushu Museum of Natural History. (in Japanese)

O. Hayashi, Record of the number of horseshoe crab spawning pairs in Sone Higata, Kokura in 2000. Kabutogani **21**, 23 (2001a). Bulletin of the Nihon Kabutogani-wo mamoru-kai. (in Japanese)

O. Hayashi, Survey of horseshoe crabs in Sone Higata in 2000. Kabutogani **21**, 26–31 (2001b). Bulletin of the Nihon Kabutogani-wo mamoru-kai. (in Japanese)

O. Hayashi, Survey of horseshoe crabs in Sone Higata in 2001. Kabutogani **22**, 32–35 (2002). Bulletin of the Nihon Kabutogani-wo mamoru-kai. (in Japanese)

O. Hayashi, K. Samoto, S. Takahashi, O. Hayakawa, Record of the number of spawning horseshoe crab pairs in 2001 for Kokura (4 sections of Sone Higata). Kabutogani **22**, 27 (2002). Bulletin of the Nihon Kabutogani-wo mamoru-kai. (in Japanese)

O. Hayashi, Record of the spawning horseshoe crab pairs in 2002 in Sone Higata. Kabutogani **23**, 15 (2003a). Bulletin of the Nihon Kabutogani-wo mamoru-kai. (in Japanese)

O. Hayashi, Survey of horseshoe crabs in Sone Higata in 2002. Kabutogani **23**, 18–22 (2003b). Bulletin of the Nihon Kabutogani-wo mamoru-kai. (in Japanese)

O. Hayashi, Observations of species in the tidal flat where horseshoe crabs live. Weekly Magazine, Japan's Natural Monuments: Animal Version, 34, Kabutogani, 32–33, (Shogakukan Publisher. Tokyo, 2003c)

O. Hayashi, Consideration of the survey methods with markings on horseshoe crabs. Kabutogani **25**, 31–36 (2005). Bulletin of the Nihon Kabutogani-wo mamoru-kai. (in Japanese)

O. Hayashi, S. Takahashi, O. Hayakawa, Record of the number of spawning horseshoe crab pairs in 2003 for Kokura (4 sections of Sone Higata). Kabutogani **24**, 32 (2004a). Bulletin of the Nihon Kabutogani-wo mamoru-kai. (in Japanese)

O. Hayashi, S. Takahashi, O. Hayakawa, M. Yasunaga, Survey of horseshoe crabs in Sone Higata in 2003. Kabutogani **24**, 52–54 (2004b). Bulletin of the Nihon Kabutogani-wo mamoru-kai. (in Japanese)

O. Hayashi, S. Takahashi, O. Hayakawa, Record of the number of spawning horseshoe crab pairs in 2004 for Kokura (4 sections of Sone Higata). Kabutogani **25**, 22 (2006). Bulletin of the Nihon Kabutogani-wo mamoru-kai. (in Japanese)

O. Hayashi, K. Shimizu, Horseshoe crab conservation activities in Sone Higata. Kabutogani **27**, 55–57 (2007). Bulletin of the Nihon Kabutogani-wo mamoru-kai. (in Japanese)

O. Hayashi, S. Takahashi, O. Hayakawa, Record of the number of spawning horseshoe crab pairs in 2006 for Kokura (4 sections of Sone Higata). Kabutogani 27, **45** (2007). Bulletin of the Nihon Kabutogani-wo mamoru-kai. (in Japanese)

# Distribution of Horseshoe Crabs, *Carcinoscorpius rotundicauda*, and *Tachypleus gigas* in Singapore, Including Coney Island

**Jasmin Lim, Yi Long Ng, Oliver Chang, Prashant Shingate, Boon Hui Tay, Qian Tang, Frank Rheindt, Hor Yee Tong, Karenne Tun, Byrappa Venkatesh, and Laura Yap**

## 1 Introduction

Horseshoe crabs can be found over a wide geographical range. *Limulus polyphemus* (Atlantic horseshoe crab) resides along the northeastern coast of the United States from the Gulf of Maine to the Gulf of Mexico. *Tachypleus tridentatus* (tri-spine horseshoe crab), *Tachypleus gigas* (coastal horseshoe crab), and *Carcinoscorpius rotundicauda* (mangrove horseshoe crab) are distributed from the southeastern coast of India to the Philippines, to the southwestern coast of Japan (Chatterji and Abidi 1993; Sekiguchi and Shuster 2009). Out of the four extant species of horseshoe crabs, only *T. gigas* and *C. rotundicauda* can be found in Singapore.

Anecdotal evidence has pointed toward horseshoe crab populations in Singapore being in decline, as similarly reported in other Asian countries (Mishra 2009; Shin et al. 2009). In Asia, horseshoe crabs are consumed as an exotic delicacy and used as fish bait and in fishmeal preparations (Christianus and Saad 2007; Faridah et al. 2015). In addition, horseshoe crabs face threats such as habitat loss caused by natural coastal erosion and anthropogenic activities such as coastal pollution and land reclamation (Jackson et al. 2005; Ngy et al. 2007; Nelson et al. 2015). In the

J. Lim (✉) · Y. L. Ng · O. Chang · L. Yap
School of Applied Science, Republic Polytechnic, Singapore
e-mail: Jasmin_lim@rp.edu.sg

P. Shingate · B. H. Tay · B. Venkatesh
Comparative Genomics Laboratory, Institute of Molecular and Cell Biology, A*STAR, Proteos, Singapore

Q. Tang · F. Rheindt
National University of Singapore, Singapore, Singapore

H. Y. Tong · K. Tun
National Parks Board, Singapore, Singapore

 
J. T. Tanacredi et al. (eds.), *International Horseshoe Crab Conservation and Research Efforts: 2007-2020*, https://doi.org/10.1007/978-3-030-82315-3_16

Singapore context, one major threat would be entanglement in abandoned fishing and prawn nets. Through the numerous rescue and coastal clean-up programs, horseshoe crabs getting caught in drift nets were a fairly common occurrence. Habitat loss is likely to be the main threat locally as rapid urbanization of the country has led to mangrove areas in Singapore decreasing from 6400 ha to 734.9 ha between 1953 and 2011 (Yang et al. 2011). As such, most of Singapore's mangrove forests are medium (100 ha) to small remnant patches (20–50 ha), and isolated strips (10 ha or less) confined to sheltered bay, tidal river, and river mouths. Notwithstanding the offshore islands of Pulau Tekong and Pulau Ubin, the largest proportion of mangrove forests on the main island is found at northwestern Singapore, in the areas of Sungei Buloh Wetland Reserve, and Mandai and Kranji mangroves (Yang et al. 2011).

Published studies on the population distribution, population structure, and breeding of horseshoe crabs in Singapore have been mainly focused at Mandai mudflats (Cartwright-Taylor et al. 2009, 2011, 2012; Cartwright-Taylor and Hsu 2012). It is the only known location with a substantial breeding population of *C. rotundicauda*, where they inhabit the brackish intertidal waters of mudflats and use the Mandai mangroves as nursery areas (Chatterji and Abidi 1993; Ng and Sivasothi 2002). In contrast, sightings of *T. gigas* have been infrequent with no known breeding sites reported for Singapore. There has only been anecdotal evidence that *T. gigas* was once abundant in Singapore, and its distribution in Singapore is unknown. Thus far, only Cartwright-Taylor et al. (2011) have published the capture of six live adult *T. gigas* females at Changi Point that had become entangled in a fisherman's net. There is an urgent need for more studies on this species, especially on the timing and spawning sites (Cartwright-Taylor 2015).

The statuses of both *C. rotundicauda* and *T. gigas* are listed as "Data Deficient" in the International Union for Conservation of Nature (IUCN) Red List, with no detailed information on their distribution and population trends, due to factors such as the absence of field data as well as inconsistent methodology leading to different outcomes (World Conservation Monitoring Centre 1996a, 1996b; Nelson et al. 2019). In the Singapore Red Data Book, the national conservation statuses of *C. rotundicauda* and *T. gigas* are listed as "Vulnerable" and "Endangered," respectively (Davidson et al. 2008). There is an urgent need to understand their distribution in Singapore and in turn devise appropriate conservation strategies due to rapid urbanization threatening their habitats.

The aim of the present study was to provide important information on the distribution of horseshoe crabs, *C. rotundicauda* and *T. gigas* in northern Singapore, including the offshore Coney Island. Coney Island is a 50-ha park connected to the mainland of Singapore via two bridges from Pasir Ris Coast Industrial Park and Punggol Promenade, and it was opened to the public in October 2015. One goal of the project was to gather baseline data on the horseshoe crabs after public access to Coney Island to enable long-term monitoring of possible impacts from anthropogenic activities on their distribution and population size/density (Cheo Pei Rong, pers. comm.). A second goal of the study was to collect information on horseshoe crabs at Sungei Buloh Wetland Reserve, Singapore's first ASEAN Heritage Park that covers 202 ha of mangroves, mudflats, ponds, and forests. Information on their distributions would help the local government in policy planning, hopefully to also achieve recovery of declining horseshoe crab populations through legislative governance and beach restorations as similarly seen in America and China (Beekey and Mattei 2015; Botton et al. 2015; Kwan et al. 2017).

## 2 Materials and Methods

### *2.1 Study Sites*

Based on literature reviews on horseshoe crab habitats, an island-wide survey was conducted to determine the spatial distribution of *C. rotundicauda* and *T. gigas* around Singapore. Five accessible sites, where the natural topography of the shore has not been destroyed by urban development, were selected for the survey (Fig. 1).

Site 1, Sungei Buloh Wetland Reserve, is a protected reserve with a large mangrove area and a muddy substrate. The sampling shore length of approximately 1 km was surveyed over the period of November 2017 to November 2018.

Site 2, Punggol Settlement, has a sandy substrate and is near to human development. The sampling shore length of approximately 800 m was surveyed from November 2017 to December 2018.

Site 3, Pasir Ris Park Beach, has a sandy substrate and is near to human development. It has a nearby mangrove forest. The sampling shore length of approximately 3.4 km was surveyed from November 2017 to December 2018.

Site 4, Changi Beach, has a sandy substrate and is near to human development. The sampling shore length of approximately 3.3 km was surveyed from May 2017 to December 2018

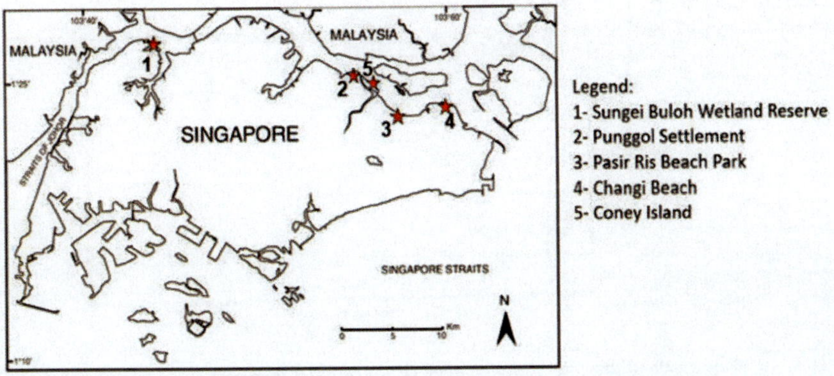


**Fig. 1** Sites of the island-wide survey of horseshoe crabs *C. rotundicauda* and *T. gigas* on northern shores of Singapore's main island and offshore Coney Island

Site 5, Coney Island, is an offshore nature park connected to the mainland of Singapore via two bridges. It has a small mangrove area that is inaccessible, but otherwise sandy substrate. The sampling shore length of approximately 2 km was surveyed from October 2015 to May 2019.

## 2.2 *Visual Search*

Prior to each survey, surveyors were briefed to walk abreast of one another in a line, at a distance of 1 m from each other, from one end of a survey site to the other covering the entire survey area. Surveys were mostly conducted at low tides of 0.8 m above Chart Datum, although there were a few occasions in which the team performed surveys at tide levels of 1.5 m and above to explore the possibility of horseshoe crab sightings in high tides.

When horseshoe crabs were sighted, information such as GPS coordinates, species, status (alive, dead, or moult), gender, weight, habitat substrate, and presence of eggs was recorded. Distinguishing morphological characteristics between the two species previously described by Sekiguchi and Nakamura (1979), such as differences in shape and size of carapace, telson cross section, as well as length of marginal spines, were used to help differentiate between the specimens found in the study. The gender of the adults was determined based on the presence of pedipalps in males as well as the shape of the first two pairs of claws. Morphometric measurements such as carapace width, carapace length, and total body length were also recorded (Fig. 2). Juveniles were classified as those <8 cm carapace width for *C. rotundicauda* (Cartwright-Taylor et al. 2012).

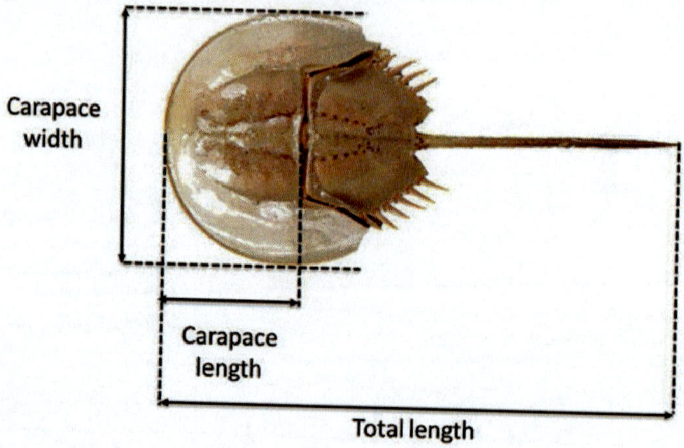


**Fig. 2** Measurements of morphological characteristics of horseshoe crab

# 3 Results and Discussion

## 3.1 Distribution of Both Species in Singapore

Survey results demonstrated that there was little overlap in distribution between both species, with coexistence occurring only on Coney Island (Fig. 3). *T. gigas* live in sandy to muddy marine habitats, while *C. rotundicauda* prefer muddy areas in brackish waters. Coney Island has a mangrove forest in close proximity to sandy beaches. Hence, it has substrate and habitat types favoring both species. Although coexistence of both species has been also previously reported at Changi Beach (Cartwright-Taylor et al. 2011), the current study only found *T. gigas* at the site. There could be a low occurrence of finding both species together at Changi Beach as the location has a natural sandy beach more suited to *T. gigas*. Incidences of coexistence for both species have also been reported in India and Indonesia (Wells et al. 1983; Chatterji 1999; Behera et al. 2015). *T. gigas* and *C. rotundicauda* coexisting with each other may be possible because of overlapping ecological niches and their ability to tolerate a wide range of environmental parameters such as temperature, salinity, and dissolved oxygen (Vestbo et al. 2018; Sekiguchi and Shuster 2009). However, distinct preferences for habitats of *T. gigas* (intertidal sandy) and *C. rotundicauda* (mangrove areas) are likely to have played a part in determining their distinctly separate distribution in Singapore.

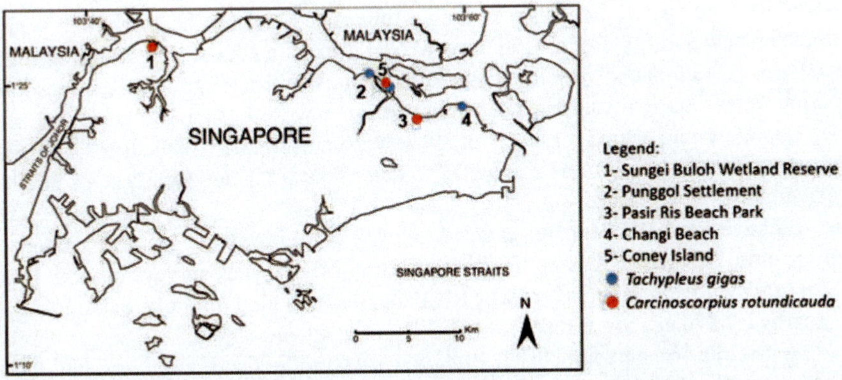


**Fig. 3** Distribution of horseshoe crabs on northern shores of Singapore main island and Coney Island

## 3.2 *Changi Beach, Pasir Ris Park Beach, and Punggol Settlement*

During the study period, low numbers of live horseshoe crabs were found at Changi Beach (30.8%), Pasir Ris Beach (7.7%), and Punggol Settlement (5.9%) (Table 1).

There was an oil spill that occurred in Johor Strait near Punggol Settlement in 2017, so that could have polluted the waters, causing mortality of the horseshoe crabs (Chan 1977; Kotwani 2017). Pollution from the oil industry has been hypothesized as a cause for the decline of horseshoe crab populations in Mexico because horseshoe crabs are benthic and burrow in sediments where dense crude oil precipitates and lingers (Zaldivar-Rae et al. 2009). In addition, exposure to oil affects their development and their diet is based on bivalves, which are known to accumulate pollutants (Strobel and Brenowitz 1981; Gold-Buchot et al. 2007; Botton and Itow 2009). As there is no previous data on horseshoe crabs populations at these locations, information from the current study would serve as baseline data to see if there would be an increase in population after the area recovers from the oil spill.

Survey of *T. gigas* distribution around the shores of the main island of Singapore suggests that isolated individuals can still be found at Changi Beach and Punggol Settlement, which are in the northeastern part of Singapore's main island (Table 1). A pair of *T. gigas* was found in amplexus at Changi Beach on May 26, 2017, which was a new moon (Fig. 4). The four female coastal horseshoe crabs at Changi Beach were also found congregated on July 13, 2018, which was a super new moon. There has been a reported correlation in the occurrence of *T. gigas* with tidal amplitude and lunar phase, with more crabs seen at higher tidal amplitudes (Chatterji and Abidi 1993). Reproductive behavior of horseshoe crabs has been observed to be periodic and synchronized with high tide, where they migrate from deeper to shallower water to spawn. In Malaysia, the nesting of *T. gigas* has been reported year-round, at new and full moons (Zaleha et al. 2012, Nelson et al. 2016). In the current study, no nesting sites were found. Factors such as sediment grain size, monsoon period, water current, and metal contamination have all been shown to impact *T. gigas* spawning behavior (Tan et al. 2012; Nelson et al. 2015, Nelson et al. 2016). More studies will need to be conducted to determine if Changi Beach is a suitable spawning site for *T. gigas*.

Although Pasir Ris Park Beach was located in close proximity to these two locations, only *T. gigas* carcasses were found during the 1-year survey. Instead, live *C. rotundicauda* were found at Pasir Ris Park Beach, which might be explained by a nearby 6-ha mangrove forest.

No juveniles or subadults of *C. rotundicauda* were seen at these three sites. The nearest mangrove settlements within the vicinity of Punggol Settlement and Changi Beach are approximately 2 km away, located at Coney Island and Pulau Ubin, respectively. *C. rotundicauda* juveniles typically remain on the surface of the wet mud of their spawning ground, hence they are unlikely to move over a far distance. Moreover, anthropogenic activities at these survey sites are high as these are recreational sandy beaches with frequent human traffic. With respect to Pasir Ris Park

**Table 1** Site characteristics and number of adult and juvenile *Carcinoscorpius rotundicauda* and *Tachypleus gigas* found at Changi, Pasir Ris Park, and Punggol Settlement beaches

| | | | | | *C. rotundicauda* | | | | *T. gigas* | | | |
|---|---|---|---|---|---|---|---|---|---|---|---|---|
| | | | | | | Alive | | | | Alive | | |
| Location | Total HSC | Alive | Dead | Habitat type | Dead | Juveniles | Females | Males | Dead | Juveniles | Females | Males |
| Changi Beach | 15 | 6 | 9 | Sandy beach | 2 | 0 | 0 | 0 | 7 | 0 | 5 | 1 |
| Pasir Ris Park Beach | 26 | 2 | 24 | Sandy beach | 11 | 0 | 1 | 1 | 13 | 0 | 0 | 0 |
| Punggol Settlement | 34 | 2 | 32 | Sandy beach | 20 | 0 | 0 | 0 | 12 | 0 | 1 | 1 |

**Fig. 4** A pair of *T. gigas* in amplexus at Changi Beach in 2017

**Table 2** Number of adult and juvenile *C. rotundicauda* and *T. gigas* found at Coney Island across 5 years

| Time period | *C. rotundicauda* | | | | *T. gigas* | | | |
|---|---|---|---|---|---|---|---|---|
| | Dead | Alive | | | Dead | Alive | | |
| | | Juveniles | Females | Males | | Juveniles | Females | Males |
| October–December 2015 | 0 | 0 | 0 | 0 | 25 | 0 | 4 | 4 |
| April–August 2016 | 0 | 0 | 0 | 0 | 16 | 0 | 0 | 0 |
| March 2017 | 5 | 0 | 0 | 0 | 2 | 0 | 0 | 0 |
| March–October 2018 | 6 | 0 | 1 | 0 | 19 | 0 | 0 | 2 |
| March–May 2019 | 9 | 0 | 0 | 0 | 2 | 0 | 0 | 0 |

Beach, the current survey did not include the mangrove area as it was inaccessible. A higher number of juveniles could possibly have been observed if the mangrove area was surveyed.

## 3.3 Coney Island

Out of 30 surveys to the island, only 10 live *T. gigas* in total were found over four surveys (Table 2). Two breeding *T. gigas* pairs were spotted on November 26 and November 27, 2015, at the boundary area between areas A and B, as well as in area B of Coney Island (Fig. 5a). Different body lengths and carapace widths indicated that the sightings on both days were of different individuals. This period coincided with the full moon on November 26, 2015.

Similarly, a female *C. rotundicauda* and a male *T. gigas* were found in amplexus on October 25, 2018, which was a full moon, at the marked location in area C (Fig. 5a and b). The amplexus pair was spotted in close proximity to the mangrove and provided interesting evidence of possible cross-species mating.

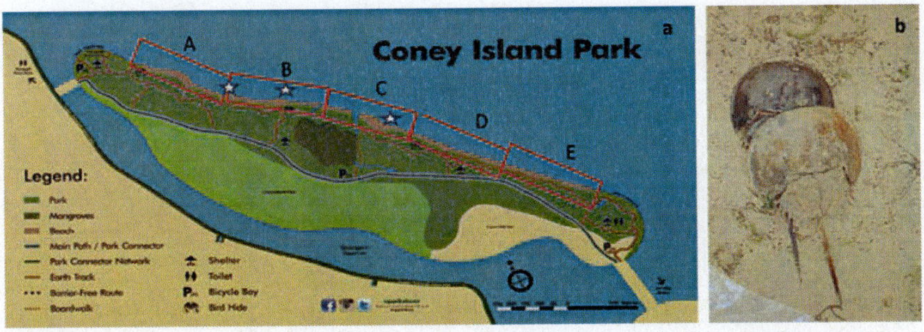


**Fig. 5** (**a**) Horseshoe crabs were found in amplexus at the locations marked with a star; (**b**) amplexus between female *C. rotundicauda* (below) and male *T. gigas* observed in 2017

This corroborates with anecdotal sightings of horseshoe crabs with hybrid morphological characteristics at Kranji mudflats (Lester Tan, pers comm.). Nonetheless, there needs to be more supporting data for this hypothesis as experimental hybridization using reciprocal fertilization of gametes between the three Indo-Pacific species and *Limulus* were nonfertile (Sugita et al. 1988).

Together with the Changi Beach data, these observations indicate that there is higher chance of finding live *T. gigas* only when they come to shore for spawning and nesting activities, which are reported to be highest during full moon and new moon periods for horseshoe crabs (Brockmann and Penn 1992; Zaleha et al. 2012).

Although there has been no reported sightings of *C. rotundicauda* previously at Coney Island (Cheo Pei Rong, pers. comm.), the presence of a mangrove indicates a possible habitat with suitable substrate for the species (see Fig. 5a). Inaccessibility to the mangrove area could be a reason why they have not been previously spotted, as unlike the other Asian horseshoe crab species, *C. rotundicauda* does not move out to sea with the receding tide.

## 3.4 *Comparison of T. gigas at Different Sites*

Many studies have reported on the variation of horseshoe crab morphometric characteristics between population and genus due to environmental conditions, including physiochemical parameters, diets, maturity, genetics, and population density (Chatterji et al. 1994; Vijayakumar et al. 2000; Graham et al. 2009; Shuster and Sekiguchi 2009; Noor Jawahir et al. 2017; Razak and Kassim 2018). Studying the morphometric characteristics of *T. gigas* individuals at Coney Island, Changi Beach, and Punggol Settlement would provide valuable baseline data on the average *T. gigas* size at the different sites. The prosomal size range for Coney Island was large, with three individuals in the 16–18 cm class size and a larger female at 34 cm. In contrast, the four females at Changi Beach were more similar in prosomal width, with the smallest at 23.2 cm and the largest at 26.7 cm (Table 3). In instances where *T. gigas* pairs were found in amplexus, the females were larger than the males.

**Table 3** Comparison of morphometric characters of *Tachypleus gigas* (mean ± SD) among several sampling locations in this study (NA = data not available)

| | Total length (cm) | | Prosoma width (cm) | |
|---|---|---|---|---|
| | Female | Male | Female | Male |
| Coney Island (*n*=10) | 34.60±11.29 | 34.37±1.51 | 21.75±7.08 | 16.23±2.78 |
| Changi Beach (*n*=4) | 43.2±4.49 | NA | 25.28±1.37 | NA |
| Punggol Settlement (*n*=2) | 42.9 | 34.3 | 23.4 | 19.5 |

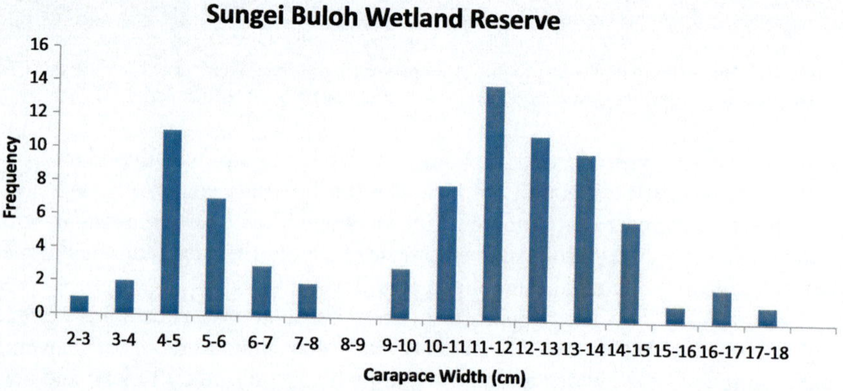


**Fig. 6** Distribution of frequency of carapace width of *C. rotundicauda* at Sungei Buloh Wetland Reserve

Further genetic analyses will be beneficial to help elucidate if the different sites are a single population of *T. gigas* due to their close proximity.

## 3.5 Sungei Buloh Wetland Reserve

At Sungei Buloh Wetland Reserve, the mud substrate and low salinity were favored only by *C. rotundicauda*, and *T. gigas* was absent from the site. The average salinity recorded for Sungei Buloh Wetland Reserve was 18.6, which is lower than the salinity range of 25–35‰ suitable for the incubation of *T. gigas* eggs (Zaleha et al. 2011).

Unlike the other sites, Sungei Buloh had healthier numbers of *C. rotundicauda*, with 83 live specimens found out of a total of 98 samples (84.7%). There were juveniles and adults present throughout the entire year of sampling (Fig. 6). Interestingly, there were only two amplexus pairs found and just in November 2017 and 2018.

Breeding of *C. rotundicauda* is believed to occur year-round as pairs in amplexus could be sighted throughout the year with no seasonal pattern in Kranji mudflats. However, a high proportion of small juveniles could be found in January, with the population shifting to be dominated by adults in July (Cartwright-Taylor & Hsu 2012). More work needs to be carried out in order to determine if similar trends exist in Sungei Buloh Wetland Reserve.

## 4 Conclusion

This is the first study that reports the distribution and morphometric characteristics of the coastal horseshoe crabs, *T. gigas*, in Singapore, especially that on offshore Coney Island. This report also gives baseline data for the *C. rotundicauda* population at Sungei Buloh Wetland Reserve, which requires further monitoring into their population density. Hence, this study addresses the data gap in both data-deficient species, and with more consistent monitoring, can be used to determine the effects of threats to the species and guide conservation strategies if action is required.

**Acknowledgments** This research was supported by the National Research Foundation, Prime Minister's Office, Singapore, under its Marine Science Research and Development program (Award No. MSRDP-P19).

## References

M.A. Beekey, J.H. Mattei, The mismanagement of *Limulus polyphemus* in Long Island Sound, USA: What are the characteristics of a population in decline? in *Changing Global Perspectives on Horseshoe Crab Biology, Conservation and Management*, ed. by R. H. Carmichael, M. L. Botton, P. K. S. Shin, S. G. Cheung, (Springer, Cham, 2015), pp. 433–461

S. Behera, B. Tripathy, K. Sivakumar, et al., Distribution and abundance of two sympatric species of horseshoe crabs along the Odisha coast, India, in *Changing Global Perspectives on Horseshoe Crab Biology, Conservation and Management*, ed. by R. H. Carmichael, M. Botton, P. K. S. Shin, S. G. Cheung, (Springer, Cham, 2015), pp. 181–191

M.L. Botton, R.H. Carmichael, P.K.S. Shin, et al., Emerging issues in horseshoe crab conservation: A perspective from the IUCN species specialist group, in *Changing Global Perspectives on Horseshoe Crab Biology, Conservation and Management*, ed. by R. H. Carmichael, M. Botton, P. K. S. Shin, S. G. Cheung, (Springer, Cham, 2015), pp. 369–381

M.L. Botton, T. Itow, The effects of water quality on horseshoe crab embryos and larvae, in *Biology and Conservation of Horseshoe Crabs*, ed. by J. T. Tanacredi, M. L. Botton, D. R. Smith, (Springer, New York, 2009), pp. 439–454

H.J. Brockmann, D. Penn, Male mating tactics in the horseshoe crab, *Limulus polyphemus*. Anim. Behav. **44**(4), 653–665 (1992)

L. Cartwright-Taylor, J. Lee, C.H. Chia, Population structure and breeding pattern of the mangrove horseshoe crab *Carcinoscorpius rotundicauda* in Singapore. Aquat. Biol. **8**, 61–69 (2009)

L. Cartwright-Taylor, V.B. Yap, C.C. Hsu, et al., Distribution and abundance of horseshoe crabs *Tachypleus gigas* and *Carcinoscorpius rotundicauda* around the main island of Singapore. Aquat. Biol. **13**(2), 127–136 (2011)

L. Cartwright-Taylor, H.H. Ng, T.Y Goh, Tracked mangrove horseshoe crab *Carcinoscorpius rotundicauda* remain resident in a tropical estuary. Aquat. Biol. **17**, 235–245 (2012)

L. Cartwright-Taylor, C.C. Hsu, Follow-up study on population structure and breeding pattern of the mangrove horseshoe crab *Carcinoscorpius rotundicauda* in Singapore. Aquat. Biol. **14**, 217–222 (2012)

L. Cartwright-Taylor, Studies of horseshoe crabs around Singapore, in *Changing Global Perspectives on Horseshoe Crab Biology, Conservation and Management*, ed. by R. H. Carmichael, M. L. Botton, P. K. S. Shin, S. G. Cheung, (Springer, Cham, 2015), pp. 193–211

E. Chan, Oil pollution and tropical littoral communities: Biological effects of the 1975 Florida Keys oil spill, in Proceedings International Oil Spill Conference. API, Washington, DC (1977)

A. Chatterji, S.A.H.J. Abidi, The Indian horseshoe crab: A living fossil. J. Indian Ocean Stud. **1**, 43–48 (1993)

A. Chatterji, J.K. Mishra, R. Vijayakumar, et al., Length-weight relationship of the Indian horseshoe crab *Tachypleus gigas* (Müller). Indian J. Fish. **41**(2), 111–113 (1994)

A. Chatterji, New record of the sympatric distribution of two Asian species of the horseshoe crab. Indian J. Mar. Sci. **77**(6), 43–48 (1999)

A. Christianus, C.R. Saad, Horseshoe crabs in Malaysia and the world. FishMail **16**, 8–9 (2007)

G.W.H. Davidson, P.K.L. Ng, H.C. Ho, *The Singapore Red Data Book: Threatened Plants and Animals of Singapore*, 2nd edn. (The Nature Society (Singapore), Singapore, 2008)

M. Faridah, N. Ismail, A.B. Ahmad, et al., The population size and movement of coastal horseshoe crab, *Tachypleus gigas* (Müller) on the east coast of Peninsular Malaysia, in *Changing Global Perspectives on Horseshoe Crab Biology, Conservation and Management*, ed. by R. H. Carmichael, M. L. Botton, P. K. S. Shin, S. G. Cheung, (Springer, Cham, 2015), pp. 213–228

G. Gold-Bouchot, O. Zapata-Pérez, V. Ceja-Moreno, et al., Biological effects of environmental pollutants in American Oyster, *Crassostrea virginica*: A field study in Laguna de Terminos, Mexico. Int. J. Env. Health **1**(2), 171–184 (2007)

L.J. Graham, M.L. Botton, D. Hata, et al., Prosomal-width-to-weight relationships in American horseshoe crabs (*Limulus polyphemus*): Examining conversion factors used to estimate landings. Fish. Bull. **107**, 235–243 (2009)

N.L. Jackson, K.F. Nordstrom, D.R. Smith, Influence of waves and horseshoe crab spawning on beach morphology and sediment grain-size characteristics on a sandy estuarine beach. Sedimentology **52**, 1097–1108 (2005)

J.K. Mishra, Horseshoe crabs, their eco-biological status along the northeast coast of India and the necessity for ecological conservation, in *Biology and Conservation of Horseshoe Crabs*, ed. by J. T. Tanacredi, M. L. Botton, D. R. Smith, (Springer, New York, 2009), pp. 89–96

B.R. Nelson, B. Satyanarayana, J.H.Z. Moh, et al., Episodic human activities and seasonal impacts on the *Tachypleus gigas* (Müller, 1795) population at Tanjung Selangor in Peninsular Malaysia. Estuar. Coast. Shelf S. **164**, 313–323 (2015)

B.R. Nelson, B. Satyanarayana, J.H.Z. Moh, et al., The final spawning ground of *Tachypleus gigas* (Müller, 1785) on the east Peninsular Malaysia is at risk: A call for action. PeerJ **4**, e2232 (2016)

B.R. Nelson, J.H.Z. Moh, B. Satyanarayana, et al., Effects of shore sedimentation to *Tachypleus gigas* (MÜLLER, 1785) spawning activity from Malaysian Waters. J. Sustain. Sci. Manag. **14**(1), 41–60 (2019)

P.K.L. Ng, N. Sivasothi, *A Guide to the Mangroves of Singapore: I (the Ecosystem and Plant Diversity)*, 2nd edn.. BP Guide to Nature Series (Singapore Science Centre, Singapore, 2002)

L. Ngy, C.F. Yu, T. Takatani, et al., Toxicity assessment for the horseshoe crab *Carcinoscorpius rotundicauda* collected from Cambodia. Toxicon **49**, 843–847 (2007)

A.R. Noor Jawahir, M. Samsur, M.L. Shabdin, et al., Morphometric allometry of horseshoe crab, *Tachypleus gigas* at west part of Sarawak waters, Borneo, East Malaysia. AACL Bioflux **10**(1), 18–24 (2017)

M. Kotwani, Part of Changi Beach temporarily closed due to oil spill. https://www.channelnewsasia.com/news/singapore/part-of-changi-beach-temporarily-closed-due-to-oil-spill-7535166 (2017). Accessed 25 Aug 2019

B.K. Kwan, J.H. Cheung, A.C. Law, et al., Conservation education program for threatened Asian horseshoe crabs: A step towards reducing community apathy to environmental conservation. J. Nat. Conserv. **35**(1), 53–65 (2017)

M.R.M. Razak, Z. Kassim, Comparison of horseshoe crabs (*Tachypleus gigas*) morphometry between different populations using allometric analysis. AACL Bioflux **11**(1), 143–157 (2018)

K. Sekiguchi, K. Nakamura, Ecology of the extant horseshoe crabs, in *Biomedical Applications of the Horseshoe Crab (Limulidae)*, ed. by E. Cohen, (Alan R. Liss, New York, 1979), pp. 37–45

K. Sekiguchi, C.N. Shuster Jr., Basic habitat requirements of the extant species of horseshoe crabs, in *Biology and Conservation of Horseshoe Crabs*, ed. by J. T. Tanacredi, M. L. Botton, D. R. Smith, (Springer, New York, 2009), pp. 5–24

P.K. Shin, H. Li, S.G. Cheung, Horseshoe crabs in Hong Kong: Current population status and human exploitation, in *Biology and Conservation of Horseshoe Crabs*, ed. by J. T. Tanacredi, M. L. Botton, D. R. Smith, (Springer, Boston, 2009), pp. 347–360

C.N. Shuster, K. Sekiguchi, Basic habitat requirements of the extant species of horseshoe crabs (Limulacea), in *Biology and Conservation of Horseshoe Crabs*, ed. by J. T. Tanacredi, M. L. Botton, D. R. Smith, (Springer, Boston, 2009), pp. 115–129

H. Sugita, Y. Yamamichi, K. Sekiguchi, Experimental hybridization, in *Biology of Horseshoe Crabs*, ed. by K. Sekiguchi, (Science House, Inc., Tokyo, 1988), pp. 288–308

C.J. Strobel, A.H. Brenowitz, Effects of Bunker C oil on juvenile horseshoe crabs *Limulus polyphemus*. Estuaries **4**, 157–159 (1981)

A.N. Tan, A. Christianus, S. Shakibazadeh, et al., Horseshoe crab, *Tachypleus gigas* (Müller, 1785) spawning population at Balok Beach, Kuantan, Pahang, Malaysia. Pak. J. Biol. Sci. **15**(13), 610–620 (2012)

S. Vestbo, M. Obst, F.J. Quevedo Fernandez, et al., Present and potential future distributions of Asian horseshoe crabs determine areas for conservation. Front. Mar. Sci. **5**(164), 1–16 (2018)

R. Vijayakumar, S. Das, A. Chatterji, A.H. Parulekar, Morphometric characteristics in the horseshoe crab *Tachypleus gigas* (Arthropoda: Merostomata). Indian J. Mar. Sci. **29**, 333–335 (2000)

S.M. Wells, R.M. Pyle, N.M. Collins, *The IUCN Invertebrate Red Data Book* (IUCN, Cham, 1983), pp. 235–243

World Conservation Monitoring Centre, *Tachypleus gigas*. The IUCN Red List of Threatened Species 1996: e.T21308A9266907 (1996a). https://doi.org/10.2305/IUCN.UK.1996.RLTS.T21308A9266907.en. Downloaded on 30 August 2019

World Conservation Monitoring Centre, *Carcinoscorpius rotundicauda*. The IUCN Red List of Threatened Species 1996: e.T3856A10123044 (1996b). https://doi.org/10.2305/IUCN.UK.1996.RLTS.T3856A10123044.en. Downloaded on 30 August 2019

S. Yang, R.F.L. Lim, C.R. Sheue, et al. The current status of mangrove forests in Singapore. Proceedings of Nature Society, Singapore's Conference on 'Nature Conservation for a Sustainable Singapore', pp. 99–120 (2011)

J. Zaldívar-Rae, R.E. Sapién-Silva, M. Rosales-Raya, et al., American horseshoe crabs, *Limulus polyphemus*, in Mexico: Open possibilities, in *Biology and Conservation of Horseshoe Crabs*, ed. by J. T. Tanacredi, M. L. Botton, D. R. Smith, (Springer, Boston, 2009), pp. 97–113

K. Zaleha, I. Hazwani, H. Siti Hamidah, et al., Effect of salinity on egg hatching and early larvae of horseshoe crab *Tachypleus gigas* (Müller 1785) in laboratory culture. J. Appl. Sci. **11**, 2620–2626 (2011)

K. Zaleha, B.A. John, H.E. Atika, et al., Spawning and nesting behaviour of *Tachypleus gigas* along the East Coast of Peninsular Malaysia. Int. J. Biol. **4**(2), 102–111 (2012)

# An Assessment of Horseshoe Crab Breeding Habitat and Trends on Long Island, New York, 2003–2017

J. Wisnewski and J. T. Tanacredi

## 1 Introduction

This investigation provides habitat factors that influence the health of the horseshoe crab population on Long Island, New York. Horseshoe crabs are true paleo survivors basically unchanged for 445 million years and continuing to exist in spite of five mass extinction events in Earth's history. These organisms fill vital niches within estuarine environments by being essential participants in the coastal near shore marine food webs. Horseshoe crab eggs provide a necessary protein source for migratory shore birds. The horseshoe crab meat is used as bait to catch eel and conch. Additionally, an exact of the blood of horseshoe crabs is an effective endotoxin detections system used to test biotechnology products, and health-care services saving countless lives each year. *Limulus polyphemus*, which inhabit the east coast of the United States, have experienced declining numbers in the past two decades. Long Island provides a preferred habitat for crabs.

J. Wisnewski (✉)
Department of Earth and Environmental Sciences, The Morrissey College of Arts and Sciences, Boston College, Chestnut Hill, MA, USA

J. T. Tanacredi
BCES Department, Molloy College Professor of Earth & Environmental Sciences, Rockville Centre, NY, USA

CERCOM -Center of Environmental Studies, Molloy College, Rockville Centre, NY, USA



J. T. Tanacredi et al. (eds.), *International Horseshoe Crab Conservation and Research Efforts: 2007-2020*, https://doi.org/10.1007/978-3-030-82315-3_17

### 1.1 Horseshoe Crab General Background

Horseshoe crabs have inhabited the earth's waters as early as 445 million years ago (mya), the Paleozoic Era. Horseshoe crabs have been some of the most resilient species on earth, surviving five mass extinctions (Tanacredi 2001). Horseshoe crabs have undergone little physiological change exhibiting that they have evolved to be well suited for their environment. Horseshoe crabs belong to the family *Limulacea*, are related to trilobites, and are merostome arthropods or organisms whose ventral mouth is surrounded by legs. The fossil record consistently shows that horseshoe crabs were composed of three parts: prosoma, opisthosoma, and telson, which is how it has derived its name of being a "living fossil" (Sekiguchi and Shuster 2009). The modern horseshoe crab still consists of all three parts, but is defined as having a fused opisthosoma. While the outer anatomy has changed little, the internal physiology has been able to adapt to different ecosystems these organisms inhabit (estuaries and nearshore ocean) allowing horseshoe crabs to be well suited for many environments (Sekiguchi and Shuster 2009). Evidence of the modern crab can be found in the Jurassic Period, and the first *Limulus polyphemus*, or the American horseshoe crab, can be found in Colorado about 60 million years ago (Shuster 2001).

Before 60 mya, most horseshoe crab populations were concentrated around what is today Europe. After 60 mya, horseshoe crabs migrated and, currently, the four species of horseshoe crabs can only be found on the eastern side of North America and Asia (Tanacredi 2001). *Limulus polyphemus* is the only North American species of horseshoe crabs, and the three other species, *Tachypleus tridentatus, Tachypleus gigas,* and *Carcinoscorpius rotundicauda,* can be found in Japan, China, and Indonesia, respectively (Shuster 2001). *L. polyphemus* branched off from the Indo-Pacific species about 135 million years ago (Sekiguchi and Shuster 2009). This North American species dwells in beaches from Nova Scotia to the Gulf of Mexico, and so *L. polyphemus* has adapted to live in a wide range of ecosystems that are composed of varying temperatures and salinity and are known as generalists because of this.

The anatomy of a horseshoe crab allows this organism to be well suited for many environments (Fig. 1). The prosoma is a horseshoe-shaped cephalic shield that protects the appendages and their vital organs, such as their tubular heart or nervous system. The opisthosoma is the midpiece and uses the carapace, which is the actual exoskeleton piece, to protect these appendages and book gills. The appendages consist of five pairs of legs with pincers, called pedipalps, and a pair of chelicerae, which are smaller feeding pincers. For males, the first pair of pedipalps are mating claws to clasp onto females during mating. For females, the first pair are normal pedipalps. The first four appendages are used for both feeding and walking while the last pair is for strictly movement. Horseshoe crabs also have five branchial appendages toward the telson also known as book gills, which aid in propulsion, reproduction, and respiration. The third part of the crab is the telson, which is the tail used to turn crabs over (Tanacredi 2001).

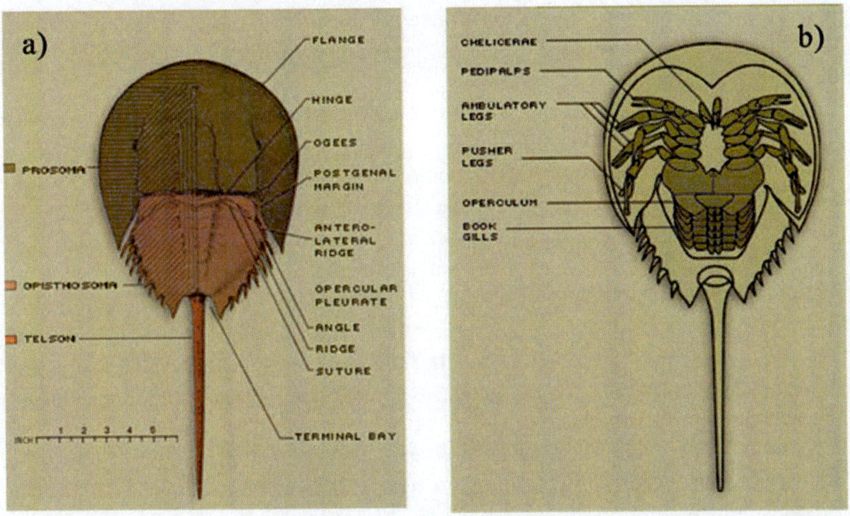


**Fig. 1** Anatomy of a horseshoe crab: (**a**) top view of a horseshoe crab including the three main sections of the external anatomy; (**b**) Bottom view of view exposing the appendages and book gills. (Source: http://www.horseshoecrab.org/anat/anat.html)

The book gills are crucial to horseshoe crab survival because they aid in respiration, propulsion, and reproduction. They consist of six pairs of gills; the first pair is called the operculum, which is protective covering over the other five pairs. Due to the amount of folding, these gills have a large surface area that increases the amount of gas exchange by taking oxygen out of the water that passes through the gills for respiration. When spawning, horseshoe crabs must leave the water in search of a sandy shore to lay their eggs or deposit their sperm on. Generally, horseshoe crabs can remain out of water for up to four days as long as their book gills are moist. By burying themselves in sand or folding in half, horseshoe crabs can ensure that their gills remain moist until the next tide brings them back into the water (U.S. Fish & Wildlife 2009). The gills aid in propulsion because they push water and move the crab forward, similar to paddles. In addition to protection, the operculum contains the genital pores that release eggs and sperm during spawning. When breeding, the gills aid in pushing the eggs and sperm from the pores and into the burrow that the female creates for her eggs (Ehlinger 2001).

For *L. polyphemus*, spawning occurs during summer months, generally May through August, because their breeding becomes inactive when temperature is below 16–20 °C. Tides are known to be a limiting factor because horseshoe crabs prefer semi-diurnal tides with a larger amplitude in order for breeding (Sekiguchi and Shuster 2009). These crabs prefer to breed on intertidal beaches because they are protected from high wave action and provide a shallow flat that is ideal for juveniles to mature on. Horseshoe crabs spawn a few days before, during, and a few days after the new and full moon high tides bringing crabs to the highest high tide mark (Brockmann and Smith 2009). In addition, this synchronization of breeding with the highest tides is thought to be linked to the amount of water flowing through the gills and changes in pressure on the carapace, which initiates the migration to the sandy shores (Chen et al. 2004).

Once crabs reach sexual maturity around 10 years old, the tide brings horseshoe crabs ashore with the males linked to the females (amplexus) to ensure fertilization. Males use their mating claws to latch onto the larger females and the females take the males ashore. Generally, one male is attached to one female but one to two other males, called satellite males, may hover around the nesting pair and deposit their sperm on the eggs as well so that their sperm may fertilize some of their eggs. Females may return to the beach a few times throughout breeding season until all of their eggs have been released. During breeding, the male will stay attached to her until she has released all of her eggs. After this, males return to the beach with new females and so males will continue to deposit their sperm and come ashore (Brockmann and Smith 2009).

On average, females lay about 20 or more egg clusters with a reported average of 3650 eggs per cluster, which can occur over several nights (Botton et al. 1994). Females burrow into the sand to create a nest about 5–10 cm deep in order to keep the eggs safe from predators and natural environmental occurrences. When the female is fully buried or buried at eye level, she will stop and remain in the burrow until she has released her eggs providing a nest for the eggs to stay in until hatched and ready to survive on the continental shelf (Brockmann and Smith 2009).

Horseshoe crabs prefer sandy beaches because they require sediment that is aerated, low in organic matter, and low in mud and silt so that the eggs can develop properly. Low wave action protects the juveniles and the mating horseshoe crabs. High wave action has been known to sweep mating horseshoe crabs out to sea, disrupting breeding (Sekiguchi & Shuster 2009). The ideal breeding beaches additionally have shallow flats for juveniles to feed and grow on before they make their way to the deep ocean. Through studies ranging from tracking to genetic analysis, it is evident that horseshoe crabs exhibit philopatry where horseshoe crabs will often return to their natal beaches. Moore and Perrin (2007) used tracking devices to show that horseshoe crabs will return to their natal beach, and Itow et al. (2004) used genetic analysis to show that there is a large genetic difference between populations, suggesting little gene flow between the populations. Additionally, larvae have limited dispersal rates because they inhabit tidal flats and have limited swimming abilities (Botton et al. 2010).

Another characteristic of the ideal *Limulus polyphemus* habitat is the presence of an adjacent continental shelf, which provides a shallow aquatic habitat because of the divergent plate margins. In the Atlantic, the shelf along New York's coast is about 190 km wide, providing enough shallow space and resources for horseshoe crabs (Kraft 1988). It has been noted that the continental shelf provides a habitat for many benthic creatures, such as bivalves, gastropods, and crustaceans, that are food for horseshoe crabs and enough space for the crabs to spread out to reduce competition (Shuster and Sekiguchi 2009). The breeding beaches and continental shelf provide a suitable habitat for each life stage for horseshoe crabs. During embryonic development, the eggs live in their protected nests on the sandy shore. After this stage, the larvae and juveniles will mature through a series of molts in the near shore, shallow waters. As the horseshoe crabs age and reach their terminal molt, they continue to migrate to deeper water that is more suitable for their mature age with abundant resources (Shuster and Sekiguchi 2009).

In the continental shelf and shoreline habitats, horseshoe crabs play a vital role in their ecosystem. Horseshoe crabs consume a large number of invertebrates, especially bivalves, which controls the invertebrate population. Juvenile horseshoe crabs provide food for a wide variety of fish and other marine organisms, such as hermit crabs. In the adult stage, horseshoe crabs have a hard carapace that protects them from most predation, but loggerhead turtles have been known to consume adult crabs (Botton 2009). When breeding crabs are stranded on the beach, different bird species, such as gulls, feed on them (Botton 2009). The adult carapace provides a habitat for epibionts, invertebrates, and algae, in an epibiosis relationship where neither the host nor the organism that grows on it is harmed or benefited (Botton 2009; Tanacredi, 2001).

One of the most significant roles of the horseshoe crab in their ecosystem is the food resource their eggs provide for migratory birds. Tispoura and Burger (1999) found that the stomach content of shorebirds found along the east coast of the United States mostly consists of horseshoe crab eggs. Migratory shorebirds stop at northeast beaches on their way to the Arctic for breeding season mainly during the end of May, which is the start of spawning for horseshoe crabs. These birds need to consume as many eggs as possible in order to build up their weight to complete their migration. Haramis et al. (2007) demonstrated that Red Knots will consume on average 18,350 eggs per day and Ruddy Turnstones will consume on average 13,300 eggs per day. Generally, the bird's feeding patterns do not limit horseshoe crab populations because peak migration occurs in May and peak spawning occurs in June. Additionally, wave action aids in the survival of eggs by reburying exposed eggs. However, the amount of horseshoe crab eggs available for consumption is vital to the survival of many migratory shorebirds (Botton 2009).

Horseshoe crabs have been harvested to be used as bait for eels and conch fisheries (Bradley, 2001; Tanacredi, 2001). Pregnant females and juveniles are seen as more productive bait and thus are worth more money than males and so are especially targeted by fishermen during breeding seasons. Horseshoe crab harvesting for use as bait is a significant impact on breeding populations primarily because the animals are removed from the breeding population permanently. Fishermen generally harvest them offshore prior to the crabs ability to reach breeding beaches (Tanacredi 2015, Roseneal).

Horseshoe crab copper-based blue blood is collected to provide an extract limulus amebocyte lysate (LAL), which is used as an efficient endotoxin detection system for such products for human health such as medication and vaccines (Ding and Ho 2010). This has made horseshoe crabs very desirable to the biotechnology industry and caused an increase in horseshoe crab collection; approximately 600,000 globally are crabs taken annually for bleeding purposes. During collection, horseshoe crabs are taken off the beach, drained of a portion of their blood, and released back into the water. However, about 10% are thought to not survive this process and, generally, many of these crabs are not returned to their natal beaches, which is crucial for productive breeding (Tanacredi 2001). Recognizing the devastating effect this may have on the individual horseshoe crab populations, at breeding sites pharma companies are currently exploring creating a synthetic LAL, which could help the horseshoe crab populations significantly (Ding and Ho 2010).

## 1.2 *Long Island Background (Fig. 2)*

Long Island is in the State of New York and is 189 km long and 32 miles wide, making it the United States' largest adjoining island to the continent. Long Island is part of the Atlantic Coastal Plain, whose topography was initially formed from glacial deposits during the Cretaceous Period. These deposits came from glacial advances and outwash from continental glaciers receding over a 19,000-year period. Five glacial stages form Long Island: the Wisconsin twice and the Manhasset, Jameo, and Mannetto Stages, as each phase retreated and left deposits the North Shore of Long Island and outwash that formed the South Shore developed geologically (Fuller 1914). Two terminal moraines, Ronkonkoma and Harbor Hill, make up the backbone of Long Island and caused the formation of the North and South forks at the eastern end of the island. South of the southern moraine, glacial outwash formed the rest of the southern shore creating a sandier shore than the hilly North Shore (Sanders and Merguerian 1995).

The North Shore formed from glacial deposits and from stream erosion. Long Island is currently separated from Connecticut by the Long Island Sound, which can be anywhere from 1 to 8 km wide. Subglacial stream erosion can account for the harbors and bays that form the North Shore. Due to its hilly nature, the North Shore mostly consists of steep sloping beaches and gravel mixed with sandy shores (Fuller 1914).

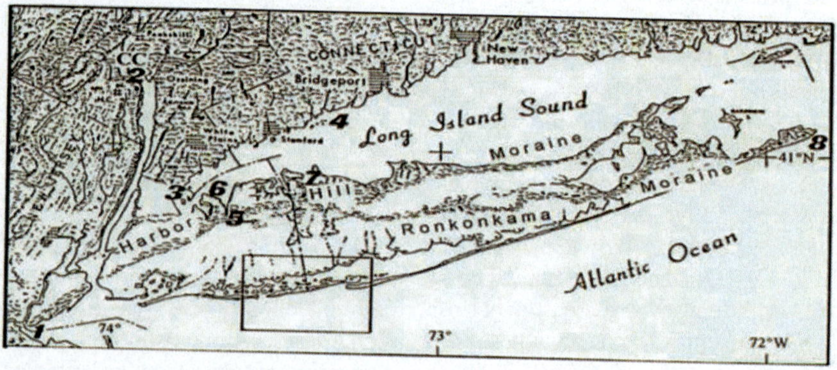


**Fig. 2** Geological map of Long Island. (Source: Sanders and Merguerian 1995)

The South Shore differs from the North Shore in that it was formed from glacial outwash due to glacial retreat and contains long stretches of gently sloping sandy shores (Fuller 1914). This glacial outwash and rise in sea level formed the barrier islands by providing a sediment source for sand deposits to the mainland. Migrating islands with rising sea level created many bays and ponds between these islands and the mainland. These barrier islands are long and continuous with inlets providing connections to the Atlantic Ocean. They also help provide protection from erosion, wave action, and storm protection for the mainland while also providing popular tourist destinations (Schwab et al. 2000). Formed by glaciers, the North and South Shore of Long Island is home to many sandy beaches that provide essential ecosystems for the rich biodiversity along these shores.

The Center for Environmental Research and Coastal Ocean Monitoring (CERCOM) at Molloy College has tracked the horseshoe crab population at 115 locations on Long Island since 2002 and conducted surveys at Long Island in abundant beaches to collect data on the presence of horseshoe crabs. Due to its geological origins, the North and South Shores both contain many inlets, bays, and estuarine systems that have lower wave action with sandy shores, resulting in these beaches being the ideal breeding ground for horseshoe crabs. The South Shore is protected by Long Island's barrier island system, and the North Shore is surrounded by the Long Island Sound. In recent years, the Long Island horseshoe crab population has declined (Tanacredi 2015), and it is due to this decline of horseshoe crabs that this investigation evaluates the sustainability of a set of Long Island horseshoe crab breeding beaches.

This investigation will evaluate the ecological condition of the *Limulus polyphemus* populations and breeding habitat on the Long Island coast.

This analysis will

1. Compare horseshoe crab population data from 2003 to 2017 from the 115 CERCOM/Molloy College Annual Horseshoe Crab Inventory beach sites over the North and South Shores of Long Island to reveal trends in the Long Island Horseshoe Crab breeding population
2. Compare sea surface data from the North Shore and South Shore of Long Island to assess these environmental factors that may have influenced horseshoe crab health over the time period that these populations were monitored
3. Do a potential breeding site analysis of a 56-beach subset with a comparison of physical beach habitat (aerial map stratigraphy) every 3 years from 2004 to 2016 to identify relative physical changes in beach geomorphology that may have affected horseshoe crab breeding

# 2 Methods

## 2.1 *Study Area*

The study area includes 56 beach sites along Long Island's North and South Shores. The North Shore contains 22 beaches grouped together into 11 beach zones based on location. The South Shore contains 25 beaches grouped into nine beach zones. These beaches were identified by CERCOM as ideal breeding habitats for horseshoe crab populations, due to the sandy and low wave action shoreline, and have had horseshoe crabs identified as breeding on them for the last 19 years (Fig. 3). These breeding sites were initially established, identified by GPS locations in 2000, the first initial extensive inventory of Horseshoe Crabs on Long Island. CERCOM's annual Horseshoe Crab Inventory is the most comprehensive large-scale monitoring program in New York State for horseshoe crabs.

## 2.2 *Horseshoe Crab Inventory*

Horseshoe crab inventory data as obtained from CERCOM of Molloy College on Long Island, New York, of this analysis, only included the initial 56 beaches are analyzed annually so as to have a complete long-term time series of horseshoe crab data.

Volunteers conduct their observations between May and August and go to each beach once during the full or new moon. Beach sites are visited between an hour before and an hour after the highest tide one day before and one day after a full or

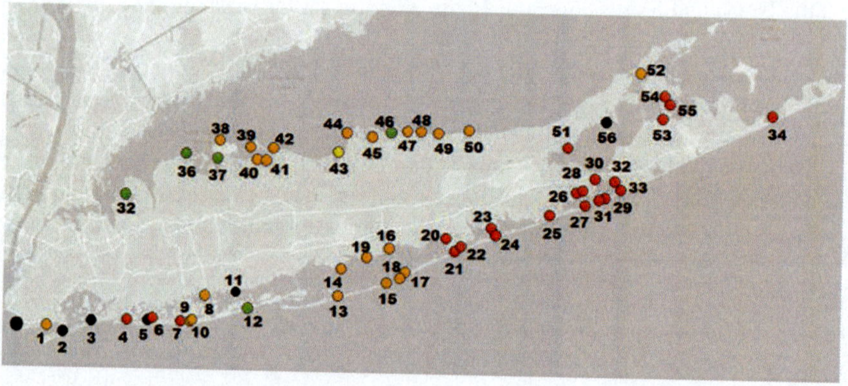


**Fig. 3** Location and classification of the 56 beach sites investigated on Long Island. Colors represent classification based on horseshoe crab population trend results from this study. Green represents active beaches, yellow represents recovering beaches, orange represents pending beaches, red represents nonbreeding beaches, and black represents incomplete data

new moon. Because horseshoe crabs come to shore to breed during the highest high tides due to the tidal signals, the horseshoe crabs respond to changes in water temperature and are present on the beaches when the surveys are conducted (Chen et al. 2004). If no horseshoe crabs were observed, the beach total was zero, indicating breeding inactivity determining the classification of the beach zones. The beach boundaries and weather conditions are noted. Horseshoe crabs that are on land and are visible within the water from the water's edge are counted, and females, males, and dead crabs are identified and reported online to CERCOM for record purposes.

## 2.3 *Horseshoe Crab Analysis*

The dataset collected from CERCOM (2003–2017) was used to analyze the horseshoe crab population. Data from the original 56 beach sites were extracted from the annual data sets (Fig. 3), and only the total #Horseshoe Crabs from each beach were used to identify breading sites. Each beach was then further grouped into 22 regions along Long Island's North (Fig. 37) and South Shores (Fig. 38). Each group was then analyzed by plotting totals against years to compare with past population and identify trends.

Each beach group was assigned a horseshoe crab health rating: nonbreeding, pending, recovering, and active breeding. A nonbreeding beach zone is any beach zone with one or less horseshoe crabs for three or more successive years in a row. A pending beach zone is defined as a beach zone with one or less horseshoe crab in 2017 in at least one beach site within the group. A recovering beach zone is any beach zone that had one or less horseshoe crabs in the past, but has since been observed as a "breeding beach." An active beach zone is any beach zone that has two or more horseshoe crabs and has remained at a consistent level of breeding activity for the last three years. These plots and ratings were used to identify problem beaches and make an assessment of the health of the horseshoe crab breeding population on Long Island.

## 2.4 *Temperature Data Collection*

Environmental factors were investigated by sea surface temperatures obtained from the National Oceanic and Atmospheric Administration's (NOAA) buoy system. Meteorological data are available from two locations off the coast of the North Shore and the South Shore of Long Island (Fig. 4).

| Site # | Site Name | 2003 | 2004 | 2005 | 2006 | 2007 | 2008 | 2009 | 2010 | 2011 | 2012 | 2013 | 2014 | 2015 | 2016 | 2017 | Longtitude | Latitude |
|---|---|---|---|---|---|---|---|---|---|---|---|---|---|---|---|---|---|---|
| | **SOUTH SHORE** | | | | | | | | | | | | | | | | | |
| | **Jamaica Bay** | 72 | 160 | 1273 | 89 | 32 | 1500 | 0 | 35 | 71 | 37 | 40 | 5 | 3 | 10 | 0 | | |
| 1 | Sheepshead Bay/Shore BLVD | 28 | 0 | 1035 | 80 | 26 | 600 | n/a | 35 | 0 | 0 | 0 | 0 | 0 | 0 | 0 | 40.580095 | -73.932225 |
| 2 | Breezy Point Roxbury, Bayside Av | 44 | 57 | 80 | 0 | 0 | n/a | n/a | 0 | 10 | 7 | 5 | 2 | 1 | 3 | 0 | 40.568362 | -73.889184 |
| 3 | Jamaica Bay/Beach Channel Brid | 0 | 103 | 158 | 9 | 6 | 900 | n/a | n/a | 61 | 30 | 35 | 3 | 2 | 7 | 0 | 40.590658 | -73.813825 |
| | **Long Beach** | 11 | 56 | 16 | 16 | 4 | 0 | 0 | 3 | 47 | 8 | 0 | 2 | 0 | 2 | 10 | | |
| 4 | Sage Pond- Reynolds Channel | 0 | 23 | 0 | 0 | 4 | 0 | 0 | 0 | 0 | 0 | 0 | 0 | 0 | 0 | 0 | 40.594103 | -73.718685 |
| 5 | Long Beach, Magnolia Blvd. | 0 | 32 | 16 | 16 | 0 | n/a | 0 | n/a | 0 | 0 | 0 | 2 | | 2 | 9 | 40.593876 | -73.663693 |
| 6 | Island Park, Seaview Ave | 0 | 1 | 0 | 0 | 0 | 0 | 0 | 3 | 47 | 0 | 0 | 0 | 0 | 0 | 1 | 40.596797 | -73.651310 |
| 7 | Pt Lookout Lido Blvd | 11 | 0 | 0 | 0 | 0 | 0 | 0 | 0 | 0 | 8 | 0 | 0 | 0 | 0 | 0 | 40.594638 | -73.576527 |
| | **Jones Beach** | 7 | 40 | 15 | 15 | 36 | 330 | 0 | 4 | 21 | 0 | 0 | 18 | 0 | 40 | 0 | | |
| 8 | Jones Beach, Short Beach Island | 0 | 40 | 15 | 15 | 33 | 200 | 0 | 4 | 0 | 0 | 0 | 7 | 0 | 40 | 0 | 40.592241 | -73.550604 |
| 9 | Jones Beach, Sloop Channel | 7 | 0 | 0 | 0 | 3 | 130 | 0 | 0 | 21 | 0 | 0 | 11 | 0 | 0 | 0 | 40.594149 | -73.544903 |
| | **Massapequa** | 0 | 10 | 0 | 26 | 0 | 0 | 0 | 0 | 0 | 1 | 6 | 54 | 12 | 220 | 0 | | |
| 10 | Massapequa, Division Ave | 0 | 10 | 0 | 26 | 0 | 0 | 0 | 0 | 0 | 1 | 6 | 54 | 12 | 220 | 0 | 40.651191 | -73.477016 |
| 11 | Unqua Pt. | 0 | 0 | 0 | 0 | 4 | 0 | 0 | 0 | 0 | | | 5 | | | | 40.650525 | -73.429928 |
| | **Ocean Beach Beaches** | 0 | 80 | 0 | 0 | 0 | 38 | 6 | 0 | 0 | 3 | 1 | 0 | 6 | 8 | 7 | | |
| 12 | Ocean Pkwy., Gilgo Island | 0 | 80 | 0 | 0 | 0 | 38 | 6 | 0 | 0 | 3 | 1 | 0 | 6 | 8 | 7 | 40.619006 | -73.396453 |
| | **Fire Island** | 0 | 909 | 6 | 6 | 0 | 5 | 0 | 1 | 1130 | 0 | 10 | 136 | 16 | 6 | 0 | | |
| 13 | Fire Island, Ocean Beach (bay sic | 0 | 4 | 0 | 0 | 0 | 0 | 0 | 1 | 0 | 0 | 10 | 1 | 0 | 1 | 0 | 40.645582 | -73.164171 |
| 14 | Heckscher St Pk | 0 | 5 | 6 | 6 | 0 | 5 | 0 | 0 | 0 | 0 | 0 | 0 | 0 | 2 | 0 | 40.698497 | -73.157259 |
| 15 | Fire Island, Barrett Beach (bay si | 0 | 900 | 0 | 0 | 0 | 0 | 0 | 0 | 1130 | 0 | 0 | 135 | 16 | 3 | 0 | 40.673111 | -73.039168 |
| | **Great South Bay** | 50 | 128 | 0 | 1 | 3 | 2 | 918 | 256 | 2494 | 922 | 994 | 1019 | 45 | 19 | 377 | | |
| 16 | Bayport/Bluepoint, Bluepoint Av | 50 | 42 | 0 | 0 | 1 | 2 | 0 | 0 | 0 | 0 | 6 | 19 | 0 | 0 | 0 | 40.738642 | -73.032793 |
| 17 | Fire Island, Watch Hil (bay side) | 0 | 6 | 0 | 0 | 0 | 0 | 0 | 23 | 1247 | 20 | 0 | 7 | 0 | 1 | 0 | 40.694007 | -72.990650 |
| 18 | Davis Park | 0 | 40 | 0 | 1 | 0 | 0 | 918 | 233 | 1002 | 902 | 988 | 989 | 45 | 15 | 377 | 40.683109 | -73.004397 |
| 19 | CERCOM | 0 | 40 | 0 | 0 | 2 | 0 | 0 | 0 | 245 | 0 | 0 | 4 | 0 | 3 | 0 | 40.721655 | -73.090335 |
| | **Mastic** | 22 | 2 | 0 | 163 | 4 | 0 | 0 | 0 | 0 | 0 | 0 | 0 | 0 | 0 | 0 | | |
| 20 | Mastic, Sandy Pt, Bellport Bay | 22 | 0 | 0 | 0 | 0 | 0 | 0 | 0 | 0 | 0 | 0 | 0 | 0 | 0 | 0 | 40.761381 | -72.885920 |
| 21 | Smith Pt (bay side) | 0 | 2 | 0 | 63 | 4 | 0 | 0 | 0 | 0 | 0 | 0 | 0 | 0 | 0 | 0 | 40.736931 | -72.863114 |
| 22 | Mastic Beach, Cranberry Rd. | 0 | 0 | 0 | 100 | 0 | 0 | 0 | 0 | 0 | 0 | 0 | 0 | 0 | 0 | 0 | 40.746441 | -72.847061 |
| | **Dune Road** | 22 | 469 | 9 | 26 | 43 | 51 | 0 | 0 | 86 | 8 | 0 | 5 | 1 | 0 | 0 | | |
| 23 | Moriches Bay, Smith St | 1 | 216 | 0 | 8 | 0 | 0 | 0 | 0 | 86 | 8 | 0 | 0 | 1 | 0 | 0 | 40.783105 | -72.768580 |
| 24 | Moriches Inlet East Cut, Cupsogu | 0 | 200 | 0 | 9 | 0 | 0 | 0 | 0 | 0 | 0 | 0 | 2 | 0 | 0 | 0 | 40.769266 | -72.756313 |
| 25 | Quogue, Meadow Lane | 21 | 3 | 9 | 9 | 43 | 0 | 0 | 0 | 0 | 0 | 0 | 0 | 0 | 0 | 0 | 40.810004 | -72.614678 |
| 26 | Tiana, Romana Dr. | 0 | 50 | 0 | 0 | 0 | 51 | 0 | 0 | 0 | 0 | 0 | 3 | 0 | 0 | 0 | 40.854014 | -72.545795 |
| 27 | Tiana, Dune Rd | 25 | 0 | 0 | 0 | 0 | 0 | 0 | 1 | 0 | 0 | 0 | 0 | 0 | 0 | 0 | 40.829114 | -72.521483 |
| 28 | Rampasture Rd | 50 | 0 | 0 | 0 | 0 | | 0 | 1 | 0 | 0 | 0 | 0 | 0 | 0 | 0 | 40.857374 | -72.528925 |
| | **Shinnecock** | 10 | 297 | 59 | 59 | 8 | 0 | 0 | 0 | 0 | 0 | 0 | 0 | 0 | 0 | 2 | | |
| 29 | Shinnecock Inlet, West Cut (bays | 0 | 40 | 0 | 0 | 0 | 0 | 0 | 0 | 0 | 0 | 0 | 0 | 0 | 0 | 1 | 40.844891 | -72.469983 |
| 30 | Shinnecock Canal, Mohawk Rd. | 0 | 7 | 1 | 1 | 8 | 0 | 0 | 0 | 0 | 0 | 0 | 0 | 0 | 0 | 1 | 40.881560 | -72.497878 |
| 31 | Shinnecock Inlet, East Cut (bay si | 8 | 50 | 0 | 0 | 0 | 0 | 0 | 0 | 0 | 0 | 0 | 0 | 0 | 0 | 0 | 40.841075 | -72.487098 |
| 32 | Shinnecock Bay, Little Neck Rd | 0 | 100 | 0 | 0 | 0 | 0 | 0 | 0 | 0 | 0 | 0 | 0 | 0 | 0 | 0 | 40.876814 | -72.445757 |
| 33 | Shinnecock Bay Meadow Lane, H | 2 | 100 | 58 | 58 | 0 | 0 | 0 | 0 | 0 | 0 | 0 | 0 | 0 | 0 | 0 | 40.860210 | -72.429112 |
| 34 | Hither Hills St. Pk., Napeague Hb | 0 | 50 | 0 | 30 | 0 | 0 | 0 | 0 | 0 | 0 | 0 | 0 | 0 | 0 | 0 | 41.009357 | -72.036907 |

**Fig. 37** North Shore horseshoe crab data. Data collected from CERCOM was organized by beach zone. This table includes site numbers and coordinates corresponding to Fig. 3

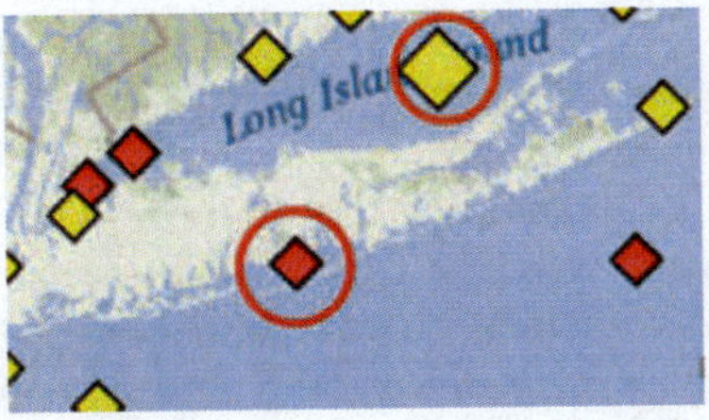


**Fig. 4** Location of NOAA buoys. Yellow square represents the North Shore station (Station 44039 Central Long Island Sound 41.138 N 72.655 W) and the red square represents the Great South Bay Station (Station 44069 Great South By 40.699 N 73.087 W). (Source: http://www.ndbc.noaa.gov)

**CERCOM?**

Temperature data were evaluated to investigate the relationship between horseshoe crab breeding and varying temperatures. For each station, sea surface and air temperature data from 2003 to 2017 were obtained from NOAA's online archives (http://www.ndbc.noaa.gov). Anomalous years were identified as those outside one standard deviation of the mean (95% confidence limits) for the entire time series. This data was then compared to anomalous years of no breeding activity in the beach zones identified.

## 2.5 *Land-Use Data*

Anthropogenic factors were identified by a map of Long Island's land use to identify human interaction with beach zones (http://www.longislandindexmaps.org; Fig. 6). This was used to identify concentrations of residential developments, agricultural development, and recreational beach use.

## 2.6 *Potential Breeding Area Collection and Analysis*

In order to understand physical changes in horseshoe crab breeding habitats, data of the 56 beach sites was analyzed to quantify changes to the Long Island shoreline. Using New York State Orthos Online, satellite images were taken from each beach site with their longitudinal and latitudinal location (https://orthos.dhses.ny.gov/?id=974149). The images are available for the years 2004, 2007, 2010, 2013, and 2016, with the 3-year time interval between images to identify site-related shoreline changes.

The images were downloaded from Orthos Online and imported into ArcGIS. Using these images, all beach locations are recorded by CERCOM as sample site "beach areas." The area of the beach went up to the high waterline mark and did not include any areas with rocks, visible human infrastructure, or any other features that would potentially hinder horseshoe crab breeding, and extended to the low tide mark that was determined by either the water's edge or the next darkest line visible on the image. This is the literal zone. Horseshoe crabs prefer to breed in undisturbed, sandy shores around the highest high tide mark, but will breed in the intertidal zone so that was included in the outlined area in the images (Tanacredi et.al 2009). The resulting area is defined here as "the potential breeding area." After this line was determined, the "total" area was calculated within the beach limits defined by the sample sites. The area data were examined for any visible physical changes or the total areas of each site at each 3-year interval and were compared to help quantify coastline changes. Horseshoe crab breeding depends on the suitability of the breeding beach site, and so any changes in the potential sustainability of the breeding site area could have observable effects on fecundity, or this animal's site fidelity.

## 3 Results

### 3.1 Horseshoe Crab Inventory and Classification

Taken from 2003 to 2017, The CERCOM Long Island horseshoe crab population/habitat inventory data indicates the presence or absence of adult horseshoe crabs at specific beaches. Each beach is classified for this investigation into the respective breeding beach groupings (Table 1). Results exhibit the distribution of nonbreeding, pending breeding, recovering, and active breeding sites with the South Shore having

**Table 1** Summary of beach classification based on CERCOM horseshoe crab inventory results. Nonbreeding sites have had no breeding activity in any consecutive three years. Pending sites have either had no breeding in the last year or include a beach that had no breeding within the last year. Recovering sites have had no breeding within the past three years, but have since recovered. Active breeding sites have seen horseshoe crab populations in the last three years (total breeding sites investigated = 20)

| Nonbreeding (NB) | Pending (P) | Recovering (R) | Active breeding (AB) |
|---|---|---|---|
| Jamesport (NS) | Greenport (NS) | Stony Brook (NS) | Manhasset (NS) |
| Long Beach (SS) | PJ Harbor (NS) | | Mt. Sinai (NS) |
| Shelter Island (NS) | Jones Beach(SS) | | Ocean Beaches (SS) |
| Dune Road (SS) | North Shore Sound (NS) | | Oyster Bay (NS) |
| Mastic Beach (SS) | Massapequa (SS) | | Cold Spring (NS) |
| Shinnecock (SS) | Northport (NS) | | |
| | Fire Island (SS) | | |
| | Great South Bay (SS) | | |
| *North Shore – 2* | *North Shore – 4* | *North Shore - 1* | *North Shore – 4* |
| *South Shore – 4* | *South Shore – 4* | *South Shore – 0* | *South Shore – 1* |
| *Total – 6* | *Total – 8* | *Total – 1* | *Total – 5* |

19 individual nonbreeding beach sites compared with the 4 nonbreeding beach sites on the North Shore.

The North Shore has two nonbreeding, four pending, one recovering, and four active breeding zones (Table 1). The South Shore has four nonbreeding, four pending, and one active breeding zones. Results show more (NB) sites on the South Shore, more (AB) zones on the North Shore, and evenly distributed (P) breeding zones through the North and South Shores.

## 3.2 *Sea Surface Water Temperature Data*

Temperature data indicate that there have been no significant changes to average sea surface temperature (SST) during the survey period. The South Shore (24.54 ± 2.52 °C) ranged from 19.18 °C to 38.90 °C. The North Shore (24.37 ± 0.91 °C) ranged from 22.59 °C to 26.09 °C. Although these four years 2003, 2004, 2005, and 2011 are anomalous with respect to SST for Long Island, they do fall within the horseshoe crab preferred temperature ranges of 16–33 °C. The North Shore is also within the appropriate breeding horseshoe crab temperature ranges experienced on Long Island (Fig. 5).

## 3.3 *Land-Use Analysis*

The land use on Long Island comprised residential, commercial, agricultural, and recreational use, with most of Long Island used for residential purposes (Fig. 6). The southwest shore and a majority of the South Shore, especially the barrier islands (Fire Island National Seashore), are categorized as community services and recreational use, respectively. Both classifications involve summer beachgoing. The upper North fork, northeastern end of Long Island, is mostly used for agricultural land. The North Shore also has some recreational use, but less than the South Shore that has long stretches of sandy beaches on the barrier islands, extending to Montauk Point.

## 3.4 *Horseshoe Crab Potential Breeding Area Results*

The change in area of potential horseshoe crab breeding habitats over time, determined from satellite image analyses, was compared with horseshoe crab breeding activity classifications (Table 1). Beaches were separated into North and South Shores to look at potential breeding areas of each shore showing that the South Shore has decreased 2% more than the North Shore (Table 2a). This section will look at each beach zone within each classification to find anomalously high and low

a)

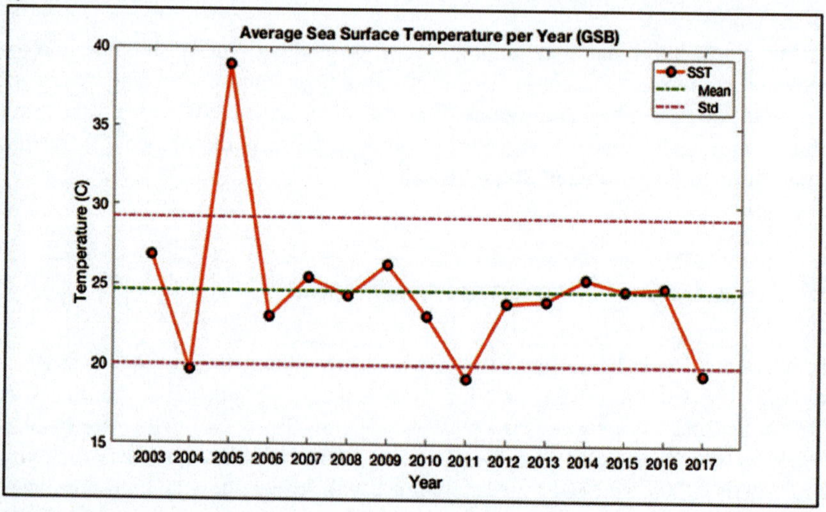


b)

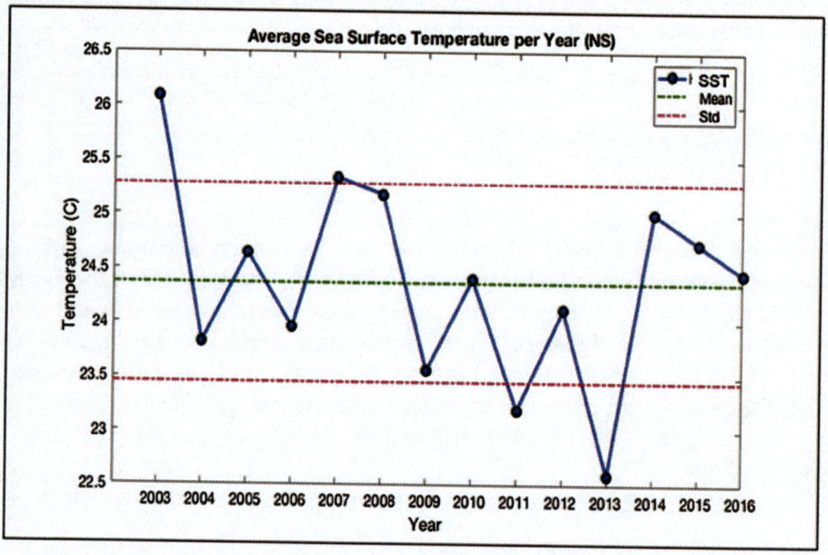


**Fig. 5** Sea surface temperature on Long Island: (**a**) average year-round SST for Station 44069; (**b**) average year-round SST for Station 44039. Data was unavailable for the North Shore in 2017

years (greater than or less than one standard deviation of the mean). Breeding areas will then be compared to inactive breeding years and to the overall percent change of potential breeding area from 2004 to 2016. This breeding beach health and level of breeding activity is important since NYSDEC, which manages the horseshoe crab population on Long Island, has been rated “poor” by the Atlantic States Marine

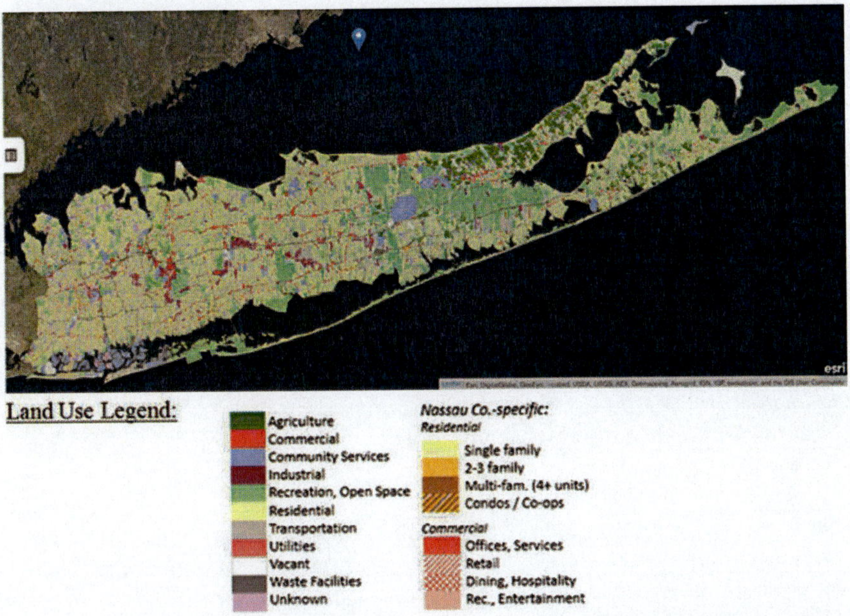


**Fig. 6** Land use of Long Island in 2017. (Source: http://www.longislandindexmaps.org)

Fisheries Commission (ASMFAC) for the last three years (2017–2020) (ASMFC Annual Report, 2020).

## Active Breeding Beach Zones (AB)

Active breeding zones are mostly in the North Shore with Ocean Beach in the South Shore being the one exception (Figs. 7, 8, 9, 10, and 11). Three out of five beaches have only one anomalous year of potential breeding area change (Table 2a). Two beaches also had anomalously high breeding areas in 2016. All active breeding beaches were active after 2013, and 2013 is the last year of anomalously low-potential beach areas for this classification. The North Shore potential breeding area of this classification increased while the South Shore's breeding area decreased (Table 3b).

## Recovering Beach Zones (RB)

There is only one beach zone in this classification and it is on the North Shore (Fig. 12). This zone had two anomalous years that do not correspond with breeding inactivity (Table 2b). Stony Brook has a 1% increase in potential breeding area from 2003 to 2016 (Table 3c).

**Table 2** Summary of potential breeding area and inactive breeding year results. (a–d) Each classification of breeding activity is broken down by beach zone, shows the mean and standard deviation, indicates the years that were above and below one standard deviation of the mean, and notes the years of inactive breeding

| *Beach zone* | *Mean ± Std ($m^2$)* | *High* | *Low* | *Inactive years* | *Figure* |
|---|---|---|---|---|---|
| (a) Active breeding | | | | | |
| Manhasset (NS) | 40,925 ± 8721 | 2016 | 2010 | 2006–2007 | 7 |
| Ocean Beach (SS) | 13,945 ± 310 | 2004 | – | 2006–2007; 2010–2011; 2013 | 8 |
| Cold Spring Harbor (NS) | 16,005 ± 5908 | 2016 | – | 2003–2010 | 9 |
| Oyster Bay (NS) | 8656.70 ± 1844 | – | 2004 | 2003–2006 | 10 |
| Mt. Sinai (NS) | 20,681 ± 2379 | 2007 | 2013 | 2006, 2011–2012 | 11 |
| (b) Recovering | | | | | |
| *Beach zone* | *Mean ± Std ($m^2$)* | *High* | *Low* | *Inactive years* | *Figure* |
| Stony Brook (NS) | 32,267 ± 8331.10 | 2013 | 2010 | 2007–2015 | 12 |
| (c) Pending | | | | | |
| *Beach zone* | *Mean ± Std ($m^2$)* | *High* | *Low* | *Inactive years* | *Figure* |
| Fire Island (SS) | 18,350 ± 2328 | 2013 | 2010 | 2003, 2007, 2009, 2012, 2017 | 13 |
| Jones Beach (SS) | 17,290 ± 4089 | 2007 | – | 2009, 2012, 2013, 2015, 2017 | 14 |
| Massapequa (SS) | 5144 ± 505 | – | 2010 | 2007–2012; 2017 | 15 |
| Greenport (SS) | 14,299 ± 550 | 2005 | 2010 | 2004, 2007–2013; 2013, 2017 | 16 |
| Northport Bay (NS) | 28,433 ± 3528 | – | 2016 | 2017 (Kirschbaum) | 17 |
| Port Jefferson Harbor (NS) | 84,208 ± 3528 | 2010 | – | 2006, 2008, 2013, 2017 | 18 |
| Great South Bay (SS) | 3352 ± 736 | 2013 | – | 2005, 2017 (all except Davis) | 19 |
| North shore sound (NS) | 20,681 ± 1752 | – | 2010, 2016 | 2003, 2005, 2009–2013; 2015, 2017 (all except Miller) | 20 |
| (d) Nonbreeding | | | | | |
| *Beach zone* | *Mean ± Std ($m^2$)* | *High* | *Low* | *Inactive years* | *Figure* |
| Jamesport (NS) | 22,558 + −1175 | – | 2010 | 2009–2010; 2014–2017 | 21 |
| Mastic (SS) | 5982 ± 2051 | 2013 | 2007 | 2004–2005; 2007–2017 | 22 |
| Long Beach (SS) | 3705 + −1011 | 2013 | 2010 | 2005–2006 | 23 |
| Dune Road (SS) | 8105 ± 1440 | 2013 | – | 2009, 2010, 2013; 2015–2017 | 24 |
| Shinnecock (SS) | 8524 ± 419 | 2010 | 2016 | 2008–2017 | 25 |
| Shelter Island | 28,859 + −3420 | 2013 | 2007 | 2008–2017 | 26 |

## Pending Beach Zones (PB)

Pending beach zones have a general downward trend around 2016 (Figs. 13, 14, 15, 16, 17, 18, 19, and 20). Beach zones where all beaches are at zero horseshoe crab population in 2017 are Fire Island, Jones Beach, Massapequa, and Greenport. Beach zones that contain at least one beach with zero horseshoe crabs are Northport Bay,

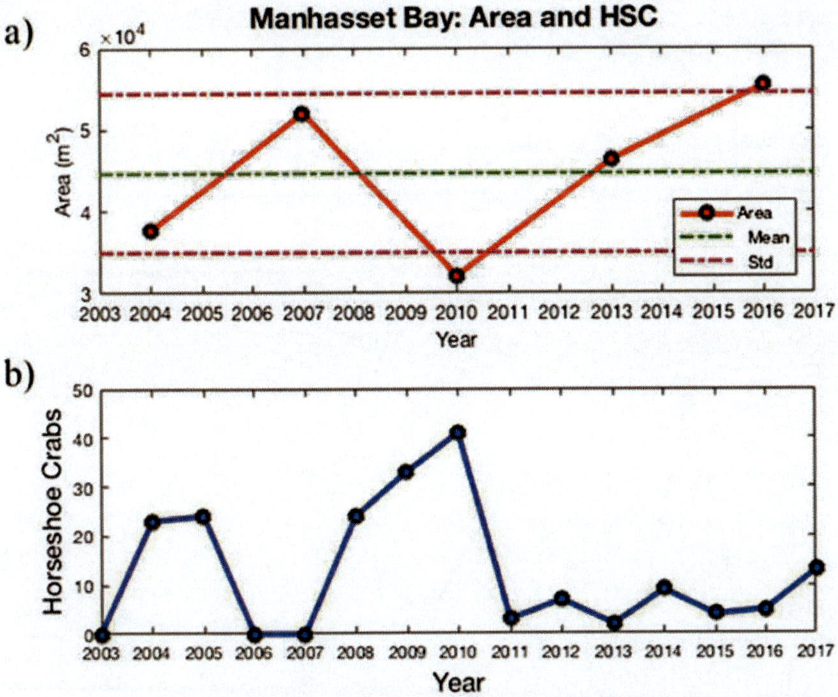


**Fig. 7** Manhasset Bay: (**a**) potential breeding area on Manhasset Bay with the mean and standard deviation; (**b**) the HSC inventory Manhasset Bay is an active breeding zone

Port Jefferson Harbor, Great South Bay, and North Shore Sound. The two zones in the North Shore had anomalously low years in 2016 and four zones had anomalously low years in 2010 (Table 2c). Both North and South Shore pending breeding zones demonstrated a 1% decrease in potential breeding area (Table 3d).

## Nonbreeding Beaches (NBB)

Nonbreeding zones consist of four South Shore zones and two North Shore zones (Fig. 21, 22, 23, 24, 25, and 26). These zones all have years of anomalously low-potential breeding areas except for Dune Road (Table 2d). Zones, such as Mastic in 2007 or Jamesport in 2010, have anomalously low-potential breeding area years while also having inactive breeding. For Mastic, once this low year was hit, active breeding was stopped. North Shore zones experienced a positive 3% change while the South decreased in potential breeding area (Table 3e).

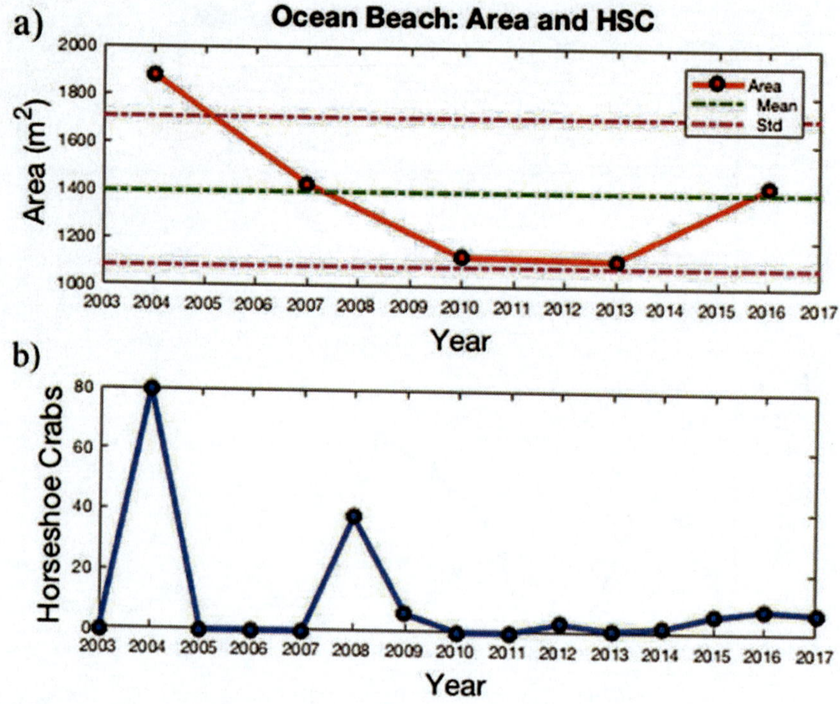


**Fig. 8** Ocean Beach: (**a**) potential breeding area on Ocean Beach with the mean and standard deviation; (**b**) the HSC inventory Ocean Beach is an active breeding zone

## 3.5 *Case Studies*

In order to further investigate the implications of potential breeding area for each classification of beaches, one beach was chosen from each to study: Manhasset, Stony Brook, Northport Bay, and Mastic. Each beach zone represents each breeding classification.

### Manhasset

Manhasset is an actively breeding beach zone consisting of Plum Beach, which is in Manhasset Bay. The areal outlines were drawn to exclude the residential area (Fig. 27). Plum Beach is secluded and generally only used by the residents with little interference to the beach. The only noticeable changes to the area were due to the presence or absence and the width of the spit in the east of the photographs that first appears in 2007, which could be the natural process of the beach.

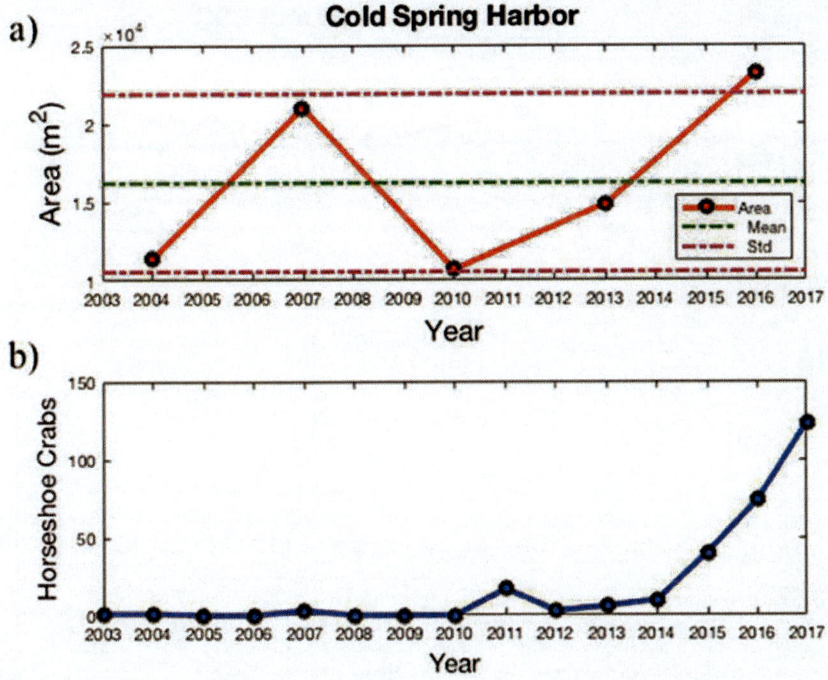


**Fig. 9** Cold Spring Harbor: (**a**) potential breeding area on Cold Spring Harbor with the mean and standard deviation; (**b**) the HSC inventory Cold Spring Harbor is an active breeding zone

## Stony Brook Zone

Stony Brook is located on the Smithtown Bay on Long Beach Road. It is adjacent to the Smithtown Bay Yacht Club, a road and a parking lot. In 2004 and 2007, there is a groin on the east edge of the beach (Fig. 28). This groin is composed of rocks and exhibits sediment build up on the west side. The area for 2016 was difficult to discern due to a large wave crashing over the shore.

## Northport Bay Zone

Northport Bay is located on the North Shore of Long Island and consists of Caumsett State Park, Huntington Bay, Centerport, Bayview Ave, and Kirchsham State Park. Caumsett State Park is on Huntington Bay and Morris Rock and is left relatively undisturbed without any residential interference (Fig. 29). Centerport is a recreational beach between Northport Harbor and Centerport Harbor (Fig. 30). There is a large parking lot next to the beach to facilitate beach going, and in 2016, there was a large reduction on shore on the tip. Huntington Bay is located across the Huntington Sound from Caumsett State Park and is on Price Bend (Fig. 31). This is a heavily

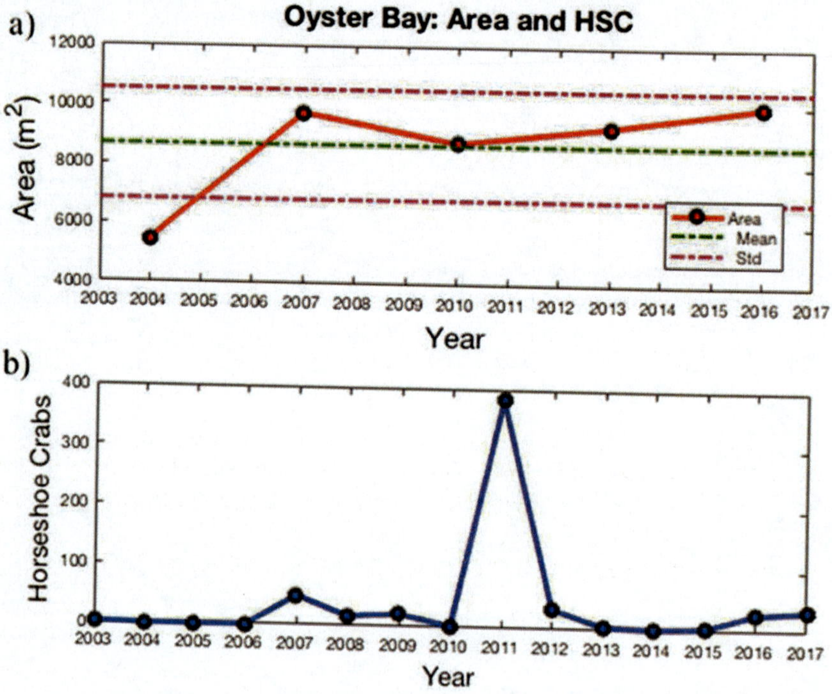


**Fig. 10** Oyster Bay: (**a**) potential breeding area on Oyster Bay with the mean and standard deviation; (**b**) the HSC inventory Oyster Bay is an active breeding zone

populated area and is a popular recreational beach and docking spot for boats. The Bayview Ave site is on Northport Harbor and contains a pier because it is another popular boat docking site (Fig. 32). Additionally, this area is highly developed. Kirchsbaum Park is another state park that is located on the Long Island Sound in Fort Salonga and is next to a manmade intel and channel (Fig. 33). There is residential area next to this park and industrialized factories next to the channel.

## Mastic Zone

The Mastic Beach zone is located on the South Shore and contains beach sites on Mastic Beach, which is on the mainland and on Smith Point, a part of the barrier island system. Cranberry Rd. site is located on Smith Point, popular summer beach, and is separated by the main island by Narrow Bay (Fig. 34). This site is located right next to a large parking lot and contains a small portion of actual shore. The Smith Point site is located on the mainland and is a part of Smith Point Park (Fig. 35). This site is adjacent to a road and a few homes. In 2013, the wharf at the point was broken, and in 2016 the pier was fixed again. Sandy Point is between the

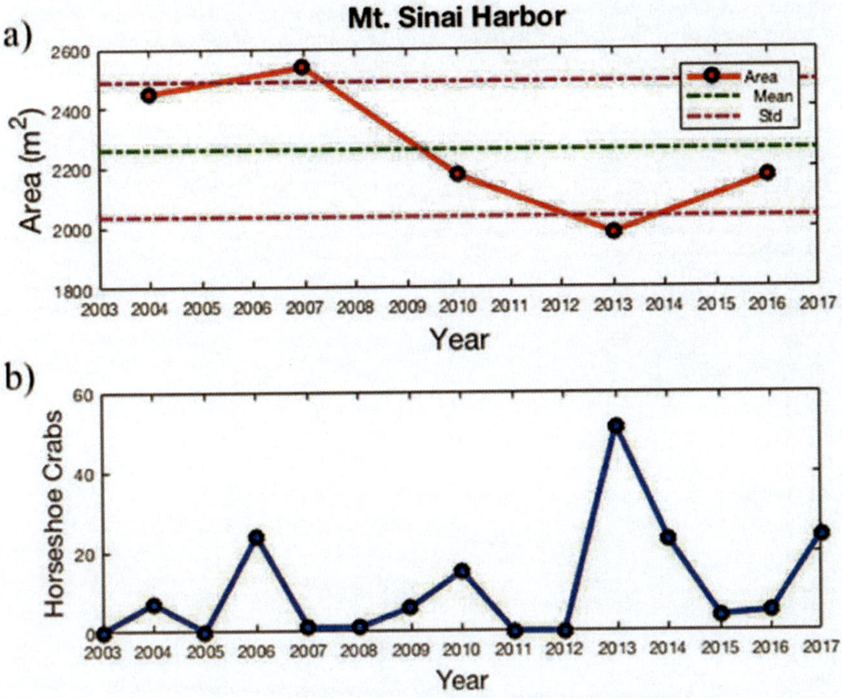


**Fig. 11** Mt. Sinai: (**a**) potential breeding area on Mt. Sinai Harbor with the mean and standard deviation; (**b**) the HSC inventory Mt. Sinai Harbor is an active breeding zone

Wertheim Wildlife National Refuge and a residential area (Fig. 36). This beach had a narrow shore until 2010.

# 4 Discussion

Horseshoe crabs are an ecologically seminal species that provide humans with essential health and ecosystem services. Horseshoe crab's contributions to human disease prevention, especially since the COVID-19 pandemic, and the pharmacological search for a successful immunization strategy, their paleo-survivability and the conservation contribution two host of other dependent species, have chronically been impacted by an alphabet of urbanizing disturbances resulting in a global decline in populations and their documented vulnerability threating extirpation at least, and extinction at worst, thus being Red Listed by the IUCN (Smith, et al.) The shallow estuarine shores and sandy beaches of Long Island provide an ideal habitat for horseshoe crabs because it supplies potential breeding area and a space for nested egg development as well as a nearshore ocean for juvenile growth and adult survival. The sustainability of the Long Island horseshoe crab population at 115

**Table 3** Summary of percent change from 2004 to 2016 for shores and beach zones: (a) divides zones by shore, mean and standard deviation and the percent change of the mean potential breeding areas from 2004 to 2016. (b–e) separates each classification by north and south shore and shows the mean and standard deviation and percent change

| (a) North vs. South Shore | | |
|---|---|---|
| *Shore* | *Mean ± Std (m²)* | *Percent change* |
| North | 22,358 ± 1906 | −2% |
| South | 8250 ± 688 | −4% |
| (b) Active breeding | | |
| *Shore* | *Mean ± Std (m²)* | *Percent Change* |
| North | 17,998 ± 4185 | 61% |
| South | 1394 ± 310 | −24% |
| (c) Recovering | | |
| *Shore* | *Mean ± Std (m²)* | *Percent change* |
| North | 32,267 ± 8331 | 1% |
| South | – | – |
| (d) Pending | | |
| *Shore* | *Mean ± Std (m²)* | *Percent change* |
| North | 17,958 ± 1742 | −18% |
| South | 11,562.62 ± 928 | −14% |
| (e) Nonbreeding | | |
| *Shore* | *Mean ± Std (m²)* | *Percent change* |
| North | 25,708 ± 1790 | 3% |
| South | 6579 ± 973 | −11% |

sites has been monitored and has been declared to be on a decline for the last 10 years (Tanacredi 2020). Results of this study investigation have revealed possible connections between environmental, anthropogenic, and physical changes to potential horseshoe crab breeding sites and the breeding activity level of the Long Island horseshoe crab population.

The Long Island horseshoe crab population is at risk for continued breeding inactivity due to the fact that only 5 out of 20 beach zones in this study are classified as active breeding (AB) beach zones (Fig. 3). The North Shore of Long Island has four active breeding (AB) zones while the South Shore only has one active breeding (AB) zone. The South Shore due to its geomorphology should be a better suited breeding area for horseshoe crabs than the North Shore. South Shore is characterized by long stretches of gently sloping, sandy beaches on barrier islands and estuarine ecosystems that buffer these areas from intensive wave actions. The North Shore is characterized by steep sloping, gravel beaches on a terminal moraine, and more relatively deeper harbors and bays.

Considering environmental factors that affect the disparity in inactive breeding (IB) North and South Shore's average yearly sea surface temperatures (SST) showed that water temperatures of both shorelines fall within the preferred temperature ranges for horseshoe crab breeding, and no significant relationship between Long Island SST and decline of horseshoe crab populations (Fig. 4). Comparing land use,

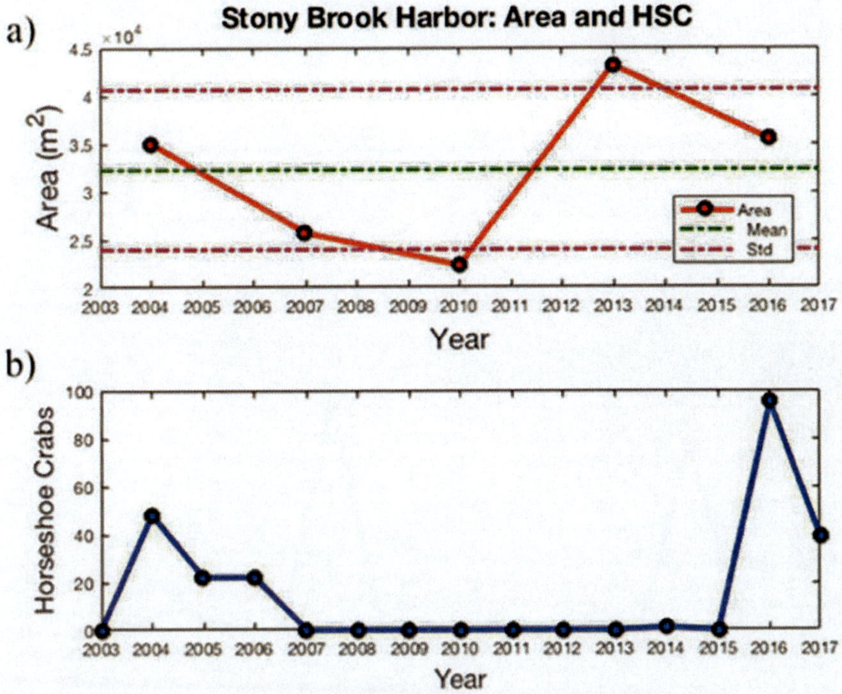


**Fig. 12** Stony Brook Harbor: (**a**) potential breeding area on Stony Brook with the mean and standard deviation; (**b**) the HSC inventory Stony Brook is a recovering breeding zone

the South Shore coastline exhibits more open recreational spaces and the north has more residential areas (Fig. 6). Considering the physical potential breeding area, the North Shore has decreased by 2% and the South Shore has decreased by 4% (Table 3a).

## *4.1 Land Use*

Areas of inactive beaches are generally concentrated around (open) zoned recreational spaces used during the summer (Tanski 2012; Fig. 6). The South Shore of Long Island is a popular tourist area for the summer containing community service and open recreational spaces such as Long Beach, Jones Beach, Fire Island National Seashore, Shinnecock, Great South Bay, and Dune Road. On the North Shore side, most areas are residential areas, especially Manhasset, Oyster Bay, and Cold Spring Harbor. On the eastern end of the North Shore, the town of Jamesport is near a large agricultural region and Shelter Island has open recreational beaches. This area is also a popular summer beach use residential area. Because horseshoe crabs come onshore to breed during the summer, their potential breeding areas are subjected to

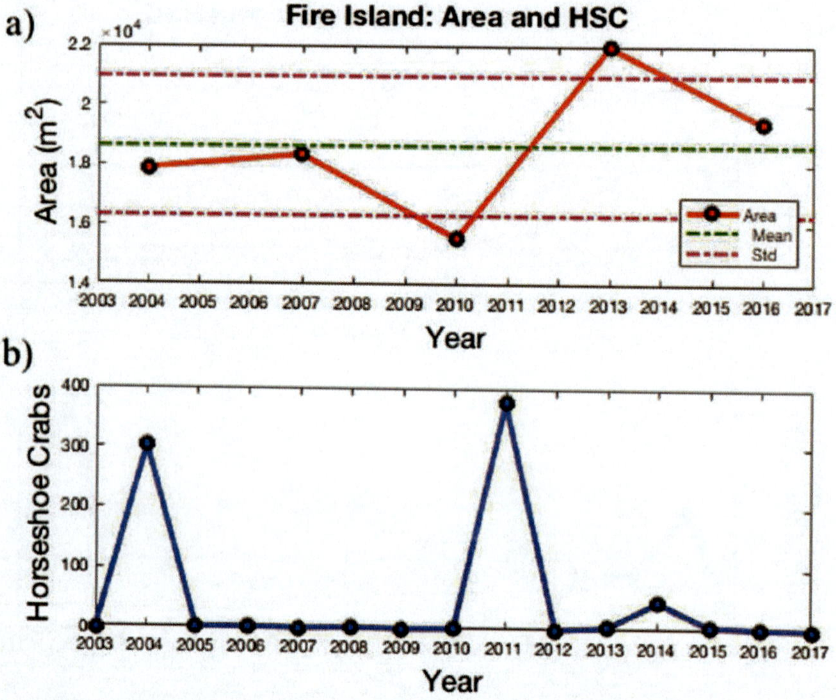


**Fig. 13** Fire Island: (**a**) potential breeding area on Fire Island with the mean and standard deviation; (**b**) the HSC inventory Fire Island is a pending breeding zone

human interference from summer beach goers, such as humans taking over the physical area or disturbing established burrows. Residential areas may still use private beaches during the summer months, but these beaches are less disturbed because they are not open to the public. The sandy and gently sloping South Shore attracts summer tourism that can lead to human disturbances to horseshoe crabs during breeding months.

These open recreation beaches are essential for horseshoe crab survival because these sites are horseshoe crabs' natal beaches. Horseshoe crabs exhibit natal breeding, meaning that these organisms have strong site fidelity and breed on their natal beach. *Limulus polyphemus* only inhabit a small range along the continental shelf providing some evidence that the crabs do not migrate. If the natal beach is disturbed, horseshoe crabs may not migrate to another beach. Even more extreme, if breeding habitat is completely lost, then horseshoe crabs will likely not breed elsewhere. The affect requires further investigation.

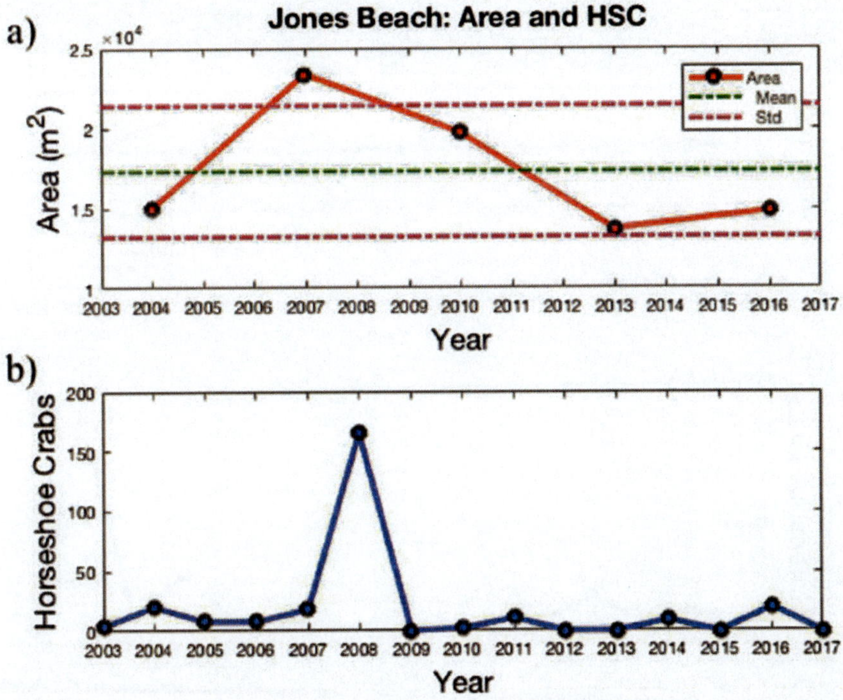


**Fig. 14** Jones Beach: (**a**) potential breeding area on Jones Beach with the mean and standard deviation; (**b**) the HSC inventory Jones Beach is a pending breeding zone

## 4.2 *Potential Breeding Area*

When the potential breeding (PB) area sites were evaluated, results show an overall decrease in potential breeding area in both the North and South Shores by 2% and 4%, respectively, suggesting that the total potential breeding areas for horseshoe crabs on Long Island are decreasing by 3% (Table 3a). Some reasons for changes in potential breeding area could include human disturbances (e.g., beach going) or loss of habitat due to developments and resulting from Hurricane Sandy in 2012. Overall, most beach zones experienced a decrease in potential breeding area, but many areas increased in 2013 (Table 2). During Superstorm Sandy, coastal sand dunes experienced overwash and thus experienced vertical dune erosion transporting sediments landward causing a flattening and elongation of the beaches (USGS 2016). This sediment transport could account for increases in beach area that are observed in 2013, such as in Cupsogue Beach, Ocean Beach, and Fire Island. Other beaches underwent beach nourishment in 2013: Jones Beach, Ocean Beach, Long Beach, and Dune Road. With Hurricane Sandy's pathway, the South Shore experienced the brunt of the storm causing not only changes in potential beach areas, but could also damage to the horseshoe crab populations themselves (Western Carolina University).

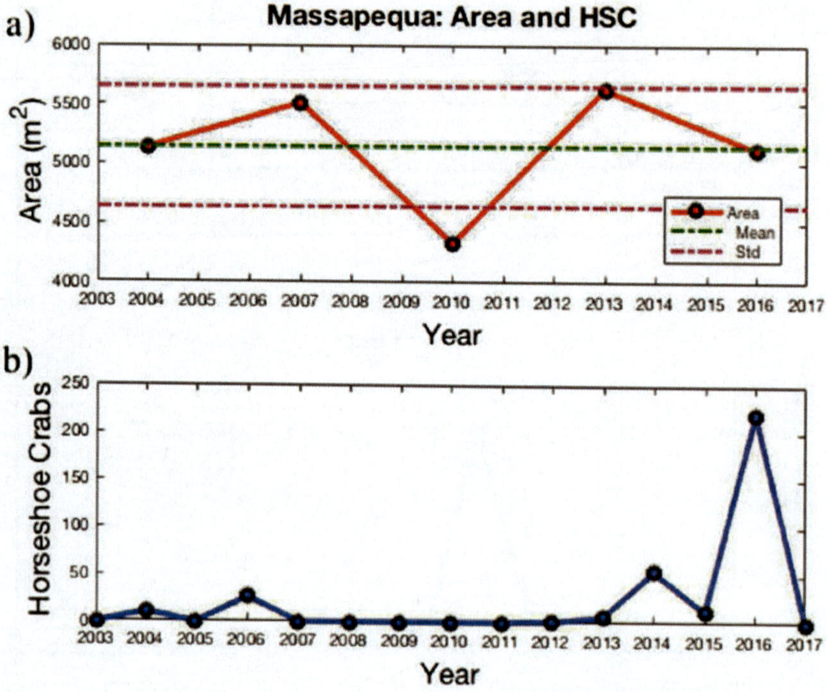


**Fig. 15** Massapequa: (**a**) potential breeding area on Massapequa with the mean and standard deviation; (**b**) the HSC inventory Massapequa is a pending breeding zone

Data collected on shorebird recovery post Sandy on New Jersey Shorelines of Delaware Bay have returned to prestorm levels, yet these are short-term and preliminary results (Wheeling 2014). Trend analysis shows the South Shore is losing more potential breeding area. The active breeding zones on the North Shore exhibited a 61% increase in potential breeding area and the South Shore active breeding zone had a 24% decrease (Table 3b). The South Shore nonbreeding beach zones have a negative 11% change while the North Shore nonbreeding beach zones have a positive 3% change (Table 3e). North Shore nonbreeding activity zones do still have a slight increase, but not as dramatic as the actively breeding zones. This suggests that the South Shore nonbreeding beach zones are losing more potential beach breeding area on natal beaches, which inhibits horseshoe crab breeding activity. To further highlight the changes in potential breeding areas, four beach zones were selected as "Case Studies" each representing a different classification condition.

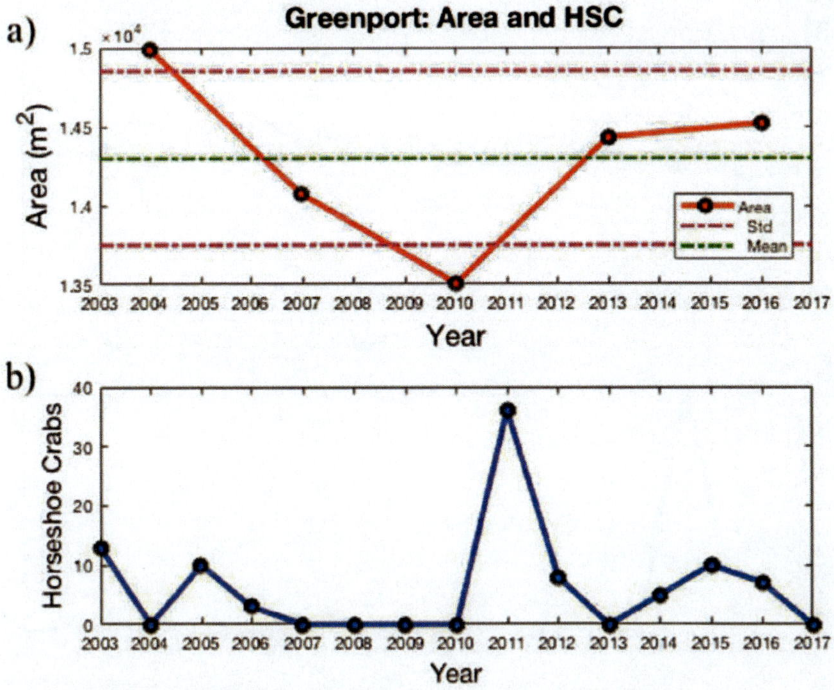


**Fig. 16** Greenport: (**a**) potential breeding area on Greenport with the mean and standard deviation; (**b**) the HSC inventory Greenport is a pending breeding zone

## 4.3 *Case Studies*

Manhasset beach zone was chosen as a representative of actively breeding beaches. This zone consists of Plum Beach, which sits on Manhasset Bay and has supported actively breeding horseshoe crabs since 2007. Plum Beach sits at the end of Plum Rd and is surrounded by a residential area (Fig. 27). The beach here is cut off from summer tourism limiting the amount of recreational use to the residents of this road. No notable changes were observed from 2004 to 2016. Plum beach is home to an actively breeding population of horseshoe crabs, which corresponds to the private beach with little potential breeding area changes. The North Shore has an increase in potential breeding area for active breeding, suggesting that horseshoe crabs are able to access their natal beaches and have enough room to breed.

Stony Brook zone is the only zone within the recovering classification. The beach site sits on Long Beach road on Smithtown Bay and is next to the Smithtown Yacht Club, which suggests that this beach may be used for recreational use (Fig. 28). However, this recreational use is most likely restricted to members of the club and thus this beach would not be heavily populated during the summer months. Additionally, two parking lots and a marina sit next to this beach, which indicate

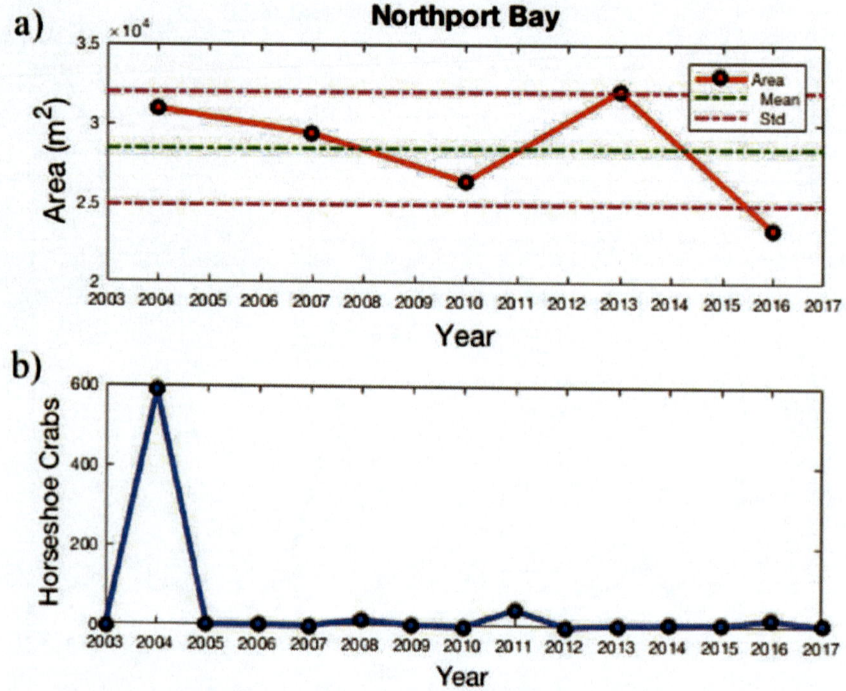


**Fig. 17** Northport Bay: (**a**) potential breeding area on Northport Bay with the mean and standard deviation; (**b**) the HSC inventory Northport Bay is a pending breeding zone

possible human impacts on the potential breeding area (Fig. 28). Horseshoe crabs were inactive from 2007 to 2015, becoming active again in 2016. Potential breeding area experienced a decrease from 2004 to 2010 until potential breeding area increased in 2013, suggesting that the lowest potential breeding area occurs at the same time as inactivity and that the highest potential breeding area could be a factor in breeding recovery.

Pending beaches are more difficult to relate to potential breeding area changes because a pending beach was actively breeding in 2016 and inactive in 2017, but potential breeding area data end in 2016 so it is important to look at factors leading up to inactivity in 2017 instead. The four South Shore pending zones decreased by 14% and the four North shore zones decreased by 18% (Table 3d). Northport was chosen as a representative of the eight beach zones in the pending breeding classification. This zone, consisting of Caumsett State Park, Huntington Bay, Centerport, Bayview Ave, and Kirschbaum State Park, is on the North Shore, but is subjected to more active recreational use. Caumsett and Kirschbaum are two parks that are more secluded and have little human interference (Figs. 29 and 33). Centerport, Huntington Bay, and Bayview beach sites are located in more residential and recreational areas (Figs. 30, 31, and 32). Centerport is a popular recreational beach; Bayview and Huntington Bay are in residential areas, but are both popular boating

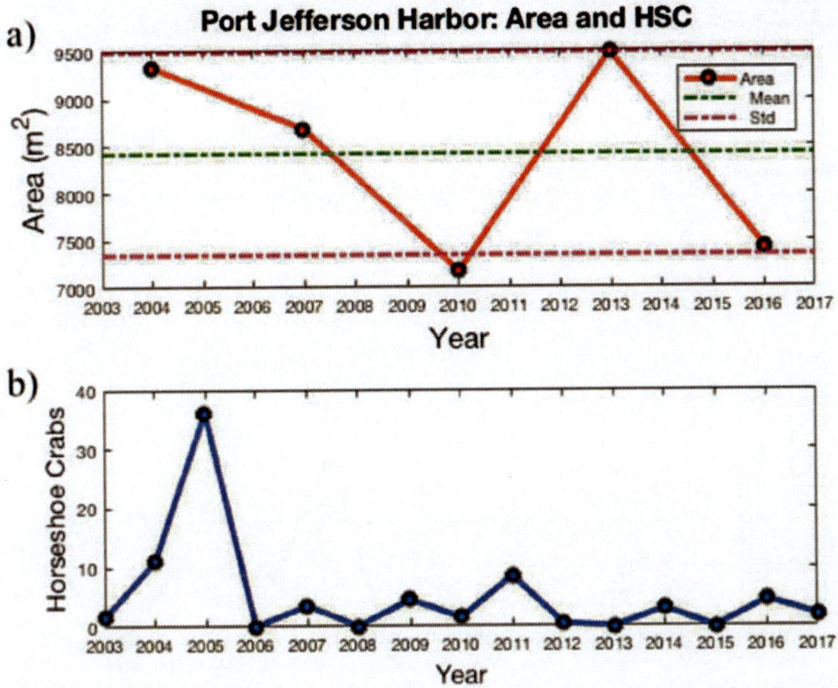


**Fig. 18** Port Jefferson Harbor: (**a**) potential breeding area on Port Jefferson Harbor with the mean and standard deviation; (**b**) the HSC inventory Port Jefferson Harbor is a pending breeding zone

areas and so have more summertime human interaction. Northport has a mixture of residential and summertime recreational use sites and had breeding inactivity in 2017. Pending beach zones are areas that need to be monitored because changes to these beaches could result in decreasing horseshoe crab breeding activity.

Mastic zone, consisting of Sandy Point on Bellport Bay, and Smith Point and Cranberry Drive Beach located on Narrow Bay, was chosen as a representative of nonbreeding beach zones. All beach sites became inactive in 2007, and this corresponds with the lowest potential breeding area (Figs. 35, 36, 37, and 38). There were no human disturbances in 2007; however, a significant narrowing of the beaches decreased the potential breeding area during the same year breeding became inactive. Contributing factors to this could be summer recreation at these beaches. For example, Cranberry Drive Beach is located right next to a road and has a large parking lot, suggesting that the public has access to this beach and that the beach is capable of holding a large volume of people due to the parking lot (Fig. 35). Nonbreeding beach zones, especially on the South Shore, have no breeding activity, an overall decrease in potential breeding, and often are used for summer recreation.

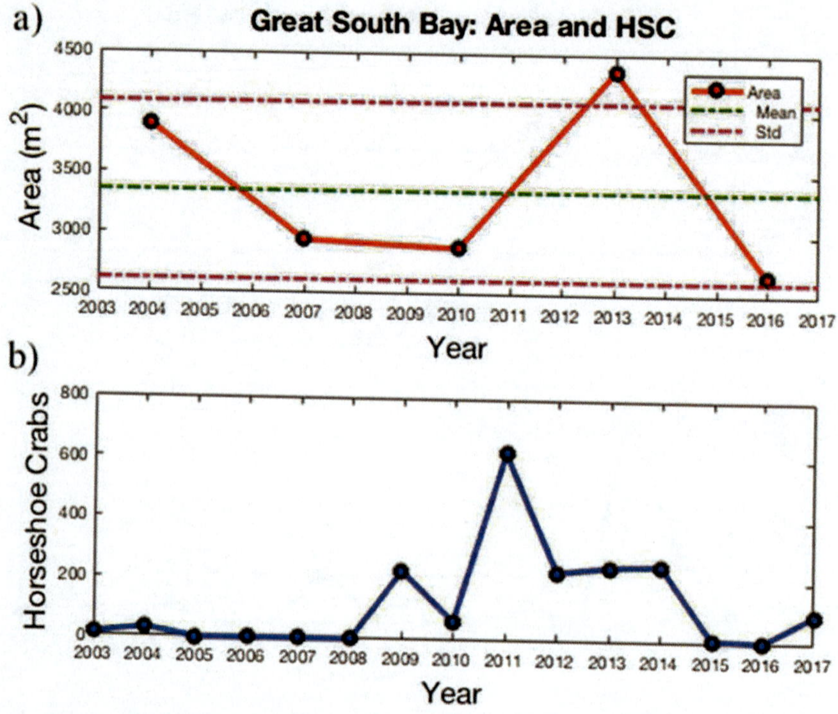


**Fig. 19** Great South Bay: (**a**) potential breeding area on Great South Bay with the mean and standard deviation; (**b**) the HSC inventory Great South Bay is a pending breeding zone

## *4.4 Conclusions*

Horseshoe crabs require specific breeding conditions as well as relief from anthropogenic threats for a successful population to be sustainable. Part of these ideal conditions are a large enough area free from disturbances to facilitate burrowing, egg laying, and egg development. Long Island exhibits an overall pattern of decreased horseshoe crab breeding activity connected with a decrease in potential breeding area, especially in areas of summer recreational use. This pattern has the potential to significantly alter the Long Island horseshoe crab to replenish their population. The South Shore of Long Island is subjected to more summer recreation use and breeding inactivity compared to the North Shore. Horseshoe crabs require natal beaches with available breeding areas undisturbed by heavy summer tourism to survive on Long Island.

As an essential keystone species in their ecosystem and an essential contributor to human quality of life, horseshoe crabs have a special niche they occupy in their habitat. If you protect horseshoe crab habitat you protect all associated species associated with these coastal estuarine ecosystems. Results of this investigation exhibiting the South Shore of Long Island being more impacted than North Shore may be

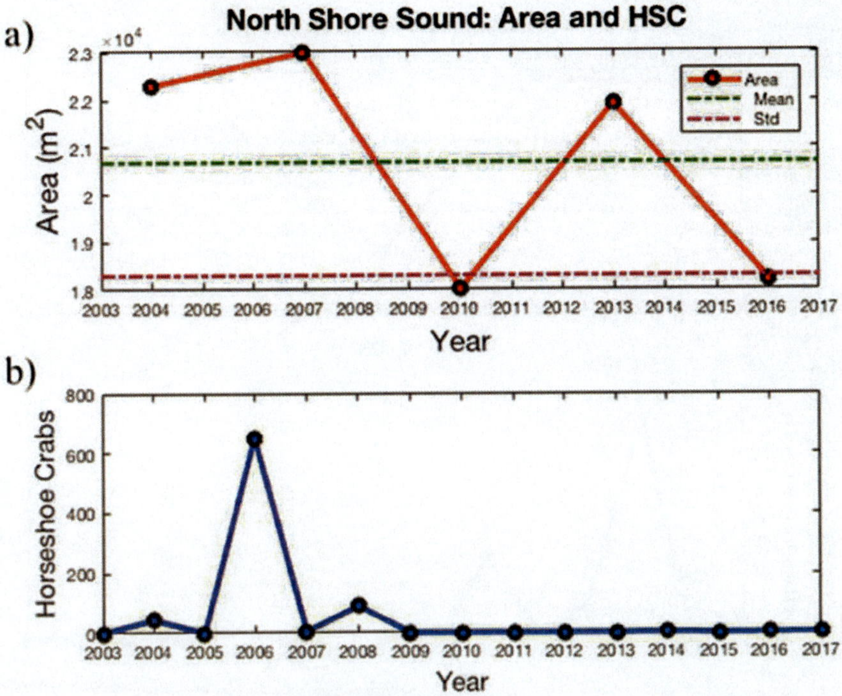


**Fig. 20** North Shore Sound: (**a**) potential breeding area on North Shore Sound with the mean and standard deviation; (**b**) the HSC inventory North Shore Sound is a pending breeding zone

attributable to growing disturbances to horseshoe crabs, such as collecting horseshoe crabs for bait (for bait, NYS removes over 135,000 (2019) from the LI Horseshoe Crab breeding population annually mostly from South Shore coastal areas). Adult horseshoe crabs are removed in fishing hauls, which prevent gravid females from nesting. The direct effect on their population numbers requires further investigation or causing changes to their natal beaches affecting their sustainability and limiting their potential breeding area.

Future monitoring work will require expanded data retrieval. Tagging adult crabs and the sites may be required. Land-use data was only available for 2017 and not for the complete time series. The potential breeding areas taken from the New York State website are available for the period 2004–2016. Results do not cover short trends in area changes.

Investigations into how various strength storms will affect potential breeding habitats and the population itself are needed. Pending beach zones should be analyzed to understand the inactivity of breeding during the same year. Pending beaches have the potential to become active again and so their potential breeding area could be understood for active breeding. With the growing human population of Long Island and the increasing trend of people interacting with the coastline, an

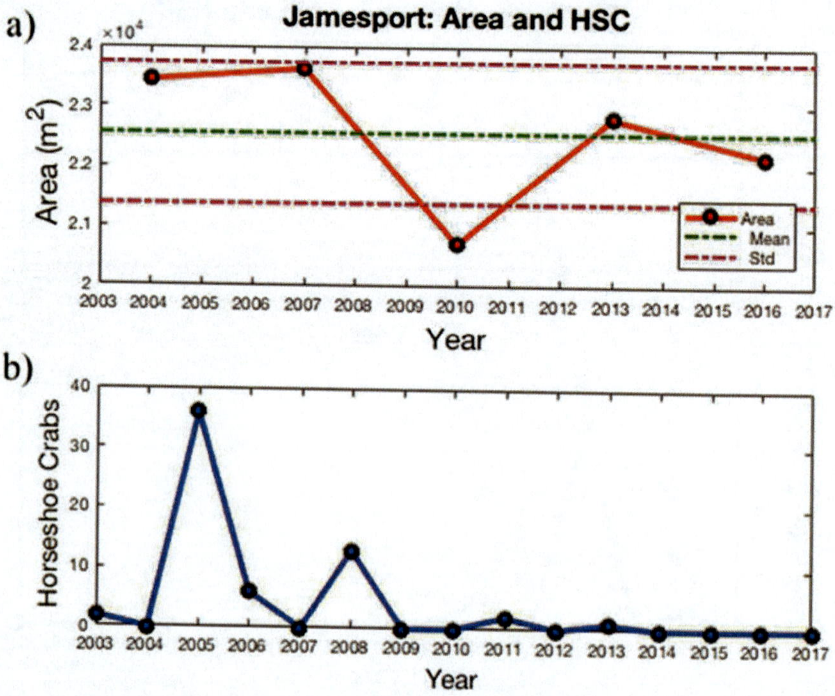


**Fig. 21** Jamesport: (**a**) potential breeding area on Jamesport with the mean and standard deviation; (**b**) the HSC inventory Jamesport is a nonbreeding zone

investigation into specific summer human recreational activities and its associated infrastructure is required.

**Acknowledgments** I would like to thank the Center for Environmental Research and Coastal Ocean Monitoring at Molloy College for a summer research internship. I would like to extend my sincerest gratitude to Dr. Gail Kineke, Boston College, for her guidance and support.

## Appendix

Appendix 1 Horseshoe crab inventory form. CERCOM volunteers fill out this form after each survey, recording horseshoe crab data.

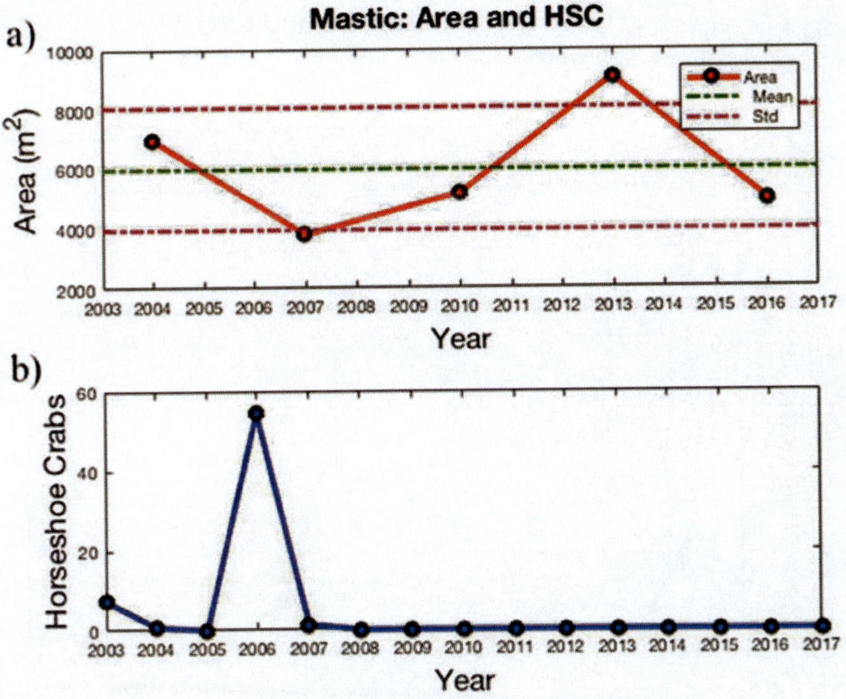


**Fig. 22** Mastic Beach: (**a**) potential breeding area on Mastic with the mean and standard deviation; (**b**) the HSC inventory Mastic is a nonbreeding zone

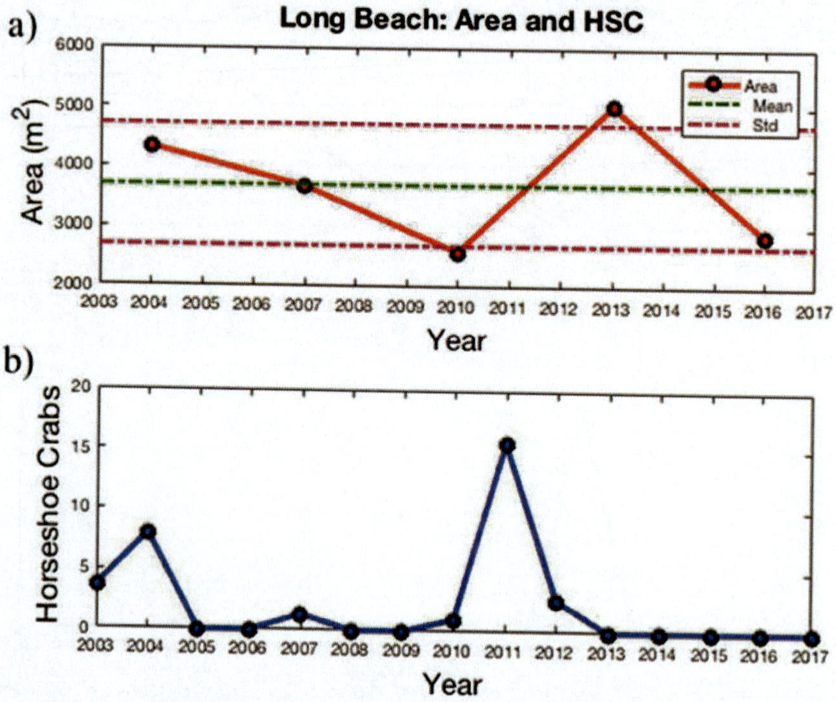


**Fig. 23** Long Beach: (**a**) potential breeding area on North Shore Sound with the mean and standard deviation; (**b**) The HSC inventory Long Beach is a non-breeding zone

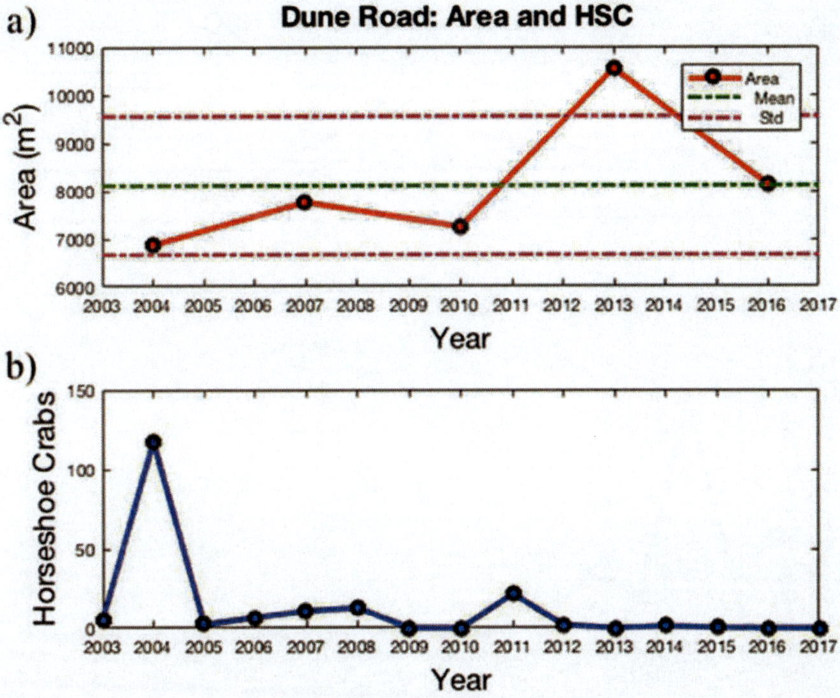


**Fig. 24** Dune Road: (**a**) Potential breeding area on Dune Road with the mean and standard deviation; (**b**) the HSC inventory Dune Road is a nonbreeding zone

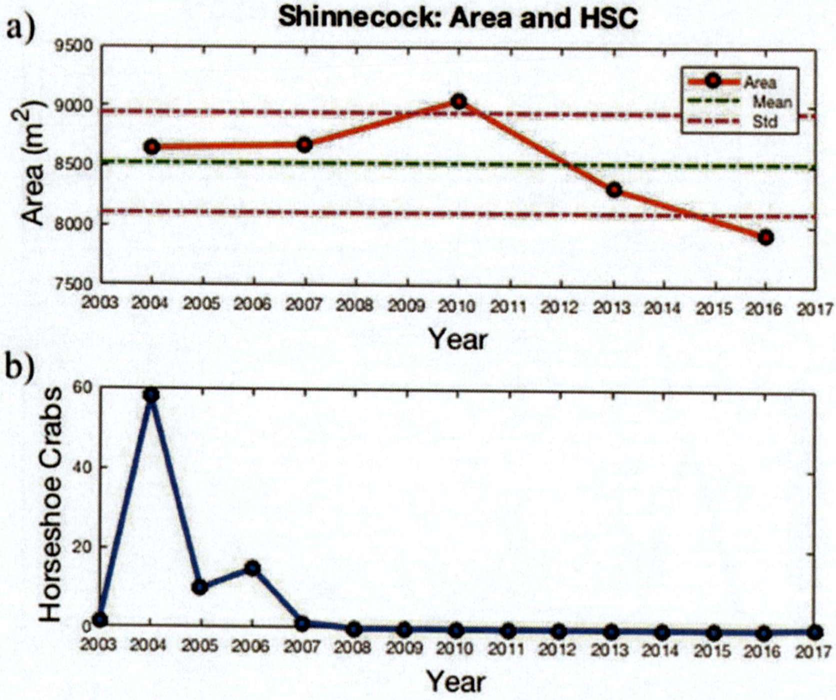


**Fig. 25** Shinnecock: (**a**) potential breeding area on Shinnecock with the mean and standard deviation; (**b**) the HSC inventory Shinnecock is a non-breeding zone

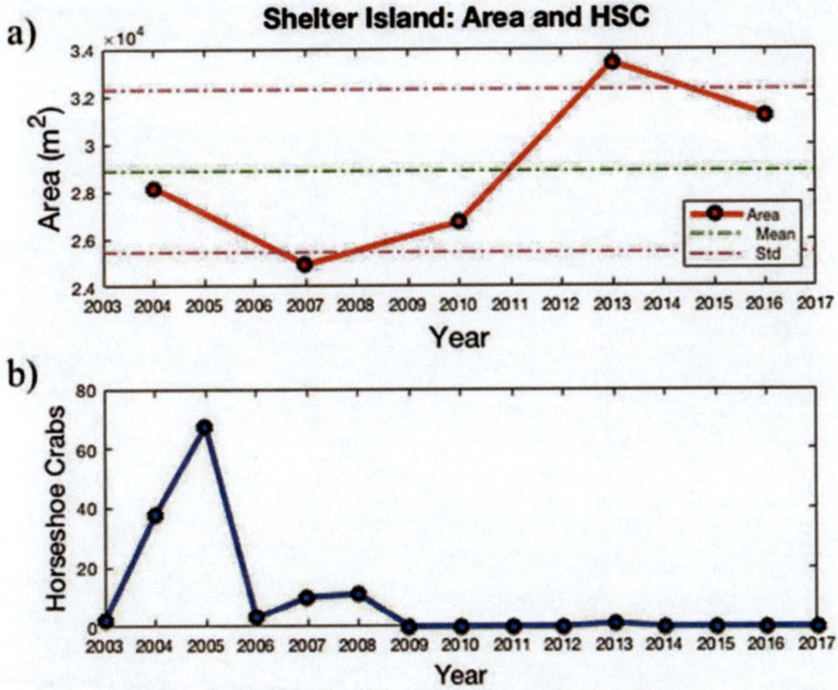


**Fig. 26** Shelter Island: (**a**) potential breeding area on Shelter Island with the mean and standard deviation; (**b**) the HSC inventory North Shore Sound is a nonbreeding zone

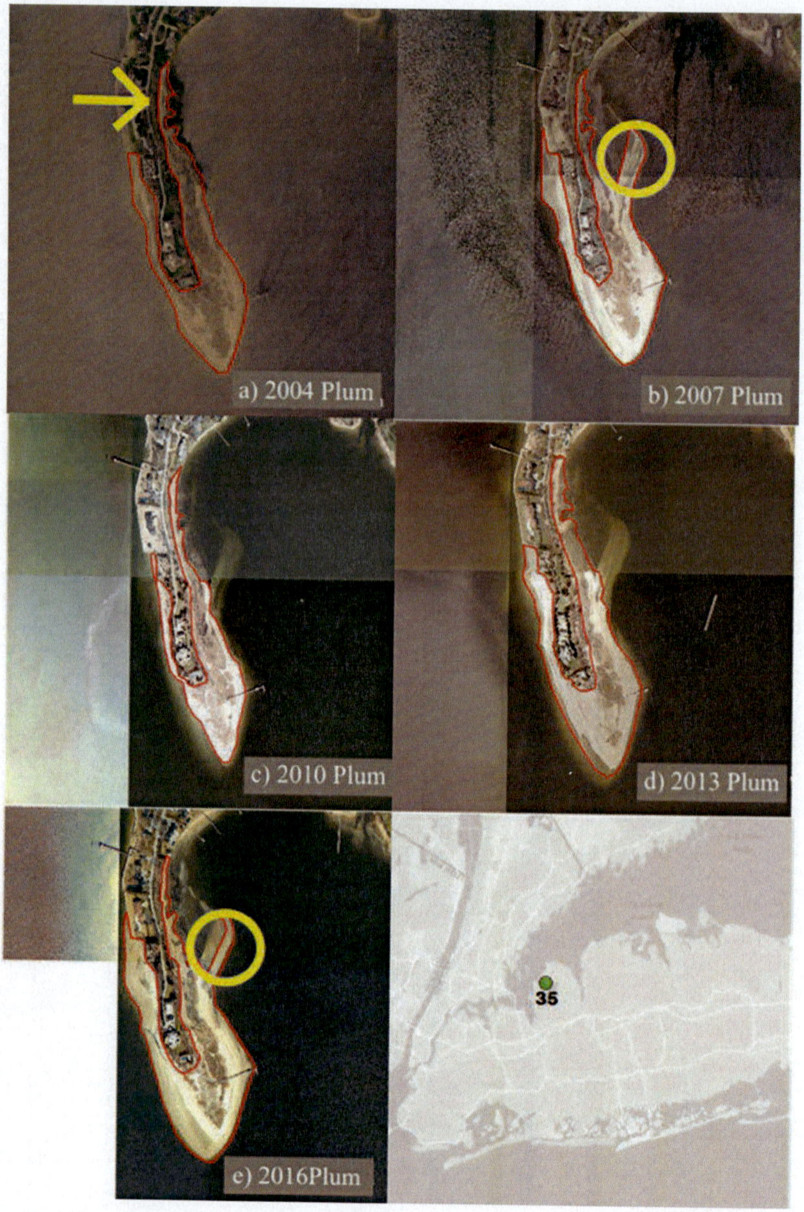


**Fig. 27** Images from Manhasset Bay, site: Plum Beach. Images (**a**–**e**) are 2004, 2007, 2010, 2013, and 2016, respectively. The arrow indicates the private road, Plum Rd. The circle indicates when the spit was developed enough for horseshoe crab breeding

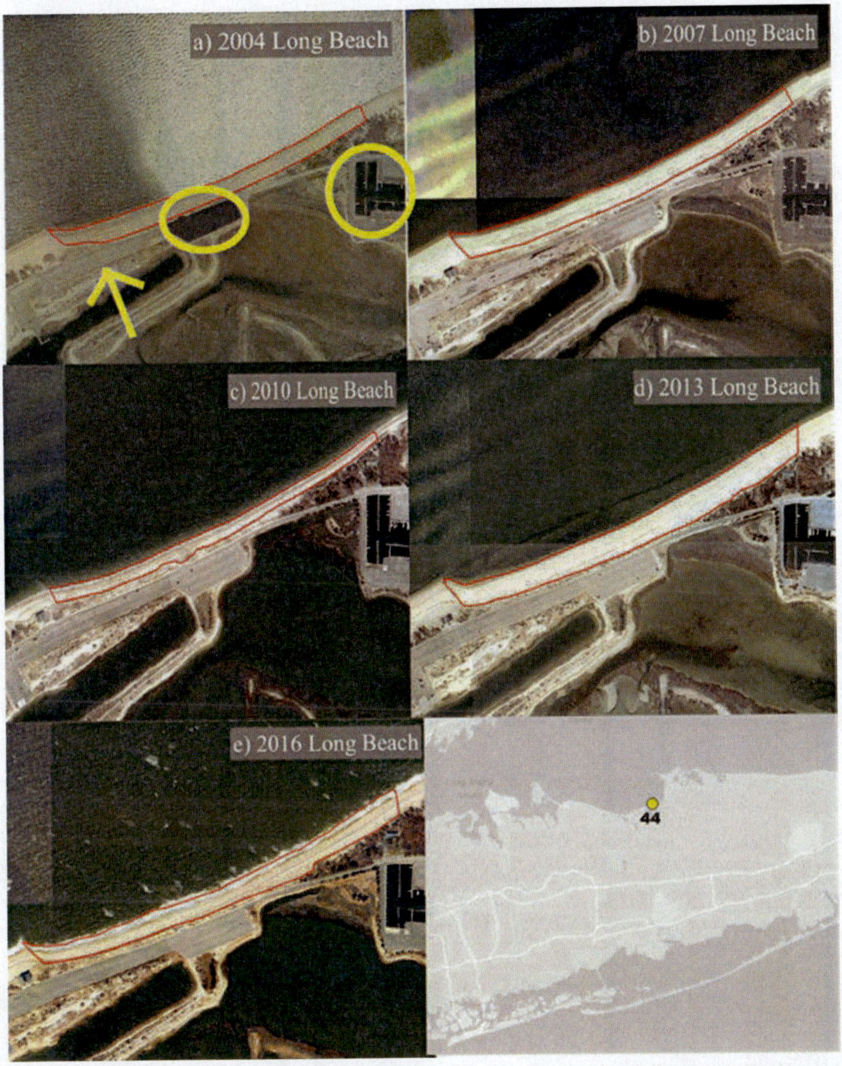


**Fig. 28** Images from Stony Brook Harbor, site: Long Beach. Images (**a–e**) are 2004, 2007, 2010, 2013, and 2016, respectively. The arrow points to Long Beach Rd. The oval encloses a parking lot. The circle indicates the marina for the Smithtown Yacht Club

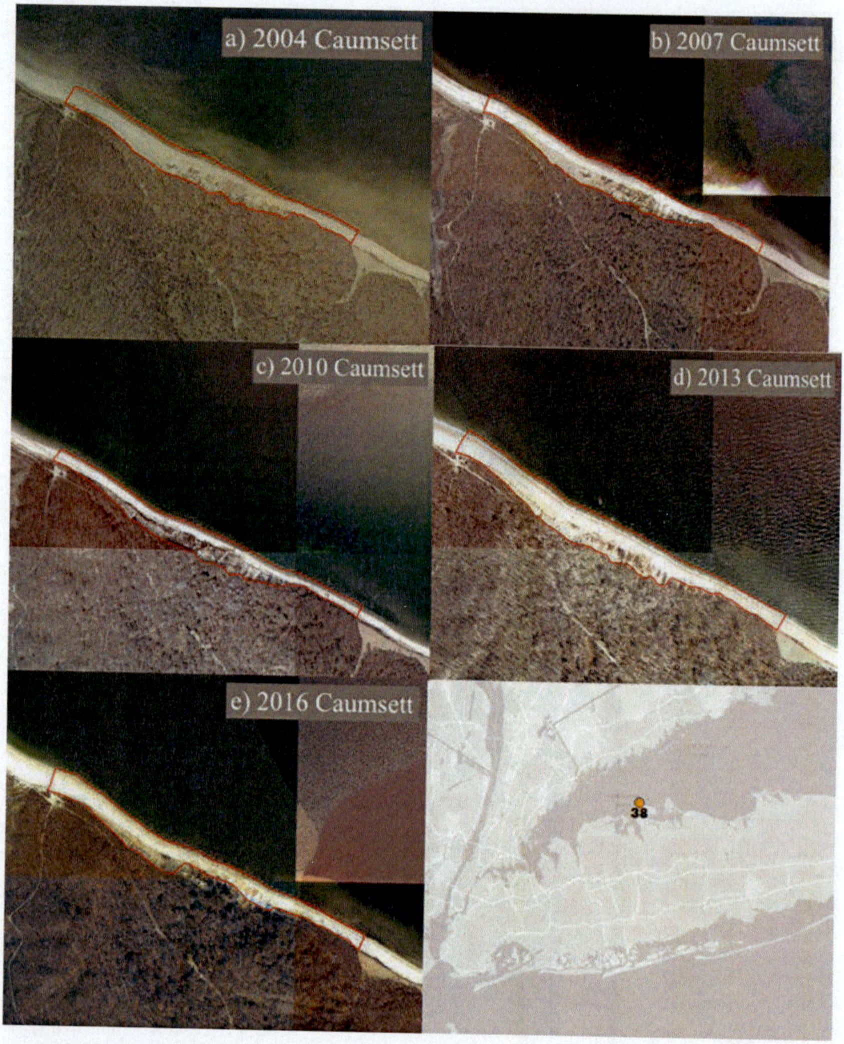


**Fig. 29** Images from Northport, site: Caumsett. Images (**a–e**) are 2004, 2007, 2010, 2013, and 2016, respectively. Located on a state park, there are no obvious changes or human interactions

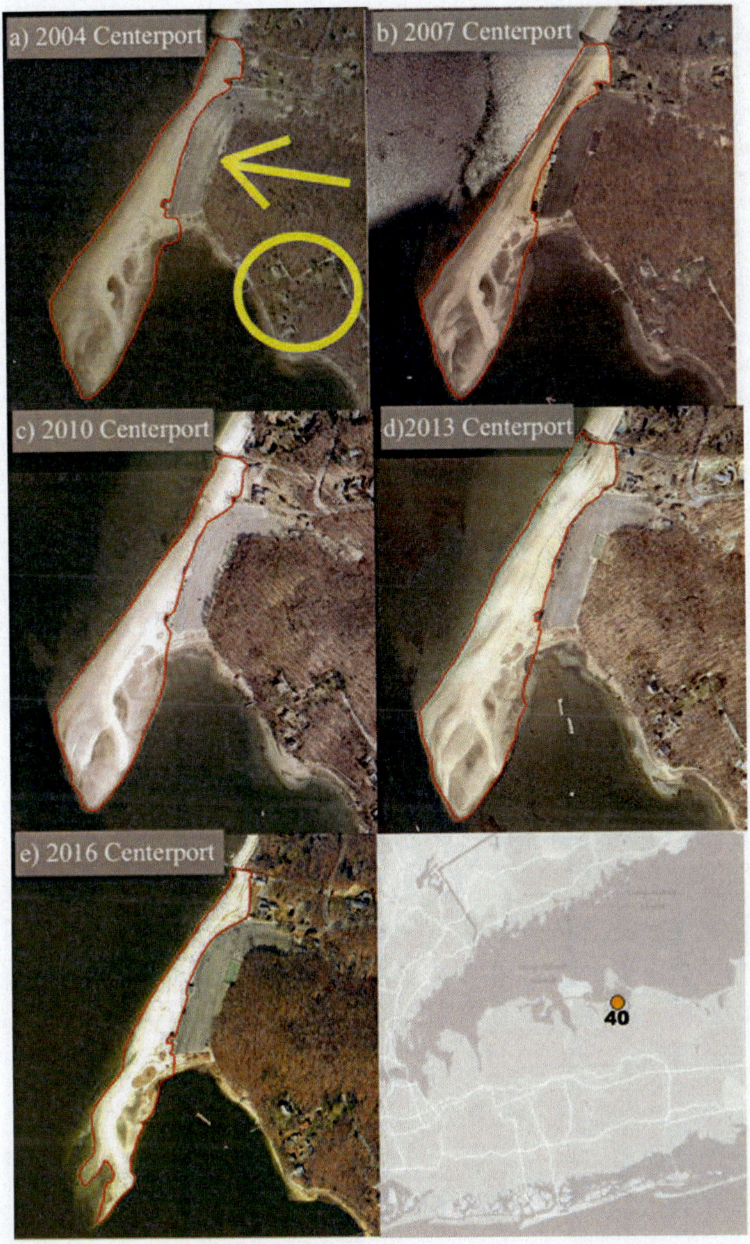


**Fig. 30** Images from Northport, site: Centerport. Images (**a**–**e**) are 2004, 2007, 2010, 2013, and 2016, respectively. The arrow indicates a parking lot. The circle encompasses a residential area

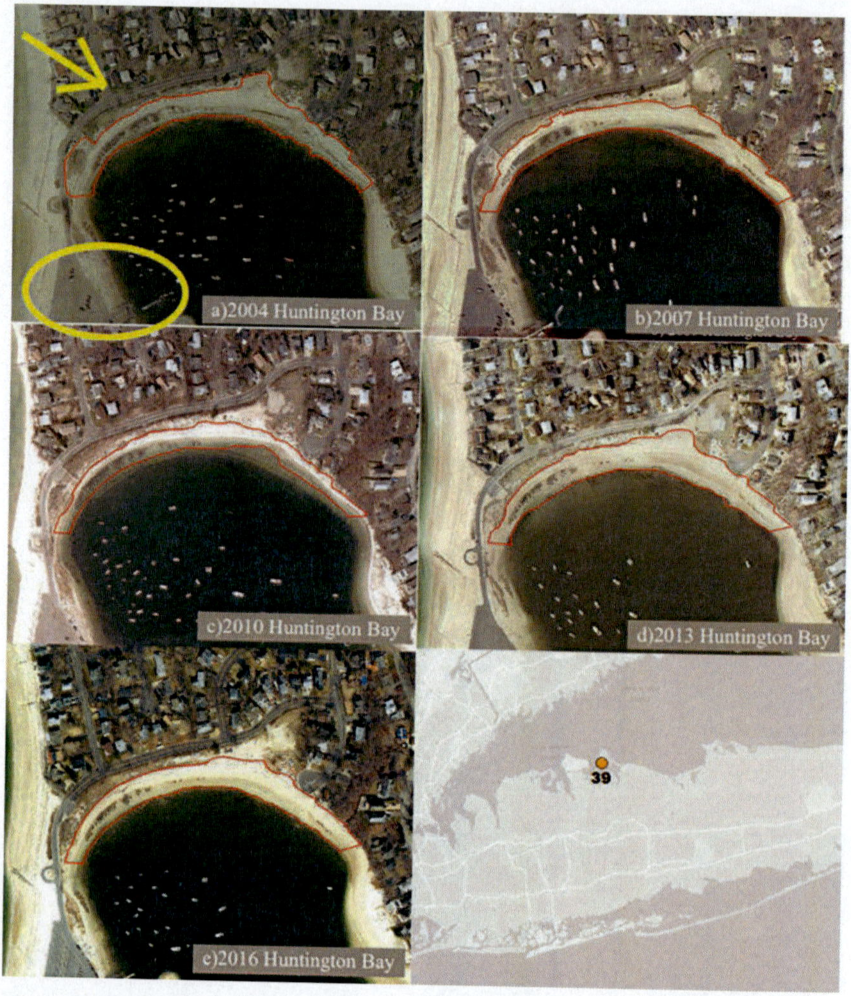


**Fig. 31** Images from Northport, site: Huntington Bay. Images (**a–e**) are 2004, 2007, 2010, 2013, and 2016, respectively. This site is located is a densely populated residential area. The arrows point to the main road. The circle shows a parking lot and pier that is present in every year

**Fig. 32** Images from Northport, site: Bayview Ave. Images (**a–e**) are 2004, 2007, 2010, 2013, and 2016, respectively. Bayview is a heavily populated residential area, and the arrow points to the main road, Bayview Ave. The two circles indicate piers that are present in every year

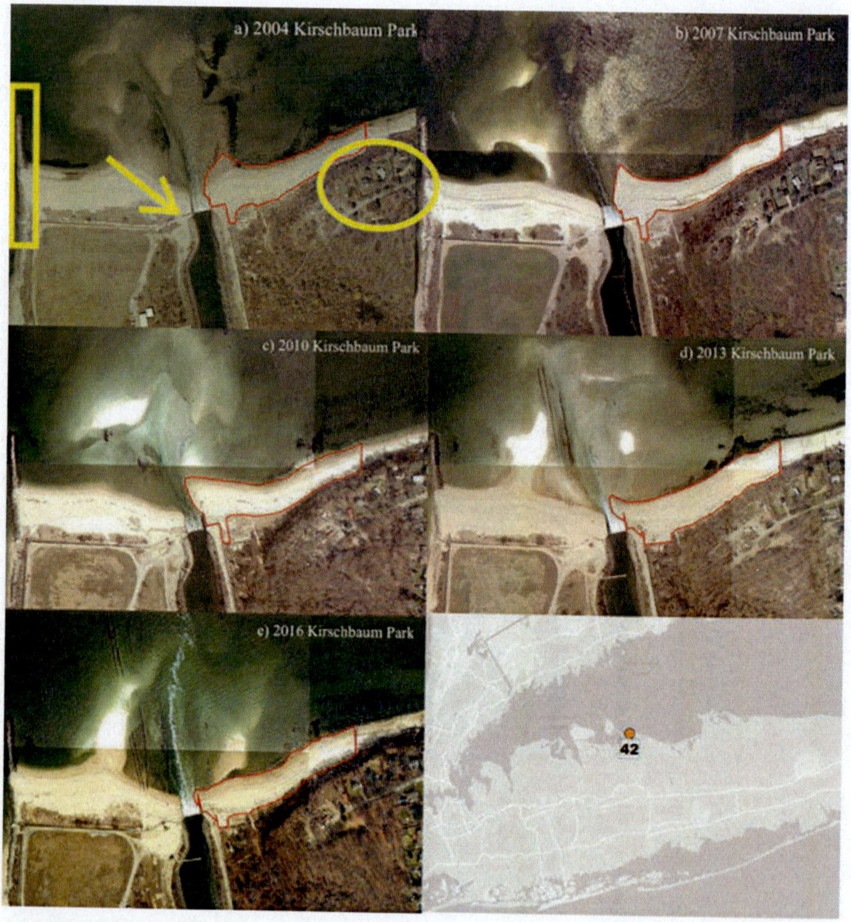


**Fig. 33** Northport. Images from Northport, site: Kirschbaum Park. Images (**a–e**) are 2004, 2007, 2010, 2013, and 2016, respectively. This site is located in a park next to a residential area, indicated by the circle. The arrow points to a manmade canal. The rectangle points out a jetty that keeps open an inlet next to this site

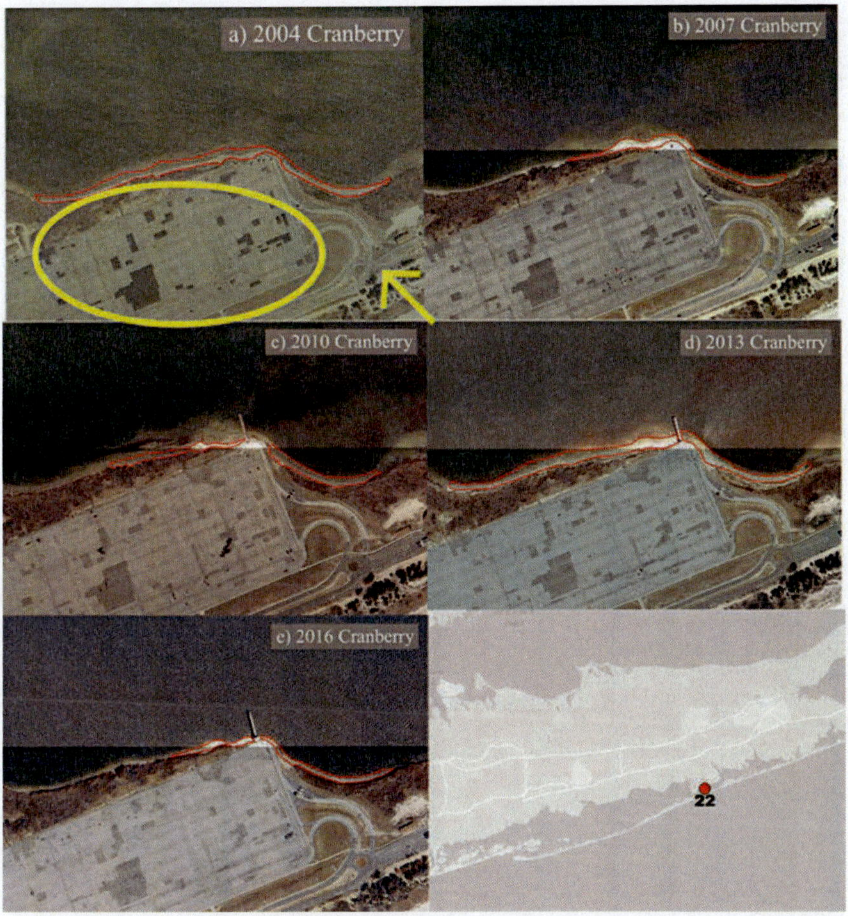


**Fig. 34** Mastic. Images from Mastic, site: Cranberry Road. Images (**a**–**e**) are 2004, 2007, 2010, 2013, and 2016, respectively. Cranberry Road site is next to a large parking lot, shown in the circle, and Cranberry Road, pointed out by the arrow

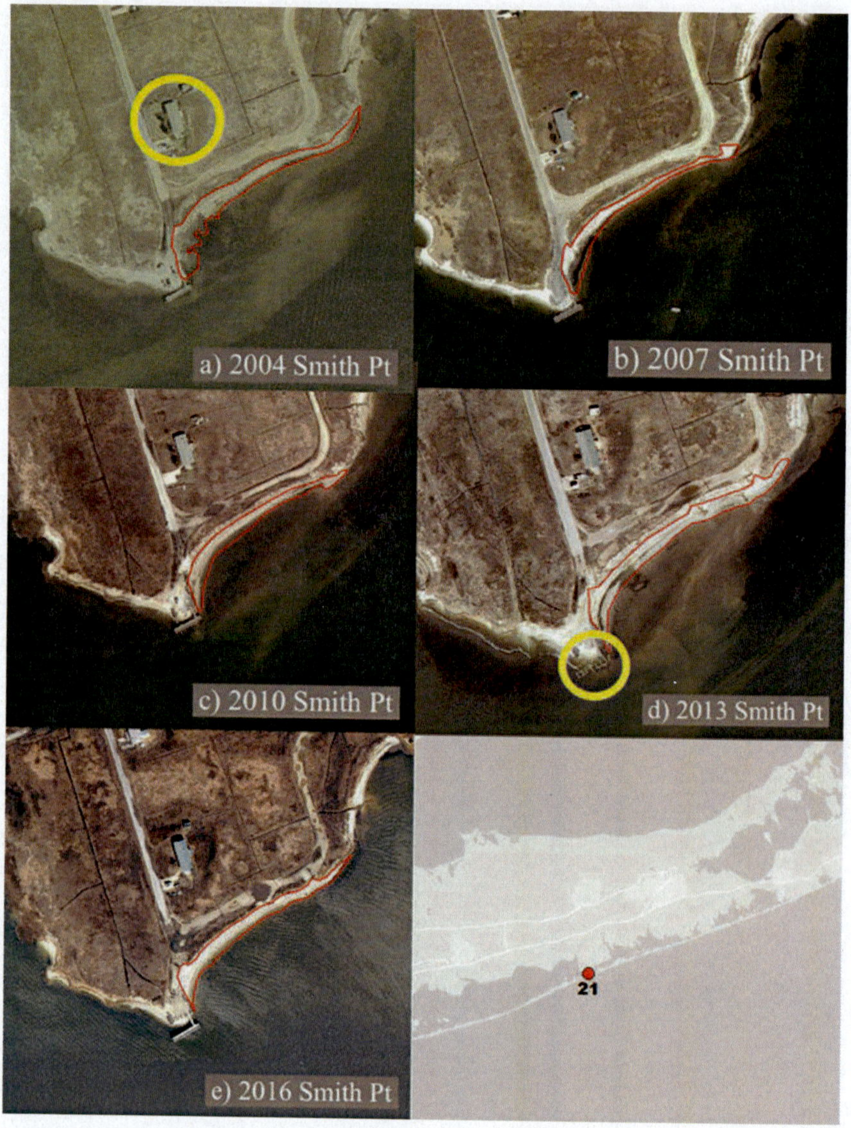


**Fig. 35** Mastic. Images from Mastic, site: Smith Point. Images (**a–e**) are 2004, 2007, 2010, 2013, and 2016, respectively. Smith Point is located in a residential area, indicated by the circle in (**e**). The circle in (**d**) notes changes to the wharf, from either Sandy or construction

**Fig. 36** Mastic. Images from Mastic, site: Sandy Point. Images (**a**–**e**) are 2004, 2007, 2010, 2013, and 2016, respectively. Sandy Point is located in a park next to a residential area indicated by the circle

| Site # | Site Name | 2003 | 2004 | 2005 | 2006 | 2007 | 2008 | 2009 | 2010 | 2011 | 2012 | 2013 | 2014 | 2015 | 2016 | 2017 | Longtitude | Latitude |
|---|---|---|---|---|---|---|---|---|---|---|---|---|---|---|---|---|---|---|
| | **NORTH SHORE** | | | | | | | | | | | | | | | | | |
| | **Manhassett Bay** | 0 | 23 | 24 | 0 | 0 | 24 | 33 | 41 | 3 | 7 | 2 | 9 | 4 | 5 | 13 | | |
| 35 | Manhassett Bay, Plum Beach Rd | 0 | 23 | 24 | 0 | 0 | 24 | 33 | 41 | 3 | 7 | 2 | 9 | 4 | 5 | 13 | 40.833167 | -73.727293 |
| | **Oyster Bay** | 2 | 0 | 0 | 0 | 50 | 17 | 21 | 2 | 383 | 32 | 5 | 0 | 4 | 28 | 33 | | |
| 36 | Oyster Bay, Baywille, Oak Point Rd | 2 | 0 | 0 | 0 | 50 | 17 | 21 | 2 | 383 | 32 | 5 | | 4 | 28 | 33 | 40.915068 | -73.568108 |
| | **Cold Spring Harbor** | 1 | 1 | 0 | 0 | 3 | 0 | 0 | 0 | 17 | 2 | 6 | 10 | 40 | 74 | 123 | | |
| 37 | Cold Spring Hbr, West Neck Rd. | 1 | 1 | 0 | 0 | 3 | 0 | 0 | 0 | 17 | 2 | 6 | 10 | 40 | 74 | 123 | 40.905261 | -73.482525 |
| | **Northport Bay** | 0 | 1762 | 4 | 4 | 0 | 47 | 7 | 0 | 125 | 0 | 12 | 15 | 20 | 58 | 16 | | |
| 38 | Caumsett St. Park | 0 | 900 | 0 | 0 | 0 | 0 | 0 | 0 | 0 | 0 | 12 | 10 | 7 | 15 | 3 | 40.941955 | -73.476927 |
| 39 | Huntington Bay, Eaton's Neck Rd. | 0 | 12 | 4 | 4 | 0 | 12 | 0 | 0 | 125 | 0 | 0 | 5 | 7 | 43 | 10 | 40.928718 | -73.398110 |
| 40 | Centerport, Little Neck Rd. | 0 | 850 | 0 | 0 | 0 | 35 | 7 | 0 | 0 | 0 | 0 | 0 | 6 | 0 | 3 | 40.904792 | -73.378824 |
| 41 | Bayview Ave | 0 | 2 | 0 | 0 | 1 | 0 | 0 | 0 | 0 | 0 | 0 | | 0 | 0 | 28 | 40.904821 | -73.355673 |
| 42 | Kirschbaum Park | 0 | 2 | 0 | 0 | 0 | 0 | 0 | 0 | 0 | 0 | 0 | | 1 | 0 | 0 | 40.927512 | -73.338782 |
| | **Stony Brook Harbor** | 0 | 48 | 22 | 22 | 0 | 0 | 0 | 0 | 0 | 0 | 0 | 1 | 0 | 95 | 39 | | |
| 43 | Smithtown Bay, Long Beach | 0 | 48 | 22 | 22 | 0 | 0 | 0 | 0 | 0 | 0 | 0 | 1 | 0 | 95 | 39 | 40.922401 | -73.169094 |
| | **PJ Harbor** | 3 | 22 | 72 | 0 | 7 | 0 | 9 | 3 | 17 | 1 | 0 | 6 | 0 | 9 | 4 | | |
| 44 | Old Field, Crane Neck Pt. | 3 | 0 | 48 | 0 | 4 | 0 | 0 | 0 | 6 | 0 | 0 | 0 | 0 | 0 | 4 | 40.959725 | -73.148393 |
| 45 | Pt. Jefferson Harbor, Chestnut Ave | 0 | 22 | 24 | 0 | 3 | 0 | 9 | 3 | 11 | 1 | 0 | 6 | 0 | 9 | 0 | 40.953040 | -73.081720 |
| | **Mt. Sinai Harbor** | 0 | 7 | 0 | 24 | 1 | 1 | 6 | 15 | 0 | 0 | 51 | 23 | 4 | 5 | 24 | | |
| 46 | Mt. Sinai Harbor, Shore Rd. | 0 | 7 | 0 | 24 | 1 | 1 | 6 | 15 | 0 | 0 | 51 | 23 | 4 | 5 | 24 | 40.963621 | -73.034535 |
| | **North Shore Sound** | 0 | 172 | 0 | 2602 | 20 | 356 | 0 | 0 | 0 | 0 | 0 | 4 | 0 | 2 | 8 | | |
| 47 | Miller Place, Seacliff Ae. | 0 | 20 | 0 | 704 | 0 | 56 | 0 | 0 | 0 | 0 | 0 | 3 | 0 | 2 | 8 | 40.966176 | -72.991552 |
| 48 | Sound Beach, Hagermen Landing | 0 | 100 | 0 | 606 | 20 | 100 | 0 | 0 | 0 | 0 | 0 | 0 | 0 | 0 | 0 | 40.966326 | -72.954490 |
| 49 | Shoreham, Beach Lane | 0 | 2 | 0 | 845 | 0 | 150 | 0 | 0 | 0 | 0 | 0 | 0 | 0 | 0 | 0 | 40.962615 | -72.911178 |
| 50 | Wading River, Herod Point Rd. | 0 | 50 | 0 | 447 | 0 | 50 | 0 | 0 | 0 | 0 | 0 | 1 | 0 | 0 | 0 | 40.970299 | -72.830218 |
| | **Jamesport** | 2 | 0 | 36 | 6 | 0 | 13 | 0 | 0 | 2 | 0 | 1 | 0 | 0 | 0 | 0 | | |
| 51 | Jamesport, South Jamesport Park | 2 | 0 | 36 | 6 | 0 | 13 | 0 | 0 | 2 | 0 | 1 | 0 | 0 | 0 | 0 | 40.940435 | -72.569864 |
| | **Greenport** | 13 | 0 | 10 | 3 | 0 | 0 | 0 | 0 | 36 | 8 | 0 | 5 | 10 | 7 | 0 | | |
| 52 | Greenport, Shore Rd. | 13 | 0 | 10 | 3 | 0 | 0 | 0 | 0 | 36 | 8 | 0 | 5 | 10 | 7 | 0 | 41.086582 | -72.383016 |
| | **Shelter Island** | 6 | 113 | 202 | 9 | 31 | 33 | 0 | 0 | 0 | 0 | 3 | 0 | 0 | 0 | 0 | | |
| 53 | Bay Point, Noyack-Long Beach Rd | 0 | 2 | 2 | 0 | 0 | 0 | 0 | 0 | 0 | 0 | 3 | 0 | 0 | 0 | 0 | 40.999747 | -72.322046 |
| 54 | Shelther Island, South Ferry Rd. | 6 | 99 | 200 | 9 | 0 | 0 | 0 | 0 | 0 | 0 | 0 | 0 | 0 | 0 | 0 | 41.044058 | -72.318467 |
| 55 | North Haven, Shelter Island Ave. | 0 | 12 | 0 | 0 | 30 | 33 | 0 | 0 | 0 | 0 | 0 | 0 | 0 | 0 | 0 | 41.027534 | -71.304978 |
| | **Cutchogue** | | | | | | | | | | | | | | | | | |
| 56 | New Suffolk | n/a | n/a | n/a | n/a | 1 | 356 | n/a | n/a | n/a | n/a | n/a | n/a | n/a | n/a | n/a | 40.991426 | -72.470719 |

**Fig. 38** South Shore horseshoe crab data. Data collected from CERCOM was organized by beach zone. This table includes site numbers and coordinates corresponding to Fig. 3

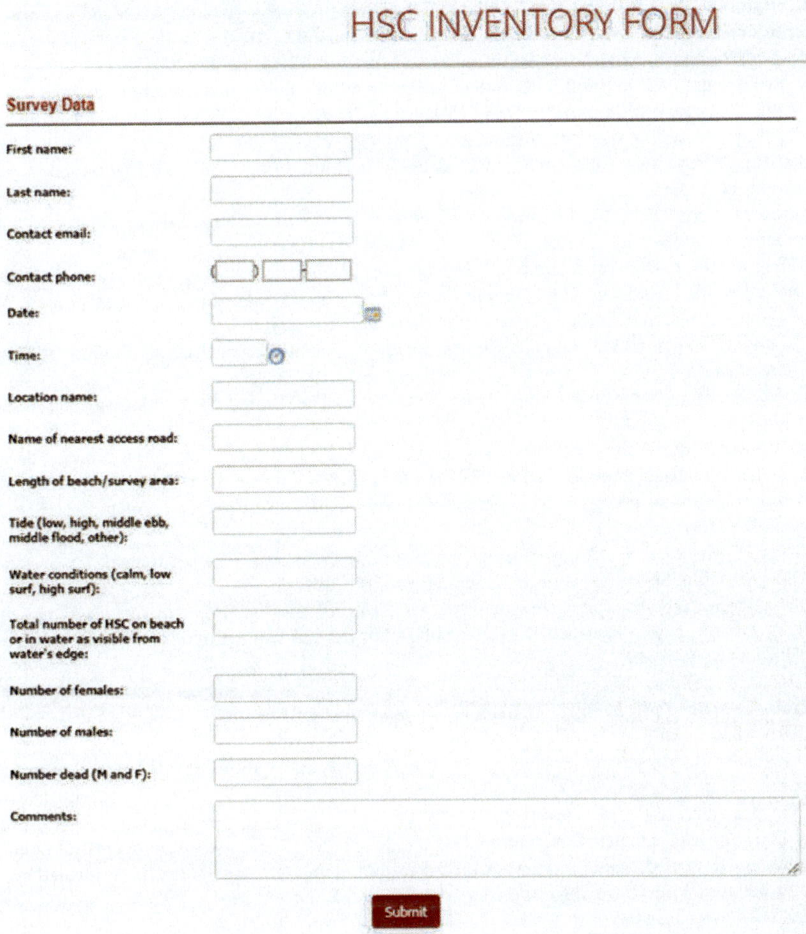

HSC INVENTORY FORM

Survey Data

First name:

Last name:

Contact email:

Contact phone:

Date:

Time:

Location name:

Name of nearest access road:

Length of beach/survey area:

Tide (low, high, middle ebb, middle flood, other):

Water conditions (calm, low surf, high surf):

Total number of HSC on beach or in water as visible from water's edge:

Number of females:

Number of males:

Number dead (M and F):

Comments:

Submit

# References


US Fish & Wildlife, The horseshoe crab: *Limulus polyphemus* a living fossil. U.S. Fish & Wildlife Service (2009)

USGS, Hurricane Sandy response – Storm impacts and vulnerability of coastal beaches. United States Geological Survey (2016)

M.L. Botton, R.E. Loveland, T.R. Jacobsen, Site selection by migratory shorebirds in Delaware Bay, and its relationship to beach characteristics and abundance of horseshoe crab (*Limulus polyphemus)* eggs. Am. Ornithol. Soc. **111**(3), 605–616 (1994)

M.L. Botton, The ecological importance of horseshoe crabs in estuarine and coastal communities: A review and speculative summary, in *Biology and Conservation of Horseshoe Crabs*, ed. by J. T. Tanacredi, M. L. Botton, D. Smith, 1st edn., (Springer, Boston, 2009), pp. 45–60

M.L. Botton, R.A. Tankersley, R.E. Loveland, Developmental ecology of the American horseshoe crab *Limulus polyphemus*. Current Zoology **56**(5), 550–562 (2010)

J. Bradley, Bait and switch. E: the Environmental Magazine **12**(4), 19–22 (2001)

J.H. Brockmann, M.D. Smith, Reproductive competition and sexual selection in horseshoe crabs, in *Biology and Conservation of Horseshoe Crabs*, ed. by J. T. Tanacredi, M. L. Botton, D. Smith, 1st edn., (Springer, Boston, 2009), pp. 199–216

Center for Environmental Research and Coastal Ocean Monitoring. 2003–2017 Horseshoe Crab Inventory Reports

C. Chen, H. Yeh, P. Lin, Conservation of the horseshoe crab at Kinmen, Taiwan: Strategies and practices. Biodiversity Conversation **13**(10), 1889–1904 (2004)

J.L. Ding, B. Ho, Endotoxin detection – From *Limulus* amebocyte lystate to recombinant factor C, in *Endotixins: Structure, Function and Recognition. Vol. 53 Subcellular Biochemistry*, ed. by X. Wang, P. Quinn, (Springer, Dordrecht, 2010), pp. 187–208

G. Ehlinger, Indian River lagoon species inventory. Smithsonian Marine Station at Fort Pierce (2001)

M.L. Fuller, The geology of Long Island, New York. The United States Geological Survey 82 (1914)

G.M. Haramis, W.A. Link, P.C. Osenton, D.B. Carter, R.G. Weber, N.A. Clark, M.A. Teece, D.S. Mizrahi, Stable isotope and pen feeding trial studies confirm the value of horseshoe crab *Limulus polyphemus* eggs to spring migrant shorebirds in Delaware Bay. J. Avian Biol. **38**(3), 367–376 (2007)

T. Itow, J.K. Mishra, A.T. Ahmed, Horseshoe crabs (king crabs) in the Bay of Bengal, South Asia. Bull. Fac. Edu. Shizuoka. Univ. Nat. Sci. **54**, 13–30 (2004)

T.L. Kraft, Geology, in *The Delaware Estuary: Rediscovering a Forgotten Resource*, ed. by T. L. Bryant, J. R. Pennock, (University of Delaware Sea Grant College Program, Newark, 1988), pp. 30–41

S. Moore, A. Perrin, Seasonal movement and resource-use patterns of resident horseshoe crab (*Limulus Polyphemus)* populations in a Maine, USA Estuary. Estuar. Coasts **30**(6), 1016–1026 (2007)

J.E. Sanders, C. Merguerian, Evidence for pre-Woodfordian ages of Long Island's terminal moraines, in *Long Island's Geologists' Abstracts Collection*, ed. by Stony Brook University, (Stony Brook, New York, 1995), pp. 91–106

W.C. Schwab, R.E. Thieler, R.A. James, D.S. Foster, A. Swift, J.F. Denny, Influence of inner-continental shelf geologic framework on the evolution and behavior of the barrier-island system between Fire Island Inlet and Shinnecock Inlet, Long Island, New York. J. Coast. Res. **16**(2), 408–422 (2000)

K. Sekiguchi, C. Shuster Jr., Limits on the global distribution of horseshoe crabs (Limulacea): Lessons learned of observations: Asia and America, in *Biology and Conservation of Horseshoe Crabs*, ed. by J. T. Tanacredi, M. L. Botton, D. Smith, 1st edn., (Springer, Boston, 2009), pp. 5–25

C.N. Shuster Jr., Two perspectives: Horseshoe crabs during 420 million years, worldwide, and the past 150 years in the Delaware Bay area, in *Limulus in the Limelight: A Species 350 Million Years in the Making and in Peril?* ed. by J. T. Tanacredi, 1st edn., (Springer, Boston, 2001), pp. 17–40

C. Shuster Jr., K. Sekiguchi, Basic habitat requirements of the extant species of horseshoe crabs (Limulacea), in *Biology and Conservation of Horseshoe Crabs*, ed. by J. T. Tanacredi, M. L. Botton, D. Smith, 1st edn., (Springer, Boston, 2009), pp. 5–25

J.T. Tanacredi, *Limulus in the Limelight: A Species 350 Million Years in the Making and in Peril?* 1st edn. (Springer, Boston, 2001), pp. 7–16

J.T. Tanacredi, Habitat Inventory Analysis of Limulus on Long Island, In: Carmichael, R., et.al. Chapt 12, Changing Perspectives on Horseshoe crab Biology, Conservation and Management, ISBN: 978-3-319-19542-1 (eBook) (2015)

J.T. Tanacredi, Long Island Horseshoe Crab network: Annual Inventory Report. Center for Environmental Research and Coastal Ocean Monitoring. (CERCOM), Molloy College (request jtanacredi@molloy.edu) (2020)

J.T. Tanacredi, et al., Conservation and Biology of the Horseshoe Crab, (Springer, 2009)

J. Tanski, Long Island's Dynamic South Shore – A Primer on the Forces and Trends Shaping our Coast. New York Sea Grant (2012)

N. Tispoura, J. Burger, Shorebird diet during spring migration stopover on Delaware Bay. Condor **101**(3), 635–644 (1999)

Wheeling, K, "Hurricane Sandy, Restoration Saves Shorebirds, living fossils they rely on" GeoSpace, ASU Blogshere Oct. 14 (2014)

# Part III
# Conservation Efforts & Managment Education

# Identifying the Major Threats to American Horseshoe Crab Populations, with Emphasis on Delaware Bay

**Mark L. Botton, Robert E. Loveland, Daphne Munroe, David Bushek, and James F. Cooper**

## 1 Introduction

The objective of this review is to examine the major threats to American horseshoe crabs, *Limulus polyphemus*, particularly in the Delaware Bay region of the United States. We begin with a brief overview of the past and present commercial exploitation of horseshoe crabs for fertilizer and bait, current management practices, and horseshoe crab population status and trends. We then discuss what we believe to be the existential threat to horseshoe crabs, which is the erosion and degradation of essential spawning habitat. The consensus of the scientific community is that overfishing and habitat loss pose the most acute threats to horseshoe crab populations, both in North America (Smith et al. 2016a, b) and Asia (Akbar John et al. 2018). This does not, however, preclude the possibility that specific local factors could stress particular populations and exacerbate the threats posed by overfishing and habitat loss.

M. L. Botton (✉)
Fordham University, Department of Natural Sciences, New York, NY, USA
e-mail: botton@fordham.edu

R. E. Loveland
Rutgers University, Department of Ecology, Evolution & Natural Resources, New Brunswick, NJ, USA

D. Munroe · D. Bushek
Haskin Shellfish Research Laboratory, Rutgers University, Port Norris, NJ, USA

J. F. Cooper
Greensboro, NC, USA



J. T. Tanacredi et al. (eds.), *International Horseshoe Crab Conservation and Research Efforts: 2007-2020*, https://doi.org/10.1007/978-3-030-82315-3_18

The public perception of these effects is driven, at least in part, by assertions made in the media, through blogs, websites, or articles published in newspapers and magazines. As one example, both biomedical bleeding and oyster aquaculture are strongly implicated in the following:

> The welk [*sic*] and eel bait industry gobbles up the lion share at a quota of more than 600,000 each year, although the actual kill has gone down over the last few years. The medical industry bleeds over 600,000, probably kills about a third, and almost certainly diminishes reproduction of the other 400,000 because they bleed only females. The aquaculture industry expanded their reach and area over the last two years taking more than a half mile of prime crab spawning habitat. (Conservewildlifenj 2017)

The same source also writes that "the oyster aquaculture industry ruins horseshoe crab breeding habitat through industrial level ATV (all-terrain vehicle) use on the intertidal flats used by crabs to feed and impeding crabs from easily reaching the shoreline to breed." In an otherwise well-sourced article, the alarming headline "Medical Labs May Be Killing Horseshoe Crabs" implicates that the biomedical bleeding of horseshoe crabs contributes to their decline (Carson 2016).

Here, we evaluate the evidence that rack-and-bag oyster culture and biomedical bleeding may be having significant negative impacts on horseshoe crab populations in comparison to habitat loss and overfishing by the bait industry.

## 1.1 Background

Horseshoe crabs are an ancient group of marine chelicerates with a deep fossil ancestry extending to at least the Late Ordovician, ca. 445 million years before present (Rudkin and Young 2009). Representatives of the genus *Limulus* have been dated to the Late Jurassic (Błażejowski 2015), but today the genus has only one extant species, the American horseshoe crab, *L. polyphemus*. These occur along the east coast of North American, ranging from the Gulf of Maine to the Yucatán Peninsula. Three other species of horseshoe crabs (*Tachypleus tridentatus*, *T. gigas*, and *Carcinoscorpius rotundicauda*) are found in Asia, ranging from southern Japan to the Bay of Bengal (Akbar John et al. 2018).

In a comprehensive review of population genetic, morphometric, and other data, Smith et al. (2016a) recognized six Atlantic coast subpopulations of American horseshoe crabs: Gulf of Maine, mid-Atlantic, southeast Atlantic, Florida Atlantic, Northeast Gulf of Mexico, and Yucatán Peninsula. Delaware Bay, in the approximate center of the mid-Atlantic region, has long been recognized as having the largest concentrations of spawning horseshoe crabs throughout its range (Shuster Jr and Botton 1985). The large size of the Delaware Bay population is believed to be a consequence of the bay's long sandy shoreline, favorable hydrographic features (e.g., moderately large tidal range, ideal temperature, and salinity), abundant areas of intertidal sand flats that serve as juvenile nursery habitats, and the richness of food resources in the bay and nearby continental shelf (Shuster Jr 2015).

## 1.2 Commercial Exploitation and Management

The high quantity of horseshoe crabs in Delaware Bay has fostered several periods of commercial fisheries and has contributed to their central ecological importance to the ecosystem. In the mid-nineteenth and early twentieth centuries, millions of horseshoe crabs were collected each year for use as fertilizer and livestock feed (Shuster Jr. 2003; Kreamer and Michels 2009). Following several decades of relatively modest fishing pressure between the 1940s and 1970s (Botton and Ropes 1987a; Kreamer and Michels 2009), the harvesting of horseshoe crabs as bait for American eel (*Anguilla rostrata*) and whelk (*Busycon carica* and *Busycotypus canaliculatus*) fisheries expanded rapidly in the 1980s and early 1990s, reaching a peak of over 6.1 million pounds (2.8 million kg, or approximately three million individuals) in 1997 (ASMFC 1998). Because egg-bearing females were preferentially targeted during the spawning season by the bait fishermen, concerns were raised about the sustainability of this fishery. The potential threat to horseshoe crabs themselves was especially heightened by the growing awareness of the importance of their eggs as food for migratory shorebirds in Delaware Bay (Myers 1986; Castro and Myers 1993; Clark et al. 1993; Botton et al. 1994; Botton and Harrington 2003; Mizrahi and Peters 2009), and the potentially catastrophic impacts that a reduced supply of eggs would have on the ability of shorebirds to gain sufficient mass during their Delaware Bay stopover. Intensive foraging on horseshoe crab eggs is a strategy used by many birds to successfully complete their migration to their Arctic breeding grounds (Baker et al. 2004; Niles et al. 2009).

The coast-wide fishery management plan (FMP) that was initiated by the Atlantic States Marine Fisheries Commission (ASMFC) in 1998 specifically addressed the size of the horseshoe crab population sufficient to meet the energetic needs of migratory shorebirds in Delaware Bay (ASMFC 1998). Among other measures, the FMP established maximum allowable horseshoe crab harvests for each member state and, through an ongoing adaptive management process, ASMFC technical committees regularly review the status and trends for both horseshoe crabs and migratory shorebirds, thus modifying the allowable harvest in accordance with these data (McGowan et al. 2011, 2015; Millard et al. 2015). A no-take area in federal waters adjacent to the mouth of Delaware Bay, known as the Carl N. Shuster Jr. Horseshoe Crab Reserve, was also established to protect horseshoe crabs that seasonally migrate between Delaware Bay and the continental shelf (Botton and Ropes 1987b). The 2019 regulations for the four states bordering the Delaware Bay region (New Jersey, Delaware, Maryland, and Virginia) stipulate that only male horseshoe crabs can be collected. States may enact even more conservative measures: for example, since 2006, New Jersey has had a moratorium on the collection of all horseshoe crabs from State waters, extending to a distance of 3 nautical miles (5.6 km) from the shoreline. Since the implementation of the FMP, harvests of horseshoe crabs have averaged about 700,000 per year, or about 1/3 of the annual harvest prior to 2000 (Fig. 1). The horseshoe crab population in Delaware Bay now appears to be stable and increasing in the most recent surveys (Smith et al. 2016a,

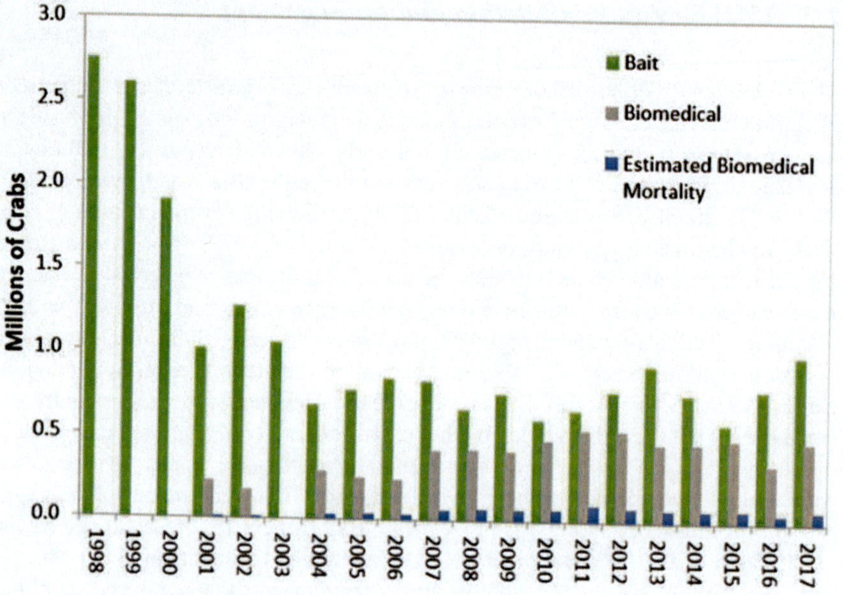


**Fig. 1** Landings of horseshoe crabs, 1988 to present. (Source: ASMFC). Biomedical mortality is estimated at 15% of the number of bled animals

b; ASMFC 2019), but there are concerns about whether the population (especially the number of females) has rebounded sufficiently to sustain migratory shorebirds (e.g., Niles et al. 2009; Smith et al. 2009; Karpanty et al. 2011). The most recent (2016) IUCN Red List classification of *Limulus polyphemus* as Vulnerable is based primarily on threats faced by the species in the extreme northern (Gulf of Maine) and southern (Yucatán) portions of its range, rather than on the mid-Atlantic and southeast Atlantic where populations appear to be more stable (Smith et al. 2016b). The *rufa* Red Knot, one of six subspecies of Red Knots globally, is acknowledged to be the shorebird that is most reliant on Delaware Bay horseshoe crab eggs and was listed as a Threatened species in 2014 (USFWS 2014). The link to declines in *rufa* Red Knots with declines in horseshoe crab eggs is confounded by increases in timing mismatches between relatively fixed Red Knot migrations and variable timing in horseshoe crab spawning driven by local weather patterns (Tucker et al. 2019).

## 2 Loss and Degradation of Spawning Habitat

Sandy estuarine beaches are the optimal spawning habitats for *Limulus polyphemus*; factors such as beach width, sediment grain size, hardness (compaction), depth of oxygen penetration, and wave energy are among the factors that contribute to the selection of particular locations for egg-laying (Botton et al. 1988, 2018; Penn and

Brockmann 1994; Smith et al. 2011). In Delaware Bay, sea level rise and beach erosion have exposed areas of intertidal peat (Botton et al. 1988), which are the remnants of several-thousand-year-old salt marshes (Knebel et al. 1988). Peat banks, or beaches with only a thin veneer of sand overlaying peat, tend to have far fewer eggs than more optimal beaches (Botton et al. 1988). The avoidance of peat is most likely a consequence of the reduced developmental success of horseshoe crab eggs in low $O_2$/high $H_2S$ environments (Vasquez et al. 2015; Funch et al. 2016). The encroaching bay has also compelled the use of various shoreline armoring practices along Delaware Bay, including the use of revetments and bulkheads, that have diminished the suitability of the habitat for horseshoe crabs (Loveland and Botton 2015). Shoreline armoring is also a major factor in the declining populations of horseshoe crabs in Asia (Akbar John et al. 2018).

As we have recently discussed the importance of sea level rise and beach erosion to horseshoe crabs in Delaware Bay (Loveland and Botton 2015), we will concentrate our discussion here on the central portion of the Cape May Peninsula (New Jersey), in the vicinity of the Rutgers Cape Shore Laboratory (Fig. 2). There are several reasons why we focus on this area. First, this beach has been the primary study site for horseshoe crabs in New Jersey, beginning with studies by Carl N. Shuster Jr. in the late 1940s (Shuster Jr 1955), followed by a series of ecological and behavioral studies by Botton and Loveland from the 1970s through the 2000s (reviewed by Botton 2009). We therefore have substantial ground truth documentation of the changes in the habitat that have occurred over these several decades of study, which we supplement with Google Earth imagery. Second, this area has historically been, and is once again, an area in Delaware Bay where intertidal culturing

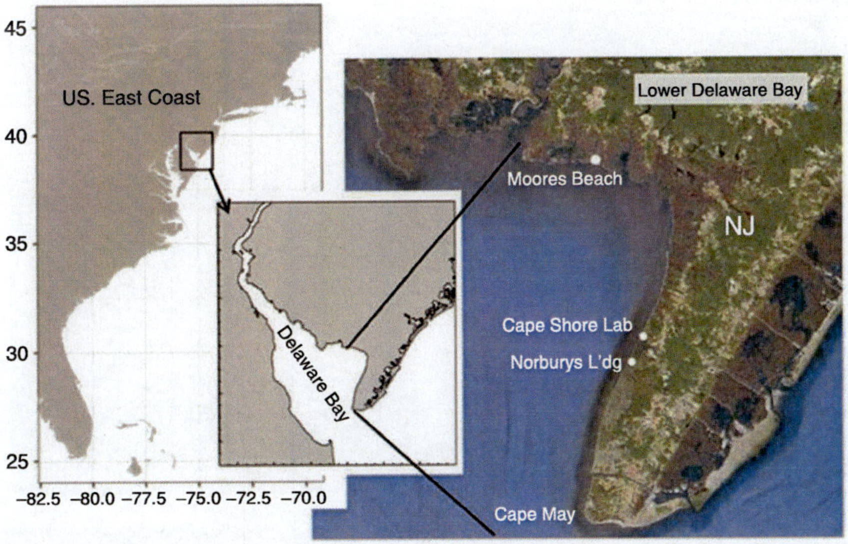


**Fig. 2** Map of lower Delaware Bay, NJ, showing locations of the beaches noted in the text

of American oysters (*Crassostrea virginica*) has taken place. The potential impacts of oyster culture on horseshoe crabs (and shorebirds) are discussed in Sect. 3. Finally, collection of horseshoe crabs for fertilizer (Shuster Jr. 2003) once took place in this region; in fact, some older residents still use the location name "King Crab Landing" (a/k/a Highs Beach) when referencing the small town just north of the Cape Shore Laboratory. In the 1980s and 1990s, horseshoe crabs were hand-collected for biomedical bleeding from this location, but the majority of the Delaware Bay area crabs are now collected offshore during the summer and fall (J. Cooper, pers. comm.). The biomedical use of horseshoe crabs, and its possible impact on the population, is discussed in Sect. 4.

## 2.1 *Habitat Loss Along the Cape May Peninsula*

Early photographs from the 1930s confirm that there was a largely uninterrupted and continuous area of sandy beach in the vicinity of the Cape Shore Laboratory (Fig. 3). Experimental culture of oysters has taken place in this area of Delaware Bay since the 1930s; this has included rack culture (Fig. 4a) and the placement of large quantities of surf clam shell on the intertidal sand flats for oyster spat collection (Fig. 4b). This region was also notable for its large horseshoe crab population that supported the local harvesting of the animals for the fertilizer industry (Shuster Jr. 2003). When Shuster Jr and Botton (1985) conducted their high tide surveys of spawning activity in the late 1970s, this region had the highest density of horseshoe crabs in all of Delaware Bay. Peak spawning counts in the range of 20,000–35,000

**Fig. 3** Photograph of the Cape Shore Laboratory beach ca. 1930s. The original Oyster Research Laboratory is at the left; in the foreground are oyster racks and a railway leading from the intertidal flats to the beach. (Photo credit: T. C. Nelson, courtesy of W. J. Canzonier)

**Fig. 4** Photographs documenting the presence of shellfish culture in the vicinity of the Cape Shore Laboratory. (**a**) Oyster conditioning racks on the intertidal flats ca. 1930s. (Photo credit: T. C. Nelson, courtesy of W. J. Canzonier). (**b**) Aggregations of surf clam (*Spisula solidissima*) shell on the intertidal flats, 1964. Shell was collected from local clam processing facilities and placed on the flats to serve as substrate for oyster spat, which were later moved to subtidal leased areas elsewhere in Delaware Bay. (Photo credit: H. Hidu, courtesy of W. J. Canzonier)

animals per km were recorded in the 1970s and 1980s (Shuster Jr and Botton 1985), and even neap tide abundance in the late 1980s often exceeded 7000 animals per km (Botton et al. 1988). Well into the 1990s, it was common to see an almost solid carpet of horseshoe crabs along the beach at high tide, with numerous satellite males jostling for position around the mated pairs. Small patches of peat began to be noticeable in the 1980s, leading to the initial observations about the avoidance of these sediments by spawning adults. Nonetheless, the vast majority of this habitat was optimal spawning beach (Shuster Jr and Botton 1985; Botton et al. 1988) because of its favorable sediment texture, width, and lack of shoreline armoring.

Google Earth images of the Cape Shore region from 1991 and 2016 (Fig. 5) show important changes in the beach and shoreline. In 1991, there was a wide band of sand along a relatively straight and uniform dune line protecting the marshes and forests behind it (Fig. 5a). In fact, the marsh was a functioning freshwater marsh with little to no saltwater intrusion (D. Bushek, personal observation). A tidal inlet at Green Creek (Fig. 5b) was constructed in 1994 to replace a former pipeline that

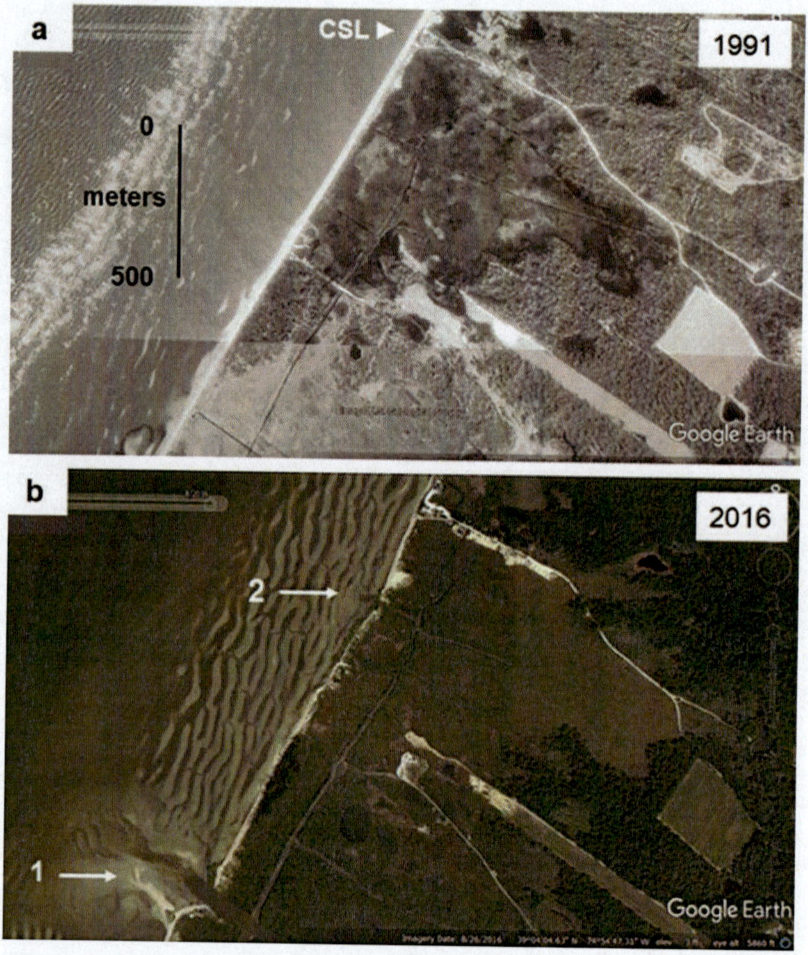


**Fig. 5** Google Earth images showing changes to the shoreline at the Cape Shore Laboratory (CSL) region between (**a**) 1991 and (**b**) 2016. The locations of the 1994 and 2010 dredged tidal creeks are shown by arrows 1 and 2, respectively. Note the overall loss of sandy beach and the increasing areas of shoreline peat, marsh, and overwashed sand. (Images are taken at an altitude of 5860 ft (1.78 km))

drained the freshwater marsh above the high tide line, thus converting it to salt marsh (Weggel 2011). The Schellenger's Creek inlet, closer to Norbury's Landing (south of the area shown on the map), was dredged in 1995 by the Cape May County Mosquito Commission and a culvert was placed closer to the Cape Shore lab to allow saltwater to convert that portion of former freshwater marsh, all as a mechanism for mosquito control. The culvert near the Cape Shore Laboratory failed and by 2010 another tidal inlet had formed (Fig. 5b). As these tidal inlets expanded, shoreline erosion accelerated as evidenced by increasing areas of overwash through the time series of images (Fig. 6). Shoreline that was formerly sandy beach has transitioned into peat banks and salt marsh, with many areas of overwashed sand (Fig. 6a–c). The beach immediately fronting the Cape Shore Laboratory has been stabilized by a gabion wall in order to protect the facilities (Fig. 6d). In brief, in less than three decades, the nearly 3 km beach from the Cape Shore Laboratory to Norbury's Landing has transformed from one of the most productive horseshoe crab habitats in the world to a locale that is marginal or unsuitable for horseshoe crab spawning.

Although the recession of the shoreline near the Cape Shore Laboratory may have been accelerated by the entrainment of sediments at the mouths of the tidal

**Fig. 6** Recent (2017) photographs of the Cape Shore Laboratory region. (**a**) A large peat outcrop to the south of the Cape Shore Laboratory (approximately the same location shown in Fig. 5). (**b**) Area of salt marsh that has developed near tidal creek 2 labeled in Fig. 5b). (**c**) One of the large sand overwash areas in this region. (**d**) A portion of the ~100 m gabion wall at the Cape Shore Laboratory beach. (Photo credit: M. Botton)

creeks (Weggel 2011), erosion is not at all atypical. To the contrary, many other Delaware Bay beaches have experienced similar alterations (Loveland and Botton 2015). Some bay shore communities (e.g., Fortescue, East Point, and Pierce's Point) have installed stone revetments, or wooden and sheet steel bulkheads to protect property, which have diminished the suitability of these beaches for horseshoe crabs (Botton et al. 1988; Jackson and Nordstrom 2009). Other communities have lost the battle against the rising sea level and have been abandoned (e.g., Moores Beach, Thompsons Beach, Bay Point). It is also the case that the effects of sea level rise along the bay were obvious long before Hurricane Sandy in late October 2012 although some narratives suggest otherwise (e.g., Palmquist 2018). We note that a closer inspection of the Google Earth images reveals a shoreline retreat of 20–50 m between 1991 and 2016 and that much of this shoreline retreat was evident in Google Earth by 2011. Under virtually all scenarios of $CO_2$ and other greenhouse gas emissions, sea level will continue to rise some 0.7–1.2 m before the year 2300 (Mengel et al. 2018). This will have the potential to inundate much of the New Jersey coast (Cooper et al. 2008) and have devastating consequences for shallow-water and shoreline ecosystems.

The ongoing erosion and degradation of essential spawning habitat thus poses a significant threat to American (Smith et al. 2016a, b) and Asian (Akbar John et al. 2018) horseshoe crabs. It is against this backdrop that we now examine two contentious issues: the effects of oyster culture and biomedical bleeding.

# 3 Horseshoe Crabs and Oyster Farms

## *3.1 Background*

The Delaware Bay has approximately 261 km of shoreline (Lathrop et al. 2013) with one third (33.0%) of that suitable for horseshoe crab spawning. Only a small portion of the suitable crab habitat (~5%) is also home to intertidal oyster farming (Munroe et al. 2017). Oyster farms currently occupy approximately 10 acres along the lower Delaware Bay of New Jersey on which they produce over 1.8 million market-sized oysters annually (Calvo 2016). During the first half of the twentieth century, large wooden intertidal racks were used to cultivate oysters over wide expanses of this region (Fig. 4). That practice stopped following the onset of MSX (multinucleated sphere unknown) disease in 1957, which killed as many as 95% of the oysters cultivated on commercial leases (Ford and Haskin 1982). In 1962, Dr. Harold Haskin began breeding oysters that survived MSX, creating the first disease-resistant lines of oysters that set the stage for the rejuvenation of oyster aquaculture in Delaware Bay and elsewhere along the mid-Atlantic coast (Haskin and Ford 1979; Ford and Haskin 1987). Shellfish aquaculture along the Cape Shore region of the Delaware Bay in New Jersey, like other forms of molluscan aquaculture, is viewed as a low-impact, sustainable food production system (Shumway et al. 2003;

Hilborn et al. 2018; van der Schatte et al. 2018). As farm production rebuilds in New Jersey and expands regionally, nationally, and globally, the industry faces challenges in assuring ecological sustainability and social license (Billing 2018). In particular, the nature of the interaction among farms and wildlife such as birds and mammals that may use habitat near to or occupied by farms is in many cases poorly understood (Price et al. 2017; Barrett et al. 2018).

Despite the historical use of the lower Delaware Bay for intertidal oyster cultivation during a period when horseshoe crabs were heavily harvested for fertilizer (Shuster Jr. 2003; Kreamer and Michels 2009), little data has been collected specifically addressing the ability of horseshoe crabs to traverse intertidal oyster farms. Modern farms use rack and bag methods, a farming activity that grows oysters in specialized cultivation bags elevated off the bottom on top of metal racks (Fig. 7). If crab migration activity is impeded or harmed by farm gear as they move past farms to reach spawning habitat, spawning may be inhibited and population-level consequences may result. Precautionary measures were recently implemented in response to a dearth of science pertaining specifically to the issue (Walsh 2016). To address

**Fig. 7** Horseshoe crabs swim among and below oyster racks as tide floods in the Delaware Bay, spawning habitat can be seen in the background of the photo. (Photo credit: D. Munroe)

this information gap, and to better understand the interactions among crabs and intertidal oyster farms in the Delaware Bay, a series of experiments were conducted in 2016 (Munroe et al. 2017) and then expanded upon in 2018 at the Rutgers University Cape Shore Laboratory and on adjacent active commercial oyster farms, co-located along the Delaware Bay in New Jersey during active horseshoe crab spawning activity from May through mid-June in each year.

The primary goal of the experiments was to assess the ability of crabs to move around and among oyster farms during low- and high-tide conditions as they migrate to and from the beach where they spawn. Experiments included surveys of the distribution of crabs on the tidal flats within and outside farms during low tide, and controlled behavioral experiments to assess the ability of crabs to move among and past farm gear both under water and in dry (low tide) conditions.

## Surveys

Low-tide crab surveys were done using paired transects on two farms in 2016 and four farms in 2018. These low-tide surveys were conducted because Delaware Bay water conditions are sufficiently turbid that crabs cannot be observed visually when water is present (high tide). The survey design assumed that crabs moving through adjacent control and farm transects were equally distributed, and those that remain during low tide are not sufficiently mobile to redistribute substantially after the tide recedes. Thus, the low-tide distribution represented the distribution (not abundance) of crabs as the tide receded, and that any differences observed between paired transects would thus be attributable to the presence of farms. At each farm site, paired, 1-meter-wide transects were oriented perpendicular to the shore, mapped, and marked with poles. All mapping and marking of transects was performed prior to the arrival of spawning horseshoe crabs. Each pair included one transect that intersected a farm and a parallel control transect passing through adjacent unfarmed intertidal habitat, ~25 m away. The locations of habitat features (e.g., sloughs and sandbars) and farm gear (e.g., racks) along each transect were noted. Here, and in subsequent surveys and experiments, controls were selected to have an equivalent bottom type, habitat features, and distance from high tide as the paired farm. The inshore edge of each farm was offset from the high-tide line by at least 91.5 m (300 feet), an area in which no farm gear is permitted due to precautionary restrictions. Complete descriptions, including schematics of transect layout, are provided in Munroe et al. (2017, 2020).

During daytime low tides, starting in early May and continuing through the end of crab spawning activity, transects on all four farms were walked and all crabs encountered along the transect were documented, and their location (inshore of, or within farm) was noted. Walks were repeated at least weekly through the duration of the spawning season. The data collected during these surveys were used to test for differences in the number of crabs observed inshore and within farm gear among paired farm and control transects.

Two of the four paired sets of transects were also surveyed during high tides over the course of 1 week between late May and early June in 2018, a period of high horseshoe crab activity. These high-tide surveys were performed using dual-frequency identification sonar (DIDSON) mounted beneath a small boat. This sonar technology is a nondestructive, nonintrusive tool that creates echograms capable of visualizing submerged habitat and organisms moving within that habitat (Moursund et al. 2003). During each survey, the boat motored slowly along each transect, following the same path that was walked during low-tide using poles located at each end of the transects as guiding markers. In addition, counts of crabs over time (15 minutes) were repeatedly made using the sonar from the boat at fixed stations within the farm and at a nearby control location with no farm gear. Only two farms were able to be surveyed by sonar in this manner because of logistical limitations due to battery and boat speed constraints.

In total, six replicate high-tide events were surveyed and the sonar videos used to count the number of single and paired horseshoe crabs along each transect. Because these sonar videos were taken from above as the boat motored over the tidal flats, oyster bags on the racks sometimes obscured the ability of the counter to see portions of the bay floor beneath the bags; therefore, videos along farm segments were corrected for area obscured. The crab counts collected during these high-tide surveys were then used to test for differences in the number of crabs along paired farm and control transects.

## Behavior Experiments

As a compliment to the surveys, agility and behavior experiments were conducted during low tide on the beach, under water in tanks in controlled conditions, and under water in natural conditions at farm and control sites.

The low-tide beach experiment was performed on hard sand habitat during a daytime low tide in late April 2016. In this experiment, described in detail in Munroe et al. (2017), the height of oyster racks were varied between 7.5 and 30.5 cm above the sand. Horseshoe crabs ranging in prosoma width from 17.5 to 23.0 cm were placed right side up approximately 1 m from the oyster rack, then observed as each walked beneath the rack. The rack height and success or failure of each crab to pass beneath or around the rack was recorded each time the experiment was repeated. The goal of this experiment was to determine at what clearance height crabs of varying sizes could pass beneath an oyster rack during low tide.

The second experiment performed in a tank filled with water aimed to test whether mature horseshoe crabs (including amplexed mating pairs) can pass beneath, around, or over oyster racks of varying heights when under water. These agility experiments were conducted in a large fiberglass tank in 2018. Twenty crabs, collected at random, were used for each trial. Before being placed into the tank, the prosomal width of each crab was measured and the sex determined. Each crab was marked with an identification number, which allowed observers to identify behaviors of individual crabs during the course of the trial. Marked animals were placed into the tank and allowed to acclimate for 15 minutes before each trial was initiated. During this acclimation period, many of the males attached to females, creating amplexed pairs and these pairings were also noted.

A total of 11 oyster gear treatments were tested in the tank. Ten farm gear treatments of varying heights and configurations were used, plus a control in which the footprint of an oyster rack was drawn on the tank bottom but no physical structure was placed in the tank. The suite of treatment types included three rack heights (7 cm, 12 cm, and 20 cm) with and without oyster bags attached, an oyster bag on the bottom (no rack), a floating oyster bag tethered to the bottom with 6.35 mm (¼ inch) braided sinking line, oyster bags leaning on the side of a rack, and a rack on its side (no bag). These various gear configurations encompassed the gear types typically used in the intertidal farms in Delaware Bay, and others currently disallowed due to the restrictions and precautionary measures to protect red knots (Walsh 2016). All of the gear used were identical to those used by farmers with the exception that all were shorter than those used on farms. All gear also had bungee cords with metal hooks attached mimicking what farmers use to hold bags onto the racks. All oyster bags used in treatments were plastic mesh with 1 cm openings, measured 8 cm × 48 cm × 90 cm (3″ × 19″ × 36″), and contained adult oyster shells to mimic live oysters in the bags (see Munroe et al. (2020) for schematic details of gear configurations).

During each agility experiment, a given treatment was placed in the tank with the 20 acclimated crabs. The crabs were observed continuously for 15 minutes as they moved about the tank and interacted with the oyster gear. A record for each individual crab was kept noting each time that an individual crab passed to the side, beneath, or over a treatment structure. Each treatment was replicated between 9 and 14 times, for a total of 139 trials. Horseshoe crabs used in the agility trials ranged in size from 16 to 28 cm prosomal width for males (mean = 20.1 cm), and from 21 to 30.5 cm for females (mean = 25.8 cm), and had sex ratios consistent with those in the spawning populations. Details of the agility trials are documented in Munroe et al. (2020).

The behavior experiments that observed crab behavior under water in natural conditions were performed using DIDSON sonar to record crabs moving among real oyster farm gear and control sites between May 13 and 23, 2019. Two sonars were used concurrently to collect paired video at two locations equidistant from the high-tide line and at the inshore edge of a commercial farm and at a comparable control location. The sonars were fixed in place, were set to image the bottom, and were tethered by cable to land where each fed live video to computers and recording devices. In total, seven high-tide events were observed and recorded, with recording starting as the tide submerged both sonars, and ending when the tide dropped below the units.

The tracks of single crabs and pairs in amplexus in these videos were analyzed to evaluate the path followed by the crabs (distance over ground), the speed of movement, and the direction. These metrics were used to compare behavior of horseshoe crabs at farm and control sites on the flats. Further details of how these videos were collected and analyzed can be found in Munroe et al. (2020).

### Survey Results

The number of horseshoe crabs observed during low tide along transects at each farm varied throughout the spawning season, among years, and among farm sites. Numbers of horseshoe crabs along the transects ranged from a low of zero at the beginning and end of the observation period, to a high of 135 per transect during the observation made on May 21, 2018; a period falling between the new and full moon in late May. In total, over all transects and across the entire 2016 observation period, 853 crabs were observed on the two farms (Munroe et al. 2017), whereas 1176 crabs were observed in 2018 on the four farms studied (Munroe et al. 2020). No difference was found among the number of crabs counted inshore of farm gear compared to controls, nor within the farm footprint compared to controls. The fact that no difference is observed among crab counts at paired farm and control survey sites at low tide, regardless of whether the observations are inshore of or within the farm footprint, indicates that crabs do not differentially use intertidal habitats in locations where farm gear is present.

High-tide surveys using DIDSON sonar occurred during the week of May 28, 2018, encompassing the full moon in late May. At the same time, high abundances of crabs were observed spawning along the entire length of beaches within the study area. In 2018, two of the paired farm transects were surveyed during high tide using DIDSON sonar. Sonar video of horseshoe crabs during high tide showed large aggregations of crabs in sloughs (muddy depressions) and sparsely distributed crabs moving independent of other crabs in seemingly random directions outside of sloughs. Crabs did not move *en masse* as a unit; rather they crossed the sand flats in all directions moving as singles or pairs with few other horseshoe crabs nearby. On many occasions, horseshoe crabs were observed to move under and out from farm gear unimpeded. On other occasions, a crab was observed to bump into the leg of a rack or another horseshoe crab, after which they would alter direction slightly and continue moving.

Similar to the results found in the low-tide transect survey, differences in numbers of crabs were observed over time (among high tides) and among farms, but no difference was observed between paired farm and control areas (Munroe et al. 2020). Within the intertidal region inshore of the farm gear, no significant difference was detected among control and farm counts of single horseshoe crabs nor amplexed pairs. Similarly, within the area of the farm gear (outer intertidal), no significant difference was detected among control and farm counts of single or amplexed horseshoe crabs.

Counts of horseshoe crabs made at fixed stations over time, standardized to crabs/minute, ranged from 0 to nearly 2 crabs per minute moving through an area of bottom approximately 10 $m^2$. No significant difference was detected among paired control and farm counts for single crabs, nor amplexus pairs. When counts were corrected for view obstruction by farm gear, no significant difference among single horseshoe crab counts at control versus farm was detected; however, significantly more amplexed pairs were observed at the farm station. In agreement with the low-tide surveys, no difference was observed among horseshoe crab counts at paired farm and control sites during high tide, indicating that crabs use these intertidal habitats consistently, regardless of the presence of farm gear.

## Behavioral Experiments: Results

The low-tide experiment on hard sand demonstrated that horseshoe crabs of all sizes tested could pass beneath racks that had a clearance above the sand of 10 cm or more (Fig. 7) (Munroe et al. 2017). Six of the 48 horseshoe crabs in the experiment did not pass beneath or around the rack, and all of those occurred when they encountered racks with only very small clearance (7.5 cm). In all six of those cases, the crabs bumped up against the rack edge, stopped, then buried slightly, and stayed in place. Four other horseshoe crabs, when bumping against the lowest rack height, changed course and continued around the rack, a behavior commonly observed in the survey experiment described above when horseshoe crabs bumped into one another underwater.

In the tank experiments, across all 11 gear configurations, male, female, and amplexed pairs were all observed moving around and under/over/through the oyster farm gear without difficulty. Interestingly, this included single female horseshoe crabs and crabs engaged in amplexus successfully passing both under and over the racks with the least clearance. In general, amplexed pairs and females tended to move along the walls of the tank when no gear was present, likely due to edge effects of the tank. When the tallest racks (greatest clearance) were used horseshoe crabs tended to pass under them, whereas when the shortest racks were used crabs tended to pass over them. This behavior is likely due to the fact that prosoma heights of spawning male and female horseshoe crabs (7.5 and 10.0 cm, respectively, Kraeuter and Fegley 1994) are greater than the shortest rack height tested, and less than the tallest rack height tested. Among all of the 128 trials performed in this experiment, no horseshoe crab was ever observed to be stuck or impeded from moving past or through the oyster farm gear (Munroe et al. 2020). This experiment demonstrates that rack and bag or floating farm gear are not obstacles that impede horseshoe crab movement whether elevated above the sediment or not.

Finally, the behavior experiments conducted in 2019 using fixed DIDSON sonars provided a novel and informative new look at horseshoe crab behavior under unmanipulated conditions (Fig. 8). In those experiments, horseshoe crabs were observed as they moved across the tidal flats at a farm site and nearby nonfarm (control) site. These concurrent observations of horseshoe crabs at a farm and control location allowed comparison of the numbers of crabs moving around, and evaluation of their behavior including speed, path straightness, and direction. No difference was observed in the number of crabs moving around at the farm compared to the control location (Munroe et al. 2020). Likewise, horseshoe crab speed (~12 cm/second) and direction (generally following tidal currents) did not change as they moved through the farm. Path straightness was slightly altered (~3% less straight) as they moved through the farms, likely due to occasionally needing to navigate around the legs of racks (Munroe et al. 2020).

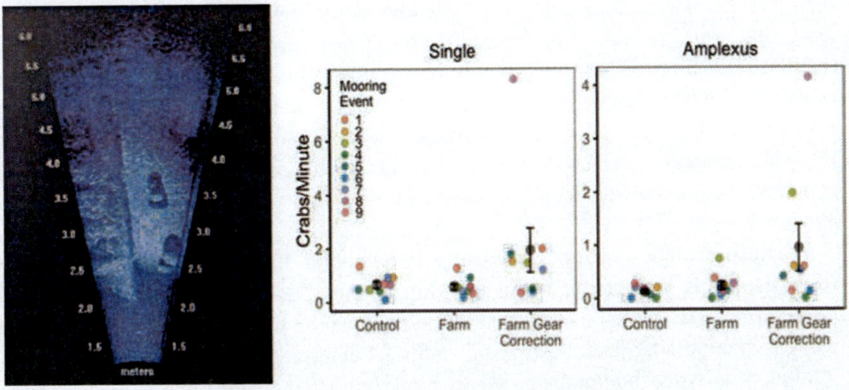


**Fig. 8** Screenshot of sonar video showing crabs in amplexus beside bags of oysters on a rack (left). Single and amplexus crab counts from all paired mooring events (right). Gray points represent the mean ± standard deviation. Colored points show average per observation event

### *3.2 Assessing the Impacts of Oyster Culture*

Across all of the surveys and experiments performed, our results indicate that horseshoe crabs can successfully traverse farms and reach spawning beaches, and that horseshoe crabs do not avoid farm gear when accessing spawning beaches. The studies discussed here show no difference in the numbers of horseshoe crabs reaching inshore spawning habitat due to farm gear, suggesting that reproductive behavior and capacity is unimpacted by oyster farms. When sonar counts of horseshoe crabs at farm sites were corrected for obstruction of the view of the bottom, there in fact appears to be more crabs moving within farms at high tide compared to control sites without gear. Attractiveness of farm structures in marine habitats, such as fish net pens and shellfish gear, to mobile fish is well documented (Callier et al. 2017). It is possible that horseshoe crabs also find oyster farm gear attractive due to increased foraging opportunities, shelter, or other cues; future research may address this possibility.

## 4 Biomedical Use of Horseshoe Crabs

### *4.1 Discovery of a Reagent for Detection of Bacterial Endotoxin*

For centuries, physicians experimented unsuccessfully with injection therapy because patients developed devastating infections and high fevers called “injection fever.” Florence Seibert (1925) proved that the fevers were caused by Gram-negative bacteria (GNB) and sought to make injectable fluids safe by eliminating GNB and their pyrogenic (fever inducing) extracts that contained endotoxin. She avoided the fever reactions caused by endotoxin (pyrogen) by making pyrogen-free water by distillation, using a rabbit fever test to verify safety, and producing sterile solutions of saline and dextrose in glass bottles by steam sterilization. Under Seibert's influence, Baxter began production of sterile IV fluids (LVPs, large volume parenterals) in 1933 at Glenview IL. LVPs were essential for managing the wounded in WWII. It is inconceivable that sterile, pyrogen-free IV solutions were first produced less than a century ago.

Horseshoe crabs and mankind have a unique bond. Physician scientists from The Johns Hopkins University, while working at the Woods Hole Marine Biological Laboratory (MBL), discovered that blood cells of the horseshoe crab had a unique way of recognizing and destroying certain bacteria. While studying the innate immunity of horseshoe crabs at MBL, Frederick Bang (1956) observed that injection of GNB or their extracts caused them to die, not by infection, but by coagulation of their hemolymph. In collaboration with hematologist Jack Levin in 1963, they observed that endotoxin caused this unexpected phenomenon by inducing the amebocytes to release an enzyme coagulation cascade that produced clotting (Levin and Bang 1964). This finding led to the creation of *Limulus* amebocyte lysate (LAL). Levin envisioned a simple test for endotoxin in septic patients, but this application never materialized.

Cooper et al. (1971) collaborated to show that LAL reagent was the optimum tool for screening injectable drugs, vaccines, and implantable devices for the presence of life-threatening endotoxin. Tens of thousands of rabbits were then used annually to test for endotoxin pyrogen as a potential contaminant in all injectable fluids. Levin's novel reagent was compared with the required rabbit pyrogen test (RPT) for endotoxin. LAL was consistently a more simple, sensitive, and specific test than the costly, variable RPT. The potential for LAL to replace the RPT was intensely studied by the parenteral drug industry and the Food and Drug Administration (FDA). The public was generally unaware of horseshoe crabs until they learned about the value of their blood to healthcare. In contrast, thousands of people today volunteer their time for horseshoe crab surveys along our coast and in Asia, and become part of citizen science events that heighten their awareness of horseshoe crabs and their important ecological relationships (Kreamer and Kreamer 2015; Nishimura and Iwaoka 2015; Liao et al. 2019; Zaukia et al. 2019).

## 4.2 FDA Elects to License and Regulate LAL as a Biological Product

The FDA became a stakeholder in LAL when Cooper collaborated with the Agency's Biologics Division to establish an LAL test capability (Cooper et al. 1972). Seligmann envisioned LAL reagent as the endotoxin test of the future and began a program in 1973 that developed regulations for the production of LAL reagents. A firm wishing to market LAL had to submit for review and approval two detailed documents, an application for a suitable facility and a submission detailing the LAL production process. FDA began licensing in the LAL industry in 1977. The regulations required a catch-and-release policy for horseshoe crabs. The FDA conducts biannual inspections of LAL firms to review compliance with drug regulations, such as Good Manufacturing Practices (GMPs), and the firms' written, FDA-approved procedures (FDA 2018). LAL was first used for drug testing in 1974 when it was required as a safety test for influenza vaccine.

Baxter Travenol, world's largest producer of injectables and medical devices, made a corporate decision to go from rabbit to LAL testing in the 1970s. A global study totaling 356,548 LAL tests and 66,594 rabbit tests proved LAL's specificity and sensitivity; this report led the FDA to approve LAL as an alternative to rabbits (Pearson and Weary 1980).

## 4.3 The LAL Industry

The first commercial LAL production was established by Cooper at Chincoteague VA in 1971. Five horseshoe crab bleeding facilities are located on the eastern US coast from Massachusetts to South Carolina that produce LAL for FDA-approved reagent. The approval by the FDA in 1987 for the use of LAL reagent as an official test for bacterial endotoxin (pyrogen) led to increased production of LAL during the 1990s to meet the growing needs of the pharmaceutical firms. Approximately 450,000 horseshoe crabs are now collected annually by US biomedical LAL firms (Fig. 1). Male and female donors are bled in about equal proportion, For example, males comprise approximately 60% of the crabs processed at two market-leading LAL firms. The worldwide market for amebocyte lysates is approximately $500 million, including TAL (*Tachypleus* amebocyte lysate). LAL and TAL firms have the crucial responsibility of providing >70 million test units annually for assuring the safety of injectable products. Horseshoe crabs bled for LAL production in the United States are returned to sea in a timely manner. Biomedical business provides livelihood for many watermen as an alternative to a bait fishery.

*Tachypleus tridentatus* used in China for TAL are most often diverted to commercial markets rather than return to sea because there is no regulatory policy for conservation (Gauvry 2015). The sharp decline of *T. tridentatus* in Asian waters led the IUCN to add them to their list of Endangered Species (Laurie et al. 2019). Eight Chinese firms produce about 15% of the amebocyte lysate global market. There will be pressure to turn to the use of LAL as the horseshoe crab population is exhausted in the South China Sea (Gauvry 2015).

The biomedical community has minimized its impact on horseshoe crab populations through 45 years of consistent conservation practices. From the outset, biomedical firms used a return-to-sea policy to minimize impact on horseshoe crabs. The FDA made this policy a condition for licensure for LAL production. In 1990, Jim Finn and Benjie Swan of Finn-Tech, a New Jersey LAL producer, introduced the Delaware Bay Spawning Survey that provides critical data on horseshoe crab population and migration (Swan 2005). The survey continues under the coordination of Swan. Initially used as an educational tool, the survey has become a management tool.

## 4.4 Fishery Management Plan (FMP) and Biomedical Uses of the Horseshoe Crab

The advent of the horseshoe crab bait industry raised the concern of LAL firms for a diminished horseshoe crab stock. At the urging of South Carolina's LAL firm, Endosafe, Inc., the State enacted legislation in 1991 for possession of horseshoe crabs, which must be collected by hand harvest, and limited their collection for biomedical and research applications. These regulations became a model for oversight of horseshoe crab use by the ASMFC. New Jersey also banned collection of horseshoe crabs for the bait industry in 2007.

The ASMFC created a FMP for limiting the horseshoe crab bait harvest (ASMFC 1998). The biomedical industry was exempted from ASMFC harvest limits because of low mortality and the critical need to assure safety of injectable medications. In anticipation of a growing LAL market, the FMP included a mortality threshold of 57,500 (not a limit); the average total estimated mortality for the past 5 years is 67,500. In response, the ASMFC sponsored a meeting of state marine resource leaders and scientists from biomedical firms to write Biomedical Best Management Practices (BMPs) for LAL firms (ASMFC 2011). All aspects of crab collection, handling, bleeding, and return-to-sea were addressed in the practices. Although LAL firms operate in diverse conditions and locations, basic operating procedures and conservation steps were identified and agreed upon. Biomedical firms use all possible conservation measures to assure the continued availability of healthy crab populations. Only healthy crabs are bled to avoid bacterial contamination of valuable LAL reagent.

## 4.5 Estimated Mortality from LAL Processes

LAL-related mortality is widely debated and estimates range from 0% to 30% (Smith et al. 2016a, b) (Table 1). The bleeding step does not result in immediate death because specimens are prescreened for health and lack of injury. However, the stress of collection and transport processes may cause mortality in horseshoe crabs that are unhealthy; death that occurs up to point of release is reported as LAL mortality and usually constitutes 2–3% of total catch.

Marine resource managers became interested in postrelease mortality to aid stock assessment of horseshoe crabs and assure the public that LAL processing was

**Table 1** Summary of estimated postbleeding mortality studies relative to LAL Biomedical Firms and Best Management Practices (BMPs)

| Author and date | Location | Number | Mortality | Relevance of methodology to biomedical practices |
|---|---|---|---|---|
| Rudloe 1983 | Florida Gulf Coast | 10,062 | 11% | 1. Release and recapture from bay. Recovery of 1415 crab with 85 dead |
| | | 80 | 2.5% | 2. Bled and unbled crab held in small pen for 30 days |
| Thompson 1998 | Charleston, SC | 40 | 15% | Bled and unbled held in open sea-water enclosure for 7 days |
| Yadon and Endosafe 1999 | Charleston | 252 | 8.3% | Bled and unbled crab held in sea pond for 2 weeks |
| Walls and Berkson 2003 | Hampton, VA | 400 | 8% | Bled and unbled crab held in replicated flow-through tanks for 2 weeks |
| Hurton and Berkson 2006 | Blacksburg, VA | 200 | 0 | 1. Bled and unbled crab held in tanks for 2 weeks; "low stress conditions" |
| | | 195 | 8.3% | 2. Bled and unbled held in tanks under "high stress conditions" |
| Leschen and Correia 2010 | Woods hole, MA | 281 | 29.8% | Crab excessively stressed and held in tanks. Methods not representative of biomedical BMP due to excessive stress. Unexplained among-tank variation |
| Anderson et al. 2013 | Durham, NH | 56 | 17.9 | Crab excessively stressed and held in various small tanks. Methods not representative of biomedical BMP due to elevated temperature and air exposure |
| Hamilton et al. 2019 | Mariculture center, Bluffton, SC | 100 | 11% | Bled and unbled crab held in seawater ponds at low densities for up to 8 weeks. Observed negative impact of heavy epibiont coverage |

not a threat to their populations. Table 1 summarizes the results of 10 estimated biomedical mortality studies and relates their methods to best management practices. Two of the ten mortality studies addressed in Table 1 reported estimated mortality rates that were conspicuously high as outliers.

Challenges to conducting simulated postrelease studies include (1) containment in marine environments that allow for prompt renourishment of donors after being bled; (2) design of simulated processing steps that are representative of LAL industry practices (BMPs); and (3) storage in a nontoxic environment that has sufficient oxygen, salinity, and other requirements.

When these and other conditions are not met, anomalously high mortality rates are observed. For example, a small study of 56 crabs reported an 18% loss (Anderson et al. 2013). The excessive stress and containment in multiple small tanks rendered the experimental conditions as being nonrepresentative of biomedical LAL practices and unaligned with BMPs. Specimens were subjected to long periods out of water and high temperatures, when kept in a barrel in mid-day sun for 4 hours. The study reported observations termed "sublethal" effects of bleeding. This simply meant that horseshoe crabs were less active for a day or two after bleeding (Smith et al. 2016a). The study stated that animals no longer spawned after bleeding, but no data to supported this conclusion. In contrast, Hamilton et al. (2019) and Swan (photo by personal communication 2018) observed spawning activity within a week of bleeding for tagged specimens (Fig. 9). The 2019 Stock Assessment presented tagging data indicating that horseshoe crabs bled, tagged, and released did not experience a reduction in long-term survival due to bleeding when compared to animals that were just caught, tagged, and released.

A study by Leschen and Correia (2010) reported the effects of two LAL treatment methods on horseshoe crabs held in saltwater tanks at the MBL. The results indicated toxicity in at least four of the holding tanks. The methods section specified that three groups of horseshoe crabs were held in six flow-through seawater tanks that contained 5 cm of sediment. Tanks differed by volume but shared a common source and flow of seawater. A similar number of animals from each treatment group and a control was assigned to each tank. Although the mortality of horseshoe crabs was similar for the two treatment groups, there was a significant difference in mortality with respect to the tanks. Mortality did not align with treatment group. Mortality by tank varied from 8.7% to 48%. The mortality rate for tanks 1 and 4 averaged 12.7%, whereas the rate for tanks 3 and 5 averaged 45%; one control died in each of the tanks. This unexplained difference in mortality indicated that there was an apparent risk factor in at least two of the tanks, such as a chemical or microbial contaminant, overcrowded conditions, or failure to maintain a required provision condition, such as oxygen, that impacted negatively on female horseshoe crabs that were stressed by bleeding. Potentially, these conditions produced anomalously high, estimated postrelease mortality.

The 2019 Horseshoe Crab Benchmark Stock Assessment and Peer Review Report (ASMFC 2019) found no discernible evidence of adverse effects upon either horseshoe crabs or migratory birds from LAL production. The most salient finding

**Fig. 9** A horseshoe crab bled and tagged June 28, 2018, and released to the south of Moores Beach, Delaware Bay, NJ. It was found spawning July 1, 2018, on Moores Beach. (Photo credit: J. Cooper)

was that the ASMFC estimated mortality is less than 1% of total estimated horseshoe crab mortality, rendering the 15% estimate a moot point. The Assessment and Peer Review Report's findings were that the natural mortality (24%), loss of habitat, bait fishing, and discards from various fisheries are the major threats to horseshoe crab sustainability. The major limitation of simulated holding studies is that the bled, stressed horseshoe crabs are not restored to their habitat for foraging and recovery, such as tidal flats. A recent study by Owings et al. (2019) found that the behavioral impacts of bleeding were short-lived, with bled crabs exhibiting similar biological rhythms and seasonal migratory behaviors to unbled crabs within 1–2 weeks after bleeding.

### 4.6 *The Future of LAL Products*

Technical advances reduce LAL needs. Charles River Labs attained FDA approval for a miniaturized LAL-cartridge-based system that reduces LAL content by 95%. Recombinant LAL products (rFC) are being evaluated for robustness, specificity, and sensitivity (e.g., Li et al. 2015; Tsuchiya 2020). The FDA has zero tolerance for endotoxin contamination and will not approve these products until they are validated as equivalent and specific as LAL for endotoxin detection, as done in the Baxter study described above. The US Pharmacopeia committed to introducing an Informational Chapter that provides a guideline for comparing recombinant LAL products with the horseshoe crab-derived LAL using reference standards and naturally occurring endotoxin in pharmaceutical water samples (Akers et al. 2020).

## 5 Summary and Conclusions

Rack-and-bag oyster culture on Delaware Bay tidal flats does not have deleterious effects on adult horseshoe crabs. Field studies (Sect. 3) indicate that the animals can successfully move in or around oyster farms and reach spawning beaches, and that crabs do not become entrained in farm gear when accessing spawning beaches. Moreover, the assertion that oyster aquaculture in the vicinity of the Cape Shore Laboratory is having harmful impacts to *Limulus* in an area of "prime horseshoe crab habitat" (Conservewildlifenj 2017) is undermined by indisputable evidence (Sect. 18.1) that this area has transitioned from the optimal sandy beach studied from the 1970s through the early 2000s to unsuitable habitat (peat banks and salt marsh) today. Some authors (e.g., Burger et al. 2015; Burger 2018) suggest that migratory shorebirds might show avoidance of oyster racks, but it should be noted that these field studies were conducted in a different portion of Delaware Bay (Reeds Beach), and deployed experimental racks that differed significantly in several ways from current aquaculture practices at the Cape Shore region. Historically, many more shorebirds have been found at Reeds Beach than at the Cape Shore

(Botton et al. 1994), and the absence of broad intertidal sand flats at Reeds Beach would, in any event, make this an unlikely area for the potential future expansion of oyster aquaculture. Interestingly, an oyster reef project at south Reeds Beach is being promoted as beneficial to both horseshoe crabs and shorebirds (e.g., Mirin 2015; Post 2015).

With regard to the effects of biomedical bleeding, simulated postbleeding mortality studies that are generally compatible with the biomedical BMPs indicate that the estimated biomedical mortality is less than 15%. The mortality caused by the biomedical industry is small in comparison to the bait fishery (Fig. 1). The most recent Horseshoe Crab Benchmark Stock Assessment and Peer Review Report (ASMFC 2019) found no credible evidence that biomedical use threatens the sustainability of the horseshoe crab or availability of their eggs for migratory birds. If approximately half of the 440,000 estimated horseshoe crabs collected for LAL production comes from the Delaware Bay stock, and half of these are female, then the number of Delaware Bay females bled would be about 110,000. Assuming 15% LAL-related mortality, then the number of females lost would be no more than 16,500 (under worst circumstances). That is about 0.2% of the 7.6 million female crabs estimated by the 2019 Assessment findings (ASMFC 2019). When one compares 0.2% with the 24% total mortality, LAL-related loss is small. Although further quantitative studies are required, discards (bycatch) of horseshoe crabs from various fisheries are likely to have a far more significant negative impact on horseshoe crab populations than biomedical bleeding (ASMFC 2019). The deleterious effects of the biomedical industry for horseshoe crab sustainability are minimized because of consistent and unique conservation efforts, such as the catch-and-release policy, support for banning the bait fishery, adherence to LAL good management practices, and coordination of the Delaware Bay horseshoe crab spawning survey.

In conclusion, we believe that the ongoing loss of high-quality spawning habitat is a greater threat to horseshoe crabs in Delaware Bay than oyster culture and biomedical mortality. We believe that future discussion needs to focus on the preservation and, if feasible, replenishment of the remaining optimal habitats. Given the inevitable landward retreat of beaches in an era of ongoing sea level rise, planners must envision where future sandy beach will likely be located – these are not necessarily the same places where beaches are found today. This will entail further studies of the sediment budget along Delaware Bay. We also believe that conversations and hard decisions should be made concerning the relative merits of shoreline armoring, beach nourishment, and buy-outs and abandonment of heavily impacted areas (Loveland and Botton 2015).

**Acknowledgments** The idea for this chapter originated from talks given by the authors at a horseshoe crab symposium at the 148th Annual Meeting of the American Fisheries Society; we thank D. R. Smith and M. J. Millard for organizing this event. D. M. and D. B. thank J. Daw and N. Cleary for their efforts collecting and analyzing data on crabs at oyster farms, along with many others that assisted from the Haskin Shellfish Research Laboratory. We are grateful for the continued cooperation of the oyster farmers of Delaware Bay who allow us access to their farms as well as the discussions about our work with the Agency Working Group and its Stakeholder Committee assigned to develop and modify relevant conservation measures. We also thank Benjie Swan and the late Carl N. Shuster for their tireless efforts in support of the horseshoe crab spawning surveys in Delaware Bay. Contributions by D. B. and D. M. were supported in part by the USDA National Institute of Food and Agriculture Hatch project accession numbers 1002345 and 1009201 through the New Jersey Agricultural Experiment Station, Hatch projects NJ32115 and NJ32114. Additional support for studies described in Sect. 3 was provided by New Jersey Sea Grant and Cape May County of New Jersey. This publication is the result of work sponsored by New Jersey Sea Grant with funds from the National Oceanic and Atmospheric Administration (NOAA) Office of Sea Grant, U.S. Department of Commerce, under NOAA grant #NA10OAR4170085 and the New Jersey Sea Grant Consortium. The statements, findings, conclusions, and recommendations are those of the author(s) and do not necessarily reflect the views of New Jersey Sea Grant or the U.S. Department of Commerce. NJSG-20-963.

## References

B. Akbar John, B.R. Nelson, H.I. Sheikh, S.G. Cheung, Y. Wardiatno, B.P. Dash, K. Tsuchiya, Y. Iwasaki, S. Pati, A review on fisheries and conservation of Asian horseshoe crabs. Biodivers. Conserv. **27**, 3573–3598 (2018). https://doi.org/10.1007/s105331-018-1633-8

J. Akers, D. Guilfoyle, D. Hussong, K. McCullough, R. Mello, D. Singer, E. Tidswell, R. Tirumalai, Functinal challenges for alternative bacterial endotoxins tests. American Pharm Review **2020**, 76–81 (2020)

R.L. Anderson, W.H. Watson, C.C. Chabot, Sublethal behavioral and physiological effects of the biomedical bleeding process on the American horseshoe crab, *Limulus polyphemus*. Biol. Bull. **225**, 137–151 (2013)

Atlantic States Marine Fisheries Commission (ASMFC), Interstate fishery management plan for horseshoe crab. Atlantic States Marine Fisheries Commission, Fishery Management Report No. 32. Washington, DC (1998)

Atlantic States Marine Fisheries Commission (ASMFC), Horseshoe crab biomedical ad-hoc working group report (2011), www.asmfc.org/uploads/file/biomedAdHocWGReport_Oct2011.pdf. Accessed May 2018

Atlantic States Marine Fisheries Commission (ASMFC), 2019 horseshoe crab benchmark stock assessment and peer review report (2019), www.asmfc.org/uploads/file/5cd5d6f1HSCAssessment_PeerReviewReport_May2019.pdf. Accessed July 2019

A.J. Baker, P.M. González, T. Piersma, L.J. Niles, I. de Lima do Nascimento, P.W. Atkinson, N.A. Clark, C.D.T. Minton, M.K. Peck, G. Aarts, Rapid population decline in red knots: Fitness consequences of decreasing refuelling rates and late arrival in Delaware Bay. Proc. Biol. Sci. **271**, 875–882 (2004)

F.B. Bang, A bacterial disease of *Limulus polyphemus*. Bull. Johns Hopkins Hosp. **98**, 325–351 (1956)

L.T. Barrett, S.E. Swearer, T. Dempster, Impacts of marine and freshwater aquaculture on wildlife: A global meta-analysis. Rev. Aquac., 1–23 (2018). https://doi.org/10.1111/raq.12277

S.L. Billing, Using public comments to gauge social licence to operate for finfish aquaculture: Lessons from Scotland. Ocean Coast. Manag. **165**, 401–415 (2018)

B. Błażejowski, The oldest species of the genus *Limulus* from the Late Jurassic of Poland, in *Changing Global Perspectives on Horseshoe Crab Biology, Conservation and Management*, ed. by R. H. Carmichael, M. L. Botton, P. K. S. Shin, S. G. Cheung, (Springer International, Switzerland, 2015), pp. 3–14

M.L. Botton, The ecological importance of horseshoe crabs in estuarine and coastal communities: A review and speculative summary, in *Biology and Conservation of Horseshoe Crabs*, ed. by J. T. Tanacredi, M. L. Botton, D. R. Smith, (Springer, New York, 2009), pp. 45–64

M.L. Botton, C.P. Colón, J. Rowden, S. Elbin, D. Kriensky, K. McKown, M. Sclafani, R. Madden, Effects of a beach nourishment project in Jamaica Bay, New York on horseshoe crab (*Limulus polyphemus*) spawning activity and egg deposition. Estuar. Coasts **41**, 974–987 (2018). https://doi.org/10.1007/s12237-017-0337-8

M.L. Botton, B.A. Harrington, Synchronies in migration: Shorebirds, horseshoe crabs, and Delaware Bay, in *The American Horseshoe Crab*, ed. by C. N. Shuster, R. B. Barlow, H. J. Brockmann, (Harvard Press, Cambridge, 2003), pp. 6–32

M.L. Botton, R.E. Loveland, T.R. Jacobsen, Beach erosion and geochemical factors: Influence on spawning success of horseshoe crabs (*Limulus polyphemus*) in Delaware Bay. Mar. Biol. **99**, 325–332 (1988)

M.L. Botton, R.E. Loveland, T.R. Jacobsen, Site selection by migratory shorebirds in Delaware Bay, and its relationship to beach characteristics and the abundance of horseshoe crab (*Limulus polyphemus*) eggs. Auk **111**, 605–611 (1994)

M.L. Botton, J.W. Ropes, The horseshoe crab, *Limulus polyphemus*, fishery and resource in the United States. Mar. Fish. Rev. **49**, 57–61 (1987a)

M. L. Botton, J. W. Ropes, Populations of horseshoe crabs, *Limulus polyphemus*, on the Northwestern Atlantic continental shelf. Fish Bull., **85**, 805–812 (1987b)

J. Burger, L.J. Niles, A.D. Dey, T. Dillingham, A.S. Gates, J. Smith, An experiment to examine how Red Knots *Calidris canutus rufa* and other shorebirds respond to oyster culture at Reeds Beach, Delaware Bay, New Jersey. Wader Stud **122**, 89–98 (2015)

J. Burger, Use of intertidal habitat by four species of shorebirds in an experimental array of oyster racks, reefs and controls on Delaware Bay, New Jersey: Avoidance of oyster racks. Sci. Total Environ. **624**, 1234–1243 (2018)

M.D. Callier, C.J. Byron, D.A. Bengtson, P.J. Cranford, S.F. Cross, U. Focken, H.M. Jansen, P. Kamermans, A. Kiessling, T. Landry, F. O'beirn, Attraction and repulsion of mobile wild organisms to finfish and shellfish aquaculture: A review. Rev Aquacult **10**(4), 924–949 (2017)

L.M. Calvo, New Jersey shellfish aquaculture situation and outlook report. 2015 Shellfish aquaculture survey results. Haskin Shellfish Research Laboratory, Port Norris, pp. 1–12 (2016), https://hsrl.rutgers.edu/outreach/aquaculture/AquacultureReports/NJAquaculture_Survey_Yr2016.pdf. Accessed 1 Dec 2018

C. Carson, Medical labs may be killing horseshoe crabs (2016), https://www.scientificamerican.com/article/medical-labs-may-be-killing-horseshoe-crabs/. Accessed 12 Jan 2019

G. Castro, J.P. Myers, Shorebird predation on eggs of horseshoe crabs during spring stopover on Delaware Bay. Auk **110**, 927–930 (1993)

K.E. Clark, L.J. Niles, J. Burger, Abundance and distribution of migrant shorebirds in Delaware Bay. Condor **95**, 694–705 (1993)

Conservewildlifenj, 20 years on Delaware Bay: Shorebird need more horseshoe crabs (2017), http://www.conservewildlifenj.org/blog/wp-content/uploads/2017/06/Del-Bay-blog-5-img-6.htm. Accessed 21 January 2019

J.F. Cooper, J. Levin, H.N. Wagner Jr., Quantitative comparison of in vitro (*Limulus*) and in vivo (rabbit) methods for detection of endotoxin. J. Lab. Clin. Med. **78**, 138–148 (1971)

J.F. Cooper, H.D. Hochstein, E.B. Seligmann Jr., The limulus test for endotoxin (pyrogen) in radiopharmaceuticals and biologicals. Bull Parenter Drug Assoc **34**, 153–162 (1972)

M.J.P. Cooper, M.D. Beevers, M. Oppenheimer, The potential impacts of sea level rise on the coastal region of New Jersey, USA. Clim. Chang. **90**, 475–492 (2008)

Food and Drug Administration (FDA), CFR Title 21 Part 210 Current good manufacturing practices (2018), https://www.accessdata.fda.gov/scripts/cdrh/cfdocs/cfcfr/CFRSearch.cfm?fr=210.1

S.E. Ford, H.H. Haskin, Infection and mortality patterns in strains of oysters *Crassostrea virginica* selected for resistance to the parasite *Haplosporidium nelsoni* (MSX). J. Parasitol. **73**, 368–376 (1987)

S.E. Ford, H.H. Haskin, History and epizootiology of *Haplosporidium nelsoni* (MSX), an oyster pathogen in Delaware Bay, 1957–1980. J. Invertebr. Pathol. **40**(1), 118–141 (1982)

P. Funch, T. Wang, C. Pertoldi, A.B. Middelbo, Low oxygen levels slow embryonic development of *Limulus polyphemus*. Biol. Bull. **231**, 113–119 (2016). https://doi.org/10.1086/690091

G. Gauvry, Current horseshoe crab harvesting practices cannot support global demand for TAL/LAL: The pharmaceutical and medical device industries' role in the sustainability of horseshoe crabs, in *Changing Global Perspectives on Horseshoe Crab Biology, Conservation and Management*, ed. by R. H. Carmichael, M. L. Botton, P. K. S. Shin, S. G. Cheung, (Springer International, Switzerland, 2015), pp. 475–500

K.L. Hamilton, L.E. Burnett, K.G. Burnett, R.E.G. Kalisperis, A.E. Fowler, Physiological impacts of time in holding ponds, biomedical bleeding, and recovery on the Atlantic horseshoe crab, *Limulus polyphemus*. Comp. Biochem. Physiol. A **239** (2019). https://doi.org/10.1016/j.cbpa.2019.110554. Accessed June 2020

H. Haskin, S.E. Ford, Development of resistance to *Minchinia nelsoni* (MSX) mortality in laboratory-reared and native oyster stocks in Delaware Bay. Mar. Fish. Rev. **41**, 54–63 (1979)

R. Hilborn, J. Banobi, S.J. Hall, T. Pucylowski, T.E. Walsworth, The environmental cost of animal source foods. Front. Ecol. Environ. **16**(6), 329–335 (2018)

L. Hurton, J. Berkson, Potential causes of mortality for horseshoe crabs (*Limulus polyphemus*) during the biomedical bleeding process. Fish. Bull. **104**, 293–298 (2006)

N.L. Jackson, K.F. Nordstrom, Strategies to conserve and enhance sandy barrier habitat for horseshoe crabs (*Limulus polyphemus*) on developed shorelines in Delaware Bay, USA, in *Biology and Conservation of Horseshoe Crabs*, ed. by J. T. Tanacredi, M. L. Botton, D. R. Smith, (Springer, New York, 2009), pp. 399–416

S.M. Karpanty, J. Cohen, J.D. Fraser, J. Berkson, Sufficiency of horseshoe crab eggs for red knots during spring migration stopover in Delaware Bay USA. J. Wildl. Manag. **75**, 984–994 (2011)

H.J. Knebel, C.H. Fletcher III, J.C. Kraft, Late Wisconsinian-Holocene paleogeography of Delaware Bay: A large coastal plain estuary. Mar. Geol. **83**, 115–133 (1988)

J.N. Kraeuter, S.R. Fegley, Vertical disturbance of sediments by horseshoe crabs (*Limulus polyphemus*) during their spawning season. Estuaries **17**, 288–294 (1994)

G. Kreamer, S. Michels, History of horseshoe crab harvest on Delaware Bay, in *Biology and Conservation of Horseshoe Crabs*, ed. by J. T. Tanacredi, M. L. Botton, D. R. Smith, (Springer, Boston, 2009), pp. 299–313

G. Kreamer, S. Kreamer, Green eggs and sand, team *Limulus*, and more: Educating for horseshoe crab conservation in the United States, in *Changing Global Perspectives on Horseshoe Crab Biology, Conservation and Management*, ed. by R. H. Carmichael, M. L. Botton, P. K. S. Shin, S. G. Cheung, (Springer International, Switzerland, 2015), pp. 557–574

R.G.J. Lathrop, L. Niles, D. Merchant, T. Farrell, C. Licitra, Mapping and assessing critical horseshoe crab spawning habitats of Delaware Bay: Update to 2007/10 and post-Sandy. Walton Center for Remote Sensing and Spatial Analysis, Rutgers University, New Brunswick (2013), Accessed online https://crssa.rutgers.edu/projects/hcrab/reports/CRSSAreport_20130916_Horseshoe_Crab_Habitat_in_DelBay_Update.pdf 17 June 2020

K. Laurie, C.-P. Chen, S.G. Cheung, V. Do, H. Hsieh, A. John, F. Mohamad, S. Seino, S. Nishida, P. Shin, M. Yang, *Tachypleus tridentatus*. The IUCN Red List of Threatened Species 2019: e.T21309A149768986 (2019), https://doi.org/10.2305/IUCN.UK.2019-1.RLTS.T21309A149768986.en. Accessed 8 Aug 2019

A.S. Leschen, S.J. Correia, Mortality in female horseshoe crabs (*Limulus polyphemus*) from biomedical bleeding and handling: Implications for fisheries management. Mar. Freshw. Behav. Physiol. **43**, 135–147 (2010)

J. Levin, F.B. Bang, The role of endotoxin in the extracellular coagulation of *Limulus* blood. Bull. Johns Hopkins Hosp. **115**(3), 265–274 (1964)

P. Li, B. Ho, J.L. Ding, Biotechnology efforts to conserve horseshoe crabs through the development of recombinant factor C-based endotoxin test, in *Changing Global Perspectives on Horseshoe Crab Biology, Conservation and Management*, ed. by R. H. Carmichael, M. L. Botton, P. K. S. Shin, S. G. Cheung, (Springer International, Switzerland, 2015), pp. 501–512

Y. Liao, H.L. Hsieh, S. Xu, Q. Zhong, J. Lei, M. Liang, H. Fang, L. Xu, W. Lin, X. Xiao, C.P. Chen, S.G. Cheung, B.K.Y. Kwan, Wisdom of crowds reveals decline of Asian horseshoe crabs in Beibu Gulf, China. Oryx **53**, 222–229 (2019)

R.E. Loveland, M.L. Botton, Sea level rise in Delaware Bay: Adaptations of spawning horseshoe crabs (*Limulus polyphemus*) to the glacial past, and the rapidly changing shoreline of the bay, in *Changing Global Perspectives on Horseshoe Crab Biology, Conservation and Managemen*, ed. by R. H. Carmichael, M. L. Botton, P. K. S. Shin, S. G. Cheung, (Springer International, Switzerland, 2015), pp. 41–63

C.P. McGowan, D.R. Smith, J.A. Sweka, J. Martin, J.D. Nichols, R. Wong, J.F. Lyons, L.J. Niles, K. Kalasz, J. Brust, Multispecies modeling for adaptive management of horseshoe crabs and red knots in the Delaware Bay. Nat. Resour. Model. **24**, 117–156 (2011)

C.P. McGowan, D.R. Smith, J.D. Nichols, J.E. Lyons, J. Sweka, K. Kalasz, L.J. Niles, R. Wong, J. Brust, M. Davis, B. Spear, Implementation of a framework for multi-species, multi-objective adaptive management in Delaware Bay. Biol. Conserv. **191**, 759–769 (2015)

M. Mengel, A. Nauels, J. Rogelj, C.-F. Schleussner, Committed sea-level rise under the Paris Agreement and the legacy of delayed mitigation action. Nat. Commun. **9**, 601 (2018). https://www.nature.com/articles/s41467-018-02985-8

M.J. Millard, J.A. Sweka, C.P. McGowan, D.R. Smith, Assessment and management of North American horseshoe crab populations, with emphasis on a multispecies framework for Delaware Bay, U.S.A. populations, in *Changing Global Perspectives on Horseshoe Crab Biology, Conservation and Management*, ed. by R. H. Carmichael, M. L. Botton, P. K. S. Shin, S. G. Cheung, (Springer International, Switzerland, 2015), pp. 407–431

B. Mirin, Red Knot rescue (2015), https://www.audubon.org/news/red-knot-rescue. Accessed 12 Jan 2019

D.S. Mizrahi, K.A. Peters, Relationships between sandpipers and horseshoe crabs in Delaware Bay: A synthesis, in *Biology and Conservation of Horseshoe Crabs*, ed. by J. T. Tanacredi, M. L. Botton, D. R. Smith, (Springer, New York, 2009), pp. 65–87

R.A. Moursund, T.J. Carlson, R.D. Peters, A fisheries application of a dual-frequency identification sonar acoustic camera. ICES J. Mar. Sci. **60**(3), 678–683 (2003)

D. Munroe, D. Bushek, P. Woodruff, L. Calvo, Intertidal rack-and-bag oyster farms have limited interaction with horseshoe crab activity in New Jersey, USA. Aquac. Environ. Interact. **9**, 205–211 (2017). https://doi.org/10.3354/aei00227

D.M. Munroe, T.M. Grothues, N.E. Cleary, J. Daw, S. Estrada, Oyster aquaculture does not impede spawning beach access for Atlantic horseshoe crabs *Limulus polyphemus*. Aquac. Environ. Interact. **12**, 81–90 (2020). https://doi.org/10.3354/aei00351

J.P. Myers, Sex and gluttony on Delaware Bay. Nat. Hist. **95**(5), 68–77 (1986)

L.J. Niles, J. Bart, H.P. Sitters, A.D. Dey, K.E. Clark, P.W. Atkinson, A.J. Baker, K.A. Bennett, K.S. Kalasz, N.A. Clark, J. Clark, S. Gillings, A.S. Gates, P.M. González, D.E. Hernandez, C.D.T. Minton, R.I.G. Morrison, R.R. Porter, R.K. Ross, C.R. Veitch, Effects of horseshoe crab harvest in Delaware Bay on Red Knots: Are harvest restrictions working? Bioscience **59**, 153–164 (2009)

C. Nishimura, C. Iwaoka, Building local community awareness of horseshoe crabs in Kujukushima, Nagasaki, Japan, in *Changing Global Perspectives on Horseshoe Crab Biology, Conservation and Management*, ed. by R. H. Carmichael, M. L. Botton, P. K. S. Shin, S. G. Cheung, (Springer International, Switzerland, 2015), pp. 575–585

M. Owings, C. Chabot, W. Watson III, Effects of the biomedical bleeding process on the behavior of the American horseshoe crab, *Limulus polyphemus*, in its natural habitat. Biol. Bull. **236**, 207–223 (2019). https://doi.org/10.1086/702917

D. Palmquist, Nature returns: Restoration brings back birds (2018), https://medium.com/usfishandwildlifeservicenortheast/nature-returns-restoration-brings-back-birds-ed715091c9b2/. Accessed 12 Jan 2019

F.C. Pearson, M. Weary, The significance of *Limulus* amebocyte lysate specificity on the pyrogen evaluation of parenteral drugs. Bull. Parenter. Drug Assoc. **34**(2), 103–108 (1980)

D. Penn, H.J. Brockmann, Nest-site selection in the horseshoe crab *Limulus polyphemus*. Biol. Bull. **187**, 373–384 (1994)

J. Post, Future of horseshoe crabs looks better (2015). Cape May County Herald, 12 Feb 2015, https://www.capemaycountyherald.com/news/environment/article_5f595461-a7e0-5537-9c15-c928f7c6ec9c.html. Accessed 12 Jan 2019

C.S. Price, E. Keane, D. Morin, C. Vaccaro, D. Bean, J.A. Morris Jr, Protected species and marine aquaculture interactions (2017). NOAA Technical Memorandum NOS NCCOS 211. Beaufort, NC. p 85, https://doi.org/10.7289/V5/TM-NOS-NCCOS-211

A. Rudloe. The effect of heavy bleeding on mortality of the horseshoe crab, Limulus polyphemus, in the natural environment. J. Invert. Pathol. **42**, 167–176 (1983)

D.M. Rudkin, G.A. Young, Horseshoe crabs – An ancient ancestry revealed, in *Biology and Conservation of Horseshoe Crabs*, ed. by J. T. Tanacredi, M. L. Botton, D. R. Smith, (Springer, New York, 2009), pp. 25–44

F.B. Seibert, The cause of many febrile reactions following intravenous injections. Am. J. Phys. **71**, 621–651 (1925)

C.N. Shuster Jr, On morphometric and serological relationships within the Limulidae, with particular reference to *Limulus polphemus* L. PhD thesis, New York University, New York. Diss Abstr 18, pp. 371–372 (1955)

C.N. Shuster Jr., King crab fertilizer: A once-thriving Delaware Bay industry, in *The American Horseshoe Crab*, ed. by C. N. Shuster Jr., R. B. Barlow, H. J. Brockmann, (Harvard University Press, Cambridge, MA, 2003), pp. 341–357

C.N. Shuster Jr., The Delaware Bay area, U.S.A.: A unique habitat of the American of horseshoe crab, *Limulus polyphemus*, in *Changing Global Perspectives on Horseshoe Crab Biology, Conservation and Management*, ed. by R. H. Carmichael, M. L. Botton, P. K. S. Shin, S. G. Cheung, (Springer International, Switzerland, 2015), pp. 15–39

C.N. Shuster Jr., M.L. Botton, A contribution to the population biology of horseshoe crabs, *Limulus polyphemus* (L.), in Delaware Bay. Estuaries **3**, 363–372 (1985)

S.E. Shumway, C. Davis, R. Downey, R. Karney, J. Kraeuter, J. Parsons, R. Rheault, G. Wikfors, Shellfish aquaculture–in praise of sustainable economies and environments. World Aquacult. **34**(4), 8–10 (2003)

D.R. Smith, E.M. Hallerman, M.J. Millard, J.A. Sweka, R.G. Weber, An incomplete analysis: Response to Niles et al. 'Are harvest restrictions working?'. Bioscience **59**, 541 (2009)

D.R. Smith, N.L. Jackson, K.F. Nordstrom, R.G. Weber, Beach characteristics mitigate effects of onshore wind on horseshoe crab spawning: Implications for matching with shorebird migration in Delaware Bay. Anim. Conserv. **14**, 575–584 (2011)

D.R. Smith, H.J. Brockmann, M.A. Beekey, T.L. King, M.J. Millard, J. Zaldivar-Rae, Conservation status of the American horseshoe crab, (*Limulus polyphemus*): A regional assessment. Rev. Fish Biol. Fish. **27**, 135–175 (2016a)

D.R. Smith, H.J. Brockmann, M.A. Beekey, T.L. King, M.J. Millard, J. Zaldivar-Rae (2016b) *Limulus polyphemus*. The IUCN Red List of Threatened Species 2016: e.T11987A80159830, https://doi.org/10.2305/IUCN.UK.2016-1.RLTS.T11987A80159830.en. Downloaded on 04 January 2019

B.L. Swan, Migrations of adult horseshoe crabs, *Limulus polyphemus*, in the middle Atlantic bight: A 17-year tagging study. Estuaries **28**, 28–40 (2005)

M. Thompson, Assessments of the population biology and critical habitat for the horseshoe crab, *Limulus polyphemus*, in the South Atlantic Bight. M.S. thesis, 136 p. Univ. Charleston, Charleston (1998)

M. Tsuchiya, Innovative mechanism of *limulus* amebocyte lysatae activation to achieve specificity and sensitivity to endotoxin; comparison with recombinant factor c reagents. Int. J. Develop. Res. **14**, 36751–36756 (2020). https://doi.org/10.37118/ijdr.19019.05.2020

A.M. Tucker, C.P. McGowan, M.J. Catalano, A. DeRose-Wilson, R.A. Robinson, J. Zimmerman, Foraging ecology mediates response to ecological mismatch during migratory stopover. Ecosphere **10**(10), e02898 (2019). https://doi.org/10.1002/ecs2.2898

US Fish and Wildlife Service (USFWS), Endangered and threatened wildlife and plants; threatened species status for the rufa Red Knot. Fed. Regist. **79**(238), 73706–73748 (2014)

O.A. van der Schatte, L. Jones, L.L. Vay, M. Christie, J. Wilson, S.K. Malham, A global review of the ecosystem services provided by bivalve aquaculture. Rev. Aquac., 1–23 (2018). https://doi.org/10.1111/raq.12301

M.C. Vasquez, A. Murillo, H.J. Brockmann, D. Julian, Multiple stressor interactions influence embryo development rate in the American horseshoe crab, *Limulus polyphemus*. J. Exp. Biol. **218**, 2355–2364 (2015)

E.A. Walls, J. Berkson, Effects of blood extraction on horseshoe crabs (*Limulus polyphemus*). Fish. Bull. **101**, 457–459 (2003)

W.L. Walsh, Biological opinion on the effects of existing and expanded structural aquaculture of native bivalves in Delaware Bay, Middle and Lower Townships, Cape May County, New Jersey on the federally listed red knot (*Calidris canutus rufa*). US Fish and Wildlife Service, New Jersey Field Office, Galloway, p. 175 (2016), Accessed online (https://www.fws.gov/northeast/njfieldoffice/pdf/AquaculturePBO_20160401.pdf) 17 June 2020

J.R. Weggel, The impact of opening an inlet in the Delaware Bay at Norbury's Landing, NJ. J Coast. Res. **59**(S1), 273–279 (2011)

D. Yadon, An evaluation of Limulus post-bleeding mortality experiment data: statistical analyses performed by South Carolina Department of Natural Resources (SCDNR) in cooperation with Endosafe, Inc. SCDNR report, 4 pp. (1999)

N.A.M. Zaukia, B. Satyanarayanaa, N. Fairuz-Fozia, B.R. Nelson, M.B. Martin, B. Akbar-John, A.J.K. Chowdhury, Citizen science frontiers horseshoe crab population regain at their spawning beach in East Peninsular Malaysia. J. Environ. Manag. **232**, 1012–1020 (2019). https://doi.org/10.1016/j.jenvman.2018.12.002

# The Need to Establish Horseshoe Crab Global Biorepository

B. Akbar John, Qian Tang, and Mike Eackles

## 1 Background

Horseshoe crabs are facing serious threats across their natural range because of overharvesting, habitat loss, and pollution. The harvest pressure on wild horseshoe crab populations in the United States and that in Asia are not the same. In the United States, harvest of *Limulus polyphemus* is for biomedical bleeding practice and eel and whelk fishery, which is well monitored by the Atlantic State Marine Fishery Commission (ASMFC), whereas in Asia, harvest pressure varies depending on the local preference/demand, species availability, and local regulative measures, and there are no standard regulative protocols to monitor. Within Asia, regional legislative measures to protect the same species vary between countries, and sometimes within the same country (John et al. 2021). For example, *Tachypleus gigas* (the coastal horseshoe crab) and *Carcinoscorpius rotundicauda* (the mangrove horseshoe crab) collected from Peninsula Malaysia can be exported nationwide and can be used for trans-border trade as these two species are considered as fishery commodity. However, such transportation is illegal in East Malaysia (Sabah and Sarawak). In some Asian countries (Bangladesh, India, Indonesia, Singapore, Vietnam, Mainland China, and Japan), horseshoe crabs are protected under law and are listed from "Vulnerable" to "Critically Endangered" under regional Red List

B. Akbar John (✉)
Institute of Oceanography and Maritime Studies (INOCEM), Kulliyyah of Science, International Islamic University Malaysia (IIUM), Kuantan, Pahang, Malaysia

Q. Tang
Department of Biological Sciences, National University of Singapore, Singapore, Singapore

M. Eackles
U.S. Geological Survey, Leetown Science Center, Aquatic Ecology Branch, Kearneysville, WV, USA

 
J. T. Tanacredi et al. (eds.), *International Horseshoe Crab Conservation and Research Efforts: 2007-2020*, https://doi.org/10.1007/978-3-030-82315-3_19

assessment (based on the species) or listed under "Species Protection Act," while in other Asian countries (Malaysia, Thailand, Cambodia, Taiwan, the Philippines, and Hong Kong), regional Red List assessment is unavailable or under preparatory phase. Therefore, to protect Asian horseshoe crab species from overharvesting is more challenging compared to the situation in the United States.

Horseshoe crab harvest for biomedical bleeding practice and local delicacy in the last two decades have tremendously increased the pressure on wild population size in some Asian countries like mainland China, Malaysia, Indonesia, and Vietnam. Female-specific catch (especially of gravid females) has triggered imbalance sex ratio in wild population. Besides, habitat destruction due to landfill projects and port constructions has triggered significant decline in the horseshoe crab population (John et al. 2018a, b; Zauki et al. 2019) to the extent that *Tachypleus tridentatus* (tri-spine horseshoe crab) is locally extinct from Haomeiliao Nature Reserve, Taiwan (only nursery ground in the west coast of Taiwan) (Hsieh and Chen 2009). Historical records revealed that up to 80–90% of horseshoe crab population has declined in mainland China from the 1950s to the 1970s (Liang 1985; WWF 2007). Rapid industrialization and pollution led to the diminishing of breeding grounds and eventually led to the local extirpation of *T. tridentatus* population in Tolo Harbour, Hong Kong (Chiu and Morton 2003). Such situation of local extinction or rare spotting of horseshoe crabs in their nesting grounds is also true in many regions in other parts of Asia.

## 2 Horseshoe Crab Biorepository

We propose to establish global horseshoe crab biorepository of tissue/bodily fluid/sperm cells/DNA/RNA/protein collected from natural populations. The repository will provide raw materials and historical references for in-depth genomic analyses in line with the technological advancement in the future. Horseshoe crabs should be collected from the nesting grounds and released back to the environment after the collection of demographic data and tissue specimens. Demographic data such as sampling date, geolocation, and species fact sheet, comprising morphometric measures (in centimeter) such as total length (TL), carapace width (CW), interocular width (IOW), tail length (TaL), weight (W in gram), epibiont attachment, and morphological abnormalities (such as shortened telson, abnormal carapace, reduced opisthosoma, etc.), should be recorded (Fig. 1). Gill tissue of 3 × 3 centimeter size excised from first or second gill flap or small pincher of claw should be preserved in absolute ethanol, while blood samples should be derived in different anticoagulant formula for further analysis. Sperm samples should be cryopreserved in liquid nitrogen before transporting it to the mid repository. Researchers at mid repository have to prepare three subset samples, isolate DNA/RNA (if needed), and measure the sample quality. Tissue subset samples ($n = 3$) should be prepared and of which one set of samples will be transported from mid repository to Lee Kong Chian Natural History Museum (LKCNHM), Singapore, for long-term storage as cryo-collection.

**Fig. 1** Some distinct morphological abnormalities such as shortened telson, abnormal carapace, and reduced opisthosoma should be documented: (**a**) shortened telson; (**b**) abnormal carapace; (**c**) reduced opisthosoma in *Tachypleus gigas* sampled from Malaysia

The tissues in the cryo-collection are available via loan request to the museum and may be subject to the loan policy. Other subset samples should be stored in absolute ethanol (tissue samples) and preserved at −80 °C (DNA/RNA samples) or using liquid nitrogen. The digital database will be an open access to public through a public domain. We propose the following framework and workflow to establish global horseshoe crab biorepository (Fig. 2).

## 3 Nagoya Protocol

In accordance with Article 15 of the Convention of Biological Diversity (CBD), the compliance to the Nagoya Protocol on access to genetic resources and equitable sharing of benefits derived from the horseshoe crab biorepository will bring up greater research collaboration. To achieve the goal of Nagoya Protocol on global multilateral benefit sharing mechanism (Article 10), prior consent and legislative permission should be derived from countries where horseshoe crabs are considered as protected species or listed under regional Red List. Consideration on approval to the transfer of genetic resource between countries could be based on two conditions: (1) the definition of "biological materials" or "species" or "derivative from the species" under each country's regional protection law and (2) mutually agreed terms on

Fig. 17.1 (continued)

the monetary and nonmonetary benefit sharing plan. Arguably, no clear definition on the "derivatives of the species" is given by regional protective laws of any country where horseshoe crabs are found. In such condition, benefits arising from the utilization of this repository are shared between countries in a fair and equitable way (Oberthür and Rosendal 2014).

## 4 Data-Sharing Plan Benefits

Horseshoe crabs are ancient species having distributed in the North American Atlantic coast (*L. polyphemus*) and throughout Southeast Asian waters (*Tachypleus gigas, T. tridentatus,* and *Carcinoscorpius rotundicauda*). Each species possesses essential and unique ecological and biomedical values. However, currently two species (*L. polyphemus* and *T. tridentatus*) are used for biomedical bleeding practice. The foundation of horseshoe crab global biorepository will provide a reliable data source to scientists or other research entities to explore how to sustainably exploit or effectively conserve the horseshoe crabs. The repository will provide raw materials and historical references for in-depth genomic analyses in line with the technological advancement in the future. It can also help in extending research collaborations among researchers around the globe. Besides, it will substantially reduce the budget spent for sampling and transportation.

Fig. 17.1 (continued)

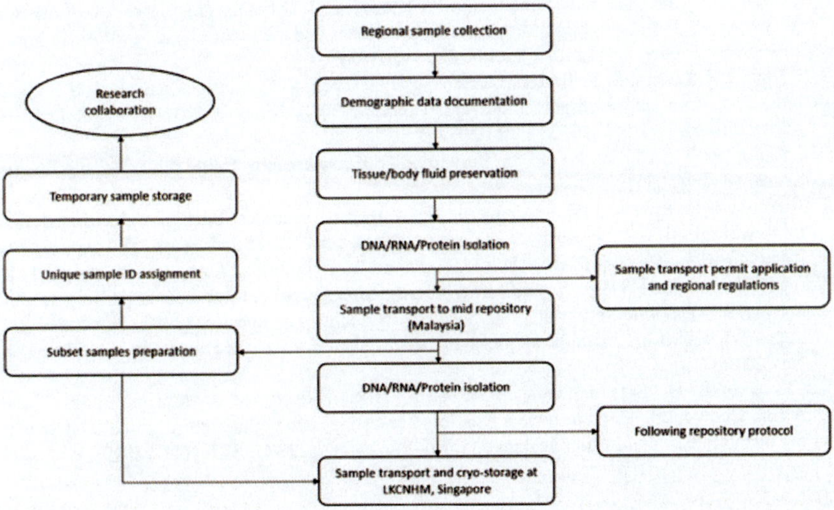


**Fig. 2** Framework on establishing horseshoe crab biorepository

## 5 Current Horseshoe Crab at Mid Repository

Presently, mid repository collection includes all four extant species of horseshoe crab sampled from Eastern United States from the state of Maine to Mexico in the Atlantic coastline and all representative counties from Southeast Asia. Over 1500 individual tissue samples in subsets are ready at the mid-repository and to be stored at LKCNHM, Singapore, as cryo-collection.

**Acknowledgments** The authors extend their sincere thanks to the contributors of this global horseshoe crab biorepository.

## References

H.M.C. Chiu, B. Morton, The status of horseshoe crabs in Hong Kong, in *Perspectives on Marine Environmental Change in Hong Kong and Southern China, Proceedings of the Hong Kong Workshops Reunion Conference on Perspectives on Marine Environmental Change in Hong Kong and Southern China (1997–2001)*, ed. by B. Morton, (Hong Kong University Press, Hong Kong, 2003), pp. 743–758

H.L. Hsieh, C.P. Chen, Conservation program for the Asian horseshoe crab, *Tachypleus tridentatus* in Taiwan: Characterizing the microhabitat of nursery grounds and restoring spawning grounds, in *Biology and Conservation of Horseshoe Crabs*, ed. by J. T. Tanacredi, M. L. Botton, D. R. Smith, (Springer, New York, 2009), p. 423

B.A. John, B.R. Nelson, H.I. Sheikh, S.G. Cheung, Y. Wardiatno, B.P. Dash, K. Tsuchiya, Y. Iwasaki, S. Pati, A review on fisheries and conservation status of Asian horseshoe crabs. Biodivers. Conserv. **29**(14), 3573–3598 (2018a)

B.A. John, B.R. Nelson, H.I. Sheikh, S.G. Cheung, Y. Wardiatno, B.P. Dash, K. Tsuchiya, Y. Iwasaki, S. Pati, Correction to: A review on fisheries and conservation status of Asian horseshoe crabs. Biodivers. Conserv. **29**(14), 3845 (2018b)

B.A. John, P.K.S. Shin, M.L. Botton, G. Gauvry, S.G. Cheung, K. Laurie. Conservation of Asian horseshoe crabs on spotlight. Biodivers. Conserv. **30**(1):253–256 (2021). Available from: https://doi.org/10.1007/s10531-020-02078-3

G.Y. Liang, Preliminary survey on the horseshoe crab resources in Beibu Gulf. Guangxi Agric. Sci. **2**, 18–20 (1985)

S. Oberthür, G.K. Rosendal, Global governance of genetic resources: Background and analytical framework, in *Global Governance of Genetic Resources: Access and Benefit Sharing after the Nagoya Protocol*, ed. by S. Oberthür, G. K. Rosendal, (Routledge, Abingdon, 2014), pp. 1–17

World Wildlife Fund (WWF), Green turtle, horseshoe crab and chinese bahaba in deep trouble reveals WWF public votes for their favourite "Ocean's 10" species. (2007, November 15). https://www.wwf.org.hk/en/news/press_release/?1333/Green-turtle-Horseshoe. Accessed Nov 2017

N.A.M. Zauki, B. Satyanarayana, F.N. Fozi, B.R. Nelson, M.B. Martin, A.J.B. John, A.J.K. Chowdhury, Citizen science frontiers horseshoe crab population regain at their spawning beach in East Peninsular Malaysia. J. Environ. Manag. **232**, 1012–1020 (2019)

# Economics of the Limulus/Tachypleus Amebocyte Lysate (LAL/TAL) Industry Relative to the Sustainability of Horseshoe Crabs Worldwide

**Thomas J. Novitsky**

## 1 Introduction

The Limulus amebocyte lysate (LAL) manufacturing industry (Novitsky 2009), including its regulation by the United States Food and Drug Administration (USFDA), has been mindful of horseshoe crab conservation from the recognition in the late 1970s that LAL could rapidly expand its application as a clinical diagnostic as well as a safety test in pharmaceutical drug quality control (Galler 1979; Rudloe 1983). However, by the mid-1990s, researchers had noted a definite decline in horseshoe crab populations in the Delaware Bay area (Shuster Jr et al. 2003). These findings seemed to correlate with a decline in migratory shorebirds, especially the red knot (Shuster Jr et al. 2003; Botton et al. 2003), that feast on horseshoe crab eggs during the birds' long northward trek to Arctic breeding grounds in the spring. It should be noted that the Delaware Bay area represents the geographic center for the Atlantic coast horseshoe crab population as well as the major stopover area for the migrating red knot. It may not be a coincidence that the birds stopped to feed at precisely the same time as horseshoe crabs were spawning (Mizahi and Peters 2009). All data considered, especially the response of the public to the plight of the red knot and its possible connection to horseshoe crabs eggs as survival food, coupled with a suspected increased harvest of horseshoe crabs for bait and biomedical use, lead to harvest regulations (Atlantic States Marine Fisheries Commission 1998).

The horseshoe crab harvest along the Atlantic coast was, and currently remains, mainly for whelk and eel bait. A smaller but significant harvest was, and currently remains, for the production of LAL. However, unlike horseshoe crabs killed for bait, horseshoe crabs used for the production of LAL (biomedical use) were returned alive to the ocean following bleeding. There is some mortality associated with

T. J. Novitsky (✉)
Trident Biosystems, Inc., Hyannis, MA, USA



J. T. Tanacredi et al. (eds.), *International Horseshoe Crab Conservation and Research Efforts: 2007-2020*, https://doi.org/10.1007/978-3-030-82315-3_20

biomedical use. This was assumed to be around 15% with a range of 5–30% (Smith et al. 2017). Due to relatively low mortality, horseshoe crabs for biomedical use were exempted from a harvest quota (Atlantic States Marine Fisheries Commission 1998). The regulations however required reporting of harvest data for both the bait and biomedical industries to the ASMFC (Smith et al. 2017). A harvest year survey in 1998 was used as the basis to set quotas and the ASMFC initiated enforcement in1999 for those states with significant harvest operations. Following the implementation of the ASMFC regulations and reporting requirements, horseshoe crab populations along the Atlantic coast, with few exceptions, have stabilized. Although specific, that is, regional harvest, numbers for biomedical use are confidential, the overall biomedical harvest has increased faster than the bait harvest over the period from 1999 to the present especially in areas where biomedical companies harvest (Novitsky 2015). While the overall harvest of horseshoe crabs for both bait and biomedical vary from year to year, over the last decade there does not seem to be a trend either upward or downward for either. Bait recorded its largest harvest in 1999 (2,605,280) (Smith et al. 2017) and then declined until 2004. It now averages about 700,000 animals per year (2004–2017), with a high in 2013 of 951,362 and a low of 612,499 in 2010 (ASMFC 2019). The biomedical harvest over the same period (including bait crabs that were bled but counted against the bait quota) was about 500,000/year, with a high in 2012 of 622,098 and a low in 2005 of 282,723. It should be noted that the number of bait crabs bled for biomedical use has steadily increased from 36,103 in 2005 to 95,231 in 2017 (ASMFC 2019).

While there does not appear to be consistent growth in the number of horseshoe crabs used to produce LAL, the global LAL market is thought to be steadily growing at a rate of about 10% per year. The discrepancy between the biomedical harvest and the LAL market may be due to differences in the amount of blood (i.e., amebocytes) needed for specific final product (LAL) formulations. It also appears that the LAL producers are becoming more efficient in the bleeding operation while LAL users are switching to the newer automated LAL assays that use far less reagent per assay.

In Asia, mainly China, the situation with the Asian horseshoe crab, *Tachypleus tridentatus*, is entirely different. There the major use of Tachypleus is for human consumption. If the horseshoe crab is used at all for bait, it is minor and localized to small fishing villages. In addition to use as human food, Tachypleus is used to make a product very similar to LAL—Tachypleus amebocyte lysate (TAL). Unlike the procedure used to make LAL, where bled animals are returned to the ocean alive, Tachypleus is returned for human consumption following processing for TAL. As TAL is a secondary market for the horseshoe crab in China, the supplier(s) of Tachypleus for the food market also control the supply of horseshoe crabs for TAL (personal observation by the author during China visits in 2012 and 2015). The system is actually quite efficient as the supplier provides the animals needed for TAL from his large supply for food (Fig. 1). Following bleeding for TAL, animals that are still quite lively are returned alive to the supplier (Figs. 2 and 3) and re-enter the live-food market supply chain. Animals that die during bleeding but are still considered "fresh" are butchered for their meat (Figs. 3 and 4). Both these activities

**Fig. 1** Crab supplier storefront Dongxing, Guangxi Province, China. Blue crates (empty stack center left) are used to hold the crabs until packed in ice for shipment in Styrofoam containers (lower right)

**Fig. 2** Bled crabs behind one of the bleeding facilities near Dongxing (2015) awaiting return to food supplier. Same blue containers used for transport both directions

could be considered comparable to the US practice of bleeding animals to produce LAL then returned for bait. A practice unique to China uses horseshoe crabs that die during bleeding but found not fit for human consumption, and those that die during harvest or storage prior to bleeding, for the production of chitin. Dead horseshoe crabs are set aside near the bleeding operation or pre-bleeding holding ponds to dry in the sun (Fig. 5) prior to their sale for chitin. Despite the fact that all harvested

**Fig. 3** Postbleeding crabs being sorted to return to food supplier in a processing area behind one of the bleeding facilities (same time and facility as shown in Fig. 2) near Dongxing. Horseshoe crabs in foreground still viable to be sold as "live." Others in background near the seated women are being butchered for their meat

**Fig. 4** Meat from one previously bled, butchered *Tachypleus tridentatus*. Note the obvious areas that correspond to where legs and telson (left end) were attached

Tachypleus in Asia end up dead, the system is quite efficient in using all of the harvest. Even the plasma (hemolymph) that is a by-product during the collection of amebocytes is sometimes saved to be reinjected into bled crabs as a clever way to increase their weight for sale in the food market (Fig. 6).

**Fig. 5** Horseshoe crabs that died during or following transport, storage, and bleeding spread out to dry in the sun on Dongbi Island, Fuqing Bay, Fuzhou Province, China (2012). This photo shows the hillside adjacent to pens where horseshoe crabs were held prior to bleeding. Dried carcasses will be sold for chitin

**Fig. 6** Typical way horseshoe crabs are displayed for sale in a Fuzhou seafood market (2012). The signs on the tank list the prices for a small animal, 48 yuan/kg (about US$7), and 318 yuan/kg (about US$47) for a large female

Thus, use for human consumption, coupled with extensive habitat destruction spanning the last 40 years, has finally gained the Asian horseshoe crab a spot on the Red List of the International Union for the Conservation of Nature (IUCN) as an endangered species (Laurie et al. 2019). It is thought the TAL market, even though much smaller than that for LAL, is expanding at least as fast if not slightly faster as

China continues to modernize its pharmaceutical industry. This puts additional pressure on the Asian horseshoe crab.

There is no published data available for the number of Tachypleus bled for TAL except for some private market studies and personal communications that could not be independently confirmed. For example, in 2012, there were a total of 60,000 horseshoe crabs processed at a bleeding facility near Fuqing in Fuzhou Province (Personal communication). The facility located there was used to bleed crabs for two of China's top three TAL manufacturers (Fig. 7). While historically there were many crabs near the coastal cities of Fuzhou and Xiamen, in 2012 all the crabs used by the two companies (Fuzhou Xinbei and Xiamen BioEndo Technology Co., respectively) located in these cities were shipped from Guangxi Provence and stored in saltwater ponds until processed for TAL. Ironically, Xiamen was the site of the first TAL manufacturer in China, the Horseshoe Crab Manufactory company (now Xiamen BioEndo Technology Co.), whose original bleeding location was ideal due to the excellent supply of Tachypleus nearby. The author visited this company in 1987. Due to dwindling supply, by 2015 both these companies had relocated their bleeding operations to Guangxi Province to be closer to the horseshoe crab supply. China's largest TAL manufacturer, Zhanjiang A & C (affiliated with Charles River Endosafe), also has a bleeding facility here.

Prior to the perceived threat to Limulus, with the extra attention it received from the migratory bird connection, the US biomedical industry was always cognizant of real and perceived threats to the source of LAL. To this end, all in the industry supported the return of horseshoe crabs to the ocean as soon as possible following bleeding and more recently have formalized best management practices (BMP) as

**Fig. 7** Bleeding facility near the Fuqing Bay where horseshoe crabs were until ready to be bled (Fig. 5). This facility was used by both Fuzhou Xinbei and Xiamen BioEndo Technology Co. to bleed crabs. Note: this building houses a bleeding laboratory on par with those of US LAL manufacturers despite the deceptive nature of the façade

part of their continuing effort to reduce mortality (ASMFC 2011). As mortality related to biomedical use depends on many factors (Berkson 2009; Smith et al. 2017; Hurton et al. 2009), the ASMFC now reports the estimated mortality assuming 4, 15, and 30% (ASMFC 2019). It is doubtful that any of the biomedical processors approach 30% and most will claim a percentage closer to 4% (personal observations).

As not all horseshoe crabs delivered for biomedical processing are bled (ASMFC 2019), it is obvious that the condition of the animal prior to bleeding is an important factor in both the quantity and quality of the blood/amebocytes obtained as well as overall mortality. No studies have been conducted looking at how collection/transport conditions relate to the quantity and quality of the final LAL product. In the current BMP, <24 hours is recommended for the time horseshoe crabs are held at the bleeding facility. No recommendation is made however for the total time, that is, from capture until return to the ocean post bleeding (ASMFC 2011). Actual times from harvest until horseshoe crabs arrive at the bleeding facility are unknown. Taking into account the time it takes to collect, either by trawl or hand, transportation time (usually, but not always, by refrigerated truck), holding by the collector (usually overnight as crabs are collected one day and delivered to the bleeding site the next morning), return of horseshoe crabs usually at the end of the bleeding day to fit with the <24 hours recommended by BMP, simple logistics suggest the entire operation could easily take up to 72 hours. Hand harvested horseshoe crabs are also often placed in floating pens ("cars") until needed. Holding times in the pens prior to unloading and transport to the bleeding facility are unknown, but during this time the animals are crowded, cannot feed, and most certainly experience low oxygen conditions. In addition, since most hand harvest occurs during spawning, captured females are further impaired as they would be prevented from normal egg laying if held in floating pens.

Notably, all the LAL manufacturers established and/or supported programs to educate the public as to the importance of conserving the horseshoe crab as an essential part of ensuring a safe drug supply. Lonza, for example, has supported a regular series of "Global Endotoxin Testing Summit" meetings where invited speakers from all stakeholder groups, including the general public, gather to freely discuss issues concerning the importance and well-being of horseshoe crabs (Lonza 2019). Many articles in the popular press, several with in-depth interviews of LAL manufacturers, have been published over the years. There are far too many of these to reference here. It should be noted, however, that nearly all of these articles misinterpret one or many of the scientific facts associated with LAL and its production.

The LAL producers (as well as independent researchers collaborating with them) led the way in studying mortality through various laboratory experiments as well as tagging and release surveys (Hurton et al. 2009; Owens et al. 2019; Leschen and Correia 2010; Hurton and Berkson 2006; Kurz and James-Pirri 2002; Walls and Berkson 2003). Most of these studies' findings generally agreed with Rudloe's (Rudloe 1983). More recent studies conducted by academic researchers not connected to the LAL industry have suggested greater mortality (Leschen and Correia

2010) and morbidity (Anderson et al. 2013; Owings et al. 2019). Mortality was also found to be greater when bled females were compared to bled males (Leschen and Correia 2010). Breeding surveys also indicate a possible adverse effect on the fecundity of bled females (Leschen et al. 2006). Even before these studies, academics and some in the LAL industry became interested in an LAL alternative, not only as a hedge against a sudden loss of horseshoe crabs as a raw material, but also to maintain market position against new competition. In 1997, researchers at the National University of Singapore succeeded in producing a recombinant factor C from a gene isolated from *Carcinoscorpius rotundicauda,* one of the three Asian species of horseshoe crabs and one easily accessible to the researchers (Ding et al. 1997; Ding and Ho 1999). This discovery is the forerunner of the current commercial rFC products (Bolden et al. 2020). Around the same time, researchers at Associates of Cape Cod, Inc. isolated an endotoxin binding protein (ENP) from *Limulus polyphemus,* which they eventually cloned in yeast (Wainwright and Novitsky 1997). On further analysis, the endotoxin binding portion of ENP was recognized as a small peptide that could then be synthesized chemically. One of the uses found for this synthetic peptide was a very specific and extremely rapid (less than 3 minutes) assay for endotoxin (Wainwright and Novitsky 2001; Sloyer and Novitsky 2000). Despite the specificity, rapid assay time, and low cost, the peptide assay was abandoned as being an impractical LAL assay alternative. This was due to low sensitivity (0.01–0.1 EU/mL with initial formulations and assay protocols), sample interference, the requirement of a special instrument, and an indication that the FDA would not consider licensing as an LAL replacement. Not surprisingly, these objections were the same ones originally attributed to rFC.

Given these reasons, it should come as no surprise that there was little interest and basically no acceptance of the commercial version of rFC by the pharmaceutical industry even for nonregulated use, for example, not for product release. Due to the conservative nature of the pharmaceutical industry, that is, reluctance to switch to something new when the existing technology was accepted and worked well, even when the USFDA decided that the industry could use rFC as an alternative if they could prove that rFC gave results similar to LAL, there was little incentive to switch. In addition, the rFC test required special instrumentation (a fluorescence spectrophotometer), was less sensitive than LAL (although now 0.005 EU/mL in water is claimed depending on the length of incubation and instrument used), took a longer time for a result compared with LAL for the same level of sensitivity, was subject to excessive sample interference, and provided no cost savings. In the last few years, beginning with the licensing of the rFC patent by a German company (Hyglos GmBH, now Hyglos-BioMérieux) to produce LAL (the original license was held by only a single company, Lonza, who marketed the product as PyroGene™ rFC Assay), and a disingenuous marketing campaign (e.g., using rFC rather than LAL was a way to "save the horseshoe crab"), did interest in rFC increase dramatically. Corporate advertising as well as popular articles (Maloney et al. 2018) provided incentive for pharmaceutical companies as well as regulatory agencies to take a closer look at rFC (Bolden et al. 2020). Recent articles in the popular press continue to provide misinformation that biomedical use is adversely impacting horseshoe crab survival. One particularly misleading report cites a potential huge increase in LAL testing for coronavirus vaccine (Fox 2020). Had the authors or publishers of these articles bothered to check, there is no regulatory requirement to test either bacterial or viral, for example, coronavirus vaccine batches with LAL (Brito and Singh 2010). While some types of vaccines contain relatively high levels of endotoxin relative to other injectable drugs, the fact that vaccines are given in very low doses (around 1 mL) and are injected into muscle tissue rather than directly into the blood stream makes them unlikely to cause any problem other than perhaps some redness/tenderness at the injection site. In any case, it is normal for all pharmaceutical manufacturers including those producing vaccines to control potential endotoxin contamination in all raw materials, especially water, with LAL. The amount of testing related to these processes, however, even for several new vaccine trials occurring simultaneously would be minimal.

Since the biomedical harvest, at least for Limulus, is currently sustainable, even with the majority of Limulus use in the United States is for bait, the misleading and emotional argument for replacing LAL with rFC should not factor in any decision to switch. More cogent arguments for "saving" the horseshoe crab would be to reduce or eliminate the bait harvest and improve the bleeding operation. Unfortunately, only token efforts have been made to produce an alternative bait (Novitsky et al. 2002; Ferrari and Targett 2003) and the current BMP's promulgated by the LAL manufacturers in concert with the ASMFC (2011) may be inadequate.

This chapter will examine in detail the scientific merits as well as practical requirements (regulatory approval, user acceptance, as well as market growth) for a synthetic LAL replacement. Given any large-scale replacement of LAL will take time, the sustainability of horseshoe crabs in both the United States and Asia will be examined with a view to steps to take to not only ensure sustainability but to increase populations. As a corollary, the development of a practical bait alternative with the goal of eliminating the bait harvest will also be discussed.

## 2 LAL Alternatives

Over the last several decades, there have been numerous attempts to produce a synthetic alternative for LAL. These include (not listed in any order of appearance or significance) (1) the In Vitro Pyrogen Test (National Toxicology Program 2020); (2) the rENP fluorescence polarization assay (Sloyer and Novitsky 2000); (3) the rFC assays, that is, EDOZYME® II rFC Test, EDOZYME® II GO, EndoLISA®, and Pyrogene™ rFC Assay (Bolden et al. 2020); (4) the EAA™ Endotoxin Activity Assay, a whole blood assay for clinical use (Spectral Diagnostics, Inc., www.specraldx.com); and (5) the Pyrosmart™ assay (Bolden et al. 2020). (Note: In this chapter, only the rFC assay will be discussed in detail.) The Pyrosmart™ assay has not yet been introduced for sale and no thorough comparative studies on it have been performed (Tamura 2016). However, this assay comprises clones of the complete suite of enzymes in the natural TAL product. As such it should be considered a "true" synthetic alternative, that is, the assay should be able to be performed in a manner very similar to, if not exactly as, LAL/TAL, that is, no special instrumentation required, little or no retraining of operators. There is a strong likelihood that this reagent will be as sensitive and exhibit sample compatibility as LAL/TAL. This later attribute is one of the key reasons for the reluctance of industry to accept rFC. Although it has been suggested that the key reason for the lack of regulatory acceptance of rFC was that it was available from only a single vendor (Bolden et. al. 2020). This seems like a poor excuse and a way to avoid addressing the far more important issues related to rFC performance. If multiple manufacturer sourcing is the prime reason for acceptance of new assays, new and potentially better assays, for example, Pyrosmart™, may never gain acceptance until patents expire or the patent owner is willing to provide multiple licenses. The multiple source argument also avoids the fact that currently there are plenty of valid (if the rFC claims of equivalency to LAL are correct) alternatives to rFC, for example, LAL and TAL, from multiple suppliers.

In order for rFC to be used as an assay, its endotoxin binding, that is, activation, must be measured. In the native LAL, the initial reaction of endotoxin with factor C results in an enzymatic cascade that ends up cleaving the soluble protein coagulogen to its insoluble form, coagulin (Novitsky and Hochstein 2003). The result of this is visual, either as a gel-clot (the original assay version), turbidity, or color, that is, the chromogenic assay (Novitsky and Hochstein 2003). The chromogenic assay employs a small peptide that is similar to the cleavage site in coagulogen but with an added para-nitroaniline group that is colorless prior to being acted upon by an activated clotting enzyme, the last enzyme in the LAL cascade, whereupon it turns yellow and can be quantitatively measured by spectrophotometry at a wavelength of 405 nm (Novitsky and Hochstein 2003). It is this chromogenic version of the LAL test that is used with Pyrosmart™. The cascade attribute of the natural LAL reaction is what imparts sensitivity (as little as 0.001 EU/mL have been reliably detected with LAL) (Novitsky and Hochstein 2003). The rFC assay, however, only contains recombinant factor C. As such there is no cascade to increase sensitivity. To compensate for the lack of the cascade, the rFC assay employs a fluorescent substrate that reacts to endotoxin-activated rFC. Generally, fluorescence provides additional sensitivity (compared to visible color) in assays where it is employed. However, this enhanced sensitivity is offset by signal quenching. Thus, while the rFC using fluorescence may approach the sensitivity of a chromogenic LAL assay, in practice, using actual samples (other than pure water), the rFC fails compared to LAL.

It is interesting to note that many claims of rFC's equivalence to LAL often use examples of nondrug samples that naturally contain endotoxin levels far in excess of those found in drugs manufactured in Good Manufacturing Practices (GMPs) qualified manufacturing facilities. Many of the examples used to show the equivalence of rFC to LAL were house dust, air, purified endotoxin preparations, and various bacteria (Bolden et. al. 2020). It has been well established and therefore should be no argument that rFC binds to endotoxin and is activated in a manner similar to factor C. The critical question however is whether or not rFC can adequately replace the LAL test for pharmaceutical quality control. House dust and endotoxin in air are not equivalent to drug products designed for injection. In addition, these samples contain such massive amounts of endotoxin that large dilutions are required to get to a measurable range. The actual sample that is measured therefore basically resembles water. In the cases where examples of actual pharmaceutical products are compared, the products selected would have been expected to work given their lack of interference at the dilutions tested. One attribute of rFC often referenced as an improvement over LAL is its specificity for endotoxin and lack of reactivity with glucan. As glucan contamination is a rare occurrence in injectable drugs, this is not a large problem for the pharmaceutical industry. Where glucan may occur as a potential contaminant in a raw ingredient, all current LAL manufacturers provide endotoxin-specific LAL assays. It is also very interesting that in the latest review of rFC to LAL (Bolden et. al. 2020), no mention whatsoever is made to the actual sensitivity and perhaps more importantly, sample interference, differences between the two assays.

It could be argued that the USFDA is the major impediment, both to improvements in the traditional LAL reagents (i.e., any changes to the currently approved LAL formulation or assay protocol requires pre-marketing approval from CBER, a process that could take months if not years until acceptance) but more importantly in the acceptance of any alternative. The very fact that the LAL reagent and its production needs to be licensed by the Center for Biologics Evaluation and Research (CBER) was a result of CBER's (formerly the Bureau of Biologics) initial involvement with the use of LAL as a replacement for the USP's Pyrogen Test (rabbit test) (Levin et al. 2003). With few exceptions, the current USFDA licenses tests for specific analytes. An excellent example is a test that actually uses LAL, that is, the Fungitell® (i.e., glucan) assay to diagnose invasive fungal infection. This reagent however was not licensed by CBER but rather by the USFDA's Center for Drug Evaluation and Research (CDER) under their 510 K criteria (FDA 2020). Fungitell® was approved with NO other LAL-based licensed product available (FDA 2020). As CDER is the FDA agency that regulates diagnostic assays, had LAL been discovered more recently, especially without the intimate involvement of the Bureau of Biologics in LAL's early research and development (Levin et al. 2003), LAL would most likely be regulated by CDER. This would be accomplished under the 510(k) approval process and be based pragmatically on evidence that the assay does what it claims. Ironically, another endotoxin assay, although not based on LAL, has been approved by CDER to detect endotoxin in blood, the EAA™ Endotoxin Activity Assay (Spectral Diagnostics, Inc., www.specraldx.com). Should other clinical LAL applications come along, even if they use the exact protocol as an LAL licensed for pharmaceutical use by CBER, they would undoubtedly be licensed by CDER. Thus, it seems incongruous that the FDA's CBER does not allow any valid test for bacterial endotoxin as long as the test could be shown to be reproducible, is able to detect pyrogenic levels of endotoxin, and shows no reaction with nonendotoxins. As a precursor to such action, the USP has had a standardized endotoxin available for many years (United States Pharmacopeia (2020). Thus, any new "endotoxin assay" that could be validated using the USP standard should be accepted. In fact, while the USFDA required a Pyrogen Test (rabbit) according to USP general chapter 151

(United States Pharmacopeia (2020), the USFDA did not license rabbit producers, but rather incorporated a requirement for licensed drug manufacturers to use, by reference, the USP Pyrogen Test (Levin et al. 2003). Other than problems related to standardization and performance, a phamacopeial monograph, especially if the various pharmacopeias are harmonized, should be the overarching control for substitution, not only for rFC but for any future "endotoxin assay" (e.g., Pyrosmart™) that can be adequately validated. The European and Japanese Pharmacopeias already have monographs for rFC (Bolden et. al. 2020). The USP plans to have a monograph sometime in 2021. Although there are still problems with harmonization, that is, insuring that the related chapters in all the major pharmacopeias agree (the various pharmacopeias already have done this for LAL in their Bacterial Endotoxin Test monographs), the USFDA, as its prescribed function, should simply require drug manufacturers to follow the appropriate USP monograph and provide supporting test results to the USFDA upon request if user complaints or other performance issues arise.

There is another more practical consideration that may play into the acceptance of rFC. This also relates to CBER's requirement that drug manufacturers use only a "licensed" LAL product. To make LAL for pharmaceutical use, LAL manufacturers need both a product as well as a facility license from CBER. Obtaining both requires time and significant capital. Seeing that the global LAL market is estimated to be only US$300–500 M, shared by four producers, the FDA requirements effectively preserve market share for these companies by barring new competition. The fact that Europe has no horseshoe crabs but was required to import LAL if they wanted to sell their drug products in the United States effectively kept them from having their own endotoxin assay. Their first attempts were with the In Vitro Pyrogen Test, but it never gained wide acceptance. In the United States, there was certainly no interest in an FDA-approved alternative for LAL, especially from the LAL manufacturers that had no alternative assay of their own, for example, Associates of Cape Cod, Inc., FujiWako, and Charles River/Endosafe. Lonza, considered to have the largest US and global LAL market share, and possessing the only alternative, rFC Pyrogene™, would benefit either way. It appears that only after Lonza had a serious European competitor for Pyrogene did they increase their marketing and regulatory approval efforts for the alternative.

One final consideration: the technical difficulty of producing any of the rFC components is daunting. While the production of LAL requires care during the collection of blood, separating amebocytes from plasma (hemolymph), washing and lysing amebocytes, and finally formulating, filling reagent vials and lyophilization, these procedures are well established. In fact, Jack Levin made perfectly good LAL at the Marine Biological Laboratory without any of the modern equipment now used. In fact, Levin worked in a non-air conditioned laboratory with open windows (no screens!) and certainly without a biological cabinet to filter out endotoxin laden airborne contaminants (Levin personal communication, Levin and Bang 1968). For rFC, production is dependent on keeping all the cell culture and purification systems as well as all raw materials perfectly clear of endotoxin. This is a much more difficult task than for LAL as it requires scrupulous attention to detail and sophisticated equipment.

## 3 Bait Alternatives

Alternatives to the horseshoe crab as bait have been considered for some time but have never gained much interest. This is surprising in light of the fact that the bait harvest has always been responsible for greater horseshoe crab mortality than when the animals are bled for LAL and released. Female horseshoe crabs were preferred as bait for the American eel, as eels were attracted to females for their eggs, a favorite food. While the eel fishery declined in the United States, the whelk fishery grew. It was found that whelks also liked horseshoe crabs, but unlike the eel, whelks liked to eat the crab. In addition, whelks did not seem to care if the bait was male or female.

Early efforts to reduce bait use employed bait bags that allowed smaller portions of the crab (as little as ¼ of a crab) to be used rather than the whole animal. This practice most likely reduced the bait demand but certainly did nothing to eliminate it (Millard et al. 2015). Using another approach, an eel and whelk attractant found in the horseshoe crab egg was suggested as part of a new artificial bait (Ferrari and Targett 2003). This idea was later incorporated in an artificial bait (Delaware Sea Grant 2013; Wakefield 2013) and commercialization was achieved by LaMonica Fine Foods in Millville, NJ (http://www.capemayfoods.com/home.html). The ASMFC decided to look more carefully at this bait alternative (Eyler et al. 2015). This product however used about 1/8 of a female horseshoe crab for each unit of bait since the chemical attractant was too complicated to duplicate synthetically. Initial field trials, however, found the bait to be a poor alternative (McKiernan 2016) for numerous reasons: (1) it was more expensive than a horseshoe crab when compared to the cost of a quarter of a horseshoe crab – the minimum needed as bait, (2) refrigeration was required prior to use, (3) product had a short shelf life, (4) product was not easily obtained from the supplier, and, perhaps most importantly for bait use, and (5) product did not last long enough in the trap to be effective.

Another attractant, employing the discarded hemolymph from LAL production, was also studied but never commercialized (Novitsky unpublished results).

Although research for an effective bait alternative is still ongoing, there seems to be little interest or support for this effort that could easily and substantially reduce horseshoe crab mortality in the United States.

## 4 Horseshoe Crab Sustainability

Horseshoe crab sustainability and perhaps even survival depends on many factors, not the least of which is habitat protection. This chapter however is focused on efforts to protect the horseshoe crab from overharvesting and mortality and morbidity related to bait and biomedical use. As mentioned in the Introduction, there are many reasons for concern that the current biomedical harvest regulations may not be sufficient. While a detailed discussion of the various ways to decrease the mortality and morbidity of horseshoe crabs for biomedical use is beyond the scope of this chapter, the following recommendations deserve careful consideration:

1. Ban harvesting during spawning[1]
2. Limit the time horseshoe crabs can be held in "cars"
3. Consider using only males for biomedical (and bait) use[1] in critical areas especially where surveys indicate the ratio of females to males to be out of balance
4. Add regulations to prohibit harvest in areas where crab populations are relatively isolated (applies especially to parts of the Atlantic northeast coast). Rotating moratoria on selected overfished areas could also be used if monitored closely
5. Improve and enforce best management practices for biomedical use
6. Increase tagging and tracking for bled crabs to get a more accurate picture of mortality associated with biomedical use
7. Lower the bait quotas by the number of bait crabs used for LAL
8. Increase support for independent research on methods to lower biomedical mortality
9. Make biomedical harvest data public to allow for independent review
10. Fund development of an artificial bait that works and is economically competitive with the horseshoe crab
11. Provide a means for horseshoe crab bait fishermen to easily transition to biomedical harvest. This should be coupled with easing restrictions on the disposition of horseshoe crab bycatch, that is, encourage use of bycatch for biomedical instead of discarding nontargeted captured animals

It should be noted, if it is not already obvious, increased use of an LAL alternative were applicable, coupled with the elimination of the bait catch would not only assure sustainability for biomedical use, but would probably result in an increase of *Limulus polyphemus* over its entire US range.

## 5 Conclusions

From currently available information, the *Limulus polyphemus* harvest for all uses in the United States is generally sustainable. The same is not true for *Tachypleus tridentatus* and possibly for the other two Asian horseshoe crab species. There is also no convincing data to show that biomedical use is adversely affecting Limulus populations especially when compared to current bait use. This essentially negates

[1] Some states already ban harvest during spawning and specify use of males only for bait.

the claim that an LAL alternative will have a significant positive impact on horseshoe crab mortality. While the goal of finding an LAL alternative is worthwhile, any substitute should stand on scientific merit alone. Currently, the evidence seems clear that rFC is an adequate replacement for LAL for many applications and the reasons given for the lack of acceptance by the USFDA and USP seem contrived.

A larger concern for the widespread acceptance of rFC and its wholesale replacement of LAL is that it will take years for change to occur. During that time, the current LAL manufacturers need to improve their processes to lower mortality. In addition, a major effort to reduce or eliminate the bait harvest should be initiated and supported by all stakeholders.

Special note should be taken of the dire situation in Asia. It is highly unlikely that TAL use in China can easily be replaced by rFC due to the cost of required equipment. Currently, much of China still uses the very simple gel clot version of the LAL assay. Although it is plausible that LAL production in the United States could be increased to provide enough reagent to cover the Asian market, once again, the cost of reagent (and possibly politics) may present a significant impediment to its acceptance.

## References


R.I. Anderson, W.H. Watson III, C.C. Chabot, Sublethal behavioral and physiological effects of the biomedical bleeding process on the American horseshoe crab. Biol. Bull. **225**,137–151 (2013)

Atlantic States Marine Fisheries Commission, Horseshoe crab biomedical ad hoc working group report, Atlantic States Marine Fisheries Commission, Arlington (2011)

Atlantic States Marine Fisheries Commission, Horseshoe crab benchmark stock assessment and peer review report, Atlantic States Marine Fisheries Commission, Arlington (2019)

Atlantic States Marine Fisheries Commission (ASMFC), Interstate fishery management plan for horseshoe crab. Atlantic States Marine Fisheries Commission, Fishery Management Report No. 32, Arlington (1998)

J. Berkson, An integrative approach to horseshoe crab multiple use and sustainability, in *Biology and Conservation of Horseshoe Crabs*, ed. by J. T. Tanacredi, M. L. Botton, D. R. Smith, (Springer, New York, 2009), pp. 387–398

J. Bolden, C. Knutsen, J. Levin, C. Milne, T. Morris, N. Mozier, I. Spreitzer, F. von Wintzingerode. Currently available recombinant alternatives to horseshoe crab blood lysates: Are they comparable for the detection of environmental endotoxins? A Review. PDA J. Pharm. Sci. Technol. **74(5)**, 602–611 (2020)

M.L. Botton, B.A. Harrington, N. Tsipora, D. Mizrahi, Synchronies in migration: Shorebirds, horseshoe crabs, and Delaware Bay, in *Changing Global Perspectives on Horseshoe Crab Biology, Conservation and Management*, ed. by R. Carmichael, M. L. Botton, P. K. S. Shin, S. G. Cheung, (Springer, New York, 2003), pp. 5–32

L.A. Brito, M. Singh, Acceptible levels in vaccine formulations during preclinical research. J. Pharma. Sci. **100**, 34–37 (2010)

Delaware Sea Grant, New artificial bait could reduce stress on horseshoe crabs. Delaware Bay Sea Grant (2013), http://www.deseagrant.org/news/new-artificial-bait-could-reduce-number-horseshoe-crabs-used-catch-eel-whelk

J.L. Ding, B. Ho, United States Patent #5858706, Expression of Carcinoscopius rotundicauda factor C in eukaryotes (1999)

J.L. Ding, C. Chai, A.W.M. Pui, B. Ho, Expression of full length and deletion homologues of *Carcinoscorpius rotundicauda* factor C in *Saccharomyces cerevesiae*: Immunogicity and endotoxin binding. J. Endotoxin Res. **4**, 33–43 (1997)

S. Eyler, S. Michels, K. Rootes-Murdy, *Review of the Atlantic States Marine Fisheries Commission Fishery Management Plan for Horseshoe Crab (Limulus Polyphemus)* (Atlantic States Marine Fisheries Commission, Arlington, 2015) http://www.asmfc.org/uploads/file/56d76a40hscFMPReview2015.pdf

FDA, 510(k) premarket notification, Fungitell STAT (2020), http://www.accessdata.fda.gov/cdrh_docs/pdf19/K191167.pdf

K. Ferrari, N. Targett, Chemical attractants in horseshoe crab, Limulus polyphemus, eggs: The potential for an artificial bait. J. Chem. Ecol. **29**, 477–496 (2003)

A. Fox, The race for a coronavirus vaccine runs on horseshoe crab blood. Smithsonian Magazine. Smartnews. Keeping you current (2020), www.smithsonianmag.com

S.R. Galler, *Limulus polyphemus*, a target of opportunity, in *Biomedical Applications of the Horseshoe Crab (Limulidae)*, ed. by E. Cohen, F. B. Bang, J. Levin, J. J. Marchalonis, T. G. Pistole, R. A. Prendergast, C. Shuster Jr., S. W. Watson, (Alan R. Liss, Inc., New York, 1979), pp. 107–108

L. Hurton, J. Berkson, Potential causes of mortality for horseshoe crabs (*Limulus polyphemus*) during the biomedical bleeding process. Fish. Bull. **104**, 293–298 (2006)

L. Hurton, J.M. Berkson, S. Smith, The effect of hemolymph extraction volume and handling stress on horseshoe crab mortality, in *Biology and Conservation of Horseshoe Crabs*, ed. by J. T. Tanacredi, M. L. Botton, D. R. Smith, (Springer, New York, 2009), pp. 331–346

W. Kurz, M.J. James-Pirri, The impact of biomedical bleeding on horseshoe crabs, *Limulus polyphemus*, movement patterns on Cape Cod, Massachusetts. Mar. Freshw. Behav. Physiol. **35**, 261–268 (2002)

K. Laurie, C.-P. Chen, S.G. Cheung, V. Do, H. Hsieh, F. Mohamed, S. Seino, S. Nishida, P. Shin, M. Yang, Tachypleus tridentatus, tri-spine horseshoe crab, errata version. The IUCN red list of threatened species (2019). https://doi.org/10.2305/IUCN.UK.2019-1.RLTS.T21309A14968986.en

A.S. Leschen, S.J. Correia, Mortality in female horseshoe crabs (*Limulus polyphemus*) from biomedical bleeding and handling: Implications for fisheries management. Mar. Freshw. Behav. Physiol. **43**, 135–147 (2010)

A.S. Leschen, S.P. Grady, I. Valiela, Fecundity and spawning of the Atlantic horseshoe crab, *Limulus polyphemus*, in Pleasant Bay, Cape Cod, Massachusetts, USA. Mar. Ecol. **27**, 54–65 (2006)

J. Levin, F.B. Bang, Clottable protein in limulus: Its localization and kinetics of coagulation by endotoxin. Thromb. Diath. Haemorrh. **19**, 186–197 (1968)

J. Levin, H.D. Hochstein, T.J. Novitsky, Clotting cells and *limulus* amebocyte lysate: An amazing analytical tool, in *The American Horseshoe Crab*, ed. by C. N. Shuster Jr., R. B. Barlow, H. J. Brockman, (Harvard University Press, Cambridge, 2003), pp. 310–340

Lonza, Bioscience solutions: 5th global endotoxin testing summit. Where science meets inspiration (2019), www.lonza.com/endosummit

T. Maloney, R. Phelan, N. Simmons, Saving the horseshoe crab: A synthetic alternative to horseshoe crab blood for endotoxin detection. PLoS Biol. **16**(10), PMC6200078 (2018). http://www.nim.nih.gov

D. McKiernan, *Memorandum: Continuing Horseshoe Crab Alternative Bait Trials* (Commonwealth of Massachusetts, Division of Marine Fisheries, Boston, 2016)

M.J. Millard, J.A. Sweka, C.P. McGowan, D.R. Smith, Assessment and management of North American horseshoe crab populations with emphasis on a multispecies framework for DelawareBay, USA populations, in *Changing Global Perspectives on Horseshoe Crab Biology, Conservation and Management*, ed. by R. Carmichael, M. L. Botton, P. K. S. Shin, S. G. Cheung, (Springer, New York, 2015), pp. 407–431

D.S. Mizahi, K.A. Peters, Relationships between sandpipers and horseshoe crab in Delaware Bay: A synthesis, in *Biology and Conservation of Horseshoe Crabs*, ed. by J. T. Tanacredi, M. L. Botton, D. R. Smith, (Springer, New York, 2009), pp. 65–87

National Toxicology Program, In vitro pyrogen test methods (2020), http://ntp.niehs.nih.gov/go/iccvam-pyrogen

T.J. Novitsky, Biomedical applications of Limulus amebocyte lysate, in *Biology and Conservation of Horseshoe Crabs*, ed. by J. T. Tanacredi, M. L. Botton, D. R. Smith, (Springer, New York, 2009), pp. 315–329

T.J. Novitsky, Biomedical implications for managing the *Limulus polyphemus* harvest alone the northeast coast of the United States, in *Changing Global Perspectives on Horseshoe Crab Biology, Conservation and Management*, ed. by R. Carmichael, M. L. Botton, P. K. S. Shin, S. G. Cheung, (Springer, New York, 2015), pp. 483–500

T.J. Novitsky, H.D. Hochstein, Limulus endotoxin test, in *Rapid Microbial Methods in the Pharmaceutical Industry*, ed. by M. C. Easter, (Interpharm/CRC, Boca Raton, 2003), pp. 187–210

T.J. Novitsky, M.E. Dawson, E.J. Paus, United States Patent #6,391,295. Artificial Bait (2002)

M. Owings, C. Chabot, W. Watson III, Effects of the biomedical bleeding process on the behavior of the American horseshoe crab, *Limulus polyphemus*, in its natural habitat. Biol. Bull. **236**, 207–223 (2019)

A. Rudloe, The effect of heavy bleeding on mortality of the horseshoe crab, *Limulus polyphemus*, in the natural environment. J. Invertebr. Pathol. **42**, 167–176 (1983)

C.N. Shuster Jr., M.L. Botton, R.E. Loveland, Horseshoe crab conservation: A coast-wide management plan, in *The American Horseshoe Crab*, ed. by C. N. Shuster Jr., R. B. Barlow, H. J. Brockman, (Harvard University Press, Cambridge, 2003), pp. 358–377

J. Sloyer, T.J. Novitsky, Use of rENP to quantitate endotoxin by fluorescence polarization. J. Endotoxin Res. **6**, 101 (2000)

D.R. Smith, H.J. Brockman, M.A. Beekey, T.L. King, M.J. Millard, J. Zaldívar-Rae, Conservation status of the American horseshoe crab, (*Limulus polyphemus*): A regional assessment. Rev. Fish Biol. Fish. **27**, 135–175 (2017)

H. Tamura, The road to next generation LAL technology. EC Bacteriol Virol Res ECO.01 (2016), https://www.ecronicon.com/eco16/pdf/ECBVR-01-ECO-03.pdf

United States Pharmacopeia 43, Bacterial endotoxins test <85> (2020), http://www.uspnf.com

United States Pharmacopeia 43, Pyrogen test <151> (2020), http://www.uspnf.com

N.R. Wainwright, T.J. Novitsky, United States Patent #5,614,369 Endotoxin Binding and Neutralizing Protein and Uses Thereof (1997)

N.R. Wainwright, T.J. Novitsky, United States Patent #6,222,021 Fragments of Endotoxin Binding Protein and Uses Thereof (2001)

K. Wakefield, Saving the horseshoe crab: designing a more sustainable bait for eel and conch fisheries. Delaware Sea Grant Report (2013), http://www.deseagrant.org/sites/default/files/attachments/SavingTheHorseshoeCrab.pdf

E.A. Walls, J. Berkson, Effects of blood extraction on horseshoe crabs (*Limulus polyphemus*). Fish. Bull. **101**, 457–459 (2003)

# LAL/TAL and Animal-Free rFC-Based Endotoxin Tests: Their Characteristics and Impact on the Horseshoe Crab Populations in the United States and Asia

Glenn A. Gauvry, Thomas Uhlig, and Karolina Heed

## 1 Introduction and History of Bacterial Endotoxin Testing

The human immune system is able to fend off a variety of organisms potentially harmful to the body. Even its comparably primordial innate immune system can detect a number of microbe- or pathogen-associated molecular patterns (MAMPs/PAMPs) without prior exposure and may initiate an appropriate immune response, including the recruitment of the adaptive immune system (Mogensen 2009). Among these MAMPs, bacterial endotoxin, a major constituent of the outer membrane of Gram-negative bacteria, acts as one of the most potent immunological stimulants (Raetz and Whitfield 2002). Even minute amounts can cause a wide spectrum of immune reactions ranging from fever – leading to the categorization of endotoxin as "pyrogen" (fever-inducing substance) – to organ dysfunction and septic shock, if they enter the blood circulation (Wolff 1973). Due to the severe physiological reactions to this potential contaminant and the abundance of Gram-negative bacteria (Gorbet and Sefton 2005), parenterally administered pharmaceuticals, i.e., ones injected into the body, and medical devices exposed to the blood circulation, are required to be tested for endotoxin (United States Pharmacopeia USP 2016a, European Pharmacopoeia Ph.Eur. 2016a, Japanese Pharmacopoeia JP 2016). Only

G. A. Gauvry (✉)
Ecological Research & Development Group Inc., Dover, DE, USA
e-mail: erdg@horseshoecrab.org

T. Uhlig
Hyglos GmbH – a bioMérieux Company, Bernried, Germany
e-mail: thomas.uhlig@biomerieux.com

K. Heed
TRIGA-S Scientific Solutions, Habach, Germany
e-mail: karolina.heed@triga-s.de



J. T. Tanacredi et al. (eds.), *International Horseshoe Crab Conservation and Research Efforts: 2007-2020*, https://doi.org/10.1007/978-3-030-82315-3_21

concentrations considered harmless are permitted, most commonly 5 endotoxin units (EU) per kilogram bodyweight (USP 2016a, Ph.Eur. 2016a, JP 2016).

The first authorized method to test for endotoxin and, indistinctively, other pyrogens was the rabbit pyrogen test (RPT), established in the 1920s (Seibert 1923) and introduced into the United States Pharmacopeia in 1942 (USP 1942). As part of this procedure, a pharmaceutical product under investigation is injected into the ear veins of several rabbits. If the product contains elevated levels of endotoxin, the rabbits display a body temperature increase beyond a predefined threshold. Consequently, the said product will be rejected and disposed. However, this method is only qualitative, as the exact pyrogen concentration remains unknown.

In the 1950s, Frederik Bang and Jack Levin found that the hemolymph of *Limulus polyphemus* clotted upon exposure to components of Gram-negative bacteria, later revealed as endotoxin (Bang 1953; Levin 1964). Specifically, the horseshoe crab's immune cells, called amebocytes, secreted coagulating agents. This discovery was decisive for the development of the *Limulus* amebocyte lysate (LAL) bacterial endotoxin test (BET). In the mid-1970s, horseshoe crabs began to be harvested for their hemolymph used in the production of LAL. Within a few years, this animal extract came into global use by the pharmaceutical and medical device industries to test for the presence of bacterial endotoxins in parenteral drugs and implantable medical devices for both humans and animals. The LAL test proved to be a simpler, more reliable, more specific, cheaper, quantitative, and more humane alternative to the RPT it replaced (Cooper et al. 1971). By the late 1970s, the Asian *Tachypleus* species began to be harvested for the same purpose, the production of equivalent *Tachypleus* amebocyte lysate (TAL) (Xiamen Bioendo Technology 2019).

In parallel to the spreading use of LAL/TAL, recombinant deoxyribonucleic acid (DNA) technology emerged, i.e., the creation of DNA sequences not found in nature (Cohen 1973; Hughes 2001; Cohen 2013). This allowed the deliberate transfer of genes between organisms and their expression into proteins in easily cultivated organisms of choice, such as bacteria, yeast, or animal cells (Brondyk 2009). Thus, a wide range of biological products began to be manufactured, starting with insulin (Goeddel et al. 1979) and expanding to many different proteins and other biomolecules.

While the said biomolecules could be processed into products such as pharmaceuticals, recombinant DNA technology also allowed for their thorough analysis, greatly facilitating research into biochemistry. As part of this development, the molecular machinery in LAL/TAL was explored and unraveled in the late 1980s and early 1990s (Nakamura et al. 1986; Muta et al. 1991; Iwanaga and Lee 2005). Based on this knowledge, the endotoxin detector protein of the lysate, Factor C, was successfully synthesized in an organism other than horseshoe crabs, namely, yeast, insect, and mammalian cells (Roopashree et al. 1996, 1997, Ding and Ho 2003, Ding et al. 1997, Pui et al. 1997). Since this recombinant horseshoe crab Factor C (rFC) fulfilled the same function as its native equivalent, it was incorporated into a bacterial endotoxin test equivalent to LAL/TAL tests, i.e., the first rFC assay from 2003 (Ding and Ho 2001, 2003, 2010, Lonza 2018).

## 2 Characteristics of LAL/TAL and rFC Tests

LAL/TAL contains a multitude of proteins produced and secreted by amebocytes. Many of the functionally notable ones include the so-called zymogens or proenzymes. These are proteins that require activation before they become enzymatically active. Here, such activity is reflected in the cleavage of other proteins at specific sites, i.e., the activation of other zymogens and otherwise functional proteins.

Upon exposure to endotoxin, the zymogen Factor C changes its structural conformation and excises a part of itself. Thus, active Factor C cleaves Factor B which will in turn cut proclotting enzyme, whose active form clotting enzyme truncates coagulogen into coagulin. Ultimately, several coagulin molecules cross-link each other into a solid network: the hemolymph or lysate coagulates (Fig. 1a). At each step of the enzymatic cascade, the reaction is amplified. Consequently, minute amounts of endotoxin can elicit a strong response, reflected in the paramount sensitivity of LAL/TAL. On the other hand, the additionally present zymogen Factor G can be activated by beta-glucans (Mikami et al. 1982; Roslansky and Novitsky 1991; Muta et al. 1995), e.g., from fungi or water filters made of cellulose (Pearson et al. 1984). Factor G cleaves proclotting enzyme equivalent to Factor B. Moreover, LAL/TAL has been shown to be reactive to substances such as thrombin, thromboplastin, nucleic acids (Elin and Wolff 1973), peptidoglycans and exotoxins from Gram-positive bacteria (Wildfeuer et al. 1974; Baek et al. 1985; Brunson and Watson 1976), and dithiothreitol (DTT; Platica et al. 1978). Accordingly, LAL/TAL can also react in the absence of endotoxin.

In its simplest and most classical format, the so-called gel clot test, LAL/TAL is simply subjected to a sample inside of a glass vial. An amount of endotoxin equal to or greater than the lysate's sensitivity will lead to coagulation within 1 hour at 37 °C, visible as turbid solid clot that stays at the bottom of the vial, when inverted.

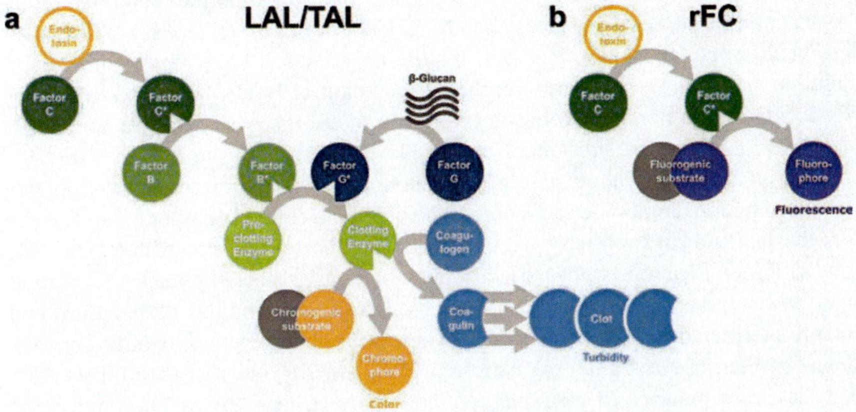


**Fig. 1** Enzyme biochemistry in (**a**) LAL/TAL and (**b**) rFC. * denotes an activated enzyme. Color and turbidity formation can be measured using an absorbance reader, fluorescence development using a fluorescence reader

In the course of coagulation, turbidity gradually increases, a process that can be measured in instruments reading absorbance. The more endotoxin, the quicker the turbidity rises. As samples and defined amounts of standard endotoxin can be compared with regard to turbidity development upon addition of LAL/TAL, the amount of endotoxin in the samples can be deduced (Urbascheck et al. 1987). This quantitative BET is known as turbidimetric LAL/TAL test. To enhance the rise in absorbance, a synthetic color-generating substrate of clotting enzyme can be added to LAL/TAL, which is the test principle of the so-called chromogenic test (Berzofsky 1994), a second quantitative BET. For convenience in handling and measurement as well as for saving lysate and sample, both quantitative test formats are usually performed in microtiter plates. These are plastic vessels commonly equipped with 96 wells, i.e., small separate reaction spaces functionally equivalent to tiny test tubes.

In contrast to LAL/TAL, rFC assays solely rely on the enzymatic function of Factor C. Instead of Factor B, a synthetic fluorescence-generating molecule is processed by rFC after activation by endotoxin (Fig. 1b). In correspondence to the chromogenic LAL/TAL test, this rFC reagent is added to samples and control standard endotoxin (CSE) inside of microtiter plates. The rise in fluorescence intensity, measured by a respective instrument, is proportional to the amount of endotoxin. While the response-amplifying enzymes of LAL/TAL are missing, even fluorescence light invisible to the human eye can be measured. Therefore, these fluorogenic rFC assays display sensitivity to endotoxin which is comparable to LAL/TAL tests. Since 2015, the entire line of Factor C, Factor B, and proclotting enzyme has been produced recombinantly and integrated into a commercially available chromogenic assay (Mizumura et al. 2017). All of these rFC-based assays do not react to beta-glucans for the lack of Factor G (Bolden 2019). Accordingly, they do not show false-positive signals in testing these polysaccharides and potentially other LAL-reactive substances (see above), i.e., signals that do not come from endotoxin.

Comparing BET, i.e., LAL/TAL and rFC tests in different formats (Williams 2007), you will not find that "one size fits all," meaning a perfect combination of robustness, speed, sensitivity, easy handling, and low price. Gel clot LAL/TAL tests are very simple and low-priced and require only modest equipment, namely, for dilution and heating. Furthermore, the high amount of lysate compared to sample lends the test rather high robustness against interference (Hughes et al. 2015). However, a variety of substances commonly interfere with the enzymes in BET (Twohy et al. 1984). The relatively high amount of lysate required for gel clot tests means that more horseshoe crabs need to be bled for a defined number of BET trials. On the downside of simplicity, the incubation time needs to be controlled manually, and clot identification is not perfectly unambiguous, especially since coagulation can be hampered by mechanical shock. Moreover, automation of handling and result interpretation is hardly feasible. These aspects appear particularly unfavorable considering ever tighter regulations on data integrity, i.e., the strive for preventing data manipulation to guarantee patient safety (United States Food and Drug Administration FDA 2018). Consequently, the gel clot test is primarily used by manufacturers with low budget and long-standing products whose change of endotoxin control would ask for considerable efforts. While the gel clot test is qualitative or semiquantitative at best – it returns ranges of endotoxin concentrations ("between," "greater than," "lower than") – all other test formats yield accurate, numeric values and are accordingly defined as quantitative BET procedures.

Turbidimetric tests are their simplest variants in the sense of only using the lysate with a few stabilizers, just as gel clot tests. Hence, they are comparably inexpensive and particularly useful for non-challenging samples such as purified water which could account for up to 70% of all bacterial endotoxin testing. Since the increase of absorbance from rising turbidity however lags behind color formation in the presence of a respective substrate, chromogenic tests tend to be more robust than their turbidimetric counterparts. In this regard, rFC assays are comparable to chromogenic LAL/TAL tests but largely remain unaffected by color in samples.

LAL/TAL is provided in a lyophilized, solid form, and a suitable solution is added to reconstitute and prepare the reagent for a test. Reconstituted LAL/TAL should be used immediately, as it starts coagulating, even if endotoxin is not added. Potentially similar to human blood (Hurley et al. 2015), horseshoe crab hemolymph might harbor a low level of inherent endotoxin that causes this phenomenon. On the other hand, rFC reagents are supplied in a liquid form and remain inactive for hours after reagent preparation for a test, i.e., the mixing of rFC, its fluorogenic substrate, and its activity-promoting buffer. This flexibility enhances test robustness and facilitates automation, as the reagents can be provided to a pipetting machine with less respect for its speed of action. Thus, the rate of errors and manual labor can be reduced.

Each test format from a specific manufacturer shows a unique level of tolerance or susceptibility to interference from a range of substances. To elaborate, whereas one BET reagent may yield accurate results in testing high amounts of a certain compound which interferes with another reagent, the relationship may be inverted in case of analyzing another compound for endotoxin (McCullough and Weidner-Loeven 1992). Likewise, several reagents may systematically return different results from a single sample (Kikuchi et al. 2017, 2018). As the amebocyte extracts and rFC preparations are not expected to show tremendous quality differences between manufacturers – the former come from large pools of horseshoe crabs alike, and the latter are structurally highly similar proteins independent from the species donating the Factor C gene – most of these peculiarities might be associated with the choice of stabilizers for both the enzymatic reagents and the co-supplied control standard endotoxin (Parenteral Drug Association 2019).

## 3 Regulation of Bacterial Endotoxin Tests

Ever since its first commercialization in 2003, rFC has faced an uphill battle to gain the regulatory and customer acceptance of LAL/TAL, independent from manufacturer and test format. When LAL emerged as alternative reagent for bacterial endotoxin testing, the advantages over the rabbit pyrogen test (RPT) were obvious (Federal Register 1977). LAL was specific to the predominantly found pyrogen endotoxin and, as only revealed later, beta-glucans. In this respect, the LAL test was more sensitive (van Noordwijk and de Jong 1976; Wachtel and Tsuji 1977; Mascoli and Weary 1979) and, after the introduction of endotoxin standards, also more quantitative (Rudbach et al. 1976).

Generally, the LAL test's convenience was striking, as it could simply be bought from reagent vendors, whereas the rabbits most commonly needed to be bred and kept in-house, resulting in considerable maintenance expenses. Likewise, the test procedure was shortened from several hours to one, allowing for quicker results. Accordingly, LAL tests offered substantial cost savings and the ability to test radio-pharmaceuticals whose short half-life requires rapid results (Hartung 2015; Liebsch 1995). Meanwhile, the influence of biological variability was reduced, as lysate produced from numerous horseshoe crabs yielded more reproducible results than the typical RPT which could only involve up to eight rabbits. Under these circumstances, the LAL test quickly gained approval by the FDA (Federal Register 1973) and adoption as compendial method into USP XX Chapter <85> (USP 1980). The European (Ph.Eur.) and Japanese Pharmacopoeias (JP) followed with their harmonized Chaps. 2.6.14. and 4.01, respectively.

Upon their introduction, rFC assays generally improved on specificity, reproducibility, and animal protection. rFC is not activated by LAL-active beta-glucans (see section above), providing a more specific bacterial endotoxin test. The biological variability was diminished even further (McKenzie et al. 2011), since rFC always bears the very same primary structure (amino acid sequence of the protein), is synthesized by genetically identical cells, and is purified in a tightly controlled biochemical environment. To compare, LAL comes from genetically different populations of horseshoe crabs that may additionally be subject to seasonal environmental influences and varying states of health (Jorgensen and Smith 1973). Particularly in relationship to different chemical environments as found in the wide spectrum of pharmaceutical samples, the enzymatic activity of rFC can thus be reproduced more easily than in case of the animal extract LAL. Consequently, preparation procedures established for specific samples can be applied more reliably.

rFC can help to take pressure off the global horseshoe crab populations that are harvested for the manufacturing of LAL/TAL. Even if *Limulus polyphemus*' mortality due to respective best-practice handling is comparably low (ASMF 2018), any horseshoe crab that can avoid bleeding maintains a higher chance of survival and reproduction. In a single 30 L production run (Fig. 2), the amount of rFC produced is equivalent to the bleeding on average of 6000 horseshoe crabs without the associated collection, transportation, bleeding, husbandry, and release associated with the production of LAL.

Whereas environmental protections – and wildlife conservation in particular – have gained momentum in the past decades, the manufacturing industry has often only contributed to this trend upon pressure from regulatory bodies, the public and governmental incentives. Accordingly, the adoption of rFC for ethical reasons has mostly been the effort of dedicated individuals such as Jay Bolden of Eli Lilly, who in 2013 began leading his company away from traditional TAL/LAL BET and

**Fig. 2** Reactor holding cell culture for manufacturing recombinant proteins such as Factor C

toward the animal-free rFC. In his 2018 progress report to the United Nations Global Compact (UNGC; Eli Lilly and Company 2018), Eli Lilly and Company CEO David A. Ricks indicated that by 2020, 90% of the company's endotoxin test will be conducted via synthetic compounds.

From its introduction in 2003, it took 9 years until the FDA explicitly mentioned rFC assays as acceptable reagents for BET, followed by the European Pharmacopoeia in 2016 (FDA 2012; Ph.Eur. 2016b). The primary validation of the first rFC assay by its manufacturer Lonza (Loverock et al. 2009) and the introduction of an rFC assay by the manufacturer Hyglos (now bioMérieux, Grallert et al. 2011) had provided rFC users with crucial validation support and the opportunity to switch between reagent suppliers. Thus, they certainly helped in driving this first step in gaining regulatory acceptance. Nonetheless, rFC assays were officially regulated as alternative BET methods, at least until Chap. 2.6.32. Test for bacterial endotoxins using recombinant Factor C of the European Pharmacopoeia becomes effective (EDQM 2019) in 2021. This potentially stimulates the chapter's adoption into USP and JP as well.

If users currently decide to use an rFC assay for release of pharmaceutical products or medical devices, they need to follow a validation adhering to USP Chapter <1225> Validation of Compendial Procedures (USP 2016b) according to the FDA (Fig. 3, FDA 2012). Namely, they show that the method is working in their laboratory as intended by demonstrating acceptable accuracy, precision, specificity, detection limit, quantitation limit, linearity, range, and robustness. While specificity is established for each kind of sample individually in the test for interfering factors, other aspects can be analyzed within a week (Williams 2018). Using comparative data between rFC assay and several LAL tests, e.g., from the manufacturer-sponsored primary validation of the method, users can even omit repeating the direct test comparison that is required by the FDA. Although rFC manufacturers have thus attempted to ease the validation, it remains a hurdle that does not apply to users of compendial LAL/TAL; some preparatory testing and the test for interfering factors are sufficient for establishing routine LAL/TAL testing. Given high financial pressure and regulatory scrutiny, many healthcare companies have therefore kept choosing LAL/TAL over rFC.

In terms of equivalence to LAL/TAL tests, rFC assays face a particular challenge. They share the biochemistry of the decisive Factor C, the handling is highly similar to quantitative LAL tests, and regulatory requirements for routine testing are the same (USP 2016a, Ph.Eur. 2016a, JP 2016). Ultimately, even the read-out, namely, endotoxin or international units (EU/IU), is identical. While these matches

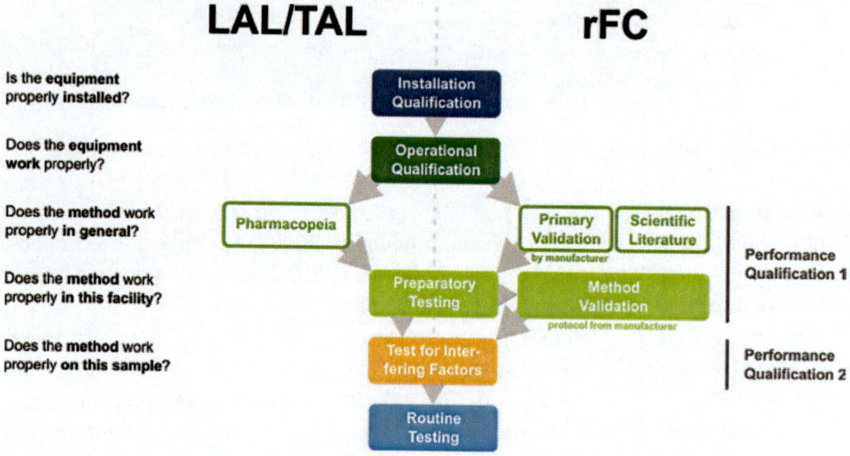


**Fig. 3** Comparison of processes for establishing bacterial endotoxin testing in pharmaceuticals or medical devices using either LAL/TAL or rFC. Colored boxes indicate activities that need to be carried out in the premises of the BET user, hile white boxes denote documents that are used as references

ease the shift from LAL/TAL to rFC in laboratory practice, they have been exploited by LAL manufacturers to compare these reagents' results and question the validity of rFC (Dubczak 2018), especially since users rarely compare different LAL/TAL reagents on samples containing endotoxin.

As indicated before, each reagent features a unique composition of enzymes, stabilizers, standard endotoxin, and additional laboratory equipment such as microtiter plates or glass vials. Hence, every BET returns individual values, i.e., each LAL/TAL and rFC test yields a specific value (Wachtel and Tsuji 1977; Kikuchi et al. 2017, 2018). At least for all LAL/TAL tests independent from the format, they are nevertheless assumed to be acceptable approximations to true endotoxin concentrations. In contrast to this categorical acceptance, differences of any rFC assay to any LAL/TAL test have been highlighted by a few LAL manufacturers and dedicated LAL users as supposedly unacceptable, particularly if an rFC assay returned lower values (Bolden et al. 2015; Dubczak 2018). Then again, the latter has also been found vice versa (Kikuchi et al. 2017, 2018; Williams 2019).

Ideally, equivalence is studied on large multicenter datasets that are robust against outliers. Most notably, the Pharmaceuticals and Medical Devices Agency of Japan (PMDA) compared three chromogenic LAL tests and three rFC-based assays in an extensive study involving up to 5 laboratories and purified endotoxin samples from 18 different bacterial strains, crude endotoxin preparations from 5 strains, and 6 environmental endotoxin samples (Kikuchi et al. 2017, 2018). They concluded that LAL and rFC detected endotoxin in all investigated samples, did not demonstrate any clear principle-specific difference, and could therefore be considered equivalent (Fig. 4).

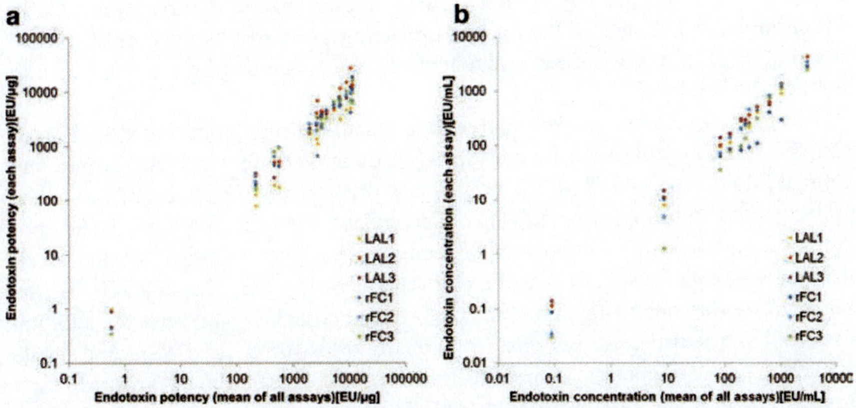


**Fig. 4** Comparison of data from collaborative study to compare kinetic-chromogenic LAL tests and rFC-based assays on (**a**) purified lipopolysaccharides from defined bacterial strains and (**b**) crudely purified endotoxin from defined bacterial strains and naturally occurring endotoxin from environmental water samples. (Adapted from Kikuchi et al. 2017)

Despite studies from the PMDA (Kikuchi et al. 2017, 2018), pharmaceutical manufacturers (Chen and Mozier 2013; Bolden 2019), academic institutions, and rFC manufacturers (Loverock et al. 2009) as well as contract research organizations, common criticism of rFC assays has focused on an ostensible lack of data proving the equivalence to LAL tests. For comparison, a Baxter Travenol study featuring thousands of parallel RPT and LAL tests has been cited (Mascoli et al. 1979; Dubczak 2018). Then again, only 37 LAL tests and 4 RPT returned positive results at the time. The study indeed indicated that endotoxin was the most prevalent pyrogen and that the absence of endotoxin might be equated with the absence of pyrogens. However, the equivalence of both tests with regard to endotoxin from different bacterial strains (see PMDA study above) was not studied. Moreover, despite other LAL-RPT comparison studies (Wachtel and Tsuji 1977), it is unclear if this data can be extrapolated from gel clot to other LAL test formats. Finally, the impartiality of the authors might be questioned, as Baxter Travenol manufactured their own LAL and therefore had vital commercial interest in promoting LAL.

Another argument against rFC has been the FDA license that is granted to each LAL reagent's manufacturing since 1977 (Federal Register 1977). When the first rFC assay was commercialized, the manufacturer submitted a request for designation in order to learn about the FDA center responsible for granting such premarket approval. Since rFC was synthetically produced, neither an animal product with its inherent variability nor intended for diagnosing diseases in humans or animals, the FDA concluded that premarket approval was not required (Berzofsky 2004). At this point, it should be noted that the FDA license is not a general indicator for the quality of LAL tests and their capability to detect endotoxin; it is simply intended to regulate and control the manufacturing process of LAL (Burgenson 2019). Whereas rFC manufacturers lack respective FDA oversight, they nonetheless adhere to stringent guidelines such as Good Manufacturing Practice (GMP) and International Organization for Standardization (ISO) requirements (Burgenson 2019). Accordingly, any change of the reagent impacting assay results must be disclosed, and rFC users may audit these manufacturers, thus encouraging the adherence to strict quality systems.

Due to their different reagent sources, the manufacturing processes of LAL/TAL and rFC actually differ substantially (Jorgensen and Smith 1973; Armstrong and Conrad 2008). For LAL/TAL, hemolymph is drawn from the hearts of horseshoe crabs, pooled and supplemented with anticoagulant. Via centrifugation, the amebocytes are separated from the rest of the hemolymph. Water is added to induce cell bursting (osmotic pressure) and thereby release of the enzymes relevant for the reagent. Another centrifugation strips them from cell debris, and after the addition of stabilizing substances, the enzymes are freeze-dried to yield solid and stable LAL/TAL. rFC, in contrast, is produced in cells that secrete the enzyme into their growth medium and that are subsequently removed using centrifugation or filtration. To purify rFC, the medium flows through chromatographic resins that selectively bind rFC or its impurities. Similar to LAL/TAL, stabilizing substances are added to allow for years of shelf lives. On the other hand, freeze-drying (lyophilization) is not necessary, as rFC does not display LAL/TAL's previously mentioned unspecific activity in the absence of endotoxin.

## 4 Effects of Horseshoe Crab Harvesting and Habitat Loss

Distributed along the Atlantic coast of the United States and the Southeast Gulf of Mexico and throughout coastal South and Southeast Asia (Fig. 5), each of the world's four extant horseshoe crab species exhibits genetic variation throughout their spawning range and is at risk of local extinction primarily due to anthropogenic activities that vary between the countries they inhabit. The primary stressors are loss of coastal habitat, both marine and terrestrial, from land reclamation, infrastructure development, coastline armoring, erosion, and pollution, bycatch, and unsustainable harvesting for bait, human consumption, amebocyte lysate production made from the hemolymph of the *Limulus* and *Tachypleus* species, and chitin, fertilizer, and traditional Chinese medicine production (Smith et al. 2017; Akbar et al. 2018; IUCN 2019). The future survival of the four horseshoe crab species will ultimately depend upon the preservation of spawning and nursery habitat and overcoming social, cultural, and corporate indifference to unsustainable harvesting practices. This is a challenging prospect in light of the ever-increasing human density along the same beaches and nearshore areas horseshoe crabs rely upon for propagation and growth (Gauvry 2015).

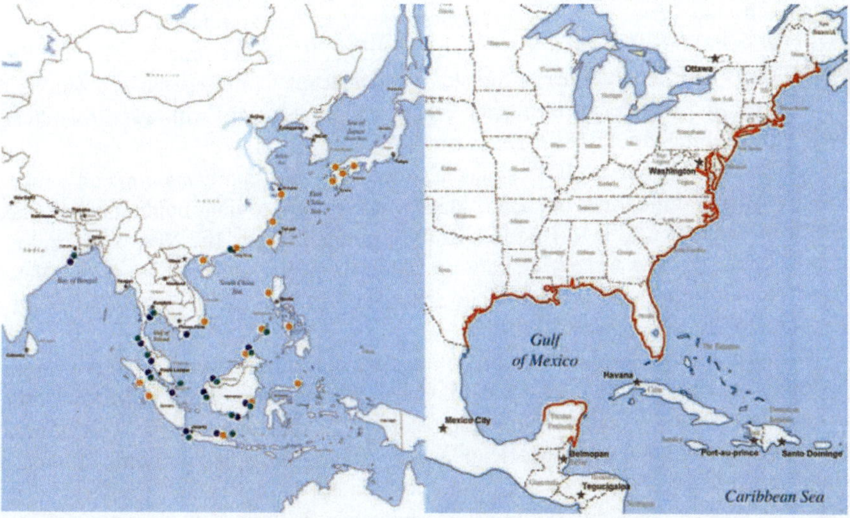


**Fig. 5** Geographic ranges of *Carcinoscorpius rotundicauda* (green), *Tachypleus tridentatus* (orange), *Tachypleus gigas* (blue), and *Limulus polyphemus* (red)

In the United States, from the late 1800s to the early 1900s, American horseshoe crabs were harvested unsustainably for fertilizer and livestock feed, depleting the resource by the 1940s (ASMFC 1998). As the fertilizer and livestock feed user groups moved away from horseshoe crabs to alternative and synthetic products, the population began to recover. By the 1960s, all commercial harvest of horseshoe crabs had ceased, and by the late 1970s, the population had substantially recovered (Shuster 1996; ASMFC 1998). Although the use of horseshoe crabs in medical research began in the early 1900s (Shuster 1962), it was not until the commercialization of LAL and the emergence of this new user group in the mid-1970s that horseshoe crabs were once again harvested in greater numbers. In the mid-1990s, yet another commercial user group emerged, who began harvesting the resource for bait in the American eel (*Anguilla rostrata*) and whelk pot fisheries (*Busycon* spp.).

Although the majority of the American horseshoe crabs harvested for LAL are returned-to-sea with a relatively low mortality of 15% (ASMFC 2018), the combined unregulated harvest from these two user groups again resulted in the population's decline (ASMFC 2018). In 1997, the States of Delaware and New Jersey, along with a ground swell of support from birding organizations, horseshoe crab advocacy groups, and the scientific community, urged the Atlantic States Marine Fisheries Commission, a US Federal Governmental body with broad fisheries management authority, to develop a horseshoe crab fisheries management plan (FMP) to monitor and regulate the harvest of the American horseshoe crab coast wide (ASMFC 1997). The purpose of the horseshoe crab FMP is to ensure a sustainable population throughout its spawning range, so as to ensure the continued role of the horseshoe crab resource in the ecology of coastal ecosystems, food resource for migratory birds and other dependent wildlife and the fishing and non-fishing public (ASMFC 2018). In March of 2001, to protect essential marine habitat in the mid-Atlantic region, home to the largest concentration of the American horseshoe crab, the National Marine Fisheries Service (NMFS), at the recommendation of the ASMFC, established the Carl N. Shuster, Jr. Horseshoe Crab Reserve. It prohibited the harvest and transfer of horseshoe crabs in federal waters within a 30 nautical mile radius of the mouth of the Delaware Bay (ASMFC 2001).

Presently, the American horseshoe crab species *Limulus polyphemus* is only marginally stable throughout its US range, with an improving population in the Northeast region (Maine to Rhode Island), still declining in the New York region (Connecticut to New Jersey), remaining stable in the Delaware Bay region (New Jersey to Virginia), and increasing in abundance in the Southeast region (North Carolina to Florida; ASMFC 2019).

The Mexican horseshoe crab population on the species extreme southern range is poorly understood and does not fall under the jurisdiction of the ASMFC. There is insufficient data to confirm population stability (Zaldivar-Rae et al. 2009; Smith et al. 2017). Although, under Mexican law, the harvest, sale, and purchase of horseshoe crabs are illegal, there is no active management of the species. Consequently, there is a growing illegal harvest of horseshoe crabs for bait in the octopus fishery with a preference for males which are most compatible with traditional drift fishing techniques (Smith et al. 2017).

While the regulated harvest of the American horseshoe crab used for bait in the American eel (*Anguilla rostrata*) and whelk pot fisheries (*Busycon* spp.) continues to decline from a high of 2.6 million in 1999 to an averaged low of 753,000 between 2004 and 2017 (ASMFC 2019), a 245% decrease, the harvest for LAL production continues to increase from a low of 130,000 crabs in 1989 (FDA 1998; ASMFC 1998) to an averaged high of 485,197 between 2004 and 2017 (ASMFC 2019), a 273% increase. Although a recent study indicates no adverse impact on the American horseshoe crab population from current LAL production harvesting levels (Smith and Newhard 2019; ASMFC 2019), it is uncertain whether LAL production, which is expected to double by 2024 (Fig.6) (Zion Market Research 2019), can continue to increase to meet its projected growth, much less absorb the Asian TAL market as it begins to decline from unsustainable harvesting practices. In an effort not to exceed their current harvesting footprint, some LAL producers have partnered with the American eel (*Anguilla rostrata*) and whelk pot fisheries (*Busycon* spp.), bleeding animals before they are used as bait (ASMF 2018). This practice may expand as these two user groups seek to share a finite resource managed for sustainability within a growing market. For as long as the harvest of the American horseshoe crab can be managed for sustainability and the availability of viable animal-free BET is available, it will be unlikely that the American eel and whelk pot fisheries will be asked to reduce their harvest in favor of biomedical use. Although often viewed as the villains in this debate by LAL producers and conservationists, who move within a different social economic class, it is important to remember that commercial fishing is one of the oldest vocations in the United States and helped provide the financial means to launch the American Revolution, thus offering the enduring status of a heritage activity (Kurlansky 1997).

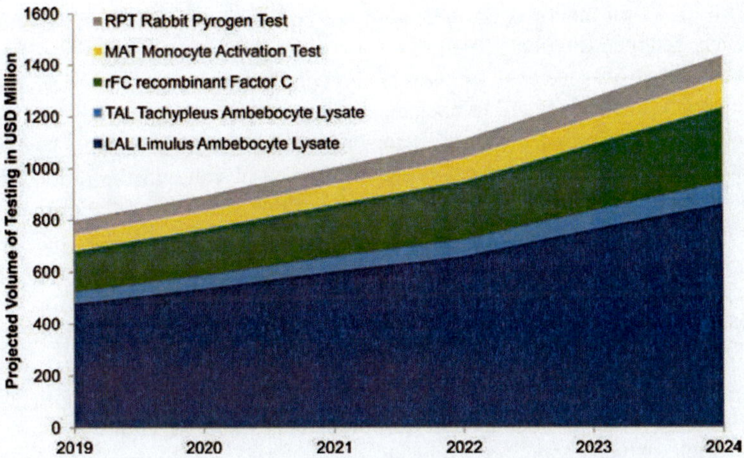


**Fig. 6** Projected volume of pyrogen and endotoxin testing by type in USD million. (Adapted from Zion Market Research 2019)

Throughout South and Southeast Asia, home to three of the world's four horseshoe crab species (*Tachypleus tridentatus*, *Tachypleus gigas*, *Carcinoscorpius rotundicauda*), horseshoe crab conservation advocacy is only beginning to take root after centuries of exploiting this marine life primarily for human consumption and, since the late 1970s, the production of the bacterial endotoxin test TAL. When human considerations overshadow environmental concerns, ignorance and indifference are often at the core of a species decline. In 2019, the IUCN Horseshoe Crab Specialist Group, in collaboration with its members throughout South and Southeast Asia, released their IUCN Red List assessment of *Tachypleus tridentatus*. The assessment classified the species as endangered throughout its spawning range and at risk of local extinction (IUCN 2019). It is believed that once the Red List assessments for the two other Asian horseshoe crab species have been completed, they will also be listed at risk. Although there is a growing trend throughout Asia to breed and release horseshoe crab larvae into areas of population decline, currently there is no matrix to assess the efficacy of these programs.

Horseshoe crabs are eaten in many South and Southeast Asian countries, with China being the largest consumer and the primary supplier of TAL (Akbar et al. 2018; IUCN 2019; Zion Market Research 2019). The by-products of these user groups supply secondary markets with chitin, fertilizer, and traditional Chinese medicine (Novitsky 2017, pers. comm). The consumption of horseshoe crabs throughout South and Southeast Asia is driven by poverty, livelihood, and social status, intertwined with broad historical and cultural trends, linked to economic growth and social stratification (Fabinyi 2011). Without action on the consumption end of the commodity trade, either through public awareness campaigns and/or cost that exceeds market viability, coupled with government intervention, it is unlikely the trade in horseshoe crabs will abate. Although there are some indications that human consumption of wildlife in China is declining, consumers with higher income and educational backgrounds are consuming wildlife at a higher rate (Zhang and Yin 2014), and there is no indication the consumption of horseshoe crabs in China has declined (Fu et al. 2019).

Some mangrove species (*Carcinoscorpius rotundicauda*) contain a tetrodotoxin that can be fatal if eaten (Kanchanapongkul 2008; Suleiman et al. 2017). Consequently, harvest of this species for human consumption remains local. It is also small and difficult to harvest in quantity to be of value in the production of amebocyte lysate. Therefore, the largest threat to this species and the primary reason for its decline are the loss of habitat throughout its range (Akbar et al. 2018). South Asia, Southeast Asia, and Asia-Pacific contain approximately 46% of the world's mangrove ecosystems, including the most biodiverse mangrove forests. On the other hand, this region also exhibits the highest global rates of mangrove loss and deforestation for alternative land uses.

Currently, there are only a few South and Southeast Asian countries with laws in place to protect their horseshoe crab resource and its habitat, and fewer still who enforce them, which encourage a substantial illegal trade (Akbar et al. 2018; IUCN 2019). The world's four horseshoe crab species are not protected by the Convention on International Trade in Endangered Species (CITES). Consequently, cooperation between countries that import horseshoe crabs from countries that ban and/or restrict their export is poorly regulated and enforced (Akbar et al. 2018; IUCN 2019). NGOs with the support of the scientific community alone cannot reverse deeply ingrained social, cultural, and corporate indifference to the decline of these species, without a government coordinated conservation and management effort, coupled with a robust law enforcement mechanism within and between countries where horseshoe crabs live and spawn.

TAL producers in China, some of whom are LAL producers in the United States, circumvent the application of sustainable harvesting practices used in the United States for *Limulus*, by implying they are beholden to poorly enforced or nonexistent government harvesting regulations and the pervasive social, cultural, and corporate indifference toward the sustainability of the Asian horseshoe crab species (Dubczak 2019a, b). However, it is the ever-increasing distance that harvesters incur to secure the horseshoe crab resource and the need to recoup associated cost that make the bleed-and-release model used in the United States untenable in Asia. This in turn strengthens the symbiotic relationship between the harvesters and importers for human consumption and TAL production. TAL producers rent their horseshoe crabs from harvesters or importers only long enough to drain their hemolymph, before returning them to be butchered for human consumption and secondary markets i.e., chitin, fertilizer, and traditional Chinese medicine (Novitsky 2017, pers. comm). These secondary markets ebb and flow with the demand for human consumption and TAL. With Asia being one of the largest and fastest growing healthcare markets (IMS 2017), the demand for TAL is expected to double by 2024 (Zion Market Research 2019), potentially offsetting decreases in human consumption through public awareness campaigns.

Since the Chinese horseshoe crab population, which was once robust, is no longer viable for this scale of enterprise, animals from countries with poorly enforced or nonexistent harvesting and/or export laws will continue to be exploited. The report that TAL only represents 20% of the world's BET market (Dubczak 2019a, b) should not be taken as an insignificant number, for the affected species is already red-listed as endangered and on the verge of extinction (IUCN 2019). Can an ethical argument to be made for why producers of TAL should continue to exploit these diminishing resources, or why the pharmaceutical, medical device, and healthcare providers should continue to use TAL when there are other equally viable BET?

## 5 Absence of Clear Ethical Directives

First introduced in 1959 by Russell and Burch, the concept of replace, reduce, and refine (3Rs) was intended to promote the humane treatment of animals used in scientific procedures (Russell and Burch, 1959). Even though the 3Rs are now widely accepted and practiced as ethical standards in Western societies (Orlans et al. 1998; NHMRC 2013; CCAC Guidelines 2017), the scope and scales of 3Rs still remain to be clarified. For example, the majority of invertebrate species, which include the four extant horseshoe crab species, are excluded from legislation regulating scientific research on animals, with the exception of cephalopods (Berry et al., 2015). However, as was the case for the four suborders of cephalopods, the growing global support behind the conservation of the horseshoe crab species has begun to deepen our understanding of what constitutes pain and suffering in other living beings as a result of human activities. There is a growing animal-free endotoxin detection trend developing in Europe (Council of Europe, 2010). It already manifested in the introduction of the Monocyte Activation Tests (MAT), a whole-pyrogen assay based on human immune cells, as a replacement for the rabbit pyrogen test (RPT) by the Ph.Eur. into chap. 2.6.30 (2010). Regarding BET, Ph.Eur. Chaps. 5.1.10 (2016b) and 2.6.32 (EDQM 2019) continue in advancing the 3Rs. The former already included rFC assays as alternative BET, and the latter will make them compendial. This will eliminate the necessity for additional method validation and thereby present them as fully accepted BET replacements of LAL/TAL (Fig. 3).

In the absence of clear ethical directives, leadership is required. There is only one user group who has the ability to transcend the geographic boundaries, government malaise, linguistics, and social, cultural, and corporate indifference that hamper the conservation of two of the three Asian horseshoe crab species (*Tachypleus tridentatus*, *Tachypleus gigas*), and that is the pharmaceutical, medical device industries and healthcare facilities around the world who are reliant upon TAL for bacterial endotoxin test. Most are unaware that their TAL supply chain will not meet future demands and, by their continued use of TAL, are complicit in the decline of the *Tachypleus* species. By implementing a supply chain management policy, which replaces their use of TAL with LAL and augments their endotoxin test requirements with rFC assays where applicable, they can reduce harvesting pressure on the two Asian *Tachypleus* species and at the same time reduce transferred harvesting pressure on the American horseshoe crab species (*Limulus polyphemus*) used in the production of LAL, which is also expected to double by 2024 (Zion Market Research 2019).

## 6 Conclusion

By excluding the presence of harmful amounts of endotoxin in pharmaceuticals and medical devices, LAL and TAL have helped in guaranteeing patient safety for decades. Established as they are, these animal extracts will likely stay in operation for the foreseeable future. Nonetheless, ample evidence has demonstrated that rFC-based assays fulfill the function of LAL/TAL equivalently, with the added advantage of providing viable animal-free BET. Presently, replacing the LAL/TAL lysate with rFC may require additional validation procedures. However, this relatively minor effort can provide a modern, reliable analysis method that can also make a significant contribution to the conservation of the horseshoe crab species.

Until there is a viable alternative bait for the American eel (*Anguilla rostrata*) and whelk pot fisheries (*Busycon* spp.) and a migration toward rFC assays or other animal-free bacterial endotoxin tests, the American horseshoe crab (*Limulus polyphemus*) will continue to be managed for the migratory shorebirds, other dependent wildlife, and the fishing and non-fishing public, regardless of the bacterial endotoxin test industry's reliance upon this species. The social elitism and distain currently shown toward the horseshoe crab bait fishery by the bacterial endotoxin test producers in the United States need to be reconciled, as both user groups must find ways to work together to share this finite and sustainably managed resource, within a growing marketplace.

In Asia, the three horseshoe crab species (*Tachypleus tridentatus*, *Tachypleus gigas*, *Carcinoscorpius rotundicauda*) will continue to decline until there is a coordinated conservation effort on the part of the governments where these animals live and spawn and a change in the social, cultural, and corporate indifference to unsustainable harvesting practices. Until then, these animals will keep losing essential habitat and genetic diversity and be harvested unsustainably for human consumption, the production of TAL, and secondary markets. The pharmaceutical, medical device industries and healthcare facilities have the ability to make a significant contribution to the conservation of the Asian horseshoe crab species by eliminating the purchase of TAL from their supply chain. Additionally, they would be wise to make BET supply chain management decisions based upon the collective knowledge of the IUCN and ASMFC and the vast global network of NGOs and scientists around the world who are dedicated to the biology and conservation of these ancient mariners. These organizations are working independent of the marketplace and stakeholders to understand and protect the world's four extant horseshoe crab species and have proven themselves to be honest brokers, versus the often biased opinions of some user groups who harvest and/or exploit them.

**Acknowledgments** *The authors would like to thank* Tina Berger, Atlantic States Marine Fisheries Commission (ASMFC) Director of Communications, for her assistance in gathering historical data on the ASMFC's management of *Limulus polyphemus* as well as Kevin Williams and Brendan Tindall, bioMérieux, for their critical reading of the manuscript.

## References

P. Armstrong, M. Conrad, *Blood Collection from the American Horseshoe Crab* (Limulus Polyphemus. JoVE, Cambridge, MA, 2008), p. 20

Atlantic States Marine Fisheries Commission (ASMFC), *Proceedings of the Atlantic States Marines Fisheries Commission 56th Annual Meeting* (American Eel Management Board Meeting, Hershey, 1997)

Atlantic States Marine Fisheries Commission (ASMFC), *Interstate Fishery Management Plan for Horseshoe Crab, Fishery Management Report No 32 of the Atlantic States Marine Fisheries Commission* (Washington, DC, 1998)

Atlantic States Marine Fisheries Commission (ASMFC), *Review of the Fishery Management Plan for Horseshoe Crab (Limulus polyphemus)*, (Washington, DC, 2001)

Atlantic States Marine Fisheries Commission (ASMFC), *Review of the Atlantic States Marine Fisheries Commission Fisheries Management Plan for Horseshoe Crab* (Washington, DC, 2018)

Atlantic States Marine Fisheries Commission (ASMFC), *Horseshoe Crab Benchmark Stock Assessment and Peer Review Report* (Washington, DC, 2019)
L. Baek, N. Høiby, J.B. Hertz, F. Espersen, Interaction between *Limulus* Amoebocyte Lysate and soluble antigens from *Pseudomonas aeruginosa* and *Staphylococcus aureus* studied by quantitative immunoelectrophoresis. J. Clin. Microbiol. **22**(2), 229–237 (1985)
A. Berry, A. Vitale, C. Carere, E. Alleva, EU guidelines for the care and welfare of an "exceptional invertebrate class" in scientific research. Annali Istituto Superiore di Sanita **51**, 267e269 (2015)
R.N. Berzofsky, Kinetic assay for endotoxin using limulus amebocyte lysate and chromogenic substrate. Patent US5310657. Priority October 30, 1989, Granted May 5, 1994 (1994)
R.N. Berzofsky, Does endotoxin testing = FDA licensing? LAL Review **1004**(1), 1–3 (2004)
J. Bolden, Recombinant Factor C: Progressive Endotoxin Detection One Year In. 5th Global Endotoxin Testing Summit. June 3, 2019 (2019)
J. Bolden, C. Platco, J. Dubczak, J.F. Cooper, K.Z. McCullough, The use of endotoxin as an analyte in biopharmaceutical product hold-time studies. Pharmacopeial Forum 41(5), Stimuli to the Revision Process (2015)
W.H. Brondyk, Chapter 11 selecting an appropriate method for expressing a recombinant protein, in *Methods in Enzymology*, vol. 463, (Academic Press, New York, 2009), pp. 131–147. https://doi.org/10.1016/S0076-6879(09)63011-1
K.W. Brunson, D.W. Watson, *Limulus* amebocyte lysate reaction with streptococcal pyrogenic exotoxin. Infect. Immun. **14**(5), 1256–1258 (1976)
A. Burgenson, Another Perspective on rFC. PDA Letter. (2019). https://www.pda.org/pda--letter-portal/home/full-article/another-perspective-on-rfc. August 05, 2019. Accessed on 30 Aug 2019.
Canadian Council on Animal Care, *CCAC Guidelines: Husbandry of Animals in Science* (Ottawa, Ontario, Canada, 2017)
L. Chen, N. Mozier, Comparison of *Limulus* Amebocyte Lysate Test Methods for Endotoxin measurement in protein solutions. J. Pharm. Biomed. Anal. **80**, 180–185 (2013)
S.N. Cohen, 2013. DNA cloning: A personal view after 40 years. Proc. Natl. Acad. Sci. U. S. A. **110**(39), 15521–15529 (2013). https://doi.org/10.1073/pnas.1313397110. Epub 2013 Sep 16
S.N. Cohen, A.C.Y. Chang, H.W. Boyer, R.B. Helling, Construction of biologically functional bacterial plasmids in vitro. Proc. Natl. Acad. Sci. U. S. A. **70**(11), 3240–3244 (1973)
J.F. Cooper, J. Levin, H.N. Wagner Jr., Quantitative comparison of in vitro and in vivo methods for the detection of endotoxin. J. Lab. Clin. Med. **78**(1), 138 (1971)
Council of Europe, Directive 2010/63/EU of the European Parliament and of the Council of 22 September 2010 on the protection of animals used for scientific purposes. Off. J. Eur. Union L **276**, 20 (2010)
J.L. Ding, B. Ho, A new era in pyrogen testing. Trends Biotechnol. **19**(8), 277–281 (2001)
J.L. Ding, B. Ho, Assays For Endotoxin. US 6645724 B1. Application April 7, 1999. Publication November 11, 2003. National University of Singapore, (2003)
J.L. Ding, B. Ho, Endotoxin detection - from Limulus amebocyte lysate to recombinant factor C. Subcell. Biochem. **53**, 187–208 (2010)
J.L. Ding, C. Chai, A.W.M. Pui, et al., Expression of full length and deletion homologues of *Carcinoscorpius rotundicauda* Factor C in *Saccharomyces cerevisiae*: Immunoreactivity and endotoxin binding. J. Endotoxin Res. **4**, 33–43 (1997)
J. Dubczak, Standing Guard. PDA Letter, September 26, 2018. (2018). https://www.pda.org/pda--letter-portal/archives/full-article/standing-guard. Accessed on August 30
J. Dubczak, Expert View: Charles River, Conservation initiatives for an invaluable partner. (2019a). europeanpharmaceuticalreview.com
J. Dubczak, *Blue-Blooded Crabs at Heart of Pharma Dilute on Drug Testing* (Reuters Health News, Zurich, 2019b)
EDQM Council of Europe, New general chapter on test for bacterial endotoxins using recombinant factor C. Give your feedback. (2019). https://www.edqm.eu/en/news/new-general-chapter--test-bacterial-endotoxins-using-recombinant-factor-c-give-your-feedback January 09, 2019. Accessed on 28 Aug 2019

Eli Lilly and Company. United Nations Global Compact Communication on Progress Report. (2018). https://assets.ctfassets.net/hadumfdtzsru/3zPl8biFFz7pvlsjTzP9OQ/ea26d-bece0a590ba58f8d7bcbb1f80f9/2018_United_Nations_Global_Compact_Communication_on_Progress_Report.pdf. Lilly Corporate Center, Indianapolis, IN (USA)

R.J. Elin, S.M. Wolff, Nonspecificity of the *Limulus* amoebocyte lysate test: Positive reactions with polynucleotides and proteins. J. Infect. Dis. **128**(3), 349–352 (1973)

European Pharmacopoeia 6.7. 2010. 2.6.30. Monocyte activation test, EDQM Council of Europe, Strasbourg, France

European Pharmacopoeia 8.8. 2016a. 2.6.14. Bacterial endotoxins, EDQM Council of Europe, Strasbourg, France. pp. 194–198

European Pharmacopoeia 8.8. 2016b. 5.1.10. Guidelines for using the test for bacterial endotoxins, EDQM Council of Europe, Strasbourg, France. pp. 5931–5934

M. Fabinyi, Historical, cultural and social perspectives on luxury seafood consumption in China. Environ. Conserv. **39**, 83–92 (2011)

Federal Register, Status of biological substances used for detecting bacterial endotoxins. **38**(8), 1404 (1973)

Federal Register, Licensing of limulus amebocyte lysate, use as an alternative for rabbit pyrogen test. **42**(213), 57749 (1977)

Y. Fu, S. Huang, Z. Wu, C.C. Wang, M. Su, X. Wang, et al., Socio-demographic drivers and public perceptions of consumption and conservation of Asian horseshoe crabs in northern Beibu Gulf, China. Aquat. Conserv. Mar. Freshwat. Ecosyst. **29**(8), 1268–1277 (2019)

G. Gauvry, Current horseshoe crab harvesting practices cannot support global demand for TAL/LAL: The pharmaceutical and medical device industries' role in the sustainability of horseshoe crabs, in *Changing Global Perspectives on Horseshoe Crab Biology, Conservation and Management*, ed. by R. H. Carmichael, M. L. Botton, P. K. S. Shin, S. G. Cheung, (Springer International Publishing, Cham, 2015), pp. 475–482

D.V. Goeddel, D.G. Kleid, F. Bolivar, H.L. Heyneker, D.G. Yansura, R. Crea, T. Hirose, A. Kraszewski, K. Itakura, A.D. Riggs, Expression in Escherichia coli of chemically synthesized genes for human insulin. Proc. Natl. Acad. Sci. U. S. A. **76**(1), 106–110 (1979)

M.B. Gorbet, M.V. Sefton, Endotoxin: The uninvited guest. Biomaterials **26**, 6811–6817 (2005)

H. Grallert, S. Leopoldseder, M. Schuett, P. Kurze, B. Buchberger, EndoLISA: A novel and reliable method for endotoxin detection. Nature methods, application notes. October 2011, iii-v (2011)

T. Hartung, The human whole blood pyrogen test – Lessons learned in twenty years. ALTEX **32**(2), 79–100 (2015)

S.S. Hughes, Making dollars out of DNA. The first major patent in biotechnology and the commercialization of molecular biology, 1974-1980. Isis **92**(3), 541–575 (2001)

P.F. Hughes, R. Candau-Chacon, C. Thomas, C. Gomez-Broughton, K. Suvarna, L. Rani Narasimhan, B. Chi, Low endotoxin recovery: An FDA perspective. BioPharma Asia March/April **2015**, 14–25 (2015)

J.C. Hurley, P. Nowak, L. Ohrmalm, C. Gogos, A. Armaganidis, E.J. Giamarellos-Bourboulis, Endotoxemia as a diagnostic tool for patients with suspected bacteremia caused by gram-negative organisms: A meta-analysis of 4 decades of studies. J. Clin. Microbiol. **53**(4), 1183–1191 (2015)

IMS Institute for Healthcare Informatics, *IMS Global Use of Medicines* (IMS Institute for Healthcare Informatics, Norwalk, 2017)

IUCN, The IUCN red list of threatened species. Version 2019-1 (2019). Available at: www.iucnredlist.org. (Accessed: 21 March 2019)

S. Iwanaga, B.L. Lee, Recent advances in the innate immunity of invertebrate animals. J. Biochem. Mol. Biol. **38**(2), 128–150 (2005)

J. H. Jorgensen, R. F. Smith, Rapid detection of contaminated intravenous fluids using the Limulus in vitro endotoxin assay. Appl. Microbiol. **26**(4), 521–524 (1973)

Akbar B. John, B. R. Nelson,·Hassan I. Sheikh,· S. G. Cheung,· Yusli Wardiatno, Bisnu Prasad Dash, Keiji Tsuchiya, Yumiko Iwasaki, Siddhartha Pati. "A review on fisheries and conservation status of Asian horseshoe crabs, Biodiversity and Conservation, Springer Nature B.V.2018. 27:3573-3598

J. Kanchanapongkul, Tetrodotoxin poisoning following ingestion of the toxic eggs of the horseshoe crab Carcinoscorpius rotundicauda, a case series from 1994 through 2006. Southeast Asian J. Trop. Med. Public Health **39**, 303–306 (2008)

Y. Kikuchi, Y. Haishima, C. Fukui, T. Murai, Y. Nakagawa, A. Ebisawa, K. Matsumura, K. Ouchi, T. Oda, M. Mukai, T. Masuda, Y. Katto, Y. Takasuga, A. Takaoka, Collaborative study on the bacterial endotoxins test using recombinant factor c-based procedure for detection of lipopolysaccharides. Pharm. Med. Device Regul. Sci. **48**(4), 252–260 (2017)

Y. Kikuchi, Y. Haishima, C. Fukui, Y. Nakagawa, A. Ebisawa, T. Morioka, K. Matsumura, K. Ouchi, K. Uchida, O. Martinez, T. Oda, M. Mukai, T. Masuda, Y. Tsukihashi, Y. Takasuga, A. Takaoka, Collaborative Study on the Bacterial Endotoxins Test Using Recombinant Factor C-based Procedure for Detection of Lipopolysaccharides, Part 2. Pharm. Med. Device Regul. Sci. **10**, 708–719 (2018)

M. Kurlansky, *Cod: A Biography of the Fish that Changed the World* (Penguin Press, New York, 1997)

M. Liebsch, Die Geschichte der Validierung des LAL-Tests. Zentralstelle zur Erfassung und Bewertung von Ersatz- und Ergänzungsmethoden zum Tierversuch, Berlin (Germany). ALTEX **12**(2), 76–80 (1995)

Lonza, FDA approves first drug using the recombinant factor C assay for endotoxin testing. News Release. Walkersville, MD (USA) / Basel (CH). November 8, 2018. (2018). http://e3.marco.ch/publish/lonza/551_871/181108_Product-News_PyroGene-Eli_Lilly_FINAL_FINAL.pdf

B. Loverock, B. Simon, A. Burgenson, A. Baines, A recombinant factor C procedure for the detection of gram-negative bacterial endotoxin. The United States Pharmacopeial Convention, Stimuli to the Revision Process. Pharmacopeial Forum **35**(6), 1613–1621 (2009)

C.C. Mascoli, M.E. Weary, Applications and advantages of the Limulus amebocyte lysate (LAL) pyrogen test for parenteral injectable products. Prog. Clin. Biol. Res. **29**, 387–402 (1979)

K.Z. McCullough, C. Weidner-Loeven, Variability in the LAL test: Comparison of three kinetic methods for the testing of pharmaceutical products. J. Parenter. Sci. Technol. **46**(3), 69–72 (1992)

J.H. McKenzie, K.U. Alwis, J.E. Sordillo, K.S. Kalluri, D.K. Milton, 2011. Evaluation of lot-to-lot repeatability and effect of assay media choice in the recombinant Factor C assay. J. Environ. Monit. **13**(6), 1739–1745 (2011). https://doi.org/10.1039/c1em10035a. Epub 2011 May 9

T. Mikami, T. Nagase, T. Matsumoto, S. Suzuki, M. Suzuki, Gelation of *Limulus* Amoebocyte Lysate by simple polysaccharides. Microbiol. Immunol. **26**(5), 403–409 (1982)

H. Mizumura, N. Ogura, J. Aketagawa, M. Aizawa, Y. Kobayashi, S.I. Kawabata, T. Oda, Genetic engineering approach to develop next-generation reagents for endotoxin quantification. Innate Immun. **23**(2), 136–146 (2017). https://doi.org/10.1177/1753425916681074

T.H. Mogensen, Pathogen recognition and inflammatory signaling in innate immune defenses. Clin. Microbiol. Rev. **22**(2), 240–273 (2009). https://doi.org/10.1128/CMR.00046-08

T. Muta, T. Miyata, Y. Misumi, F. Tokunaga, T. Nakamura, Y. Toh, Y. Ikehara, S. Iwanaga, Limulus factor C. an endotoxin-sensitive serine protease zymogen with a mosaic structure of complement-like, epidermal growth factor-like, and lectin-like domains. J. Biol. Chem. **266**(10), 6554–6561 (1991)

T. Muta, N. Seki, Y. Takaki, et al., Purified horseshoe crab factor G. 1995. Reconstitution and characterization of the (1,3)-β-D-glucan-sensitive serine protease cascade. J. Biol. Chem. **270**, 892–897 (1995)

T. Nakamura, T. Morita, S. Iwanaga, Lipopolysaccharide-sensitive serine-protease zymogen (factor C) found in Limulus hemocytes. Eur. J. Biochem. **154**, 511–521 (1986). https://doi.org/10.1111/j.1432-1033.1986.tb09427.x

National Health and Medical Research Council, Australian Research Council, Australian code for the care and use of animals for scientific purposes, 8th edition, (2013)

Novitsky, (2017), personal communication

F.B. Orlans, T.M. Beauchamp, R. Dresser, D.B. Morton, J.P. Gluck, *The Human Use of Animals: Case Studies in Ethical Choice* (Oxford University Press, Oxford, 1998)

Parenteral Drug Association, Technical Report No. 82. Low Endotoxin Recovery. Bethesda, MD (USA) (2019)

F.C. Pearson, J. Bohon, W. Lee, G. Bruszer, M. Sagona, G. Jakubowski, R. Dawe, D. Morrison, C. Dinarello, Characterization of *limulus* Amoebocyte lysate-reactive material from hollow-fiber dialyzers. Appl. Environ. Microbiol. **48**, 1189–1196 (1984)

M. Platica, W. Harding, V.P. Hollander, Dithiols simulate endotoxin in the *limulus* reaction. Experientia **34**(9), 1154–1155 (1978)

A.W.M. Pui, B. Ho, J.L. Ding, Yeast recombinant Factor C from horseshoe crab binds endotoxin and causes bacteriostasis. J. Endotoxin Res. **4**, 391–400 (1997)

C.R. Raetz, C. Whitfield, Lipopolysaccharide endotoxins. Annu. Rev. Biochem. **71**, 635–700 (2002)

S.D. Roopashree, B. Ho, J.L. Ding, Expression of *Carcinoscorpius rotundicauda* Factor C in *Pichia pastoris*. Mol. Mar. Biol. Biotechnol. **5**, 334–343 (1996)

S.D. Roopashree, B. Ho, J.L. Ding, Recombinant COS-1 cells express *Carcinoscorpius rotundicauda* Factor C. Biotechnol. Lett. **19**, 357–361 (1997)

P.F. Roslansky, T.J. Novitsky, Sensitivity of Limulus amebocyte lysate (LAL) to LAL-reactive glucans. J. Clin. Microbiol. **29**(11), 2477–2483 (1991)

J.A. Rudbach, F.I. Akiya, R.J. Elin, H.D. Hoch-stein, M.K. Luoma, E.C.B. Milner, K.C. Milner, K.R. Thomas, Preparation and properties of a National Reference Endotoxin. J.Clin.Microbiol. **3**, 21–25 (1976)

W.M.S. Russell, R.L. Burch, *Principles of Humane Experimental Technique* (Methuen & Co, London, 1959)

F.B. Seibert, The cause of many febrile reactions following intravenous injections. Am. J. Phys. **71**(3), 621–651 (1923)

C.N. Shuster Jr., Serological correspondences among horseshoe 'crabs' (Limulidae). Zoologica **47**(1), 1–9 (1962)

C.N. Shuster Jr., *The Delaware Bay Area - An Ideal Habitat for Horseshoe Crabs* (Public Service Electric and Gas Company, Hancocks Bridge, 1996), p. 26. + appendices

D.R. Smith, J.J. Newhard, *The Long-Term Effect of Bleeding for Limulus Amebocyte Lysate on Annual Survival: An Analysis of Tagging Data* (Fourth International Workshop on the Science and Conservation of Horseshoe Crabs, Qinzhou City, 2019)

D.R. Smith, H.J. Brockmann, M.A. Beekey, T.L. King, M.J. Millard, J.A. Zaldívar-Rae, Conservation status of the American horseshoe crab, (*Limulus polyphemus*): A regional assessment. Rev. Fish Biol. Fish. **27**, 135–175 (2017)

M. Suleiman, J. Muhammad, J. Jelip, T. William, T.H. Chua, An outbreak of tetrodotoxin poisoning from consuming horseshoe crabs in Sabah. Southeast Asian J. Trop. Med. Public Health **48**(1), 197–203 (2017)

The Japanese Pharmacopoeia 4.01 Bacterial Endotoxins Test, Seventeenth Edition (JP17), *The Ministry of Health, Labour and Welfare, Pharmaceuticals and Medical Devices Agency* (Tokyo, Japan, 2016) pp. 110–114

C.W. Twohy, A.P. Duran, T.E. Munson, Endotoxin contamination of parenteral drugs and radiopharmaceuticals as determined by the *Limulus* amebocyte lysate method. J. Parenter. Sci. Technol. **38**, 190–201 (1984)

United States Food and Drug Administration, *Guidance for Industry - Pyrogen and Endotoxins Testing: Questions and Answers* (U.S. Department of Health and Human Services, Rockville, 2012)

United States Food and Drug Administration, *Guidance for Industry - Data Integrity and Compliance With Drug CGMP: Questions and Answers* (U.S. Department of Health and Human Services, Rockville, 2018)

United States Pharmacopeia, *Pyrogen Test, in the United States Pharmacopeia XII* (The United States Pharmacopeial Convention, Rockville, 1942)

United States Pharmacopeia, *<85> Bacterial Endotoxins Test, in the United States Pharmacopeia XX and the National Formulary XV* (The United States Pharmacopeial Convention, Rockville, MD, USA, 1980)

United States Pharmacopeia, Bacterial endotoxins test, in *The United States Pharmacopeia 39 and the National Formulary 34*, (The United States Pharmacopeial Convention, Rockville, 2016a), pp. 161–166

United States Pharmacopeia, *Validation of Compendial Procedures, in the United States Pharmacopeia 39 and the National Formulary 34* (The United States Pharmacopeial Convention, Rockville, 2016b)

B. Urbascheck, K.P. Becker, B. Ditter, Method for determining endotoxin concentrations. Patent US4663298A. Priority October 31, 1984. Granted May 5, 1987 (1987)

J. van Noordwijk, Y. de Jong, Comparison of the Limulus test for endotoxin with the rabbit test for pyrogens of the European Pharmacopoeia. J. Biol. Stand. **4**(2), 131–139 (1976)

R.E. Wachtel, K. Tsuji, Comparison of limulus amebocyte lysates and correlation with the United States pharmacopeial pyrogen test. Appl. Environ. Microbiol **33**, 1265–1269 (1977)

A. Wildfeuer, B. Heymer, K.H. Schleifer, O. Haferkamp, Investigations on the specificity of the *Limulus* test for the detection of endotoxin. Appl. Microbiol. **28**(5), 867–891 (1974)

K.L. Williams, Limulus Amebocyte Lysate discovery, mechanism, and application, in *Endotoxins, Pyrogens, LAL Testing and Depyrogenation*, 3rd edn., (CRC Press, London, 2007), pp. 191–219

K.L. Williams, Set-up, performance and alternative validation of recombinant Factor C assays for endotoxin testing. Eur. Pharm. Rev. **23**(3), 53–55 (2018)

K.L. Williams, Response to "Standing Guard". PDA Letter, February 2019, 17–19 (2019)

S.M. Wolff, Biological effects of bacterial endotoxins in man. J. Infect. Dis. **128**, 259–264 (1973)

Xiamen Bioendo Technology Co., Ltd, About Us. (2019). https://www.lalendotoxin.com/about-us_d1. Accessed on 29 Aug 2019

J. Zald´ıvar-Rae, R.E. Sapie´n-Silva, M. Rosales-Raya, H.J. Brockmann, American horseshoe crabs, Limulus polyphemus, in Mexico: Open possibilities, in *Biology and Conservation of horseshoe Crabs*, ed. by J. T. Tanacredi, M. L. Botton, D. R. Smith, (Springer, New York, 2009), pp. 97–113

L. Zhang, F. Yin, Wildlife consumption and conservation awareness in China: A long way to go. Biodivers. Conserv. **23**, 2371–2381 (2014)

Zion Market Research, *Pyrogen Testing Market* (New York, NY, 2019)

# The Power of Citizen Science: 20 Years of Horseshoe Crab Community Research Merging Conservation, Education, and Management

**Jennifer H. Mattei, Jo-Marie E. Kasinak, Samah Senbel, and Kirk Bartholomew**

## 1 Introduction

Citizen science, the active public involvement in scientific research, has its origins in ancient China and can be traced back over several millennia where community members tracked locust outbreaks conceivably to look for patterns that could have predictive value to help prepare for or prevent future crop loss (Irwin 2018). More recently, the power of citizen science was revealed by Tian et al. (2011) when they concluded the analysis of this nearly 2000-year-long data set, changes in climate, that is, dryer and cooler periods, were tied to locust outbreaks. Long-term population data, which are rare, can give us better information on how climate change will impact organisms, including humans, in the future. Today, citizen scientists help generate very large data sets that are not possible to obtain by other means on thousands of projects worldwide (Irwin 2018). Researchers who employ citizen science methodology can also produce large data sets that cover vast geographic areas and with leadership, commitment, and perseverance produce a temporally expansive data set covering several decades. The power of citizen science can lead to policy changes as well. For example, within the United States, in the early 1960s scientists

J. H. Mattei (✉) · K. Bartholomew
Department of Biology, College of Arts & Sciences, Sacred Heart University, Fairfield, CT, USA
e-mail: matteij@sacredheart.edu

J.-M. E. Kasinak
Department of Biology, College of Arts & Sciences, Sacred Heart University, Fairfield, CT, USA

Department of Marine Sciences, University of Connecticut, Storrs, CT, USA

S. Senbel
School of Computer Science and Engineering, Sacred Heart University, Fairfield, CT, USA

 
J. T. Tanacredi et al. (eds.), *International Horseshoe Crab Conservation and Research Efforts: 2007-2020*, https://doi.org/10.1007/978-3-030-82315-3_22

enlisted the aid of local residents to help collect poisoned and dead birds as part of a study on the effects of dichloro-diphenyl-trichloroethane (DDT) on bird mortality, finding a 70% reduction in the resident population of robins after several years of using DDT to control Dutch elm disease (Wurster et al. 1965). After it became obvious that DDT was toxic to more than pest insects, the Environmental Protection Agency (1972) banned the use of DDT in the United States. Soon after, the Endangered Species Act (1973) was passed to protect many species' declining populations or at least prevent their extinctions.

All species have intrinsic value and conservation of species is ethically important, yet some species have greater ecosystem effects than others. As Sean Carroll states it in *The Serengeti Rules* (2016), "some species are more equal than others," and the details of their interactions and effects can take a lifetime of study to unravel. If a once abundant species becomes rare, it is very difficult to predict what interactions or impacts on ecosystem processes the species had when it was abundant. For example, the gray wolf (*Canis lupus*) was once considered vermin and nearly extirpated from the lower 48 states of the United States by the 1970s (Hoy et al. 2019) was never considered to be an important species for regulating large ungulates. After its reintroduction to Yellowstone National Park, the entire ecosystem has been transformed. This grand experiment has clearly demonstrated how important "keystone" species are for maintaining biodiversity and changing many species interactions within the Yellowstone ecosystem that supply us with renewable natural resources (Farquhar 2019). The ecological definition of a keystone species is one whose impacts on an ecosystem are large relative to their low abundance/biomass; starfish and sea otters are additional examples in North America of well-documented keystone species (Paine 1966; Reisewitz et al. 2006; Mittelbach and McGill 2019).

Other species significantly impact ecosystem function by their sheer abundance and can be permanent or temporal within the ecosystem. These "dominant" species (Grime 1987; Mittelbach and McGill 2019) may be common or restricted in their range, but their abundance is proportionate to their effects on community diversity and/or ecosystem function (Avolio et al. 2019). For example, saltmarsh grasses (*Spartina alternaflora*, Bortolus et al. 2019) sequester carbon, filter pollutants from the water, hold sediments, abate wave energy, and act as nursery habitat for many species (Smee 2019). As saltmarsh grasses are filled in or removed, all of these services and biodiversity are lost. The eastern oyster (*Crassostrea virginica*) with very similar ecosystem function as *Spartina* was historically a dominant species and once constructed massive oyster reefs along the eastern coast of the United States. However, today oyster reefs have become one of the most imperiled habitats in our temperate estuarine ecosystems due to human harvest/consumption and shell removal (Ermgassen et al. 2012). Coastal States have only recently begun oyster management programs for habitat restoration and to enhance ecosystem services (i.e., storm protection, water filtration, and fish habitat) rather than for commercial harvest (Grabowski et al. 2012).

There is a debate in the community ecology literature surrounding the definitions of the terms "foundation species" vs. "dominant species." We will adopt the specific three-part definition proposed by Ellison (2019) that foundation species are usually abundant primary producers, nontrophic, and species that physically modify the environment where they are found. Thus, Avolio et al. (2019) relegate "foundation species" as a restricted subset of "dominant" species. S. *alternaflora* exemplifies a dominant species that is also a foundation species based on evidence described in detail by Crotty et al. (2019).

Based on the discussion above, spawning American horseshoe crabs (*Limulus polyphemus*) are not keystone species as they are erroneously referred to in the popular literature and review articles (Degnarain 2020; Krisfalusi-Gannon et al. 2018), but historically they were a dominant species, super abundant along the east coast of the United States and responsible for increased biodiversity of our estuarine ecosystem (Walls et al. 2002). Historically, their importance to ecosystem function during their massive spawning events was not recognized. They were harvested by the millions through the 1990s often viewed as a nuisance species (Berkson and Shuster 1999). In one geographic location, Delaware Bay, they remain the dominant species of coastal sandy beaches within estuaries. When spawning horseshoe crab numbers are high (over a million individuals) during their 4–6 weeks spawning period (May–June), this species alone substantially increases biodiversity and biomass on coastal beach habitats (Walls et al. 2002; Botton et al. 2003; Sweka et al. 2007; Botton 2009; Niles et al. 2009). Karpanty et al. (2006) documented unequivocally that "abundance of horseshoe crab eggs on sandy beaches is driving movement and distribution of red knots and that there is little alternative food during the migratory stopover in Delaware Bay."

Spawning horseshoe crabs are a temporally dominant species based on the sheer number of dependent species and over a million individual migratory birds that visit the Bay every spring feeding on their eggs, larvae, and adult carcasses (Fig. 1, Tables 1, 2, and 3). Fourteen species of shorebirds were reported to be feeding on *Limulus* eggs during a 5-day observation period reported on eBird (May 25–31, 2019). These observations were recorded by researchers and citizen scientists (Table 1A & B, www.ebird.com) with individual birds numbering in the thousands during that 6-day period. A significant tourist industry has built up around this phenomenon so that Delaware Bay is the first stopover site ranked with "hemispheric importance" and has been referred to as the "Serengeti of bird watching" (Niles et al. 2009; Degnarain 2020). Overharvest and habitat loss during the late 1990s through early 2000s led to substantial spawning horseshoe crab decline and populations of dependent species declined with them (Niles et al. 2009; Faurby et al. 2010). As adult horseshoe crabs decline, so too their eggs to consume, predators are forced to switch to alternate food sources, for example, loggerhead sea turtles (*Caretta caretta*) were observed to feed mainly on *Limulus* in the 1980s and switched to blue crab and finfish in the mid-1990s to 2000s (Seney and Musick 2007). If there are little to no food substitutes (Karpanty et al. 2006), the decline of a dominant species alters local food webs and may result in a loss of local biomass (Dobson et al. 2006; Pálinkás 2018; Sanchez-Bayo and Wyckhuys 2019; Wilson 2016). This occurred in the 2000s in Delaware Bay with the decline of spawning horseshoe crabs the population crash of red knots was documented by Baker et al. (2004). By restricting the horseshoe crab harvest to males only in Delaware and with a

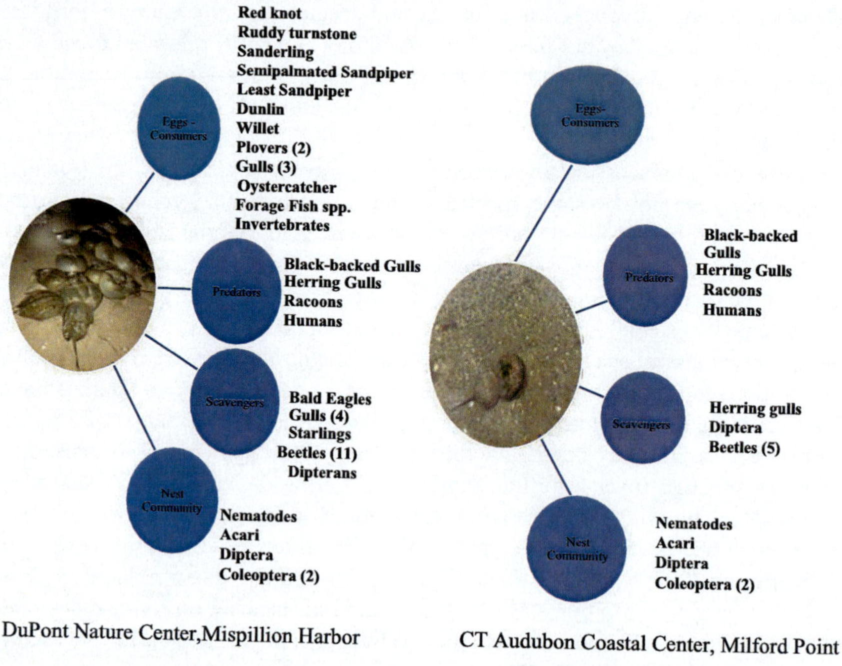


**Fig. 1** Number of species associated with spawning communities of horseshoe crabs in Delaware Bay vs. Connecticut shoreline. This includes observed predation on eggs and live adults, scavengers of horseshoe crab carcasses, and biota surrounding nests buried in the sand approximately 10 cm deep. Number of species identified in parentheses

**Table 1** List of shorebird species observed to be opportunistic consumers of horseshoe crab eggs on the surface of sand May 25–31, 2019 (www.ebird.org)

| Species | Common name |
|---|---|
| *Calidris canutus rufa* | Red knot |
| *Arenaria interpres* | Ruddy turnstone |
| *Calidris alba* | Sanderling |
| *Calidris pusilla* | Semipalmated sandpiper |
| *Calidris minutilla* | Least sandpiper |
| *Calidris alpina* | Dunlin |
| *Tringa semipalmata* | Willet |
| *Pluvialis squatarola* | Black-bellied plover |
| *Charadrius semipalmaus* | Semipalmated plover |
| *Limnodromus griseus* | Short-billed dowitcher |
| *Leucophaeus atricilla* | Laughing gull |
| *Larus argentatus* | Herring gull |
| *Larus delawarensis* | Ring-billed gull |
| *Haematopus palliatus* | American oystercatcher |

**Table 2** Scavengers of dead or dying adult horseshoe crabs found on beaches at (**A**) DuPont Nature Center, Mispillion Harbor, Delaware, and (**B**) CT Audubon Coastal Center, Milford

| A) | |
|---|---|
| **Bird species** | **Common name** |
| *Leucophaeus atricilla* | Laughing gull |
| *Larus argentatus* | Herring gull |
| *Larus delawarensis* | Ring-billed gull |
| *Larus marinus* | Great black-backed gull |
| *Haliaeetus leucocephalus* | Bald eagle |
| *Turnus vulgaris* | European starling |
| **Insect species** | **Family** |
| *Baeckmanniolus dimidiatippenis* | Histeridae |
| *Hypocaccus fraternus* | Histeridae |
| *Saprinus pensylvanicus* | Histeridae |
| *Omorgus suberosus* | Trogidae |
| *Phaleria testacea* | Tenebrionidae |
| *Blapstinus metallicus* | Tenebrionidae |
| *Collops nigriceps* | Melyridae |
| *Coccinella septempunctata* | Coccinellidae |
| *Cycloneda munda* | Coccinellidae |
| *Astylopsis sexguttata* | Cerambycidae |
| *Dermestes carnivorus* | Dermestidae |
| *Diptera, spp.* | Numerous fly larvae |

| B) | |
|---|---|
| **Bird species** | **Common name** |
| *Larus argentatus* | Herring gull |
| **Insect species** | **Family** |
| *Baeckmanniolus dimidiatippenis* | Histeridae |
| *Hypocaccus fraternus* | Histeridae |
| *Saprinus pensylvanicus* | Histeridae |
| *Phaleria testacea* | Tenebrionidae |
| *Dermestes frischii* | Dermestidae |
| *Diptera, spp.* | Fly larvae |

Source: Mattei pers. obs., eBird; Insect species, Conover et al. (2015), Pupedis (2019)

complete ban on harvest in New Jersey, the Delaware Bay population has stabilized (Smith et al. 2017; ASMFC 2019).

Avolio et al. (2019) suggests a method of testing the dominant status of a species is by examining the community diversity and biomass where it has been removed or has declined. In LIS, the population is low and has continued to decline due to over-harvest and habitat loss (Mattei et al. 2010; Atlantic States Marine Fisheries Commission (ASMFC) 2019). This provides an opportunity to comparatively test if horseshoe crabs are a dominant species when abundant and not dominant when the

**Table 3** Invertebrate taxa found in or near horseshoe crab nests (depth 5–10 cm) in both DuPont Nature Center, Mispillion Harbor, Delaware, and CT Audubon Coastal Center, Milford (Conover et al. 2015; Pupedis 2019)

| Order | Family |
|---|---|
| Coleoptera | Histeridae (*Baeckmanniolus dimidiatippenis*) larvae and adults<br>Tenebrionidae spp. |
| Diptera | Unknown species |
| Acari | Unknown species |
| Nematodes | Unknown species |

population is sparse. If horseshoe crabs are found in low densities, the surrounding ecosystem will be depauperate because of dramatically lower egg and spawning adult densities. We examined data collected from the DuPont Nature Center, Mispillion Harbor, Delaware, and CT Audubon Coastal Center, Milford Point, during visits in the spring of 2019. Figure 1 illustrates the list of species found and their trophic status. Horseshoe crabs no longer function as a dominant species on the beaches of LIS, no shorebirds were observed feeding on horseshoe crab eggs, there were fewer shorebird species, and those present have comparably lower densities compared to Delaware Bay. This is most likely due to loss of abundant horseshoe crab eggs to feed on during spring migration (Table 1, Beekey et al. 2013). In fact, the only known bird species that has been documented consuming horseshoe crabs eggs annually on a few beaches in LIS is the Canada goose (*Branta canadensis*). In Rye, NY, and Guilford, CT, geese were observed floating over nesting areas before the tide went out, moving their webbed feet rapidly and causing the *Limulus* eggs to float to the sandy surface. They would then consume all the eggs (Mattei, unpublished data). This was later confirmed at low tide by collecting goose scat and easily finding the chorion of many *Limulus* eggs in the scat (Mattei, unpublished data). In addition, there were fewer insect species and numbers of individual insects associated with fewer decaying adults and within sparse horseshoe crab nests based on a preliminary study conducted by staff at the Peabody Museum (Table 2, Pupedis 2019).

The LIS horseshoe crab population is in need of conservation, not to prevent extirpation but to increase the population density to restore its dominant status. One of the first steps to properly manage an ecologically important dominant species that is in decline is to assess its movement patterns and breeding range. Do spawning horseshoe crabs on the coast of CT migrate to neighboring States and return again? Are there separate breeding populations in CT and NY? Does the NY region contain a genetically homogeneous population or is it divided into subpopulations? In 2006, the Connecticut Department of Energy and Environmental Conservation (CTDEEP) set aside three no harvest areas, one at Sandy Point in West Haven and one at Milford Point, Milford, and one on a small island near Westbrook. Now, more than 10 years later, are these refugia increasing the number of spawning horseshoe crabs in LIS?

In this chapter, we report on *Limulus* movement patterns by using US Fish & Wildlife Service (USFWS) white disc tags and citizen scientist participants to tag and collect recapture information on Long Island Sound horseshoe crabs over a

20-year period, covering 110 miles of coastline. We examine evidence for change in the spawning population of horseshoe crabs in CT. We also report on microsatellite analysis of five sample populations distributed across the Sound to test if the genetic population data matches the tag movement patterns. It is important to know if *Limulus* that cross the Sound from CT to NY, subsequently spawn in NY and if individuals originally tagged in Rye, NY, spawn as far away as Groton, CT (~140 kms). This information will inform State fisheries regulators of the importance of better management practices for these populations to allow for greater biodiversity to return to the New York region. The methods and results from this work may help with the study and management of the three Asian species of horseshoe crab that have also experienced population declines throughout their distribution. The authors have encouraged colleagues in Hong Kong, Malaysia, and India to try these methods by sending specially made tags for deployment on horseshoe crabs in their study areas.

# 2 Methods

## *2.1 Mark-Recapture Study*

Project *Limulus* researchers and volunteers have been tagging horseshoe crabs in LIS since 1997. From 1997 to 1998, we used a three-inch blue plastic T-bar-type tag often used in fish research. We found this method difficult to use and the T-bars were easily pulled out and lost. During the 1999 spawning season through 2006, we tagged horseshoe crabs using yellow Floy Cinch-tags (model FT-4, 8″: http://www.floytag.com). In 2007, we switched to white disc tags issued by the U.S. Fish and Wildlife Service (USFWS). For more detailed methods, see Beekey and Mattei et al. (2015). The tag number, sex, mating behavior, and shell condition were recorded (Mattei et al. 2010). Copies of the data sheets we use with our citizen scientists are available at www.projectlimulus.org. Recapture data (i.e., re-sightings) were obtained by researchers and volunteers who reported tag numbers to the USFWS phone number on the tag or via email to info@projectlimulus.org. We placed signs at pavilions near beaches and coastal nature centers so people who encounter tagged horseshoe crabs learned how to report the re-sighting. Of the 21,245 recapture records, approximately 15,655 horseshoe crabs recaptured from 1998 through 2018 were analyzed and modeled to discover movement patterns. For data cleaning, visualization, and data modeling details, see Youssef et al. (2019). An app was created to help engage the public and inform NY and CT State conservation staff of the movement patterns of adult horseshoe crabs throughout the region (https://senbel.shinyapps.io/RCrab/). Shiny is an R package that allows the user to build interactive web apps straight from R. Once the data were cleaned, maps were created for the movement patterns from where the horseshoe crabs were originally tagged to where they were recaptured. The majority of the recaptures were reported by citizen scientist volunteers (horseshoe crabs were released after the tag was reported) and some individual crabs were reported multiple times. The beaches of Long Island Sound were divided into five regions labeled (A) Rye, New York, (B) Milford, (C) West Haven, (D) Groton in Connecticut, and (E) North Shore, Long Island, New York. These regions correlate with the areas where we collected tissue samples for genetic analysis (Table 4).

## 2.2 *Spawning Census Data*

To estimate the density of spawning horseshoe crabs in a defined area, we trained citizen scientist to survey spawning beaches near their homes. The reported average annual spawning index includes data from all CT beaches surveyed from 2008 to 2018. Four beaches were surveyed every year: Milford Point in Milford, Short Beach in Branford, Bluff Point in Groton, and Calf Pasture Beach in Norwalk. Citizen scientists and even researchers come and go during a 20-year period so the number of beaches included in the annual index varies (Table 5). From 2008 to 2010, surveys were conducted as described in Smith et al. (2002) except 5 $m^2$ quadrats were used to count horseshoe crabs every 10 meters of the length of a beach. After 2010, the method was simplified by counting every horseshoe crab found along a premeasured length of beach within 3 m of the waterline starting at high tide. CT has many pocket beaches and, in some cases, the entire length of the beach was surveyed, and in others the majority of a beach was surveyed. While one person walked the waterline starting the survey at high tide, the other team member held a 3 m rope taught and walked three meters into the water. The team walked together the length of the premeasured survey area, which resulted in a rectangle of known area for calculating spawning females/$m^2$. All horseshoe crabs encountered were counted and listed by their sex and mating behavior (single, pairs, or female plus the number of males around her). Solitary females were not included in the spawning index. Yearly average spawning indices were calculated by averaging the spawning female density for each site (night surveys) within each lunar cycle from May through June (e.g., May new/full moon, June full/new moon) each year.

**Table 4** Total count of recaptured horseshoe crabs included in the analysis was 15,655: (A) Each region listed below includes the number of horseshoe crabs originally tagged in that area; (B) number of horseshoe crabs that were recaptured in each region

| A | | | | | |
|---|---|---|---|---|---|
| Rye Beach (A) | Milford Point (B) | Sandy Point (C) | Bluff Point (D) | Long Island (E) | Other |
| 1430 | 6155 | 3946 | 3815 | 137 | 172 |
| B | | | | | |
| Rye Beach (A) | Milford Point (B) | Sandy Point (C) | Bluff Point (D) | Long Island (E) | Other |
| 1198 | 5883 | 3985 | 3447 | 657 | 485 |

**Table 5** Number of night-time horseshoe crab spawning surveys conducted on Connecticut beaches from 2008 to 2018

| Town | Beach name | 2008 | 2009 | 2010 | 2011 | 2012 | 2013 | 2014 | 2015 | 2016 | 2017 | 2018 | Total |
|---|---|---|---|---|---|---|---|---|---|---|---|---|---|
| Branford | | | | | | | | | | | | | |
| | Bayberry Beach | 4 | 15 | | | | | | | | | | 19 |
| | Montgomery Beach | 3 | 13 | | | | | | | | | | 16 |
| | Short Beach | 6 | 14 | 12 | 10 | 15 | 10 | 17 | 2 | 3 | 5 | 7 | 101 |
| Bridgeport | | | | | | | | | | | | | |
| | Arthur Street | 3 | | | | | | | | | | | 3 |
| | Ash Creek | 3 | | | | | | | | | | | 3 |
| | Ferris Street | 1 | | | | | | | | | | | 1 |
| | Seabright Beach | 3 | | | | | | | | | | | 3 |
| Clinton | | | | | | | | | | | | | |
| | Cedar Island | | | 4 | 8 | 10 | 2 | | | | | | 24 |
| | Town Beach | 6 | | 3 | | | | | | | | | 9 |
| Fairfield | | | | | | | | | | | | | |
| | Penfield 1 | | | | 4 | | | | | | | | 4 |
| | Penfield 2 | | | | 4 | | | | | | | | 4 |
| | Penfield 3 | | | | 3 | | | | | | | | 3 |
| | Penfield 4 | | | | 3 | | | | | | | | 3 |
| | Penfield Beach | 5 | 11 | 7 | 2 | 2 | 9 | 6 | | | | | 42 |
| | Penfield Reef | | 5 | 1 | | | | | | | | | 6 |
| Greenwich | | | | | | | | | | | | | |
| | Elias Point | | | 3 | | | | | | | | | 3 |
| | Greenwich Point I | 2 | | | | | | | | | | | 2 |
| | Greenwich Point II | 3 | | | | | | | | | | | 3 |
| | Greenwich Point III | 11 | 1 | | | | | | | | 2 | | 14 |
| | Greenwich Point IV | 5 | 1 | | | | | | | | 2 | | 8 |
| | Greenwich Point Park | | | | | | | | | | | 1 | 1 |
| | Greenwich Point V | 6 | 1 | | | | | | | | 2 | | 9 |
| | Shell Island | | | 1 | | | | | | | | | 1 |
| | Tod's Point | | | | | | | | | | | 8 | 8 |
| Groton | | | | | | | | | | | | | |
| | Bluff Point | 2 | 6 | 2 | 6 | 7 | 11 | 5 | 3 | 4 | 5 | 3 | 54 |
| | Esker Point | 1 | 6 | 1 | 5 | 3 | 1 | 1 | 4 | | | 2 | 24 |
| | Esker Point-GLP Kiddie | | | | | | | | | | 1 | 2 | 3 |
| | Jupiter Point | | | | | 11 | | | | | | | 11 |

(continued)

**Table 5** (continued)

| Town | Beach name | 2008 | 2009 | 2010 | 2011 | 2012 | 2013 | 2014 | 2015 | 2016 | 2017 | 2018 | Total |
|---|---|---|---|---|---|---|---|---|---|---|---|---|---|
| | Kiddie Beach | | 1 | 1 | | 1 | | 1 | | | | | 4 |
| | Kiddie Beach- GLP | | | | | | | | | | | 1 | 1 |
| Guilford | | | | | | | | | | | | | |
| | Indian Cove Private | | 3 | 12 | 13 | 11 | 7 | 12 | 6 | 5 | 8 | 7 | 84 |
| | Indian Cove Public | | 3 | 13 | 11 | 11 | 7 | 9 | 6 | 5 | 8 | 9 | 82 |
| | Joshua Cove | | 3 | 8 | | | | | | | | | 11 |
| Milford | | | | | | | | | | | | | |
| | Milford Point – Front | 15 | 15 | 13 | 7 | 8 | 11 | 8 | 5 | 6 | 7 | 12 | 107 |
| | Milford-Pointside | | 11 | 10 | | | | | | | | | 21 |
| | Smith's Point | | | | | | | | 1 | | | | 1 |
| Mystic | | | | | | | | | | | | | |
| | Enders Island | | 3 | | 1 | 1 | | | 1 | | | | 6 |
| | Masons Island | | 2 | | | | | | | | | | 2 |
| | Williams (YMCA) | | 5 | | 8 | 4 | 2 | 5 | 4 | | 1 | | 29 |
| Niantic | | | | | | | | | | | | | |
| | Giant's Neck | | | | 11 | 3 | 4 | | | | | | 18 |
| Noank | | | | | | | | | | | | | |
| | Mumford Cove | | | | 1 | | | 1 | | | | 1 | 3 |
| Norwalk | | | | | | | | | | | | | |
| | Calf Pasture Beach | 9 | 13 | 9 | 2 | 4 | 1 | 5 | 3 | 3 | 2 | 3 | 54 |
| Old Lyme | | | | | | | | | | | | | |
| | Griswold Point | | | | | | 1 | | | | | | 1 |
| Old Saybrook | | | | | | | | | | | | | |
| | Harvey's Beach | | | 4 | 1 | | | | | | | | 5 |
| | Cove Island East | 3 | | 3 | | | | | | | | | 6 |
| | Cove Island West | 3 | | 3 | | | | | | | | | 6 |
| | Horseshoe Beach | | | 4 | | | | | | | | | 4 |
| | Southfield Beach | 1 | | | | | | | | | | | 1 |
| Stonington | | | | | | | | | | | | | |
| | Big Dock | | 4 | | | | | | | | | | 4 |
| | East Neck | | | | 1 | | | | | | | | 1 |
| | Hopkins Beach | | | | | | | | | | | 1 | 1 |

(continued)

**Table 5** (continued)

| Town | Beach name | 2008 | 2009 | 2010 | 2011 | 2012 | 2013 | 2014 | 2015 | 2016 | 2017 | 2018 | Total |
|---|---|---|---|---|---|---|---|---|---|---|---|---|---|
| | Lords PT-Hopkins | | | | 4 | 5 | | 1 | 4 | 2 | | 3 | 19 |
| | Open Way | | 3 | | | | | | | | | | 3 |
| | Sandy Point | | 2 | | 3 | 5 | | | 2 | | 1 | 2 | 15 |
| | Trestle Beach | | 4 | | | 1 | 1 | | | | | | 6 |
| | Wamphassuc | | | | | 6 | | | | | | | 6 |
| Stratford | | | | | | | | | | | | | |
| | Stratford PT | | | | 3 | 12 | 11 | 6 | 5 | 5 | 7 | 12 | 61 |
| Waterford | | | | | | | | | | | | | |
| | Jordan Cove | 2 | | 3 | 4 | 2 | | | | | | | 11 |
| West Haven | | | | | | | | | | | | | |
| | Sandy Point | 12 | 17 | 12 | 6 | 11 | 8 | 11 | 9 | 6 | | | 92 |
| Westbrook | | | | | | | | | | | | | |
| | Menunketesuck Island | | | | | | | | 1 | | | | 1 |
| | West Beach | | | | | | | | 1 | | | | 1 |
| Total | | 109 | 162 | 129 | 121 | 133 | 86 | 88 | 57 | 39 | 51 | 74 | 1049 |

Total number of surveys at each beach and for each year included. Zeros omitted for clarity

## *2.3 Genetic Characterization of LIS Horseshoe Crab Population*

Horseshoe crab leg muscle tissue samples were collected from

A. Rye Playland Beach, Rye, NY – 38 samples (RB)
B. Milford Point Beach, Milford, CT – 38 samples (MP)
C. Sand Point Beach, West Haven, CT – 35 samples (SP)
D. Bluff Point Beach, Groton, CT – 38 samples (BP)
E. Cedar Beach, Mt. Sinai, NY – 37 samples (CB)

The distal segment of the third walking leg of each individual was excised, stored on ice in the field, and subsequently frozen at −80 °C until further processing. Frozen muscle tissue dissected from stored samples was ground in liquid nitrogen without thawing and stored at −80 °C. DNA was isolated from ground tissue with the Gentra Puregene Cell Kit (Quiagen, Valencia, CA) following the manufacturer's protocols. Eleven Microsatellite Loci (A315, D6, A40, D3, A67, A37, A42, A64, A5, A52, and A38) were genotyped by fluorescent primer PCR using primer pairs and amplification conditions previously described in King et al. (2005) and Kasinak et al. (2011). Amplified microsatellite loci were sized by capillary electrophoresis at the DNA Analysis Facility on Science Hill (Yale University, New Haven, CT) using an Applied Biosystems 3730xl 96-Capillary Genetic Analyzer (DS31 5 Color Dye Family, Liz-500 size standard). Electrophoretogram output was analyzed for determination of fragment size and microsatellite allele assignment using GeneMarker

(Softgenetics, State College, PA). Data were analyzed for inconsistent allele calls and the presence of null alleles using Micro-Checker (Van Oosterhout et al. 2004). Genotype data were analyzed for linkage disequilibrium and Hardy–Weinberg equilibrium using the GenePop (Raymond and Rousset 1995; Rousset 2008) version 4.2 as implemented on the website http://genepop.curtin.edu.au/. http:// using the default Markov chain parameters. The following analyses of the genotype data were calculated using Genalex version 6.503 (Peakall and Smouse 2012): (1) AMOVA (CoDom-Allelic-$F_{ST}$, Standard Permute) for pairwise comparison of genetic differentiation between sample sites (Michalakis and Excoffier 1996), (2) G-statistics (Codominant, 999 permutations) for estimation of global population genetic parameters (number of alleles, heterozygosity, $G''_{ST}$, etc.) and pairwise comparison of genetic differentiation between sample sites (Meirmans and Hedrick 2011), and (3) population assignment (Codominant, Assign All Pops, Leave One Out) for genotype-based assignment of samples individuals to sample sites (Paetkau et al. 2004).

# 3 Results

## 3.1 *Mark-Recapture Study*

Citizen scientists (trained volunteers) tagged and gathered data on 62,214 horseshoe crabs over 20 years of the study. Sacred Heart University faculty and students tagged 33,353 horseshoe crabs during the same period for a total of 95,567 tags deployed between 1999 and 2019. Citizen scientist volunteers greatly increased the tagging effort of horseshoe crabs in LIS and tagged up to 76% of animals in a given season (Fig. 2).

The majority (82%) of tagged/recaptured horseshoe crabs are found within less than 10 km from where they were originally tagged and 79% of the recaptured horseshoe crabs were only found once. Three horsecrabs were recaptured 7, 8, and 9 times, respectively. There was no correlation found between the time that past between recaptures and the distance traveled (see Youssef et al. 2019 for details).

Within LIS, horseshoe crabs exhibit moderate site fidelity, 18% of the horseshoe crab recaptures were more than 10 km away from where they were initially tagged. Horseshoe crabs were tagged and found to cross LIS to Long Island (NY) beaches from all regions of this study. In Rye, NY region (A), 1,430 spawning adult crabs were tagged and recaptured of which 164 (11.47%) were recaptured on Long Island region beaches (E) (Fig. 3a). From the Milford region (B) 6,155 recaptured tagged adults resulted in 256 (4.16%) recaptured on Long Island (E) (Fig. 3b). In West Haven, CT region (C), out of the 3,946 tagged and recaptured horseshoe crabs, 89 (2.26%) were recaptured on Long Island (E) (Fig. 3c). It was found in Groton, CT (D), that of the 3,815 horseshoe crabs tagged and recaptured, 38 (1%) were

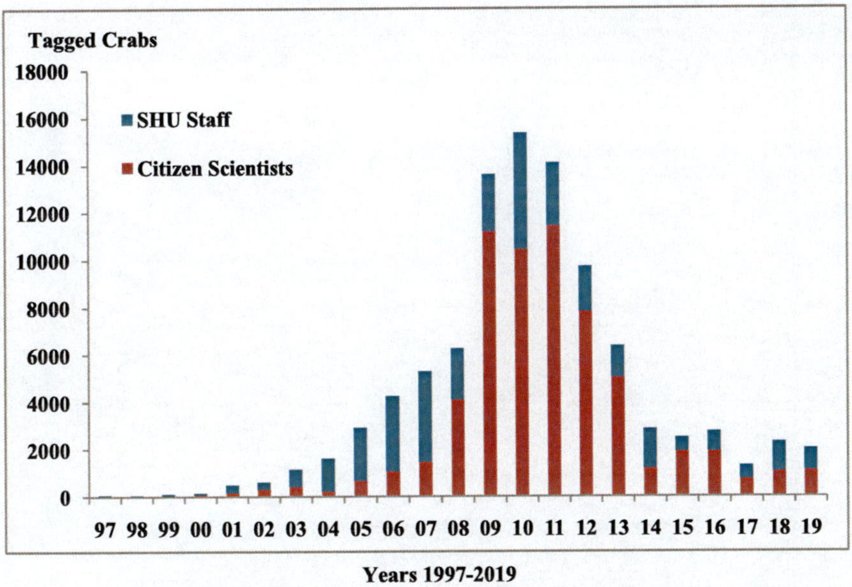


**Fig. 2** Number of horseshoe crabs tagged by research team at Sacred Heart University vs. citizen scientists

recaptured on Long Island (E) (Fig. 3d). Finally, of the 137 tagged and recaptured horseshoe crabs on Long Island (E), 30 (21.89%) crossed the Sound and were recaptured in CT and Rye, NY regions (A–D) (see Fig. 3e).

Analysis of the tag data reveals that over the past 10 years the average number of spawning females paired with one male was 72%, single females remained high at 25%, and females found with multiple males remained low on average less than 3%. There were no significant changes in number of single females, pairs, or spawning groups (Fig. 4).

## *3.2 Spawning Census Results*

From 2008 through 2018, 19 different CT beaches have been surveyed at night for spawning horseshoe crabs (Table 5) for a total of 1,049 surveys. Some beaches are only surveyed once a season, or only surveyed for 1 year, but several beaches have been surveyed multiple times every year since 2008 (Table 5). Ten years of spawning census data indicate that the spawning index fluctuates from year to year and has remained very low (0.009–0.036 females/m$^2$, Fig. 5).

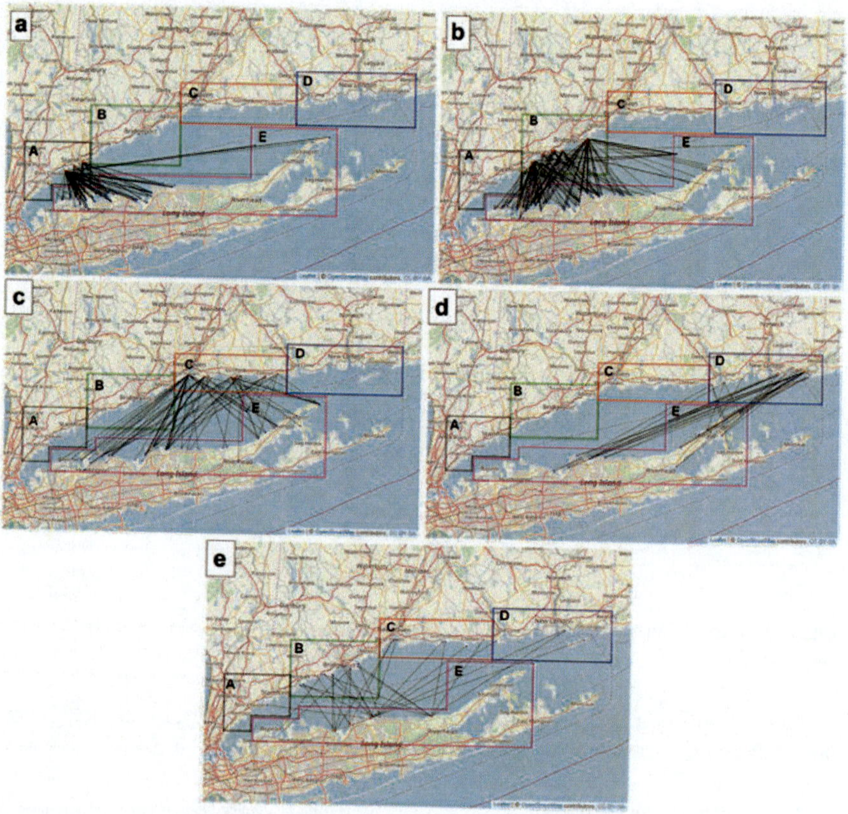


**Fig. 3** Map with lines depicting the shortest distance traveled by horseshoe crabs that were originally tagged on beaches (circles) of (**a**) Rye, New York region (A), and were re-sighted on other regions on beaches of Long Island (E, arrowheads), (**b**) originally tagged on beaches of Milford region (B) and were re-sighted on beaches of Long Island (E, arrowheads), (**c**) West Haven region (C) and were re-sighted on beaches of Long Island (E, arrowheads), (**d**) Groton region (D) and were re-sighted on beaches of Long Island (E, arrowheads), (**e**) Long Island (E, circles) and were re-sighted on other regions in Connecticut (B, C, and D, arrowheads) and in the Rye, New York region (A)

## 3.3 *Genetic Characterization of LIS Horseshoe Crab Population*

To date, a total of 186 individual horseshoe crabs from five different beach locations within and just adjacent to LIS have been genotyped at 11 different microsatellite loci (Fig. 6) in order to assess the genetic health of the LIS population and test for evidence of regional population differentiation within LIS. No significant deviations from Hardy–Weinberg equilibrium were detected, and there was no evidence of significant inbreeding with a near zero $F_{IS}$ (Wright's inbreeding coefficient) value

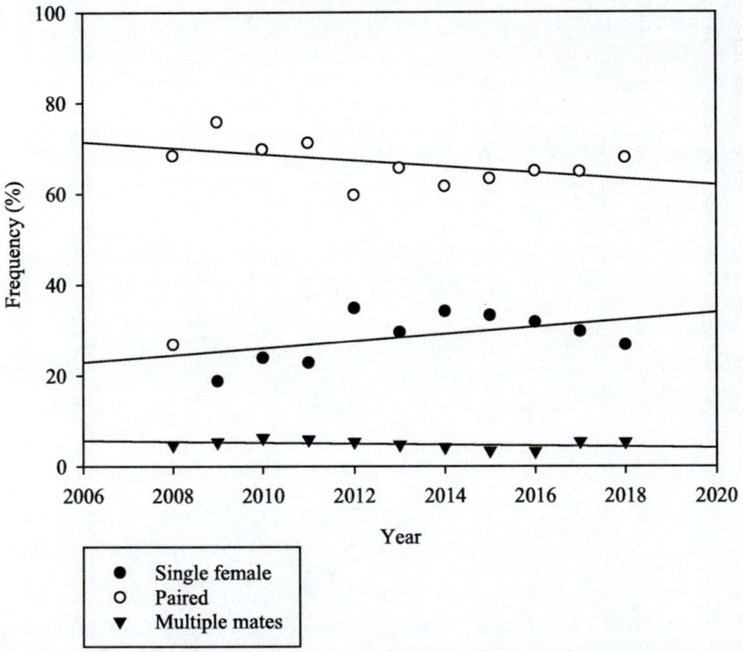


**Fig. 4** Mating patterns for all female horseshoe crabs tagged from 2008 to 2018
Paired females (P) open circles, single females (SF) closed circle, and females with multiple (MM) mates triangles are plotted. ($P$: $y = -0.6769x + 1429.2$, $R^2 = 0.2396$, $p = 0.13$), (SF: $y = 0.7891x - 1560$, $R^2 = 0.2623$, $p = 0.11$), (MM: $y = -0.1122x + 230.79$, $R^2 = 0.1463$, $p = 0.25$)

($F_{IS}$-0.010 +/– 0.011). Overall number of alleles, number of effective alleles, observed heterozygosity, and F-statistics (data not shown) were similar to those reported previously for mid-Atlantic and southern New England (King et al. 2005).

Analysis of regional genetic differentiation within LIS based on the microsatellite data described above indicates no significant genetic differentiation between samples sites. Pairwise comparisons based on $G''_{ST}$, chosen due to the low number of sample sites (Meirmans and Hedrick 2011), result in $G''_{ST}$ values very close to zero, indicating little genetic differentiation between samples sites. Furthermore, the high probabilities of greater than or equal to $G''_{ST}$ values when comparing random groups of sampled individuals are in support of a genetically homogenous population (Fig. 6). The regional genetic differentiation within Long Island Sound, AMOVA analysis based on $F_{ST}$ indicates 0% of the observed molecular variance among populations (sample sites) and population assignment tests assign individuals to sample sites correctly only 18% of the time. This is slightly worse than would be expected based on random assignment with five populations and again supports that this is a genetically homogenous population (20%, data not shown).

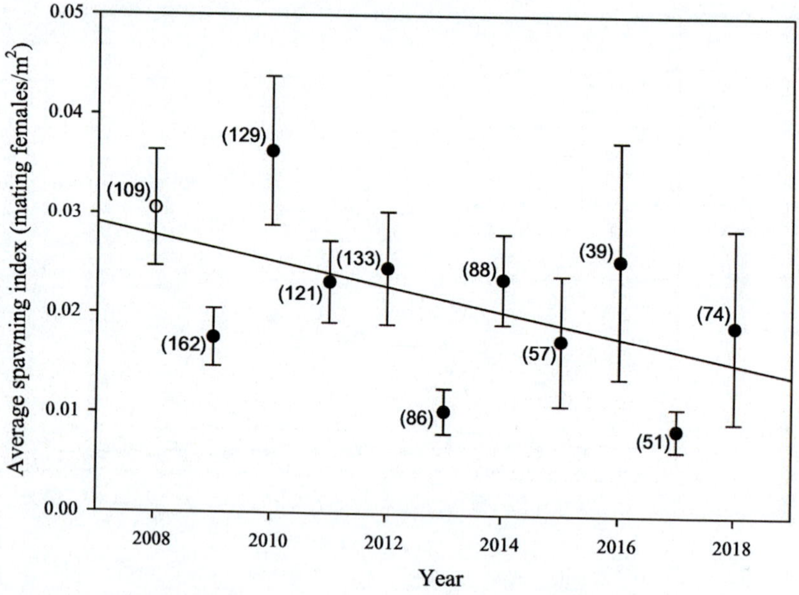


**Fig. 5** Average nighttime spawning index/m$^2$ ± SE for all surveyed beaches from 2008 to 2018. Parentheses indicate the number of surveys included in the average ($Y = -0.0013x + 0.0291$, $R^2 = 0.2609$, $p = 0.11$)

## 4 Discussion

By involving the public in this research (see www.projectlimulus.org), not only were we able to gather a very large data set both over time (20 years) and over space (>160 km), but we also gathered support for the conservation of a dominant species in the LIS ecosystem that prior to the start of the study was not recognized as important. Many of our citizen scientists are now advocating for a moratorium on harvesting horseshoe crabs in LIS. Community members from the Town of Stratford that participated as citizen scientist became invested in horseshoe crab conservation. They successfully pressured their State representatives to pass a bill to protect spawning horseshoe crabs from harvest on the Town's beaches (Dixon 2017). This was a rare political event given that the CT Department of Energy and Environmental Protection (CTDEEP) regulates all harvesting that occurs off CT waters and individual towns are usually discouraged from making their own harvest regulations. Since 2001, Connecticut's annual quota has been set by ASMFC at 48,689 crabs. The reported annual harvest since 2001 has generally varied between 15,000 and 30,000 crabs and has never approached the state quota; however, the catch may be underreported and there may also be illegal harvest. Licensed harvesters may take horseshoe crabs by hand off of spawning beaches (both males and females) or catch them in trawl nets (CTDEEP 2020). CTDEEP noted the decline in

| $G''_{st}$ | CB | SP | BP | RB | MP |
|---|---|---|---|---|---|
| CB | 0.000 | 0.717 | 0.629 | 0.229 | 0.591 |
| SP | -0.008 | 0.000 | 0.910 | 0.430 | 0.887 |
| BP | -0.005 | -0.015 | 0.000 | 0.139 | 0.996 |
| RB | 0.008 | 0.002 | 0.013 | 0.000 | 0.412 |
| MP | -0.004 | -0.014 | -0.027 | 0.001 | 0.000 |

**Fig. 6** Sampling locations/sampled individuals and analysis of genetic differentiation between sample sites. Sampling locations: RB (Rye Beach, Rye, NY), CB (Cedar Beach, Mt. Sinai, NY), MP (Milford Point, Milford, CT), SP (Sandy Point, West Haven, CT), and BP (Bluff Point, Groton, CT). Pairwise comparisons of genetic differentiation between sample sites is reported as $G''_{ST}$ (below diagonal). Probability of $G''_{ST}$ for random selection of subpopulations from all sampled individuals being greater than or equal to value is reported above table diagonal

the horseshoe crab population from their annual trawl data and in 2006 set aside three area beaches as no harvest zones for *Limulus* during the spawning season, Milford, West Haven, and Westbrook. These no harvest areas were expected to allow more females to deposit eggs and allow for an increase in the adult population within 10 years time. Both the tag data and the census data (Figs. 4 and 5) clearly show no significant increase in the adult spawning population after 13 years with the three established no harvest zones. This conservation measure is important; however, it is simply not enough to make a difference in the surviving adult population.

The phenomenon of adult female horseshoe crabs coming to shore during spawning events but without a male attached has been reported in other low-density populations in Massachusetts and Florida (James-Pirri 2010; Brockmann and Johnson 2011). However, over the past 13 years, out of all the females that have been tagged on shore, an average of 25% are without a mate attached. Our working hypothesis is the density of adults is so low that even though males make up 60% of the adult population they may be having difficulty finding the females. We also found in previous work that the population is aging (based on shell condition, Beekey and Mattei 2015) and a small proportion of adults may have injuries to their shells that prevent amplexus (Brockmann and Johnson 2011), but this does not account for the large number of single females documented in LIS. If the density of adult males and females were increasing due to conservation measures, single female numbers would be expected to significantly decline. Figure 4 (closed circles) clearly demonstrates that this is not the case, even with the skewed male sex ratio over 20 years that has ranged from 1:2 to 1:1.5 (females:males, Beekey and Mattei 2015) 20–30% of the females tagged on spawning beaches are alone and not reproducing at their maximum rate.

In general, the majority (82%) of tagged and recaptured horseshoe crabs that reside in LIS stay within the Sound and maintain a moderate home range of under 10 km. Horseshoe crabs that were recaptured 5–10 years after they were initially tagged were not significantly further away from their initial tagging location than those recaptured 1–2 years later (Youssef et al. 2019). However, the horseshoe crabs that did travel were found to be capable of crossing the LIS, whether they had been tagged in the west (Rye, NY) or in the east (Groton, CT) (see Figs. 3a–d). In addition, horseshoe crabs tagged on Long Island can also cross over LIS to CT (Fig. 3e). These types of movement patterns or migration events do not guarantee that the animals are interbreeding and for years NY and CT fisheries management departments have managed their horseshoe crab populations as if they were genetically distinct populations. The genetic analysis we conducted shows no significant difference between the five sample populations tested. The horseshoe crab population living within the LIS is panmictic with no population substructure between Rye, NY, and Groton, CT, and between these northern sites and across LIS to Long Island (Fig. 6). Inbreeding depression was not detected from the alleles we sampled, and this population of horseshoe crab residing in LIS was found to be in relatively good health based on general population genetic parameters. Horseshoe crabs in LIS are not genetically distinct from other mid-Atlantic populations including populations from Delaware Bay (King et al. 2015), and it has been suggested that the entire region south of New Hampshire to and including North Carolina can be treated as a broad management unit (Smith et al. 2017). Recently, the ASMFC (2019) declared the horseshoe crab stock status in CT and NY as poor. By 2021, CTDEEP and NY State Department of Environmental Protection must amend their management plan to increase the population of horseshoe crabs.

The spawning indices observed in the Delaware Bay (Smith and Robinson 2015) are 25–100 times greater than spawning indices currently observed on CT beaches (0.009–0.036 females/$m^2$). The horseshoe crab's important role as a dominant species cannot be fulfilled at these low population levels. Current management strategies, including the four no harvest zones in CT (the fourth includes Stratford beaches) and the current catch limits, are not maintaining the population of horseshoe crabs within the LIS to provide a sustainable fishery into the future (Smith et al. 2017). The extremely low density of spawning horseshoe crabs in the LIS calls for changes in management to allow the population time to recover from decades of overharvest and habitat loss. A moratorium on adult horseshoe crab harvest will build the population back up to harvestable levels over time. In addition, coastal habitat restoration is imperative, not only for horseshoe crabs but for many of the species we harvest that generates revenue for CT, especially those that rely on *Limulus* eggs as a resource. There are no areas of LIS that are left untouched by human activities. Establishing a marine protected area (MPA) in LIS would bring back a healthy and diverse benthic zone that juvenile and adult horseshoe crabs need to survive.

Given their genetic and behavioral similarities, the three Asian horseshoe crab species (*Tachypleus tridentatus, T. gigas,* and *Carcinoscorpius rotundicauda*) were potentially dominant species in the not too distant past (Laurie et al. 2018; Liao et al. 2019). Recently, the tri-spine horseshoe crab (*T. tridentatus*) has been given the status of Endangered on the IUCN Red List (2019) (https://www.iucnredlist.org/species/21309/133299524) due to major population declines throughout its range and soon the other two species will also be listed (conclusion of the 4th International Workshop on the Science and Conservation of Horseshoe Crabs, Qinzhou, China, June, 2019). The three Asian species' populations have been greatly reduced for the same reasons many *Limulus* populations have declined: overharvest, loss of habitat from seawall construction, land reclamation, climate change, entrapment by intake pipes of coastal power plants, toxic pollution, plastic and ghost nets/traps entanglement, marsh reclamation, and intense human development along shorelines, particularly in estuaries (Tan et al. 2012; Faurby et al. 2010; Beekey and Mattei 2015; Mattei et al. 2015; Smith et al. 2017; Botton et al. 2018; John et al. 2018; Vestbo et al. 2018), and in Asia, the addition of being sold for human consumption (John et al. 2018; Laurie et al. 2018). The biodiversity and biomass lost by dominant Asian horseshoe crab species being reduced to Endangered status will never be known for sure. The worldwide decline of horseshoe crabs could very well be linked to the worldwide decline of shorebirds (Weidensaul 2018) among other environmental factors.

All four species of horseshoe crab require a reduction in harvest and coastal habitat restoration. Nature-based solutions have been developed to abate waves and arrest coastal erosion to slow the impact of sea level rise (Mattei et al. chapter 10, this volume) and would be beneficial to horseshoe crabs and the world's human population whose health depends on them (Arnold 2020).

**Acknowledgments** We thank the hundreds of Project *Limulus* volunteers of all ages that have donated thousands of hours of their time to help understand the ecology and conservation of the horseshoe crab in Long Island Sound. Numerous conservation organizations have helped organize volunteers, conduct tagging sessions, collate data, and assist with our educational outreach. In particular, we would like to thank the staff and members of The Maritime Aquarium, CT Audubon Society, Audubon Connecticut, and many others. The tagging aspect of the project would not have been possible without the assistance of Josh Newhard and the staff at the US Fish & Wildlife Service, Maryland Fishery Resources Office. We would like to particularly recognize our undergraduate students who have worked on these projects and kept the energy level high as well as the quality of data collected in excellent condition. Finally, we thank Dr. S.G. Cheung and anonymous reviewers for feedback that greatly improved this manuscript.

## References

C. Arnold, Horseshoe crab blood is key to making a COVID-19 vaccine-but the ecosystem may suffer. National Geographic, July 2, 2020 (2020)

ASMFC, 2019 Horseshoe Crab Benchmark Stock Assessment and Peer Review Report, Atlantic States Marine Fisheries Commission, Arlington (2019)

M.L. Avolio, E.J. Forrestel, C.C. Chang, K.J. La Pierre, K.T. Burghardt, M.D. Smith, Demystifying dominant species. New Phytol. **223**, 1106–1126 (2019)

A.J. Baker, P.M. Gonzalez, T. Piersma, et al., Rapid population decline in red knots: Fitness consequences of decreased refueling rates and late arrival in Delaware Bay. Proc. R. Soc. B Biol. Sci. **271**(1541), 875–882 (2004)

M.A. Beekey, J.H. Mattei, The mismanagement of *Limulus polyphemus* in Long Island Sound: What are the characteristics of a population in decline? in *Changing Global Perspectives on Horseshoe Crab Biology, Conservation and Management*, ed. by R. H. Carmichael, M. L. Botton, P. K. S. Shin, S. G. Cheung, (Springer, New York, 2015), pp. 433–461

M.A. Beekey, J.H. Mattei, B.J. Pierce, Horseshoe crab eggs: A rare resource for predators in Long Island Sound. J. Exp. Mar. Biol. Ecol. **439**, 152–159 (2013)

J. Berkson, C.N. Shuster Jr., The horseshoe crab: The battle for a true multiple-use resource. Fisheries **24**(11), 6–10 (1999)

A. Bortolus, P. Adam, J.B. Adams, M.L. Ainouche, et al., Supporting *Spartina*: Interdisciplinary perspective shows *Spartina* as a distinct solid genus. Ecology **100**, e02863 (2019)

M.L. Botton, The ecological importance of horseshoe crabs in estuarine and coastal communities: A review and speculative summary, in *Biology and Conservation of Horseshoe Crabs*, ed. by J. T. Tanacredi, M. L. Botton, D. R. Smith, (Springer, New York, 2009), pp. 45–63

M.L. Botton, C.N. Shuster, J.A. Keinath, Horseshoe crabs in a food web: Who eats whom? in *The American Horseshoe Crab*, ed. by C. N. Shuster Jr., R. B. Barlow, H. J. Brockmann, (Harvard University Press, Cambridge, MA, 2003), pp. 133–153

M.L. Botton, C.P. Colón, J. Rowden, S. Elbin, D. Kriensky, K. McKown, M. Sclafani, R. Madden, Effects of a beach nourishment project in Jamaica Bay, New York, on horseshoe crab (*Limulus polyphemus*) spawning activity and egg deposition. Estuar. Coasts **41**(4), 974–987 (2018)

H.J. Brockmann, S.L. Johnson, A long-term study of spawning activity in a Florida Gulf Coast population of horseshoe crabs (*Limulus polyphemus*). Estuar. Coast **34**, 1049–1067 (2011)

S.B. Carroll, *The Serengeti Rules: The Quest to Discover How Life Works and Why It Matters* (Princeton University Press, Princeton, 2016)

D.R. Conover, R. Pupedis, J.H. Mattei, Colonization of *Limulus polyphemus* nests by invertebrates on sandy beaches of Long Island Sound. Professional Science Master's Thesis, Sacred Heart University, Fairfield (2015)

S.M. Crotty, A.H. Altieri, J.F. Bruno, H. Fischman, M.D. Bertness, The foundation for building the conservation capacity of community ecology. Front. Mar. Sci. **6**, 1–8 (2019)

CTDEEP (CT Department of Energy and Environmental Protection), Public information document, Horseshoe Crab Public Informational Meeting CT DEEP Marine Headquarters, Old Lyme, CT, https://portal.ct.gov/-/media/DEEP/fishing/NoticesFishermen/HorseshoeCrabMeeting.pdf?la=en (2020)

N. Degnarain, Big Pharma's U-turn threatens 450-million-year-old species on front line of coronavirus defense, Forbes, https://www.forbes.com/sites/nishandegnarain/2020/06/28/big-pharmas-u-turn-threatens-450-million-year-old-species-on-front-line-of-coronavirus-defense/#1f92e77625d5 (June 20, 2020)

K. Dixon, Horseshoe crab ban okayed for Stratford. CT Post, May 10th, https://www.ctpost.com/local/article/Horseshoe-crab-ban-okayed-for-Stratford-11135901.php (2017)

A. Dobson, D. Lodge, J. Alder, G.S. Cumming, J. Keymer, J. McGlade, H. Mooney, Habitat loss, trophic collapse, and the decline of ecosystem services. Ecology **87**(8), 1915–1924 (2006)

A.M. Ellison, Foundation species, non-trophic interactions, and the value of being common. iScience **13**, 254–268 (2019)

Environmental Protection Agency, DDT ban take effect December 31, 1972 Press Release, https://archive.epa.gov/epa/aboutepa/ddt-ban-takes-effect.html (1972)

P.S.E.Z. Ermgassen, M.D. Spalding, B. Blake, et al., Historical ecology with real numbers: Past and present extent and biomass of an imperiled estuarine habitat. Proc. R. Soc. B **279**, 3393–3400 (2012)

B. Farquhar, Wolf reintroduction changes ecosystem in Yellowstone, https://www.yellowstonepark.com/things-to-do/wolf-reintroduction-changes-ecosystem (2019)

S. Faurby, T.L. King, M. Obst, E.M. Hallerman, C. Pertoldi, P. Funch, Population dynamics of American horseshoe crabs- historic climatic events and recent anthropogenic pressures. Mol. Ecol. **19**, 3088–3100 (2010)

J.H. Grabowski, R.D. Brumbaugh, R.F. Conrad, A.G. Keeler, J.J. Opaluch, C.H. Peterson, M.F. Piehler, S.P. Powers, A.R. Smyth, Economic valuation of ecosystem services provided by oyster reefs. Bioscience **62**(10), 900–909 (2012)

J.P. Grime, Dominant and subordinate components of plant communities: Implications for succession, stability, and diversity, in *Colonization, Succession and Stability: 26th Symposium of the British Ecological Society*, ed. by A. J. Gray, M. J. Crawley, P. J. Edwards, (Blackwell, Oxford, 1987), pp. 413–428

S.R.R. Hoy, O. Peterson, J.A. Vucetich, Ecological studies of wolves on Isle Royale. Annual report 2018–2019, School of Forest Resources and Environmental Science, Michigan Technological University, Houghton, Michigan (2019)

A. Irwin, Citizen science comes of age. Nature **562**, 480–482 (2018)

M.J. James-Pirri, Seasonal movement of the American horseshoe crab *Limulus polyphemus* in a semi-enclosed bay on Cape Cod, Massachusetts (USA) as determined by acoustic telemetry. Curr Zool **56**, 575–586 (2010)

B.A. John, B.R. Nelson, H.I. Sheikh, S.G. Cheung, Y. Wardiatno, B.P. Dash, K. Tsuchiya, Y. Iwasaki, S. Pati, A review on fisheries and conservation status of Asian horseshoe crabs. Biodivers. Conserv. **27**, 3573–3598 (2018)

S. Karpanty, J.D. Fraser, J. Berkson, L.J. Niles, A.D. Dey, E. Smith, Horseshoe crab eggs determine red knot distribution in Delaware Bay habitats. J Wildl Manag **70**(6), 1704–1710 (2006)

J. Kasinak, K. Bartholomew, M. Beekey, J.H. Mattei, Movement patterns and population genetics of the American horseshoe crab in relation to Long Island Sound conservation strategies. Proceedings of the 2010 Biennial Long Island Sound Research Conference. University of Connecticut, Stamford, CT, pp. 20–25 (2011)

T.L. King, M. Eackles, A.P. Spidle, H.J. Brockmann, Regional differentiation and sex-biased dispersal among populations of the horseshoe crab *Limulus polyphemus*. Trans. Am. Fish. Soc. **134**, 441–465 (2005)

T.L. King, M.S. Eackles, A.W. Aunins, H.J. Brockmann, E. Hallerman, B.L. Beaver, Conservation genetics of the American horseshoe crab (*Limulus polyphemus*): Allelic diversity, zones of genetic discontinuity, and regional differentiation, in *Changing Global Perspectives on Horseshoe Crab Biology, Conservation and Management*, ed. by R. H. Carmichael, M. L. Botton, P. K. S. Shin, S. G. Cheung, (Springer, New York, 2015), pp. 65–96

J. Krisfalusi-Gannon, W. Ali, K. Dellinger, L. Robertson, T.E. Brady, M.K.M. Goddard, R. Tinker-Kulberg, C.L. Kepley, A.L. Dellinger, The role of horseshoe crabs in the biomedical industry and recent trends impacting species sustainability. Front. Mar. Sci. **5**, 1–13 (2018)

K. Laurie, C.P. Chen, S.G. Cheung, V. Do, H. Hsieh, A. John, F. Mohamad, S. Seino, N. Shin, P. Shin, M. Yang, *Tachypleus tridentatus*. The IUCN Red List of Threatened Species e.T21309A133299524 (2018)

Y. Liao, H. Hsieh, S. Xu, Q. Zhong, J. Lei, M. Liang, H. Fang, L. Xu, W. Lin, X. Xiao, C. Chen, S.G. Cheung, B.K.Y. Kwan, Wisdom of crowds reveals decline of Asian horseshoe crabs in Beibu Gulf, China. Oryx **53**(2), 222–229 (2019)

J.H. Mattei, M.A. Beekey, A. Rudman, A. Woronik, Reproductive behavior in horseshoe crabs: Does density matter? Curr Zool **56**(5), 634–642 (2010)

J.H. Mattei, M. Botton, M. Beekey, C. Colón, Horseshoe crab research in urban estuaries: Challenges and opportunities, in *Changing Global Perspectives on Biology, Conservation, and Management of Horseshoe Crabs*, ed. by R. Carmichael, M. L. Botton, P. K. S. Shin, S. G. Cheung, (Springer Publishing, New York, 2015), pp. 537–555

P. Meirmans, S. Hedrick, Assessing population structure: FST and related measures. Mol. Ecol. Resour. **11**, 5–18 (2011)

Y. Michalakis, L. Excoffier, A generic estimation of population subdivision using distances between alleles with special reference for microsatellite loci. Genetics **142**, 1061–1064 (1996)

G.G. Mittelbach, B.J. McGill, *Community Ecology*, 2nd edn. (Oxford University Press, New York, 2019)
L.J. Niles, J. Bart, H.P. Sitters, A.D. Dey, K.E. Clark, et al., Effects of horseshoe crab harvest in Delaware Bay on red knots: Are harvest restrictions working? Bioscience **59**(2), 153–164 (2009)
D. Paetkau, R. Slade, M. Burden, A. Estoup, Genetic assignment methods for the direct, real-time estimation of migration rate: A simulation-based exploration of accuracy and power. Mol. Ecol. **13**, 55–65 (2004)
R.T. Paine, Food web complexity and species diversity. Am. Nat. **100**, 65–75 (1966)
M. Pálinkás, Characteristics of collapsing ecosystems and main factors of collapses, in *Ecosystem Services and Global Ecology*, ed. by L. Hufnagel, (Intechopen, London, 2018), pp. p17–p38
R. Peakall, P.E. Smouse, GenAlEx 6.5: genetic analysis in Excel. Population genetic software for teaching and research--an update. Bioinformatics **28**(19), 2537–2539 (2012)
R. Pupedis, Notes on invertebrate biota found on dead horseshoe crabs on the beach and surrounding nests of horseshoe crabs (i.e. buried eggs) in Connecticut and Delaware beaches. Collections held at the Peabody Museum, Yale University, New Haven, CT (2019)
M. Raymond, F. Rousset, GENEPOP (version 1.2): Population genetics software for exact tests and ecumenicism. J. Hered. **86**, 248–249 (1995)
S.E. Reisewitz, J.A. Estes, C.A. Simenstad, Indirect food web interactions: Sea otters and kelp forest fishes in the Aleutian archipelago. Oecologia **146**(4), 623–631 (2006)
F. Rousset, Genepop'007: A complete reimplementation of the Genepop software for Windows and Linux. Mol Ecol Resour **8**, 103–106 (2008)
F. Sánchez-Bayo, K.A.G. Wyckhuys, Worldwide decline of the entomofauna: A review of its drivers. Biol. Conserv. **232**, 8–27 (2019)
E.E. Seney, J.A. Musick, Historical diet analysis of loggerhead sea turtles (*Caretta caretta*) in Virginia. Copea **2**, 478–489 (2007)
D.L. Smee, Coastal ecology: Living shorelines reduce coastal erosion. Curr. Biol. **29**, R411–R413 (2019)
D.R. Smith, T.J. Robinson, Horseshoe crab spawning activity in Delaware Bay, USA, after harvest reduction: A mixed-model analysis. Estuar Coast **38**(6), 2345–2354 (2015)
D.R. Smith, P.S. Pooler, B.L. Swan, S.F. Michels, W.R. Hall, P.J. Himchak, M.J. Millard, Spatial and temporal distribution of horseshoe crabs (*Limulus polyphemus*) spawning in Delaware Bay: Implications for monitoring. Estuaries **25**, 115–125 (2002)
D.R. Smith, H.J. Brockman, M.A. Beekey, T.L. King, M.J. Millard, J. Zaldívar-Rae, Conservation status of the American horseshoe crab, (*Limulus polyphemus*): A regional assessment. Rev. Fish Biol. Fish. **27**(1), 135–175 (2017)
J.A. Sweka, D.R. Smith, M. Millard, An age-structured population model for horseshoe crabs in the Delaware Bay area to assess harvest and egg availability for shorebirds. Estuar. Coast **30**(2), 277–286 (2007)
A.N. Tan, A. Chistianus, S. Shakibazadeh, P. Hajeb, Horseshoe Crab, *Tachypleus gigas* (Müller, 1785) spawning population at Balok Beach, Kuantan, Pahang, Malaysia. Pak. J. Biol. Sci. **15**(13), 610–620 (2012)
H. Tian, L. Stige, B. Cazelles, K. Kausrud, R. Svarverud, N. Stenseth, Z. Zhang, Reconstruction of a 1,910-y-long locust series reveals consistent associations with climate fluctuations in China. Proc. Natl. Acad. Sci. **108**(35), 14521–14526 (2011)
C. Van Oosterhout, W.F. Hutchinson, D.P. Wills, P. Shipley, Micro-checker: Software for identifying and correcting genotyping errors in microsatellite data. Mol. Ecol. Notes **4**, 535–538 (2004)
S. Vestbo, M. Obst, F.J.Q. Fernandez, I. Intanai, P. Funch, Present and potential future distributions of Asian horseshoe crabs determine areas for conservation. Front. Mar. Sci. **5**(164), 1–16 (2018)
E. Walls, J. Berkson, S. Smith, The horseshoe crab, *Limulus polyphemus*: 200 million years of existence, 100 years of study. Rev. Fish. Sci. **10**, 39–73 (2002)
S. Weidensaul, Losing ground: What's behind the worldwide decline of shorebirds? Living Bird, Cornell Lab of Ornithology, https://www.allaboutbirds.org/losing-ground-whats-behind-the-worldwide-decline-of-shorebirds/ (2018)

E.O. Wilson, *Half Earth: Our Planets Fight for Life*, vol 978 (Liveright, NY, 2016)
D.H. Wurster, C.F. Wurster, W.N. Strickland, Bird mortality following DDT spray for Dutch elm disease. Ecology **46**(4), 488–499 (1965)
I. Youssef, S. Senbel, J. Kasinak, J.H. Mattei, RCrab: An R analytics tool to visualize and analyze the movement of horseshoe crabs in Long Island Sound. Institute of Electrical and Electronics Engineers (IEEE) Long Island Systems, Applications, and Technology Conference (LISAT) Proceedings, May 3rd, https://ieeexplore.ieee.org/document/8817346 (2019)

# Relative Abundance of Horseshoe Crabs in the Delaware Bay Region: A Critical Factor for Adaptive Resource Management

**David N. Hata and Eric M. Hallerman**

## 1 Introduction

Horseshoe crabs, *Limulus polyphemus*, occur along the Atlantic coast of the United States and in the eastern Gulf of Mexico, with the center of abundance in the mid-Atlantic region, from New Jersey to Virginia (Shuster 1982; Botton and Ropes 1987). The largest spawning aggregation occurs in Delaware Bay, where horseshoe crabs deposit their eggs in intertidal beach sand in May and June (Shuster and Botton 1985). While at Delaware Bay, migrating shorebirds rely heavily on those eggs as a major energy source to fuel their spring migration (Castro and Myers 1993; Tsipoura and Burger 1999). Biomedical companies catch horseshoe crabs for their blood, from which they produce *Limulus* amebocyte lysate (LAL), which is used to detect contamination of injectable drugs and implantable devices by Gram-negative bacteria, and which is the most sensitive means available for detecting their endotoxins (Novitsky 1984). Horseshoe crabs also support a commercial fishery which harvests them for use as American eel (*Anguilla rostrata*) and whelk (family Melongenidae) bait (ASMFC 1998). The magnitude of the commercial harvest in the 1990s raised concerns that a decreasing horseshoe crab population contributed to serious declines in dependent shorebird populations (Berkson and Shuster 1999).

The Atlantic States Marine Fisheries Commission (ASMFC 1998) instituted an Interstate Fishery Management Plan in order to maintain horseshoe crab populations for sustainable use by the fishing and non-fishing public, including the biomedical industry, and dependent fish and wildlife species. The ASMFC in 2012 adopted the Adaptive Resource Management (ARM) framework (McGowan et al.

D. N. Hata · E. M. Hallerman (✉)
Department of Fish and Wildlife Conservation, Virginia Tech University,
Blacksburg, VA, USA
e-mail: ehallerm@vt.edu



J. T. Tanacredi et al. (eds.), *International Horseshoe Crab Conservation and Research Efforts: 2007-2020*, https://doi.org/10.1007/978-3-030-82315-3_23

2011a) to guide managers to set defensible horseshoe crab harvest regulations by accounting for multiple species effects, focusing on the demographic rebuilding of the threatened red knot (*Calidris canutus rufa*) population (U.S. Fish and Wildlife Service 2014) that feeds on horseshoe crab eggs during its northward spring migration through the Delaware Bay. Implementation of the ARM framework is dependent on annual indices of horseshoe crab abundance in the Delaware Bay region that are not dependent upon catch in the fishery, as well as on red knot abundance estimates. Female horseshoe crab spawning population size and spawning population sex ratios are critical trigger points in the model. Monitoring of both horseshoe crab and red knot populations is critical to (1) assess the state of the ARM model systems, (2) evaluate model prediction against observation, (3) assess whether management objectives are achieved, and (4) provide feedback into the model to refine it. We have been conducting annual benthic trawl surveys of relative abundance of horseshoe crabs in the Delaware Bay region to provide input data necessary for successful implementation of the management model and thereby to support the ASMFC fishery management plan. Here, we summarize our findings with reference to horseshoe crab and shorebird conservation in the Delaware Bay area.

## 2 Methods

The horseshoe crab trawl surveys were conducted in two areas (Fig. 1). The coastal Delaware Bay area (CDBA) survey extended in the Atlantic Ocean from the shore out to 22.2 km (12 nautical miles, nm) in order to encompass the distribution of horseshoe crabs (Botton and Haskin 1984; Botton and Ropes 1987; Hata and Berkson 2004) and from 39° 20' N (Atlantic City, New Jersey) to 37° 40' N (slightly north of Wachapreague, Virginia). This area was sampled from 2002 to 2011 and again in 2016 and 2017. The Delaware Bay (DB) survey area extended from the Bay mouth to a line between Egg Island Point, New Jersey, and Kitts Hummock, Delaware. The DB was sampled from 2010 to 2012 and in 2016 and 2017. The surveys were conducted in the fall from September to November in order to obtain information on primiparous crabs, i.e., those who are mature, but have not yet spawned. Attainment of this life stage requires that the crabs have undergone a terminal molt, which occurs in late summer or fall in this region (Hata and Berkson 2004).

The CDBA survey area was stratified by distance from the shore (0–3 nm = 0–5.6 km, 3–12 nm = 5.6–22.2 km) and bottom topography (trough, non-trough), factors which significantly influence catch sizes (Hata and Berkson 2004). The DB survey area was stratified by bottom topography only. Sampling was conducted aboard a 16.8-m chartered commercial fishing vessel. We used a two-seam flounder trawl with an 18.3-m headrope and 24.4-m footrope, rigged with a Texas Sweep of 13-mm link chain and a tickler chain (Hata and Berkson 2004). The net body consisted of 15.2-cm (6-inch) stretched mesh, and the bag consisted of 14.3-cm (5 5/8-inch) stretched mesh. Tows were usually 15-minute bottom time but were occasionally shorter to avoid fishing gear (e.g., gill nets, crab, and whelk pots) or

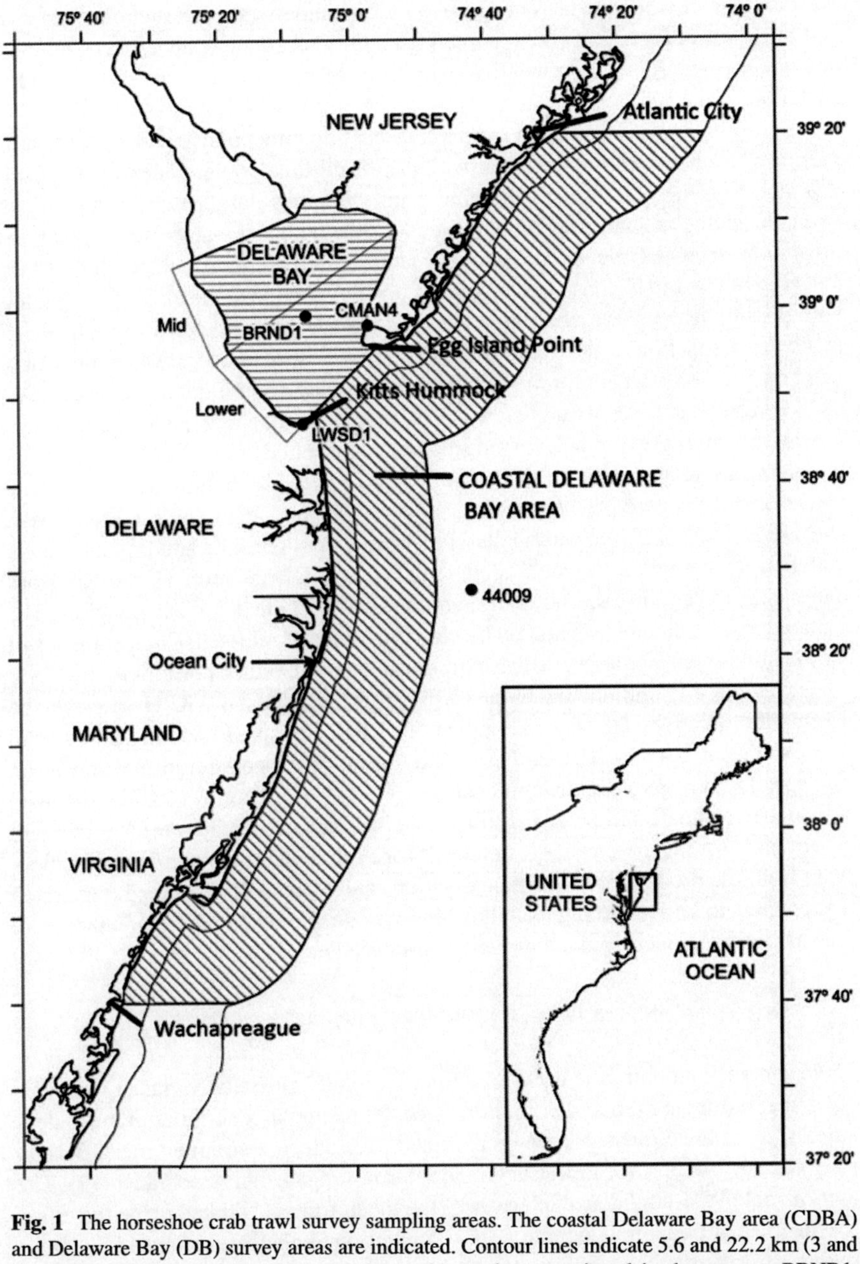


**Fig. 1** The horseshoe crab trawl survey sampling areas. The coastal Delaware Bay area (CDBA) and Delaware Bay (DB) survey areas are indicated. Contour lines indicate 5.6 and 22.2 km (3 and 12 nautical miles) from the shore. Hydrographic stations mentioned in the text are BRND1, Brandywine Shoal, DE; CMAN4, Cape May, NJ; LWSD1, Lewes, DE; and offshore buoy 44,009

vessel traffic or when the net unexpectedly hit an underwater obstruction. Start and end positions of each tow were recorded from GPS when the winches were stopped and when retrieval began, respectively. Bottom water temperature was recorded for each tow.

Horseshoe crabs were culled from the catch, and either all individuals or a subsample was examined for prosomal width (PW, millimeters) and identified for sex and maturity (Fig. 2). Maturity classifications were immature; newly mature, those that are capable of spawning but have not yet spawned; and mature, those that have previously spawned. Newly mature and mature males are morphologically distinct and are believed to be classifiable without error. Although some error is associated with distinguishing newly mature from immature females, the effort was made to make the distinction in the field, as the data provide a leading indicator of recruitment of female spawners for the following spring and are important for setting target harvest levels. All examined females that were not obviously mature (i.e., bearing rub marks) or immature (too small or soft-shelled) were probed with an awl to determine the presence or absence of eggs. Females with eggs but without rub marks were considered newly mature. Females with both eggs and rub marks were considered mature. Initial sorting classifications were presumed adult males (newly mature and mature), presumed adult females, and all immature. Up to 25 adult males, 25 adult females, and 50 immatures were retained for examination. The remainder were counted separately by classification and released. Characteristics of the examined subsamples were then extrapolated to the counted portions of the catch.

The average 15-minute tow in the CDBA was 1.22 km at 4.9 KPH. The average 15-minute tow in the DB was 1.26 km at 5.0 KPH. Net spread was measured from 113 tows from 2011 to 2016 and averaged 9.6 m. The regression of net spread on tow speed was negative and significant ($r = -0.245$; $t = 2.66$; $p = 0.009$), so speed was used to estimate net spread for collections in which net spread was not measured.

For each tow, catch density (catch/km$^2$) for each demographic group was calculated from the product of tow distance (in km) and estimated net spread (converted from meters to km) assuming that all fishing was done only by the net and that there was no herding effect from the ground gear (sweeps):

$$\text{catch} / \text{km}^2 = \text{catch} / \left[\text{tow distance}(\text{km}) \times \text{net} - \text{spread}(\text{km})\right]$$

Within each stratum, the mean catch per km$^2$ and associated variance were calculated assuming a normal distribution catch-frequency model. Stratum mean densities and variance estimates were combined to produce a stratified mean density ($\bar{X}_{\text{st}}$) using formulas for a stratified random sampling design (Cochran 1977). The approximate 95% confidence intervals of the mean were calculated using the effective degrees of freedom (Cochran 1977). Population totals were estimated by multiplying stratified mean density ($\bar{X}_{\text{st}}$) by survey area (CDBA = 5127.1 km$^2$; DB = 528.4 km$^2$):

$$\text{Population total} = \bar{X}_{\text{st}} \times \left(5{,}127.1 \,\text{or}\, 528.4 \,\text{km}^2\right).$$

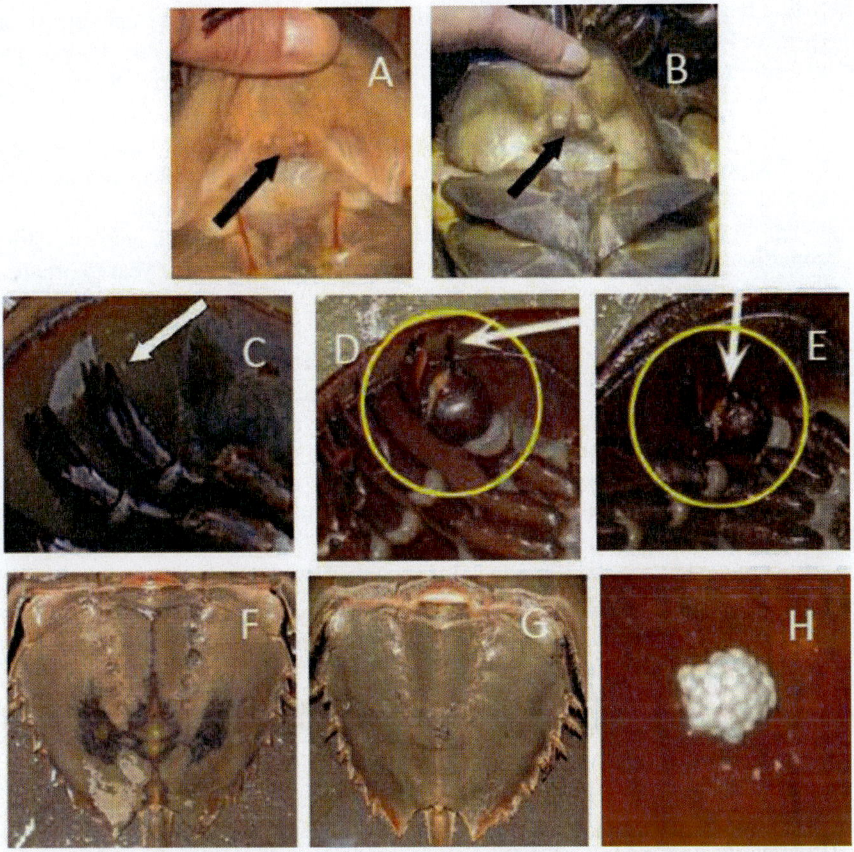


**Fig. 2** Classification of horseshoe crabs by sex and maturity stage as immature, newly mature (mature but has not yet spawned), and multiparous (mature and has spawned). Identification of sex and maturity may require examination of the rear body segment (opisthosoma), front legs (pedipalps), and genital operculum, i.e., the first "flap" before the book gills. (**a**) Even in small immature crabs, males can be identified by the genital papillae on the underside of the genital operculum. The genital papillae are raised nodes, easily visible in larger individuals, and easily felt in even small individuals as hard bumps. (**b**) In both juveniles and adults, females can be identified by the gonopores on the underside of the genital operculum. The gonopores appear as soft bumps and are distinguishable by touch from the hard genital papillae of males even at sizes as small as 75 millimeters prosomal width. (**c**) The pedipalps (the first set of legs) of crabs of both sexes are undifferentiated, looking the other, "true" legs. The pedipalps are undifferentiated in mature females as well. (**d**) Newly mature males are capable of spawning, but have not yet spawned. The claws on the pedipalps, the chelae, are modified into a "boxing glove" shape. The fixed finger of the chela is thin and somewhat fragile, while the movable finger is much stouter. The chelae are used to latch onto the female during spawning, termed amplexus. (**e**) During the male's first amplexus, the fixed finger is broken off, thereby identifying a multiparous male that has spawned. (**f**) During amplexus, the contact of the male on top of the female rubs the opisthosoma – the rear body section – of the female. This rubbing results in worn, often dark patches. Females with these marks are classified as multiparous, indicating they have previously spawned. (**g**) Females that lack rub marks may be immature or newly mature. The state of maturity for these females can only be determined by the presence of eggs. (**h**) The subfrontal area of the prosoma, the main part of the body, is relatively thin and easily punctured with an awl; eggs will often exude from the hole. If eggs are present but there are no rub marks, the female is classified as newly mature

Life histories for the sexes differ, with males maturing at age 11 and females at age 12 or later in this region (Smith et al. 2009). Thereafter, spatial distributions and survival rates for the sexes differ. Because numbers of reproductive females may limit the potential for population growth, female spawning population size and spawning population sex ratio are critical triggers within the ARM model. Annual size-frequency distributions, in intervals of 10-mm PW, were calculated for each sex/maturity category by pooling size-frequency distributions of all stations in a stratum in a year to calculate the relative proportions for each size interval. Those proportions then were multiplied by the stratum mean density that year to produce a stratum size-frequency distribution. Stratum size-frequency distributions then were multiplied by the stratum weights and added in the same manner as calculating the stratified mean density. Areas under the distribution curves then represented the stratified mean density at each size interval. The regression of stratified mean densities of each demographic group was assessed using a two-tailed $t$ test (Zar 1984). Delaware Bay size-frequency distributions also were subdivided into the lower DB, 0–20 km from the Bay mouth, and the mid-DB, >20 km from the Bay mouth (Fig. 1), to examine spatial distribution by size. Size-at-age designations followed Smith et al. (2009).

Recruitment of horseshoe crabs seems to be strongly affected by first-year survival. To assess the hypothesis that temperature affects critical first year survival, we examined the relative mean catch density of a putative year class with average January subsurface water temperatures during their first winter – e.g., the abundance of horseshoe crabs believed to have been spawned in May–June 2000 was related to January 2001 water temperature. We examined immature crabs 150–180 mm PW believed to be age 10 (see Results) and average January water temperatures recorded at Cape May, New Jersey, from 1998 to 2008 (Center for Operational Oceanographic Products and Services, National Oceanic and Atmospheric Administration, NOAA; https://tidesandcurrents.noaa.gov) and from hydrographic buoy 44,009 near the Delaware-Maryland border from 1994 to 2008 (National Data Buoy Center, NOAA; http://www.ndbc.noaa.gov/) (Fig. 1). Water temperatures at the hydrographic buoy are measured 2.07 m below mean lower low water (MLLW) and at Cape May 1.04 m below MLLW, where MLLW is the lower of the two low tides per day averaged over a 19-year period. Water temperature data were included only if at least 95% of the 744 possible hourly measurements (24 hrs × 31 days) were recorded. Water temperatures at Cape May were strongly correlated with temperatures at the Lewes, Delaware, and Brandywine Shoal, Delaware stations (Lewes, $n = 10$, $r = 0.937$, paired $t = 7.59$, two-tailed $p < 0.001$; Brandywine Shoal, $n = 7$, $r = 0.964$, $t = 8.10$, $p < 0.001$), but the Cape May data were more complete. Cape May January temperatures also were correlated with buoy 44,009 temperatures ($n = 17$; $r = 0.784$, $t = 4.89$; $p < 0.001$). Although coastal water temperatures observed at buoy 44,009 are likely not encountered by first year horseshoe crabs, those temperatures may be indicative of the temperature variability within Delaware Bay and extend the data range earlier than those available for the Bay hydrographic sites.

# 3 Results

## 3.1 *Size*

Size-frequency distributions of immature horseshoe crabs in the CDBA survey displayed considerable interannual variability, and modal groups were generally indistinct. Size-frequency distributions from the DB were likewise indistinct. Size-frequency modal groups became more apparent when the annual distributions were averaged (Fig. 3). Following the size-at-age designations provided by Smith

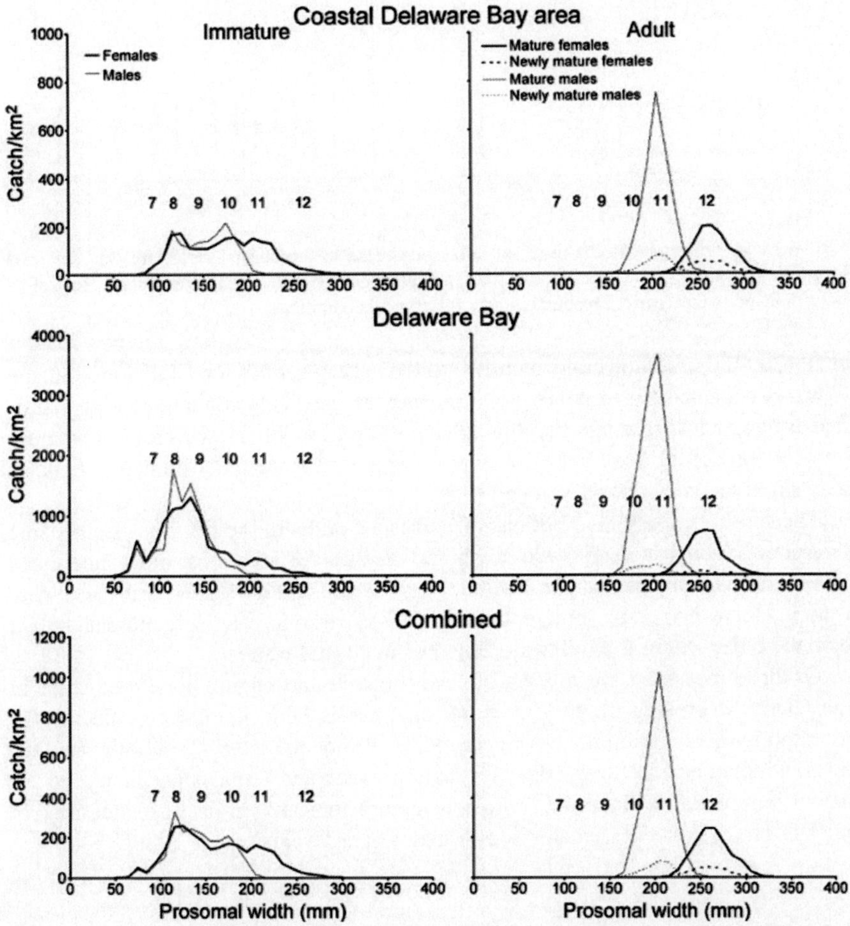


**Fig. 3** Relative size-frequency distributions of horseshoe crabs, by demographic group in the coastal Delaware Bay area and Delaware Bay trawl surveys, all years averaged, and surveys combined. Surveys were combined by weighting by survey areas. Putative age designations 7 through 12 are from Smith et al. (2009). Note differences in *y*-axis scales

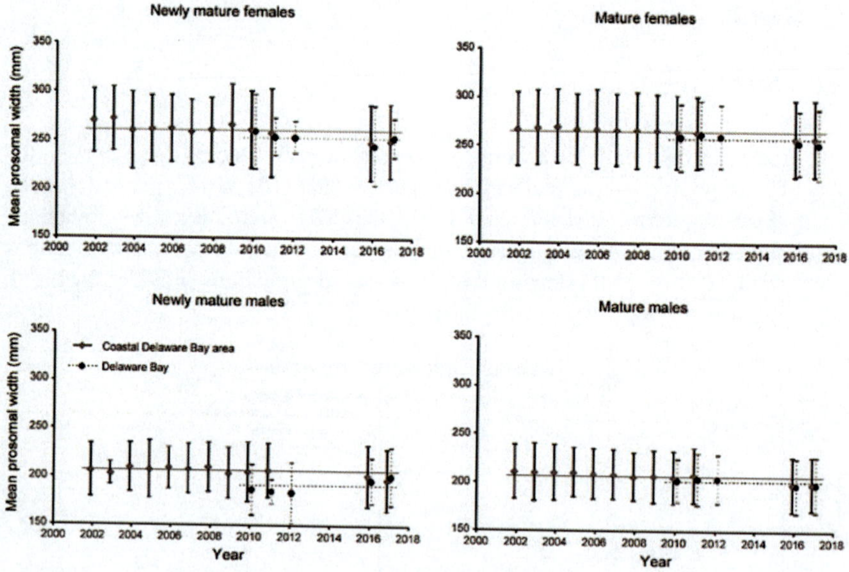


**Fig. 4** Mean prosomal widths (mm) of horseshoe crabs in the coastal Delaware Bay area and Delaware Bay surveys, by year. Vertical lines indicate means ± two standard deviations. Horizontal lines indicate overall mean prosomal widths for all years combined

et al. (2009), horseshoe crabs recruited to the survey gear in the CDBA around age 8. Males became newly mature predominantly at age 11. Most females were classified as newly mature at age 12, although a portion were not. However, newly mature females were identified by the presence of eggs, and those older immature females may simply have not yet developed eggs.

The size-frequency distributions of horseshoe crabs in the DB likewise became clearer when survey years were combined. Within the DB, most horseshoe crabs also recruited to the gear at age 8, although some younger horseshoe crabs also were caught. Horseshoe crabs younger than age 10 appear to be relatively more abundant than the older within the Delaware Bay than in coastal waters.

Of those measured, mean PWs of newly mature and mature horseshoe crabs in the CDBA decreased slightly over the time series (Fig. 4), not significantly as assessed by observation of overlapping 95% confidence intervals. Within the DB, mean widths decreased over time for mature males and females but increased for newly mature males. However, few newly mature horseshoe crabs were measured in the DB surveys, just 46 females and 47 males over 5 years.

## 3.2 *Spatial Distribution*

Within DB, the smallest immature horseshoe crabs, modally around 75 mm PW, were most abundant in the mid-Delaware Bay, farthest from the Bay mouth (Fig. 5). Immature crabs larger than 100 mm PW were more evenly distributed in the mid-DB and lower DB. Average densities of newly mature horseshoe crabs were generally low throughout the Bay, whereas densities of mature horseshoe crabs were higher in the lower DB.

The smallest juveniles in the CDBA tended to be distributed in the inshore (< 5.6 km from shore) strata, while larger juveniles were more evenly distributed inshore and offshore (Fig. 5). On average, about 43% of all immature females and 46% of immature males in the CDBA occurred in the inshore strata during the fall, although the inshore strata comprised only about 25% of the total survey area. About 61% and 59% of immature females and males <150 mm, respectively, occurred inshore in the fall. Mature females were more evenly distributed, with 29% occurring inshore, while 40% of mature males were inshore.

Immature and newly mature horseshoe crabs in the CDBA were distributed primarily south of Delaware Bay. Average catches increased rapidly south of the Bay mouth and remained consistently high to the southern limit of the survey area (Fig. 6). Average catches of mature males and females were low at the northern and southern limits of the survey area, with notable peaks around 38.3° N and 38.8° N, respectively, corresponding to the vicinity of Ocean City Inlet, Maryland, and around the Delaware Shelf Valley, which is a deep trench outside of DB. Average catch density generally decreased with distance from the shore but increased with depth. This is somewhat deceiving though, as the deepest sites were associated with the Delaware Shelf Valley at intermediate distances: only six collections were deeper than 30 m, all from 10.9 to 14.0 km from shore.

## 3.3 *Year Classes*

Catch densities of horseshoe crabs in coastal waters were correlated within putative sex/size/age groups within particular survey years, but not between survey years. Mean densities of immature females <150 mm in the CDBA were correlated with mean densities of immature males <150 mm over the time series ($n = 12$; $r = 0.997$; paired $t = 41.72$; two-tailed $p < 0.001$). Mean densities of small immature females 150–180 mm ($FI_{150–180}$) and approximately age 10 (Fig. 2) were correlated with mean densities of immature males 150–180 mm ($MI_{150–180}$) ($n = 12$; $r = 0.908$; $t = 6.87$; $p < 0.001$). Larger immature females 180–240 mm ($FI_{180–240}$) and age 11 increased with mean densities of newly mature males, which were of similar size and age ($n = 12$; $r = 0.932$; $t = 8.15$; $p < 0.001$). However, mean immature male $MI_{150–180}$ densities (age 10) were not correlated with mean densities of newly mature males (age 11) the following year ($n = 10$; $t = 0.12$; $p = 0.906$). Mean immature female $FI_{180–240}$ densities were not correlated with densities of newly mature females age 12 the following year ($n = 10$; $t = 1.84$; $p = 0.104$). Similarly, densities of newly mature males were not correlated with densities of newly mature females the following year, which were of the same year class ($n = 10$; $t = 2.30$; $p = 0.051$).

Water temperatures encountered during their first winter may influence demographic recruitment and subsequent horseshoe crab year-class strength in the

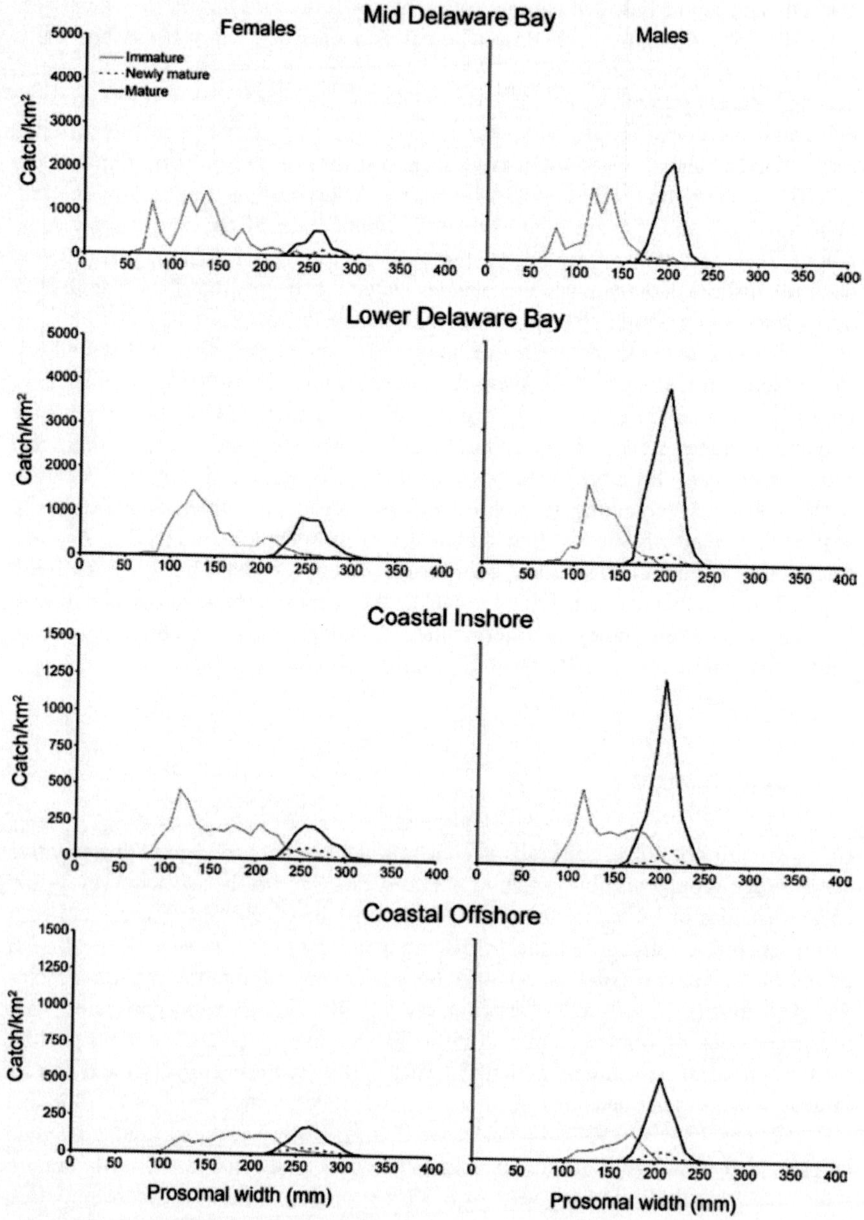


**Fig. 5** Relative size-frequency distributions of horseshoe crabs, by demographic group in the mid and lower sections of Delaware Bay and inshore and offshore strata of the coastal Delaware Bay area surveys, all years averaged. Note differences in *y*-axis scales

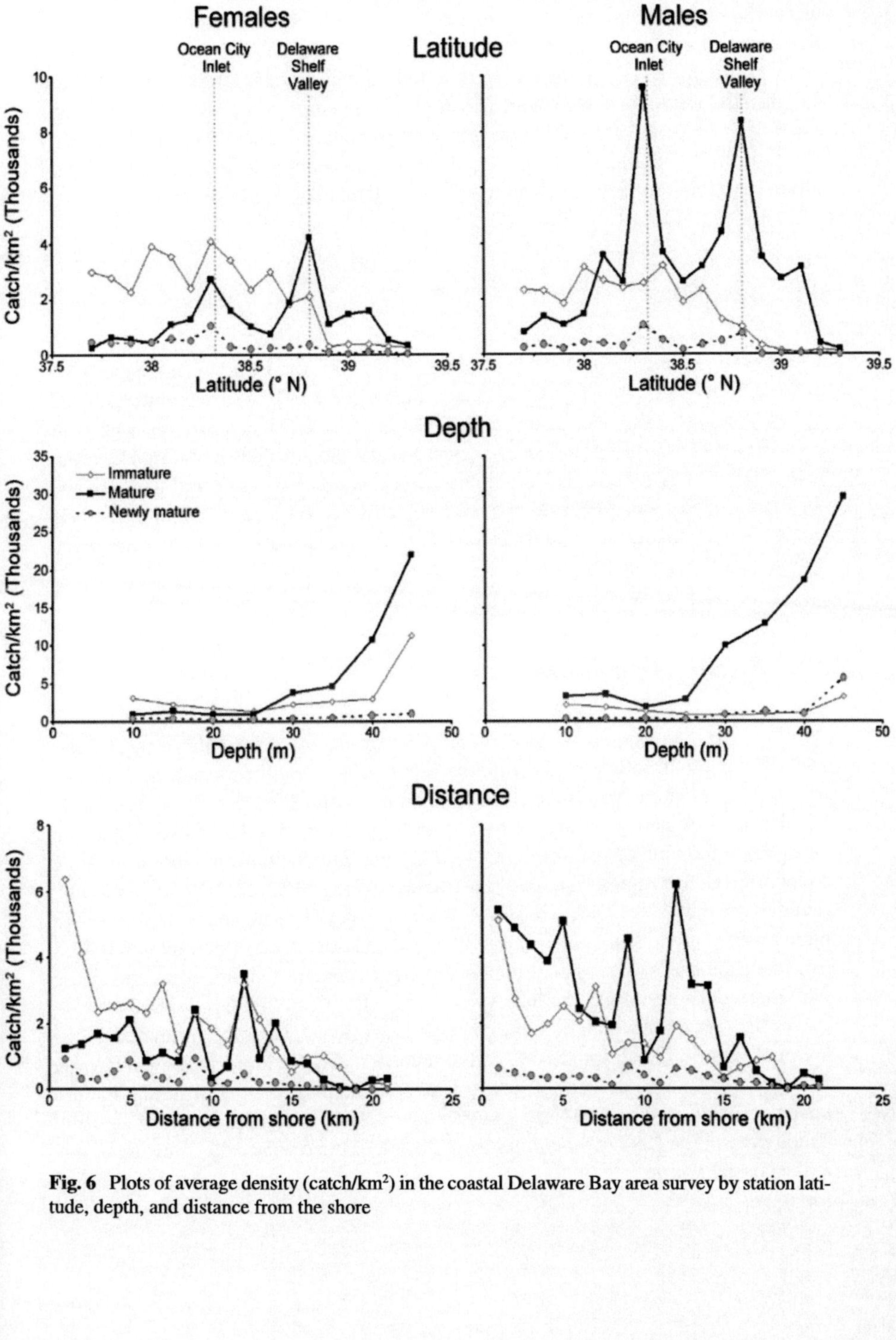


**Fig. 6** Plots of average density (catch/km²) in the coastal Delaware Bay area survey by station latitude, depth, and distance from the shore

Delaware Bay area. CDBA catch densities of immature males 150–180 mm PW and approximately 10 years of age were correlated with average January Cape May water temperatures in their first winter ($n = 7$; $r = 0.769$; $t = 2.69$; $p = 0.043$). However, densities of immature females $FI_{150–180}$ were not correlated with water temperature ($n = 7$, $t = 1.86$; $p = 0.122$). Coastal January water temperatures at buoy 44,009 also were correlated with densities of immature males $MI_{150–180}$ ($n = 11$, $r = 0.655$; $t = 2.60$; $p = 0.029$) but not with immature females $FI_{150–180}$ ($t = 2.09$; $p = 0.067$). Although not definitive, these relationships suggest possibilities for future research into factors affecting recruitment and year-class strength.

## 3.4 Sex Ratios

The sex ratio of horseshoe crabs less than 150 mm PW was 1.02:1.00 males/females in DB and 0.99:1.00 in the CDBA. Mature males in the CDBA waters outnumbered mature females 2.46:1.00 in the fall. The sex ratio of newly mature crabs was 0.98:1.00, despite males generally being a year younger than females. However, within DB, newly mature males outnumbered females 3.17:1.00, and the sex ratio for mature crabs was 3.85:1.00. For the Bay and coastal waters combined, newly mature males outnumbered females 1.28:1.00, and mature males outnumbered mature females 2.97:1.00.

## 3.5 Relative Abundance

Stratified mean densities for all demographic groups were variable in the CDBA and DB regions throughout the time series (Fig. 7). Stratified mean densities of immature and mature horseshoe crabs were higher within the DB than in the CDBA, but densities of newly mature females and males were similar between the two areas. Over the time series from 2002 to 2017, the population of mature horseshoe crabs in the CDBA appeared to increase (males, $n = 12$, $r = 0.804$, $t = 4.28$, $p = 0.002$; females, $r = 0.594$, $t = 2.34$, $p = 0.042$). However, there was no apparent increase in numbers of newly mature males ($t = 1.37$; $p = 0.201$) or females ($t = 0.69$; $p = 0.505$) over the time series. The annual mean densities of mature females were correlated with mean densities of mature males ($n = 12$; $r = 0.807$; $t = 4.32$; $p = 0.002$).

For the 4 years in which both regions were sampled, 76% of the estimated population of mature females and 71% of the mature males occurred in the CDBA, on average. Eighty-nine percent of newly mature females and 84% of newly mature males occurred in the CDBA region, and 65% and 62% of immature females and males were in the CDBA. In comparison, 90.7% of the combined survey areas was in the coastal region.

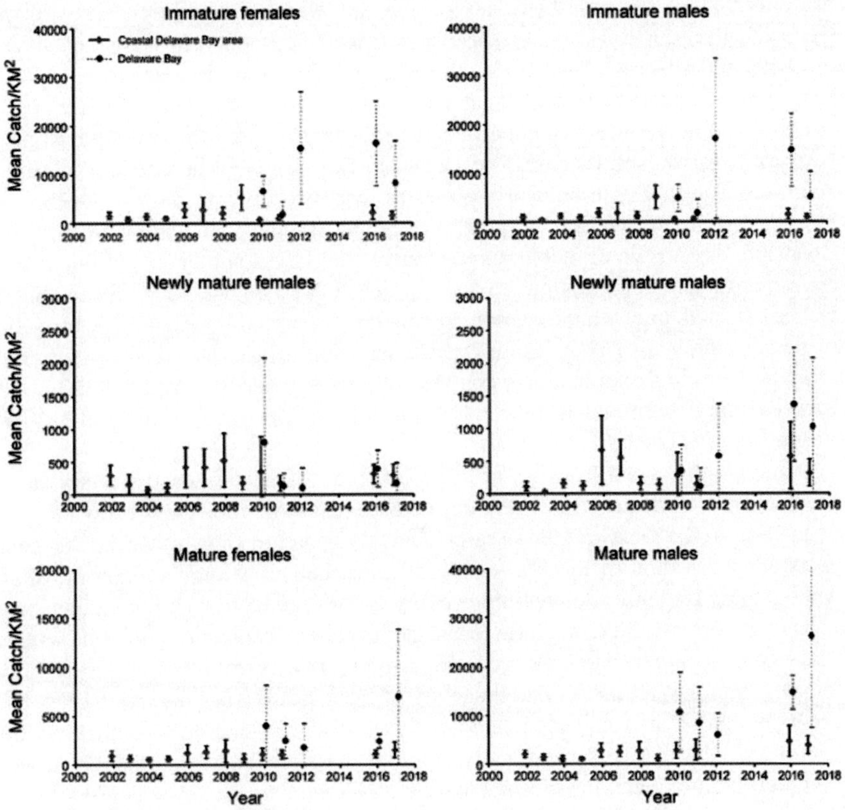


**Fig. 7** Stratified mean density (catch/km$^2$) of horseshoe crabs in the coastal Delaware Bay area and Delaware Bay surveys by demographic group. Vertical lines indicate 95% confidence limits. Note differences in *y*-axis scales

# 4 Discussion

Once dramatically more abundant, horseshoe crab populations along the mid-Atlantic coast of North America declined through the twentieth century, a decline that became more apparent in the 1990s with the impact of the biomedical and bait fisheries. The Interstate Fishery Management Plan (IFMP) (ASMFC 1998) was established with the goal of managing horseshoe crab populations for the continued use by the fishing and non-fishing public, migrating shorebirds, and other dependent wildlife. With this goal in mind, management actions have been adopted for all coastal states collectively by the ASMFC or unilaterally by the individual states, including harvest limits and moratoria, closed areas, closed seasons, and female harvest restrictions. The Carl N. Shuster, Jr. Horseshoe Crab Reserve, located in federally managed waters off the mouth of the Delaware Bay estuary, was created in 2001 to protect the spawning population of horseshoe crabs (Marine Conservation

Institute 2020). Our horseshoe crab trawl survey was designed in the early 2000s (Hata and Berkson 2004) and has been conducted since then, informing the harvest management process. Since 2013, the Adaptive Resource Management (ARM) framework – established through Addendum VII of the IFMP – incorporating both red knot and horseshoe crab abundance levels has been used to set optimized harvest levels for horseshoe crabs "of Delaware Bay origin" (Hallerman et al., this volume). The ARM framework regulates harvest in New Jersey, Delaware, Maryland, and the coastal portion of Virginia. Tagging studies show that the Delaware Bay spawning population is part of the larger population of horseshoe crabs across the mid-Atlantic region (Walls 2001; Grogan 2004; Swan 2005; Hallerman et al. this volume), a finding which is supported by screening of genetic markers (King et al. 2015). Using genetic markers, Hallerman et al. (this volume) examined the composition of five commercial harvests and estimated that Delaware Bay spawning assemblages on average contributed 51% of Maryland and 35% of Virginia coastal catches.

Although estimated landings of horseshoe crabs have decreased steadily since the IFMP was established in 1998, our data and those of colleagues show that the population in the Delaware Bay region has only begun to show recovery in recent years. Consideration of the life history of horseshoe crabs may explain the slow demographic response. Population recovery is constrained by egg production, survival in the pre-exploitation phase (eggs and juveniles), generation time (the time it takes those eggs to develop into spawning adults, 11 years for males and over 12 for females, Smith et al. 2009), and longevity of the spawning phase (and thereby the accumulation of spawners over time). A review of the literature shows that the life history includes extremely high mortality in the first year and then moderate levels of mortality throughout the juvenile and adult phases. Carmichael et al. (2003) estimated that only 0.001% of Cape Cod, Massachusetts, eggs survived through the first year. Botton et al. (2003) estimated a 0.003% survival rate through the first year in Delaware Bay. Application of an age-structured population model (Sweka et al. 2007) estimated that the probability of a population increase was less than 50% at high egg mortality rates, regardless of female harvest levels, and that population growth was most sensitive to first year survival rates. Results of our simple analyses suggested that first year winter temperature may influence year-class strength. In contrast, Grady and Valiela (2006) suggested that population growth in Cape Cod was sensitive to mortality of older juveniles. Carmichael et al. (2003) estimated that 78% of Cape Cod juveniles that survived their first year ultimately attained adulthood. Horseshoe crabs do not mature until they are 11–12 years of age (Smith et al. 2009) and may live an additional 8–10 years (Botton and Ropes 1988; Swan 2005). Hence, year-class strength may not be highly dependent on the spawning population size, but small changes in first year mortality may have a substantial influence on year-class strength. Our results showed the ontogenetic movement of horseshoe crabs from Delaware Bay and nearshore waters to locations farther offshore. Our results showed change of the sex ratio with age, suggesting higher mortality of females after maturation, which could affect the potential for demographic recovery. The sex ratios that we observed largely parallel those seen in horseshoe crab spawning surveys (e.g., Hallerman et al., this volume). Horseshoe crab fecundity increases with size (Leschen et al. 2006), so the apparent decrease in mean prosomal widths of mature and newly mature females that we observed poses implications for annual egg production, which may affect future year-class strength as well as the quantity of eggs available to foraging shorebirds.

Uncertainty about the factors affecting the dynamics of horseshoe crab populations affects understanding of the impacts of exploitation and management actions. Sweka et al. (2007) predicted that management actions reducing early life-stage mortality would have a greater effect on horseshoe crab and red knot population dynamics than adjusting harvest levels alone. Although harvest is not directed at intermediate juvenile stages, they are susceptible to an unknown, though probably low, level of fishing-related mortality. Horseshoe crabs are harvested by biomedical companies that extract blood from them and then return them to the water. As blood volume is related to horseshoe crab size (Hurton et al. 2005), larger crabs, including females, are preferred. Estimated mortality from harvest, handling, and bleeding range from 8% to 15% in males (Walls and Berkson 2003; Hurton and Berkson 2006) and 10–29% in females (Hurton and Berkson 2006; Leschen and Correia 2010; Anderson et al. 2013), depending in part on harvest methods. There also are sublethal behavioral and physiological effects of harvest, handling, and bleeding (Anderson et al. 2013). Biomedical harvest for the entire coast has increased since our monitoring program was initiated in 2004 (ASMFC 2020), likely leading to proportionate changes in the level of industry-associated mortality. However, the estimated mortality rate in the fishery management plan does not include mortality due to the entire biomedical harvest process (ASMFC 2010). For horseshoe crabs harvested by trawl, juvenile horseshoe crabs and a small percentage of adult horseshoe crabs captured are culled at sea due to damage (Hata pers. observ.), and a number of both juvenile and adult horseshoe crabs may be damaged but not captured or retained by the gear, as we have seen in our trawl survey and experimental work (Hata and Hallerman, unpublished data). Damage and associated mortality are also applicable to the trawl bait fishery, as well as to bycatch in other fisheries, such as for flounder. Mortality from dredges, as used in the whelk fishery, may be higher (Smith et al. 2017). These respective sources of mortality are operating upon a depleted population with a reduced number of larger, generally mature, crabs.

Adult life stages are subject to both fishing and natural mortality. They are the primary targets of the bait and biomedical fisheries and suffer spawning-related mortality such as stranding. Botton and Loveland (1989) estimated 10% mortality within the Delaware Bay spawning population from stranding each year; older adults strand at higher frequencies than younger adults (Penn and Brockmann 1995; Smith et al. 2010). Annual variability in first year mortality may manifest as strong or weak year classes which eventually enter the mature stage, with substantial or insignificant contributions to the standing spawning stock. Weak year classes in the spawning population may be mitigated by strong ones in a multiple age-class spawning population such as horseshoe crabs, but that effect might be lessened if the spawning population is reduced by high natural and fishing mortality.

Demographic rebuilding in marine species is relatively difficult for those that are long-lived and have late maturity, and often with limited geographic range, regardless of fecundity (Sadovy 2001). A population's rate of natural increase – its replacement rate – is more important than fecundity itself (Sadovy 2001). Slight changes in first year mortality would have a greater impact on horseshoe crab population growth than any other life history parameter, including harvest rate (Sweka et al. 2007). Davis et al. (2006) estimated that population recovery could take less than 5 years, but possibly more than 15 years. Even with low egg mortality and no female harvest, doubling the female spawning population would take more than 15 years (Sweka et al. 2007). Depletion of the horseshoe crab resource was recognized in the late 1990s, and coast-wide management began in earnest in the early- to mid-2000s. Recovery of a depleted horseshoe crab stock would not be expected until at least one generation had recovered – in the case of horseshoe crabs, 11–12 years. Rebuilding the spawning population may require a further 8 years, assuming that they survive an additional 8 years after maturing (Botton and Ropes 1988). We are only now entering the timeframe where horseshoe crab population recovery may become evident. Females from eggs produced by mature females in 2004, at the time of the lowest mean density that we observed, became newly mature only in fall 2016 and spawned for the first time in spring 2017. We have not yet been able to examine relationships between spawning population size and year-class strength because of the long time to recruit to the gear, long generation time, and the hiatus in the trawl survey time series in 2013–2015.

Horseshoe crab population recovery may be adversely impacted by other ecological processes. That is, a reduced population creates an ecological void which may be occupied by expansion of other species; recovering stocks then may suffer from heightened competition for resources such as available prey. For example, such a process may be hampering recovery efforts for Scotian Shelf Atlantic cod (*Gadus morhua*) (Bundy and Fanning 2005); in contrast, Barents Sea Atlantic cod, which are not subject to similar levels of trophic competition, do show signs of recovery (Dolgov 2009). Even in the case of a recovered population, previous abundance levels might not be attained due to shifts in the ecosystem (Lotze et al. 2011). The possibility that such community-level ecological processes may be impeding recovery of Delaware Bay horseshoe crabs has not been investigated.

The demographic recovery of the Delaware Bay horseshoe crab population is of interest for its potential to contribute to the recovery of the *rufa* red knot population that depends upon its spawning to provide a critical food resource during its spring northward migration to its reproductive range on northern Canada. McGowan et al. (2011b) presented a model linking (1) red knot spring migration stopover mass gain to horseshoe crab spawning abundance and (2) subsequent apparent annual survival to 180-gram mass state at departure from the Delaware Bay stopover area. Model selection statistics showed support for a positive relationship between horseshoe crab spawning abundance during the migration stopover and the probability of red knots gaining critical mass of 180 grams. McGowan et al. (2011a) used the model to compare different simulated horseshoe crab harvest strategies and found that restricting crab harvest can benefit red knot populations. While the models were initially weighted based on stakeholder confidence in these hypotheses, weights will be adjusted based on monitoring and updating of Bayesian model weights. McGowan et al. (2015) used a dynamic optimization routine to identify the optimal harvest policy for horseshoe crabs, given possible actions, stated objectives, and competing hypotheses about system function. The Adaptive Resource Management plan was reviewed, accepted, and implemented by the Atlantic States Marine Fisheries Commission in 2012 and 2013. The current adaptive framework represents a true multispecies management effort where additional data over time are employed to improve the predictive models and reduce parametric uncertainty (Millard et al. 2015). Continued monitoring of the horseshoe crab resource by us and by colleagues will be critical to effective application of the Adaptive Resource Management model.

**Acknowledgments** Our research has been supported by the US Department of Commerce, National Marine Fisheries Service, the National Fish and Wildlife Foundation, and the Atlantic States Marine Fisheries Commission and commercial entities including Associates of Cape Cod, Bernie's Conchs, Chesapeake Bay Packing, LaMonica Fine Foods, Lonza, Sea King Corporation, Southern Connection Seafood, and Spots Fish Company.

## References

R.L. Anderson, W.H. Watson III, C.C. Chabot, Sublethal behavioral and physiological effects of the biomedical bleeding process on the American horseshoe crab, *Limulus polyphemus*. Biol. Bull. **225**(3), 137–151 (2013)

ASMFC (Atlantic States Marine Fisheries Commission), *Interstate fishery management plan for horseshoe crab. Fishery management report no. 32* (Atlantic States Marine Fisheries Commission, Arlington, VA, 1998)

ASMFC (Atlantic States Marine Fisheries Commission), A framework for adaptive management of horseshoe crab harvest in the Delaware Bay constrained by red knot conservation. Stock Assessment Report No. 09–02 (Suppl. B), (Atlantic States Marine Fisheries Commission, Arlington, VA, 2009)

ASMFC (Atlantic States Marine Fisheries Commission), *Review of the Fishery Management Plan in 2009 for Horseshoe Crab (Limulus Polyphemus)* (Atlantic States Marine Fisheries Commission, Arlington, VA, 2010)

ASMFC (Atlantic States Marine Fisheries Commission), *2013 Horseshoe Crab Stock Assessment Update* (Atlantic States Marine Fisheries Commission, Arlington, VA, 2013)

ASMFC (Atlantic States Marine Fisheries Commission), *Horseshoe Crab* (Atlantic States Marine Fisheries Commission, Arlington, VA, 2020)

J. Berkson, C.N. Shuster Jr., The horseshoe crab: The battle for a true multiple-use resource. Fisheries **24**(11), 6–10 (1999)

M.L. Botton, H.H. Haskin, Distribution and feeding of the horseshoe crab, *Limulus polyphemus*, on the continental shelf off New Jersey. Fish. Bull. **82**, 383–389 (1984)

M.L. Botton, R.E. Loveland, Reproductive risk: High mortality associated with spawning by horseshoe crabs (*Limulus polyphemus*) in Delaware Bay, USA. Mar. Biol. **101**, 143–151 (1989)

M.L. Botton, J.W. Ropes, Populations of horseshoe crabs, *Limulus polyphemus*, on the northwestern Atlantic continental shelf. Fish. Bull. **85**, 805–812 (1987)

M.L. Botton, J.W. Ropes, An indirect method for estimating longevity of the horseshoe crab (*Limulus polyphemus*) based on epifaunal slipper shells (*Crepidula fornicata*). J. Shellfish Res. **7**, 407–412 (1988)

M.L. Botton, R.E. Loveland, A. Tiwari, Distribution, abundance, and survivorship of young-of-the-year in a commercially exploited population of horseshoe crabs *Limulus polyphemus*. Mar. Ecol. Prog. Ser. **265**, 175–184 (2003)

A. Bundy, L.P. Fanning, Can Atlantic cod (*Gadus morhua*) recover? Exploring trophic explanations for the non-recovery of the cod stock on the eastern Scotian Shelf, Canada. Can. J. Fish Aquat. Sci. **62**, 1474–1489 (2005)

R.H. Carmichael, D. Rutecki, I. Valiela, Abundance and population structure of the Atlantic horseshoe crab *Limulus polyphemus* in Pleasant Bay, Cape Cod. Mar. Ecol. Prog. Ser. **246**, 225–239 (2003)

G. Castro, J.P. Myers, Shorebird predation on eggs of horseshoe crabs during spring stopover on Delaware Bay. Auk **110**, 927–930 (1993)

W.G. Cochran, *Sampling techniques*, 3rd edn. (John Wiley and Sons, Inc., New York, 1977), 428 p

M.L. Davis, J. Berkson, M. Kelly, A production modeling approach to the assessment of the horseshoe crab (*Limulus polyphemus*) population in Delaware Bay. Fish. Bull. **104**, 215–225 (2006)

A.V. Dolgov, Trophic structure of the Barents Sea fish assemblage with special reference to the cod stock recoverability. Prog Oceanogr **81**, 165–173 (2009)

S.P. Grady, I. Valiela, Stage-structured matrix modeling and suggestions for management of Atlantic horseshoe crab, *Limulus polyphemus*, populations on Cape Cod, Massachusetts. Estuar. Coasts **29**, 685–698 (2006)

W.N. Grogan, *A Mid-Atlantic Study of the Movement Patterns and Population Distribution of the American Horseshoe Crab (Limulus Polyphemus). M.Sc. thesis* (Virginia Polytechnic Institute and State University, Blacksburg, 2004)

E.M. Hallerman, D.N. Hata, M.S. Eackles, T.L. King, Population genetics and movement show metapopulation dynamics of Delaware Bay region horseshoe crabs, (this volume)

D. Hata, J. Berkson, Factors affecting horseshoe crab *Limulus polyphemus* trawl survey design. Trans. Am. Fish Soc. **133**, 292–299 (2004)

L. Hurton, J. Berkson, Potential causers of mortality for horseshoe crabs (*Limulus polyphemus*) during the biomedical bleeding process. Fish. Bull. **104**, 293–298 (2006)

L. Hurton, J. Berkson, S. Smith, Estimation of total hemolymph volume in the horseshoe crab *Limulus polyphemus*. Mar. Freshw. Behav. Physiol. **38**, 139–147 (2005)

T.L. King, M.S. Eackles, A.W. Aunins, H.J. Brockmann, E. Hallerman, B.B. Brown, Conservation genetics of the horseshoe crab (*Limulus polyphemus*): Allelic diversity, zones of genetic discontinuity, and regional differentiation, in *Changing Global Perspectives on Biology, Conservation and Management of Horseshoe Crabs*, ed. by R. Carmichael, M. L. Botton, P. Shin, S. G. Cheung, (Springer International Publishing, Berlin, 2015), pp. 65–96

A.S. Leschen, S.J. Correia, Mortality in female horseshoe crabs (*Limulus polyphemus*) from biomedical bleeding and handling: Implications for fisheries management. Mar. Freshw. Behav. Physiol. **43**, 135–147 (2010)

A.S. Leschen, S.P. Grady, I. Valiela, Fecundity and spawning of the Atlantic horseshoe crab, *Limulus polyphemus*, in Pleasant Bay, Cape Cod, Massachusetts, USA. Mar. Ecol. **27**, 54–65 (2006)

H.K. Lotze, M. Coll, A.M. Magera, C. Ward-Paige, L. Airoldi, Recovery of marine animal populations and ecosystems. Trends Ecol. Evol. **26**, 595–605 (2011)

Marine Conservation Institute, Carl N. Shuster, Jr. Horseshoe Crab Reserve. Atlas of Marine Protection. (2020). http://www.mpatlas.org/mpa/sites/8591/

C.P. McGowan, D.R. Smith, J.A. Sweka, J. Martin, J.D. Nichols, R. Wong, J.E. Lyons, L.J. Niles, K. Kalasz, J. Brust, M. Klopfer, B. Spear, Multispecies modeling for adaptive management of horseshoe crabs and red knots in the Delaware Bay. Nat. Resour. Model **24**, 117–156 (2011a)

C.P. McGowan, J.E. Hines, J.D. Nichols, J.E. Lyons, D.R. Smith, K.S. Kalasz, L.J. Niles, A.D. Dey, N.A. Clark, P.W. Atkinson, C.D. Minton, Demographic consequences of migratory stopover: Linking red knot survival to horseshoe crab spawning abundance. Ecosphere **2**, 1–22 (2011b)

C.P. McGowan, D.R. Smith, J.D. Nichols, J.E. Lyons, J. Sweka, K. Kalasz, L.J. Niles, R. Wong, J. Brust, M. Davis, B. Spear, Implementation of a framework for multi-species, multi-objective adaptive management in Delaware Bay. Biol. Conserv. **191**, 759–769 (2015)

M.J. Millard, J.A. Sweka, C.P. McGowan, D.R. Smith, Assessment and management of north American horseshoe crab populations, with emphasis on a multispecies framework for Delaware Bay, USA populations, in *Changing Global Perspectives on Biology, Conservation and Management of Horseshoe Crabs*, ed. by R. Carmichael, M. L. Botton, P. Shin, S. G. Cheung, (Springer International Publishing, Berlin, 2015), pp. 407–431

T.J. Novitsky, Discovery to commercialization: The blood of the horseshoe crab. Oceanus **27**, 13–18 (1984)

D. Penn, H.J. Brockmann, Age-biased stranding and righting in male horseshoe crabs, *Limulus polyphemus*. Anim. Behav. **49**, 1531–1539 (1995)

Y. Sadovy, The threat of fishing to highly fecund fishes. J Fish Biol **59**(Suppl. A), 90–108 (2001)

C.N. Shuster Jr., M.L. Botton, A contribution to the population biology of horseshoe crabs, *Limulus polyphemus* (L.), in Delaware Bay. Estuaries **8**, 363–372 (1985)

C.N. Shuster Jr., A pictorial review of the natural history and ecology of the horseshoe crab, *Limulus polyphemus*, with reference to other Limulidae, in *Physiology and Biology of Horseshoe Crabs: Studies on Normal and Environmentally Stressed Animals*, ed. by J. Bonaventura, C. Bonaventura, S. Tesh, (Alan R. Liss, Inc., New York, 1982), pp. 1–52

D.R. Smith, M.T. Mandt, P.D.M. MacDonald, Proximate causes of sexual size dimorphism in horseshoe crabs (*Limulus polyphemus*) of the Delaware Bay. J. Shellfish Res. **28**, 405–417 (2009)

D.R. Smith, L.J. Brousseau, M.T. Mandt, M.J. Millard, Age and sex specific timing, frequency, and spatial distribution of horseshoe crab spawning in Delaware Bay: Insights from a large-scale radio telemetry array. Curr. Zool. **56**, 563–574 (2010)

D.R. Smith, H.J. Brockmann, M.A. Beekey, T.L. King, M.J. Millard, J. Zaldivar-Rae, Conservation status of the American horseshoe crab, (*Limulus polyphemus*): A regional assessment. Rev. Fish Biol. Fish. **27**, 135–175 (2017)

B.L. Swan, Migrations of adult horseshoe crabs, *Limulus polyphemus*, in the Middle Atlantic Bight: A 17-year tagging study. Estuaries **28**, 28–40 (2005)

J.A. Sweka, D.R. Smith, M.J. Millard, An age-structured population model for horseshoe crabs in the Delaware Bay area to assess harvest and egg availability for shorebirds. Estuar. Coasts **30**, 277–286 (2007)

N. Tsipoura, J. Burger, Shorebird diet during spring migration stopover on Delaware Bay. Condor **101**, 635–644 (1999)

U.S. Fish and Wildlife Service. Endangered and threatened wildlife and plants; final Threatened status for the rufa red knot (*Calidris canutus rufa*). Docket No. FWS–R5–ES–2013–0097; RIN AY17, (U.S. Fish and Wildlife Service, Northeast Region, New Jersey Field Office, Pleasantville, New Jersey, 2014, 383 p)

E.A. Walls, *A Horseshoe Crab (Limulus Polyphemus) Demographic Study. M.Sc. thesis* (Virginia Polytechnic Institute and State University, Blacksburg, 2001)

E.A. Walls, J. Berkson, Effects of blood extraction on horseshoe crabs (*Limulus polyphemus*). Fish. Bull. **101**, 457–459 (2003)

J.H. Zar, *Biostatistical Analysis*, 2nd edn. (Prentice Hall, Englewood Cliffs, 1984)

# Detrimental Coastline Alteration at Horseshoe Crab Spawning Grounds in Pahang, South China Sea

**Nurul Ashikin Mat Zauki, Shantika Maylana Sastraprawira, Mohammad Yahya Ibrahim Alkhalili, Nur Fairuz-Fozi, Siddhartha Pati, Salwa Shahimi, Chong Ju Lian, Hisham Atan Edinur, Akbar John, Ahmed Jalal Khan Chowdhury, Behara Satyanarayana, and Bryan Raveen Nelson**

## 1 Introduction

The coastal zone represents a dynamic interaction between land and ocean processes where high energetic tidal regime, monsoonal wind, oceanic current, and fluvial discharge influence the topography of coastal landscapes (Yaacob et al. 2018; Mentaschi et al. 2018; Ariffin et al. 2018; Lee et al. 2011; Suhaila et al. 2010).

N. A. M. Zauki (✉) · N. Fairuz-Fozi · B. Satyanarayana
Mangrove Research Unit (MARU), Institute of Oceanography and Environment, Universiti Malaysia Terengganu, Terengganu, Malaysia

S. M. Sastraprawira · M. Y. I. Alkhalili · B. R. Nelson (✉)
Institute of Tropical Biodiversity and Sustainable Development, Universiti Malaysia Terengganu, Terengganu, Malaysia
e-mail: bryan.nelson@umt.edu.my

S. Pati
SIAN Institute, Association for Biodiversity Conservation and Research (ABC), Odisha, India

S. Shahimi · C. J. Lian
Faculty of Marine and Environment Sciences, Universiti Malaysia Terengganu, Terengganu, Malaysia

H. A. Edinur
Forensic Science Programme, School of Health Sciences, Universiti Sains Malaysia, Kelantan, Malaysia

A. John
Institute of Oceanography and Maritime Studies (INOCEM), Kulliyyah of Science, IIUM, Pahang, Malaysia

A. J. K. Chowdhury
Department of Marine Science, Kulliyyah of Science, International Islamic University Malaysia Kuantan, Jalan Sultan Ahmad Shah, Kuantan, Malaysia

 
J. T. Tanacredi et al. (eds.), *International Horseshoe Crab Conservation and Research Efforts: 2007-2020*, https://doi.org/10.1007/978-3-030-82315-3_24

Seasonal transitions between rough and calm weather either sweep out (erosion) or restore (accretion) land masses that lay adjacent to water bodies (Nelson et al. 2019; Ariffin et al. 2018; Fairuz-Fozi et al. 2018; Nelson et al. 2016a) and demand for manmade developments (e.g., seawalls, breakwaters, jetties, etc.) as intervention against active manmade activities like sand mining, construction of dams, or irrigation system (Mentaschi et al. 2018; Yaacob et al. 2018; Nelson et al. 2016b; Nelson et al. 2015). The National Coastal Erosion Study (1984–1986) that was implemented by the Ministry of Environment and Water, Malaysia, identified a total of 29% shorelines suffering from erosion (Ghazali 2006). Since shore erosion became a national concern, the National Coastal Vulnerability Index Study (NCVI) derived the following:

- *Category 1*: Shorelines currently in a state of erosion where shore-based facilities or infrastructure are in immediate danger of collapse or damage.
- *Category 2*: Shorelines eroding at a rate whereby public property and agriculture land of value will become threatened within 5–10 years unless remedial action is taken.
- *Category 3*: Undeveloped shorelines experiencing erosion but with no or minor consequent economic loss if left unchecked.

Reconstruction of shore morphology (i.e., hotels, resorts, sports, etc.) by converting vegetated (e.g., mangrove forest) areas gathered opinions and interventions that used revetments and beach nourishment strategies through the *General Administrative Circular No. 5/1987* that is managed by the Department of Irrigation and Drainage (Ghazali 2006). Public awareness is raised by the Guidelines for Erosion Control for Development Projects in the Coastal Zone (JPS 1/97), whereas erosion control strategies use the Integrated Coastal Zone Management and Integrated Shoreline Management Plan. Reclamation projects <50 ha are supervised by the Environmental Quality (Amendment) Act (1985), whereas projects >50 ha are assessed under the Environmental Assessment Order 1987 (Ghazali 2006).

Unfortunately, horseshoe crabs are not placed in Wildlife Protection Act 2010 nor in Fisheries Act 1985, which means that the coastline modifications in horseshoe crab spawning grounds can easily take place in Malaysia. For instance, a 1-km groyne-shaped airport runway extension toward the South China Sea developed in 2010 (c.f. Ariffin et al. 2018) made Setiu Wetland (some 65 km away) impacted from shore erosion. Eventually, it led to the disappearance of *C. rotundicauda* and their spawning grounds. Similarly, construction of fishing jetty in Balok and Tanjung Selangor (Pahang) altered the shore morphology of horseshoe crab breeding grounds and caused momentary reductions in *T. gigas* spawning activity (Nelson et al. 2016a, 2015). Elsewhere, the conversion of mangrove forests into agricultural (paddy) lands has limited the spatial-temporal distribution of aquatic life in West Peninsular Malaysia (Ghazali et al. 2018). With this, the present study aims to communicate on impact of both monsoon seasons (northeast, southwest, and intermonsoon seasons) and manmade structures in Balok and Pahang Tua River estuaries, areas associated with coexisting *T. gigas* and *C. rotundicauda* populations.

## 2 Changing of Sediment Properties in Light of Coastal Disturbances

The sand-like shore at Balok River appears similar to shores at the open sea and is considered suitable for fishing activities (i.e., boat docking and fishing jetty; Fig. 1). In 2012, the coastal development continued with the construction of parking lot and ~ 50 m riprap was built fringing the wet market and parking lot areas. It was this period that fine sediment dominated the surface of Balok Beach and reduction in horseshoe crab spawning activity is witnessed (*c.f.* Nelson et al. 2016a; Zauki et al. 2019a,b).

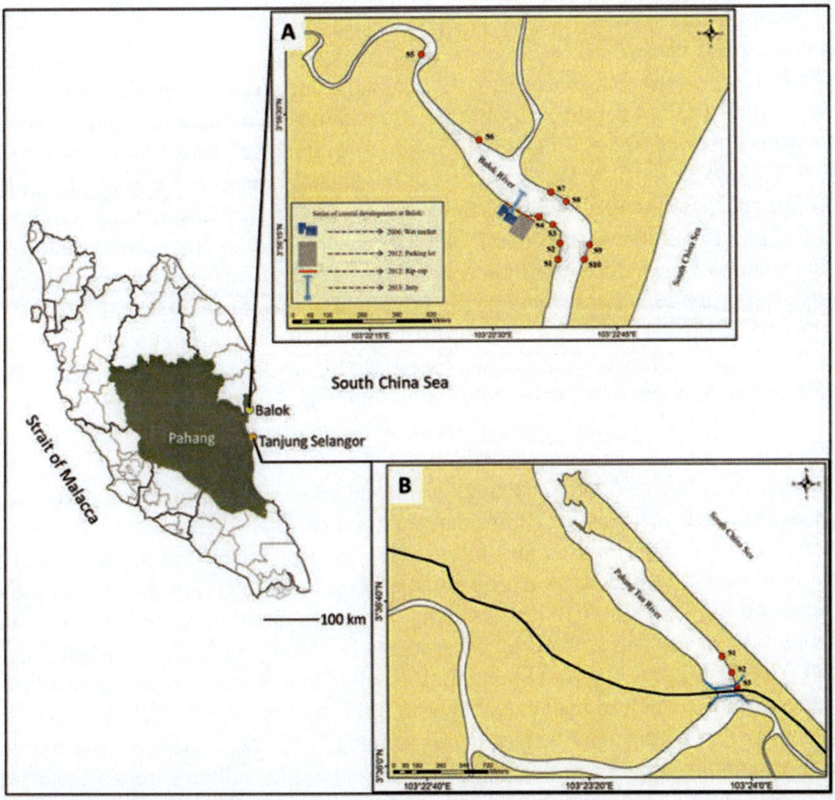


**Fig. 1** The location of Balok and Tanjung Selangor in Peninsular Malaysia. (**a**) The 10 study sites (S1–S10) in Balok River vicinity that were visited between 2010 and 2017 for sediment collection. (**b**) The three study sites (S1–S3) in Pahang Tua River vicinity that were visited between 2010 and 2011. A series of constructions near Balok Beach started with a wet market in 2006, parking lot and riprap in 2012 and fishing jetty in 2013. Similarly, a road bridge construction is responsible for the land reclamation and mangrove clearing at Tanjung Selangor. These depictions are represented using ArcGIS v.10.1 (Esri Inc., USA)

The coastline protection measures at Balok via riprap construction were implemented to reduce beach erosion, but it became the source of sediment disturbance later on. The improvement of the wooden fishing jetty a year later was aimed at safety of local fishermen and eased the collection of fin- and shellfish from vessels. Observations in 2010 indicated that the sediment grain size as medium sand (1.2–1.5 Xφ) got shifted to fine sand (2.5–2.6 Xφ) in 2013 and medium-fine sand (1.8–2.7 Xφ) in 2016–2017 (Table 1). This sediment transition was taught to be detrimental since horseshoe crab counts were decreased in our 3 days catch effort data between 2009 and 2013 (Nelson et al. 2016a, b). However, our perception changed with the citizen scientist initiative (c.f. Zauki et al. 2019a, b) after witnessing the presence of 239 *T. gigas* and 80 *C. rotundicauda* during 2012, 263 *T. gigas* and 91 *C. rotundicauda* during 2014, and 312 *T. gigas* and 119 *C. rotundicauda* during 2016 in Balok environment.

Sampling sites were arranged based on *T. gigas* presence for the 2010 and 2013 sampling and included *C. rotundicauda* environment for the 2016–2017 sampling (Table 1). The sediment sorting in 2010 that ranged between moderately and poorly sorted (0.9–1.0 σφ) nature was gradually transitioned to moderately well sorted to moderately sorted (0.5–0.7 σφ) in 2013 and moderately well sorted to poorly sorted (0.5–1.3 σφ) in 2016–2017. In relation to the annual monsoonal impact, the early (2010) analysis revealed medium sand grain size in the northeast monsoon (NEM) (1.9 Xφ) and southwest monsoon (SWM) (1.6 Xφ) period while coarser sand (0.8 Xφ) covered Balok microenvironment in inter-monsoon (IM) season, whereas the sediment properties changed to fine sand throughout the year in both 2013 (2.5–2.6 Xφ) and 2016–2017 (2.1–2.4 Xφ) observations (Table 2). Throughout the time of 2010, the sediment was poorly sorted by NEM (1.0 π) and SWM (1.1 π), while IM (0.8 π) was dominated by moderately sorted sediment. Throughout the annual monsoon season, the sediment sorting retained as reasonably well-sorted sediment (0.5–0.6 π).

Tanjung Selangor (Pekan, Pahang) is situated ~35 km south of Balok (Kuantan, Pahang) on the east coast of Peninsular Malaysia (Fig. 1). The anthropic activities started off with sand mining and followed by the construction of a jetty in 2010 before the road bridge construction commenced a year later. An investigation by Nelson et al. (2015) in 2010 disclosed that the sediment at those sampling sites in Tanjung Selangor was influenced by moderately sorted to poorly sorted (0.9–1.1 σφ) coarse sand (0.6–0.7 Xφ) (Table 3). It resulted in a severe decline of horseshoe crab abundance between the years 2009 and 2011 until the site is no longer feasible for *T. gigas* spawning activity. Due to this, Pahang Tua River estuary is inaccessible to the public because the entire land mass is reserved for military training onward year 2011.

The follow-up visit in 2013 revealed the conversion of sediment nature into poorly sorted (1.1 σφ) fine sand (2.0–2.1 Xφ). The similar sediment characteristic pattern was observed in terms of annual monsoonal differences where coarse sediment grains (0.5–0.7 Xφ) were found throughout the year 2010 with moderately sorted (0.9 σφ) sediment (0.5–0.7 σφ) representing SWM and IM environment while poorly sorted (1.1 σφ) sediment characterized NEM (Table 4). The sediment

**Table 1** Sediment nomenclature at Balok between 2010 and 2017

| | 2010 | | | 2013 | | | 2016–2017 | | | | | | | | | |
|---|---|---|---|---|---|---|---|---|---|---|---|---|---|---|---|---|
| Criteria | S1 | S2 | S3 | S1 | S2 | S3 | S1 | S2 | S3 | S4 | S5 | S6 | S7 | S8 | S9 | S10 |
| Mean (Xφ) | $1.2^{M}$ | $1.5^{M}$ | $1.5^{M}$ | 2.5<br>F | $2.5^{F}$ | 2.6<br>F | 2.4<br>F | $2.0^{F}$ | 2.3<br>F | 2.5<br>F | $1.8^{M}$ | $2.4^{F}$ | $2.6^{F}$ | 2.7<br>F | 2.6<br>F | 2.5<br>F |
| Sorting (σφ) | $1.0^{Ps}$ | $0.9^{Ms}$ | $1.0^{Ps}$ | $0.6^{Mw}$ | $0.7^{Ms}$ | $0.5^{Mw}$ | $0.6^{Mw}$ | $0.7^{Ms}$ | $0.6^{Mw}$ | $0.5^{Mw}$ | $1.3^{Ps}$ | $1.3^{Ps}$ | $0.7^{Ms}$ | $0.5^{Mw}$ | $0.5^{Mw}$ | $0.5^{Mw}$ |

Note: Sites are abbreviated as S1 = site 1, S2 = site 2, S3 = site 3, … and S10 = site 10 (Fig. 1). In 2010, allocated sites were visited during March, April, June, July, August, and October, while observations for 2013 were done from February to May (Nelson et al. 2016a). Period of 11 months sampling were carried out during 2016, whereas only one-time visit (February) was done during 2017. S1–S4 and S7–S10 were observed in 2017 while analyses of sediment at S5–S7 were done in 2016. Logarithmic method of moments (Blott and Pye 2001) is referred where the coefficients were expressed as *M* medium sand, *F* fine sand, *Ms*. moderately sorted, *Mw* moderately well sorted, and *Ps* poorly sorted sand

**Table 2** Sediment nomenclature presented by different monsoon seasons at Balok

| | 2010 | | | 2013 | | | 2016–2017 | | |
|---|---|---|---|---|---|---|---|---|---|
| Criteria | NEM | SWM | IM | NEM | SWM | IM | NEM | SWM | IM |
| Mean (Xφ) | 1.9 M | 1.6 M | 0.8 C | 2.5 F | 2.6 F | 2.6 F | 2.4 F | 2.1 F | 2.4 F |
| Sorting (σφ) | 1.0 Ps | 1.1 Ps | 0.8 Ms | 0.6 Mw | 0.5 Mw | 0.5 Mw | 0.8 Ms | 1.3 Ps | 1.2 Ps |

Note: Monsoon seasons are abbreviated as *NEM* northeast monsoon, *SWM* southwest monsoon, and *IM* inter-monsoon. Also, 6 months (March, April, June, July, August, and October) in 2010, 4 months (February–May) in 2013 recorded by Nelson et al. (2016a), 11 months (except March) in 2016, and 1 month (February) in 2017 were pooled according to each monsoon seasons. Logarithmic method of moments (Blott and Pye 2001) is referred where the coefficient is expressed as *C* coarse, *M* medium sand, *F* fine sand, *Ms.* moderately sorted, *Mw* moderately well sorted, and *Ps* poorly sorted sand

**Table 3** Sediment nomenclature at Tanjung Selangor between 2010 and 2013

| | 2010 | | | 2013 | | |
|---|---|---|---|---|---|---|
| Criteria | S1 | S2 | S3 | S1 | S2 | S3 |
| Mean (Xφ) | 0.7 C | 0.6 C | 0.7 C | 2.0 F | 2.1 F | 2.1 F |
| Sorting (σφ) | 1.0 Ps | 1.1 Ps | 0.9 Ms | 1.1 Ps | 1.1 Ps | 1.1 Ps |

Note: Sites are abbreviated as S1 = site 1, S2 = site 2, and S3 = site 3. Sediment data were retrieved from Nelson et al. (2015) where 12 months of sampling were carried out in 2010 and 2 months (April and October) observation were done in 2013. Logarithmic method of moments (Blott and Pye 2001) is referred where the coefficient is expressed as *C* coarse, *F* fine sand, *Ms.* moderately sorted, and *Ps* poorly sorted sand

**Table 4** Sediment nomenclature presented by different monsoon seasons at Tanjung Selangor

| | 2010 | | | 2013 |
|---|---|---|---|---|
| Criteria | SW | NE | IM | IM |
| Mean (Xφ) | 0.7 C | 0.6 C | 0.5 C | 2.0 F |
| Sorting (σφ) | 0.9 Ms | 1.1 Ps | 0.9 Ms | 1.1 Ps |

Note: Monsoon seasons are abbreviated as *NEM* northeast monsoon, *SWM* southwest monsoon, and *IM* inter-monsoon. Sediment data were retrieved from Nelson et al. (2015) where 12 months sampling was carried out in 2010 and 2 months (April and October) observation was done in 2013. Logarithmic method of moments (Blott and Pye 2001) is referred where the coefficient is expressed as *C* coarse, *F* fine sand, *Ms.* moderately sorted, and *Ps* poorly sorted sand

maintained the features of poorly sorted (1.1 σφ) fine sand (2.0 Xφ) in IM period during the visit in 2013.

## 3 Shoreline Topography Changes Assessment by Geospatial Data

In light of the coastal disturbances taking place at Balok and Tanjung Selangor, we utilized satellite images from Google Earth Pro to communicate on shoreline changes in terms of landmass loss (erosion) and gain (accretion). Altogether 10 selected points were plotted in satellite the images of 2010, 2013, and 2017, ranging from the river mouth to upstream (Figs. 2 and 3). The comparison of boundary changes was categorized into short-term period (2010–2013) and long-term period (2010–2017) (Tables 5 and 6). Overall, Balok River mouth/downstream area (P1 and P2) underwent dynamic changes. Within the short term (four years), high intensity of erosion (−42.20 m at P1) and deposition (+75 m at P2) creating the nature shaped the sediment appearances in the area. Long-term (8 years) investigation suggested that several events of land mass gains and losses have occurred. Perturbation of river mouth sediment throughout 2010–2017 resulted in a narrower river mouth width in the year 2017. The land erosion at the disturbed site (developments) (P3 and P4) was severe at first four years of assessment (−27.55 m at P3 and −7.51 m at P4) and gradually decreased by 2017 (~1–3 m differences). Further upstream (P5–P8) showed minute alternating deposition and erosion events except for the meander (P9 and P10) section where greater land mass changes (accretion and erosion) were observed. A similar pattern of dynamic land mass changes was observed shifting the morphology of river mouth of Tanjung Selangor. Massive changes of land mass were observed (+250. 04 m at P1 and −254.95 m at P2) in 2010–2013 time frame while the longer periods of time have influenced natural repairing process where smaller land changes were measured between 2010 and 2017. Moving upstream, the paired points (P3–P4 and P5–P6) undergo minute landmass differences in terms of deposition and erosion in either point while P7 and P8 showed expansion of river width (erosion) from 2010 to 2017. At P9 and P10, the river width was narrower during 2010 but loss of sediment at P10 had widened the river width during 2017.

# 4 Discussion

## *4.1 Natural Determinants of Morphodynamics in the Coastal Area*

Peninsular Malaysia experiences tropical monsoonal storms (winter storms) that interrelate with hydrodynamic interactions of tides, wind, wave action, and bed load movement (Ariffin et al. 2018; Razak et al. 2018; Mirzaei et al. 2013). Flooding by seawater creates perturbation zones that erode land masses at the shoreline, but the effects of wind (Ekman spiral) for energy transfer influence the surge trajectory (Mirzaei et al. 2013). Therefore, winds of the northeast monsoon (November–March) cause East Peninsular Malaysia to experience rough sea conditions aside from heavy rainfall (Razak et al. 2018; Suhaila et al. 2010). Comparatively, the Titiwangsa range does not permit wind movements into the West Coast and winds and rain brought by the southwest monsoon (May–September) remains on the West Coast. Instead, either coasts experience an inverse situation during the monsoon seasons where humidity decreases while hot and arid conditions prevail. Shielding by Sumatera and short distance between Malaysian and Indonesian land masses in

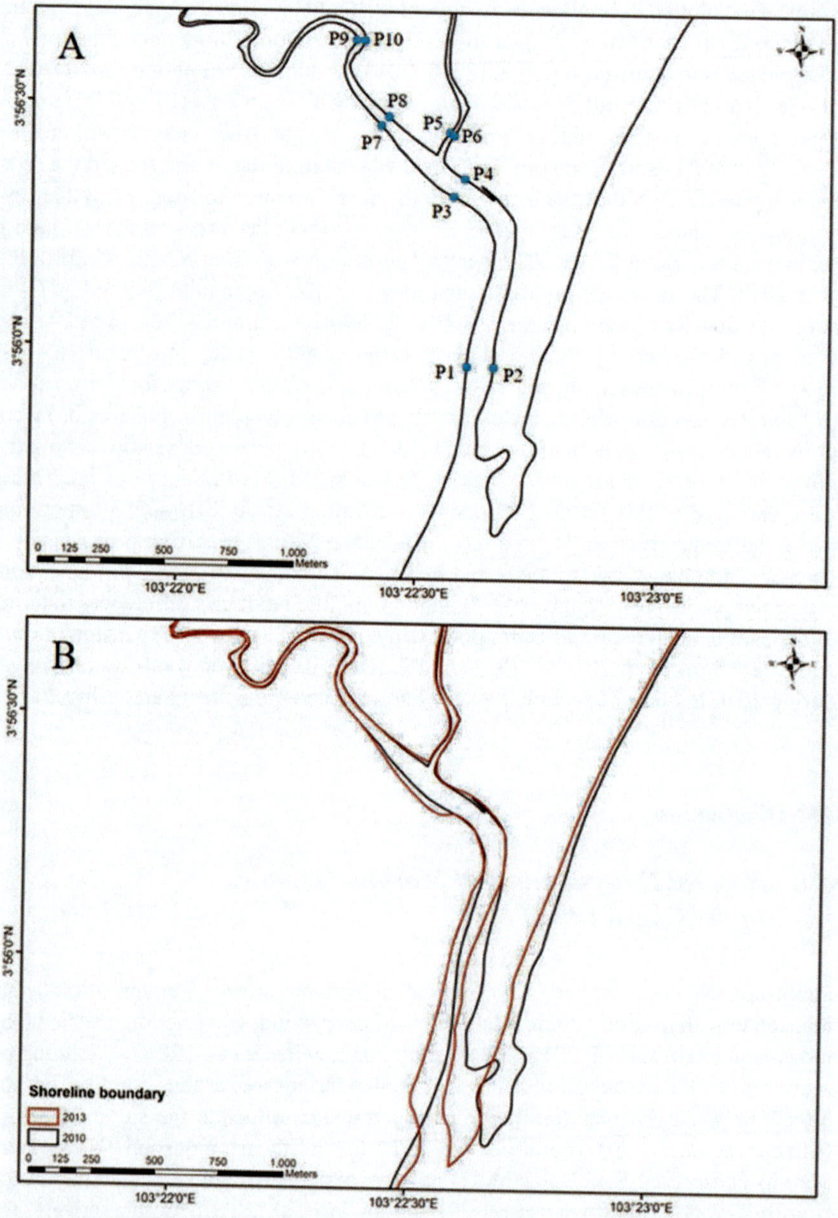


**Fig. 2** The borderline changes of Balok Beach from 2010 to 2017 digitized from satellite images retrieved from Google Earth. (**a**) 10 plots are selected as point to measure land mass gain or loss along Balok River. (**b**) The borderlines shifting between 2010 and 2013 (4 years changes) within the construction period. (**c**) The borderlines shifting between 2010 and 2017 (8 years changes) after the constructions end. (**d**) The shoreline boundary changes of the years 2010, 2013, and 2017

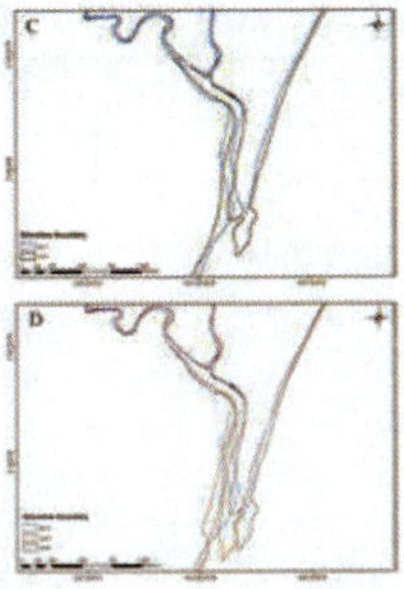

**Fig. 20.2** (continued)

the Straits of Malacca brings upon low wave activity and the resulting mud-dominant shores on the West Coast in comparison to the tidal amplitude and currents brought by South China Sea to give eastern shores sand-dominant beaches in Peninsular Malaysia (Razak et al. 2018; Ghazali 2006).

## 4.2 *Anthropic Activities and Morphodynamic Changes to the Horseshoe Crab*

In recent years, coastal encroachment changed the morphodynamic footprints particularly after structure installations (i.e., building, facilities developments), nearshore transportation (i.e., longshore and cross-shore transportation as it contributes to wave height intensity), and indirect anthropic activities like mining, irrigation systems, dam constructions, and clearing mangrove forests, the natural buffer vegetations (Mentaschi et al. 2018; Morris et al. 2018; Razak et al. 2018). The natural pressure by changing climate and manmade shoreline modifications transformed shore sediment properties in Balok and Tanjung Selangor, but fortunately the *T. gigas* and *C. rotundicauda* continue to select these natal beaches for their spawning (Nelson et al. 2019; Ariffin et al. 2018; Nelson et al. 2015, 2016a). It was different for Kuala Nerus (Terengganu) where land reclamation to produce a 1 km tarmac airport runway adjacent to open sea caused severe erosion to Batu Rakit, Tok Jembal, Cherating, and Setiu onward the year 2010 (Ariffin et al. 2016, 2018). Aside from horseshoe crab habitat loss in Setiu Wetland, property damage/loss (house damage, land loss, and complete destruction of a research station owned by Universiti Malaysia Terengganu) and livelihood impacts (fisher vessels unsafe to dock in open sea) urged the call for facilities (i.e., jetty, market, road bridge, riprap, and revetments) as intervention. Unfortunately, the same practice is repeated in Pahang where sea sand mining and coastal land reclamation in Balok and Tanjung Selangor transformed sediment of its shores from medium and medium-coarse into fine sand (Nelson et al. 2015, 2016a). Erosion (Tables 5 and 6) occurred at both areas although *T. gigas* and *C. rotundicauda* emergence was taking place at their

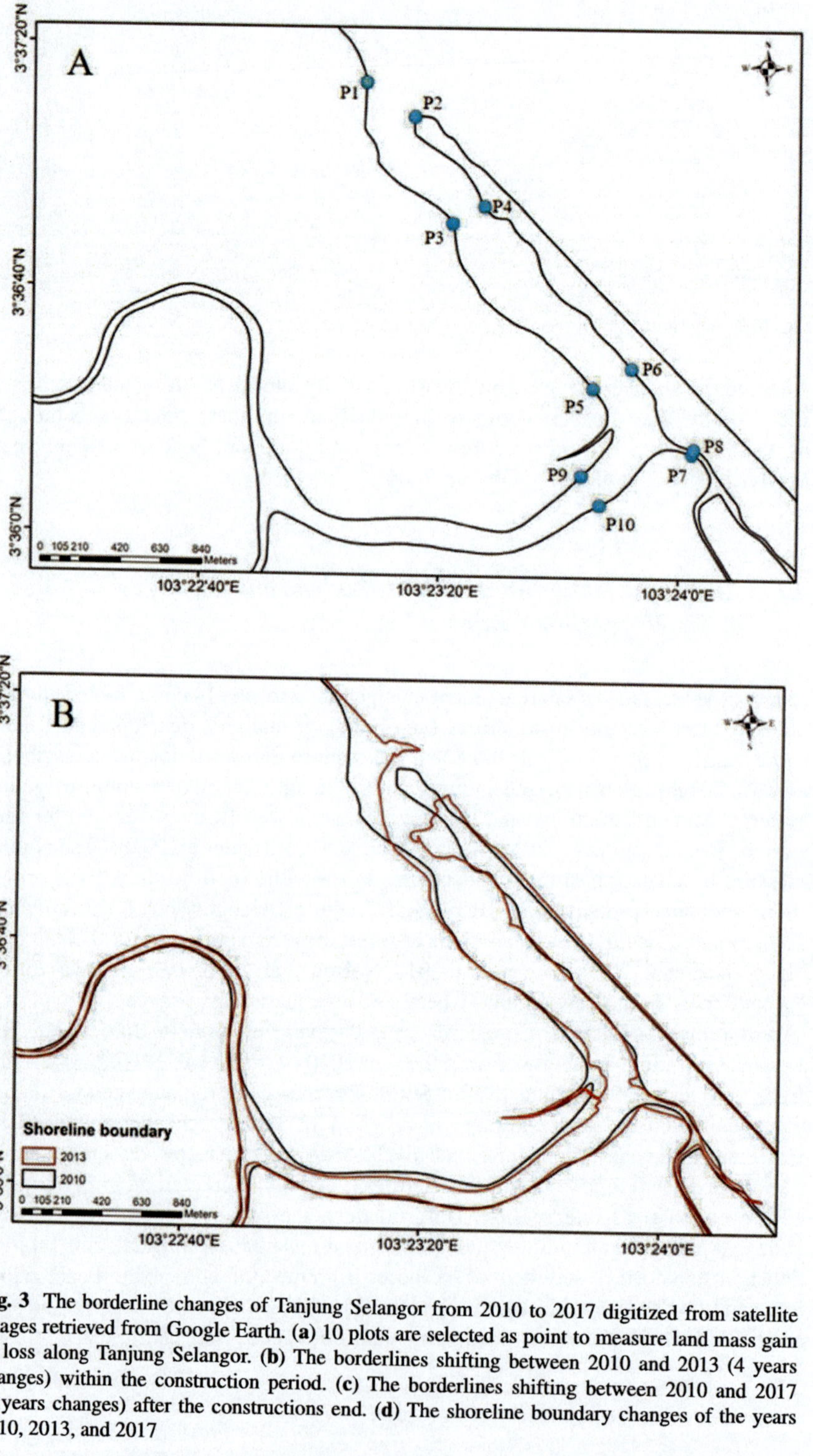


**Fig. 3** The borderline changes of Tanjung Selangor from 2010 to 2017 digitized from satellite images retrieved from Google Earth. **(a)** 10 plots are selected as point to measure land mass gain or loss along Tanjung Selangor. **(b)** The borderlines shifting between 2010 and 2013 (4 years changes) within the construction period. **(c)** The borderlines shifting between 2010 and 2017 (8 years changes) after the constructions end. **(d)** The shoreline boundary changes of the years 2010, 2013, and 2017

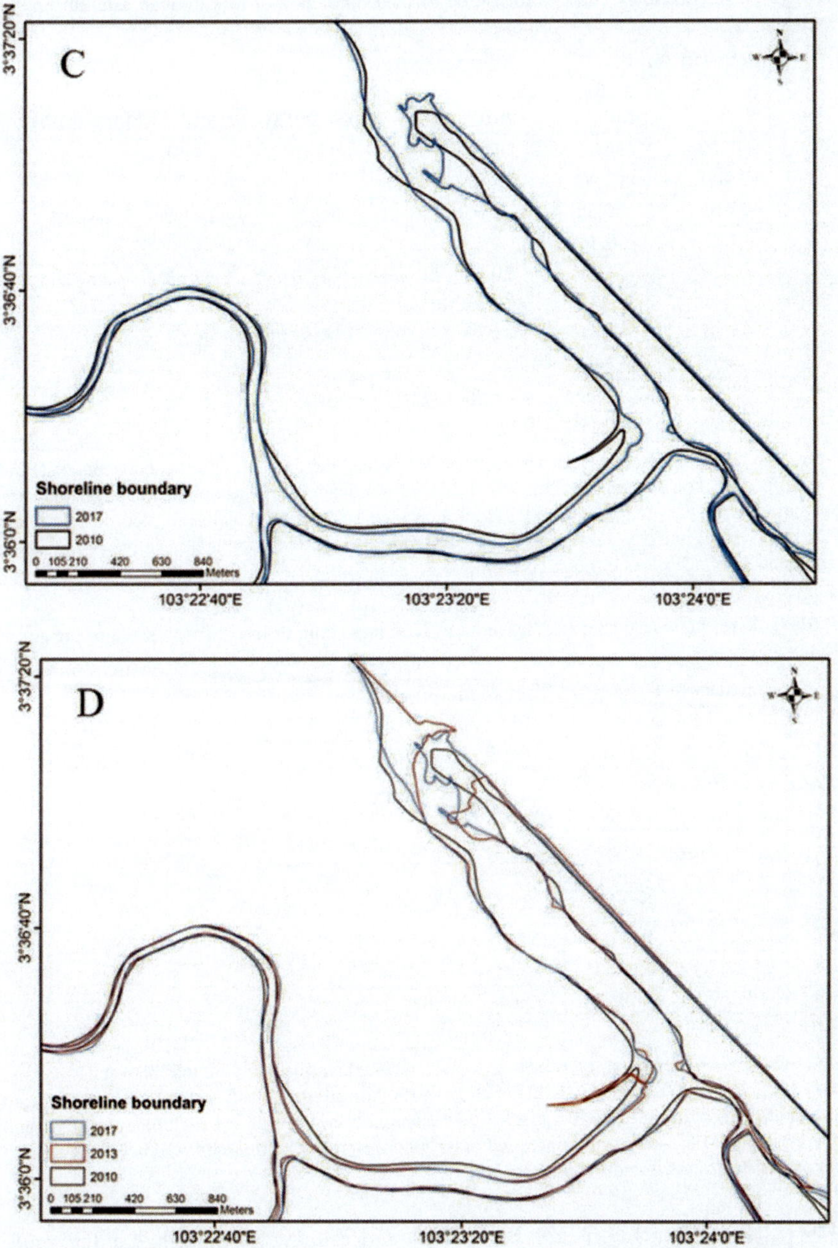


**Fig. 20.3** (continued)

natal beaches (Manca et al. 2016; Nelson et al. 2015, 2016a; John et al. 2012; Tan

**Table 5** River boundary changes measured as land mass gain or loss through satellite images at Balok

| | Boundary (m) | | River width (m) | | | | |
|---|---|---|---|---|---|---|---|
| Points | 2010 to 2013 | 2010 to 2017 | 2010 | 2013 | Diff. (2010–2013) | 2017 | Diff. (2013–2017) |
| P1 | −42.20 | +10.82 | 104.43 | 70.97 | +33.46 | 79.51 | −8.54 |
| P2 | +75.00 | +13.96 | | | | | |
| P3 | −27.55 | −30.65 | 79.66 | 116.4 | −36.74 | 120.25 | −3.85 |
| P4 | −7.51 | −9.46 | | | | | |
| P5 | −2.86 | +2.80 | 20.1 | 21.46 | −1.36 | 18.63 | +2.83 |
| P6 | +1.51 | −1.31 | | | | | |
| P7 | −5.14 | −7.64 | 44.12 | 49.67 | −5.55 | 49.2 | +0.47 |
| P8 | −0.67 | +2.13 | | | | | |
| P9 | −7.57 | −0.98 | 32.56 | 37.01 | −4.45 | 44.78 | −7.77 |
| P10 | +3.39 | −1.25 | | | | | |

Note: The points selected at river considering the visible land form or land loss between 2010 and 2017 (Fig. 3a). The differences (Diff.) of river width indicate the river boundary changes between the superscript years (e.g., 2010–2013) and measured between odd and even plots (e.g., P1 and P2). Positive (+) marking represents land mass gain through accretion and negative (−) marking represents land mass loss through corrosion

**Table 6** River boundary changes measured as land mass gain or loss through satellite images at Tanjung Selangor

| | Boundary (m) | | River width (m) | | | | |
|---|---|---|---|---|---|---|---|
| Plots | 2010–2013 | 2010–2017 | 2010 | 2013 | Diff. (2010–2013) | 2017 | Diff. (2013–2017) |
| P1 | +250.04 | −25.26 | 301.82 | 306.21 | −4.39 | 293.02 | +13.19 |
| P2 | −254.95 | +60.44 | | | | | |
| P3 | −32.62 | −39.38 | 185.6 | 202.51 | −16.91 | 220.39 | −17.88 |
| P4 | +9.67 | +11.69 | | | | | |
| P5 | +43.57 | +37.58 | 219.83 | 178.7 | +41.13 | 187.1 | −8.4 |
| P6 | −4.32 | −6.28 | | | | | |
| P7 | −28.64 | −31.76 | 26.94 | 64.31 | −37.37 | 71.76 | −7.45 |
| P8 | −8.47 | −13.15 | | | | | |
| P9 | +31.57 | +31.05 | 173.64 | 141.84 | +31.8 | 151.17 | −9.33 |
| P10 | +0.39 | −9.39 | | | | | |

Note: The points selected at river considering the visible land form or land loss between 2010 and 2017 (Fig. 4a). The differences (Diff.) of river width indicate the river boundary changes between the superscript years (e.g., 2010 – 2013) and measured between odd and even plots (e.g., P1 and P2). Positive (+) marking represents land mass gain through accretion and negative (−) marking represents land mass loss through corrosion

et al. 2012; Zaleha et al. 2012). Both Balok and Tanjung Selangor had unfavorable shores with high sediment compaction (fine sand, no foot depressions) compared to 2010 because the *T. gigas* spawning pairs reduced to <5 and nest recoveries were < 10 during full moon tides (Nelson et al. 2015, 2016a). Eventually by 2011, Setiu Wetland and Tanjung Selangor could no longer support the spawning activity of horseshoe crabs (ca. *T. gigas* and *C. rotundicauda*) because salinity (>33 ppt), pH

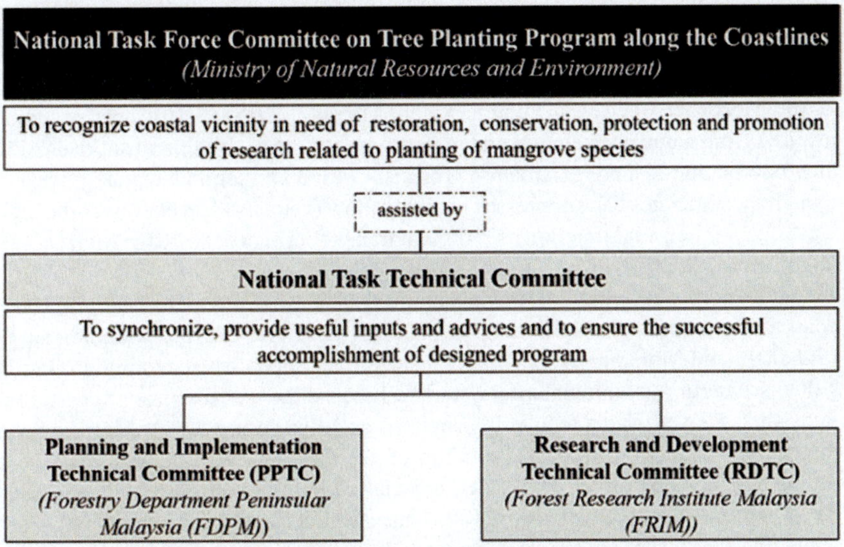


**Fig. 4** The National Task Force Committee on Tree Planting Program along the Coastline was established by the Malaysian government after the severe effects of tsunamis struck in 2004

(<6), tidal amplitude (>3.0 m), and water currents (0–2 m = 5–10 kph) and sediment composition (>2 Xφ) exceeded the edaphic perquisites of this arthropod (Nelson pers. obs.; Nelson et al. 2015, 2019).

## 4.3 *Suggestions on Mitigation Efforts to Minimize the Coastal Impacts*

Despite the mitigation efforts (installments of ripraps and revetments) to restore the coastal morphology, stakeholder participation must take place from the initiations of ideas (Ariffin et al. 2016, 2018). The involvement of several groups must consider sustainable opinions and not personal interests (Reed et al. 2009). This includes knowledge and opinions of local people because they reside in the project vicinity (Zauki et al. 2019a; Ariffin et al. 2018). Furthermore, scientific technical studies like environmental impact assessment and management plans must be considered by the stakeholders for any (new) development plans (Ghazali et al. 2018). For instance, location of the wind research station in Universiti Malaysia Terengganu was 9.7 km from the project site and communities occupy the shores in between both points (Ariffin et al. 2018). Urgency to use hard structures (riprap) as low-budget means was unsuitable, and this sort of actions can be avoided during round-table discussions on protection schemes.

Studies consider seawalls, wave breakers, revetments, and ripraps as immediate response to mitigate the erosion (Ariffin et al. 2018; Morris et al. 2018). However, these heavy structures displace seabed sediment away from the shore and make the entire shore deep. Maintenance is required to fill the displaced sediment behind the barriers and depending biodiversity can no longer use these shores as habitat (Nelson et al. 2019; John et al. 2018; Ariffin et al. 2018; Ghazali et al. 2018; Jumain et al. 2018; Morris et al. 2018; Sulaiman and Mohidin 2018). Therefore, solutions inspired from nature (vegetation and reefs) are preferred as solution to shore erosion after completion of land reclamation projects. Coastal vegetation like mangrove and mangrove associates can suppresses the water flow (velocity) and energy of the high waves using roots that function as sediment traps (Toorman et al. 2018; Morris et al. 2018; Verhagen 2018).

While northwest coast of Peninsular Malaysia was hit by tsunami in 2004, coastal vegetation in Larut Matang was successful in breaking the pounding surge (Sulaiman and Mohidin 2018). Unfortunately, horseshoe crab spawning ground in Teluk Senangin was momentarily disturbed because the tsunami brought perturbations that delayed shore recovery (Chatterji and Shaharom 2009). Moreover, an inundation model predicted 0.25–0.52 m of average sea level rise with maximum 0.5 m coastal flooding while 20% of adjacent agriculture (paddy) and mangrove forest will be fully inundated in 2100 (Ghazali et al. 2018). With mounting concerns, the Malaysian government developed the "National Task Force Committee on Tree Planting Programme along the Coastlines" that involved a "Planning and Implementation Technical Committee (PPTC)" and the "Research and Development Technical Committee (RDTC)" (Sulaiman and Mohidin 2018; Fig. 4).

Since mangrove vegetations were sought by the committee, they selected species that regenerate and colonize rapidly while offering drag forces and hydraulic resistance against high waves (Sulaiman and Mohidin 2018; Verhagen 2018). Additional research by RDTC includes sediment (mudflats) stabilization (i.e., utilization of woven geotextile tube as breakwater) and application of innovative mangrove techniques (Comp–mat, Com–pillow, and Bamboo Encasement Method) after the planting of mangrove vegetation (*Rhizophora apiculata, R. mucronata,* and *Avicennia alba*) in Kuala Bernam Forest Reserve, Selangor (Sulaiman and Mohidin 2018). However, their efforts were challenged by root anchorage of the young mangrove plants in the presence of harsh tidal action. We observe this situation in Balok where young mangrove vegetation dominated upstream Balok River, whereas due to erosion, mangrove vegetation was pushed inside the river banks. Perhaps implementing engineering structures (i.e., geo-textile tubes, brushwood dams, and rubble mounts) could maximize colonization spaces (mud bank deposition) for young mangrove plants or introducing salt marsh species (*Spartina brasiliensis*) could stabilize the substrate and support accretion in the area (ca. Toorman et al. 2018 in Guyana, South America) while wind and sediment models are used to organize the planting plots and to minimize sediment displacement during transport (Nelson et al. 2019; Goessens et al. 2014).

Community-based involvements with proper environmental management have secured a shielding layer over the ecosystem and provide a better living environment to the wildlife within (Pati 2019; Pati 2017; Pati et al. 2017). The invaluable local knowledge has become more practical and reliable in enforcing management plans. In fact, the local people are witnessing daily changes in their area and they would be able to provide precise information on the scale and magnitude of coastal impacts (Zauki et al. 2019a, b; Datta et al. 2012). Local communities are self sufficient in their management and take their actions seriously especially if it involves their resource pools. After the fishermen suffered loses from shore modifications that led to severe sediment circulation near the river mouth (Zauki et al. 2019a; Nelson et al. 2016b), they decided to organize their boat anchorage onto wooden poles at the shore, implement a centralized site for waste (i.e., old nets and rubbish) disposal, and scheduled fisheries activity (daily catch landing at 10 am and 4 pm daily except for Friday). Altogether, their actions were successful to reinstate the spawning and feeding activities of horseshoe crabs in Pantai Balok, Pahang (Zauki et al. 2019a).

## 5 Conclusion

The natural monsoon storms have significantly affected the coastal morphodynamics, especially the tropical-region of Asia. The construction of artificial structures have interfered and hindered the natural repairing process of the coastal areas in Malaysia. These problems are made worse by the climate change that also introduced hazards (storms and coastal flooding) from the shoreline mark. To mitigate all the coastal impacts, solid coastal defense structures were installed to provide shoreline protection. Nature-based protection schemes (replanting of coastal vegetation) is perhaps the best option for the management of coastal erosion after development. In addition, it is recommended that hydrological and soil models should be part of coastal development so that predictions could prepare stakeholders for the consequences.

**Acknowledgments** We extend our gratitude to the Malaysian Ministry of Higher Education for funding this project under the FRGS grant (FRGS-59380, 2015) and Ocean Park Conservation Fund, Hong Kong (OT01.2021). We would like to thank the authorities Institute of Oceanography and Environment (INOS) and School of Marine and Environmental Sciences (PPSMS) for the facilities provided.

## References

E.H. Ariffin, M. Sedrati, M.F. Akhir, N.R. Daud, R. Yaacob, M.L. Husain, Beach morphodynamics and evolution of monsoon-dominated coasts in Kuala Terengganu, Malaysia: Perspectives for integrated management. Ocean Coast. Manag. **163**, 498–514 (2018)

E.H. Ariffin, M. Sedrati, M.F. Akhir, R. Yaacob, M.L. Husain, Open sandy beach morphology and morphodynamic as response to seasonal monsoon in Kuala Terengganu, Malaysia. J. Coast. Res. **75**, 1032–1036 (2016)

S.J. Blott, K. Pye, GRADISTAT: A grain size distribution and statistics package for the analysis of unconsolidated sediments. Earth Surf. Process. Landf. **26**, 1237–1248 (2001)

A. Chatterji, F. Shaharom, Unusual spawning behaviour of the horseshoe crab (*Tachypleus gigus*, Müller) after the tsunami along Orissa Coast, India. Pertanika J. Sci. Technol. **17**(2), 263–268 (2009)

D. Datta, R.N. Chattopadhyay, P. Guha, Community based mangrove management: A review on status and sustainability. J. Environ. Manage. **107**, 84–95 (2012)

N. Fairuz-Fozi, B. Satyanarayana, N.A.M. Zauki, A.M. Muslim, M.L. Husain, S. Ibrahim, B.R. Nelson, *Carcinoscorpius rotundicauda* (Latreille, 1802) population status and spawning behaviour at Pendas coast, Peninsular Malays. Glob. Ecol. Conserv. **15**, e00422 (2018)

N.H.M. Ghazali, Coastal erosion and reclamation in Malaysia. Aquat. Ecosyst. Health Manage. **9**, 237–247 (2006)

N.H.M. Ghazali, N.A. Awang, M. Mahmud, A. Mokhtar, Impact of sea level rise and tsunami on coastal areas of North-West Peninsular Malaysia. Irrig Drain (2018)

A. Goessens, B. Satyanarayana, T. Van der Stocken, M.Q. Zuniga, H. Mohd-Lokman, I. Sulong, F. Dahdouh-Guebas, Is Matang mangrove forest in Malaysia sustainably rejuvenating after more than a century of conservation and harvesting management? PLoS One **9**, e105069 (2014)

B.A. John, B.R. Nelson, H.I. Sheikh, S.G. Cheung, Y. Wardiatno, B.P. Dash, K. Tsuchiya, Y. Iwasaki, S. Pati, A review on fisheries and conservation status of Asian horseshoe crabs. Biodivers. Conserv. **27**(14), 3573–3598 (2018)

B.A. John, B.Y. Kamaruzzaman, K.C.A. Jalal, K. Zaleha, Feeding ecology and food preferences of *Carcinoscorpius rotundicauda* collected from the Pahang nesting grounds. Sains Malays **41**, 855–861 (2012)

N.A. Jumain, N.H. Kamarulzaman, I.A. Latif, F.M. Yusoff, Impact of reclamation on fishery activities in Malaysia. Int. J. Agric. Environ. Bio-res. **4**, 87–97 (2018)

Y.K. Lee, J.H. Ryu, J.K. Choi, J.G. Soh, J.A. Eom, J.S. Won, A study of decadal sedimentation trend changes by waterline comparisons within the Ganghwa tidal flats initiated by human activities. J. Coast. Res. **27**, 857–869 (2011)

L. Mentaschi, M.I. Vousdoukas, J.F. Pekel, E. Voukouvalas, L. Feyen, Global long-term observations of coastal erosion and accretion. Sci. Rep. **8**, 12876 (2018)

A. Mirzaei, F. Tangang, L. Juneng, M.A. Mustapha, M.L. Husain, M.F. Akhir, Wave climate simulation for southern region of the South China Sea. Ocean Dyn. **63**, 961–977 (2013)

B.R. Nelson, J.H.Z. Moh, N.A.M. Zauki, B. Satyanarayana, A.J.K. Chowdhury, Effects of shore sedimentation to *Tachypleus gigas* (Müller, 1785) spawning activity from Malaysian waters. J. Sustain. Sci. Manag. **14**, 1–27 (2019)

B.R. Nelson, B. Satyanarayana, J.H.Z. Moh, M. Ikhwanuddin, A. Chatterji, F. Shaharom, The final spawning ground of *Tachypleus gigas* (Müller, 1785) on the east Peninsular Malaysia is at risk: A call for action. Peer J. **4**, e2232 (2016a)

B.R. Nelson, B. Satyanarayana, J.H. Moh, F. Shaharom, Does human infringement at the spawning grounds challenge horseshoe crab eggs and their embryogenesis? Journal of Sustainability Science and Management Special Issue 1: The International Seminar on the Straits of Malacca and the South China Sea (2016b), p. 1–10

B.R. Nelson, B. Satyanarayana, J.M.H. Zhong, F. Shaharom, M. Sukumaran, A. Chatterji, Episodic human activities and seasonal impacts on the *Tachypleus gigas* (Müller, 1785) population at Tanjung Selangor in Peninsular Malaysia. Estuar. Coast. Shelf Sci. **164**, 313–323 (2015)

A. Manca, F. Mohamad, B.R. Nelson, M.F.A. Mohd Sofa, A.A. Alia'm, A. Ismail, Trailing the spawning horseshoe crab *Tachypleus gigas* (Müller, 1785) at designated natal beaches on the East Coast of Peninsular Malaysia. Cell Dev. Biol. **5**, 171 (2016)

R.L. Morris, T.M. Konlechner, M. Ghisalberti, S.E. Swearer, From grey to green: Efficacy of eco-engineering solutions for nature-based coastal defence. Glob. Chang. Biol. **24**(5), 1827–1842 (2018)

S. Pati, Cleaning the breeding beaches and awareness event for horseshoe crab conservation along Chandipur coast and nearby estuaries in Odisha. Zoo's Print **32**, 46–48 (2017)

S. Pati, S. Tudu, A. Rajesh, G.C. Biswal, A. Chatterji, B.P. Dash, R. Samantaray, Manmade activities affected the breeding ground of horseshoe crab (*Tachypleus gigas*) along Balasore coast: Call for immediate conservation. E-Planet **15**, 145–154 (2017)

S. Pati, Trade and by-catch assessment of Indian horseshoe crab along with its conservation by integrating education and awareness among community along Balasore Coast. Final Report produced for Rufford Foundation. https://www.rufford.org/files/24630-1%20Final%20Report.pdf (2019). Accessed 31 Aug 2019

M.S.A. Razak, F.X. Suryadi, N. Jamaluddin, N.A.Z.M. Nor, Shoreline planform stability of embayed beaches along the Malaysian peninsular coast. J. Coast. Res. **85**(sp1), 631–635 (2018)

M.S. Reed, A. Graves, N. Dandy, H. Posthumus, K. Hubacek, J. Morris, C. Prell, C.H. Quinn, L.C. Stringer, Who's in and why? A typology of stakeholder analysis methods for naturalresource management. J. Environ. Manage. **90**, 1933–1949 (2009)

J. Suhaila, S.M. Deni, W.Z.W. Zin, A.A. Jemain, Trends in Peninsular Malaysia rainfall data during the southwest monsoon and northeast monsoon seasons: 1975–2004. Sains. Malays. **39**, 533–542 (2010)

R.B.R. Sulaiman, F.S.M. Mohidin, Establishment of shoreline buffer zone through rehabilitation of degraded coastal mangroves. MATEC Web Conf. **203**, 01019 (2018)

A.N. Tan, A. Christianus, S. Shakubazadeh, P. Hajeb, Horseshoe crab, *Tachypleus gigas* (Muller, 1785) spawning population at Balok Beach, Kuantan, Pahang, Malaysia. Pak. J. Biol. Sci. **15**, 610–620 (2012)

E.A. Toorman, E. Anthony, P.G. Augustinus, A. Gardel, N. Gratiot, O. Homenauth, N. Huybrechts, J. Monbaliu, K. Moseley, S. Naipal, Interaction of mangroves, coastal hydrodynamics, and morphodynamics along the coastal fringes of the Guianas, in *Threats to Mangrove forests*, (Springer, Cham, 2018), pp. 429–473

Verhagen HJ (2018) The beneficial effects of mangrove forest to sea defence structures. In Makowski C, Finkl CW Threats to Mangrove Forests. Springer, Cham, pp. 475–495

R. Yaacob, H. Shaari, N. Sapon, M.F. Ahmad, E.H. Arifin, R. Zakariya, M.L. Hussain, Annual changes of beach profile and nearshore sediment distribution off Dungun-Kemaman, Terengganu, Malaysia. J. Teknol. **80**, 57–66 (2018)

K. Zaleha, B.A. John, H. ErniAtika, B.Y. Kamaruzzaman, K.C.A. Jalal, Spawning and nesting behaviour of *Tachypleus gigas* along the East Coast of Peninsular Malaysia. Int. J. Biol. **4**, 102–111 (2012)

N.A.M. Zauki, B. Satyanarayana, N. Fairuz-Fozi, B.R. Nelson, M.B. Martin, B.A. John, A.J.K. Chowdhury, Citizen science frontiers horseshoe crab population regain at their spawningbeach in East Peninsular Malaysia. J. Environ. Manage. **232**, 1012–1020 (2019a)

N.A.M. Zauki, B. Satyanarayana, N. Fairuz-Fozi, B.R. Nelson, M.B. Martin, B. Akbar-John, A.J.K. Chowdhury, Horseshoe crab bio-ecological data from Balok, East coast Peninsular Malaysia. Data Brief **22**, 458–463 (2019b)

# Assessing the Use of Acoustic Sampling for Locating Horseshoe Crabs During Migrations Using Side Scan Sonar

Stephen D. Nagiewicz, Vince Capone, Peter F. Straub, Steven P. Evert, and Nate Robinson

## 1 Overview

Background information about the species is appropriate and necessary to be able to understand the movement and spawning processes observed and documented previously by tagging and observations made by scientists. This will help define the parameters of how to use acoustic methods to find and identify aspects of crab behavior. While sonar can see fish underwater, for example, tracking their random movement is an exercise better quantified in time versus dollars spent. What sonar can do if the species can be seen and verified by acoustic sampling is understand their environment and any associated behaviors while reaching a beach for nesting and laying eggs.

The horseshoe crab's name is somewhat misleading. Although it is shaped like a horseshoe, it is not crab. The horseshoe crab is an arachnid, a class of arthropods that also includes scorpions, spiders, mites, and ticks (Ballesteros & Sharma 2019). With two main eyes, two simple (light sensing only) eyes, and a mouth on the bottom, the horseshoe crab is well suited to life on the bottom. A brownish segmented shell offers protection and a pointed tail helps the animal right itself; it is not used for attacking or even self-defense.

Often called "living fossils," horseshoe crab ancestors can be traced back through the geological record (Shuster and Shuster 2003) to around 445 million years ago, 200 million years before dinosaurs existed. Only one species, *Limulus polyphemus*, is found in North America along the Atlantic and Gulf coasts from Maine to Mexico (Tanacredi et al. 2009).

S. D. Nagiewicz (✉) · P. F. Straub · S. P. Evert · N. Robinson
Stockton University, Galloway, NJ, USA
e-mail: Stephen.Nagiewicz@stockton.edu

V. Capone
Black Laser Learning, Hockessin, DE, USA



J. T. Tanacredi et al. (eds.), *International Horseshoe Crab Conservation and Research Efforts: 2007-2020*, https://doi.org/10.1007/978-3-030-82315-3_25

Horseshoe crabs are known for their large nesting aggregations, or groups, and often spawn during neap tides (Penn and Brockmann 1994; Cavanaugh 1975; Barlow et al. 1986) on beaches particularly in mid-Atlantic states such as Delaware, New Jersey, and Maryland. This process is known as spawning. It appears that Crabs start their inshore movement from the deep bay and coastal waters and it appears to be triggered by the lengthening of daylight hours. This happens in late May and can coincide with high tide during full or new moons.

When mating, the smaller male crab attaches itself to the top of the larger female's shell by using its specialized front claws, and together they crawl to the beach. The male fertilizes the eggs as the female lays them in a nest in the sand. Some males (called satellite males) do not attach to females but still have success in fertilizing the female's eggs as they crowd around the attached pair.

The mechanism by which horseshoe crabs locate preferred spawning habitat is not completely understood. While horseshoe crabs spawn in greater numbers and with greater fecundity along sandy beaches, horseshoe crabs can tolerate a wide range of physical and chemical environmental conditions, and they will spawn in less suitable habitats if ideal conditions are not encountered. Therefore, the presence of large numbers of horseshoe crabs on a beach is not necessarily an indicator of habitat suitability (Shuster 1994).

In addition to the intertidal zone used for spawning, horseshoe crabs also use shallow water areas such as intertidal flats in their juvenile life stages. Horseshoe crabs may congregate on intertidal flats to wait for full moon high tides because these flats provide protection from wave energy. Scientists identified that preferentially selected spawning sites were located adjacent to large intertidal sand flat areas. In addition to providing protection from wave energy, sand flats typically provide an abundance of available food for juvenile horseshoe crabs. Since several tidal cycles may be required to complete spawning, offshore intertidal flats may provide safe areas to rest between tide cycles (Thompson 1999).

According to the National Marine Fisheries Service's Northeast Fishery Center during the spawning season, adults typically inhabit bay areas adjacent to spawning beaches and feed on bivalves. Horseshoe crabs swim or crawl as their primary means of locomotion. Both larvae and juveniles are more active at night than during the day (Rudloe 1979; Shuster 1982; Thompson 1999). Juveniles typically feed prior to the daytime low tide, then burrow into the sand, remaining inactive for the remainder of the day (Rudloe 1981; Thompson 1999).

Horseshoe crabs are an important part of the ecology of coastal communities (Tanacredi et al. 2009; Shuster and Sekiguchi 2009; Rudloe and Herrnkind 1976). Their eggs are the major food source for northward-migrating shorebirds, including the federally threatened red knot. These shorebirds have evolved to time their migrations to coincide with peak horseshoe crab spawning activity, especially in the Delaware and Chesapeake Bay areas. The horseshoe crab, an important keystone species of the Delaware Bay, is an animal that is very much depended upon by many other species participating in the ecosystem. Shorebirds such as the Red Knot (*Calidris canutus*), Ruddy Turnstone (*Arenaria interpres*), and the Sanderling (*Calidris alba*) depend upon horseshoe crab eggs deposited along the banks of the Delaware Bay for their own nutritional welfare.

## 2 Methods

Research has not found published evidence that any scientist or organization has used side scan sonar to acoustically survey a species like horseshoe crabs in shallow intertidal coastal waters, so there are no known criteria for evaluating data marine biology data acquired acoustically. Dr. Daphne Monroe of Rutgers University's Cape Shore Laboratory also conducted a sonar survey in 2018–2019 in the same survey area as Stockton University but used the DIDSON 300, which is a forward looking 1.1 MHz sonar that delivers real-time, near video-quality data to study the behavior of marine life and horseshoe crabs (Monroe 2020). We believe that Rutgers using the DIDSON sonar and with Stockton using both the Klein 3900 side scan sonar and the Humminbird Helix Mega 1200 MHz is the first to use acoustic imaging to understand habitat and migration of a bottom-dwelling marine species although using different sonar platforms.

Imaging sonars produce overhead reproductions of the sea floor created with sound. Unlike photography, sound does not require a light source, is not affected by turbidity, and has significantly more range. Side scan sonar is the most common imaging sonar with the ability to generate high-resolution bottom imagery over relatively long ranges. These systems emit sound pulses with very narrow horizontal beam angles imaging narrow slices of the sea floor, which are combined to produce complete bottom imagery.

Each sonar pulse, commonly referred to as a "ping," travels across the sea floor perpendicular to the transducer array. Range and distance to features are determined by the two-way travel time from the transducer. Reflections from each ping are geographically referenced via a position and heading, resulting in an accurate depiction of the sea floor.

Sonar pulses transmitted from the towfish over the bottom out to a predetermined range produce the overhead imagery. Ensonified sediments and objects reflect different amounts of sonar energy depending upon their acoustic properties, resulting in varying color intensities in the imagery.

Objects on the sea floor can be accurately measured. Contacts that rise above the flat bottom will block the sonar pulse from reaching the bottom behind the object creating an acoustic shadow. The result is an overhead picture analogous to an aerial photograph taken with a low sun angle with extended shadows. Shadow length can be accurately measured determining the height above bottom of the object casting the shadow (Mazel, 1985).

High-frequency side scan sonar had previously been tested on scallop shells to determine the reflectivity of the shells (Capone 2008). The reflectivity of the scallop shell was visible on a medium sand bottom. While the amount of reflected sonar energy is a function of the acoustic impedance of the material being imaged, its reflection must be different than that of the background to be visible in the sonar display. Side scan sonar sediment reflectivity is based upon its grain size, which determines the strength of the acoustic backscatter. This is the first important question to be answered.

Crab presence, absence, movement, and potentially numbers of crabs will be assessed acoustically using a Klein 3900 dual-frequency side scan sonar and a Humminbird Mega 1200 MHz side imaging sonar. These sonar systems will utilize survey-grade positioning and will be operated from a 24′ survey boat. These side scan sonars are capable of resolving small targets while covering 20–80 meter swaths (depending on conditions) with image detection of individual crabs possible and masses of crab's probable. The ability to detect the crab masses will be dependent on density, bottom types, bathymetry, and other possible factors.

Two sonars were selected for this survey. The Klein 3900 side scan sonar is a high-frequency 455/900 kHz and is a towed array sonar. It is the usual tool for many commercial and academic sonar mapping surveys, but it requires advanced operating knowledge and postprocessing software for data interpretation. Humminbird Helix MEGA is a recreational fishing sonar also with high resolution but with shorter ranger capability utilizing side imaging frequencies ranging from 455/800 kHz and 1.2 MHz.

It is simple to use and less expensive and can be setup quickly and easily on most any type of vessel.

Each sonar has advantages for this application in the data they can collect; ultimately we chose to use the Humminbird Helix due to its adaptability for small vessel operation in shallow water. The towed sonar while producing high-resolution images was problematic in shallow intertidal waters near the beach in 3 feet of water or less. The Humminbird was the simplest and most effective solution for this shallow water survey (Fig. 2).

While snapshots can be saved from screen captures, the best analysis of sonar data is through postprocessing software for both the Klein and Humminbird sonars. Stockton used Sonar Pro for data collected by the Klein 3900 and SAR HAWK software for the Humminbird. SAR HAWK was developed for the many search and recovery rescue squads and police departments. It is easy to use and inexpensive and is getting wider uses in academic environments.

Transect lines were developed along a north–south line along the shoreline between Rutgers Cape Laboratory and Green Creek. This comprises approximately 1 nm of shoreline that is a mix of aquaculture and open flats. The surveys were designed in direct consultation with the project biologists following discussion on anticipated data outcomes and sampling strategies. Survey speeds will average 4–6 knots, allowing 3 nm transect lines to be run in approximately 30 minutes, or as seen in (Fig. 1) 4 nm lines in approximately 40 minutes. Four hours of survey time around the high tide will provide over 20 nm of transect line data for analysis (Fig. 3).

Running longshore survey lines just outside of the fixed gear oyster farms and north and south of those areas will provide data on crab mass approaches to those areas (presence/absence as seen in the imaging). Stockton researchers are confident that crab masses will be detectable with the Klein 3900 and with the HB sonar, but it is noted that this has not been shown and this initial investigation is a pilot study.

Acoustic image interpretations were groundtruthed by the deployment of a small oyster trawl dredge after test survey lines. It is anticipated that areas of crab masses can be delineated in postprocessing software much like bottom classification work.

**Fig. 1** Strongly reflecting bottoms such as sand or rock are bright gold while lessor reflecting sediments such as mud are darker. (Courtesy Klein and Associates)

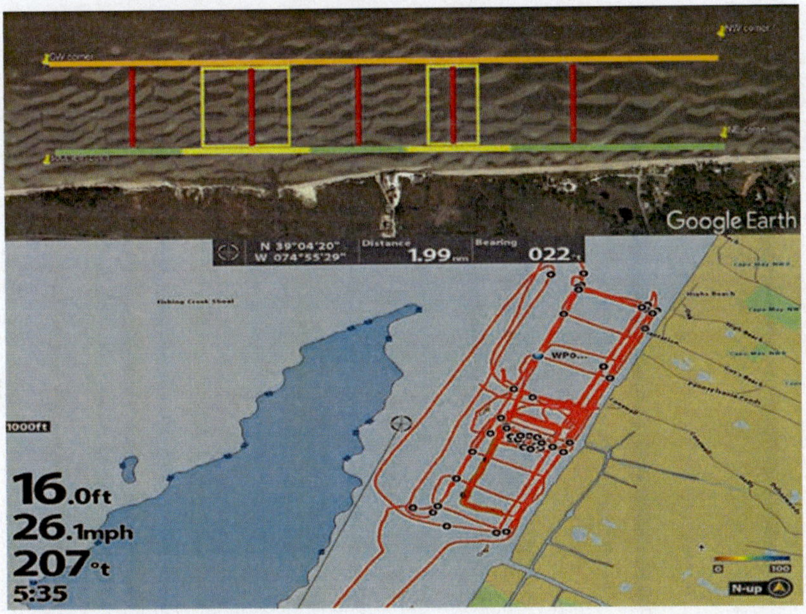


**Fig. 3** Survey area planned vs. actual. (Image courtesy of Stockton University)

Delineation of crab mass sizes may be used to suggest the number of crabs in each event if data exists or is attainable for an average of crabs per square meter when amassed. This crab number versus mass data would need to be provided by the project biologists and is considered secondary to the acoustic detection of presence/

**Fig. 2** Klein 3900 side scan sonar and Humminbird Helix Mega. (Images Courtesy of Klein & Associates and Johnson Outdoors)

**Fig. 4** Trawl groundtruth. (Image courtesy of Stockton University)

absence and movement in relation to shoreline features. Wherever possible, side-by-side comparisons of shoreline stretches with and without gear that are immediately adjacent to each other will be made.

Key research questions were to determine relative abundance and ingress/egress movement timing to/from beaches of horseshoe crabs in areas with aquaculture gear and without. Using side scan sonar, horseshoe crab abundance/density will be measured in predefined areas that are inshore and offshore of control and aquaculture gear areas that will help determine whether crabs primarily move through gear or around it. A secondary goal will be to measure the rate of movement; for example, on a rising tide do horseshoe crab arrive inshore of gear at the same time as they

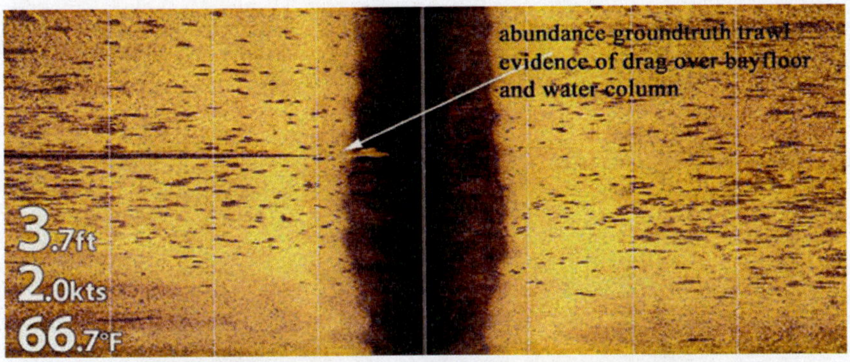


**Fig. 5** Klein sonar image of ground truth trawl. In this image, individual and mating crab pairs are visible as bright spot followed by a dark shadow trail, evidence of the sonar movement from bottom to top in this image. (Courtesy of Stockton University)

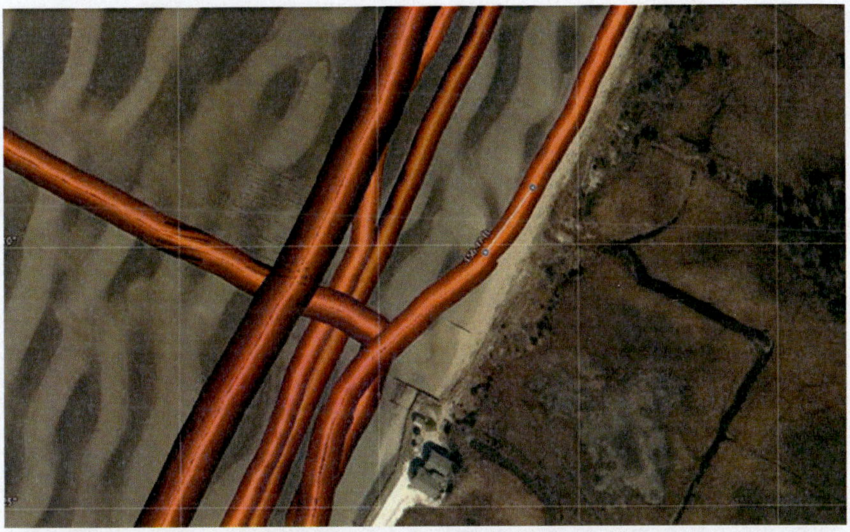

**Fig. 6** It shows about one-quarter nautical mile of survey area. In the image visible is Rutgers Cape Shore Laboratory (the end point of the survey area), the faint lines at the crossing of sonar mosaic lines are oyster gear of some of transect lines, and finally the 156 ft section of transect enlarged below

arriving in control areas? And the same goes for a falling tide on the offshore side of gear.

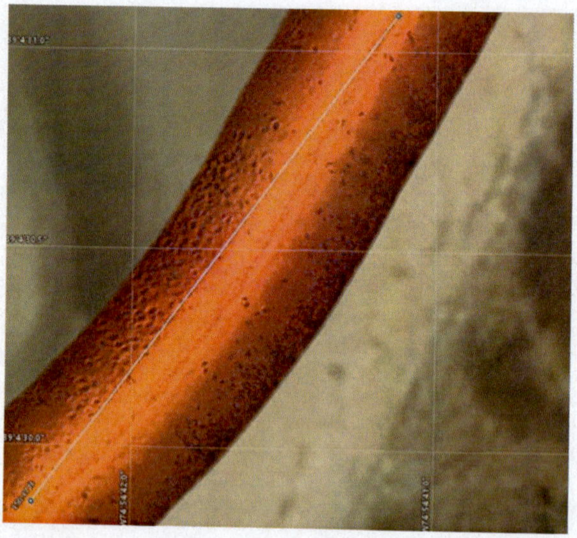

Fig. 7 SAR HAWK software mosaic of sonar run. (Courtesy of Stockton University)

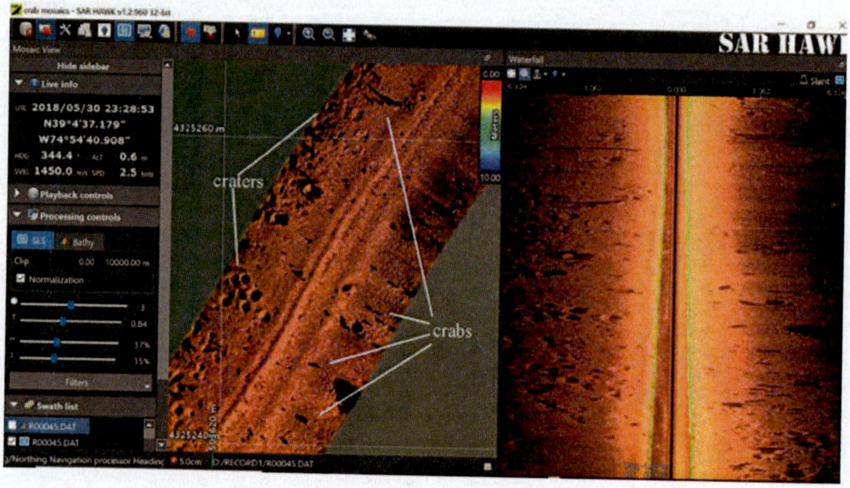


Fig. 8 Screen grab from SAR HAWK software screen showing crabs and craters or burrows. (Courtesy of S S. Nagiewicz)

## 3 Results

The survey area was described by researchers and local fishermen to be mostly flat with very little relief from natural objects (rocks, coral, or manmade detritus). Nautical charts indicate mostly sandy bottom with some area of slight depressions that had some sediment fill (mud). The site was groundtruthed during dead low tides

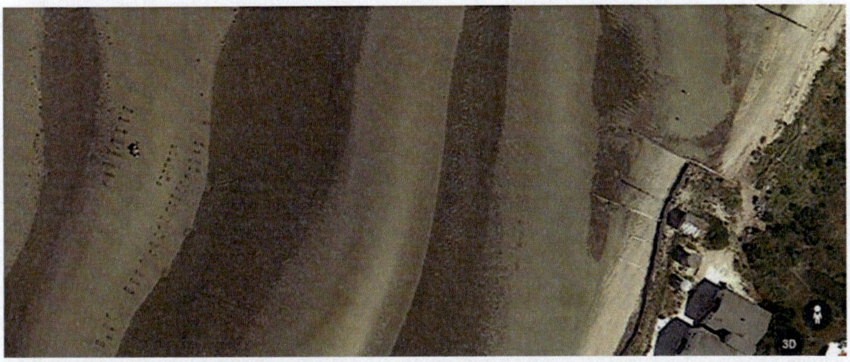


**Fig. 9** Google Earth image 5/29/20 showing a large part of the survey area just offshore of the Rutgers Cape Shore Laboratory

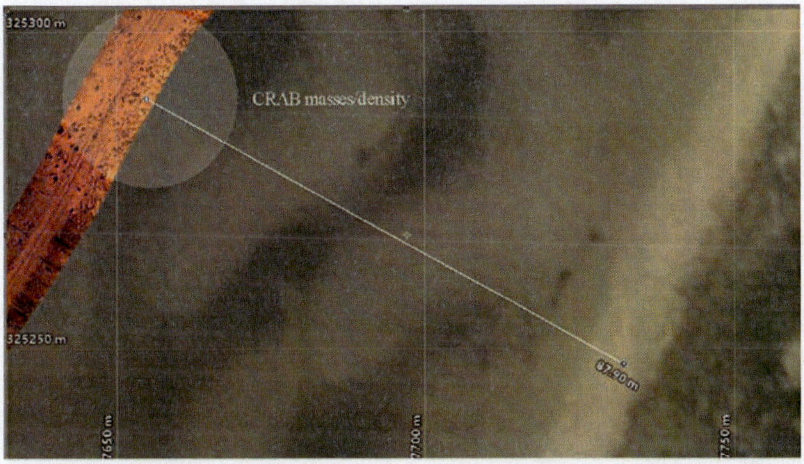


**Fig. 10** Sonar mosaic showing crab density along an inshore transect 89 meters from the beach SAR HAWK software automatically grabs Google Earth or any preloaded chart to the background in the sonar mosaic. (Courtesy of S. Nagiewicz)

as fishermen and researchers can walk out to gear or use all-terrain vehicles for oyster farming gear maintenance. There is one section of the survey area that was known to have oyster farming gear setup, which was geographically known and marked. This area would be part of the survey so carefully planned lines needed to be calculated so as not to hit and possibly adversely affect research or fishermen gear. This area needed to be included to understand how crab movement might be affected for the in-place gear.

While the targets exhibit characteristics of a sonar refection, having some idea of habitat conditions is necessary. A rock, metal debris or crab will all reflect differently based upon the contacts acoustic reflectivity relative to the bottom sediment

**Fig. 11** Cart in a crater or burrow in the survey area just one day after an evening sonar run. (Courtesy of S. Nagiewicz)

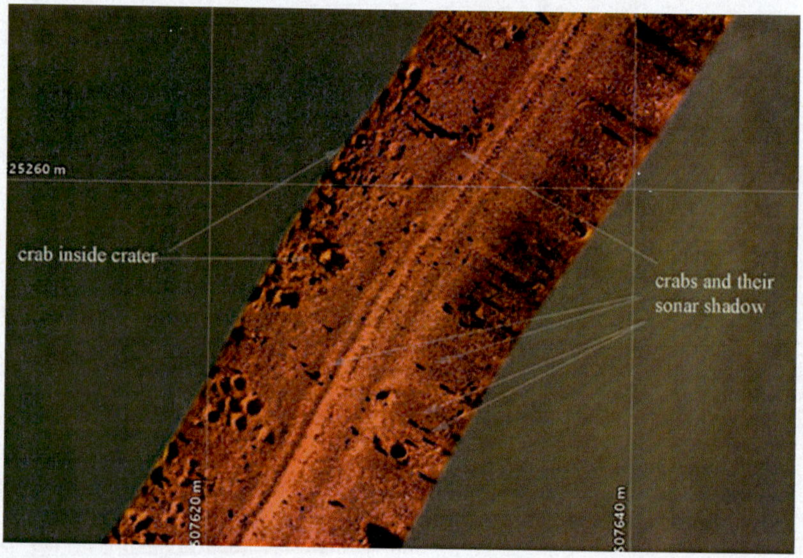


**Fig. 12** Humminbird sonar mosaic of craters or burrows that are empty and possibly have crabs still in them. (Courtesy of S. Nagiewicz)

acoustic backscatter. Whether stronger or weaker the difference in reflectivity to the sediment makes objects identifiable. Knowing the terrain and sediment composition beforehand simplifies identification. With this information, Stockton survey team had high confidence that any "targets" should then be marine life or seafloor

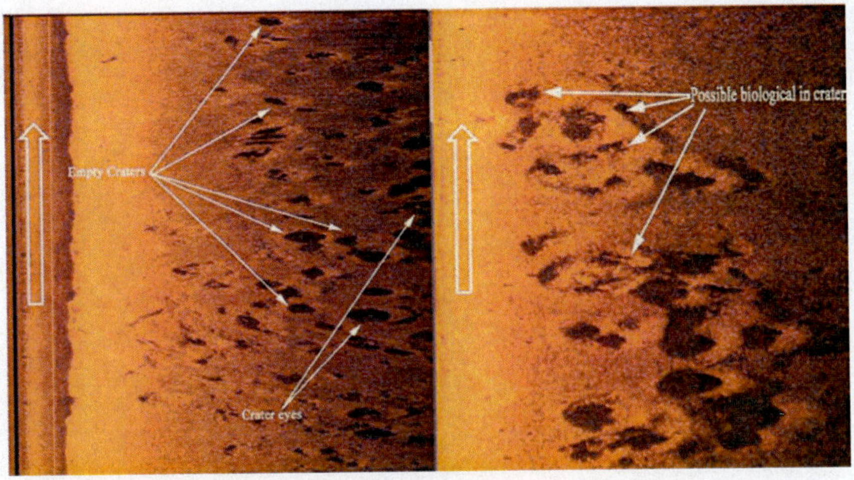


**Fig. 13** Humminbird 1200 side scan sonar images of bay bottom. (Courtesy of S. Nagiewicz)

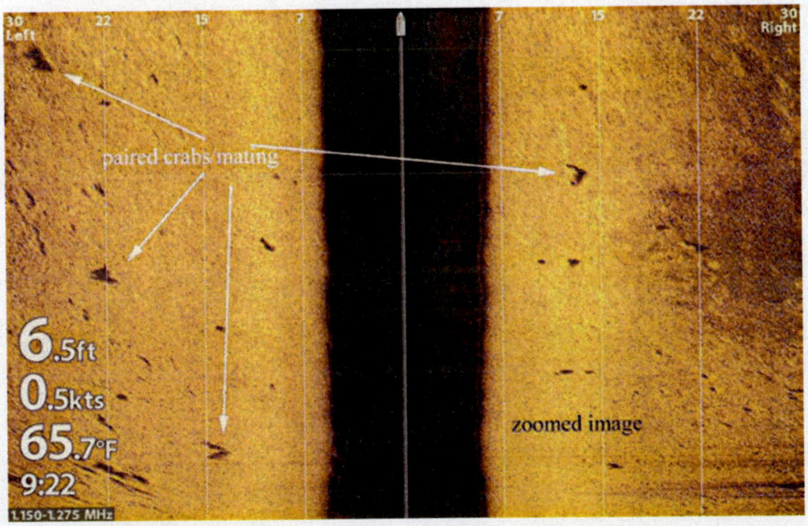


**Fig. 14** Humminbird 1200 kHz sonar screen grab that seems to show pairs of crabs mating. (Courtesy of S. Nagiewicz)

biological of some type. Also considered was the background science of horseshoe crab behavior during spawning, their apparent preferences for gradual sloping approaches to beaches for egg-laying, and abundance of food and hard bottom for easier movement by the crabs for an arduous trek.

Three sonar surveys were made during the month of May and one in June based upon high tides and changed for weather conditions. The survey area is inaccessible by beach for easy boat deployments, and this forced launching from the Cape May

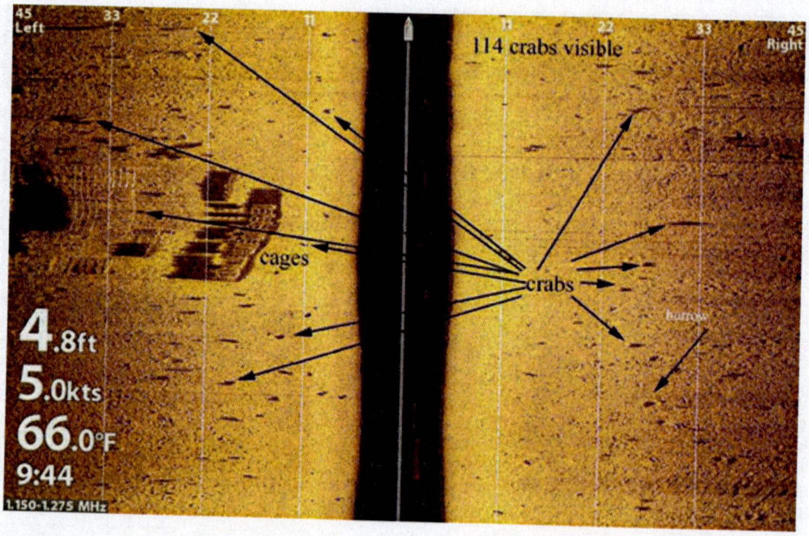


**Fig. 15** Humminbird 1200 kHz sonar image of oyster cage farm with numerous possible crab targets. (Courtesy of S. Nagiewicz)

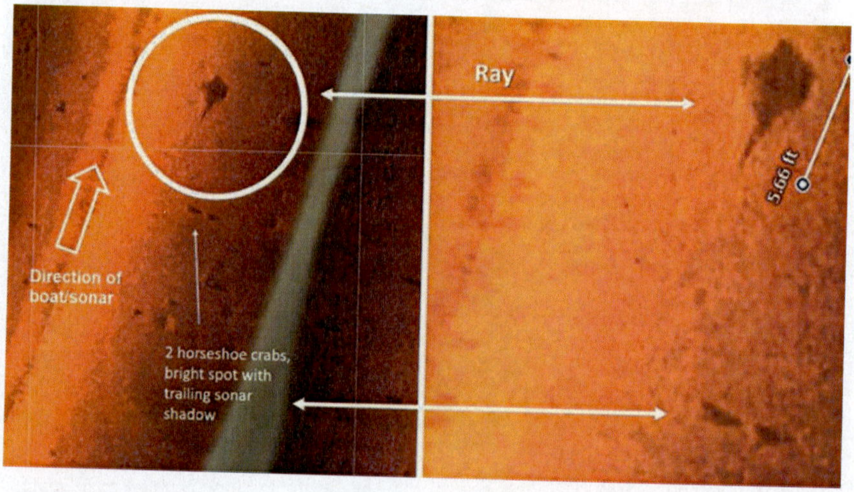


**Fig. 16** Image of some species of ray approximately 5 feet in size. Rays are predator of the horseshoe crab. (Courtesy of S. Nagiewicz)

Canal and a 30-minute transit to site each day. The initial survey took place on late evening of May 18, 2018, on the RV Osprey. This was the only trip where both Klein sonar and the Humminbird were used jointly. Groundtruthing was done along two transects, one where many targets were imaged and one where there were none. In both instances, the trawl proved the sonar data correct (Figs. 4 and 5). Surveying in shallow water with manmade gear made towing sonar potentially hazardous to both gear and sonar and the Humminbird was delivering the same acceptable results, so the decision was made to forego towed sonar and run exclusively with the transom-mounted Humminbird side imaging sonar.

Our initial swath runs were 20-40 meters, which later changed to as shallow water and wind-driven sea conditions limited sonar swaths to 10–20 meters. Wave height affects the sonar returns and can distort data, so slow and low with shorter range helped to better collect useable data.

May 24, 2018, was the second day of survey. Surface conditions were the worse on this survey date than all the others, very windy and choppy seas, all of which affected data collection. Factors besides wind and seas that alter returns and data recording can be a combination of the depth of water, boat speed, transducer placement, angle of beam, and frequency of the sonar beam. All of these played into the data recording, while images were useful as the amount of interference does make the data suspect. We could see the same numbers of potential targets but the sea conditions made measurements difficult. High tide was 6 o'clock, so the survey runs would start at 2:30 PM and end at 6:00 PM and cover all transects. The survey started with the north to south offshore transects to determine if more crabs were incoming on the rising tide. No crabs were recorded on the first offshore transect. The sonar survey finished with an inshore transect running from south to north. Many crabs were spotted along the beach. This was when running our transect lines many craters or divots were recorded as seen in Figure ##. We were not sure what these meant; however, going back to the crab research data, we understood that crabs might have completed their inshore migration and waited near the beach lying dormant and dug into the bottom to feed and wait until the right time to make their run onto the beaches to lay eggs.

On May 30, the third survey started at 6:20 PM with high tide at 9:00 PM, with light easterly winds, calm seas. As per our plan, we ran the offshore transect lines and gradually worked out way inshore as high tide approached. Like last week, many crabs were spotted on the beach.

The last survey was on June 12 . The survey started at 5:00 PM with high tide again at 9:00 PM, with light easterly winds, calm seas. As per our plan, we ran the offshore transect lines and gradually worked out way inshore as high tide approached. Like last week, many crabs were spotted on the beach, but in this survey many more holes or craters were recorded along the inshore transects, which can be seen in Fig. 10. To provide true ground coverage, the water column at the center of the sonar record was removed in Figs. 6, 7, and 8.

We established that the carapace of the crabs reflects sonar just as a rock or shipwreck would. We further established that the survey area was devoid of rocks or other human debris other than gear cages for oyster farmers, so any sonar return would have to be from a marine organism or crab. Crab targets on the left side of the image are identified by long shadow "behind" the lighter target. Craters or divots seem to suggest adaptions of the crabs to burrow into sandy bottom. The horseshoe crabs are one of the largest organisms that burrow in sandy environments so that it can feed on organisms in the seafloor and lay its eggs in the sediment.

Inshore sonar transects showing 156 feet of crab abundance (along the thin white line in the image) to the right side of the sonar mosaic close to the shoreline as the crabs move to nest. The sonar swath width is 40 meters. To the left of the sonar center line are numerous craters as shown by circular areas of no reflection. We surmised that the crabs stayed in the depressions, feeding and resting in-between visits to nesting areas on the beach. Previously, we already described how shadows were created. The crabs produce both a reflection and a small shadow. Depressions with a reflector inside have a crab present. While transiting across the sediment, crabs produce shadows as their carapace blocks the sonar energy. By reviewing the data using both reflections and shadows, sonar analysis can result in a numerical estimate of crab density.

Using the SAR HAWK postprocessing software, the target and its shadow can be accurately measured. It is interesting that the left side of the sonar track shows only craters while the right side shows large abundance of crabs quite near the surf line. Supposition is the crabs are leaving their dug in resting area and are moving to spawn on the beach.

## 4 Discussion

The primary objective of this survey was to see if horseshoe crabs could be imaged using side scan sonar and be able to track their migration to the beach egg-laying areas. A side issue would be to determine if oyster cage gear deployed in the flat sand off the beaches would affect the crab migration. N.J. State law says that cage gear needs to be supported to a height of 6–12 inches so that animals can move underneath the equipment (NJDEP-Proposed Amendments, new rules and repeals: N.J.A.25A, Oysters). Any impediment for the crabs reaching nesting areas along the beach could have an impact on shore bird feeding on the beaches after horseshoe crab egg-laying.

What we needed to understand going into this project? One, will sonar be reflected or absorbed by the crab? If sonar pulses are reflected, then the crabs would appear as targets by the sonar and could presumably be tracked as they move along the bay bottom. If the sound pulses were being absorbed, the crabs might be undetectable by sonar.

Two, we understood by chart and observation that the bay bottom in our survey area was flat, comprising both hard sand and some muddy sedimented depression areas with a gentle slope from the deep waters of the bay to the beach, presumably, ideal habitat. We also knew that little to no rock formations or human debris other than oyster cages for farming and research were found. At a dead low tide, researchers and oyster farmers could walk or ride all-terrain vehicles out to research locations in the near shore bay. Figure 9 shows a recent Google Earth image of our survey area. The images confirm most of what local researchers and fisherman have stated about conditions in this area. Hard sand bottom some contours but very little in the way of rocky out cropping's or human debris. A convenient location to test the theory that horseshoe crabs can be images using acoustic technologies for research and data gathering.

Three, what would a "migration of horseshoe crabs" actually look like on a sonar screen? How could one crab be distinguished from another and what type of density mass of crabs would be present? Initially, our expectations were that we could catch or image the crabs entering the area from deeper waters offshore in the bay.

Figure 10 is a very good example an image of a side scan sonar transect line. The range of this line is 40 meters, 20 meters on either side of the sonars path. In the highlighted circle, on the right side (the inshore side) of the center path of the sonar are targets or crabs. Their density can be approximated by taking a square meter in SAR HAWK software and counting the crabs. This data can then be extrapolated to make a better estimate over a larger area. On the left side of the path or the offshore side, numerous craters or burrows can be seen. Our hypothesis is that the crabs burrowed in for a time and at the time and time extracted themselves for the walk to the nesting areas on the beach. Figure 11 shows an image of a cart that sits in what was left of crater or burrow.

As we completed more survey lines and gained some experience analyzing the data, we began to see more crab behavior and habitat that helped us understand the migration and movement along the bay bottom during mating and nesting. Figures 12 and 13 show sections of our sonar mosaics (transect lines). Figure 12 shows particularly there are many empty burrows, but what we began to notice was some still have animals. As indicated previously, depressions are depicted as dark or nonreflective areas of the sonar record. There are times the dark area of the depression surrounds a strong reflector indicating the presence of a crab or other object inside. Does this mean it is a crab? Not necessarily, it just means that there is something in the burrow other than an empty hole. Our hypothesis is that it is a crab in the process of digging itself out. The data says there is something there, and we feel the only thing that fits is a horseshoe crab. Of course, it can also be a mound of sand or sediment from the crab's self-extraction. But it fits our theory that these "crater eyes" have a biological animal inside.

On the right and left sides of the transect line, there are crabs. Notice the reflection, followed by a shadow as the sonar reflects off the crab and passes around creating the trailing shadow. Sonar operators measure both the reflection and shadow to determine the length, width, and height of the contact. The arrows in Fig. 13 indicate the direction of the boat's travel and the sonar transducer. Note that both the Humminbird and Klein have captured similar data and imaging.

In Fig. 14, three arrows indicate larger than normal targets, which by this time we believe are horseshoe crabs. In each of the three targets, a split shadow can be seen. We feel that it indicates mating pairs, but it could also simply be two crabs traveling close together. However, the mating theory has more potential. During more than one trawl, we did capture mated pairs of crabs: the larger female and the small male.

Another aspect of the survey was to evaluate if the oyster cage gear placed by fishermen and researchers impeded the crab's movements. In Fig. 15, the cages are clearly visible with arrows pointing to crabs all around, very few traveling under or through the cages. Burrows are also noted along with crabs and trailing shadows. Once again, prior knowledge of the farming area indicated no structure other than cages, so any black spot can be considered a target (crab) (Fig. 16).

Sonar can also image schooling fish, individual fish, and other marine organisms. Tracking them is not feasible, but they do occasionally show up and are a good indicator that as a predator, there is prey in the vicinity.

We believe that using acoustic imaging technology (side scan sonar) can be another tool for marine researchers to use when evaluating in this case horseshoe crabs movement in an environment. We feel that our imaging results not only visualized how the crabs traveled through the habitat but also showed that crabs can and do move around obstacles like oyster racks to reach nesting areas. With the proper training and a small investment anyone can survey an area to understand visually understand what is happening. Can we track crab movement? Not an individually certainly, although we can see individual crabs and even mating pairs but we use this technology map an area and compute and sample population density using the side scan sonar swaths. Capturing the data is the easiest part. Much more challenging is the data analysis and a rubric for the search area that eliminates any other possibly other than your targeted species.

## References

J.A. Ballesteros, P.P. Sharma, A critical appraisal of the placement of xiphosura (chelicerata) with account of known sources of phylogenetic error. Syst. Biol. **68**(6), 896–917 (2019). https://doi.org/10.1093/sysbio/syz011

R.M. Barlow, K. Powers, H. Howard, L. Kass, Migration of Limulus for mating: Relation to lunar phase, tide height, and sunlight. Biol Bull **171**, 310–329 (1986)

V. Capone, Personal Communication with Dr. William Phoel, NOAA Fisheries (2008)

J.T. Tanacredi, M.L. Botton, D.R. Smith, editors, Biology and Conservation of Horseshoe Crabs, (Springer Book, 2009)

C. Mazel, Side Scan Sonar Record Interpretation, (Klein Associates, Inc., 1985)

D. Monroe, Oyster aquaculture does not impede spawning each access for Atlantic horseshoe crabs, *Limulus polyphemus*. Aquac. Environ. Interact. **12**, 81–90 (2020). https://doi.org/10.3354/aei00351

D. Penn, H.J. Brockmann, Nest-site selection in the horseshoe crab, Limulus polyphemus. Biol. Bull. **187**, 373–384 (1994)

A. Rudloe, Aspects of the biology of juvenile horseshoe crabs, *Limulus polyphemus*. Bull. Mar. Sci. **31**, 125–133 (1981)

A. Rudloe, W.F. Herrnkind, Orientation of Limulus polyphemus in the vicinity of breeding beaches. Mar. Behav. Physiol. **4**, 75–89 (1976)

A.E. Rudloe, Locomotor and light responses of larvae of the horseshoe crab, *Limulus polyphemus*. Biol Bull **157**, 494–505 (1979)

C.N. Shuster Jr., A pictorial review of the natural history and ecology of the horseshoe crab *Limulus polyphemus*, with reference to other Limulidae, in *Physiology and Biology of Horseshoe Crabs: Studies on Normal and Environmentally Stressed Animals*, ed. by J. Bonaventura, C. Bonaventura, S. Tesh, (Alan R Liss Inc, New York, 1982), pp. 1–52

C.N. Shuster Jr., K. Sekiguchi, Basic habitat requirements of the extant species of horseshoe crabs (Limulacea), in *Biology and Conservation of Horseshoe Crabs*, ed. by J. T. Tanacredi, M. Botton, D. R. Smith, (Springer, New York, 2009), pp. 115–129

C.N. Shuster, G,N. Shuster, Professor of Zoology University of Florida Gainesville, ed. by H.J. Brockmann, R.B. Barlow, H.J. Brockmann (The American Horseshoe Crab Harvard University Press, 2003)

M. Thompson, Assessments of the population biology and critical habitat for the horseshoe crab, *Limulus polyphemus* in South Carolina.. Medical University of South Carolina. (Unpublished - Thesis, report-manuscripts) (1999)

# Asian Horseshoe Crab Conservation: Knowledge, Attitudes, and Intentions of Local College Students in Northern Beibu Gulf, China

**Zhou Wu, Shuyan Huang, Yijian Fu, Junfu He, Wenquan Zhen, Xiaoyong Xie, Chun-Chieh Wang, Junhua Zhu, and Kit Yue Kwan**

## 1 Introduction

The Beibu Gulf is the coastal area that accommodates the highest abundance of tri-spine horseshoe crabs, *Tachypleus tridentatus* (Brockmann and Smith 2009; Liao et al. 2019; Xie et al. 2020). Horseshoe crabs play a critical ecological role in

Z. Wu
Guangxi Key Laboratory of Beibu Gulf Marine Biodiversity Conservation, College of Marine Sciences, Beibu Gulf University, Qinzhou, China

National Engineering Research Centre for Marine Aquaculture, Zhejiang Ocean University, Zhejiang, China

S. Huang · Y. Fu
Guangxi Key Laboratory of Beibu Gulf Marine Biodiversity Conservation, College of Marine Sciences, Beibu Gulf University, Qinzhou, China

Futian-CityU Mangrove R&D Center, City University of Hong Kong Shenzhen Research Institute, Shenzhen, China

J. He · W. Zhen · J. Zhu · K. Y. Kwan (✉)
Guangxi Key Laboratory of Beibu Gulf Marine Biodiversity Conservation, College of Marine Sciences, Beibu Gulf University, Qinzhou, China
e-mail: kityuekwan@bbgu.edu.cn

X. Xie
South China Sea Fisheries Research Institute, Chinese Academy of Fishery Sciences, Guangzhou, China

C.-C. Wang
Guangxi Key Laboratory of Marine Environmental Science, Guangxi Beibu Gulf Marine Research Center, Guangxi Academy of Sciences, Nanning, China



J. T. Tanacredi et al. (eds.), *International Horseshoe Crab Conservation and Research Efforts: 2007-2020*, https://doi.org/10.1007/978-3-030-82315-3_26

coastal food chains as both prey and consumers (Kwan et al. 2015; Fan et al. 2017). They are also marine substrate engineers, nutrient transporters, and substrates for epibionts in estuarine environments (Botton 2009). The juveniles can act as sentinel species for coastal health since the changes in hemolymph composition patterns reflect heavy metal and nutrient concentrations of the intertidal habitats (Kwan et al. 2018). Being ecologically important in the coastal water, local populations of horseshoe crabs suffer from multiple threats. *Tachypleus tridentatus*, the largest of the four extant horseshoe crab species, is listed as "Endangered" on the IUCN Red List (Laurie et al. 2019) and is vulnerable to exploitation. The population was harvested for food (Fu et al. 2019) and for the manufacture of *Tachypleus* amebocyte lysate (TAL, Liao and Li 2001; Gauvry 2015). Recent community interviews with local fishers in the region indicated that the mean daily harvest rate of adult *T. tridentatus* has decreased from 50–1000 individuals per day in the 1990s to 0–30 adults per day during 2011–2016 (Liao et al. 2019). In addition to excessive resource exploitation, habitat destruction due to coastal development has caused considerable population decline (Hsieh and Chen 2015). Despite the species being listed as a Key Protected Aquatic Wildlife in Guangxi region and an Endangered Species on the China Species Red List, the declining trend of Asian horseshoe crabs in the Beibu Gulf is unlikely to be reversed. This can be related to ambiguous wildlife laws, lack of scientific monitoring and evaluation systems, limited enforcement power, and an established cultural practice of consuming horseshoe crab as food. Other than wildlife laws and regulations, alternative conservation strategies are necessary to protect the remaining Asian horseshoe crab resources in the region long term.

Previous studies indicate that community-based conservation programs are one of the most promising ways to support local conservation measures (van der Ploeg et al. 2011; Şekercioğlu 2012). For example, raising laboratory-cultured juvenile horseshoe crabs at local secondary schools in Hong Kong improved students' attitudes and behaviors toward horseshoe crabs and their conservation (Kwan et al. 2017). The success of biological conservation initiatives is typically associated with a better understanding of biological and ecological information. However, a lack of knowledge of the community's values and socioeconomic factors compromises the relevant conservation programs (Aipanjiguly et al. 2003; Vincenot et al. 2015). Environmental attitude has been shown to predict support for wildlife conservation and successful policies (Kellert 1991; Fu et al. 2019). Attitudes are tendencies of people when they view an entity, which can be abstract or concrete, with some degree of favorability or unfavorability (Ajzen 2001). Positive attitudes toward entities tend to produce favorable responses, and vice versa (Bohner and Wanke 2002). Information of the attitude of local communities is thus critical for planning, implementing, and assessing conservation actions on the threatened horseshoe crabs.

Community opinion about horseshoe crabs and their corresponding conservation measures has not adequately been assessed along the horseshoe crab population range throughout Asia, even though the information is considered essential. The participants of public conservation programs in the northern Beibu Gulf are mostly local college students. The objective of this study was to gain a better understanding of local college students' attitudes and intentions toward horseshoe crabs and their conservation. We aimed to use the collected data to enhance community participation in future horseshoe crab conservation awareness programs in the Beibu Gulf region. Such programs may offer alternatives and promote positive environmental attitudes toward horseshoe crab conservation.

## 2 Materials and Methods

### *2.1 Study Area*

Located in the northern South China Sea between China and Vietnam, the Beibu Gulf is a semi-closed gulf with a total coastline of 1595 km and sea area of 128,000 $km^2$ (Chen et al. 2009). It has extensive estuarine ecosystems, including mangrove wetlands (92 $km^2$), coral reefs (31 $km^2$), and seagrass beds (9 $km^2$), which are under persistent influence of freshwater input from seven major rivers (Fangcheng, Qin, Dafeng, Nanliu, Red, Beilun, Changhua). The rivers carry a considerable amount of nutrients from the land, which in turn supports the rich biodiversity and productive fisheries in the northern South China Sea (Yu and Mu 2006). Asian horseshoe crabs, *T. tridentatus* and *Carcinoscorpius rotundicauda,* are frequently found in mangrove-dominated estuaries with sheltered sandy mudflats along the coastline of the northern Beibu Gulf (Chen et al. 2015; Liao et al. 2019). Local communities encounter adult *T. tridentatus* in aquarium tanks at seafood markets and restaurants. For *C. rotundicauda*, the mating pairs are generally found burrowing in the mud along tidal creeks or along the outer edges of mangrove forests.

### *2.2 Survey Method*

Community interviews were conducted at five tertiary education institutions during July 2017–January 2018 in the three coastal cities (Fangchenggang, Qinzhou, and Beihai) along the northern Beibu Gulf of the Guangxi Zhuang Autonomous Region, China. Interviews were performed by student helpers who are all native Mandarin Chinese speakers. Participants were intercepted at random at the campus gates and when the survey team walked through each college/university campus. They were informed of the study's objectives and told that the collected data would be anonymous. After giving verbal consent, all respondents were interviewed one-to-one using a standard anonymous questionnaire in simplified Chinese characters. To improve the validity of the data, responses were only included in the analysis if the participant grew up in Guangxi region and could name the animal (i.e., horseshoe crab) from the images printed on the questionnaire (Supplementary Material). The questionnaire contains descriptive and multiple-choice questions in three sections:

background information of respondents, basic ecological knowledge, and attitudes of and behavioral intentions toward horseshoe crab conservation. We constructed 11 statements to measure concepts and knowledge of Asian horseshoe crabs. The factual questions involved horseshoe crab taxonomy, their ecological and utilitarian values, and conservation status (Supplementary Material). Three options were provided for each question, including "correct," "wrong," and "not sure." The last part of the questionnaire tested students' attitudes and behavioral intentions toward horseshoe crab conservation, including their general support for awareness programs, and whether they discourage their parents and friends from eating horseshoe crabs. A five-point Likert scale was used for rating items agree through disagree. To calculate the scores of concepts and knowledge, positive statements with "correct" responses scored 1, "incorrect" and "not sure" scored 0. For attitudes/behavior scores, positive statements with "strongly agree" to "strongly disagree" responses scored from 5 to 1. The scores of negative statements were reversed. Both positive and negative wordings were used in the attitudes/behavior questionnaire items to minimize any acquiescent bias. The option "not sure" was included to eliminate the occurrence of random responses.

### 2.3 *Statistical Analyses*

Statistical analyses were conducted using SPSS Statistics 22 (IBM, New York, USA) after the data had been examined for normality and homogeneity of variance using Shapiro–Wilk and Levene's tests, respectively. The five-point Likert scale data measuring attitudes/behavioral intentions were treated as interval-level data. Since most of the data failed to meet the requirements of homogeneity of variance, nonparametric Mann–Whitney *U* and Kruskal–Wallis tests were used to identify significant differences in mean scores between/among different sociodemographic variables. Spearman's rank correlation coefficient was performed to measure the effect of age and academic performance [Grade Point Average (GPA)] in relation to the scores for knowledge and attitudes/behavior.

## 3 Results

The survey provided 391 completed questionnaires (Table 1). The number of respondents from each institution varied (27–176 individuals), depending on students' willingness to participate in the interviews.

Of the 11 knowledge questions, students averaged 4.77 correct responses (SD = 2.46). The students showed a good understanding that horseshoe crabs are "ancient marine organisms" (item 4; 71% correctly answered), are "ecologically important to the survival of migratory birds and other marine organisms" (item 10; 57%), and "their blood can be used to produce TAL" (item 2; 68%) (Table 2).

**Table 1** Sociodemographic features of respondents

| Variables | N | Percentage |
|---|---|---|
| Institution | 391 | |
| Beibu Gulf University | 176 | 45.01 |
| Guangxi Talent International College | 27 | 6.91 |
| Guilin University of Electronic Technology | 27 | 6.91 |
| Beihai College of Art and Design | 37 | 9.46 |
| Guangxi University of Finance and Economics | 112 | 28.64 |
| Other colleges in the northern Beibu Gulf | 12 | 3.07 |
| Gender | 391 | |
| Male | 177 | 45.41 |
| Female | 214 | 54.59 |
| Education | 391 | |
| Diploma | 71 | 18.11 |
| Bachelor | 319 | 81.63 |
| Master or above | 1 | 0.26 |
| Father's occupation | 391 | |
| Managers (administrative and commercial managers, senior officials) | 12 | 3.07 |
| Professionals (teaching professionals, business and administration professionals, civil services) | 16 | 4.09 |
| Workers (services and sales, related trade) | 108 | 27.62 |
| Fishers and farmers | 140 | 35.81 |
| Self-employed | 99 | 25.32 |
| Environmental conservation and education-related | 5 | 1.28 |
| Retired/jobless | 11 | 2.81 |
| Mother's occupation | 391 | |
| Managers (administrative and commercial managers, senior officials) | 8 | 2.05 |
| Professionals (teaching professionals, business and administration professionals, civil services) | 10 | 2.56 |
| Workers (services and sales, related trade) | 96 | 24.55 |
| Fishers and farmers | 153 | 39.13 |
| Self-employed | 105 | 26.85 |
| Environmental conservation and education-related | 2 | 0.51 |
| Retired/jobless | 17 | 4.35 |
| Awareness campaign experience | 391 | |
| Yes | 30 | 7.67 |
| No | 361 | 92.33 |

However, most students mistakenly believed that the horseshoe crab "is a kind of crab" (item 1; 16%), "its value as a traditional Chinese medicinal material has been scientifically proven" (item 3; 17%), and "is listed as a State-Protected Category II Wildlife Species" (item 11; 24%) (Table 2).

The mean (± SD) value of the attitudes/intentions scores on the five-point Likert scale (out of 5) was 3.68 ± 0.32. Most students perceived that horseshoe crab conservation is demanding, but disagreed that the horseshoe crab "has low economic

**Table 2** The frequency of response for concepts and knowledge of horseshoe crabs. Three choices of responses were provided to the college students for each question, including correct (C), incorrect (IC), and not sure (NS)

| No. | Concepts and knowledge | % Frequency of responses on Likert scale | | | Correct answers (%) |
|---|---|---|---|---|---|
| | | C | W | NS | |
| 1 | Horseshoe crab is a kind of crab | 61.64 | 15.60 | 22.76 | 15.60 |
| 2 | Blood of the horseshoe crab can be used to produce TAL for endotoxin detection | 68.03 | 11.51 | 20.46 | 68.03 |
| 3 | Horseshoe crab's value as traditional medicinal material has been proven by research | 67.27 | 16.62 | 16.11 | 16.62 |
| 4 | Horseshoe crab is an ancient marine organism, known as a "living fossil" | 71.10 | 17.39 | 11.51 | 71.10 |
| 5 | A high density of horseshoe crabs deteriorates coastal water quality | 44.24 | 39.39 | 16.37 | 39.39 |
| 6 | The increase of horseshoe crab individuals lowers the harvest of fishermen | 34.02 | 49.36 | 16.62 | 49.36 |
| 7 | Great consumption of horseshoe crab resources has little effect on coastal ecosystems | 28.90 | 52.69 | 18.41 | 52.69 |
| 8 | Horseshoe crab feeds on mollusks, polychaetes, and others; therefore, horseshoe crabs decrease marine biodiversity | 34.52 | 48.60 | 16.88 | 48.59 |
| 9 | Recent rearing techniques are capable of culturing horseshoe crab larvae to reach maturity, which can effectively enhance the natural population | 49.62 | 35.55 | 14.83 | 35.55 |
| 10 | Horseshoe crab eggs serve as important food sources for migratory birds | 56.77 | 28.65 | 14.58 | 56.77 |
| 11 | Tri-spine horseshoe crab (*Tachypleus tridentatus*) is listed as a State-Protected Category II Wildlife Species and under protection by national laws and regulations | 62.66 | 23.79 | 13.55 | 23.78 |

value" (item 8; reverse score: 4.01), "has very few natural predators" (item 9; reverse score: 3.96), and "has high reproduction ability" (item 10; reverse score: 3.97) (Table 3). A high proportion of students also opposed that "the number of horseshoe crabs consumed by themselves would not contribute to the overall declining trend of horseshoe crab population" (item 12; reverse score: 3.98), "I would not discourage my parents/friends to consume horseshoe crabs as everyone is eating them" (item 11; reverse score: 3.85), and "horseshoe crab conservation is the responsibility of government staff and ecologists. I can only do small things" (item 7; reverse score: 3.98). The students tended to respond neutrally when they were asked whether they have contributed to horseshoe crab conservation (item 14; 3.01).

Students with previous experiences of joining environmental awareness campaigns had significantly higher mean scores of both conceptual ($z = 5.483, p = 0.000$) and attitudes/behavioral ($z = 5.525, p = 0.000$) questionnaire items (Table 4). Gender ($z = -3.288$, $p = 0.001$) and parents' occupations (father: chi-square = 22.347, $p = 0.001$; mother: chi-square = 31.957, $p = 0.000$) were significantly related to

**Table 3** The frequency of response for attitudes and intentions towards horseshoe crab conservation. Five choices of responses were provided to the college students for each question, including strongly agree (SA), agree (A), neutral (NE), disagree (D), and strongly disagree (SD)

| No. | Type | Attitudes and behavior | % Frequency of responses on Likert scale | | | | | Mean score (out of 5) |
|---|---|---|---|---|---|---|---|---|
| | | | SA | A | NE | D | SD | |
| 1 | – | Horseshoe crabs are highly abundant in the northern Beibu Gulf as horseshoe crabs can be widely found in seafood markets and restaurants | 2.81 | 11.25 | 27.11 | 17.65 | 41.18 | 3.60 ± 1.00 |
| 2 | + | Citizens have the mission to conserve horseshoe crabs and other threatened species | 30.95 | 38.36 | 14.58 | 8.18 | 7.93 | 3.76 ± 1.21 |
| 3 | – | Economic benefits derived from horseshoe crabs such as food consumption and TAL production exceed the ecological values of horseshoe crabs themselves. Therefore, consumption of horseshoe crab resources should be encouraged | 6.39 | 20.20 | 27.62 | 16.62 | 29.16 | 3.29 ± 1.16 |
| 4 | – | The declining number of horseshoe crabs in the northern Beibu Gulf is not evident | 3.07 | 13.81 | 28.90 | 18.41 | 35.81 | 3.53 ± 1.04 |
| 5 | – | A decrease of horseshoe crabs would not affect my life | 0.77 | 11.76 | 29.41 | 28.90 | 29.16 | 3.74 ± 1.03 |
| 6 | + | Anthropogenic activities and unsustainable harvest for food are the major drivers of horseshoe crab decline | 13.04 | 33.76 | 28.64 | 9.46 | 15.09 | 3.25 ± 1.15 |
| 7 | – | Horseshoe crab conservation is the responsibility of government staff and ecologists. I can only do "small" things | 4.09 | 8.70 | 16.62 | 44.25 | 26.34 | 3.98 ± 1.15 |
| 8 | – | Horseshoe crab has low economic value; therefore, conservation effort is not required | 3.32 | 6.91 | 18.16 | 42.71 | 28.90 | 4.01 ± 1.09 |
| 9 | – | Adult horseshoe crab has very few natural predators. Thus, conservation efforts for horseshoe crabs are not required | 3.07 | 7.93 | 18.67 | 39.90 | 30.43 | 3.96 ± 1.09 |
| 10 | – | Horseshoe crab has high reproduction ability. The extinction risk is low, thus protection is not required | 3.58 | 7.16 | 17.65 | 39.64 | 31.97 | 3.97 ± 1.09 |
| 11 | – | I would not discourage my parents/friends from consuming horseshoe crabs as everyone is eating them | 1.79 | 7.42 | 24.30 | 30.43 | 36.06 | 3.86 ± 1.00 |
| 12 | – | The number of horseshoe crabs consumed by me would not contribute to the overall declining horseshoe crab population | 2.30 | 7.42 | 18.41 | 38.36 | 33.50 | 3.98 ± 1.04 |
| 13 | – | I do not have time to join horseshoe crab conservation educational activities | 1.53 | 11.00 | 32.48 | 22.25 | 32.74 | 3.63 ± 1.00 |
| 14 | + | I have contributed to horseshoe crab conservation | 11.76 | 19.18 | 39.39 | 11.51 | 18.16 | 3.02 ± 1.14 |

**Table 4** Statistical results of Mann–Whitney *U* (gender, education level, whether has previously joined an awareness campaign) and Kruskal–Wallis tests (education and parents' occupations) showing the effects of respondents' sociodemographic variables on their mean scores of knowledge and attitudes/behavior toward horseshoe crab conservation. Significant *p* value <0.05 is highlighted in bold

| | Mean scores of concepts and knowledge | | | Mean scores of attitudes and behavior | | |
|---|---|---|---|---|---|---|
| Variables | z value | $\chi^2$ value | Sig. | z value | $\chi^2$ value | Sig. |
| Gender | −1.747 | | 0.081 | **−3.288** | | **0.001** |
| Education | −1.300 | | 0.194 | −1.655 | | 0.098 |
| Father's occupation | | 0.661 | 0.722 | | **22.347** | **0.001** |
| Mother's occupation | | 0.166 | 0.986 | | **31.957** | **0.000** |
| Awareness campaign | **5.483** | | **0.000** | **5.525** | | **0.000** |

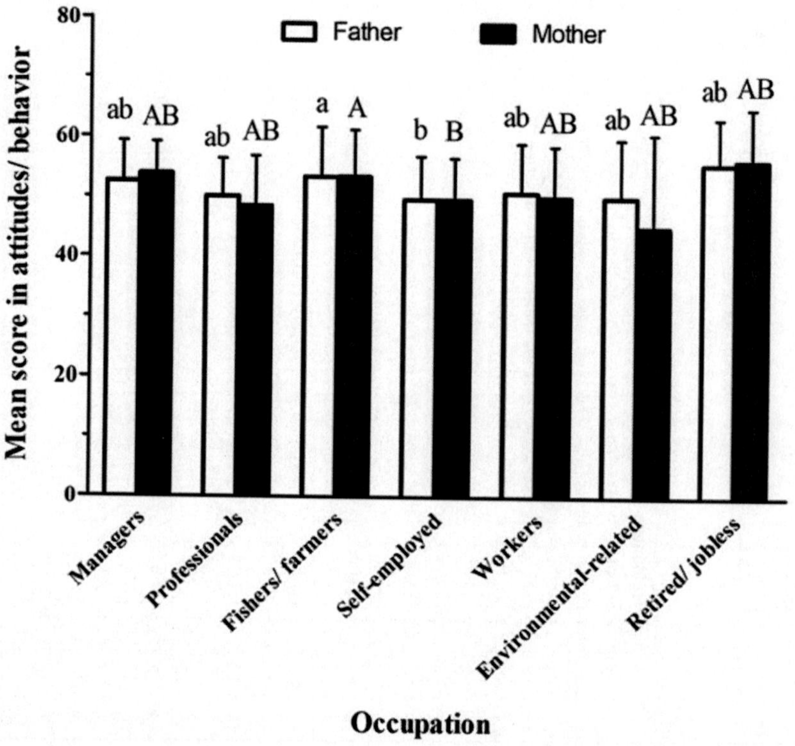


**Fig. 1** Influence of parents' occupations on the students' mean score of attitudes and behavioral intentions toward horseshoe crab conservation. Values represent the mean + standard deviation. The data were analyzed by nonparametric Kruskal–Wallis tests. Multiple Mann–Whitney *U* with Bonferroni correction with statistical difference among father's occupation is presented by different lowercase letters, whereas difference among mother's occupation is indicated by different uppercase letters

students' mean scores of attitudes/behavior toward horseshoe crab conservation, but not those for concepts and knowledge (Table 4; Fig. 1). Female students had significantly higher attitudes/behavior scores. Multiple Mann–Whitney *U* tests with Bonferroni correction showed that students whose parents are fishers and farmers had more positive attitudes and were more motivated to contribute to horseshoe crab conservation. There was a positive correlation between the students' mean scores of knowledge and attitudes/behavior ($r = 0.179$, $p = 0.000$) (Table 5). Students' age and academic performance (GPA) did not correlate to their mean scores of either knowledge or attitudes/behavior ($p > 0.05$; Table 5).

## 4 Discussion

Increased scientific knowledge is believed to result in pro-environmental attitudes and behavior (Lo et al. 2012; Tsoi et al. 2016). Positive attitudes are essential in gaining support for wildlife conservation initiatives and successful policies (Bandara and Tisdell 2003; Senko et al. 2011). Makashvili et al. (2014) demonstrated the potential role of knowledge in shaping a positive image of snakes among undergraduates in Georgia. Tsoi et al. (2016) also indicated that misconceptions about sharks among 11- to 12-year-old students led to a more negative attitude. However, Lo et al. (2012) argued that mere knowledge was ineffective in the Chinese community in facilitating wildlife conservation efforts because the implementation of conservation education in China heavily relies on passive forms of information transfer. Several previous studies pointed out that subjective norms involving social and community influences, nevertheless, were the most reliable predictor of conservation behavior (Zhang et al. 2008; Lo et al. 2012). Our findings demonstrate that students' ecological knowledge of horseshoe crabs, but not their academic performance, positively correlated to their attitudes of and behavioral intentions toward horseshoe crab conservation. Previous experience of participating in relevant conservation awareness programs also enhanced students' attitudes and behaviors toward horseshoe crabs and their conservation. Such findings are consistent with Kwan et al. (2017), which showed that students displayed emotional attachment to the horseshoe crabs when they were engaged in community programs. Students

**Table 5** Statistical results of nonparametric Spearman's correlation coefficient showing the effects of students' age, academic performance (GPA), and mean knowledge score on their knowledge of horseshoe crabs and attitudes and behavior toward horseshoe crab conservation. Significant *p* value <0.05 is highlighted in bold

| | Mean scores of concepts and knowledge | | Mean scores of attitudes and behavior | |
|---|---|---|---|---|
| Variables | r value | Sig. | r value | Sig. |
| Age | –0.016 | 0.746 | 0.047 | 0.359 |
| GPA | 0.057 | 0.266 | 0.025 | 0.624 |
| Mean knowledge score | – | – | **0.179** | **0.000** |

whose parents are fishers or farmers tended to have more positive attitudes and behavioral intentions toward horseshoe crab conservation. One possible explanation for this is the family influence since most fishers in the northern Beibu Gulf perceived horseshoe crab decline during 2011–2016, and they agreed that immediate conservation of the remaining population is needed (Liao et al. 2019). This may imply that the attitude of stakeholders, including fisherman, can play an important role in conservation management of horseshoe crabs.

College students in the region lacked ecological knowledge about horseshoe crabs, particularly their taxonomy, conservation status, and utilitarian value. These may be related to the misconceptions and inaccurate normative messages spread widely among the community, including that *T. tridentatus* is listed as a State-Protected Category II Wildlife Species, and that the horseshoe crab has high value as a traditional Chinese medicine. Most students in the community also typically perceived that current rearing techniques are capable of culturing horseshoe crab larvae to reach maturity, a false claim frequently used by fishers and seafood sellers to convince people to purchase their adult horseshoe crabs. The students exhibited generally neutral to positive attitudes and intentions toward horseshoe crab conservation. Horseshoe crabs were viewed from a utilitarian, ecological, and scientific point of view. Most students perceived that horseshoe crabs deserve more protection and agreed to take some action, such as discouraging their family and peers from eating horseshoe crabs, to conserving the population. However, the students mostly showed neutral responses when asked whether they had contributed to horseshoe crab conservation and whether the commercial value of horseshoe crabs exceeded their ecological value.

A better understanding of the knowledge, perceptions, and behavioral intentions of college students in the Beibu Gulf region serves as a basis for recommendations on how to enhance community participation in future horseshoe crab conservation awareness programs. Recommendations include the use of targeted media, normative messages, and visual displays on campuses to address students' gaps in knowledge about horseshoe crabs, especially their protection status as well as ecological and biomedical value. Integration of ecological studies on food-web models may enhance students' understanding of the ecological roles horseshoe crabs play in maintaining biodiversity in coastal and estuarine ecosystems. Since the species' utilitarian value has been reported as one of the driving factors determining attitudes toward wildlife conservation (Kellert 1991; Vincenot et al. 2015), the awareness of the contribution of *T. tridentatus* in the biomedical industry can be strengthened. In the long term, overharvesting of horseshoe crabs could be averted by reducing market demand for them as food and improving bleeding practices for TAL production. The application of animal-free synthetic TAL should be approved at the level of the government to reduce the horseshoe crab harvests for this use (Maloney et al. 2018). The community should be explicitly informed of the possible actions that they could take to facilitate local conservation efforts, such as discouraging their family and peers from purchasing horseshoe crabs for food, and not going to any restaurants that sell horseshoe crabs. Local fishers and their families in the region should be engaged in future awareness programs as they can report the recent status and drivers of horseshoe crab decline. Conservation awareness campaigns in the region should offer alternative activities to horseshoe crab exploitation and enhance positive environmental perceptions of horseshoe crabs. Future studies should examine the role of environmental attitude in bridging knowledge-action gaps and the underlying sociodemographic factors in reducing ingrained apathy to environmental conservation, particularly among youth communities.

**Acknowledgments** The work was funded by the Guangxi BaGui Youth Scholars Programme and Guangxi Recruitment Program of Hundred Global Experts. We appreciate the student volunteers from the College of Marine Sciences, Beibu Gulf University, for their help in conducting community interviews.

## Supplementary Material

Interview survey questionnaire

### *Section A: Background Information of Respondents*

1. Institution
   - ☐ Beibu Gulf University
   - ☐ Guangxi Talent International College
   - ☐ Guilin University of Electronic Technology
   - ☐ Beihai College of Art and Design
   - ☐ Guangxi University of Finance and Economics
   - ☐ Other colleges in the northern Beibu Gulf
2. Education level
   - ☐ Diploma
   - ☐ Bachelor

3. Recent Grade Point Average (GPA): ____________
4. Gender: ☐ Male ☐ Female
5. Age: ___________
6. Hometown: ___________
7. Father's occupation

| |
|---|
| ☐ Managers (administrative and commercial managers, senior officials) |
| ☐ Professionals (teaching professionals, business and administration professionals, civil services) |
| ☐ Workers (services and sales, related trade) |
| ☐ Fishers and farmers |
| ☐ Self-employed |
| ☐ Environmental conservation and education-related |
| ☐ Retired/jobless |

8. Mother's occupation

| |
|---|
| ☐ Managers (administrative and commercial managers, senior officials) |
| ☐ Professionals (teaching professionals, business and administration professionals, civil services) |
| ☐ Workers (services and sales, related trade) |
| ☐ Fishers and farmers |
| ☐ Self-employed |
| ☐ Environmental conservation and education-related |
| ☐ Retired/jobless |

9. Have you ever joined any awareness campaigns related to environmental conservation?

## *Section B: Concepts and Knowledge of Horseshoe Crabs*

| No. | Questions | Correct | Incorrect | Not sure |
|---|---|---|---|---|
| 1 | Horseshoe crab is a kind of crab | | | |
| 2 | Blood of the horseshoe crab can be used to produce TAL for endotoxin detection | | | |
| 3 | Horseshoe crab's value as traditional medicinal material has been proven by research | | | |
| 4 | Horseshoe crab is an ancient marine organism, known as a "living fossil" | | | |
| 5 | A high density of horseshoe crabs deteriorates coastal water quality | | | |
| 6 | The increase of horseshoe crab individuals lowers the harvest of fishermen | | | |
| 7 | Great consumption of horseshoe crab resources has little effect on coastal ecosystems | | | |

| No. | Questions | Correct | Incorrect | Not sure |
|---|---|---|---|---|
| 8 | Horseshoe crab feeds on mollusks, polychaetes, and others; therefore, horseshoe crabs decrease marine biodiversity | | | |
| 9 | Recent rearing techniques are capable of culturing horseshoe crab larvae to reach maturity, which can effectively enhance the natural population | | | |
| 10 | Horseshoe crab eggs serve as important food sources for migratory birds | | | |
| 11 | Tri-spine horseshoe crab (*Tachypleus tridentatus*) is listed as a State-Protected Category II Wildlife Species and under protection by national laws and regulations | | | |

## *Section C: Conservation Awareness of Horseshoe Crabs*

| No. | Questions | Strongly agree | Agree | Neutral | Disagree | Strongly disagree |
|---|---|---|---|---|---|---|
| 1 | Horseshoe crabs are highly abundant in the Beibu Gulf as horseshoe crabs can be widely found in seafood markets and restaurants | | | | | |
| 2 | Citizens have the mission to conserve horseshoe crabs and other threatened species | | | | | |
| 3 | Economic benefits derived from horseshoe crabs such as food consumption and TAL production exceed the ecological values of horseshoe crabs themselves. Therefore, consumption of horseshoe crab resources should be encouraged | | | | | |
| 4 | The declining number of horseshoe crabs in the northern Beibu Gulf is not evident | | | | | |
| 5 | A decrease of horseshoe crabs would not affect my life | | | | | |
| 6 | Anthropogenic activities and unsustainable harvest for food are the major drivers of horseshoe crab decline | | | | | |
| 7 | Horseshoe crab conservation is the responsibility of government staff and ecologists. I can only do "small" things | | | | | |
| 8 | Horseshoe crab has low economic value; therefore, conservation effort is not required | | | | | |

| No. | Questions | Strongly agree | Agree | Neutral | Disagree | Strongly disagree |
|---|---|---|---|---|---|---|
| 9 | Adult horseshoe crab has very few natural predators. Thus, conservation efforts for horseshoe crabs are not required | | | | | |
| 10 | Horseshoe crab has high reproduction ability. The extinction risk is low, thus protection is not required | | | | | |
| 11 | I would not discourage my parents/ friends from consuming horseshoe crabs as everyone is eating them | | | | | |
| 12 | The number of horseshoe crabs consumed by me would not contribute to the overall declining horseshoe crab population | | | | | |
| 13 | I do not have time to join horseshoe crab conservation educational activities | | | | | |
| 14 | I have contributed to horseshoe crab conservation | | | | | |

## References

S. Aipanjiguly, S.K. Jacobson, R. Flamm, Conserving manatees: Knowledge, attitudes, and intentions of boaters in Tampa Bay, Florida. Conserv. Biol. **17**, 1098–1105 (2003)

I. Ajzen, Nature and operation of attitudes. Annu. Rev. Psychol. **52**, 27–58 (2001)

R. Bandara, C. Tisdell, Comparison of rural and urban attitudes to the conservation of Asian elephants in Sri Lanka: Empirical evidence. Biol. Conserv. **110**, 327–342 (2003)

G. Bohner, M. Wänke, *Attitudes and Attitude Change: Social Psychology* (Hove, UK, 2002)

M.L. Botton, The ecological importance of horseshoe crabs in estuarine and coastal communities: A review and speculative summary, in *Biology and Conservation of Horseshoe Crabs*, ed. by J. T. Tanacredi, M. L. Botton, D. R. Smith, (Springer, 2009), pp. 45–63

H.J. Brockmann, M.D. Smith, Reproductive competition and sexual selection in horseshoe crabs, in *Biology and Conservation of Horseshoe Crabs*, ed. by J. T. Tanacredi, M. L. Botton, D. R. Smith, (Springer, New York, 2009), pp. 199–221

C.P. Chen, M.C. Yang, L.F. Fan, G. Qiu, Y.Y. Liao, H.L. Hsieh, Co-occurrence of juvenile horseshoe crabs *Tachypleus tridentatus* and *Carcinoscorpius rotundicauda* in an estuarine bay, southwestern China. Aquat. Biol. **24**, 117–126 (2015)

Z. Chen, S. Xu, Y. Qiu, Z. Lin, X. Jia, Modeling the effects of fishery management and marine protected areas on the Beibu Gulf using spatial ecosystem simulation. Fish. Res. **100**, 222–229 (2009)

L.-F. Fan, C.-P. Chen, M. Yang, G.L. Qiu, Y.-Y. Liao, H.-L. Hsieh, Ontogenetic changes in dietary carbon sources and trophic position of two co-occurring horseshoe crab species in southwestern China. Aquat. Biol. **26**, 15–16 (2017)

Y.J. Fu, S.Y. Huang, Z. Wu, C.C. Wang, M.D. Su, et al., Socio-demographic drivers and public perceptions of consumption and conservation of Asian horseshoe crabs in northern Beibu Gulf, China. Aquat. Conserv. Mar. Freshw. Ecosyst. **29**, 1268–1277 (2019)

G. Gauvry, Current Horseshoe crab harvesting practices cannot support global demand for TAL/LAL: The pharmaceutical and medical device industries' role in the sustainability of horseshoe crabs, in *Changing Global Perspectives on Horseshoe Crab Biology, Conservation and Management*, ed. by J. T. Tanacredi, M. L. Botton, D. R. Smith, (Springer International Publishing, Cham, 2015), pp. 475–482

H.L. Hsieh, C.P. Chen, Current status of *Tachypleus tridentatus* in Taiwan for red list assessment, in *Changing Global Perspectives on Horseshoe Crab Biology, Conservation and Management*, ed. by J. T. Tanacredi, M. L. Botton, D. R. Smith, (Springer International Publishing, Cham, 2015), pp. 383–396

S.R. Kellert, Public attitudes toward bears and their conservation, in *Bears: Their Biology and Management*, ed. by M. R. Pelton, (International Association for Bear Research and Management, Missoula, 1991), pp. 43–50

B.K.Y. Kwan, S.G. Cheung, P.K.S. Shin, A dual stable isotope study for diet composition of juvenile Chinese horseshoe crab *Tachypleus tridentatus* (Xiphosura) on a seagrass-covered intertidal mudflat. Mar. Biol. **162**, 1137–1143 (2015)

B.K.Y. Kwan, J.H.Y. Cheung, A.C.K. Law, S.G. Cheung, P.K.S. Shin, Conservation education program for threatened Asian horseshoe crabs: A step towards reducing community apathy to environmental conservation. J. Nat. Conserv. **35**, 53–65 (2017)

B.K.Y. Kwan, V.K.Y. Un, S.F. Cheung, P.K.S. Shin, Horseshoe crabs as potential sentinel species for coastal health: Juvenile haemolymph quality and relationship to habitat conditions. Mar. Freshw. Res. **69**, 894–905 (2018)

K. Laurie, C.P. Chen, S.G. Cheung, V. Do, H.L. Hsieh, A. John, F. Mohamad, S. Seino, S. Nishida, P. Shin, M.C. Yang, Tachypleus tridentatus. The IUCN Red List of Threatened Species 2019: E.T21309A133299524 (2019). https://doi.org/10.2305/IUCN.UK.2019-1.RLTS.T21309A133299524.en

Y.Y. Liao, H.L. Hsieh, S.Q. Xu, Q.P. Zhong, L. Juan, et al., Wisdom of Crowds reveals decline of Asian horseshoe crabs in Beibu Gulf, China. Oryx **53**, 222–229 (2019)

Y. Liao, X. Li, Present situation of horseshoe crab resources in the sea area of China and tactics of preservation. Resour. Sci. **23**, 53–57 (2001)

A.Y. Lo, A.T. Chow, S.M. Cheung, Significance of perceived social expectation and implications to conservation education: Turtle conservation as a case study. Environ. Manag. **50**, 900–913 (2012)

M. Makashvili, N. Kaishauri, T. Azmaiparashvili, The role of knowledge in overcoming snake fear. Proc. Soc. Behav. Sci. **152**, 184–187 (2014)

T. Maloney, R. Phelan, N. Simmons, Saving the horseshoe crab: A synthetic alternative to horseshoe crab blood for endotoxin detection. PLoS Biol. **16**(10), e2006607 (2018)

Ç.H. Şekercioğlu, Promoting community-based bird monitoring in the tropics: Conservation, research, environmental education, capacity-building, and local incomes. Biol. Conserv. **151**, 69–73 (2012)

J. Senko, A.J. Schneller, J. Solis, F. Ollervides, W.J. Nichols, People helping turtles, turtles helping people: Understanding resident attitudes towards sea turtle conservation and opportunities for enhanced community participation in Bahia Magdalena, Mexico. Ocean Coast Manage. **54**, 148–157 (2011)

K.H. Tsoi, S.Y. Chan, Y.C. Lee, B.H.Y. Ip, C.C. Cheang, Shark conservation: An educational approach based on children's knowledge and perceptions toward sharks. PLoS One **11**, e0163406 (2016)

J. van der Ploeg, M. Cauilan-Cureg, M. van Weerd, W.T. De Groot, Assessing the effectiveness of environmental education: Mobilizing public support for Philippine crocodile conservation. Conserv. Lett. **4**, 313–323 (2011)

C.E. Vincenot, A.M. Collazo, K. Wallmo, L. Koyama, Public awareness and perceptual factors in the conservation of elusive species: The case of the endangered Ryukyu flying fox. Glob. Ecol. Conserv. **3**, 526–540 (2015)

X.Y. Xie, Z. Wu, C.C. Wang, Y.J. Fu, X.P. Wang, et al., Nursery habitat for Asian horseshoe crabs along the northern Beibu Gulf, China: Implications for conservation management under baseline gaps. Aquat Conserv Mar Freshw Ecosyst **30**, 260–272 (2020)

Y. Yu, Y. Mu, The new institutional arrangements for fisheries management in Beibu Gulf. Mar. Policy **30**, 249–260 (2006)

L. Zhang, N. Hua, S. Sun, Wildlife trade, consumption and conservation awareness in southwest China. Biodivers. Conserv. **17**, 1493–1516 (2008)

# Fishery Bycatch Jeopardizes Indian Horseshoe Crab *Tachypleus gigas* Along the Northeast Coast of Odisha, India

**Siddhartha Pati, Sanatan Tudu, Apratim Sai Rajesh, Satyabrata Meher, Bryan Raveen Nelson, and Bisnu Prasad Dash**

## 1 Introduction

The mega-fauna of the marine ecosystem including predatory fish, seabirds, and marine mammals are facing decline worldwide due to anthropogenic threats such as overfishing, bycatch (unintended catch), infectious diseases, and habitat loss. Horseshoe crabs are no exception to this unfortunate development (Jackson et al. 2001; Myers and Worm 2003; Aguirre and Tabor 2004; Lewison et al. 2004; Pati et al. 2015; Pati et al. 2020a; 2021a).

The populations of horseshoe crabs, particularly the Asian species, have been dwindling under increasing human threats over the years, and there is hardly any improvement in this scenario (Chatterji and Shaharom 2009; Shin et al. 2009; Hu

S. Pati (✉) · S. Tudu · A. Sai Rajesh · S. Meher
Department of Bioscience & Biotechnology, Fakir Mohan University, Balasore, Odisha, India

SIAN Institute, Association for Biodiversity Conservation and Research (ABC), Balasore, Odisha, India

B. R. Nelson (✉)
Institute of Tropical Biodiversity and Sustainable Development, University Malaysia Terengganu, Terengganu, Malaysia

SIAN Institute, Association for Biodiversity Conservation and Research (ABC), Balasore, Odisha, India
e-mail: bryan.nelson@umt.edu.my

B. P. Dash
Department of Bioscience & Biotechnology, Fakir Mohan University, Balasore, Odisha, India

SIAN Institute, Association for Biodiversity Conservation and Research (ABC), Balasore, Odisha, India



J. T. Tanacredi et al. (eds.), *International Horseshoe Crab Conservation and Research Efforts: 2007-2020*, https://doi.org/10.1007/978-3-030-82315-3_27

et al. 2010; Cartwright-Taylor et al. 2011, Nelson et al. 2015, 2016b). No single factor is sufficient enough to explain these declining trends, but it seems most likely that horseshoe crab populations are adversely affected, resulting from pollution, degradation, and loss of estuarine spawning habitats, and commercial fishing activities (Berkson and Shuster 1999; Botton 2001; Nelson et al. 2016a; Pati et al. 2017; John et al. 2018; Vestbo et al. 2018; Mohamed et al. 2021; Tudu et al., 2021). Among these threats, a major contributing factor is the continuous and increasing destruction of horseshoe crab breeding grounds. This persistent problem has been the focus of conservation effort to prevent further decline of horseshoe crab populations (Berkson et al. 2009; Pati et al. 2015; Pati and Dash 2016; John et al. 2018).

In addition to threats to breeding sites, there have been increasing reports and growing concern regarding bycatch of horseshoe crabs in various fishing gear. This is in particular reference to commercial and small-scale fishing firms, local fishing communities, as well as their potential impacts on horseshoe crab populations. Though this is a growing and alarming factor of threat to horseshoe crabs, information on these operations and their damage assessment is scarcely available in the Indian context. While many scientific investigations, ranging from molecular and biotechnological approaches to population dynamics, breeding biology, and spawning behavior, are available in India (Chatterji et al. 2004; Ghaskadbi et al. 2008; Chatterji and Shaharom 2009; Srijaya et al. 2013; Alam et al. 2015; Chinnary et al. 2015; Pati et al. 2020b; 2021b; Nong et al., 2021), studies and documentation of these issues related to anthropogenic and bycatch activities of horseshoe crabs are limited. There are two species of Asian horseshoe crabs, *Tachypleus gigas* and *Carcinoscorpius routundicauda*, present in the Indian waters from historical records. However, it is only recently that they are beginning to gain considerable scientific attention. They are yet to receive a national priority for their protection and conservation.

In India, gillnets, set nets, and bottom trawling nets are used in continental shelf, open ocean (beyond the Exclusive Economic Zone), and nearshore estuaries to catch different types of pelagic fishes (Luther et al. 1997). Horseshoe crabs are regularly caught as bycatch species in fishing nets in many parts of the world (Kai and Morikawa 1999; Iwaoka and Okayama 2009; Smith et al. 2009; Pati et al. 2017; John et al. 2018). However, the numbers of horseshoe crabs that are entangled in gillnet fisheries in India have not been assessed. The objective of this chapter is to make available new information regarding incidental bycatch of *T. gigas* along the Balasore coast. We aimed at documenting horseshoe crab bycatch information in fishing nets using observer data provided by trained skippers from March 2017 to February 2019. Additionally, this study highlights an ongoing program that is being implemented to ensure the safe release of entangled horseshoe crabs from gillnet operations.

## 2 Materials and Methods

### 2.1 Study Sites

The coast of Balasore stretches along the Bay of Bengal in an area of around 81 km. This region has many rivers, canals, and rivulets that merge with the Bay of Bengal. The development of the Balasore coastline has been continually marked by incidents of sea erosion, groundwater depletion, water, air, and solid waste pollution, unregulated tourism, and industrialization – all fairly recent occurrences in history with significant implications for the local fishing communities. There are four major estuaries that serve as breeding grounds for *T. gigas* (Fig. 1).

Balaramgadi estuary (21°28′07.54″N and 87°04′08.52″E) in the Balasore district and its surrounding area are located 6 km to the north of Chandipur, a well-known, picturesque site. Mahisali estuary is the second estuary with the highest sightings of horseshoe crabs. It is situated around 11.3 km away from Balaramgadi, toward the south direction with the coastal beach located at 21°24′32.92S″N and 86°58′47.89″E. Khapra estuary (21°22′30.69″N and 86°56′50.78″E) is 16.5 km away from Balaramgadi estuary. The highest numbers of horseshoe crabs are found at Khandia estuary, located at 21°20′15.69″N and 86°54′46.42″E in Balasore. It is about 22.17 km away from Balaramgadi. All these estuaries are being highly exploited and used as commercial fishery landing sites, with a large number of

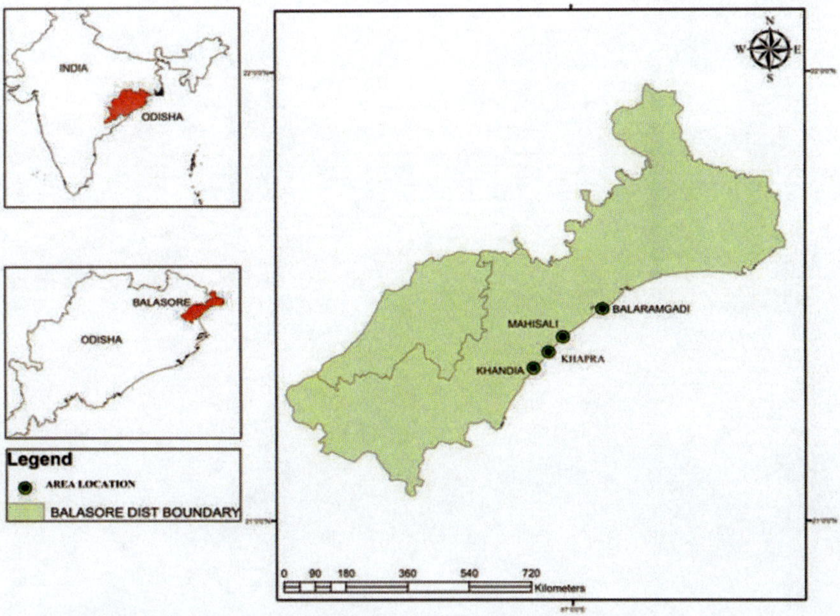


**Fig. 1** Location of study sites where bycatch assessment was carried out in Balasore, India

mechanized fishing boats, point-source irrigation, jetty construction, and other human disturbances. All these activities have led to beach degradation and heavy pollution along with a high rate of unnatural and mechanical interferences.

## 2.2 Survey Observations

This study used denotations E (entangled), D (dead), and L (alive) to indicate the state when horseshoe crabs are detangled from the nets. Such information was disclosed to fishermen in a training program that emphasized on horseshoe crab bycatch and stranding assessment. All participants were taught on bycatch assessment method including labeling observer's name, stranding date, species, and horseshoe crab number by the day, location, condition of the carcass (decomposition state), gender of carcass (as externally observed), and field notes. The study team also participated with fishermen together twice monthly, and discarded nets on the shores were exhumed, if needed, before removal between March 2017 and February 2019 (Fig. 2). Fishermen were also taught about horseshoe crab rehabilitation after freeing the live individuals from the discarded fishing nets. The crabs were

**Fig. 2** Entangled *T. gigas* in discarded fishing nets at different places on Balasore coast, India

stocked at >1 $m^2$ per individual in seawater tank until they were active before releasing them into the sea to ensure their fitness to survive in the wild.

### 2.3 Data Analysis

The horseshoe crab survey data collected from March 2017 to February 2019 were grouped based on three major seasons experienced by the coastal belt of Odisha: summer (S), monsoon rains (R), and winter (W). All entangled horseshoe crabs were also separated into live (after release) and dead (on site). Box plot of number of horseshoe crabs depicted as entangled, alive, or dead in these three seasons was presented using SPSS version 23.0 (Statistical Package for Social Sciences, Inc., Chicago, IL, USA). Student's t-test was also performed to compare 2017–2018 and 2018–2019 survey data according to the above status of horseshoe crabs in different seasons using GraphPad InStat Version 3.0 (GraphPad Software, Inc., San Doego, CA, USA). Level of significance was at $p < 0.05$.

## 3 Results

Over the 2-year survey period, a total of 6546 entangled individuals of *T. gigas* were recorded. Of these, 1408 were recovered alive, while 5138 were found dead either floating on the water or in entanglement with fishing nets. Of the dead individuals, 703 were severely decomposed or at juvenile age, in which their sex could not be determined. For the remaining live and dead individuals, a total of 3496 (59.8%) females and 2347 (40.2%) males were recorded.

Year-wise, 2377 individuals of entangled horseshoe crabs were recorded from March 2017 to February 2018, in which 483 were alive and released back into the sea (Fig. 3), whereas from March 2018 to February 2019, a total of 4169 individuals were found, in which 925 were alive and safely rescued (Fig. 4). During the 2-year survey period, the highest number of entangled animals was observed in 2018–2019. The entangled *T. gigas* generally ranged from juveniles to adults. However, based on the observations on the size of individuals obtained, most individuals recorded were predominantly adult specimens, rather than young ones.

From our extensive personal communications with community leaders and local people of the fishermen community, we noted that different types of nets were used along the Odisha coast, including Kabala jaal, Koni/Ghagra jaal, Ilish jaal, Chandi jaal, Bhetki jaal, Disco jaal – all colloquial and local names for different types of gillnets. Their lengths ranged from 2000 to 7000 m with variable widths of net sizes (Table 1). They are made of monofilament nylon and/or multifilament nylon and other PVC (vinyl) composites. These gillnets are placed at the surface and have a depth of 10–14 m from the surface with a stretched mesh size of 13–17 cm. The net is usually set in early morning and retrieved after 12 hours. It takes about 2–3 hours

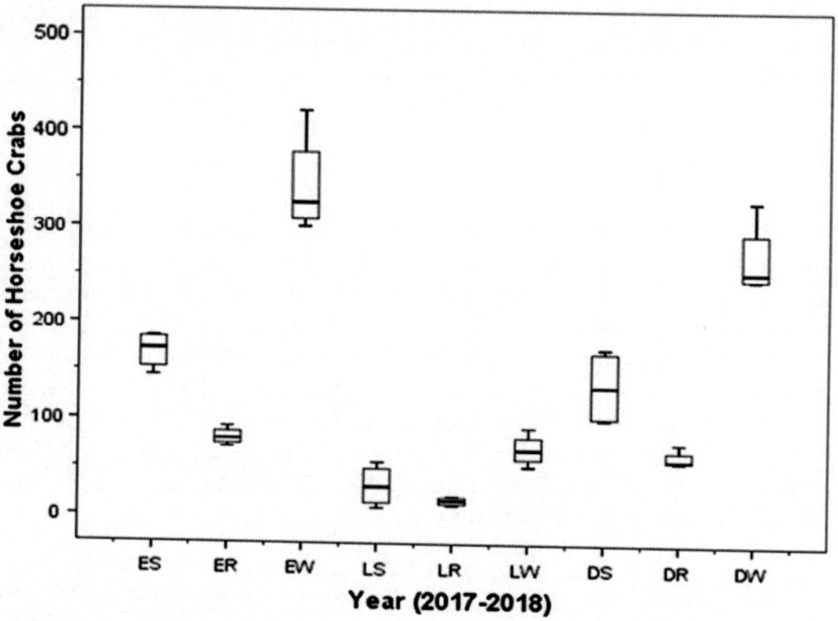


**Fig. 3** Boxplot of *T. gigas* stranded in fishing nets during 2017–18. Boxes represent the 25% and 75% quartiles, the horizontal line is the median (50% quartile), and the bars are 95% confidence intervals. E entangled, L live, D dead, S summer, R monsoon rains, W winter

on average to haul the net. Over our field observations, we found that high numbers of horseshoe crabs tended to entangle or strand in the discarded gillnets.

By comparing the 2-year survey data, highest numbers of stranded *T. gigas* were recorded in both winter times (Figs. 3 and 4). Significantly higher numbers of total entangled horseshoe crabs and dead specimens ($p < 0.001$) were also found in summer and winter in 2018–2019 than 2017–2018, whereas there was no statistically significant difference ($p > 0.05$) in data obtained in monsoon rains and in numbers of live horseshoe crabs released in summer and winter between the 2-year survey period (Fig. 5).

## 4 Discussion and Conservation Implications

In recent years, community-driven horseshoe crab conservation was carried out in India because community knowledge and their position make them directly involved with the horseshoe crab resource. For instance, the present study used colloquial names assigned to gillnets so that local fisherman communities are easy to understand what to do with respect to the survey requirements. The different names of nets used by fishermen have no relation to the catch, but are common terms within

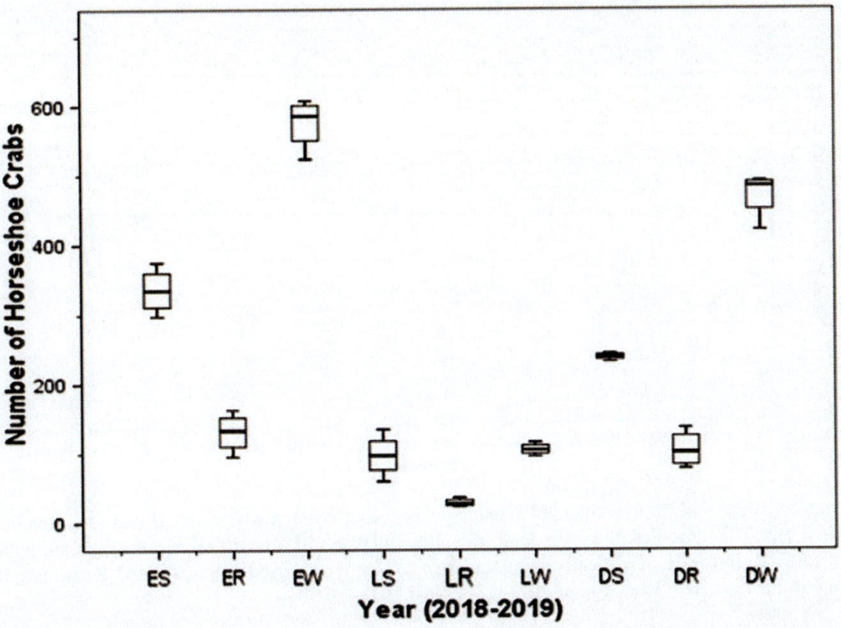


**Fig. 4** Boxplot of *T. gigas* stranded in fishing nets during 2018–19. Boxes represent the 25% and 75% quartiles, the horizontal line is the median (50% quartile), and the bars are 95% confidence intervals. E entangled, L live, D dead, S summer, R monsoon rains, W winter

**Table 1** Types of fishing nets in which entangled horseshoe crabs were observed in the present study

| Net type in local name | Mesh size (cm) | Mesh type and material |
|---|---|---|
| Kabala jaal | 2 | Monofilament, nylon |
| Koni/Ghagrajaal | 13.5 | Multifilament, high-density polyethylene |
| Ilishjaal | 6.5 | Monofilament, nylon |
| Chandijaal | 13.5 | Monofilament, nylon |
| Bhetkijaal | 15–20 | Multifilament, nylon |
| Disco jaal | 14 | Multifilament, cotton |

the community. With this, the fisherman community anticipates fin-fish as the target catch commodity, and by far, we are yet to identify horseshoe crab fishermen in Balasore district. Horseshoe crabs are passively exploited in Orissa either by wild animals or habitat loss through coastal interventions (Mishra et al. 2015; Behera et al. 2015; Pati and Dash 2016). Data collected in this study were purely accidental catch, which forces fishermen to either remove entangled horseshoe crabs or discard the entire net. Since horseshoe crabs have no use to community folks in Balasore, they are considered trash or bycatch. This implies that the 6546 horseshoe crabs

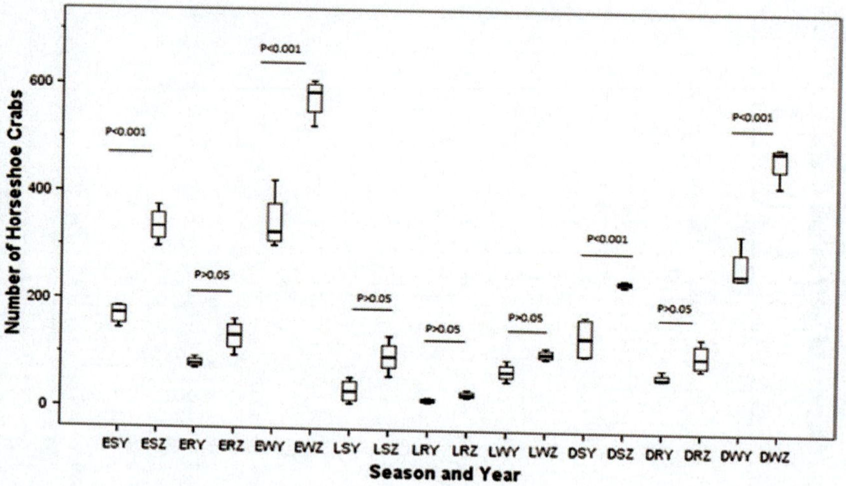


**Fig. 5** Boxplot of *T. gigas* stranded in fishing nets during 2-year study period. Boxes represent the 25% and 75% quartiles, the horizontal line is the median (50% quartile), and the bars are 95% confidence intervals. Statistically significant at $p < 0.05$. E entangled, L live, D dead, S summer, R monsoon rains, W winter, Y year 2017–18, Z year 2018–19

recorded in this study were left to die by suffocation, injury, or starvation in their own habitat, if they were not rescued and released from the entanglement.

Fishing practices are responsible for horseshoe crab decline in Southeast Asia (Fairuz-Fozi et al. 2018; Zauki et al. 2019), where consumption, biased harvest of female crabs, indiscriminate removal of entangled individuals from nets, and sale of horseshoe crabs in wet markets are considered legal. Previously, >4000 live and >10,000 dead horseshoe crabs were recovered from Orissa and Calcutta coasts between 2005 and 2007 (Basudev et al. 2013), whereas another 6050 live *T. gigas* were recovered from "Mala Jaala" nets in Ekakula (180 km south of Balaramgadi), further confirming that horseshoe crabs are threatened by fishing practices (Behera et al. 2015). Unfortunately, these horseshoe crabs became entangled in discarded nets on the shore as subsequent rise and fall of tides can gather sediments to make the nets illusive so that horseshoe crabs can be trapped in these abandoned nets when they move along the shore. In some cases, the delayed departure of horseshoe crabs from the shore to the sea due to longshore currents also causes them to strand in shallow waters (Ayaskanta et al. 2021; Fairuz-Fozi et al. 2018; John et al. 2018; Nelson et al. 2019).

The present data indicated reduced incidences of trapped horseshoe crabs during summer and monsoon rain season when compared to winter months. Fishing practices in India are governed by the Wildlife Protection Act 1972, which restricts fishing activity throughout the year except for the winter. Also, entangled horseshoe crabs were reported to increase from year 2017 to 2019 due to an increase in participants who were involved in the horseshoe crab rescue network and funding support from the Conservation Leadership Program and Wildlife Trust of India. Fisherman communities have been nurtured with basic science to enhance their traditional ecological knowledge. With such information, fishermen explored the availability and use of different types of nets and changed their procedures in setting up of fishing nets, such as the location and period of net deployment with regard to current condition, so as to reduce the entanglement of horseshoe crabs during their fishing operations (Pati 2017). Fisherman communities were also invited to share their knowledge with school children and to allow children to be involved with the release of entangled horseshoe crabs from their nets. By doing so, parents and teachers successfully converted the "touch-and-feel" experience into pledge (Pati 2019) that assures the survival of horseshoe crabs.

In Orissa, poems, idioms, songs, Oriya language, and art have elements of horseshoe crabs. Because horseshoe crabs are assimilated with culture, media coverage can be used as a strategy during regional festivals, parades, and art competitions to raise public compassion in horseshoe crab conservation. Media coverage is also responsible for recruiting members into our horseshoe crab rescue network. Since community folks are driven by compassion, they can encourage the fishing net manufacturers to adopt composite materials instead of nylon or redesign the net shapes and dimensions to allow horseshoe crabs to escape from entanglement. Overall, our horseshoe crab rescue attempts have reached four horseshoe crab spawning beaches in Balasore, three of which remained undocumented until recently. Since horseshoe crab bycatch and stranding assessments are ongoing, we anticipate extension projects that enable a wider coverage of *T. gigas* population protection throughout Orissa coastlines.

**Acknowledgments** This work was funded by Arcadia through the Conservation Leadership Programme (CLP) Rufford Foundation, UK, and Wildlife Trust of India (WTI). It is a pleasure to acknowledge our colleagues Gobinda Chandra Biswal and Shaik Rahimuddin for their invaluable assistance. Special thanks are extended to Anil Chatterji for his valuable suggestions. We greatly appreciate the time and effort of all involved.

## References

A.A. Aguirre, G.M. Tabor, Introduction: Marine vertebrates as sentinels of marine ecosystem health. Eco Health **1**, 236–238 (2004)

H. Alam, S. Chinnari, S. Pati, et al., A horseshoe crab peri-vitelline fluid triggers the human bone marrow stem cell differentiation into cardiomyocyte in vitro. Cell Dev. Biol. **4**, 3–8 (2015)

P. Ayaskanta, A.S. Rajesh, T. Sanatan, et al., Review on biology, distribution and conservation challenges for horseshoe crabs in India. J. Sustain. Sci. Manag. **16**, 332–346 (2021)

T. Basudev, K. Sivakumar, J. Sajan, et al., Assessment of the population status and threats to the horseshoe crabs along the northern east coast of India, in *Ecology and Conservation of Tropical Marine Faunal Communities*, ed. by C. Raghunathan, K. Venkataraman, (Springer, Berlin, Heidelberg, 2013), pp. 137–146

S. Behera, B. Tripathy, K. Sivakumar, et al., Distribution and abundance of two sympatric species of horseshoe crabs along the Odisha Coast, India, in *Changing Global Perspectives on Horseshoe Crab Biology, Conservation and Management*, ed. by R. H. Carmichael, M. L. Botton, P. K. S. Shin, S. G. Cheung, (Springer, Cham, 2015), pp. 181–191

J. Berkson, C.N. Shuster Jr., The horseshoe crab: The battle for a true multiple-use resource. Fisheries **24**, 6–12 (1999)

J. Berkson, C.P. Chen, J. Mishra, et al., A discussion of horseshoe crab management in five countries: Taiwan, India, China, United States, and Mexico, in *Biology and Conservation of Horseshoe Crabs*, ed. by J. T. Tanacredi, M. L. Botton, D. R. Smith, (Springer, New York, 2009), pp. 465–474

M.L. Botton, The conservation of horseshoe crabs: What can we learn from the Japanese experience? in *Limulus in the Limelight*, ed. by J. T. Tanacredi, (Kluwer Academic/Plenum Publishers, New York, 2001), pp. 41–51

A. Chatterji, S. Kotnala, R. Mathew, Effect of salinity on larval growth of horseshoe crab, *Tachypleus gigas* (Müller). Curr. Sci. **87**, 248–250 (2004)

A. Chatterji, F. Shaharom, Unusual spawning behaviour of the horseshoe crab (*Tachypleus gigas,* Müller) after the Tsunami along Orissa Coast, India. Pertanika J. Sci. Technol. **17**, 263–268 (2009)

S. Chinnari, S. Pati, B.P. Dash, et al., A new record on inducement of differentiation of dendritic cells by the peri-vitelline fluid of the fertilized eggs of Indian horseshoe crab (*Tachypleus gigas*, Müller). Bio. Forum. **7**, 678–681 (2015)

L. Cartwright-Taylor, V.B. Yap, C.C. Hsu, et al., Distribution and abundance of horseshoe crabs *Tachypleus gigas* and *Carcinoscorpius rotundicauda* around the main island of Singapore. Aquat. Biol. **13**, 127–136 (2011)

N. Fairuz-Fozi, B. Satyanarayana, N.A.M. Zauki, et al., *Carcinoscorpius rotundicauda* (Latreille, 1802) population status and spawning behaviour at Pendas coast, Peninsular Malaysia. Glob. Ecol. Conserv. **15**, e00422 (2018)

S. Ghaskadbi, V. Patwardhan, M. Chakraborthy, et al., Enhancement of vertebrate cardiogenesis by a lectin from perivitelline fluid of horseshoe crab embryo. Cell. Mol. Life Sci. **65**, 3312–3324 (2008)

M.H. Hu, Y.J. Wang, S.T. Tsang, et al., Effect of prolonged starvation on body weight and blood-chemistry in two horseshoe crab species: *Tachypleus tridentatus* and *Carcinoscorpius rotundicauda* (Chelicerata: Xiphosura, Limulidae). J. Exp. Mar. Biol. Ecol. **395**, 112–119 (2010)

C. Iwaoka, T. Okayama, Public awareness and community-based conservation for the horseshoe crab at Saikai National Park in Nagasaki Prefecture, Japan, in *Biology and Conservation of Horseshoe Crabs*, ed. by J. T. Tanacredi, M. L. Botton, D. R. Smith, (Springer, New York, 2009), pp. 571–583

J.B.C. Jackson, M.X. Kirby, W.H. Berger, et al., Historical overfishing and the recent collapse of coastal ecosystems. Science **293**, 629–638 (2001)

B.A. John, B.R. Nelson, H.I. Sheikh, et al., A review on fisheries and conservation status of Asian horseshoe crabs. Biodivers. Conserv. **27**, 3573–3598 (2018)

S. Kai, J. Morikawa, The horseshoe crab, *Tachypleus tridentatus*, collected from the coast of Nagasaki Prefecture and the East China Sea. Trans. Nagasaki Biol. Soc. **61**, 91–94 (1999)

R.L. Lewison, L.B. Crowder, A.J. Read, et al., Understanding impacts of fisheries bycatch on marine megafauna. Trends Ecol. Evol. **19**, 598–604 (2004)

G. Luther, P.P. Pillai, A.A. Jayaprakash, et al., Gillnet fisheries of India. Mar. Fisk. Infor. Serv. **150**, 1–24 (1997)

R. Mohamed, A.P. Nur, S.I. Nor, et al., Using applied statistics for accurate size classification of the endangered Tachypleus tridentatus horseshoe crab." J. Ecol. Eng. **22**, 273–282 (2021)

R. Myers, B. Worm, Rapid worldwide depletion of predatory fish communities. Nature **423**, 280–283 (2003)

S. Tudu, T. Sarkar, B. Satyanarayana, et al., Is sex reversal common phenomenon in horseshoe crabs too?. Ecol. Eng. Environ. Technol. **22**,45–52 (2021)

J.K. Mishra, A. Mishra, Yasmin, Status and threat perceptions of the Indian horseshoe crabs along the northeast coast of Bay of Bengal, India, in *Changing Global Perspectives on Horseshoe Crab Biology, Conservation and Management*, ed. by R. H. Carmichael, M. L. Botton, P. K. S. Shin, S. G. Cheung, (Springer, Cham, 2015), pp. 397–406

B.R. Nelson, J.M.H. Zhong, N.A.M. Zauki, et al., Effects of shore sedimentation to *Tachypleus gigas* (Müller, 1785) spawning activity from Malaysian waters. J. Sustain. Sci. Manage. **14**, 41–60 (2019)

B.R. Nelson, B. Satyanarayana, J.H. Moh, et al., The final spawning ground of *Tachypleus gigas* (Müller, 1785) on the east Peninsular Malaysia is at risk: A call for action. PeerJ **4**, e2232 (2016a)

B.R. Nelson, B. Satyanarayana, J.H. Moh et al., Does human infringement at the spawning grounds challenge horseshoe crab eggs and their embryogenesis? J Sustain Sci Manag Special Issue 1: The International Seminar on the Straits of Malacca and the South China Sea, 1–10 (2016b)

B.R. Nelson, B. Satyanarayana, J.M.H. Zhong, et al., Episodic human activities and seasonal impacts on the *Tachypleus gigas* (Müller, 1795) population at Tanjung Selangor in Peninsular Malaysia. Estuar. Coast. Shelf Sci. **164**, 313–323 (2015)

W. Nong, Z. Qu, Y. Li, et al., Horseshoe crab genomes reveal the evolution of genes and microRNAs after three rounds of whole genome duplication. Commun. Biol. 4, 83 (2021)

S. Pati, G.C. Biswal, B.P. Dash, Availability of *Tachypleus gigas* (Müller) along the river estuaries of Balasore district, Odisha, India. Int. J. Fish. Aquat. Stud. **2**, 334–336 (2015)

S. Pati, B.P. Dash, Horseshoe crab (*Tachypleus gigas*) as prey of domestic pig (*Susdomesticus*) in Khandia estuary, Balasore, Odisha, India. Zool. Print **31**, 14–15 (2016)

S. Pati, Cleaning the breeding beaches and awareness event for horseshoe crab conservation along Chandipur coast and nearby estuaries in Odisha. Zoo's Print **32**, 46–48 (2017)

S. Pati, S. Tudu, A. Rajesh, et al., Manmade activities affected the breeding ground of horseshoe crab (*Tachypleus gigas*) along Balasore coast: Call for immediate conservation. E-Planet **15**, 145–154 (2017)

S. Pati, Trade and by-catch assessment of Indian horseshoe crab along with its conservation by integrating education and awareness among community along Balasore Coast. Final Report for Rufford Foundation (2019). https://www.rufford.org/files/24630-1%20Final%20Report.pdf. Accessed 31 August 2019

S. Pati, S. Shahimi, H. A. Edinur, et al., Extraction of people's perception toward horseshoe crab existence in northeast coast of India. Front. Mar. Sci. **7**, 587335 (2020a)

S. Pati, A. Chatterji, B.P. Dash, et al., Structural characterization and antioxidant potential of chitosan by γ-irradiation from the carapace of horseshoe crab. Polymers **12**, 2361 (2020b)

S. Pati, S. Shahimi, N. Debabrata, et al., Predicting Tachypleus gigas spawning distribution with climate change in Northeast Coast of India. J. Ecol. Eng. **22**, 211–220 (2021a)

S. Pati, S. Tanmay, S. I. Hassan, et. al., γ-Irradiated chitosan from Carcinoscorpius rotundicauda (Latreille, 1802) improves the shelf life of refrigerated aquatic products. Front. Mar. Sci. **8**, 664961 (2021b)

P.K.S. Shin, H.Y. Li, S.G. Cheung, Horseshoe crabs in Hong Kong: Current population status and human exploitation, in *Biology and Conservation of Horseshoe Crabs*, ed. by J. T. Tanacredi, M. L. Botton, D. R. Smith, (Springer, New York, 2009), pp. 347–360

D.R. Smith, M.J. Millard, R.H. Carmichael, Comparative status and assessment of *Limulus polyphemus* with emphasis on the New England and Delaware Bay populations, in *Biology and Conservation of Horseshoe Crabs*, ed. by J. T. Tanacredi, M. L. Botton, D. R. Smith, (Springer, New York, 2009), pp. 361–386

T.C. Srijaya, P.J. Pradeep, A. Hassan, et al., Effect of peri-vitelline fluid from horseshoe crab embryo in enhancing the early gonadal development in red tilapia. Biol. Forum **5**, 1–7 (2013)

S. Vestbo, M. Obst, F.J. Quevedo Fernandez, et al., Present and potential future distributions of Asian horseshoe crabs determine areas for conservation. Front. Mar. Sci. **5**, 164 (2018)

N.A.M. Zauki, B. Satyanarayana, N. Fairuz-Fozi, et al., Citizen science frontiers horseshoe crab population regain at their spawning beach in East Peninsular Malaysia. J. Environ. Manage. **232**, 1012–1020 (2019)

# Citizen Science Approach for Indonesian Horseshoe Crab Conservation

**Yusli Wardiatno, Naila Khuril Aini, Ali Mashar, Zairion Zairion, Peter Funch, Yuni Puji Hastuti, and Untung Sangaji**

## 1 Introduction

Horseshoe crabs are one of the animals in the family Limulidae, order Xiphosura. The horseshoe crab is known as a living fossil because it has survived nearly 500 million years and is called a stabilomorph because it has a highly stable morphology (Eldredge and Stanley 1984; Kin and Błażejowski 2014). Ancient horseshoe crab fossils have been found since Ordovician times (Rudkin et al. 2008; Van Roy et al. 2010), and the existence of horseshoe crabs with the same form emerged in the Jurassic period (Sekiguchi and Sugita 1980; Briggs et al. 2005). At present, there are four species of horseshoe crab spread on the coasts of North America and Asia (Sekiguchi and Shuster 2009). Three of them are found in Indonesia, namely, tri-spine horseshoe crab (*Tachypleus tridentatus*), coastal horseshoe crab (*T. gigas*),

Y. Wardiatno (✉)
Department of Aquatic Resources Management, Faculty of Fisheries and Marine Sciences, IPB University, Kampus IPB Darmaga, Bogor, Indonesia

Environmental Research Center – IPB University, Bogor, Indonesia

Center for Coastal and Marine Resources Studies – IPB University, Bogor, Indonesia

N. K. Aini
Doctoral Program in Aquatic Resources Management, Graduate School of IPB University, Kampus IPB Darmaga, Bogor, Indonesia

A. Mashar
Department of Aquatic Resources Management, Faculty of Fisheries and Marine Sciences, IPB University, Kampus IPB Darmaga, Bogor, Indonesia

Z. Zairion
Department of Aquatic Resources Management, Faculty of Fisheries and Marine Sciences, IPB University, Kampus IPB Darmaga, Bogor, Indonesia

Center for Coastal and Marine Resources Studies – IPB University, Bogor, Indonesia



J. T. Tanacredi et al. (eds.), *International Horseshoe Crab Conservation and Research Efforts: 2007-2020*, https://doi.org/10.1007/978-3-030-82315-3_28

and mangrove horseshoe crab (*Carcinoscorpius rotundicauda*), which are found around the coast of Asia. *Limulus polyphemus*, commonly called the American horseshoe crab, is only found on the Atlantic coast of North America (Walls et al. 2002).

Horseshoe crab blood has a high economic value, and the price of blood is very expensive. *Limulus* yields a blood product, *Limulus* amebocyte lysate (LAL), which is used for detection of endotoxin contaminants in drugs and intravenous devices (Novitsky 1984; John et al. 2010; Eyler 2015). *Tachypleus* is used to produce a similar product, commonly referred to as *Tachypleus* amebocyte lysate (TAL) (John et al. 2018a, b), which is used by biomedical industries, and they are also harvested for human consumption (Manca et al. 2017). In addition, horseshoe crabs are used as eel (*Anguilla rostrata*) and whelk fishing bait in the USA (Kreamer and Michels 2009; ASMFC 2013), while in Kuala Tungkal, Jambi, Indonesia, the horseshoe crab is used as fish bait in eeltail catfishes (*Euristhmus microceps*) (Rubiyanto 2012). Other reports have revealed that Asian horseshoe crabs were previously imported to North America to be used as bait (Botton et al. 2015; Smith et al. 2016). *Limulus polyphemus* is ecologically important as a bioturbator and has important predator-prey roles in the benthic invertebrate community (Botton and Haskin 1984; Botton and Ropes 1989; Botton 2009). Their eggs are a source of protein for various types of shorebirds, and adult carapaces host many types of epibionts (Walls et al. 2002; Botton 2009). Unfortunately, such ecological role of Asian horseshoe crabs has not been reported yet.

## 2 Ecology and Value of Horseshoe Crabs in Indonesia

Ecologically and economically, horseshoe crab has very important benefits, but in Indonesia, it is a protected animal, so their utilization would be limited. On the IUCN Red List of Threatened Species, *L. polyphemus* is a vulnerable species (Smith et al. 2016). The conservation status of *T. tridentatus* was recently updated in April 2019, from the initial data-deficient category to endangered (Laurie et al. 2019). Meanwhile, the conservation status of *T. gigas* and *C. rotundicauda* is lacking data or data-deficient (DD). *C. rotundicauda*, *T gigas*, and *T. tridentatus* in Indonesia are protected animals under the Ministry of Environment and Forestry Regulation No. 20 of 2018 concerning protected plant and animal species.

P. Funch
Department of Biology – Genetics, Ecology and Evolution, Aarhus University, Aarhus, Denmark

Y. P. Hastuti
Department of Aquaculture, Faculty of Fisheries and Marine Sciences, IPB University, Kampus IPB Darmaga, West Java, Indonesia

U. Sangaji
Indonesian Police Head Quarter, Jakarta, Indonesia

At present, horseshoe crabs around the world, including Indonesia, are experiencing an increased risk of extinction. The risk of extinction is related to extreme weather changes, high capture, pollution, reclamation, and habitat destruction in coastal areas (Meilana et al. 2016; Razali and Zaleha 2017). Evidence from various studies and opinions from fishermen suggests that the number of horseshoe crab populations has decreased for all species. A significant decline in horseshoe crab populations is noted in many Asian regions including Japan (Itow 1993), China (Liao et al. 2001), Taiwan (Chen et al. 2004), the Philippines (Almendral and Schoppe 2005), and Hong Kong (Shin et al. 2009). The decline in the number of horseshoe crab is caused by several factors, such as beach reclamation which drive the habitat loss, degradation in water quality because of industrial pollution (Botton and Itow 2009), and destruction of spawning area (Nelson et al. 2016).

Regarding the ecobiological information about horseshoe crab in Indonesia, there has been an investigation of *C. rotundicauda* and *T. gigas* in Banyuasin, Sumatera Island, Indonesia. The distribution started from the waters near mangroves up to 4.2 km to the sea (Fauziyah et al. 2019). *C. rotundicauda* and *T. gigas* were widely distributed in Banyuasin waters. The prosoma width of both species was 100–150 mm (for *C. rotundicauda*) and 115–210 mm (for *T. gigas*), respectively, while the body weights were 72–285 g for *C. rotundicauda* and 99–918 g for *T. gigas*. Sari et al. (2020) described the environmental conditions in the habitat of *C. rotundicauda* in Banyuasin, Sumatera. The depth ranged from 0.55 to 6.25 m, water salinity from 17.6 to 20.8 psu, dissolved oxygen from 5.9 to 7.5 mg/L, pH between 7.6 and 8.1, and bottom temperature of 28 °C to 31.2 °C. Sediments ranged from 9.49% to 38.2% sand, 2.85% to 12.68% clay, and around 49.3% to 84.64% silt.

In a different place but on the same island, Anggraini et al. (2017) described the habitat of *C. rotundicauda* at Gisi Village, Bintan. In Bintan, *C. rotundicauda* was closely allied with sandy mudflats, brackish salinity, and the presence of mangrove areas. Anggraini et al. (2017) stated that *C. rotundicauda* in Bintan were found at temperatures between 31 and 33.8 °C, salinities between 27 and 31 psu, dissolved oxygen ranging from 5.20 to 7.40 mg/L, current velocities 0.10 to 1.10 m/s, and pH between 7.50 and 8.20. The basic sediment characteristics are sandy muddy area with a small amount of gravel. Furthermore, regarding the length-weight relationship, growth patterns of males were positive allometric, and females were negative allometric.

In addition to Sumatera Island, horseshoe crabs have also been found on Java Island. Mashar et al. (2017) found that horseshoe crabs, especially *C. rotundicauda* and *T. gigas*, inhabit the coast of northern Java, including Subang, Indramayu, Semarang, Demak, Rembang, Tuban, Lamongan, and Surabaya. Mashar et al. (2017) stated that *C. rotundicauda* was found almost in all locations, except in Tuban coast. This is not an uncommon phenomenon. *T. gigas* commonly lives in shallow waters with soft, sandy bottoms or extensive mud flats or the mangrove-mudflat ecosystem. Furthermore, Meilana et al. (2016) also found horseshoe crabs, especially *T. gigas*, in the waters of northern Java. Regarding the genetic distance, *T. gigas* in Subang is close to *T. gigas* in Banten, while *T. gigas* in Demak is close to *T. gigas* in Semarang. *C. rotundicauda* and *T. gigas* population in Demak, Madura, and Balikpapan based on pairwise matrix comparison (Fst) were significantly different ($p < 0.05$) (Aini et al. 2020a, b). These differences are due to the limited movements of horseshoe crabs between sites. Using RAPD (random amplified polymorphic DNA) marker, the highest polymorphism in *C. rotundicauda* is in Demak (74.67%), while the lowest polymorphic is in Madura (58.67%). However, the highest heterozygosity in *C. rotundicauda* is in Demak (0.2669), while the lowest heterozygosity is in Balikpapan (0.1885). Contrary to *C. rotundicauda, T. gigas* in Madura show the highest polymorphism (74.36%) and heterozygosity (0.2551), respectively, whereas the lowest polymorphism is in Demak (51.28%), but the lowest heterozygosity is Balikpapan around 0.1864. Using mark and recapture methods, Erwyansyah (2018) estimated a population of 18 0.2866 *T. tridentatus* in Balikpapan.

Nuraisah et al. (2020) have studied the diet of *Carcinoscorpius rotundicauda* and *Tachypleus gigas* in Balikpapan coastal waters, East Kalimantan, Indonesia. They found eight categories of food items in the gut of the horseshoe crabs, namely, bivalves, gastropods, scaphopods, polychaetes, echinoderms, crustaceans, leaf litter, and others. Based on the preponderance index, gastropods were the main food item of the two horseshoe crab species. In accordance with the food composition, the horseshoe crabs inhabiting Balikpapan coastal waters are categorized as benthivores.

Some regions in Indonesia consider that horseshoe crabs have economic value besides the blood content. Some people in places in northern Java such as Surabaya, Madura, and Pati sell this crab in traditional markets, and the price, depending on the size of the crab, is $1.5–$2 USD. Moreover, many people hunt for horseshoe crab eggs to be used as a food. In Kendal, Central Java, this animal is used as an icon in *dugderan* festival. The festival always starts in the beginning of fasting month (the period month depends on the lunar cycle). The eggs of horseshoe crabs are one of the most popular items. In this festival, eggs are sold at a price range of $1–$1.5 USD. Horseshoe crabs are also a delicacy in several exclusive restaurants in Thailand and Malaysia (Christianus and Saad 2007).

Some fisherman in Madura, Demak, Subang, Ujung Kulon, and Balikpapan consider that horseshoe crabs are pests because they damage their nets. Fishermen tend to kill horseshoe crab for easier release from the net. Few fishermen know that this animal is endangered according to the IUCN and protected by the Indonesian government. They tend to be indifferent about the existence of this animal. This poses a threat to the stability of horseshoe crab population in nature.

## 3 Citizen Science Approach for Conservation

To know more about the biology, ecology, and sustainability of horseshoe crab, we need additional cooperation and improved horseshoe crab research. The existence of horseshoe crabs is important, and the need to conserve horseshoe crabs is a major concern (Luo et al. 2020). The limited scientific information on the biology and ecology of horseshoe crabs in Indonesia is one of the problems in conservation efforts. Information about the population of horseshoe crab in Indonesia is still limited. Research on the three species of Asian horseshoe crab in Indonesia has not been done often, and data on population size are sparse. The database needed to estimate the status population of this animal is inadequate, compared with research on Asian horseshoe crab in other countries, including Japan, China, Malaysia, and India, which have continued to increase research in the last decade. On the other hand, research about American horseshoe crab (*L. polyphemus*) has been carried out extensively. Updating the data about horseshoe crab in Indonesia can be done through voluntary monitoring programs based on citizen science approach. The specific goals in citizen science are to monitor and learn more about the population size of horseshoe crab in Indonesia.

A citizen science approach will provide impact on the process of monitoring horseshoe crab population in Indonesia. The process of collecting information will be easier if working with citizen science groups. The group will help to collect the scientific data on all three species of horseshoe crabs, *C. rotundicauda*, *T. gigas*, and *T. tridentatus*. The groups will identify the species of horseshoe crab, count individuals, determine sex, and describe the habitat. To identify the species, the group will use the leaflet that has explained for practical identification. Some of the citizen science programs are done in some several locations. For example, in the USA, there are networks of volunteers and students that have helped to count and tag *Limulus* and save animals that have become stranded while spawning (Mattei et al. 2015). In Singapore, the Nature Society (Singapore) has been active in promoting citizen science conservation of horseshoe crabs. Since 2007, the Nature Society (Singapore) started conservation efforts to rescue horseshoe crabs entrapped in fishing nets at the Mandai mudflats. Citizen science participation is also carried out in Malaysia for monitoring horseshoe crab populations in spawning areas (Zauki et al. 2019). Citizen scientists will include a group of fishermen who have basic knowledge about a target resource and whose daily lives intensely associate with the sea and the animal.

Because there are three co-occurring species of horseshoe crabs in Indonesia, accurate species identification using morphological characteristics is vital to support this citizen science effort. The morphological characters of the three Asian horseshoe crabs are different. *C. rotundicauda* has key morphological characteristics, in which the telson in cross section is round and the body size is relatively smaller than other species. Morphologically, key characters of *T. gigas* are a telson that is triangular in cross section and a single spine on the posterior of the opisthosoma. *T. tridentatus* differs in having three spines at the posterior of opisthosoma, hence its common name tri-spine horseshoe crab (Yamasaki et al. 1988; Chiu and Morton 2003; Yang and Ko 2015).

Practical tests for identifying three species of horseshoe crabs were conducted on 27 university students who had not previously seen the animals. The students were shown specimens of adult horseshoe crabs as dead shells. All of them said that the morphological characters to distinguish species were easy to understand. The students were also successful in identifying all specimens which were given to them.

However, three of them failed to determine sex of the tested specimen (Wardiatno, unpublished data). The determination of sex is based on the different shape in male and female pedipalp.

Although the above test showed that morphological characters among the three horseshoe crab species in Indonesia are easy to understand for identification, we have found morphological variation among *T. tridentatus* in the Balikpapan coastal area that can lead to confusion. The two morphotypes were found in both males and females of *T. tridentatus*. Some individuals have three opisthosomal spines, while others have one spine at the posterior of opisthosoma (Fig. 1) with several tiny spines on dorsal opisthosoma (Fig. 2).

There were two types of variability in *T. tridentatus* (in certain Indonesian waters), and citizen scientists need to be aware of this to avoid possible misidentification of *T. tridentatus* with *T. gigas*. Aini et al. (in review) found 13.33% had one immovable spine, while 86.67% had three immovable spines on the opisthosomatic carapace. Alignment results of the samples from sequences with CO1 primers obtained nucleotide base pair length of 551–593 bp, and BLAST-n results obtained proximity of 99% with the type *T. tridentatus* in NCBI database access codes JQ739210.1, FJ860267.1, and U09387.1. The number of three spines in this section is usually used as an easy-to-see of *T. tridentatus* (Yamasaki et al. 1988; Chiu and Morton 2003; Yang and Ko 2015). Blood sample analysis from two-phenotype

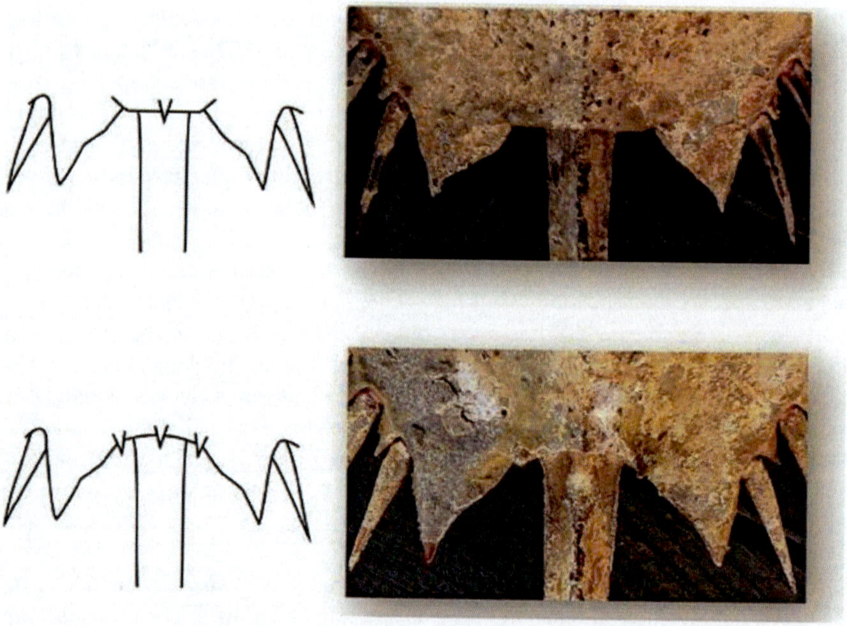

**Fig. 1** Two types of phenotype on *Tachypleus tridentatus* in Balikpapan, East Kalimantan. The dorsal posteriormost opisthosoma carries either one or three spines. (Taken from Aini et al. 2020a, b)

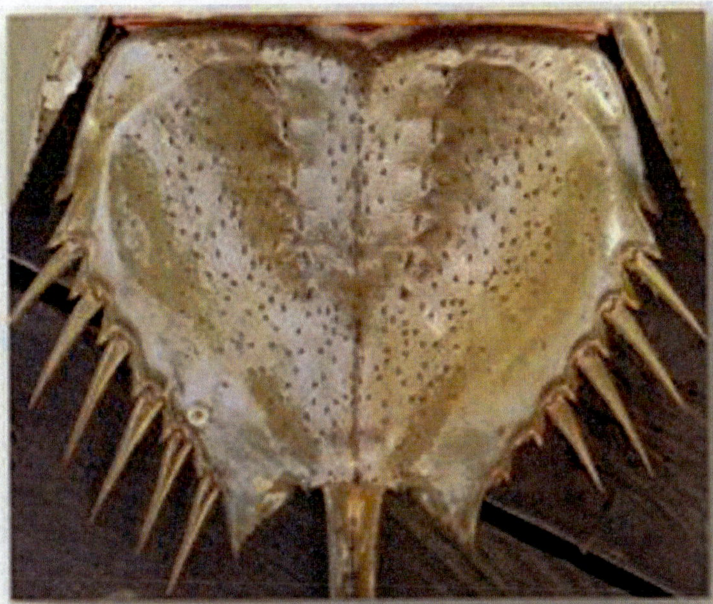

**Fig. 2** Several small spines on the dorsal opisthosoma *Tachypleus tridentatus* in Balikpapan, East Kalimantan. (Taken from Aini et al. 2020a, b)

*T. tridentatus* using COI gene sequence and aligning them with a database on the NCBI website provides validation of taxonomy. It indicates that the two phenotypes are genetically *T. tridentatus*. A 1% difference for individuals in the arthropod phylum still shows the same type (Hebert et al. 2003). Thus, it will be easier and more practical to identify *T. tridentatus* using the presence of several tiny spines on the dorsal of opisthosoma as seen in Fig. 2. However, *T. gigas* does not have the tiny spine scattered in opisthosoma except in cardiac area, or the opisthosoma of *T. gigas* is smooth (Fig. 3).

In Indonesia, horseshoe crabs are primarily a fishery bycatch. Some people who will be part of the volunteer group of citizen scientists may assist in untangling the live crab, collecting data on size and sex, and then releasing the animals. Implementation of a citizen science approach in Indonesia will greatly assist the process of monitoring population of horseshoe crab (Fig. 4). However, our study in Demak coastal area, Central Java, and Balikpapan coastal area, Kalimantan, showed that there were some obstacles to run this approach. In general, fishermen in Indonesia tend to prioritize economically valuable species, and because horseshoe crabs are of minor value, there is little regard for them. In Indonesia, horseshoe crabs are mostly a bycatch in other fisheries, and that is why the level of awareness to save these protected animals is still low. Moreover, some fishermen consider horseshoe crabs as pests, because the crabs cause damage to fishing nets. In contrast, the success of citizen science in Indonesia has been carried out in Belawan,

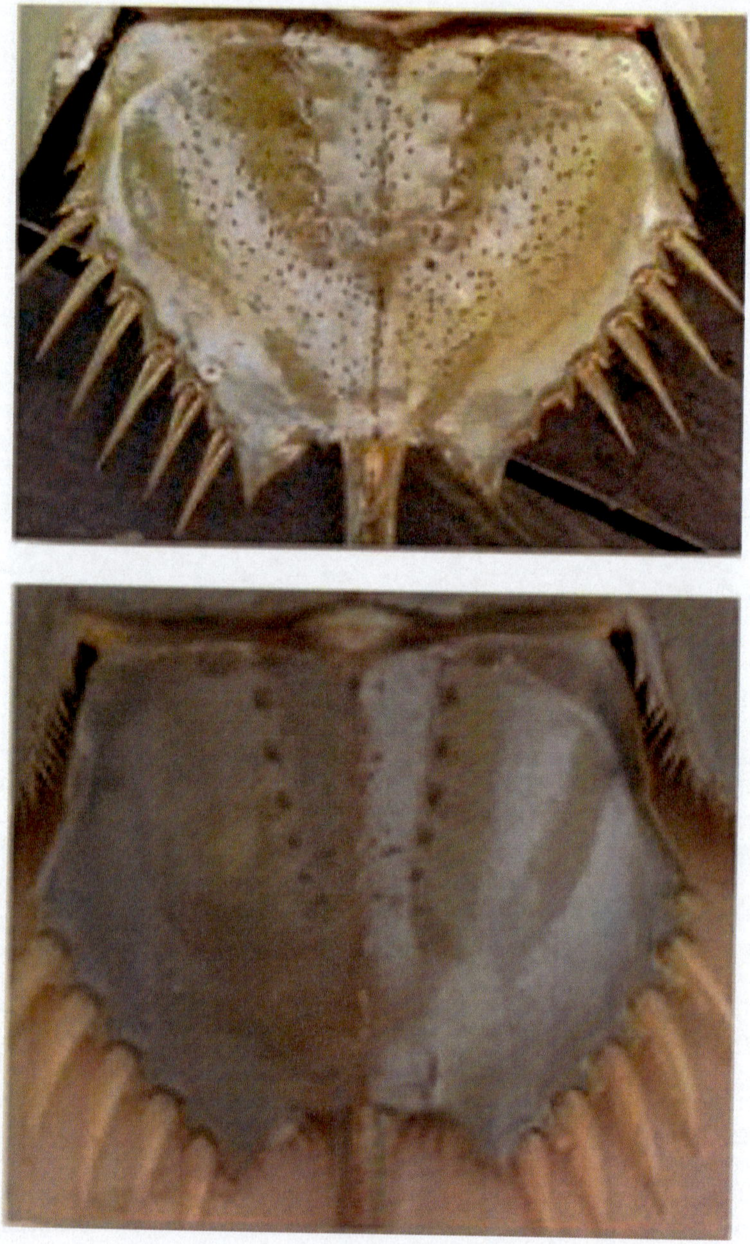

**Fig. 3** Differences of *Tachypleus tridentatus* (upper) and *Tachypleus gigas* (lower) regarding the presence of several spines. (Taken from Aini et al. 2020a, b)

**Fig. 4** Initiation to develop citizen science in Balikpapan, East Kalimantan, Indonesia

**Fig. 5** Release process of horseshoe crab in Belawan, North Sumatera, Indonesia

**Fig. 6** The hatchery of horseshoe crab managed by the Water Unit Police in Belawan, North Sumatera, Indonesia

North Sumatra, by the water police. This group initiative started as a conservation effort for sustainability of horseshoe crabs. The group actively promotes the conservation effort of horseshoe crabs, by introducing the crabs to students from secondary schools and promoting the release of the animals back to the natural environment (Fig. 5). In addition, this group also has a hatchery for horseshoe crabs, especially *T. gigas* (Fig. 6). They are also involved and assist with data collection related to the enrichment of horseshoe crab populations.

## 4 Summary

In Indonesia, the three horseshoe crab species are protected by regulations. In most distribution area, anthropogenic threats are faced by horseshoe crabs, especially when their habitat is overlapped with the fishing grounds. In such areas, they are primarily caught as a fishery bycatch and would be just thrown away as dead animals. A citizen science concept could be a help for future conservation efforts. However, there are challenges to implementing the concept in Demak coastal area, Central Java, and Balikpapan coastal area, Kalimantan, because the fisher community showed relatively low respect to noneconomic species. Yet, a successful story about citizen science implementation has been recorded in coastal area of Belawan, North Sumatra. So, horseshoe crab conservation involving local communities in Indonesia is not impossible in the future.

**Acknowledgments** Our research team activities are funded by the Research Center for Oceanography, Indonesian Institute of Sciences (PPO-LIPI) through Demand-Driven Research Fund Contract No. B-5063/IPK.2/KS.02/III/2019 to YW and IPB University through PMDSU Scheme Contract No. 6623/IT3.L1/PN/2019 to YW (following contract between IPB University and the Ministry of Research, Technology and Higher Education No. 261/SP2H/LT/DRPM/20190). The authors want to thank those who helped in research activities: Heri Seputro, Rani Nuraisah, and Ananingtyas Septia Darmarini.

## References

N.K. Aini, E. Erwyansyah, R. Kurnia, N.A. Butet, A. Mashar, Z. Zairion, P. Funch, H. Maduppa, Y. Wardiatno, Two morphotypes of *Tachypleus tridentatus* in Indonesia and implications for species identification. Biodiv. J. **11**(2), 535–540 (2020a)

N.K. Aini, A. Mashar, H.H. Maduppa, Y. Wardiatno, Genetic diversity of horseshoe crabs (*Carcinoscorpius rotundicauda* and *Tachypleus gigas*) in northern Java and Balikpapan waters based on Random Amplified Polymorphic DNA marker. J. Nat. Resourc. Environ. Manage. **10**(1), 124–137 (2020b)

M.A. Almendral, S. Schoppe, Population structure of *Tachypleus tridentatus* (Chelicerata: Merostomata) at a nursery beach in Puerto Princesa City, Palawan, Philippines. J. Nat. Hist. **39**(25), 2319–2232 (2005)

R. Anggraini, D.G. Bengen, N.M.N. Natih, Population structure and morphometry of horseshoe crab *Carcinoscorpius rotundicauda*, Latreille 1802 in Kampung Gisi coastal area of Bintan Bay of Riau Islands Province. J. Ilmu. Teknol. Kelaut. Tropis. **9**, 211–220 (2017)

ASMFC (Atlantic States Marine Fisheries Commission), *Horseshoe Crab Stock Assessment Update* (Atlantic States Marine Fisheries Commission, Washington DC, 2013). http://www.asmfc.org/uploads/file/2013HorseshoeCrabStockAssessment Report.pdf

M.L. Botton, H.H. Haskin, Distribution and feeding of the horseshoe crab, *Limulus polyphemus*, on the continental shelf off New Jersey. Fish. Bull. **82**, 383–389 (1984)

M.L. Botton, J.W. Ropes, Feeding ecology of horseshoe crabs on the continental shelf, New Jersey to North Carolina. Bull. Mar. Sci. **45**, 637–647 (1989)

M.L. Botton, The ecological importance of horseshoe crabs in estuarine and coastal communities: A review and speculative summary, in *Biology and Conservation of Horseshoe Crabs*, ed. by J. T. Tanacredi, M. L. Botton, D. R. Smith, (Springer, New York, 2009), pp. 45–63

M.L. Botton, R.H. Carmichael, P.K.S. Shin, S.G. Cheung, Emerging issues in horseshoe crab conservation: A perspective from the IUCN species specialist group, in *Changing Global Perspectives on Horseshoe Crab Biology, Conservation and Management*, ed. by R. H. Carmichael, M. L. Botton, P. K. S. Shin, S. G. Cheung, (Springer, Cham, 2015), pp. 369–381

M.L. Botton, T. Itow, The effects of water quality on horseshoe crab embryos and larvae, in *Biology and Conservation of Horseshoe Crabs*, ed. by J. T. Tanacredi, M. L. Botton, D. R. Smith, (Springer, New York, 2009), pp. 439–454

D.E.G. Briggs, R.A. Moore, J.W. Shultz, G. Schweigert, Mineralization of soft–part anatomy and invading microbes in the horseshoe crab *Mesolimulus* from the Upper Jurassic Lagerstätte of Nusplingen, Germany. Proc. R. Soc. B **272**, 627–632 (2005)

C.P. Chen, H.Y. Yeh, P.F. Lin, Conservation of the horseshoe crab at Kinmen, Taiwan: Strategies and practices. Biodivers. Conserv. **13**, 1889–1904 (2004)

A. Christianus, C.R. Saad, Horseshoe crabs in Malaysia and the world. FishMail **16**, 8–9 (2007)

H.M.C. Chiu, B. Morton, The distribution of horseshoe crabs (*Tachypleus tridentatus* and *Carcinoscorpius rotundicauda*) in Hong Kong. Asian Mar. Biol. **16**, 185–196 (1999)

H.M.C. Chiu, B. Morton, The sediment and hydrographic characteristics of three horseshoe crab nursery beaches in Hong Kong. J. Ocean Univ. Qingdao **2**, 35–43 (2003)

N. Eldredge, S.M. Stanley, *Living Fossils* (Springer, Berlin, 1984)

Erwyansyah, Populasi dan struktur genetik belangkas *Tachypleus tridentatus* Leach, 1819 sebagai dasar pengelolaan di perairan Kota Balikpapan. Master Thesis. IPB University (2018)

S. Eyler, *Review of the Atlantic States Marine Fisheries Commission Fishery Management Plan for Horseshoe Crab* (Atlantic States Marine Fisheries Commission, Arlington, 2015)

F. Fauziyah, A.I.S. Purwiyanto, W.A.E. Putri, F. Agustriani, A.Z. Mustopa, F. Fatimah, The first investigation record of threatened horseshoe crabs in the Banyuasin estuarine, South Sumatra, Indonesia. Ecol. Montenegrina **24**, 17–22 (2019)

P.D.N. Hebert, S. Ratnasingham, J.R. De Waard, Barcoding animal life: Cytochrome c oxidase subunit 1 divergences among closely related species. Proc. R. Soc. B **270**, 96–99 (2003)

T. Itow, Crisis in the Seto Inland Sea: The decimation of the horseshoe crab. EMECS Newslett. **3**, 10–11 (1993)

B.A. John, K.C.A. Jalal, B.Y. Kamaruzzaman, K. Zaleha, Mechanism in the clot formation of horseshoe crab blood during bacterial endotoxin invasion. J. Appl. Sci. **10**(17), 1930–1936 (2010)

B.A. John, B.Y. Kamaruzzaman, K.C.A. Jalal, K. Zaleha, TAL-a source of bacterial endotoxin detector in liquid biological samples. Int. Food Res. J. **19**(2), 423–425 (2012)

B.A. John, B.R. Nelson, I.S. Hassan, S.G. Cheung, Y. Wardiatno, B.P. Dash, K. Tsuchiya, Y. Iwasaki, S. Pati, A review on fisheries and conservation status of Asian horseshoe crabs. Biodivers. Conserv. **27**, 3573–3598 (2018a)

B.A. John, B.R. Nelson, I.S. Hassan, S.G. Cheung, Y. Wardiatno, B.P. Dash, K. Tsuchiya, Y. Iwasaki, S. Pati, Correction to: A review on fisheries and conservation status of Asian horseshoe crabs. Biodivers. Conserv. **27**, 3845 (2018b)

A. Kin, B. Błażejowski, The horseshoe crab of the genus *Limulus*: living fossil or stabilomorph? PLoS One **9**(10), e108036 (2014)

G. Kreamer, S. Michels, History of horseshoe crab harvest on Delaware Bay, in *Biology and Conservation of Horseshoe Crabs*, ed. by J. T. Tanacredi, M. L. Botton, D. R. Smith, (Springer, New York, 2009), pp. 299–313

K. Laurie, C.P. Chen, S.G. Cheung, V. Do, H. Hsieh, B.A. John, F. Mohamad, S. Seino, S. Nishida, P. Shin, M. Yang, *Tachypleus tridentatus* (errata version published in 2019). The IUCN Red List of Threatened Species 2019: e.T21309A149768986 (2019). https://doi.org/10.2305/IUCN.UK.20191.RLTS.T21309A149768986.en

Z. Luo, F. Miao, M. Hu, Y. Wang, Research development on horseshoe crab: A 30-year bibliometric analysis. Front. Mar. Sci. **7**, 41 (2020). https://doi.org/10.3389/fmars.2020.00041

A. Manca, F. Mohamad, A. Ahmad, M.F.A.M. Sofa, N. Ismail, Tri-spine horseshoe crab, *Tachypleus tridentatus* (L.) in Sabah, Malaysia: The adult body sizes and population estimate. J. Asia-Pac. Biodivers. **10**, 355–361 (2017)

A. Mashar, N.A. Butet, B. Juliandi, Y. Qonita, A.A. Hakim, Y. Wardiatno, Biodiversity and distribution of horseshoe crabs in northern coast of Java and southern coast of Madura. IOP Conf. Ser.: Earth Environ. Sci. **54**, 012076 (2017)

J.H. Mattei, M.L. Botton, M.A. Beekey, C.P. Colón, Horseshoe crab research in urban estuaries: Challenges and opportunities, in *Changing Global Perspectives on Horseshoe Crab Biology, Conservation and Management*, ed. by R. H. Carmichael, M. L. Botton, P. K. S. Shin, S. G. Cheung, (Springer, Cham, 2015), pp. 537–555

L. Meilana, Y. Wardiatno, N.A. Butet, M. Krisanti, Karakter morfologi dan identifkasi molekuler dengan marka gen CO1 pada mimi (*Tachypleus gigas*) di perairan utara Pulau Jawa. J. Ilmu. Teknol. Kelaut. Tropis. **8**, 145–158 (2016)

B.R. Nelson, B. Satyanarayana, J.H. Moh, M. Ikhwanuddin, A. Chatterji, F. Shaharom, The fnal spawning ground of *Tachypleus gigas* (Müller, 1785) on the east Peninsular Malaysia is at risk: A call for action. PeerJ **4**, e2232 (2016)

R. Nuraisah, N.K. Aini, A. Mashar, Z. Zairion, Y.P. Hastuti, P. Funch, Y. Wardiatno, Food habit of horseshoe crabs, *Tachypleus gigas* (Müller, 1785) and *Carcinoscorpius rotundicauda* (Latreille, 1802) in Balikpapan coastal waters, East Kalimantan. J. Nat. Resourc. Environ. Manage. **10**(2), 153–162 (2020)

C.N. Lee, B. Morton, Experimentally derived estimates of growth by juvenile *Tachypleus tridentatus* and *Carcinoscorpius rotundicauda* (Xiphosura) from nursery beaches in Hong Kong. J. Exp. Mar. Biol. Ecol. **318**, 39–49 (2005)

Y.Y. Liao, S.G. Hong, X.M. Li, A survey on the horseshoe crabs in north of South China Sea. Acta Zool. Sin. **47**, 108–111 (2001) (in Chinese, with English abstract)

T.J. Novitsky, Discovery to commercialization: The blood of the horseshoe crab. Oceanus **27**, 13–18 (1984)

K. Sekiguchi, C.N. Shuster, Limits on the global distribution of horseshoe crabs (Limulacea): Lessons learned from two lifetimes of observations: Asia and America, in *Biology and Conservation of Horseshoe Crabs*, ed. by J. T. Tanacredi, M. L. Botton, D. R. Smith, (Springer, New York, 2009), pp. 5–24

P.K.S. Shin, H.Y. Li, S.G. Cheung, Horseshoe crabs in Hong Kong: Current population status and human exploitation, in *The Biology and Conservation of Horseshoe Crabs*, ed. by J. T. Tanacredi, M. L. Botton, D. R. Smith, (Springer, New York, 2009), pp. 347–360

D.R. Smith, M.A. Beekey, H.J. Brockmann, T.L. King, M.J. Millard, J.A. Zaldívar-Rae. Limulus polyphemus. The IUCN Red List of Threatened Species 2016. e.T11987A80159830 (2016). Available online at: https://doi.org/10.2305/IUCN.UK.2016-1.RLTS.T11987A80159830.en

M. Razali, K. Zaleha, Fishery aspect of horseshoe crab (*Tachypleus gigas* Muller, 1785) in the Peninsular Malaysia: Exploitation status. Universal J. Appl. Sci. **5**(2), 11–15 (2017). https://doi.org/10.13189/ujas.2017.050202

E. Rubiyanto, Studi Populasi Mimi (Xiphosura) di Perairan Kuala Tungkal, Kabupaten Tanjung Jabung Barat, Jambi. [Master Thesis, University of Indonesia] (2012)

D.M. Rudkin, G.A. Young, G.S. Nowlan, The oldest horseshoe crab: A new xiphosurid from late Ordovician konservat-lagerstätten deposits, Manitoba, Canada. Palaeontology **51**, 1–9 (2008)

N.P. Sari, N. Gulo, M. Najah, Fauziyah, Influence of environmental variability on the body condition of the mangrove horseshoe crab *Carcinoscorpius rotundicauda* from Banyuasin Estuarine, South Sumatra, Indonesia. Ecol. Montenegrina **30**, 37–46 (2020)

K. Sekiguchi, H. Sugita, Systematics and hybridization in the four living species of horseshoe crabs. Evolution **34**, 712–718 (1980)

P. Van Roy, P.J. Orr, J.P. Botting, et al., Ordovician faunas of Burgess shale type. Nature **465**, 215–218 (2010)

E.A. Walls, J. Berkson, S.A. Smith, The horseshoe crab, *Limulus Polyphemus*: 200 million years of existence, 100 years of study. Rev. Fish. Sci. **10**, 39–73 (2002)

T. Yamasaki, T. Makioka, J. Saito, External morphology, in *The Biology of Horseshoe Crabs*, ed. by K. Sekiguchi, (Science House, Tokyo, 1988), pp. 69–132

K.C. Yang, K.S. Ko, First Record of Tri-spine Horseshoe Crab, *Tachypleus tridentatus* (Merostomata: Xiphosurida: Limulidae) from Korean Waters. Anim. Syst. Evol. Divers. **31**(1), 42–45 (2015)

N.A.M. Zauki, B. Satyanarayana, N. Fairuz-Fozi, B.B. Nelson, M.B. Martin, B.A. John, A.J.K. Chowdhury, Citizen science frontiers horseshoe crab population regain at their spawning beach in East Peninsular Malaysia. J. Environ. Manage. **232**, 1012–1020 (2019)

# Cumulative Modifications of the Endangered Horseshoe Crab *Tachypleus tridentatus* Habitat by Small-Scale Developments at Tsuyazaki Cove in Fukuoka, Japan

**Shinji Itaya, Toshifumi Wada, and Mari Shuuno**

## 1 Introduction

The tri-spine horseshoe crab, *Tachypleus tridentatus*, is a well-known living fossil, which has not changed its morphologic characteristics since approximately 200 million years ago, and is one of the four extant species of horseshoe crabs found along the coastal zones in East and South East Asia (Sekiguchi 1988; Sekiguchi 1989). *T. tridentatus* is the largest of the four species of horseshoe crabs, with males averaging 500 mm and females 600 mm in their body length (Sekiguchi 1989; Souji 1992).

The species requires different habitat types (namely, tidal flats for juveniles, sandy beaches within calm enclosed bays for spawning, and seaweed beds for subadults) through its life cycle (Sekiguchi 1988; Sekiguchi 1989; Seino et al. 1998a). Therefore, *T. tridentatus* is considered as a flagship species for monitoring the health of coastal zones (Seino and Uda 2002; Hsieh and Chen 2009).

The spawning of *T. tridentatus* in Japan takes place mostly from late June to August. They lay their eggs at the shallow slopes of sandy beaches near the high tide line. The larvae hatch in the sandy beaches and usually overwinter in the same nest and then move to adjacent mud flats in the next spring (Sekiguchi 1988; Wada et al. 2010). As with other arthropods, they grow from the larval stage to sexual maturity by molting. At each ecdysis stage, their body size increases by a factor of

S. Itaya (✉)
Graduate School of Engineering, Kyushu University, Fukuoka, Japan
e-mail: ecological-identity@nethome.ne.jp

T. Wada
University of Hyogo, Museum of Nature and Human Activities, Kobe, Japan

M. Shuuno
Tsuyazaki Seaside Nature School, Fukutsu, Japan



J. T. Tanacredi et al. (eds.), *International Horseshoe Crab Conservation and Research Efforts: 2007-2020*, https://doi.org/10.1007/978-3-030-82315-3_29

approximately 1.28. For horseshoe crabs raised in captivity from eggs, it takes at least 13 molts over 14 years for males to reach sexual maturity and 14 molts in 15 years for females (Sekiguchi 1988).

In Japan, *T. tridentatus* is listed as critically endangered (Ministry of the Environment 2006). The decline of *T. tridentatus* is mainly related to anthropogenic activities, resulting in the loss of tidal flats and spawning beaches. After researching the four species of horseshoe crabs throughout the world, Sekiguchi (1989: p.17) concluded that "there are no other horseshoe crabs' habitats in the world which have been devastatingly destroyed as those in Japan as a result of development." Habitat loss particularly caused by the development of coastal zones for economic development to increase industrial and agricultural production has resulted in the decreasing population. Although other factors (water pollution and changes in substratum condition due to dam constructions that restricted river discharges) are responsible for the species decline, habitat loss is the major problem in many cases, not only in Japan but also in other Asian countries (Sekiguchi 1988, 1989; John et al. 2018). In 2019, *T. tridentatus* was assessed as endangered by the International Union for Conservation of Nature (Laurie et al. 2019). Thus, the species as a whole is facing a threat of extinction, and therefore, urgent conservation practices are crucial.

The recent distribution and breeding locations for *T. tridentatus* in Japan are shown in Figure 1. *T. tridentatus* was still abundant along the whole area of the Seto Inland Sea and northern part of Kyushu Island from the 1920s to the 1930s. However, many of these habitats disappeared during Japan's economic development, and existing ones have become disjunct from each other. Currently, only small and fragmented areas in Okayama, Hiroshima, Yamaguchi, Oita, Fukuoka, Saga, and Nagasaki are known as the existing breeding habitats of the species. The habitats in

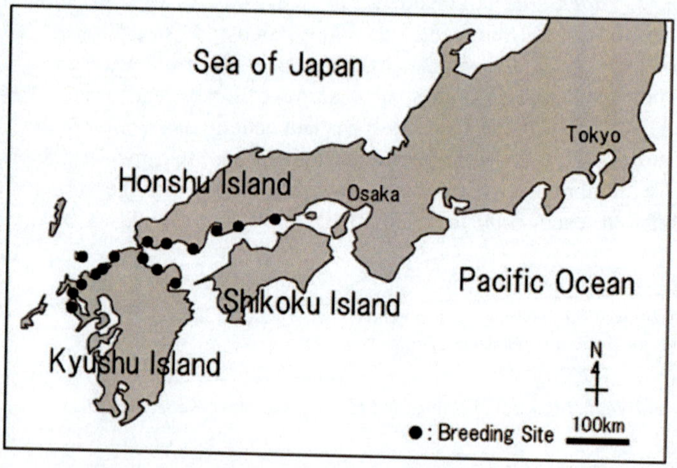


**Fig. 1** Current distribution of *T. tridentatus* in Japan, indicated by breeding sites (marked by dots). The map shows the western portion of Japan, from Tokyo region to Kyushu. (Sources: Sekiguchi 1988; Sekiguchi 1989; Hino 2008; Takeishi 2016)

Kasaoka in Okayama Prefecture and Imari in Saga Prefecture are designated as a natural monument of the nation. In Ehime, the breeding site in Saijo City is also designated as a natural monument of the prefecture, although no breeding activities have been observed in this area for decades. In Nagasaki, the species is protected under a prefectural regulation. In the meantime, no conservational regulations and laws are applied in Hiroshima, Yamaguchi, Oita, and Fukuoka (Ohtsuka et al. 2017).

Tidal flats along the coastal zones are the favorite spots for spawning *T. tridentatus*. However, it is these places where many anthropogenic activities such as commercial fishing are concentrated, making them susceptible to numerous human impacts. Recently, with the growing interest in the importance of coastal conservation, many conventional civil engineering and commercial constructions have been evaluated, and development alternatives are now required to provide for our coexistence with rare species such as *T. tridentatus* (Seino et al. 2000a). The conservation and restoration of sandy beaches necessary for *T. tridentatus* to spawn are particularly important for the survival of the species, which continues to suffer a rapid decline in numbers. In Tsuyazaki, one of the habitats of *T. tridentatus* in Japan, the number of visiting pairs for spawning decreased every year to 139 pairs in 2005, 70 pairs in 2006, 45 pairs in 2007, and 40 pairs in 2008[1](Wada et al. 2010). It is therefore essential to detect the cause of this decline in relation to coastal developments for the conservation of this rare taxon (Wada et al. 2010; Itaya et al. 2019a,b).

The aim of our research was to examine possible causes of the decline of *T. tridentatus* in relation to cumulative developments of its habitat in Tsuyazaki, Fukuoka, Japan.

## 2 Methods and Materials

### *2.1 Study Site*

The field investigation was conducted at Tsuyazaki Cove (33°47’N, 130°27′E) in Fukuoka, Japan, from 2005 to 2013. We carried out the spawning survey at the sites from site 1 to 12 (Fig. 2). The characteristics of the sites are shown in Table 1. The cove is located on a sand dune that was formed during the Holocene Epoch. The cove used to be a vast lagoon; however, it was reclaimed for salt farming during the seventeenth and the eighteenth centuries (Hirowatari and Shimoyama 1999) (Fig. 3). The salt farming area is currently used for agriculture and housing. A mud flat developed at the north end of the cove is known as the habitat of *T. tridentatus* juveniles (Wada et al. 2008). No large rivers are flowing into the cove, but only a few small creeks supply fresh water. Therefore, the influence of the supply of fresh water to the cove is minimal – the salinity of the water at Tsuyazaki Cove ranges

[1] Wada et al. (2010) included nonbreeding pairs such as moving pairs as well as breeding pairs in their study.

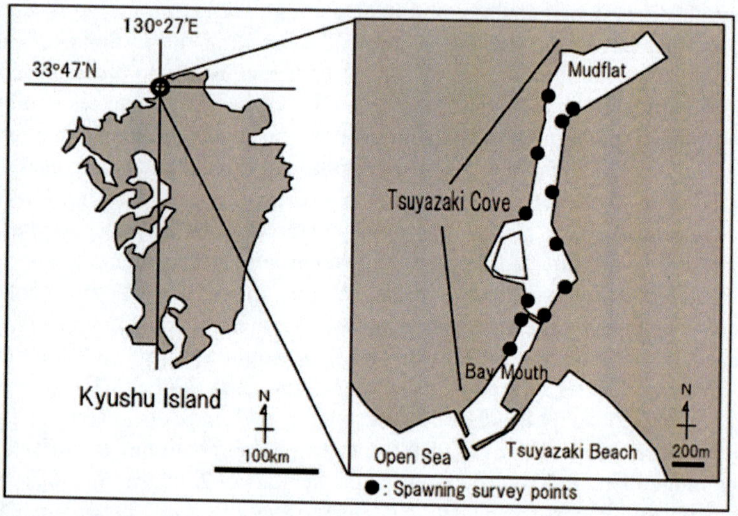


**Fig. 2** Location of the study site in Tsuyazaki, Fukuoka, Japan. The dots indicate the breeding pair counting sites (sites 1–12) at Tsuyazaki Cove

from 32.4% to 34.5% (Wada et al. 2010), and this is unique for a *T. tridentatus* habitat (Itaya et al. 2019b).

## 2.2 General Methods and Analysis

The field investigation was conducted from 2005 to 2013. During every high tide (i.e., twice a day) from late June to early August each year, we counted the number of *T. tridentatus* pairs visiting the shallow water of the cove to spawn. Counting of pairs at each high tide continued for 3 or 4 days on each occasion. However, we aborted the survey under adverse weather conditions for the safety of investigators. Thus, the average number of investigation days was 9.8 ± 1.6 days per year. We investigated for 2–3 hours each side of the high tide. In addition to the authors, the students from Fukuoka Eco Communication College were mainly involved in the field survey. Local residents also joined the survey from 2007. The average number of investigators during the period was 8.9 people (max. = 18, min. = 1). At the time of the survey, the authors conducted a pre-lecture on the significance of the study and the survey method for the participants to standardize their field skills. Each adult *T. tridentatus* was numbered with a combination of plastic tags and crayon marking for identification. We defined "breeding pair" as one spawning at the beach during our field investigation. The pairs for those visiting the beach with no observation of spawning activities and unaccompanied individuals were excluded from the data analyses.

**Table 1** Information on the approximate areas in square meters, dominant plant species, the sediment characteristics at each spawning site in Tsuyazaki, Fukuoka

| Site | ca. Area (m2) | Plant species | Sediment characteristics[a] Grain size | Median diameter (mm) | Sorting coeff. (sφ) | Mud content (%) |
|---|---|---|---|---|---|---|
| 1 | 91 | N/A | Fine sand | 0.200 | 0.369 | 0.395 |
| 2 | 114 | Brassicaceae sp. Poaceae sp. | Medium sand | 0.254 | 0.573 | 0.000 |
| 3 | 18 | N/A | Fine sand | 0.206 | 0.432 | 0.139 |
| 4 | 17 | N/A | Fine sand | 0.246 | 0.618 | 0.118 |
| 5 | 35 | N/A | Pebble | 5.004 | 0.843 | 0.815 |
| 6 | 10 | N/A | Coarse sand | 0.673 | 1.240 | 1.358 |
| 7 | 16 | N/A | Medium sand | 0.406 | 0.600 | 0.995 |
| 8 | 206 | *Angelica japonica* | Coarse sand | 0.733 | 1.093 | 5.435 |
| 9 | 18 | Ulmaceae sp. Poaceae sp. | Medium sand | 0.383 | 1.123 | 6.119 |
| 10 | 95 | N/A | Very coarse sand | 1.737 | 1.978 | 4.038 |
| 11 | 78 | N/A | Coarse sand | 0.603 | 1.306 | 4.381 |
| 12 | 34 | *Angelica japonica, Phragmites australis* | Coarse sand | 0.757 | 0.826 | 2.694 |

[a]Sediment characteristics are adapted from Wada et al. (2010)

Changes in potential spawning sites, or sandbanks, from 1948 to 2010 were estimated by interpretation of aerial photographs and QGIS ver.3.0. Suitable spawning habitats for *T. tridentatus* are reported as sand banks within the highest tidal lines of flood and neap tides (Sekiguchi 1988) and within mean higher high water and mean tide level (Seino et al. 1998b). We also perceived that *T. tridentatus* laid their eggs around the upper part of the intertidal zone during our field investigation. We gained a survey map of the highest tidal and the lowest tidal lines of the flood tide on 3 May 2010 at Tsuyazaki Cove, which was measured by a local surveyor. Based on this map together with the previous studies and our field observation, we estimated sand banks as the basement of potential spawning sites for aerial photograph interpretation in this study.

Data were analyzed using R i386 3.4.4 version and Microsoft Excel 2007. The relationship between the amount of the potential spawning sites and the area of development was tested by Pearson's coefficient of linear correlation. Pearson's chi-square test was performed for the comparison of the habitat preference between the sites. For the comparison of the changes in the number of *T. tridentatus* breeding pairs from 2005 to 2013, we divided the number of breeding pairs by sampling days and the mean number of investigators involved in each year. A Wilcoxon rank sum test was performed to evaluate a statistical difference in the number of breeding pairs between two periods from 2005 to 2008 and from 2009 to 2013.

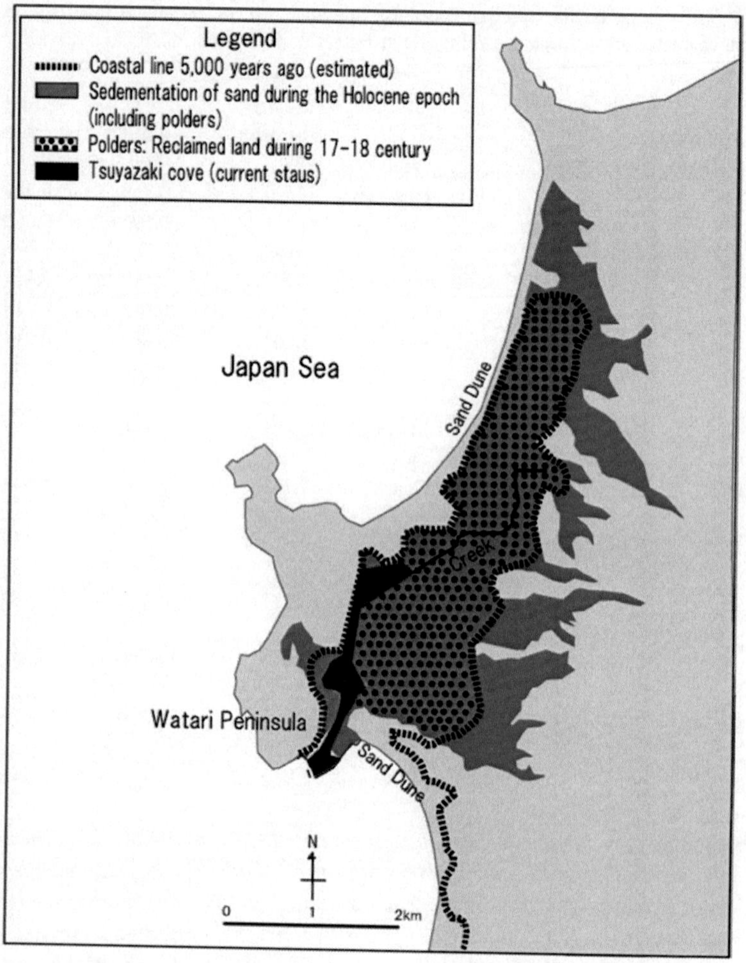


**Fig. 3** Geographical feature of Tsuyazaki Cove vicinity. (Modified from Hirowatari and Shimoyama 1999). Dotted line shows an estimated coastal line approximately 5000 years ago. Reclaimed area for salt farming during the seventeenth and the eighteenth centuries is indicated by dots

# 3 Results

## 3.1 *Changes in the Potential Spawning Sites from 1948 to 2010*

Figures 4 and 5 show the changes in the potential spawning sites of *T. tridentatus* from 1948 to 2010 in Tsuyazaki. The development of the bay mouth increased along with Japan's economic development after the Second World War, extending from

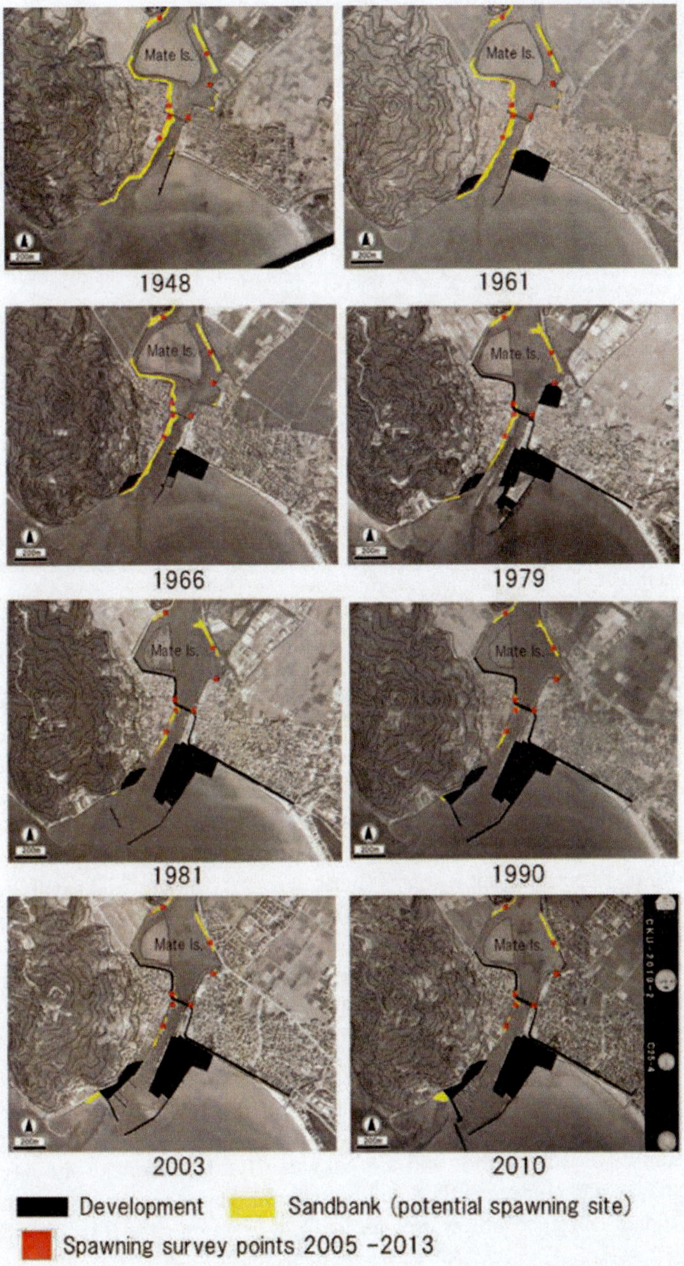


**Fig. 4** Aerial photos of the bay mouth of Tsuyazaki Cove from 1948 to 2010, showing the relationship between development, sandbank, and spawning survey points. Black, yellow, and red colors indicate the area of development, sandbank (potential spawning site), and spawning survey sites, respectively. Mate Is. indicates a deposition of sand which is locally called "Mate Island" which is submerged at high tide. Aerial photo source: Geospatial Information Authority of Japan

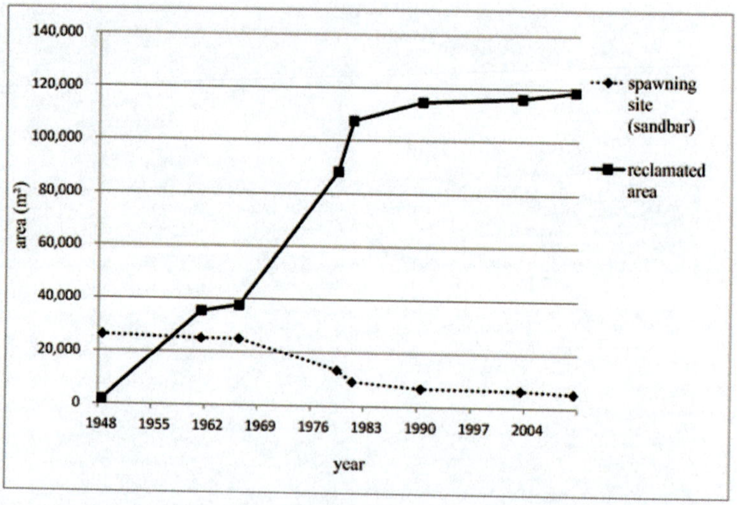


**Fig. 5** Comparison between the area of the potential spawning sites of *T. tridentatus* and area of the development from 1948 to 2010 in Tsuyazaki. Solid and dotted lines indicate the area of development and that of potential spawning site, respectively

1954 $m^2$ in 1948 to 118,681 $m^2$ in 2010. During the same period, there was a contingent decrease in the area of potential spawning sites from 26,303 $m^2$ to 5376 $m^2$ as a result of the development. A significant negative relationship was found between the amount of the potential spawning sites and the area of development (Pearson's correlation $r = -0.96$; df 6, $p < 0.05$).

In 1948, no large development was found at the bay mouth, and sandbanks were continuously formed on the West Bank. Mate Island was excavated in the late 1960s, resulting in the linearization of the flow channel. Following this, in the late 1970s, the sandbank at the south of Mate Island was disappeared as a result of the construction of coastal roads around the cove. This resulted in the reduction of a large amount of potential spawning area from 25,038 $m^2$ in 1966 to 13,547 $m^2$ in 1979 (Figs. 4 and 5). The open space of the bay mouth, which was the only area for tidal flow to exchange, was reduced due to the development during the 1980s and 1990s (Fig. 4). Finally, by 2010, the mouth of the bay was almost totally closed by the construction of a yacht harbor and a breakwater lying in front of the bay mouth (Figs. 4 and 6), and consequently a continuous sandbank (potential spawning site), which is indicated by the yellow area in Fig. 4, was fragmented (Fig. 4).

## 3.2 *Deterioration of the Most Significant Spawning Site from 2005 to 2013*

*T. tridentatus* showed a significant habitat preference for spawning over the sampling period between 2005 and 2013 (Pearson's chi-square test $x^2 = 273.9$; df = 11, $p < 0.01$). Of all spawning sites, site 2 was the most concentrated area where 34% of breeding pairs was found (75 pairs out of 223 pairs) and 22% at site 1 (48 pairs out of 223 pairs). Thus, more than half of the pairs used site 1 and site 2 for their spawning activities from 2005 to 2013 ($n = 223$) (Fig. 7).

Figure 8 is the comparison of sands accumulated at site 2 (the most concentrated spawning site) over the observed period between 2007 and 2018. A birds-eye view of the site with two essential spawning grounds is shown as quadrangles (b) and (c) (Fig. 8a). Using the stone wall as an indicator, the sand accumulation line is shown in Fig. 8b, representing the topography of accumulated sand in 2007. It is evident that the sand accumulation in 2018 was reduced less than half of that in 2007, indicating the erosion of the area together with the part of the vegetation zone, which is above the high tidal zone, took place over the years. Also, most of the smaller sand particles were washed away, and only pebble-sized particles remained as a result of the reduction of sand supply (Fig. 8c).

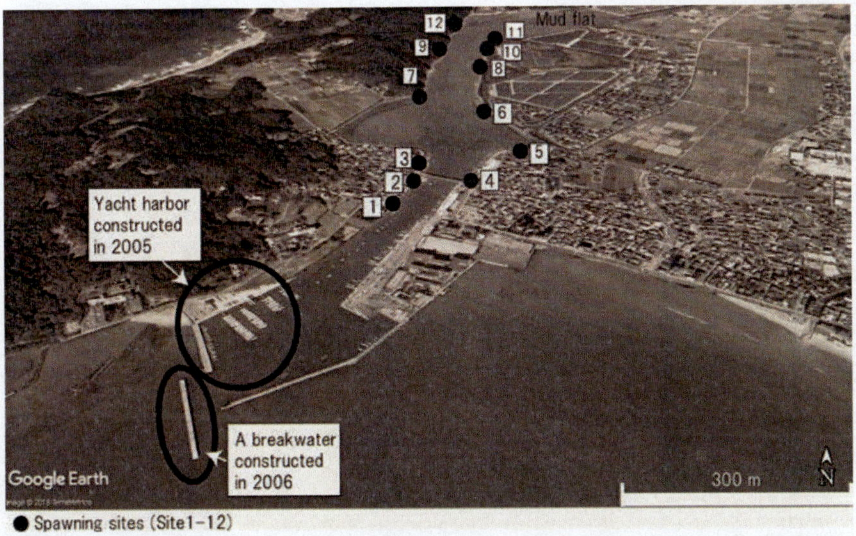


**Fig. 6** A closer look of the bay mouth of Tsuyazaki Cove in 2017. The yacht harbor constructed in 2005 and a breakwater established in 2006 can be seen at the bay mouth. Spawning sites (sites 1–12) are indicated by dots. (Photo source: Google Earth)

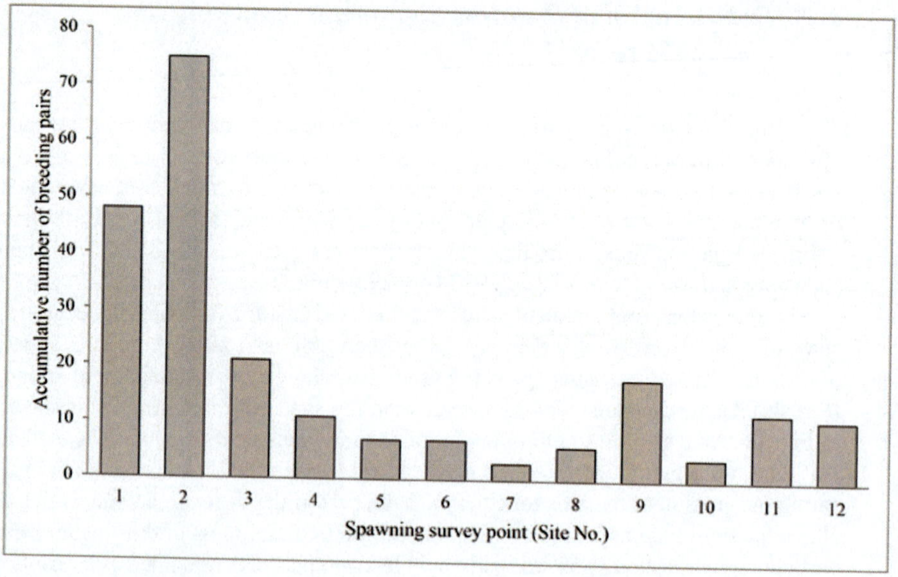


**Fig. 7** Accumulative number of breeding pairs at each survey point from 2005 to 2013. The locations of the survey points are shown in Fig. 2

## 3.3 *Decline of Breeding Pairs from 2005 to 2013*

The field survey revealed that the breeding pairs of *T. tridentatus* declined rapidly during the study period from 2005 to 2013. This decreasing trend was clear especially after 2009 since when the breeding pairs of *T. tridentatus* did not exceed four pairs (2005–2013: $\bar{x}$ =23.6 ± 29.1, max. = 83, min. = 1, $n$ = 9). A Wilcoxon rank sum test performed to evaluate a statistical difference in the number of breeding pairs per capita/day between two periods from 2005 to 2008 ($\bar{x}$ =0.36 ± 0.14, $n$ = 4) and from 2009 to 2013 ($\bar{x}$ =0.06 ± 0.03, $n$ = 5) also indicated that *T. tridentatus* declined significantly since 2009 ($W$ = 20, $p$ = 0.016) (Fig. 9).

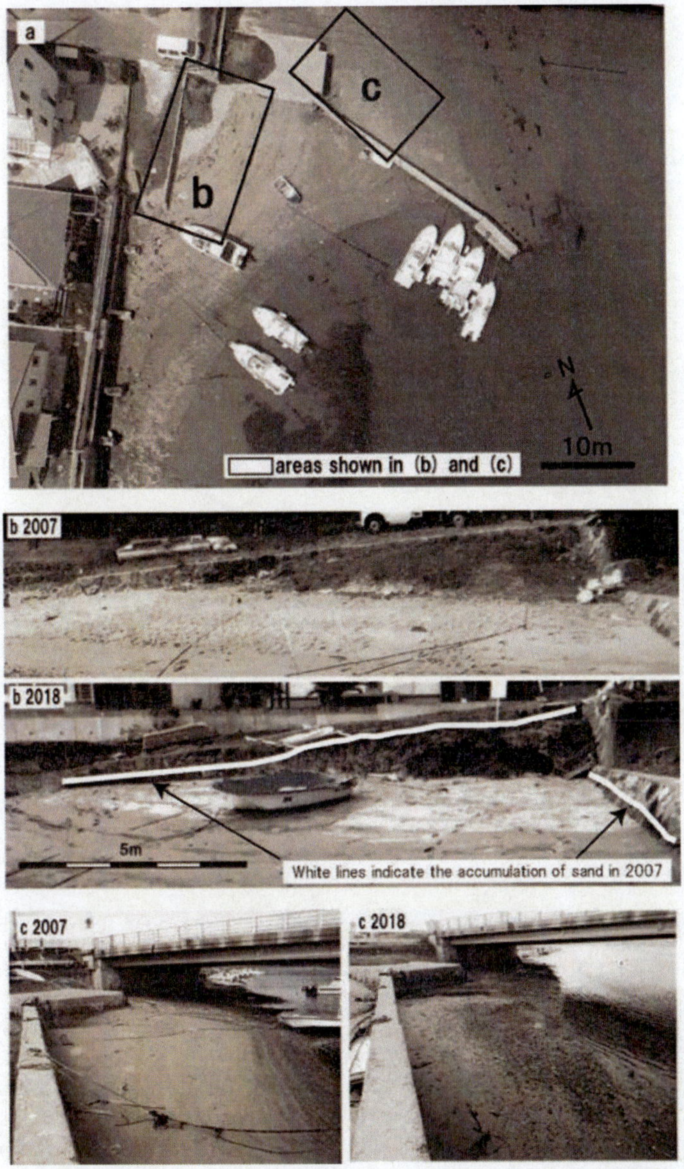


**Fig. 8** Comparison of beach morphology and condition at site 2 between 2007 and 2018. (**a**) Birds-eye view, showing the two studied areas, b and c. (**b**) Comparison of the sand accumulation at b area in 2007 (above) and in 2018 (below). (**c**) Comparison of the sand accumulation at c area in 2007 (left) and in 2018 (right). Photos by Shinji Itaya, (**a**), (**b**) 2018, (**c**) 2018, and Masami Sadanaga, (**b**) 2007, (**c**) 2007

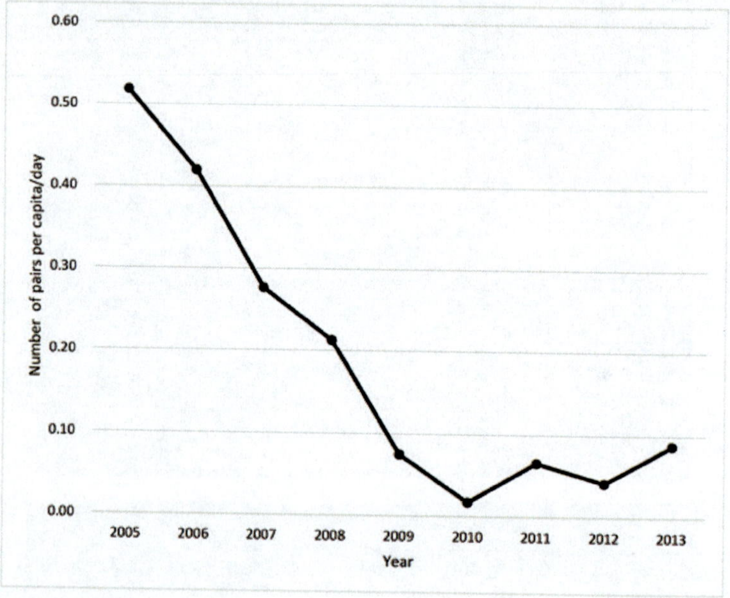


**Fig. 9** The number of *T. tridentatus* breeding pairs per capita/day from 2005 to 2013 in Tsuyazaki, Fukuoka. The number of breeding pairs per capita /day (y-axis) was estimated by dividing the number of breeding pairs by sampling days and the mean number of investigators involved in each year

# 4 Discussion

## *4.1 Changes in the Potential Spawning Sites from 1948 to 2010*

The results of this study allow us to suggest that the decline of *T. tridentatus* in Tsuyazaki could be related to the loss of sand supply to potential spawning sites caused by the development of the bay mouth. This phenomenon was confirmed by the significant negative relationship between the amount of the potential spawning sites and the area of development from 1948 to 2010 (Figs. 4 and 5).

Almost all of *T. tridentatus* spawning sites in Japan are located at river mouths where sediments are supplied from the rivers to form spawning grounds (Sekiguchi 1989). However, this is not the case in our study area. The spawning sites of *T. tridentatus* in Tsuyazaki are located on the Holocene Epoch sand dune (Hirowatari and Shimoyama 1999). In other words, the current spawning sites are on the surface of the residual sands deposited approximately 10,000 years ago. A lack of sand supplies from small creeks, which were formed for salt farming in the seventeenth century, adjacent to Tsuyazaki Cove means that the coastal and tidal currents may play a significant role in supplying sands to the spawning sites (Itaya et al. 2019b). Also, Tsuyazaki Cove is relatively protected from waves generated by the westerlies due to the presence of Watari Peninsula protruding on the southeast side of the cove (Fig. 3). It is these geological and geographical features that have provided a stable spawning habitat for *T. tridentatus* over a long period (Itaya et al. 2019b).

However, this subtle balance in nature appears to have been devastated during the economic development in Japan since the late 1960s. Our results suggest that the development of the bay mouth might have a significant effect on habitat degradation by changing the incoming and outgoing tidal current movements, which caused the reduction in the supply of sand from the open sea.

The most substantial development, which included the construction of coastal roads, was implemented in the late 1970s. Following this, the development of the West Bank at the bay mouth started since the 1980s. As a result of this development, sands flowing into the bay declined, and this, in turn, may have led to the reduction in the number and extent of potential spawning sites in the area. This current change is revealed by the buildup of sands at the corner of the yacht harbor, which otherwise could have accumulated onto the bay to provide spawning grounds for the horseshoe crabs. After these developments, each sandbank (i.e., spawning site) became completely isolated from one another (Fig. 4).

Wakamiya (1989) reported that the number of *T. tridentatus* in Tsuyazaki was relatively low compared to other areas in Fukuoka between 1980 and 1983 and that remaining spawning sites diminished in size. Although his study was not as detailed as the present one, it may be possible that the spawning sites of *T. tridentatus* had become already vulnerable due to the development even before the embarkation of our field survey in 2005.

## 4.2 *Deterioration of the Most Significant Spawning Site Resulting in the Decline of Breeding Pairs from 2005 to 2013*

The result of our field investigation indicated that more than half of the breeding pairs came to site 1 and site 2 for spawning (Fig. 7). Wada et al. (2010) speculated that there was a significant positive relationship between the median diameter of sediments and the number of reproductive visiting pairs in the same area. Although *T. tridentatus* in Tsuyazaki laid their eggs in a broader range of grain sizes from gravel (−2.32 ϕ, about 4 mm in diameter) to fine sand (2.32, about 0.25 mm in diameter) compared to other spawning habitats in Japan (Wada et al. 2010; Iida et al. 2017), approximately 80% of spawning activities were found at the sites with medium-to-fine sands including site 1 (fine sand) and site 2 (medium sand) (Wada et al. 2010). Thus, these sites are crucial for the local population in their reproduction. For the survival of *T. tridentatus*, it would appear to be necessary to conserve these areas, especially site 1 and site 2, which was used by more than 50% of the breeding pairs (Fig. 7).

However, large amounts of sand at site 2 (the most preferred spawning site) were seen to have been lost over the sampling period (Fig. 8b and c). Site 1 and site 2 are located on the West Bank of the bay mouth, where currents from the open sea could flow smoothly into the cove, enabling a sequence of sandbar formation (potential spawning sites) from the 1940s to the 1960s. The East Bank, on the other hand, was subject to the direct landfilling from the early 1960s, so no such continuous sedimentation was found even in the 1960s (Fig. 4). It is possible that the West Bank of the bay was crucial for the reproduction of the species during the sampling period (even though the spawning sites were already degraded to some extent when we started the study as described). Nevertheless, the West Bank was almost closed in 2005 by the completion of the yacht harbor. In addition, in 2006, a breakwater was established at the bay mouth to protect the yacht harbor from the surf (Fig. 6).

Although we are not sure how much these constructions (yacht harbor and breakwater) are responsible for the rapid decline of the breeding pairs in the area, in some cases, coastal developments lead to the physical obstacles for the movement of *T. tridentatus* going out/into the bays and malfunction of the process of dispersal of the larvae due to changes in current movements (Maeda et al. 2000; Seino et al. 2000a). In our surveys of horseshoe crab juveniles from 2003 to 2016 at Tsuyazaki Cove, approximately 80% of juveniles have disappeared since 2009 when comparing the average number of juveniles per survey between 2003–2008 (2003–2008: $\bar{x} = 81.4 \pm 37.5$, max. = 145, min. = 33, $n = 6$) and 2009–2016 (2009–2016: $\bar{x} = 17.9 \pm 7.8$, max. = 31, min. = 7, $n = 8$) (Itaya and Shuuno unpublished data). This decrease of the juveniles together with the decline of the breeding pairs during the same period in the area could be evidence of the impact of the constructions of the yacht harbor and breakwater in the early 2000s.

In Sone tidal flat, which is one of the largest habitats for *T. tridentatus* in Japan, an artificial island-type airport was constructed about 3 km offshore from the spawning ground in 2006 (Hara et al. 2007), and other developments are still continuing (Takahashi 2018). Although the number of spawning pairs in Sone tidal flat decreased from 2006 to 2012, it has recovered since 2013 (Hayashi 2019). However, in Tsuyazaki, the number of spawning pairs has not recovered yet remaining on the declining trend (Fig. 9). In spite of these ongoing developments, one of the reasons for Sone tidal flat to keep maintaining relatively abundant horseshoe crab numbers could be explained in terms of the size of the habitat. Sone tidal flat has an area of about 517 ha, which is ten times larger than Tsuyazaki Cove. In addition to being a small habitat, Tsuyazaki Cove had become already vulnerable due to the development when we started the survey, as revealed in this study. Under these circumstances, even small-scale developments such as the construction of the yacht harbors and the breakwater could cause serious damage to the local population.

Sandbanks used as spawning grounds by *T. tridentatus* are formed under limited conditions requiring subtle balances in nature. Therefore, it is vital to consider the ecological consequences of *T. tridentatus* habitat when changing these geomorphological features by developments (Seino et al. 2001). It seems likely that coastal developments with a poor understanding of ecological knowledge have caused the corruption of natural mechanisms to form the sandbars in the area, which led to the rapid decline of *T. tridentatus* breeding pairs in Tsuyazaki.

### 4.3 Importance of *T. tridentatus* Conservation

The Tsuyazaki tidal flat with its surrounding agricultural area is designated as an important wetland for biodiversity in Japan by the Ministry of the Environment (ME n.d.). Also, as a part of the Munakata Region, the part of the area is a UNESCO World Cultural Heritage Site (UNESCO 2017). These achievements should be highly regarded by many. However, the situation of the natural environment has become worse within the past decade. Most of the agricultural areas were turned into concrete irrigation creeks as a part of farmland consolidation in 2010, resulting in the rapid decline of another iconic endangered species, Japanese rosy bitterling (*Rhodeus ocellatus kurumeus*), a rare freshwater fish known as a biological indicator of floodplain viability. Besides, more than half of the area of the previous salt farm became mega solar power plants in 2013. Thus, both of these environmentally important components (i.e., the agricultural areas and the former site of the salt farm) forming the back marsh environment of Tsuyazaki tidal flat have crucially degraded. It is concerned that this degradation might lead to the decline of freshwater and nutrient discharge into the tidal flat and aridification in this important wetland (Itaya et al. 2018). Considering these serious changes in the natural environments, we should re-examine the value of embracing *T. tridentatus*, a living fossil, in this area and make our best efforts to fit the requirements for "biologically important wetland" and "World Cultural Heritage."

According to a recent genetic study, the northernmost populations of *T. tridentatus*, including those in Yamaguchi, Oita, Fukuoka, Saga, and Nagasaki, show genetic differentiations reflecting a unique population history for each local group (Nishida and Koike 2009). As almost all of *T. tridentatus* populations in Japan are facing the risk of extinction due to human activities, including habitat fragmentation (Sekiguchi 1988, 1989), it is imperative to conserve their remaining habitats in order to protect their genetic diversity.

Although *T. tridentatus* is listed as critically endangered in Japan (Minister of Environment 2006), these listings have no legal force, and there are no commitments imposed by their inclusion in Japan's Red Data List and Red Data Book and no specific conservation strategies are applied for those species listed (Hayama and Sekine 2003). To arrest the decline of *T. tridentatus* populations in Japan, it is necessary to establish legal bases for the protection of listed marine areas and to use the results of scientific research to conserve the species when evaluating development proposals.

### 4.4 Recommendations for Restoring the Local Population

In Tsuyazaki, most of the suitable spawning grounds have been lost as a result of the recent development. This has led to some breeding pairs appear to be forced to lay their eggs at less suitable spawning grounds (Itaya et al. 2019a). Considering the deterioration of the spawning ground and the decrease in the number of breeding pairs revealed in this study, the natural recovery of the population would not be possible under the current conditions. Therefore, in this section, we propose some suggestions to restore the spawning ground in Tsuyazaki.

The most preferable one is to restore the sand supply cycle to the spawning ground. It is highly likely that the yacht harbor is blocking the inflow of sand from the open sea (Fig. 4). Therefore, it is desirable to remove the yacht harbor, but if that is difficult, we recommend reconstructing the yacht harbor with a floor which is raised above the sea level, so that tidal currents can pass through below it. These high floor-type structures can be seen in Japan to avoid direct landfills around tidal flats, so it is technically possible.

In addition, as the number of breeding pairs has decreased significantly since 2009 (Fig. 9), it is necessary to remove the breakwater and its surrounding structures that were built in the 2000s. At the time of the construction of the breakwater, we were becoming worried about its impact on the local horseshoe crabs. Therefore, we protested to the administration in charge, which is the main body of the construction. However, the administrative officers did not know that the area was a horseshoe crab habitat, and they said that the breakwater is a slit-type through which tidal currents pass, so there should be no problem. Nevertheless, according to our observation, the basement of the breakwater and the entrance of the slits were filled with tetra pots with poor tidal flows. The administration's response to our protest was that if the impact of the breakwater and its surrounding constructions on the horseshoe crabs would be revealed in surveys, they would consider measures against the impact. Since some of the causal relationships between the development and the decline of the horseshoe crabs have been identified in this study, the administration in charge should act responsibly and take conservative measures immediately before it is too late.

Meanwhile, it would be necessary to manually restore the spawning ground as a temporary measure until the administration in charge takes measures as described above. Beach nourishments of *T. tridentatus* spawning grounds have been carried out in various places (Seino et al. 2000b; Ohtsubo et al. 2005; Hsieh and Chen 2009), and it has been reported that spawning activities increased in the nourished site compared to the non-nourished site in Imazu tidal flat which is within the same metropolitan area of Tsuyazaki Cove (Ohtsubo et al. 2005). To nourish the beach, it is necessary to consider the dynamism of drifting sand (Ishikawa et al. 2015). In our study site, artificially nourished spawning environments should be maintained by transporting sand from the sandbar that was divided by the development and supplying it to the current spawning ground. The current spawning sites, namely, sites 1, 2 and 3, and the sandy beach on the yacht harbor were originally one continuous sandbar (Fig. 4). Therefore, the sand deposited on the yacht harbor should be used for the restoration of the spawning ground.

As the number of juveniles has been decreasing as well as the spawning pairs in Tsuyazaki, a population enhancement by releasing laboratory-cultivated juveniles to their natural habitat might be necessary. Although the effectiveness of such release programs has not been evaluated, the measurement can increase the juvenile survival rate (Hsieh and Chen 2015).

## 4.5 Conclusion

The impact of coastal development upon the degradation of the habitat of *T. tridentatus* in Tsuyazaki has been confirmed in this study. Coastal development made with a poor understanding of ecological knowledge, and lacking formal protection mechanisms for listed areas, has caused not only the decline of the spawning sites but also the corruption of natural processes previously involved in the formation of sandbars in the area. Further study would be desirable to investigate the effect of the development on the changes in the water movement and the amount of sand being supplied to form the sandy beach suitable for horseshoe crab to reproduce in the area.

**Acknowledgments** We thank the students from Fukuoka Eco Communication College who involved actively in the field survey. We would like to thank Dr. Graeme K. Ward, Dr. John A. Harris, and Dr. Yumiko Iwasaki for proofreading, support, and encouragement.

We dedicate this paper to Dr. Toshifumi Wada, who passed away in November 2018. An enthusiastic ecologist and conservationist, he started the field survey of *T. tridentatus* in Tsuyazaki.

## References

Y. Hara, H. Kojima, S. Yan, K. Tanaka, H. Okumura, A. Nishioka, A fundamental study about a physical characteristics of horseshoe crab larva habitat environment in Sone tidal flat. Ann. J. Coast. Dev. **23**, 447–452 (2007) (In Japanese with English abstract)

S. Hayama, T. Sekine, Suggestions for avoiding extinctions, in *The Nature Conservation Society of Japan (ed) Protection and Laws from Ecological Perspectives*, (Koudansha, Tokyo, 2003), pp. 60–71

O. Hayashi, Spawning patterns of horseshoe crabs in Sone tidal flat. Kabutogani **39**, 16–20 (2019) (In Japanese)

A. Hino, Some aspects of conservation of an endangered marine species in the case of local activities for conservation of horseshoe crab. Ocean Policy Stud. **6**, 49–68 (2008) (In Japanese with English abstract)

F. Hirowatari, S. Shimoyama, Geological history of Tsuyazaki-town, in *History of Tsuyazaki-town –Overview History*, ed. by T. H. C. Committee, (Tsuyazaki-town, Fukuoka, 1999), pp. 107–144

H. Hsieh, C. Chen, Conservation program for the Asian horseshoe crab *Tachypleus tridentatus* in Taiwan: Characterizing the microhabitat of nursery grounds and restoring spawning grounds, in *Biology and Conservation of Horseshoe Crabs*, ed. by J. T. Tanacredi, M. L. Botton, D. R. Smith, (Springer, New York, 2009), pp. 417–438

H. Hsieh, C. Chen, Current status of *Tachypleus tridentatus* in Taiwan for red list assessment, in *Changing Global Perspectives on Horseshoe Crab Biology, Conservation and Management*, ed. by R. H. Carmichael, M. L. Botton, P. K. S. Shin, S. G. Cheung, (Springer, New York, 2015), pp. 383–396

K. Iida, M. Yonetani, R. Nakamura, Y. Kondo, O. Hayashi, S. Takahashi, S. Ohtsuka, Density of juvenile of tri-spine horseshoe crab *Tachypleus tridentatus* in the Sone Estuary, Kitakyushu, Japan, with notes on sediment particle sizes in habitats and breeding areas. Hiroshima University FSC report **15**, 1–7 (2017) (In Japanese with English abstract)

M. Ishikawa, S. Seino, T. Anai, H. Suzaki, Dynamism and related improvement to the spawning beach through integrated watershed sediment management at Imazu tidal flat, Fukuoka city. In: Committee on Environmental Systems Japan Society of Civil Engineers (ed) Proceedings of 43th Annual Meeting of Environmental Systems Research 2015, (Tokyo, 2015) (In Japanese with English abstract)

S. Itaya, M. Shuuno, S. Seino, Decline of endangered Rosy bitterling, *Rhodeus ocellatus kurumeus*, as a biological indicator of the backswamp of Tsuyazaki tidal flat and a suggestion for the conservation of the species in Tsuyazaki, Fukuoka, Japan. In: Committee on Environmental Systems Japan Society of Civil Engineers (ed) Proceedings of 46th Annual Meeting of Environmental Systems Research 2018, Tokyo, (2018) (In Japanese with English abstract)

S. Itaya, S. Seino, M. Shuuno, A. Sakurada, R. Koshiguchi, Spawning site selection of the endangered horseshoe crab *Tachyplenus tridentatus* at Tsuyazaki Cove in Fukuoka, Japan, in *Proceedings of the 10th International Conference on Asian and Pacific Coasts, Hanoi*, ed. by N. T. Viet, T. T. Tung, D. Xiping, (Springer, Singapore, 2019a), pp. 959–963. https://doi.org/10.1007/978-981-15-0291-0_131

S. Itaya, S. Seino, T. Wada, M. Shuuno, Rapid decline in the number of breeding pairs of *Tachypleus tridentatus* and the degradation of sandbanks for spawning due to cumulative bay mouth modifications at the Tsuyazaki Cove in Fukuoka, Japan. Jpn J Conserv Ecol **24**, 157–170 (2019b) (In Japanese with English abstract)

B.A. John, B.R. Nelson, H.I. Sheikh, S.G. Cheung, Y. Wardiatno, B.P. Dash, K. Tsuchiya, Y. Iwasaki, S. Pati, A review on fisheries and conservation status of Asian horseshoe crabs. Biodivers. Conserv. **27**, 3573–3598 (2018). https://doi.org/10.1007/s10531-018-1633-8

K. Laurie, C-P. Chen, S.G. Cheung, V. Do, H. Hsieh, A. John, F. Mohamad, S. Seino, S. Nishida, P. Shin, M. Yang, *Tachypleus tridentatus* (errata version published in 2019). The IUCN Red List of Threatened Species 2019: e.T21309A149768986 (2019) https://doi.org/10.2305/IUCN.UK.2019-1.RLTS.T21309A149768986.en

K. Maeda, S. Seino, S. Nishihara, A. Hino, Ecology of hatched larvae of the horseshoe crab *Tachypleus tridentatus* (Leach) in relation to the physical environment. Jpn. J. Benthol. **55**, 12–24 (2000). https://doi.org/10.5179/benthos.55.15. (In Japanese with English abstract)

Ministry of the Environment (ed.), *Threatened Wildlife of Japan-Red Data Book*, vol 7, 2nd edn. (Arthropods, Japan Wildlife Research Center, Tokyo, 2006) (In Japanese)

Ministry of the Environment, Important Wetlands in Japan for Biodiversity Conservation. (n.d.). http://www.env.go.jp/nature/important_wetland/ index.html. Accessed 7 Jan. 2019

S. Nishida, H. Koike, Genetic structure of Japanese populations of *Tachypleus tridentatus* by mtDNA AT-rich region sequence analysis, in *Biology and Conservation of Horseshoe Crabs*, ed. by J. T. Tanacredi, M. L. Botton, D. R. Smith, (Springer, New York, 2009), pp. 183–196

M. Ohtsubo, H. Ishida, H. Minei, S. Yamaoka, T. Higashi, Physical and chemical properties of the nest-site beach of the horseshoe crab rehabilitated by sand placement. J. Jpn. Soc. Soil Phys. **99**, 55–63 (2005) (In Japanese with English abstract)

S. Ohtsuka, N. Nishihara, R. Hirayama, H. Tanaka, Y. Kondo, H. Saito, N. Shimizu, K. Tomikawa, K. Iida, M. Yonetani, Occurrence of endangered tri-spine horseshoe crab in Etajima and Takehara cities of Hiroshima prefecture, western Japan. Jpn. J. Benthol. **72**, 16–26 (2017) (In Japanese with English abstract)

S. Seino, T. Uda, Environmental changes and restoration of endangered horseshoe crab habitat in Morie bay, Oita prefecture. Bull. Coast. Oceanogr. **39**(2), 117–124 (2002) (In Japanese with English abstract)

S. Seino, T. Uda, S. Mama, T. Minami, M. Serizawa, T. Furuike, K. Maeda, A. Hino, Geographical and coastal surf features of Morie bay and tidal flat as a habitat of the endangered horseshoe crab *Tachypleus tridentatus*. Ann. J. Coast. Eng. **45**, 1096–1100 (1998a) (In Japanese)

S. Seino, K. Maeda, A. Hino, T. Uda, S. Mama, N. Yamada, Why do horseshoe crabs spawn on that shore? –The geographical features, sediments and flow characteristics. Ann. J. Coast. Eng. **45**, 1091–1095 (1998b) (In Japanese)

S. Seino, T. Uda, K. Meda, K. Souji, Dispersion mechanism of hatchling of horseshow crab, *Tachypleus tridentatus* at tidal flat off the Yasaka river mouth in Morie bay. Ann. J. Coast. Eng. **44**, 1209–1214 (2000a) (In Japanese with English abstract)

S. Seino, T. Uda, Y. Tsuchiya, S. Mama, N. Yamada, S. Watanabe, Field observation of natural and artificial spawning sites of horseshoe crab *Tachypleus tridentatus* on Sone tidal flat in Kitakyushu city. Ann. J. Coast. Eng. **16**, 333–338 (2000b) (In Japanese with English abstract)

S. Seino, T. Uda, T. Shiozaki, T. Goto, T. Kuroki, T. Nakamura, Field observation of river mouth bars in Nakatsu tidal flat in view of aptitude of spawning site of horseshoe crab *Tachypleus tridentatus*. Ann. J. Coast. Eng. **17**, 375–380 (2001) (In Japanese with English abstract)

K. Sekiguchi, Distribution and ecology, in *Biology of Horseshoe Crabs*, ed. by K. Sekiguchi, (Science House, Tokyo, 1988), pp. 21–51. (In Japanese)

K. Sekiguchi (ed.), *Current Stats of Japanese Horseshoe Crab* (Horseshoe Crab Preservation Group of Japan, Tokyo, 1989) (In Japanese)

N. Souji, *Horseshoe Crabs* (Okayama Bunko, Okayama, 1992) (In Japanese)

S. Takahashi, Sone tidal flat and a wind power generation plan. In: Ramnet-J News letter vol 31 (2018). http://www.ramnet-j.org/2018/04/05/ramjnews31.pdf. Accessed 28 June 2020 (In Japanese)

Z. Takeishi, Sone tidal flat in Kitakyushu city, Fukuoka -one of the largest spawning sites of tri-spine horseshoe crab in Japan. Our Nat. Hist. **135**, 1–6 (2016) (In Japanese)

UNESCO, Sacred Island of Okinoshima and Associated Sites in the Munakata Region (2017). http://whc.unesco.org/en/list/1535. Accessed 7 Jan. 2019

T. Wada, H.D. YoneyamaT, S. Nomura, S. Itaya, M. Shuuno, Growth and development of a monitoring method for the growth of the juvenile horseshoe crab *Tachypleus tridentatus* at the Tsuyazaki mudflat in Fukuoka, Japan. Jpn. J. Conserv. Ecol. **13**, 199–205 (2008) (In Japanese with English abstract)

T. Wada, S. Itaya, M. Shuuno, Spawning sites and annual variability of the number of reproductive visiting pairs of the horseshoe crab *Tachypleus tridentatus* along the Tsuyazaki coast in Fukuoka, Japan. Jpn. J. Conserv. Ecol. **15**, 163–171 (2010) (In Japanese with English abstract)

Y. Wakamiya, Current status of *Tachypleus tridentatus* along Hakata bay and Karatsu bay in Fukuoka, in *Current Stats of Japanese Horseshoe Crab*, ed. by K. Sekiguchi, (Horseshoe Crab Preservation Group of Japan, Tokyo, 1989), pp. 105–138. (In Japanese)

# Exploitation Status of Two Species of Horseshoe Crabs from Subang, West Java, Indonesia

**Lusita Meilana, Qinhua Fang, Lukman Guam Hakim, Okoli Moses Ugochukwu, Ali Mashar, and Yusli Wardiatno**

## 1 Introduction

There are three extant species of Asian horseshoe crabs and all can be found in Indonesian waters. *Tachypleus tridentatus* has recently been declared endangered on the IUCN Red List (Laurie et al. 2019); *Tachypleus gigas* and *Carcinoscorpius rotundicauda* are still listed in the data deficient category by IUCN (2018). Despite insufficient data, recent studies have shown that human activities are seriously affecting horseshoe crab populations worldwide. They are threatened by land reclamation, habitat degradation, and pollution (Mishra 2009; Lee and Morton 2016; Jawahir et al. 2017). Horseshoe crabs are exploited for medicinal uses (Hurton 2003), and large numbers are harvested as food (Shin et al. 2009; Faridah et al. 2015). Some are used as bait for catfish (*Euristhmus microceps*) in Kuala Tungkal waters (Rubiyanto 2012), for eel (*Anguilla rostrata*), and whelk (*Busycon carica* and *Busycotypus canaliculatus*) (Novitsky et al. 2002). According to Christianus and Saad (2007), hundreds of horseshoe crabs were caught every week in Small Sadeli and Johor on Malaysia's east coast and exported to Thailand. Also, Meilana (2015) reported that horseshoe crabs from Subang in Indonesia were exported to Malaysia and Thailand for food consumption.

Clearly, sustaining horseshoe crab populations is ecologically essential, as they play key roles in their ecosystem, as bioturbators or in balancing the food chain, and

L. Meilana (✉) · Q. Fang · O. M. Ugochukwu
Coastal and Ocean Management Institute, Xiamen University, Xiamen, China
e-mail: lusitameilana@stu.xmu.edu.cn

L. G. Hakim
Department of Fisheries Resources Utilization, IPB University, Bogor, Indonesia

A. Mashar · Y. Wardiatno
Department of Aquatic Resources Management, IPB University, Bogor, Indonesia



J. T. Tanacredi et al. (eds.), *International Horseshoe Crab Conservation and Research Efforts: 2007-2020*, https://doi.org/10.1007/978-3-030-82315-3_30

as controllers of the population of many benthic invertebrates (John et al. 2012). They also serve as a food source for a multitude of marine animals, serving at least 20 species of migrating shorebirds (Beekey et al. 2013) and the mangrove monkey (*Macaca fascicularis*) (Rubiyanto 2012).

Data on horseshoe crabs is very important to understand the population dynamics in their changing habitat, especially with regard to the ongoing climate change and the various threats they are facing nowadays (John et al. 2018). Fishermen in Indonesia are closely connected to horseshoe crabs and their habitat, particularly those using gillnets as fishing gear to catch crabs as target species or bycatch (Graham 2007, Prasetyo 2017). It was even reported that horseshoe crabs became the dominant discarded bycatch in the use of gillnet fishing gears at Paloh Waters in Indonesia (Prasetyo 2017). However, no data are available for the coastal waters of the western Java Sea, where gillnet fishing is very common.

The present study, therefore is aimed at investigating the population size of horseshoe crabs entangled in gillnets set up by fishermen at different habitats in West Java, Indonesia between August and October 2016. Furthermore, morphological characters of the entangled horseshoe crabs were measured to estimate the mortality rate and the exploitation status of horseshoe crabs in the study area. These data can be used to identify the distribution pattern and the composition of horseshoe crabs spatially. Due to the high number of horseshoe crabs at specific sites, a restriction policy can be implemented to reduce the amount of bycatch. This would inform the local fishers to reduce bycatch, minimize effort, and optimized catch.

## 2 Methods

### 2.1 Sampling Sites and Techniques

From August to October 2016, the sampling was carried out to collect horseshoe crabs caught by fishermen from the coastal villages of Mayangan and Legon Wetan, Subang District, West Java, Indonesia. The study area was located in (the coastal waters of) West Java Sea between 107°43′–106°48′E and 6°10′–6°13′S (Fig. 1). Based on the activities and the fishing grounds of the cooperating fishermen, three stations (ST) with a total of 11 substations (SB) were chosen as study sites (Table 1). Station 1 was located in semi-enclosed waters surrounded by mangroves (Segara Menyan and Pondok Bali Bay) and composed of substations 1–4. Station 2, comprising of substations 5–9, was located at the open sea, while station 3 was a coastal area and contained substations 10–11. Each substation was the fishing ground of a cooperating fisherman. Detailed information on the stations and substations are given in Table 1.

The sampling was conducted using accidental sampling method, wherein fishermen who allowed us to join their fishing boat during their work time were selected (Etikan and Bala 2017). The fishermen used bottom-set gillnets to catch blue

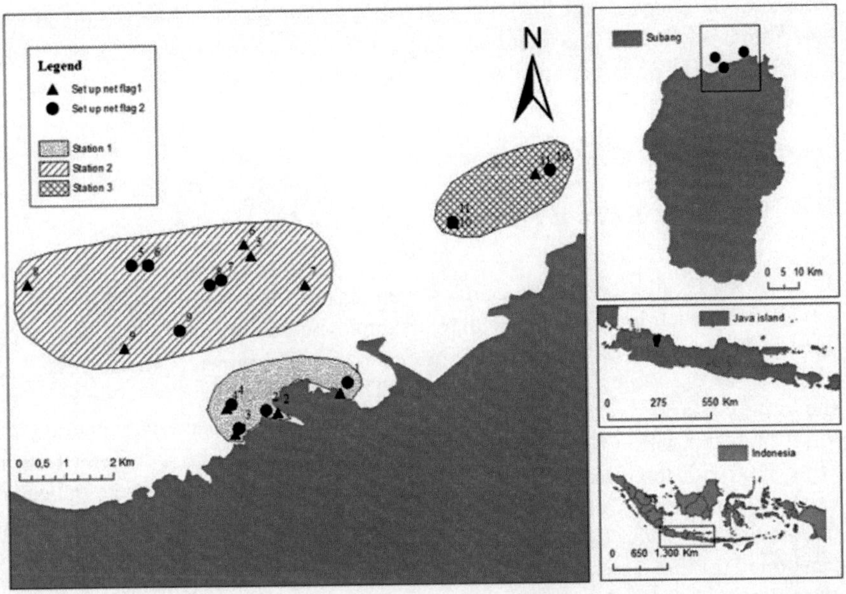


**Fig. 1** The three sampling sites ST1 = station 1, ST2 = station 2, and ST3 = station 3 in Subang Waters, West Java, Indonesia

**Table 1** The description of stations and substations

| Station | Substations (Fisherman) | Net length of bottom-set gillnets (m) | Types | Description of the surrounding habitat |
|---|---|---|---|---|
| ST1 | SB1 | 300 | Semi-enclosed waters (Segara Menyan and Pondok Bali Bay) | Natural mangrove forest, sand dune |
| | SB2 | 300 | | |
| | SB3 | 150 | | |
| | SB4 | 150 | | |
| ST2 | SB5 | 2600 | Open sea | Sandy-muddy ground |
| | SB6 | 2160 | | |
| | SB7 | 1800 | | |
| | SB8 | 3825 | | |
| | SB9 | 1200 | | |
| ST3 | SB10 | 2400 | Coastal area | Abandoned paddy fields and fish ponds |
| | SB11 | 2040 | | |

swimming crabs (*Portunus spp.*). They used nets of various sizes, mainly depending on the geology, topography, and the tidal range of the respective intertidal sampling site (Table 1). The net length varied between 150 and 3825 m (Table 1). All gillnets had a height of 95 cm and a mesh size of 10 cm. The nets were set up during the afternoon and checked the next morning. We joined the fishermen every day during the three months study period, and all entangled horseshoe crabs were collected and counted. The species were distinguished based on Sekiguchi and Nakamura (1979) and Yamasaki (1988). The total body length (TL), carapace length (CL), and telson length (TEL) of each individual were measured with a Vernier caliper to the nearest millimeter (Fig. 2). The body weight was measured to the nearest gram.

## 2.2 *Physicochemical Parameters and Sediment Samples*

In each study site, sediment samples were taken, and water parameters were assessed. At each location, the sediment samples were collected in triplicates. As seen in Kwan et al. (2015), several physicochemical parameters were selected to characterize the habitat of horseshoe crabs. The water was analyzed by measuring the salinity (psu) using a refractometer, the dissolved oxygen and temperature (°C) were analyzed with a precision of 0.01 mg (mg $L^{-1}$) using the DO meter Lutron DO-5510, with three replicates in each station. Sediment samples were collected using a Van Veen grab (opening size 25 × 45 cm, 21 cm depth) and analyzed to determine grain size distribution and organic content. All laboratory analyses were performed at the Department of Aquatic Resources Management, Faculty of Fisheries and Marine Science, Bogor Agricultural University (Indonesia). The fractions were analyzed by dry sieving and clay by hydrometer method. The particle size distribution was then determined following the Wentworth scale (Gray and

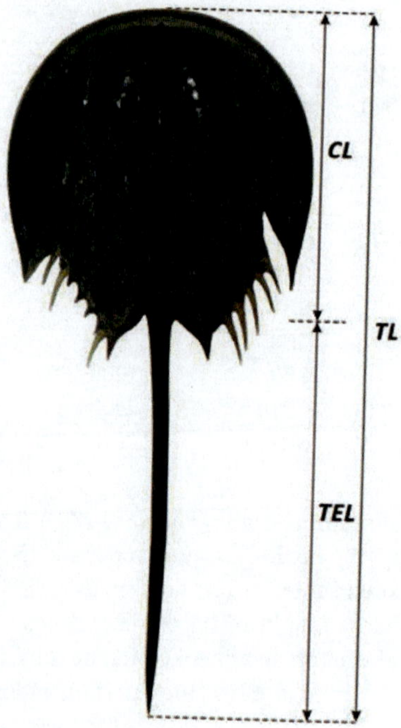


**Fig. 2** The assessed morphological parameters of horseshoe crabs: CL carapace length, TEL telson length, and TL total length. The total length equals CL + TEL

Elliott 2009). The total organic carbon content (TOC) of the sediment was determined using the Walkley–Black chromic acid wet oxidation method (Walkley and Black 1934).

## 2.3 Data Analysis

Data regarding the number of horseshoe crabs and the physicochemical parameters of the sediments and water measured at each study site were compared descriptively. Also, the relationship between the average population size and physicochemical habitat parameters of each station was analyzed using Principal Component Analysis (PCA) and BIO-ENV analysis in the Primer v.6 software.

The growth curve was analyzed according to the von Bertalanffy equation using the ELEFAN I package in the Fish Stock Assessment Tool FiSAT II (Gayanilo et al. 2005) to estimate the mortality and exploitation status of horseshoe crab species.

The growth curve was estimated using the von Bertalanffy growth model following Sparre and Venema (1998):

$$L_t = L_\infty\left[1 - e^{-K(t-t_0)}\right]$$

The growth parameters K and $L_\infty$ were estimated using the ELEFAN II package in the FiSAT II program following Ford-Walford derived from the von Bertalanffy model:

$$K = -\ln b$$

$$L_\infty = a/(1-b)$$

The $T_0$ value was estimated using the equation of Pauly (1983) as follows:

$$\log(-t_0) = -0.3922 - 0.2752(\log L_\infty) - 1.0380(\log K)$$

Where:

$K$ = the growth rate coefficient (year$^{-1}$)
$L_\infty$ = asymptotic length (mm)
$t_0$ = the age when the length equals zero (year)
$L_t$ = the horseshoe crabs length at the age t (mm)
$b$ = angle/slope, which is a form of value

Length–weight relationships (LWRs) were estimated using equation $W = aL^b$ in which $W$ is the weight (g) of the horseshoe crab and L is the total length (mm). While "a" and "b" values are from linear forms of equations above (Pauly 1983):

$$\log(W) = \log(a) + b\log(L)$$

The parameter “a” is representing the intercept, and “b” is the relationship slope. The body lengths of all individuals belonging to the same species were grouped into specified size intervals, and size-frequency histograms were plotted for each sampling month to establish LWRs that can affect “b” (Zargar et al. 2012). The size interval was 30 mm. When “b” is equal to 3, it represents an isometric growth, while allometric growth may be positive if it is more than 3, and negative if less than 3 (Ricker and Carter 1958).

The total mortality (*Z*) was estimated using the linearized catch curve procedure of ELFAN II based on the length-frequency data (Pauly 1983), where:

$$\ln \frac{C(L_1,L_2)}{\Delta t(L_1,L_2)} = h - Zt\left(\frac{L_1 + L_2}{2}\right)$$

The above equation is estimated through simple linear regression equations $y = b_0 + b_{1x}$ with $y = \ln \frac{C(L_1,L_2)}{\Delta t(L_1,L_2)} = h - Zt\left(\frac{L_1 + L_2}{2}\right)$ as ordinate, $x = t\left(\frac{L_1 + L_2}{2}\right)$ as absis, and $Z = -b$ (further derived formula can be seen in Sparre and Venema 1998).

The natural mortality (*M*) was estimated using an empirical formula following Pauly (1983):

$$\ln M = -0.0152 - 0.279 \ln L_\infty + 0.6543 \ln K + 0.463 \ln T$$

Where:

*T* = Average temperature of the water (°C)

The fishing mortality (F) was obtained by subtracting the natural mortality from the total mortality (Pauly 1983):

$$F = Z - M$$

The exploitation rate (*E*) was determined in Microsoft Excel by dividing the fishing mortality with the total mortality (*E* = fraction of *Z* caused by *F*) (Pauly 1983):

$$E = \frac{F}{F + M} = \frac{F}{Z}$$

# 3 Results

## 3.1 Number of Horseshoe Crabs and Physicochemical Habitat Parameters

This study identified two species of horseshoe crabs in Subang waters: *C. rotundicauda* and *T. gigas*. The highest number of horseshoe crabs was found at ST2 with 432 individuals, while the lowest tally was at ST1 with 146 individuals (Table 2). At the species level, the highest number found in substations during the survey period when harmonizing all sampling sites was recorded for *T. gigas* with 59.0 ± 37.2 individuals, while *C. rotundicauda's* number stood at 16.7 ± 10.7 individuals. The percentage of *C. rotundicauda* and *T. gigas* specimens collected at ST1 were equal at 50% each. At ST2, 74.1% of the collected specimens were dominated by *T. gigas*, and 25.9% belonged to *C. rotundicauda*. With 85.4% of the collected specimens, *T. gigas* was dominant again at ST3, while *C. rotundicauda* has 14.6% (Table 2).

The physicochemical analysis results characterizing the habitat are shown in Table 3 and Fig. 3. The average salinity of the surface water measured at the 11 substations varied between 26.8 and 31.7 psu, the average temperature was 29.9–30.7 °C, and the average content of dissolved oxygen of the seawater varied between 4.90 and 7.42 mg/l (Table 3). Silt was the dominant grain size in all study sites. The highest content of silt was found in the sediment of ST1 (87.22–94.60%). In general, the sediment type in the study sites was dominated by silty loam. The concentrations of TOC in the sediment differed between the sampling stations. The

**Table 2** The average number of individuals ± standard deviation and total number (N) of horseshoe crabs collected between August, September, and October 2016 shown per species and sampling site

| | | Number of individuals | | *N* |
|---|---|---|---|---|
| Station | Substation | *C. rotundicauda* | *T. gigas* | |
| ST1 | SB1 | 14.0 ± 11.0 | 8.0 ± 3.6 | 66 |
| | SB2 | 3.7 ± 1.5 | 3.0 ± 1.7 | 20 |
| | SB3 | 5.0 ± 5.0 | 4.7 ± 2.1 | 29 |
| | SB4 | 1.7 ± 1.2 | 8.7 ± 6.4 | 31 |
| | | **18.3 ± 16.2** | **18.0 ± 8.1** | **146** |
| ST2 | SB5 | 6.3 ± 6.7 | 13.3 ± 9.5 | 59 |
| | SB6 | 4.3 ± 4.2 | 15.7 ± 10.5 | 60 |
| | SB7 | 16.7 ± 10.7 | 37.0 ± 13.0 | 161 |
| | SB8 | 6.3 ± 0.6 | 29.7 ± 14.2 | 108 |
| | SB9 | 3.7 ± 2.5 | 11.0 ± 6.0 | 44 |
| | | **22.4 ± 15.8** | **64.0 ± 34.1** | **432** |
| ST3 | SB10 | 7.3 ± 4.9 | 58.3 ± 49.1 | 197 |
| | SB11 | 12.7 ± 8.3 | 59.0 ± 37.2 | 215 |
| | | **30.0 ± 11.3** | **176.0 ± 1.4** | **412** |

*ST* Stations, *SB* Substations, *N* total individual number

**Table 3** Physicochemical parameters of the water (mean value ± standard deviation); the depth as well as the mean fractions of sand, silt, and clay (%); and the total content of soil organic carbon (TOC) of the sediment of each sampling site

| Station | Substations | Surface water Salinity (psu) | Temperature (°C) | DO (mg $L^{-1}$) | Water depth (m) | Mean particle size (%) Sand | Silt | Clay | TOC (%) |
|---|---|---|---|---|---|---|---|---|---|
| ST1 | SB1 | 31.5 ± 1.1 | 30.7 ± 0.4 | 5.7 ± 3.1 | 1.2 ± 0.2 | 3.3 | 93.0 | 3.7 | 1.1 |
| | SB2 | 29.0 ± 0.8 | 29.8 ± 0.9 | 7.4 ± 2.1 | 0.8 ± 0.1 | 1.4 | 87.2 | 11.4 | 1.2 |
| | SB3 | 26.8 ± 3.4 | 29.9 ± 0.9 | 5.4 ± 0.3 | 1.2 ± 0.1 | 1.4 | 93.6 | 4.9 | 1.1 |
| | SB4 | 30.8 ± 0.7 | 29.9 ± 0.7 | 6.1 ± 0.5 | 3.0 ± 0.0 | 1.3 | 94.6 | 4.1 | 0.3 |
| | | **28.5 ± 2.9** | **30.1 ± 0.9** | **5.9 ± 2.0** | **1.2 ± 0.3** | **1.9** | **92.1** | **6.0** | **0.9** |
| ST2 | SB5 | 30.6 ± 0.5 | 30.1 ± 0.1 | 6.1 ± 0.1 | 10.2 ± 0.0 | 1.8 | 86.6 | 11.6 | 0.8 |
| | SB6 | 30.6 ± 0.5 | 30.6 ± 0.0 | 6.3 ± 0.2 | 10.9 ± 0.0 | 0.9 | 87.5 | 11.6 | 0.7 |
| | SB7 | 31.0 ± 0.0 | 30.7 ± 0.2 | 5.6 ± 0.1 | 11.6 ± 0.0 | 1.2 | 83.6 | 15.2 | 0.6 |
| | SB8 | 31.6 ± 0.5 | 30.6 ± 0.1 | 5.5 ± 0.1 | 10.5 ± 0.0 | 3.3 | 83.6 | 13.1 | 0.7 |
| | SB9 | 30.6 ± 0.5 | 29.8 ± 0.1 | 5.3 ± 0.1 | 8.4 ± 0.0 | 2.3 | 85.7 | 11.9 | 0.5 |
| | | **28.7 ± 5.6** | **30.2 ± 0.8** | **5.9 ± 0.4** | **7.6 ± 7.1** | **1.9** | **85.4** | **12.7** | **0.7** |
| ST3 | SB10 | 31.0 ± 0.0 | 30.1 ± 0.0 | 4.9 ± 0.2 | 7.7 ± 0.0 | 1.0 | 88.1 | 10.8 | 0.6 |
| | SB11 | 31.0 ± 0.0 | 30.0 ± 0.0 | 4.9 ± 0.2 | 7.7 ± 0.0 | 1.1 | 88.2 | 10.8 | 0.6 |
| | | **30.8 ± 0.4** | **30.0 ± 0.1** | **5.1 ± 0.2** | **8.1 ± 0.4** | **1.1** | **88.2** | **10.8** | **0.6** |

*ST* station, *SB* substation, *DO* dissolved oxygen

highest TOC content ranging between 0.3% and 1.23% was measured at ST1, while the lowest concentration ranging between 0.59% and 0.6% was found in the sediment of ST3 (Fig. 4). The depth at ST1 (0.88–3.02 m) was shallow compared with the depth at ST2 (8.42–11.60 m) and ST 3 (7.70–7.75 m). The relationship between environmental conditions and horseshoe crabs population size could be expected through the variance of 41.7% along axis-1 (eigenvalue: 3.32) and 20.2% along axis-2 (eigenvalue: 1.62), supported by BIO-ENV analysis. Depth, DO, and silty loam are highly correlated ($\rho = 0.68$) with the average number of the two horseshoe crab species (Fig. 3).

## 3.2 Length–Weight Relationship

The measurements revealed the body length of the entangled *C. rotundicauda* and *T. gigas* individuals (Table 4). The largest and heaviest specimen caught in the present study belonged to *T. gigas*. The highest frequency of *C. rotundicauda* was at the size range of 250–270 mm (68 individuals) (Fig. 4a). *T. gigas* had the highest frequency between 330 and 350 mm (147 Individuals) (Fig. 4b).

The following equations describe the allometric relationships between the weight and the body length presented in Fig. 5 for *C. rotundicauda* and *T. gigas*:

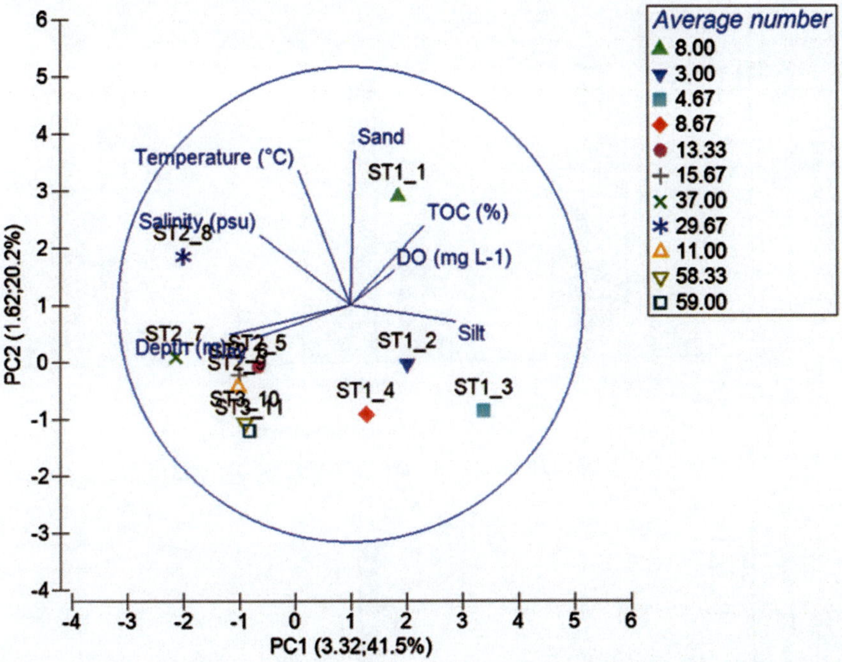


**Fig. 3** The relationship between horseshoe crabs' average number and physicochemical habitat parameters in each station. ST = Station (ST1–ST3) and substation is indicated by the number after an underscore (1–11)

| | | |
|---|---|---|
| *C. rotundicauda* | ➔ | $W = 0.00006L^{2,6355}$ with $r^2 = 0.8532$ |
| *T. gigas* | ➔ | $W = 0.00002L^{2,8897}$ with $r^2 = 0.8926$ |

The analysis of the length–weight relationship revealed that *C. rotundicauda* and *T. gigas* exhibited a negative allometric growth for the weight ($b < 3$).

## 3.3 Growth Models

Table 5 presents the results of the growth parameter analysis. According to the model, *T. gigas* had a greater asymptotic length than *C. rotundicauda* based on the body length ($L_\infty$). Incidentally, *C. rotundicauda* took a faster rate to achieve maximum length than *T. gigas* based on the average growth rate ($K$). The age at the initial condition parameter ($t_0$) of *T. gigas* was greater than *C. rotundicauda*, which may explain why *T. gigas* grew larger than *C. rotundicauda*. Therefore, in the present study, *T. gigas* grew faster and weighed more than *C. rotundicauda*. The growth curves can be seen in (Fig. 6a) and (Fig. 6b). The parameters $L_\infty$, $K$, and $t_0$ from the von Bertalanffy growth equation of the two species was estimated as follows

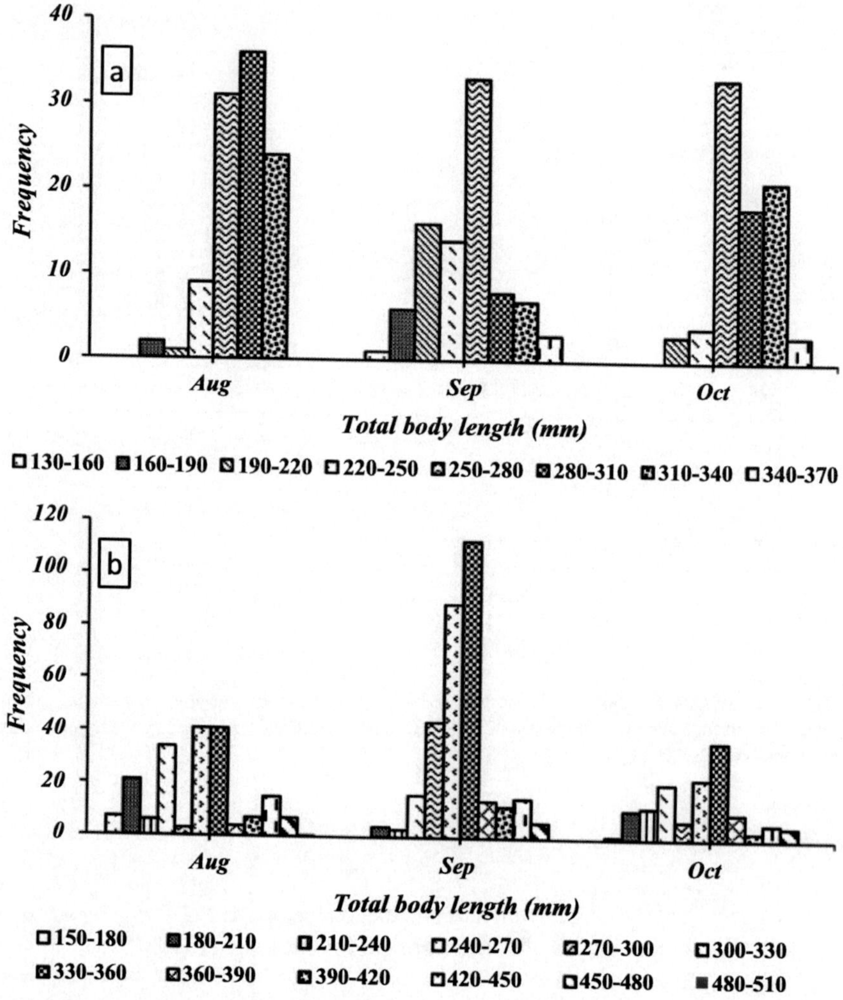


**Fig. 4** Size-frequency histograms with an interval size of 30 mm separated by sampling months in 2016 for (**a**) *C. rotundicauda,* and (**b**) *T. gigas*

**Table 4** Minimum and maximum of the body length and weight of two Asian horseshoe crab species in West Java, Indonesia

| Species | Number of individuals | Total body length (mm) | Total body length (g) | | |
|---|---|---|---|---|---|
| | | Min | Max | Min | Max |
| *C. rotundicauda* | 273 | 130 | 360 | 39 | 445 |
| *T. gigas* | 717 | 151 | 490 | 40 | 1106 |

| | | |
|---|---|---|
| *C. rotundicauda* | ➔ | $L_t = 511[1 - e^{-0.40(t-(-0.76))}]$ |

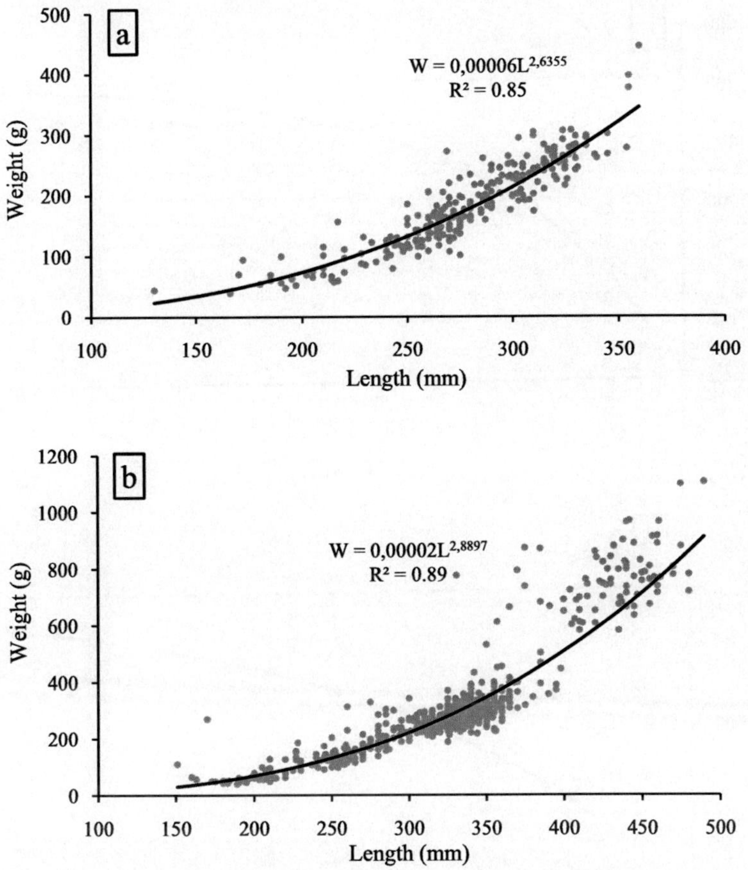


**Fig. 5** Length–weight relationship of (**a**) *C. rotundicauda,* and (**b**) *T. gigas*

**Table 5** Growth estimation of the two species *C. rotundicauda and T. gigas*

| | Growth parameters | | |
|---|---|---|---|
| Species | $L_\infty$ (mm) | $K$ (year$^{-1}$) | $t_0$ |
| *C. rotundicauda* | 511 | 0.40 | −0.76 |
| *T. gigas* | 709 | 0.23 | −1.05 |

| | | |
|---|---|---|
| *T. gigas* | ➔ | $L_t = 709[1 - e^{-0.23(t-(-1.05))}]$ |

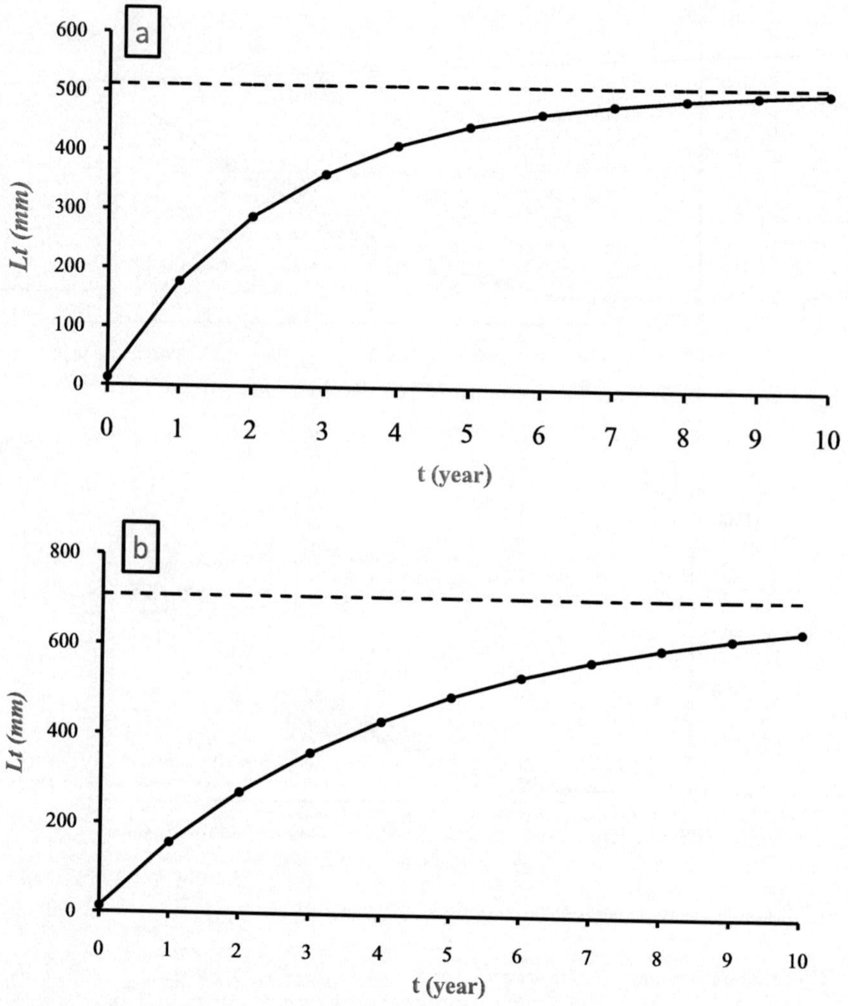


**Fig. 6** Von Bertalanffy growth curve for (**a**) *C. rotundicauda* and (**b**) *T. gigas*

## 3.4 Mortality and Exploitation Rates

Mayangan waters had an average temperature (*T*) of 30 °C. The value of catch mortality (*F*) was obtained from the difference between total mortality (*Z*) and natural mortality (*M*). Based on the calculated result presented in Table 6, the value of the total mortality coefficient (*Z*) of *T. gigas* was 1.94 per year, the catch mortality (*F*) was 1.71 per year, and the natural mortality (*M*) was 0.23 per year. It can be explained that *T. gigas* had greater catch mortality than natural mortality. In a similar pattern, *C. rotundicauda* also had greater fishing mortality than natural mortality.

**Table 6** Total mortality rate (*Z*), natural mortality rate (*M*), fishing mortality rate (*F*), and optimal exploitation rate (*E*) of *C. rotundicauda,* and *T. gigas*

| | Parameters (year$^{-1}$) | | | |
|---|---|---|---|---|
| Species | Total mortality rate (Z) | Natural mortality rate (M) | Fishing mortality rate (F) | The rate of exploitation (E) |
| *C. rotundicauda* | 2.71 | 0.37 | 2.34 | 0.86 |
| *T. gigas* | 1.94 | 0.23 | 1.71 | 0.88 |

The estimated exploitation value (*E*) of *C. rotundicauda* was 0.86 per year and *T. gigas* was 0.88 per year, meaning that the population of both species can be estimated as overexploited.

# 4 Discussion

## 4.1 *The Number of Horseshoe Crabs and Physicochemical Habitat Parameters*

There are no data on the adult horseshoe crab species in Subang waters before now. Therefore, to investigate the population size and exploitation status of horseshoe crabs entangled in gillnets of fishermen in different habitats of Western Java, Indonesia, this study used an accidental sampling method that lasted for three months. Our result revealed an overlapping presence of the two species investigated in all study sites.

It is noteworthy that the transect methods used in analyzing the abundance of horseshoe crabs as seen in Shin et al. (2009), Morton and Lee (2010), and Kwan et al. (2016) cannot be applied in the present study because horseshoe crabs aggregated in small numbers on the beaches of Subang unlike what is obtained in other countries. Hence, we obtained data by counting all entangled horseshoe crabs in fishing nets from 11 fishermen in this study. In total, 990 individual horseshoe crabs were recorded during the sampling out of which 365 individuals were found in August, 405 individuals were in September, and 220 in October. As per the knowledge of the fishermen, these high numbers coincided with the spawning season for horseshoe crabs in Subang. This is consistent with Jawahir et al. (2017), which collected higher numbers of crabs during the spawning season of horseshoe crabs in Sarawak Malaysia from March to September. They also mentioned that the population of horseshoe crabs was influenced by habitat characteristics such as water quality, sediment size, and percentage of organic carbon. In this study, there were more *C. rotundicauda* recorded in ST1 than *T. gigas* with a higher composition of silt and a lower clay content. The sites were made up of mangrove areas and was consistent with Jawahir et al. (2017). Other sediment types where *C. rotundicauda* can be found include muddy sediments, shallow waters, and swampy estuaries. Also, the number of *T. gigas* was greater in ST3 with low silt content and higher clay composition. However, both species were present in all the three sites, a fascinating discovery and a testament to the result of our PCA where depth, DO, and silty loam was highly correlated with the population size of the two horseshoe crab species. Jawahir et al. (2017) had observed a different preference habitat for the different species, a contradiction to our study.

The salinity of most sites in our study location ranges from 26.83 to 31.50 psu. Botton and Itow (2009) had pointed out that the four species they investigated have embryos and larvae, which are highly tolerant of a wide range of salinities. Also, studies in the Indian River Lagoon, Florida, showed that trilobite larvae survived in the highly variable and occasionally hypersaline (between 10 and 70 psu) water conditions (Ehlinger and Tankersley 2004). It is widely acknowledged that temperature exerts a great influence on marine fauna reproduction. Consequently, water temperature during our study period ranged from 29.87 to 30.73 °C and agrees with Chen et al. (2004) who stated that the most suitable water temperature for incubating horseshoe crab eggs and rearing of juveniles ranges from 28 to 31 °C. We, therefore, state that Subang's water temperature is conducive for horseshoe crab's growth. In Luk Keng and Yi O, Hong Kong, the higher densities of juvenile *C. rotundicauda* were found on mudflats (0.08–0.14 mm), which generally had low dissolved oxygen level (3.8–5.5 mg/l), while large juvenile populations of *T. tridentatus* were located on sandy flat or sandy-mudflat (median particle size: 0.10–0.92 mm) with higher dissolved oxygen (6.8–7.7 mg/l) (Kwan et al. 2016). In our study, ST1 has the highest concentrations of TOC in sediment. According to Zhou and Morton (2004), Kwan et al. (2015), Meilana et al. (2021), and Kwan et al. (2016), the high total organic content of sediments provides food sources for feeding and growth of juvenile horseshoe crabs.

## 4.2 Growth Rate, Mortality, and Exploitation

In this study, the morphological parameters of entangled horseshoe crabs were used to estimate the growth rate, mortality, and their exploitation rate. The growth rate model of von Bertalanffy has been used to study benthic invertebrates such as the red king crab (*Paralithodes camtschaticus*) in Norwegian waters (Windsland et al. 2013) and mangrove crab (*Ucides cordatus*) in Brazil (Pinheiro et al. 2005). Relative to our study on horseshoe crabs, the result shows that the growth rate coefficients of both species in this study are smaller than what has been studied by Sekiguchi et al. (1988), due to the absence of different juvenile and adult stage data.

Factors that may affect mortality rates include egg and larval development phases of the horseshoe crabs, environmental factors such as temperature and salinity, predation, starvation, disease, physiological changes (Geffen et al. 2007), and population density (Jørgensen and Holt 2013). Another factor from human activities is bycatch, which is an unused or unmanaged non-target catch from fishing activities (Davies et al. 2009), that has become a major issue in global fisheries (Soykan et al.

2008; Casale 2011). They pose a serious threat to marine megafauna populations and are, in part, responsible for their decline especially horseshoe crabs (Graham 2007; Meilana 2015; Prasetyo 2017). This is particularly true in Subang waters where the use of gillnets as fishing gear is common. Consequently, the mortality rate of horseshoe crabs from fishing activities is higher than mortalities resulting from natural factors in Subang. Also, the exploitation rate of horseshoe crabs in Subang is higher than 0.5. Gulland (1983) opines that exploitation rates exceeding an optimum rate of 0.5 indicates an overexploited resource. This means that horseshoe crab species are already overexploited in our study area. Retrospectively, Sparre and Venema (1998) had argued that more young fish are caught during fishing than old ones. This would lead to mortality from fishing activities being greater than mortalities from natural factors, a similar pattern with our study area as observed from the size of the horseshoe crabs caught by the gillnets. Even though horseshoe crabs are not targeted by the gillnet fishermen, they often end up as bycatch, most of which are not returned to the water. In certain localities in Indonesia, mature females are sometimes cooked or illegally exported (Meilana and Fang 2020). The exported crabs find their way to other countries where they serve different purposes. In the United States, horseshoe crabs are harvested in large numbers for biomedical and bait purposes (Smith et al. 2017). In Mainland China (Liao and Li 2001), Taiwan (Hsieh and Chen 2015), Japan (Itow 1993), and Hong Kong (Shin et al. 2009), habitat loss and degradation due to coastal development and high harvest pressure for biochemical application and food consumption have caused horseshoe crab population to decline considerably. In our study, interviews with local fishermen along the north coast of Java revealed a declining catch. A similar problem is experienced in Hau Hok Wann, Hong Kong, where only a few juvenile or adult horseshoe crabs have recently been observed (Kwan et al. 2016).

### 4.3 Management Directions

This study on the population size and habitat condition of horseshoe crabs is the first information bank when pursuing further research aimed at identifying population changes and management measures for horseshoe crab conservation in Subang. A unique finding from this study is the overlapping presence of the two species at all sites but with different relative numbers. This study also found that the population of horseshoe crabs in Subang waters is becoming depleted due to the use of gillnets by the fishermen in Subang waters.

Our study also agrees with Jawahir et al. (2017), which states that the number of horseshoe crabs can be influenced by the ecological factors of an environment. With regard to the estimation of the overexploited species, fishing nets contributed to the decline in horseshoe crab's population. Although these crabs are not the primary target, they are frequently entangled in the nets during fishing operations. Precautionary management measures and approaches are needed to control and restrict the number of entangled horseshoe crabs landed by fishermen to reduce and achieve optimum exploitation rate of 0.5 or below. Therefore, fishermen are encouraged to release back the entangled horseshoe crabs to their habitat and not to consume the grown female crabs. Minimizing effort and reducing bycatch should be conducted especially at ST1, which is a feeding and spawning ground for the crabs. It is pertinent that one protected area for horseshoe crabs should be set up especially in the shore or in shallow waters and protected from any fishing activity to help preserve their habitat and conserve the horseshoe crabs. Monthly observation should be repeated to know the spawning ground, nursery ground, and the population changes of the horseshoe crab species. These numbers, ecological parameters, growth, and exploitation data will help guide further research to identify any population changes of horseshoe crabs and conservation management in Subang waters, West Java, Indonesia. Additionally, effective policing and law enforcement are needed to stem the tide of illegal exportation and indiscriminate discard of the horseshoe crabs.

## 4.4 Limitations of the Study

This study was limited by the timeframe afforded for its research. We believed that annual data collection would have been better than the three months used for this research. Other parts of the limitation include the absence of both small-sized horseshoe crabs (<100 mm) in the two species and large-sized horseshoe crab (>365 mm) in *C. rotundicauda* individuals unlike the study by Sekiguchi et al. (1988). It can be seen from Table 7 that the total length data used in this study composed of only adult horseshoe crabs. Therefore, the presumed size (Table 8) result of this study is larger than what Sekiguchi et al. (1988) have obtained. Sekiguchi et al. (1988) reiterated the difficulty of confirming the growth history of horseshoe crabs in their natural habitats based on field observations from hatching to maturity. This is because it takes several years and molts to reach maturity and huge resources and time would be needed for this purpose. The only option left would be to rear the horseshoe crabs in a biological laboratory (Sekiguchi et al. 1988) of which the funding and time are not available to this study. We believe that field observations will not provide the same biological insight as that which laboratory observations will provide, but acknowledge the likelihood of bias since the sampling was not done all year round for a comprehensive time series of samples.

**Table 7** The total length data (average ± SD) to estimate the growth rate

| Species | Sekiguchi et al. (1988) (mm) | This study (mm) |
|---|---|---|
| *Limulus polyphemus* | 4.05 ± 0.48 – 437.9 ± 54.7 | |
| *Tachypleus tridentatus* | 6.30 ± 0.19 – 600.4 ± 33.4 | |
| *T. gigas* | 7.27 – 31 | 151 – 490 (322.07 ± 62.81) |
| *C. rotundicauda* | 4.65 – 20.14 | 130 – 360 (276.44 ± 39.49) |

**Table 8** The presumed maximum total length

| Species | Sekiguchi et al. (1988) (mm) | This study (mm) |
|---|---|---|
| *Limulus polyphemus* | 437.9 ± 54.7 | |
| *Tachypleus tridentatus* | 600.4 ± 33.4 | |
| *T. gigas* | 422.2 ± 29.1 | 709 |
| *C. rotundicauda* | 296.6 ± 30.9 | 511 |

Furthermore, the growth rate coefficient of this study was 0.23–0.40, meanwhile, the rate from Sekiguchi et al. (1988) was 1.03–1.83. This also made the $L_\infty$ for our study lie between 65% and 75% of the asymptotic length against Pauly (1983) who obtained 95% of $L_\infty$ on the oldest fish. Therefore, this presented a challenge in the use of von Bertalanffy growth models to estimate horseshoe crab growth in this study.

**Acknowledgments** The authors express heartfelt gratitude to everyone who contributed in one way or the other to make this work a success. We appreciate many fisheries experts in the Subang region for their co-operation. We would also like to use this opportunity to thank reviewers for their constructive comments in improving this paper.

## References

M.A. Beekey, J.H. Mattei, B.J. Pierce, Horseshoe crab eggs: A rare resource for predators in Long Island Sound. J. Exp. Mar. Biol. Ecol. **439**, 152–159 (2013)

M.L. Botton, T. Itow, The effects of water quality on horseshoe crab embryos and larvae, in *Biology and Conservation of Horseshoe Crabs*, (Springer, Boston, 2009), pp. 439–454

C.P. Chen, H.Y. Yeh, P.F. Lin, Conservation of the horseshoe crab at Kinmen, Taiwan: Strategies and practices. Biodivers. Conserv. **13**(10), 1889–1904 (2004)

A. Christianus, C.R. Saad, Horseshoe crabs in Malaysia and the world. Fish. Mail **16**, 8–9 (2007)

P. Casale, Sea turtle bycatch in the Mediterranean. J. Fish Fish. **12**, 299–316 (2011)

R.W.D. Davies, S.J. Cripps, A. Nickson, G. Porter, Defining and estimating global marine fisheries bycatch. J. Marine Policy **33**, 661–672 (2009)

G.S. Ehlinger, R.A. Tankersley, Survival and development of horseshoe crab (*Limulus polyphemus*) embryos and larvae in hypersaline conditions. Biol. Bull. **206**(2), 87–94 (2004)

I. Etikan, K. Bala, Sampling and sampling methods. Biom. Biostat. Int. J. **5**(6), 215–217 (2017)

M. Faridah, N. Ismail, A.B. Ahmad, A. Manca, M.Z.F.A. Rahman, M.F.S. Bahri, M.F.A.M. Sofa, I.H.A. Ghaffar, A.A. Alia'm, N.H. Abdullah, M.M.M. Kasturi, The population size and movement of coastal horseshoe crab, *Tachypleus gigas* (Müller) on the East Coast of Peninsular Malaysia, in *Changing Global Perspectives on Horseshoe Crab Biology, Conservation and Management*, (Springer, Cham, 2015), pp. 213–228

F.C. Gayanilo Jr., P. Sparre, D. Pauly, FAO-ICLARM Stock Assessment Tools II (FiSAT II). Revised version. User's guide. FAO Computerized Information Series (Fisheries). Revised version. Rome, FAO. 8 (2005), p.168

A.J. Geffen, H.W. Van der Veer, R.D.M. Nash, The cost of metamorphosis in flatfishes. J. Sea Res. **58**(1), 35–45 (2007)

J.S. Gray, M. Elliott, *Ecology of Marine Sediments: From Science to Management* (Oxford University Press, 2009)

L.J. Graham, Bycatch associated with a horseshoe crab (*Limulus polyphemus*) trawl survey: identifying species composition and distribution (Doctoral dissertation, Virginia Tech) (2007)

J.A. Gulland, *Fish Stock Assessment: A Manual of Basic Methods*, vol 425 (Wiley, New York, 1983)

L. Hurton, Reducing post-bleeding mortality of horseshoe crabs (*Limulus polyphemus*) used in the biomedical industry (Doctoral dissertation, Virginia Tech) (2003)

H.L. Hsieh, C.P. Chen, Current status of *Tachypleus tridentatus* in Taiwan for Red List assessment, in *Changing Global Perspectives on Horseshoe Crab Biology, Conservation and Management*, (Springer, Cham, 2015), pp. 383–396

T. Itow, Crisis in the Seto Inland Sea: Decimation of the horseshoe crab. EMECS Newsletter **3**, 10–11 (1993)

[IUCN] International Union for Conservation of Nature, Red List of Threatened Species [OL]. Available in http://www.iucnredlist.org/search (2018). Accessed 16 April 2018

A.N. Jawahir, M. Samsur, M.L. Shabdin, A.K. Adha, Distribution of two species of Asian horseshoe crabs at west coast of Sarawak's waters, East Malaysia. Egypt. J. Aquatic Res. **43**(2), 135–140 (2017)

B.A. John, B.R. Nelson, H.I. Sheikh, S.G. Cheung, Y. Wardiatno, B.P. Dash, K. Tsuchiya, Y. Iwasaki, S. Pati, A review on fisheries and conservation status of Asian horseshoe crabs. Biodivers. Conserv., 1–26 (2018)

C. Jørgensen, R.E. Holt, Natural mortality: Its ecology, how it shapes fish life histories, and why it may be increased by fishing. J. Sea Res. **75**, 8–18 (2013)

B. A. John, B. Y. Kamaruzzaman, K. C. A. Jalal, K. Zaleha, Feeding ecology and food preferences of *Carcinoscorpius rotundicauda* collected from the Pahang nesting grounds. Sains Malaysiana **41**(7), 855–861 (2012)

B.K. Kwan, P.K. Shin, S.G. Cheung, Preliminary home range study of juvenile Chinese horseshoe crabs, *Tachypleus tridentatus* (Xiphosura), using passive tracking methods, in *Changing global perspectives on horseshoe crab biology, conservation and management*, (Springer, Cham, 2015), pp. 149–166

B.K. Kwan, H.L. Hsieh, S.G. Cheung, P.K. Shin, Present population and habitat status of potentially threatened Asian horseshoe crabs *Tachypleus tridentatus* and *Carcinoscorpius rotundicauda* in Hong Kong: A proposal for marine protected areas. Biodivers. Conserv. **25**(4), 673–692 (2016)

K. Laurie, C. P. Chen, S.G. Cheung, V. Do, H. Hsieh, A. John, F. Mohamad, S. Seino, S. Nishida, P. Shin, M. Yang, 2019. *Tachypleus Tridentatus* (Errata Version Published in 2019). e.T21309A149768986. IUCN, Gland, Switzerland (2019)

Y.Y. Liao, X.M. Li, Present situation of horseshoe crab resources in the sea area of China and tactics of preservation. Resour. Sci. **23**, 53–57 (2001)

C.N.W. Lee, B. Morton, Changes in the distributions of juvenile horseshoe crabs (Arthropoda: Chelicerata)(2002–2014) related to environmental perturbations at Pak Nai and Ha Pak Nai, Deep Bay, Hong Kong SAR, China. Mar. Pollut. Bull. **108**(1–2), 134–146 (2016)

L. Meilana, Assessment of Horseshoe Crabs (Xiphosura, Limulidae) Morphology and Genetic as Basis for Conservation and Management in the Java Island Coast. (Master Thesis, Bogor Agricultural University) (2015)

L. Meilana, Q. Fang, Local knowledge-based study on the status of horseshoe crabs along the Indonesian coast. Reg. Stud. Mar. Sci. **36**, 1–9 (2020)

L. Meilana, A.A. Hakim, Q. Fang, Nursery habitat of three species of juvenile Asian horseshoe crabs in Teritip Beach, East Kalimantan, Indonesia: Characterization and implication. Glob. Ecol. Conserv. **26**, e01453 (2021)

J.K. Mishra, Horseshoe crabs, their eco-biological status along the northeast coast of India and the necessity for ecological conservation, in *Biology and Conservation of Horseshoe Crabs*, (Springer, Boston, 2009), pp. 89–96

B. Morton, C.N. Lee, Spatial and temporal distributions of juvenile horseshoe crabs (Arthropoda: Chelicerata) approaching extirpation along the northwestern shoreline of the New Territories of Hong Kong SAR, China. J. Nat. Hist. **45**(3-4), 227–251 (2010)

T.J. Novitsky, M.E. Dawson, E.J. Paus, Associates of Cape Cod Inc, Artificial bait. U.S. Patent 6,391,295 (2002)

D. Pauly, Some simple methods for tropical fish stock. FAO Fish. Tech. Pap. (243) (1983), p. 52

G.D. Prasetyo, Used Green LED (Light Emitting Diode) Light to Reduce Sea Turtle Bycatch in Gillnet Fisheries off Paloh Coast. (Master Thesis, Bogor Agricultural University) (2017)

M.A.A. Pinheiro, A.G. Fiscarelli, G.Y. Hattori, Growth of the mangrove crab *ucides cordatus* (brachyura, ocypodidae). J. Crustac. Biol. **25**(2), 293–301 (2005)

W.E. Ricker, N.M. Carter, Handbook of computations for biological statistics of fish populations, No. 119. The Fisheries Research Board of Canada. Queen's printer and controller of stationary, Ottawa (1958)

E. Rubiyanto, Study Population of horseshoe crabs (Xiphosura) in Peninsular Kuala Tungkal, The District of Tanjung Jabung Barat, Jambi. (Master Thesis, Indonesia University) (2012)

K. Sekiguchi, K. Nakamura, Ecology of the extant horseshoe crabs, in *Bio-Medical Applications of the Horseshoe Crab (Limulidae)*, ed. by E. Cohen, (Alan R. Liss, New York, 1979), pp. 37–45

K. Sekiguchi, H. Seshimo, H. Sugita, Post-embryonic development of the horseshoe crab. Biol. Bull. **174**(3), 337–345 (1988)

P.K. Shin, H. Li, S.G. Cheung, Horseshoe crabs in Hong Kong: Current population status and human exploitation, in *Biology and Conservation of Horseshoe Crabs*, (Springer, Boston, 2009), pp. 347–360

D.R. Smith, H.J. Brockmann, M.A. Beekey, T.L. King, M.J. Millard, J. Zaldivar-Rae, Conservation status of the American horseshoe crab (*Limulus polyphemus*): A regional assessment. Rev. Fish Biol. Fish. **27**, 135–175 (2017)

C.U. Soykan, J.E. Moore, R. Zydelis, L.B. Crowder, C. Safina, R.L. Lewison, Why study bycatch? An Introduction to the Theme Section on fisheries bycatch. J. Endangered Species Res. **5**, 91–102 (2008)

P. Sparre, S.C. Venema, Introduction to tropical fish stock assessment. Part 1: Manual. FAO Fish. Tech. Pap. **306**(1), 1–407 (1998)

A.J. Walkley, I.A. Black, Estimation of soil organic carbon by the chromic acid titration method. Soil Sci. **37**(1), 29–38 (1934)

K. Windsland, C. Hvingel, E.M. Nilssen, J.H. Sundet, Evaluation of von bertalanffy growth curves for the introduced red king crab (*Paralithodes camtschaticus*) in Norwegian waters. Fish. Res. **145**, 15–21 (2013)

T. Yamasaki, Taxonomy of horseshoe crabs, in *Biology of Horseshoe Crabs*, ed. by K. Sekiguchi, (Science House, Tokyo, 1988), pp. 419–421

U.R. Zargar, A.R. Yousuf, B. Mushtaq, D. Jan, Length–weight relationship of the Crucian carp, Carassius carassius in relation to water quality, sex and season in some lentic water bodies of Kashmir Himalayas. Turk. J. Fish. Aquat. Sci. **12**, 683–689 (2012)

H. Zhou, B. Morton, The diets of juvenile horseshoe crabs, *Tachypleus tridentatus* and *Carcinoscorpius rotundicauda* (Xiphosura), from nursery beaches proposed for conservation in Hong Kong. J. Nat. Hist. **38**(15), 1915–1925 (2004)

# Captive Breeding Success and Pitfalls for Horseshoe Crab, *Limulus polyphemus*, at CERCOM, Molloy College

**K. Maurelli**

CERCOM, Molloy College, is permitted annually by New York State DEC to collect 60 adult horseshoe crabs per site either by hand or otter trawl for the study of population monitoring, migratory behavior, and or genetic analysis from removed tissue. Horseshoe crab eggs are also permitted for collection using sediment cores. In general, all crabs are collected from waters surrounding Long Island, from Brooklyn to Montauk, and follow our 115 HSC inventory beach locations for collection opportunities (Fig. 1).

During the spring and summer months when HSCs are readily available on local beaches for CERCOM collection, the bay water temperature ranges from 64.4 °F to 83.5 °F. In support of captive adults, CERCOM lab tanks are heated to match in site conditions, and the tank water temperatures are consistently maintained at 76 °F. When new crabs are collected from the warm environment, they are received in either of the two open system tanks (flow-through well water systems) or a separate closed system tanks for acclimation and quarantine. Over the course of approximately 2 weeks, crabs that successfully acclimate will be disbursed throughout other lab tanks as breeding pairs for potential breeding and spawning. Despite this acclimation technique, CERCOM still experiences some unexplainable HSC mortality. Mortality generally follows long periods of suppressed appetite and low mobility, but no consistent patterns are shown to support "random" mortality in the aquaculture maintained animals.

K. Maurelli (✉)
Cercom, Molloy College, Rockville Centre, NY, USA
e-mail: kmaurelli@molloy.edu



J. T. Tanacredi et al. (eds.), *International Horseshoe Crab Conservation and Research Efforts: 2007-2020*, https://doi.org/10.1007/978-3-030-82315-3_31

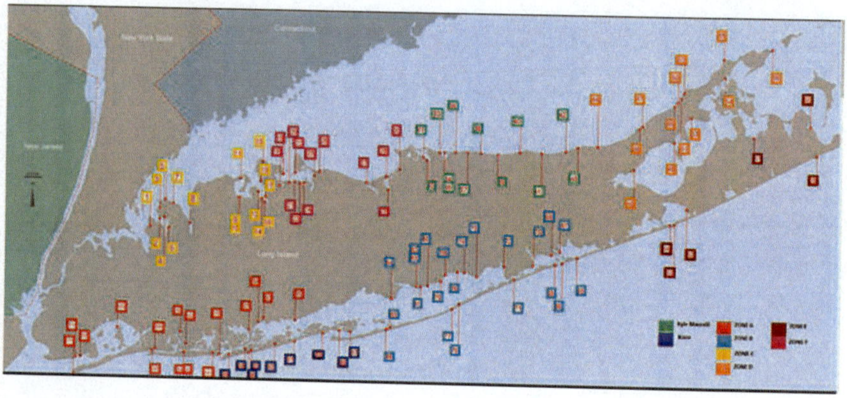

**Fig. 1** CERCOM annual HSC habitat/inventory sites (115 locations)

## 1 Salt Water Well Conditions and Infrastructure

CERCOM, Molloy College maintains two natural saltwater wells (one in each lab) to supply its HSC captive breeding aquaculture needs. It is tapped from an aquifer that runs approximately 80–100 ft. deep to the south with both oceanic and freshwater infiltration. The water is readily available and disbursed as desired into aquaculture tanks. Long-term maintenance of juvenile and adult horseshoe crabs in aquaculture settings have shown success with an optimal temperature between 59 °F and 69.8 °F and salinity around 27 ppth (Smith and Berkson 2005). Salinity is constant at 26 ppth, and the temperature is consistently 55.4 °F out of the well. The seawater pumped from the Great South Bay into the aquaculture system is the same water as the natural spawning effort for HSC in this area. By using the same water, it may limit any negative biological effects from a nonnatural source of water (Landau et al. 2015, p. 515). There have been no observed examples of abnormal HSC larvae neither from eggs spawned and developed in CERCOM well water nor from eggs spawned in the natural environment and developed in CERCOM well water. Hatch rates for spawned eggs from both CERCOM heated well water and the natural environment have been high, from 70% to 90%, respectively.

## 2 Laboratory's Bay Water Conditions and Infrastructure

There have been marine systems maintained at CERCOM, which are supplied by direct feeds of estuarine water from the Great South Bay in West Sayville, Long Island. Water temperature in these systems is influenced by natural fluctuations in bay water temperature, but, especially during winter months, they are considerably warmer due to the influence of ambient room temperature in the aquaculture lab. Bay water systems are employed in the transition/laboratory acclimation period of newly acquired adult HSCs from the natural environment. New acquisitions will be isolated for two weeks in these systems, while their systems slowly adapt to laboratory (well water) conditions. Bay water systems have also been used in housing lab-reared juveniles that have successfully made it in development to the fourth and fifth instar, 2–3 years old in the lab as seen in Fig. 2.

## 3 Closed Aquaculture Systems

CERCOM has 15 large tanks that range from 350 to 500 gallons that can house up to approx. 70 adult HSCs at one given time; tanks are set up with one to two females in each, with a minimum of two males per female to encourage healthy breeding ratios. Unlike females who are determinate spawners, males frequently return to spawning areas in nature causing the operational sex ratio to be male-biased (Brockmann et al. 2015, p. 322). All tanks are set up to be environmentally controlled; however, some tanks are kept at isothermal temperatures to prevent stress from constantly changing environmental temperatures (two open systems), and most tanks are kept at warmer temperatures to increase chances for year-round spawning. The cooler temperature, open systems have had no spawning success since 2017. The lab's closed system tanks are kept at a constant warmer temperature to encourage breeding season water parameters to cue spawning. These closed systems also allow for all water parameter adjustments such as pH, salinity, etc. Well water drawn from the 100 ft. saltwater well at CERCOM is a constant 55.4 °F with no seasonal fluctuation and is then heated within the closed systems. Dissolved oxygen (DO) in well waters are at or near 2.5 mg/L. This is enough to foster a good appetite and high survival. Adding aeration to the tanks improves DO up to 3.5–5.0 mg/L in the flow-through systems but leads to increased manganese

**Fig. 2** Lab-reared juvenile HSCs, approx. 3 inches wide by 6 inches long

precipitation on the tank and HSC carapace surface area. Based on a study from Tanacredi and Portilla (2015) optimal pH levels in successfully keeping HSCs are between 7.8 and 8.4. The investigation observed the effect pH changes had on *Limulus polyphemus* by measuring the carapace length of larvae and juveniles. For four weeks, sets of 100 fertilized eggs were exposed to five isolated systems with pH values ranging from 6.0 to 8.0. In systems with pH 6.0, 6.5, and 7.0, no juveniles survived for measurement. These results demonstrate that low pH may obstruct the progression from larva to second instar juvenile stage. In systems with pH 7.5 and 8.0, the survivorship may reflect survivorship to the juvenile stage as observed in situ pH ranges (Tanacredi and Portilla 2015). On average adult horseshoe crabs collected from local waters thrive in a captive setting for approx. 2–3 years before unexplained mortality. Water changes to these systems are preformed minimally on a weekly basis removing approx. ¼ of the tanks water and replacing this volume directly using saltwater well water. During water changes, water is always removed from the sump to ensure that the animals are not disturbed.

## 4 Captive Diet and Feeding Routines for Adult HSCs

All adults are given a diet of Atlantic surf clams *(Spisula solidissima)*, Atlantic squid *(Loligo pealei)*, silver sides *(Menidia)*, herring (*Clupea bentincki)*, and mackerel *(S. colias)*. Food is taken out 24 hours before feeding and defrosted. The average consumption is 27 g/kg HSC biomass per week. Feeding is performed 3× per week (Monday, Wednesday, and Friday), and when HSCs do not consume the food they are provided by the next feeding, the excess food is removed and the amount is reduced. When food is completely finished, the amount at next feeding is increased by approximately 1 g/kg HSC biomass. To ensure appropriate individual consumption, crabs can also be hand-fed using tongs, as seen in Fig. 3.

## 5 Captive Spawning

Adjustments to water temperature, sediment conditions, and photoperiod have been made to HSC tank environments to initiate spawning at CERCOM. Closed system tank temperatures have been raised from normal well water temperature (55.4 °F) to average mid-May to September temperatures of Great South Bay (+ - 76 °F) to allow for year-round spawning. At the same time, the photoperiod has been increased from 12 hr/12 hr (light/dark) to 16 hr/8 hr (to encourage optimal parameters for year-round spawning) concurrent with temperature adjustment (using LED marine lighting). To date, these efforts have increased spawning success. However, the same individual females subject to these environmental changes have spawned at various points throughout the year following no confirmed pattern. Since 2017, there have been over a dozen spawns, averaging approximately 10,000 to 20,000

**Fig. 3** 1800+ fertilized eggs from a single spawn, 2018

juvenile HSCs continuously being reared in the CERCOM labs each year. When a spawn occurs in the lab, the eggs are kept in the parent tank for 24–48 hours to maximize fertilization probability if there is a pair of amplexus. Eggs are then removed and placed into a ten-gallon tank with sponge filtration or placed into a down-weller system where eggs are held in warm, highly aerated, shallow water in a recirculating system. Water is pumped into each down-weller bucket at 2–4 liters per minute. After about 2–3 weeks, fertilized eggs will begin to hatch and be moved to ten-gallon tanks with sediment, lights, and sponge filtration. Occasionally, eggs have been removed prior to hatching in order to observe the larvae development under a stereo microscope where pictures and measurement data can be taken and recorded. As juvenile HSCs continue to grow and develop, they are gradually moved to larger closed systems where it is easier to observe, track, and record their molting patterns and behavior (Fig. 4).

The majority of captive spawns have been in August and September from females newly acquired to the lab earlier that spawning season. These have yielded healthy larva and good development up to the fifth instar. The later spawns have been from females well established to the captive environmental conditions of the lab, with a slower and more sporadic spawning rate. All spawns in the lab have yielded high hatch rates (< 90%) and high survival rates up to 1–2 years in lab or fifth instar (Fig. 5).

**Fig. 4** Sexually mature HSC breeding pair in amplexus, 2019

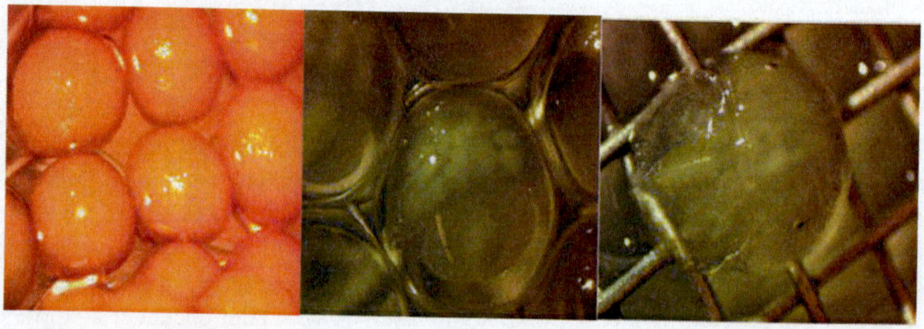

**Fig. 5** Fertilized HSC eggs, approx. 2.58 μm » HSC larvae development day 40, approx. 3.00 μm » HSC hatched juvenile, day 52, approx. 3.48 μm

## 6 Juvenile Rearing and Care

Small juveniles (second to fifth instar) are fed a mixture of Hikari frozen brine shrimp, Mysis shrimp, blood worms, clam on the half shell, oceanic plankton, First Bites fish pellets, and blended clams. This combination has yielded good growth in laboratory experimentation (presentation by Paul K. S. Shin, Ph. D., International HSC Conference, Sasebo, Japan, 2015). One consideration with juveniles is the design of their tanks (10–30 gallons). These juvenile tanks are set up with sediment

**Fig. 6** Biological sponge filter with lab reared juvenile HSCs approx. 0.5 cm long

to allow crabs to naturally borough, a heater, biological sponge filter (creating minimal to no current), LED marine light, and native algae. All food sinks to the bottom and decays slowing to allow crabs to find food successfully. Water quality in the juvenile system has been optimal (low NH3, NO2, and NO3) since sediments have matured. Sediments host denitrifying bacteria that can handle a moderate feeding load without the need to change water each week. Previous research was conducted to determine the effects of sediment type on HSC growth and survival in culture further explored the idea that sediment type affected growth by affecting nutrition. Results showed that sediment-derived organic matter made a substantial contribution to the HSC diets in natural sediment (Hieb et al. 2015).

Laboratory-reared juveniles are the primary source for experimental ocean acidification projects and Molloy College's Crab Club program with local high school science classes. Increasing public awareness and educating students about the ecological and biomedical importance of horseshoe crabs can be invaluable in facilitating their survival in urban estuaries (Mattei et al. 2015). Crabs hatched and reared at CERCOM have not made it larger than the size of an American quarter since 2017, or the fifth instar. Mortality rates are high after 1–2 years reared in the lab; however, lab-reared crabs have occasionally made it to 2–3 years of age (Figs. 6, 7, and 8).

**Fig. 7** Ten-gallon aquarium heater, lab-reared juvenile HSCs in petri dish, approx. 0.8 cm long

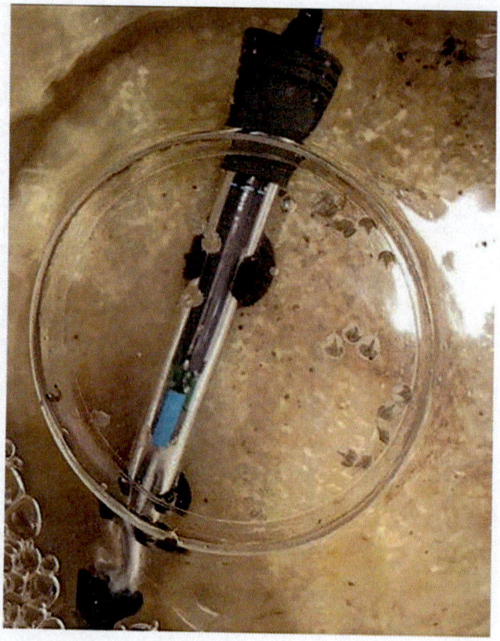

**Fig. 8** 600+ lab-reared juvenile HSCs approx. 2 inches long

## 7 Conclusion

CERCOM, Molloy College, continues to work to explore ways to improve lab rearing success. CERCOM plans to explore creating an artificial tide cycle within an aquaculture tank, developing lighting frequency that resembles the lunar cycle and its phases, and increasing breeding tank water depth. Further work on diet chemistry and microbiome is anticipated.

## References

H.J. Brockmann, S.L. Johnson, M.D. Smith, D. Sasson, Mating Tactics of the American Horseshoe Crab (Limulus polyphemus), in *Changing Global Perspectives on Horseshoe Crab Biology, Conservation and Management*, (Springer, Cham, 2015), pp. 321–351. https://doi.org/10.1007/978-3-319-19542-1_19

E.E. Hieb, J.D. Baggett, A.M. Aven, R.H. Carmichael, Effects of Sediment Type and Tank Shape on Horseshoe Crab (Limulus polyphemus) Growth and Survival in Culture, in *Changing Global Perspectives on Horseshoe Crab Biology, Conservation and Management*, (Springer, Cham, 2015), pp. 289–302. https://doi.org/10.1007/978-3-319-19542-1_16

B.J. Landau, D.R. Jones, C.B. Zarnoch, M.L. Botton, Development of aquaculture methods to enhance horseshoe crab populations: An example from Delaware Bay, U.S.A., in *Changing Global Perspectives on Horseshoe Crab Biology, Conservation and Management*, (Springer, Cham, 2015), pp. 513–536. https://doi.org/10.1007/978-3-319-19542-1_30

J.H. Mattei, M.L. Botton, M.A. Beekey, C.P. Colón, Horseshoe crab research in urban estuaries: challenges and opportunities, in *Changing Global Perspectives on Horseshoe Crab Biology, Conservation and Management*, (Springer, Cham, 2015), pp. 537–555. https://doi.org/10.1007/978-3-319-19542-1_31

S.A. Smith, J. Berkson, Laboratory culture and maintenance of the horseshoe crab (Limulus polyphemus). Lab Anim. **34**(7), 27–34 (2005). https://doi.org/10.1038/laban0705-27

J.T. Tanacredi, S. Portilla, Preliminary investigation of effects of seawater ph on early development of Limulus polyphemus, in *Changing Global Perspectives on Horseshoe Crab Biology, Conservation and Management*, (Springer, Cham, 2015), pp. 281–287. https://doi.org/10.1007/978-3-319-19542-1_15

# City-Wide Efforts in the Conservation of Horseshoe Crab (*Tachypleus tridentatus*) in Imari City, Saga, Japan

**Koyo Funai**

## 1 Conservation of Horseshoe Crabs in Imari

### *1.1 Regulations to Protect Horseshoe Crabs in Imari Bay*

There are two kinds of regulations involved in the conservation of horseshoe crabs in Imari Bay. One is related to the environmental protection, such as the "Act on Conservation and Creation of Environment in Saga Prefecture" (hereinafter, "Saga Prefecture Act") and "Act on Protection and Nurture of Environment Leading to the Future of Nagasaki Prefecture" (hereinafter, "Nagasaki Prefecture Act"). The other kind is "Act on Protection of Cultural Properties."

### *1.2 Regulations Related to Environment*

Imari Bay is a bay opening toward North West. Its south and northeast sides belong to Saga, and its west side is Nagasaki. Since these two prefectures govern certain areas of the bay, the application of their laws also is restricted, but either clearly states the protection of rare wildlife. In this paper, Saga Prefecture Act will be introduced.

Saga Prefecture Act states that "the governor can designate rare species of plants and animals in Saga Prefecture (hereinafter, rare species), which the governor decided to be protected urgently." Horseshoe crab is designated as "the species most

K. Funai (✉)
Imari City Board of Education, Imari, Saga, Japan
e-mail: funai-kouyou@city.imari.lg.jp

 
J. T. Tanacredi et al. (eds.), *International Horseshoe Crab Conservation and Research Efforts: 2007-2020*, https://doi.org/10.1007/978-3-030-82315-3_32

effected by the development work and such among rare species in Saga, with its rapidly decreasing number which needs protection."

Because of this statement, it is prohibited to capture, collect, injure, or kill live individuals (including eggs), and anyone who violates it will be sentenced to imprisonment of up to 1 year, or a fine up to 500,000JPY. Capturing horseshoe crab for the purpose of academic study or breeding requires a permission from the governor (in Nagasaki, accidental capture or injury to animals will not be applicable to the law).

Additionally, the Saga Prefecture Act states that a person who owns a property[1] where rare species exist needs pay attention to protect them when using the property. The law does not enforce such a person to protect the species, but asks to be attentive.

## 1.3 Regarding Act on Protection of Cultural Properties

According the Act on Protection of Cultural Properties, cultural properties are defined as tangible properties such as buildings, arts and crafts, and historical material and intangible properties like performing arts and craft techniques. In addition, monuments such as archeological sites, scenic spots, plants and animals, and geological minerals are also included in the cultural properties.

Among monuments, habitats inhabited, visited, or used for breeding by important species can be designated as national monuments by Minister of Education, Culture, Sports, Science and Technology.

Within Imari Bay, areas indicated in Figs. 1 and 2 are designated as national natural monument, "Imari Bay Horseshoe Crab Breeding Site." It also states, "one needs a permission from Commissioner of the Agency for Cultural Affairs when changing the current condition of the property, or modifying it, or give an influence on its preservation." This means that any act like capturing and injuring horseshoe crabs, including eggs, and changing the current state of the property such as landfill must have a permission first.

Based on the above regulation, anyone who modifies the current condition of the designated natural monument or does any harm on the conservation of it or destroys it will be sentenced to imprisonment of up to 5 years, or a fine up to 500,000JPY.

Saga Prefecture Act regulates individual protection of horseshoe crabs and other species, but as for their habitat environment, only provides cautions. Act on Protection of Cultural Properties gives a protection for the individuals and environment within the designated areas only, and horseshoe crabs living outside of the designated area are not protected at all.

Based on the past inhabitation survey, it is confirmed that adult and subadult horseshoe crabs in Imari Bay have been changing their habitat within the bay. If spawning beaches for adults and nurturing tidal flat for juveniles can definitely be

[1] The sea (public sea surface) is owned by the nation, but managed by the prefecture.

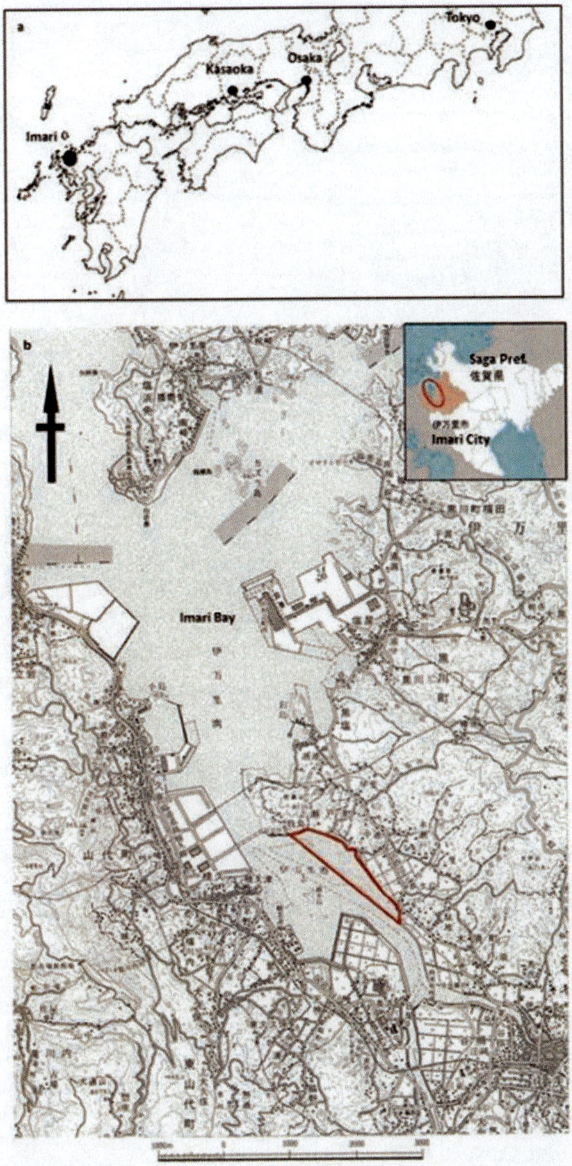


**Fig. 1** Maps showing the location of Imari Bay, Saga, Japan. a. A map of western Japan, indicating the relative location of Imari to Tokyo, Osaka, and Kasaoka. b. A detail map of Imari. The inset is Saga Prefecture with the circled area represents Imari Bay. Within Imari Bay, the area with a bold line is the national natural monument designated in 2015

protected within the designated area, then the breeding of horseshoe crabs will be secured. The adults and subadults will move outside of the designated conservation

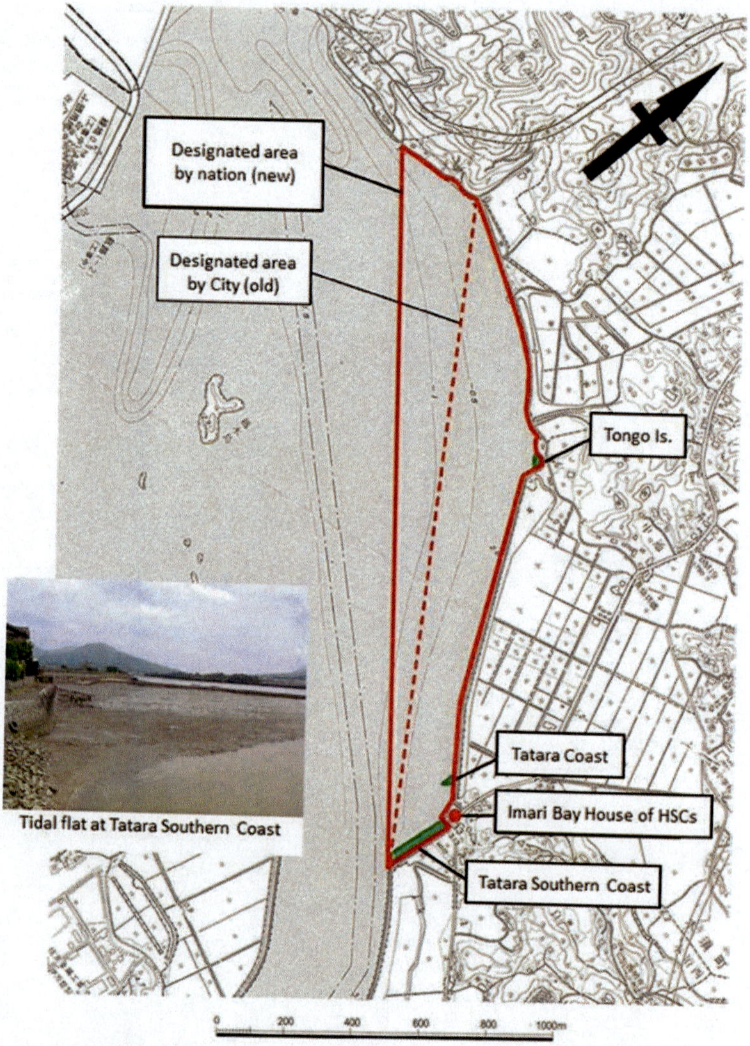


**Fig. 2** Designated area in Imari Bay for a national natural monument of Japan. From the shore to the dotted line is the area designated for protected by the City of Imari, to the solid line is the newly designated area as a national natural monument. The inset picture shows the scenery of the tidal flat at Tatara Southern Coast

area, but in that case, each individual will be protected either by Saga Prefecture Act or Nagasaki Prefecture Act.

Because different regulations are prescribed to protect either habitats or individuals of horseshoe crabs in Imari Bay, together they will conserve the wellness of this species, and this is very characteristic of the conservation method in Imari Bay.

### 1.4 Protection of Cultural Properties as Preservation and Utilization

Protection in terms of Protection of Cultural Properties under the Act means preservation or conservation and utilization. Preservation is meant for transmitting the values of cultural properties to the next generations without any modifications. For this reason, one must need to understand and evaluate the potential value of the property, and no change or modification is allowed to it.

Utilization means that the value of the property should be made available to the public. It also means that a place to learn about local culture and nature is provided and acts as a tool for community development and cultural tourism resources.

The following section explains how the City of Imari took initiatives based on the Act on Protection of Cultural Properties.

## 2 About the Preservation

### 2.1 The Necessity of Preservation

Horseshoe crabs haven't changed their morphology for over 200 million years and are called "living fossils." They are an important species in order to investigate evolutionary processes and phylogenetic relationships with fossil species. Moreover, it is urgent to protect this species as their biological distribution on a global scale is very peculiar, and therefore, further studies on them become important.

In the past, horseshoe crabs were abundant in almost all areas in Seto Inland Sea and along the northern coastline of Kyushu. However, due to the influence of seawater pollution, landfill projects, and such during the rapid economic growth period, the number of horseshoe crabs declined and their distribution became smaller as we see it today. Currently, horseshoe crab in Japan is enlisted as "Endangered Category I" in Japanese Ministry of the Environment Red List. Even from the point of view of species diversity, it is necessary to protect this endangered species from extinction.

### 2.2 Awareness among Local Citizens

Horseshoe crab conservation in Imari is characterized by a cooperation among four bodies, citizen group, school, local cooperation, and administration, that have been working together over years. Each is explained below.

## Society for the Protection of Horseshoe Crabs, Imari City (Hereinafter, City Society)

The City Society was established in November 1979. They regularly initiate cleanup activities on spawning beaches and events to observe spawning horseshoe crabs for the purpose of protecting horseshoe crabs (Fig. 3 a and b). Additionally, as an effort to increase awareness, they set up a monument of horseshoe crabs in front of Imari Railway Station. Through publicizing activities like this, they encourage people to take care of the environment where horseshoe crabs inhabit.

## Saga Prefectural High School, Science and Biology Club (Hereinafter, Club)

Since 1963, the Biology Club has been surveying the biology and spawning sites of horseshoe crabs. Especially, they have been counting the number of spawning pairs every year since 1986, creating an important dataset to show the fluctuation in population number in Imari Bay. For a long time, they have also continued to breed horseshoe crab juveniles and been teaching their breeding technique to elementary and junior high schools. (The Club started as Biology Club in 1963, then changed its name to Science and Biology Club.)

**Fig. 3** **(a)** and **(b)** Cleanup activity by Imari City Society for the Conservation of Horseshoe Crabs and Imari High School students. Every year, the Conservation groups and Imari High School students clean up the beaches before horseshoe spawning season begins

**Fig. 26.3** (continued)

## Makishima Society for Breeding Horseshoe Crabs and Fireflies (Hereinafter, Makishima Society)

Makishima Society was established in March 2006 by residents of Makishima District and has been participating at beach cleanup activities and horseshoe crab spawning observation events. In July 2009, they opened "Imari Bay House of Horseshoe Crabs" and manage it and its exhibits to observe live horseshoe crabs during spawning off-seasons. In 2014, they renewed the facility.

## The City of Imari

In order to protect horseshoe crabs, the entire area around Tatara Coast was designated as Imari City natural monument in August 1986 (Figs. 2 and 4). (This area is now included into the newly designated national natural monument.)[2] In addition, the City also provides funds for protection activities to each group.

A designation of cultural properties is determined by the city, prefecture, and nation. Any object can be designated as City Cultural Properties if it is found to

[2] Horseshoe crab spawning area was designated as city's natural monument between 1986 and 2015 (296,250m$^2$). Since 2015, the area of 581,766m$^2$ was designated as a national natural monument, which incorporates the existing designated area. The entire designated area is now governed by the nation.

**Fig. 4** A view of the tidal flat at the Tatara southern coast in front of the designated natural monument within Imari Bay

represent City's history, culture, custom, etc. Similarly, if they are important for the prefecture, they can be designated as prefectural cultural properties and for the nation as national cultural properties. These designations by either city, prefecture, or nation do not correspond to the importance of the objects, but to the geographic scale.

## Saga Prefecture

Saga Prefecture practices the protection of horseshoe crab spawning area by putting sand, beach cleanup, and provides funds for conservation groups dedicated for the breeding of this species.

## 2.3 Survey for Understanding the Actual Habitat

### Spawning Pair Number Survey

Since 1986, surveys on the count of spawning horseshoe crab pairs have been conducted by the Club every year at Tatara Coast, Tatara Southern Coast, and on Tongo Island. Between 2004 and 2006, together with the Club, graduates of the Club and volunteers did a simultaneous survey of the spawning areas in Imari Bay. For this, seemingly adequate beaches for spawning activities were selected within the Bay, and the pairs visiting the areas were counted on the day most appropriate for them to spawn. The result confirmed that the main spawning areas were Tatara Coast, Tatara Southern Coast, and on Tongo Island, where the Club has been investigating.

### Habitat Distribution Survey

Fishing industry thrives at the towns of Hatatsu and Fukushima. During the spawning season, some horseshoe crabs can get caught in the fishing net. The fishermen supply information on when and where such individuals were trapped and the fishing method they used (Fig. 5). Then Imari City Board of Education collects trapped horseshoe crabs. Safeguarded individuals were then given a marking, then were released to the sea. From these marked horseshoe crabs that come to spawn and are recaptured, their distribution and habitat extension will be clear. The fishermen who

**Fig. 5** Horseshoe crabs entangled by fishing net were rescued. After receiving a marking on the prosoma, they will be released into the bay

collaborate to protect horseshoe crabs are rewarded[3]. The City Society is involved in such a rewarding process.

The accumulation of material and data from both long- and short-term investigations is valuable to determine the past trend and predict future change in environment and ecology of horseshoe crabs. Years of survey data, which enabled Imari Bay to become designated national natural monument, will serve as an evaluation material for future conservation activities.

## 2.4 For Breeding

Since 1986, the Club has been breeding horseshoe crabs and established its method on the subject. Since 2004, the breeding program has been conducted at elementary and junior high schools in Imari City in order to increase the number of horseshoe crabs and to raise the awareness of conservation. The Club teaches its know-how.

The horseshoe crab eggs are collected from beaches between late August and early September, then hatched juveniles are released to the sea in the following late July. At the time of release, they are grown to be third to fourth instars.

Currently, any change in the number of juveniles hasn't been confirmed, but the awareness for the conservation of horseshoe crabs has increased due to the fact that young students have experienced to breed and raise juveniles on their own hands. For the purpose of encouraging conservation activities, teaching how to breed horseshoe crabs from eggs is proven effective. These activities are supported by funds from Saga Prefecture.

## 2.5 Protection of Environment

Every year before the horseshoe crab spawning season, beach cleanup is conducted on the spawning areas. In addition to City Society and Makishima Society, volunteers from Imari High School and Imari Marine Boy Scouts have been participating in the cleanup for a long time.

---

[3] The reward is not for the price of horseshoe crabs, but for the information such as the place where they were caught. As explained in "b. Laws related to environment," Imari Bay is governed by both Acts of Saga Prefecture and Nagasaki Prefecture, and the treatment of rare species is stated within. Anything protected by law is not the subject of money exchange, which is inappropriate.

**Fig. 6** Lecture by science and biology students at the event to observe horseshoe crab spawning. Such an event is held every year for citizens to encourage them to be aware of and understand the horseshoe crab conservation efforts

## 2.6 Educational Activities

One of the characteristics of educational activities in Imari is to publicly share "Calendar for Horseshoe Crab Spawning Observation," which is used to observe spawning pairs on Tatara Coast. The calendar lists spring tide season and high tide time and the degree of expectation for predicted spawning based on the past survey data.

In addition, "Event for Observing Horseshoe Crab Spawning" is held every year for the general public during the spawning season. At this event, students from the Club give lectures on site about horseshoe crab biology and their conservation activities (Fig. 6).

Through the actual observations on site, out in the nature, the participants experience the wonder of nature, and deepen their understanding of conservation.

## 2.7 Coexistence of Development and Conservation

The designated area as a national natural monument is located along the eastern coast of Imari Bay. Its coastline has become mainly reclaimed and landfilled in recent years and is now a straight line with almost no natural geographic outline.

In 1953, Saga Prefecture became the port manager of Imari Bay, and in 1955, a port planning was drawn up and a port development was started. According to the local people, many horseshoe crabs were observed before the development. After the development, the beaches were lost to landfill and revetment works, and horseshoe crabs disappeared in some areas.

Upon constructing an industrial park in Nanatsu-jima District as a part of Imari port development plan, Saga Prefecture and Imari Bay Fishery Cooperative negotiated to give the latter a fishery compensation, which was agreed in 1972. Hatatsu Fishery Cooperative also negotiated with the Prefecture and got an agreement in 1972 for the disappeared fishing area and associated fishing rights. The fishing rights within the designated area were given up, and any fishing within it has been abolished. Because of this, even though the number of horseshoe crabs has been in decline, they have become protected in the designated area, and the conservation effort is now supported by these fishermen.

For the conservation of horseshoe crabs in other areas, it is recommended that the fishermen and development business to have a good partnership.

# 3 Utilization

## *3.1 Maintenance*

Toward any changes that might happen in the future, a preservation maintenance planning is to be established, and further, it is to draw a maintenance plan, which indicates a basic concept and plan to protect and utilize the spawning area.

## *3.2 A Resource for Local Town Planning*

### Town Planning and Revitalization

Cultural properties will be adopted for the revitalization of the town planning and town itself. Town planning means that the local residents take initiatives to improve the living conditions of their town.

Having horseshoe crabs living in their region is the identity of the town, and for it to be retained for the future generations, it becomes necessary to build a better environment now.

In order to better enhance the horseshoe crab habitat, the first step in the community to do is to improve the community environment, which relates to a better town planning. Improvement of the local environment by local people can revitalize the community.

In Imari, the local conservation group, City Society, leads activities for a town planning, which adopts horseshoe crabs. Every spawning season in summer, a summer festival “Evening to Welcome Spawning Horseshoe Crabs” is held at the House

**Fig. 7** Imari Bay House of Horseshoe Crabs. This institution is available at any time of the year to learn about horseshoe crab biology and deepen understanding of horseshoe crab conservation through exhibits

of Horseshoe Crabs (Fig. 7). At the festival, Makishima Elementary School pupils present their studies on horseshoe crabs, and the embankment around spawning area is lighted up with bamboo lanterns to create a fantastic landscape.

## *3.3 Research Institutions and Exhibit Facility*

To reach out as many people as possible for the understanding of horseshoe crab conservation, an exhibit facility to observe and learn about horseshoe crabs at any time of the year, the City Society, with a support by the City of Imari, established the “Imari Bay House of Horseshoe Crabs.” In 2014, the House was renewed. Inside it, there is one pair of horseshoe crabs alive in a tank, explanation panels, molts of various growth stages, and other supporting material. The space is about 37m$^2$.

Facilities like this can be used as a resource for a cultural tourism, and in the future, a research facility together with exhibit facility will be crucial to be a focal point of conservation activities.

### 3.4 *Resource as Economic Activities*

The relationship between the conservation of cultural properties and any development can be seen conflicting. However today, because of the increasing awareness of environmental issues, people who are to develop land tend to pay a careful attention to environment. Even though it costs more for a construction with a consideration toward environment, it will boost a positive image for the firm while fulfilling the need of the conservation group to preserve the environment.

Moreover, as a strategy to keep a good relationship with the fishermen, branding of an "image" is considered. In general, the environment in which the so-called living fossil, horseshoe crab, inhabits gives an image of a clean, pristine ocean. Fish from this pristine water gives an impression of safety, artificial-free, and appetizing. By branding an image of "natural and delightful fish from a pristine sea where horseshoe crabs live," the existence of horseshoe crabs can be utilized as a resource to encourage economic activities.

## 4 Becoming a Designation as a National Natural Monument

For the conservation history of horseshoe crabs in Imari City, it started with a survey research by Biology Club at Imari High School in 1963. In November 1979, Society for the Protection of Horseshoe Crabs, Imari City was launched. Then in August 1986, the area around Tatara Coast was designated as natural monument of the City to conserve horseshoe crab spawning beaches. In March 2006, Makishima Society for Breeding Horseshoe Crabs and Fireflies was founded. In 2009, an exhibit facility, Imari Bay House of Horseshoe Crabs was opened, and then it was renewed in 2014. Each of these groups has been actively working for the conservation of horseshoe crabs.

In March 2013, an investigator from the Ministry of Culture visited Saga Prefecture and was given an explanation about the status of conservation activities at the actual survey site at Tatara Coast, followed by a consultation about a designation of the Coast as a national natural monument. Saga Prefecture prepared documents and submitted them to the Ministry in January 2015. On June 19 of the same year, the committee at the Ministry reported to the Minister of Education, Culture, Sports, Science and Technology, and on October 7, Tatara Coast area was officially designated (Fig. 2).

## 5 Records of the Survey Studies at Imari High School

As explained above, horseshoe crab conservation efforts in Imari have been conducted by citizen group, school, local cooperation, and administration that have been collaborating together in activities for years. Among them, the surveys by Imari High School Biology Club made a basis for horseshoe crab conservation in Imari Bay. It is noteworthy that their study has still been going today. They started their research on 21 spawning sites at the coast of Imari Bay in 1972 and 1973 and have almost annually published 37 reports from the Club Magazine at school since 1981.

## 6 Breeding

a. Method to Raise Juveniles.

**Fig. 8** Releasing of horseshoe crab juveniles. Juveniles hatched from eggs collected the previous year are reared at a few elementary and junior high schools, then released by students who raised them. Through these activities since childhood, people can develop a better understanding of the conservation

**Fig. 9** Second instars raised at Imari High School. These instars were hatched from eggs collected at Tatara Coast in 2015 and then were released to the same place the next year

Imari High School Science and Biology Club has published a guide to breed horseshoe crab juveniles and executes the practice. The manual is distributed to elementary and junior high schools that breed and raise juvenile horseshoe crabs.

## 6.1 Method to Release Juveniles

Horseshoe crab eggs are collected from late August to early September. The hatchlings are raised in a lab or a classroom and then can be released to the sea in late July the following year (Figs. 8 and 9). By the time of release, they grow third to fourth instars.

Deliverance of juveniles takes place in tide pools that appear at low tide. Juveniles are observed until they can dig and hide into the substrate by themselves.

Method of Release

1. Juveniles and seawater are put in a cup and handed to participants who came to release juveniles.
2. Protect eggs that were deposited on the sandy beach. Then to walk on the tidal flat, a row of planks is provided parallel to the shore to avoid being trapped into the soft substrate. Walking is permitted only on the planks.
3. Using the planks as footage, find a tide pool, and then release juveniles. Observe until they dig into the substrate by themselves.

## 7 Future Tasks

In order to sustain the existence of horseshoe crabs, our future tasks would be: (1) preservation of the entire habitat, (2) security of spawning beaches for adults and nurturing tidal flats for juvenile growth, and (3) protection efforts by the conservation groups, especially by the local residents.

As for (1), large-scale environmental pollutions caused by corporations during the period of rapid economic growth may not repeat due to environmental laws and positively changed attitude toward environment by corporations. However, miscellaneous wastewater from outdated sewer maintenance is more problematic, updating it is required as a duty of an administration.

For (2), there are three spawning locations within the nationally designated area (i.e., Tatara Coast, Tatara Southern Coast, and Tongo Island). In addition to these spawning sites, additional beaches for horseshoe crab spawning need to be established in the future.

If sand is put into the Coast, it can be washed away by the waves, and such an overflow of sand might cover the tidal flat for juveniles, it becomes necessary to consider the direction of currents and figure out how to keep the sand. Once these points are studied, they will be stated clearly on the preservation maintenance planning, which will be drawn up from now on.

For (3), four groups previously mentioned are the key point of horseshoe crab conservation in Imari City, and now to have their activities to be continued, teaching the next generations and fundraising method need to be considered.

Especially for fundraising, it is necessary to investigate ways to use horseshoe crabs as a resource of cultural tourism to attract funds.

Even though the area in Imari Bay was designated as a national natural monument of Japan, conservation and protection of horseshoe crabs are not ensured. Working with conservation groups within Japan and those abroad, these valuable animals must be protected and their safety should be guarded.

# Part IV
# Biology & Physiology

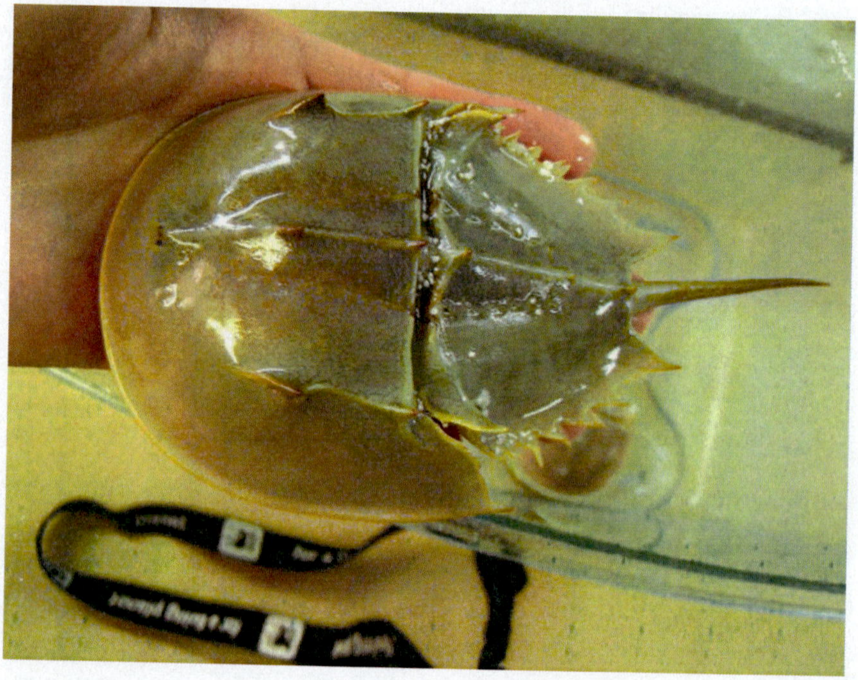

# Temperature and Salinity Preferences of Adult American Horseshoe Crabs (*Limulus polyphemus*) in the Great Bay Estuary, New Hampshire U.S.A

**Helen Cheng, Vilma Vaattovaara, Meghan Connelly, Brianna Looney, Christopher C. Chabot, and Winsor H. Watson III**

## 1 Introduction

During the spring and summer months, sexually-mature American horseshoe crabs (*Limulus polyphemus*) become active and migrate into estuaries and embayments along the United States Atlantic and Gulf of Mexico coastlines. The physical and chemical cues they use to locate suitable beaches for spawning have been previously documented. For example, physical characteristics of a beach, such as wave energy, sediment type, beach morphology, and high tide inundation, influence where horseshoe crabs spawn and lay eggs for developmental success (Rudloe and Herrnkind 1976, 1980, Botton et al. 1988, Smith et al. 2002a, b, Swan, 2005, Vaquez et al. 2015a, Cheng et al. 2016), as well as chemical features such as the amount of peat, sediment redox potential, ambient oxygen, and pore water hydrogen sulfide (Botton et al. 1988, Saunders et al. 2010, Vasquez et al. 2015a, b, 2017). Horseshoe crabs have likely evolved to prefer beaches with the right combination of these characteristics in order to optimize egg development and larval survival. Furthermore, horseshoe crabs use chemical cues emanating from conspecifics and nests with eggs to locate mates and nests (Patten 1894, Cohen and Brockmann 1983, Hassler and Brockman 2001, Ferrari and Targett 2003, Schwab and Brockmann 2007, Saunders et al. 2010).

H. Cheng (✉)
Department of Marine and Environmental Sciences, Northeastern University, Boston, MA, USA

Department of Biological Sciences, University of New Hampshire, Durham, NH, USA

V. Vaattovaara · M. Connelly · B. Looney · W. H. Watson III
Department of Biological Sciences, University of New Hampshire, Durham, NH, USA

C. C. Chabot
Department of Biological Sciences, Plymouth State University, Plymouth, NH, USA

 
J. T. Tanacredi et al. (eds.), *International Horseshoe Crab Conservation and Research Efforts: 2007-2020*, https://doi.org/10.1007/978-3-030-82315-3_33

While the cues that most influence the choice of a particular beach for spawning have been extensively investigated, little is known about the environmental factors that influence the large-scale movements of horseshoe crabs to the vicinity of these beaches when the mating season commences. Moore and Perrin (2007), James-Pirri (2010), and Schaller et al. (2010) used acoustic telemetry techniques to track the seasonal movements of horseshoe crabs throughout Taunton Bay, Maine, USA; Pleasant Bay, Massachusetts, USA; and the Great Bay Estuary, New Hampshire, USA, respectively. In these three New England bays and estuaries, horseshoe crabs tended to overwinter in deeper waters and then, just prior to the spawning season, move into areas further from the coast (for example, up into an estuary) to the vicinity of known spawning beaches.

The Great Bay Estuary is a large semi-enclosed tidal estuary composed of Little Bay, the area closest to the coast, and Great Bay proper, the upper reaches of the estuary. Great Bay and Little Bay have very large shallow regions and a deeper, central channel that runs the length of both of them. In the Great Bay estuary, after being sedentary during the winter, horseshoe crabs typically moved ~3 km up into the head of the Great Bay Estuary in the spring, and the initiation of this migration appeared to be triggered by an increase in water temperature (Schaller et al., 2010; Watson et al. 2016). This was most obvious in 2012 when horseshoe crabs initiated their migration much earlier than usual because of an unusually early, warm spring (Watson et al. 2016). A parallel study showed that, as a result, horseshoe crabs were observed on spawning beaches 2–3 weeks earlier than reported in other years (Cheng et al. 2016). These observations, in part, led to the hypothesis that horseshoe crabs can sense thermal gradients and they use them to help guide their migrations up into the estuary where optimal beaches for spawning are located. Previously, Reynolds and Casterlin (1979a) demonstrated in that horseshoe crabs will avoid extreme cold or hot water, choosing to occupy water at temperatures from 15 to 40 °C, when given a potential range of 0–50 °C. However, in the Great Bay Estuary, during the time period when horseshoe crabs typically migrate from where they overwinter to the beaches where they spawn, the water temperature is typically between 12 and 20 °C, with increasing mean surface water temperature from the open coast to the inner estuary (Short 1992; Cheng et al. 2016). Therefore, in this study, we aimed to determine if they could detect, and exhibit preferences for smaller changes in temperature than previously demonstrated by Reynolds and Casterlin (1979a).

Along with temperature differences, salinity gradients are also common in estuaries. For example, the salinity in the Great Bay Estuary has been observed to be as low as 4 psu at the head of the estuary during strong storms or large spring runoff events from three major rivers. In general, every spring in this New England estuary, the salinity is much lower in the upper reaches of the estuary than towards the ocean. As a result, the salinity near the outlet of the estuary compared to areas of the inner estuary where horseshoe crabs are most abundant can differ by about –15 psu in April to ~10 psu in June (Short 1992; Howell et al. 1999; Watson et al. 1999; Fulton et al. 2013). Therefore, it is possible that a salinity gradient could be guiding the migration of horseshoe crabs further up into the estuary in the spring.

Based on the aforementioned temperature and salinity gradients in the Great Bay Estuary, and the very directed large-scale movements of horseshoe crabs further into the estuary in the spring just prior to mating season, the first goal of this study was to test the hypothesis that horseshoe crabs detect temperature and salinity gradients and prefer warmer water and/or lower salinity water. The second goal of the study was to determine how sensitive horseshoe crabs are to changes in temperature (we did not test salinity because the results of the Y-maze turned out to be negative).

# 2 Materials and Methods

## 2.1 *Horseshoe Crabs*

All horseshoe crabs used in the behavioral and cardiac assay experiments were collected from the Great Bay Estuary, New Hampshire U.S.A. and were held in an outdoor flow-through tank at the University of New Hampshire (UNH)'s Jackson Estuarine Laboratory. Horseshoe crabs used for the Y-Maze experiments were collected from May to August of 2012 and 2013, during the spawning and post-spawning season, and were used within a week of when they were collected. Some horseshoe crabs were used for both salinity preference and temperature preference experiments and when this was done, the second experiment took place 3–4 weeks after the first experiment.

The cardiac assay experiments were conducted in the spring of 2016 and 2017 using horseshoe crabs that had been collected the previous fall and were either held in a flow-through seawater system at the UNH Jackson Estuarine Laboratory, or in recirculating chilled aquaria in a laboratory. These horseshoe crabs were fed mussels, frozen fish, or frozen shrimp once/per week, and they were acclimated to 12 °C water for at least three days prior to each experiment.

## 2.2 *Y-Maze Experimental Design*

A 2 m long x 1 m wide, two-channel "Y-maze," that included a "neutral" zone, was used to quantify the behavior of horseshoe crabs when exposed to either temperature or salinity gradients (Fig. 1a). At the end of each channel of the Y-maze was a modified 19 liter bucket (header tank) that was used to mix fresh and salt water, or create a reservoir of warm or cool water, prior to having it flow into the Y-maze. The source of water for these experiments was from the Jackson Estuarine Laboratory flow-through estuarine water system, and thus mimicked ambient water conditions in the Great Bay Estuary. Flow out of the header tanks and into the channels was controlled by a valve that could be adjusted so that each channel received a similar flow. At the other end of the tank was a wall, or a divider, that was shorter than the

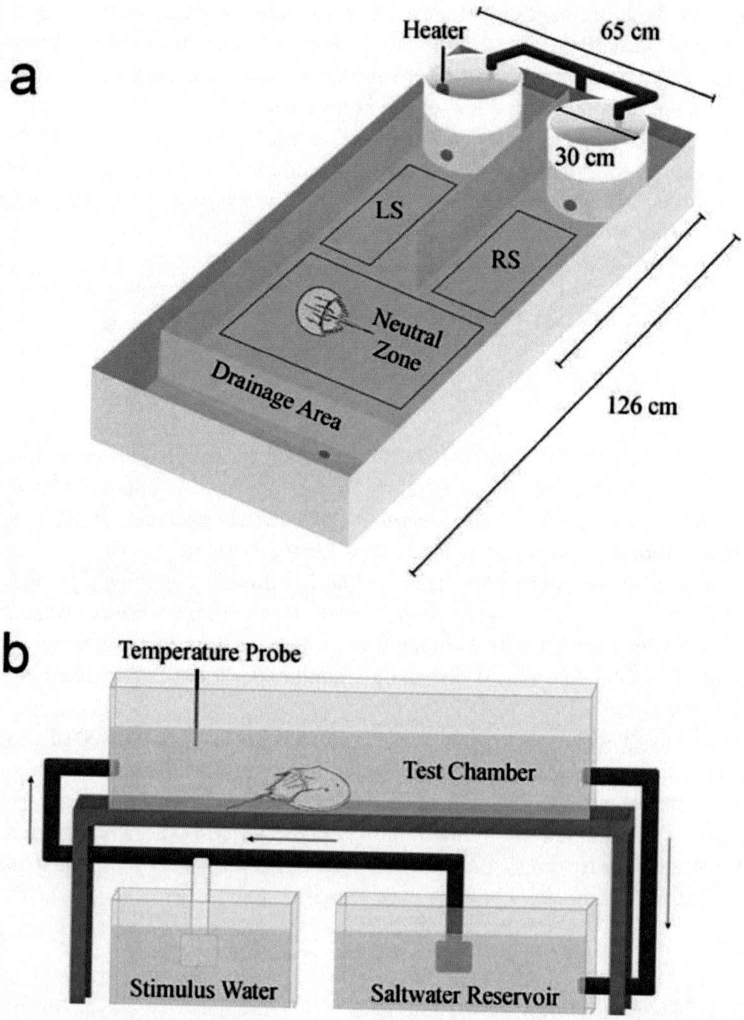


**Fig. 1** Illustrations of (**a**) the Y-maze used for the preference experiments, and (**b**) the chambers used for the cardiac assay experiments. During Y-maze experiments, conditions were changed in one or both of the header tanks (addition of freshwater, heaters, heaters and frozen ice packs); at the start of the control period and the experiment period, a horseshoe crab was held in the neutral zone for approximately 15–30 s and then was released to explore the tank. During the cardiac assay experiments, horseshoe crabs with heart electrodes attached were placed inside a light-tight recording chamber

height of the tank. Water flowed over this short barrier into a drainage area so that water flowed straight down each channel of the maze and out the other end. This reduced mixing and helped maintain a temperature or salinity difference between the two channels of the Y-maze. A color-dye test verified there was a constant flow of water at ~1.43 cm/s down each channel of the Y-maze and that there was little mixing of the water between each of the channels, except in the neutral area.

At the start of each trial, a horseshoe crab was positioned in the middle of the neutral zone for approximately 30 seconds, then released to explore the Y-maze for a 1-hour control period, during which only ambient estuarine seawater flowed into both header tanks and then into the Y-maze. When transitioning from the control period of the trial to the experimental period, the horseshoe crab was temporarily removed from the experimental arena. After adjusting the header tanks so that the salinity or temperature of one channel of the Y-maze was different than the other, the experimental period commenced the same way as the control period. The choices to which horseshoe crabs were exposed to were: (1) estuarine water diluted with fresh water versus ambient estuarine water, (2) warm water versus ambient temperature water, and (3) warm water versus cold water. The channel of the maze used for the different stimuli was randomized for each trial.

During the experimental period of the salinity choice experiment, fresh water was added to one of the header tanks and allowed to drain into one channel of the maze. A salinity probe (Vernier Software and Technology Salinity Sensor) was placed in each channel of the maze to monitor salinity changes throughout the experiment. Salinity in the channel with flowing fresh water averaged 14.8 ± 1.5 psu (range of 11–19), while the salinity in the other channel of the Y-maze averaged 16.6 ± 1.4 psu (range of 13–20). There was an average difference of 1.72 ± 0.5 psu between channels.

Two types of temperature preference experiments were carried out in the same manner as the salinity experiment described above. In each type of temperature experiment, a Vernier temperature probe was placed in each arm of the maze to monitor temperature changes throughout the experiment. The first type of experiment compared their preference for warm water (water warmed by aquarium heaters in one of the header tanks; 24.4 ± 0.4 °C) versus ambient water (~22 °C). In the second study, horseshoe crabs were given a choice between warm water (23.8 ± 1.3 °C) versus slightly colder water (21.1 ± 1.7 °C). In this second experiment, aquarium heaters were placed in one header tank and frozen ice packs were placed in the other, creating a larger temperature gradient between the two channels than in the first type of temperature experiment.

Horseshoe crab movements were recorded using a black and white camera (PC-222, SuperCircuits, Austin, TX) suspended directly above the Y-maze. The output of the camera was digitized (Canopus ® ADVC-110, Grass Valley, San Francisco, CA), time-stamped, and recorded on a computer using video capture software (Gawker, v. 0.8.3, Seattle, WA), which captured one frame every 25 seconds. Videos were subsequently analyzed either by eye (by one person to maintain consistency), or with video tracking software (Noldus EthoVision® XT Video Tracking Software, Noldus Information Technology, Wageningen, the Netherlands). In both cases, the position of the horseshoe crab for each one-second of video playback was scored as in the RS Zone, LS Zone, or neutral zone.

During the control periods of trials, when the same type of water was flowing through the entire Y-maze, we determined how often horseshoe crabs changed zones in the Y-maze in order to verify that they were exploring the Y-maze and, as a result, had ample opportunity to sample the water in each area. Horseshoe crabs that spent less than 10% of the control period in either the LS zone or the RS zone, or had a bias for one of the zones throughout the control period were not used for subsequent analyses because of an apparent bias for part of the Y-maze.

## 2.3 Cardiac Assay Experimental Design

A cardiac assay, similar to one used to test the ability of American lobster to detect changes in temperature or salinity by Dufort et al. (2001) and Jury and Watson (2000), was used to determine if horseshoe crabs (n=22) could detect a change in water temperature. Heart rate was monitored with either an impedance converter (UFI model 2991) or a modified infrared sensor (IR) that was based on the devices described in Depledge (1984) and Fedotov et al. (2000). The advantage of the IR sensor was that it was not as sensitive to changes in water temperature as the impedance-based method. When using the impedance device, coated wires with about 2 mm of exposed wire at the tips were inserted through small holes in the dorsal carapace of the horseshoe crab, on both sides of the heart. When using the IR system, a 2 cm diameter threaded cap of a plastic falcon tube that was painted black was glued to the dorsal carapace, over the middle of the heart, and then the IR sensor was screwed into the PVC cap after the glue had set. This arrangement allowed for the use the same IR sensor with numerous horseshoe crabs. The outputs of both sensors were connected to an ADInstruments (Sydney, Australia) analog to a digital converter (PowerLab 4/25) and were displayed and recorded using LabChart 7 software (ADInstruments, Sydney, Australia).

After the electrodes were in place, horseshoe crabs were secured to a plexi-glass frame with elastic bands and placed within a large, light-tight, flow-through chamber (designated as the "recording chamber"). Chilled seawater (12 °C) from an aquarium that was used as a seawater reservoir was recirculated through the recording chamber at a constant rate (Fig. 1b). A temperature probe (Vernier Software and Technology Sensors) was placed in the recording chamber and its output was recorded throughout each experiment using Vernier Software (LoggerPro 3.8.7). For some experiments, a small thermistor (custom order from Vernier Software) was inserted inside of the horseshoe crab along the lateral edge of the prosoma in order to monitor the internal body temperature; this was done for 5 horseshoe crabs out of the total of 22 horseshoe crabs that were used for the cardiac assay experiments. The output of the thermistor was also recorded using the Vernier System.

After allowing a horseshoe crab to adjust to the recording chamber for at least two hours, and during a time when their heart rate was very consistent and had not changed by more than five beats per minute in at least ten minutes, it was subjected to an increase in water temperature. This was accomplished by moving a small submersible pump from the seawater reservoir into a 4-liter container holding warm seawater (range 15–25 °C). This water then entered the recording chamber and mixed with the water already in the chamber, causing a moderate change in the temperature of the water surrounding the horseshoe crab (Fig. 1b). This method ensured that there was no change in the flow rate or other parameters that might cause the horseshoe crab to sense changes in its surroundings, other than temperature. Control experiments were carried out the same way, but the pump was moved into a container holding water that was at the same temperature as the water in the reservoir/aquarium. After exposing the horseshoe crab to a change in temperature for at least one minute, the recirculating pump was placed back in the aquarium and the horseshoe crab was once again perfused with ~12 °C seawater.

A disruption in their heart rate, typically consisting of a short slowing (bradycardia) followed by a longer increase in heart rate that lasted >35 seconds after exposing horseshoe crabs to a change in water temperature, was used as an indication that the horseshoe crab detected the change in water temperature (Jury and Watson 2000, Dufort et al. 2001). The water temperature reading at the time when the heart rate first changed was designated as the absolute temperature sensitivity threshold. The difference between this temperature and the ambient temperature in the tank was designated as the amount of change necessary to cause a response, or their temperature change threshold. If they responded to a given stimulus, then after waiting for ~1–2 hours, they were exposed to a smaller change in temperature. This was repeated until they no longer responded. Moreover, if they did not respond, after a recovery period, they were exposed to a larger change in temperature until they did respond. In this manner, it was possible to determine the minimum amount of temperature change each horseshoe crab could detect.

## 2.4 Data Analyses

In the Y-maze experiments, paired *t*-tests were used to compare the number of times horseshoe crabs switched zones and the time spent in the experimental side of the Y-maze during the control period versus the experiment period. Unpaired *t*-tests were used to compare the total time spent in both zones (experimental side versus ambient condition side) during the experiment period. Repeated Measures ANOVAs were used to compare the time spent in the experiment side of the Y-maze during each 15-minute interval of the 60-minute experiment period.

In the cardiac assay experiments, the smallest amount of temperature change required to elicit a clear cardiac response was considered the temperature change threshold for each horseshoe crab. These values were averaged for the 17 horseshoe crabs that responded to determine the mean change in temperature horseshoe crabs, as a group, could detect. To determine if horseshoe crabs had external or internal temperature receptors, the 5 other horseshoe crabs' behaviors and responses were observed.

## 3 Results

### 3.1 Overall Movements in the Y-Maze

Based on a random sample of 9 horseshoe crabs from the total number of horseshoe crabs used in all temperature experiments (n=13), horseshoe crabs changed zones an average of 37.1 ±6.6 times during the 1-hour control period, demonstrating that they were fully exploring the Y-maze and had ample opportunity to sample the water in each zone. During the experiment period when temperature changed to one or both of the header tanks, the horseshoe crabs switched zones an average of 26.2 ±6.9 times. While there was not quite a significant decrease between the number of zone changes between the control and the experimental periods ($p = 0.097$, Paired *t*-test), there was a significant decrease in zone changes over the course of the 1-hour experimental period ($p = 0.003$; Friedman Test nonparametric repeated measures ANOVA). This tendency of changing zones was likely due to the horseshoe crabs choosing to spend more time in their preferred side of the Y-maze. The same analyses were done for horseshoe crabs used in salinity experiments. Based on a random sample of 16 horseshoe crabs from the total number of horseshoe crabs used in salinity experiments (n=30), they changed zones an average of 50.0 ± 9.6 times during the 1-hour control period, again demonstrating that the horseshoe crabs were fully exploring the Y-maze and had ample opportunity to sample the water in each zone. There was a significant decrease in the amount of times horseshoe crabs switched zones (average total of 34.3 ±8.0 times) during the experiment period when the water conditions were changed on the experimental side of the Y-maze, compared to the control period ($p = 0.025$; Wilcoxon matched-pairs signed-ranks test). This change is probably due to the preference for one side versus the other, and thus, horseshoe crabs spent more time there and switched zones less.

### 3.2 Y-Maze Experiments: Temperature

#### Experiment 1: Warm Water Versus Ambient-Temperature Water

During the 1 h control period, when both sides of the Y-maze were the same temperature, horseshoe crabs ($n = 6$) spent 29.1 ± 5.0% of their time on the side of the Y-maze where warm water was going to be added during the experimental period. During the 1 h experimental period, when warm water was added to a designated channel of the Y-maze, horseshoe crabs spent almost twice as much time there (53.1 ± 11.5%) than they did before warm water was added; however, this difference was not significant ($p = 0.1642$; Paired *t*-test). But, when we compared the time spent on one side of the Y-maze versus the other during the 1 h experiment period, horseshoe crabs spent significantly more time in the warmed channel than the channel with ambient water ($p = 0.043$, unpaired *t*-test; Fig. 2a). As the experiment

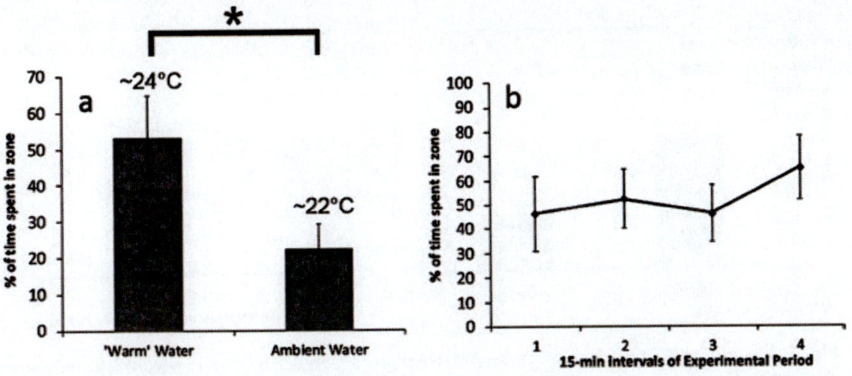


**Fig. 2** Responses of horseshoe crabs ($n = 6$) when given a choice between areas of a Y-maze with warm versus ambient temperature estuarine water. **(a)** The amount of time horseshoe crabs spent in each channel of Y-maze during the 1-hour experimental period. Asterisks indicate statistical significance. **(b)** The amount of time horseshoe crabs spent in the experimental zone (the warm side of the Y-maze) during each of the 15-minute intervals of the 1-hour experimental period

progressed, there was also a tendency for the horseshoe crabs to spend more time in the channel of the maze that was warmer (Fig. 2b), but this was not significant ($p = 0.313$; Repeated Measures ANOVA).

**Experiment 2: Warm Versus Cold Water**

When one channel of the Y-maze was heated and the other was cooled to create a larger difference in temperatures, than used in Experiment 1, horseshoe crabs ($n = 7$) spent significantly more time in the warmer side of the Y-maze ($p = 0.023$, unpaired *t*-test; Fig. 3a). They also exhibited a trend to spend more time in the experimental side of the Y-maze during the time that it was being warmed versus the control period ($p = 0.062$; Paired *t*-test). While horseshoe crabs initially spent a similar amount of time on both the 'cold' and 'warm' sides of the Y-maze during the first 15 minutes of the experiment period, they exhibited a non-significant trend to progressively express a preference for the warm water, spending 71.6 ± 14.5% of the final interval of the experiment in the 'warm' zone ($p = 0.172$; Nonparametric Repeated Measures ANOVA; Fig. 3b).

### *3.3 Y-Maze Experiments: Salinity*

The time horseshoe crabs ($n = 30$) spent in the channel of the Y-maze where freshwater was being added during the experimental period did not vary significantly from the time spent on that same side of the Y-maze during the control period (prior

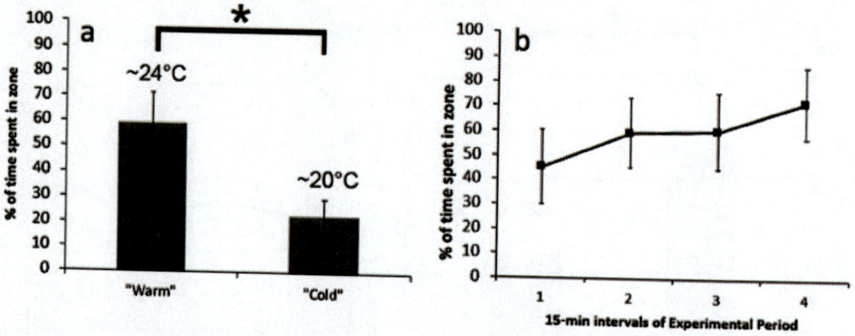


**Fig. 3** Responses of horseshoe crabs ($n = 7$) when given a choice between areas of a Y-maze with warm estuarine water versus cold estuarine water. (**a**) The amount of time horseshoe crabs spent on the "warm" or "cold" side of the Y-maze during the 1-hour experimental period. Asterisks indicate statistical significance. (**b**) The amount of time horseshoe crabs spent on the "warm" side of the Y-maze during each of the 15-minute intervals of the 1-hour experimental period

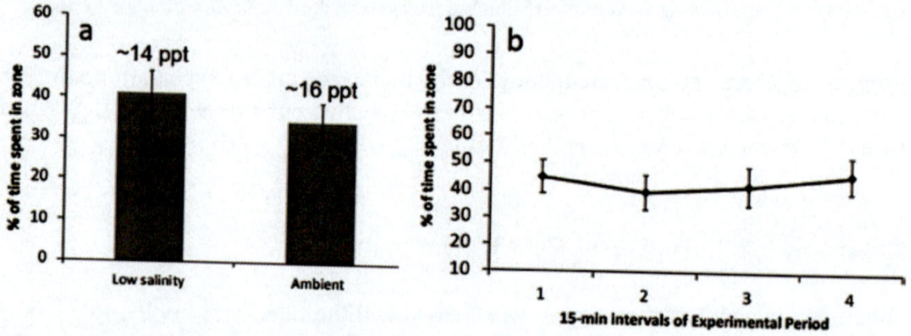


**Fig. 4** Responses of horseshoe crabs ($n = 30$), when given a choice between areas of a Y-maze with low salinity (receiving freshwater input) versus ambient estuarine water. (**a**) The amount of time horseshoe crabs spent on each side of the Y-maze during the 1-hour experimental period. Asterisks indicate statistical significance. (**b**) The amount of time horseshoe crabs spent in the experimental zone (where the salinity is lower) during each of the 15-minute intervals of the 1-hour experimental period

to freshwater being added; $p = 0.593$; Wilcoxon matched-pairs signed-ranks test). Horseshoe crabs also did not prefer one side or the other when given a choice between ambient estuarine water and lower salinity water ($p = 0.395$; Mann-Whitney Unpaired *t*-test; Fig. 4a). They also did not change their behavior much during the experimental period, spending 44.9 ± 6.2% of time during the first 15-min interval of the experimental period in the low salinity zone, and 46.4 ± 6.7% of time during the last interval ($p = 0.521$; nonparametric Repeated Measures ANOVA; Fig. 4b). While females ($n = 13$) appeared to spend slightly more time in the lower salinity area (46.3 ± 7.3%) than males ($n = 17$; 37.2 ± 8.3%), collectively, they did not prefer either side of the Y-maze throughout the experiment. Therefore, it did not

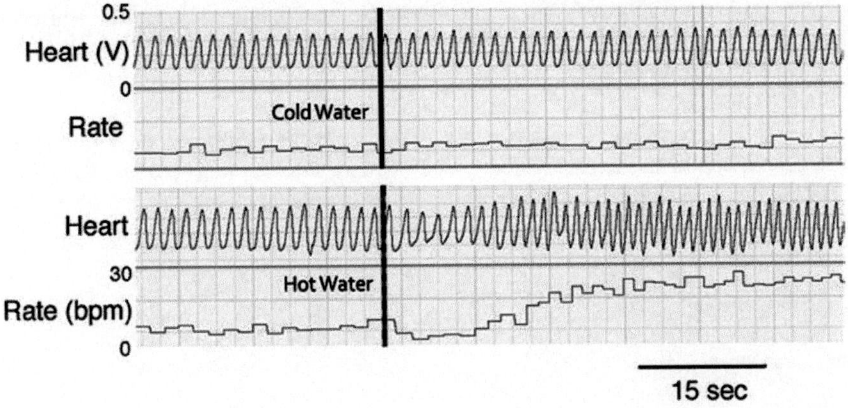


**Fig. 5** A sample record showing the cardiac response of a horseshoe crab to water that was 3 °C colder than the water in the experimental chamber (12 °C) (top), and then to water that was 3 °C warmer (bottom). The stimuli were applied at the solid or dashed lines. Note that while there was no change in the heart rate in response to cold water, but the addition of hot water caused the heart rate to slow for a brief period (bradycardia), and then speed up for a longer period of time. This "startle response" indicated that the horseshoe crabs sensed the change in water temperature

appear, under the circumstances provided to them, that horseshoe crabs preferred water that was either at a higher or lower salinity.

## 3.4 *Cardiac Assay Results*

A typical cardiac response to a change in water temperature is shown in Fig. 5. When at rest in the dark experimental chamber, horseshoe crab heart rates tended to be very consistent, varying less than 5 bpm for several minutes at a time, as seen in the beginning of Fig. 5. Nevertheless, we always waited to add stimulus water until heart rate had been very consistent for at least 10 minutes. At that point, when warm water was added, there was a brief slowing of the heart (bradycardia), followed by an increase in heart rate (Fig. 5). This rapid response suggests that horseshoe crabs sensed the change in temperature using receptors located externally. Moreover, the fact that they did not respond to water that was at the same temperature as the experimental chamber, or slightly colder (Fig. 5), further demonstrates that horseshoe crabs were responding to an increase in water temperature rather than some other stimulus that was associated with the procedure.

All seventeen horseshoe crabs tested responded to a temperature increase that was more than 2.6 °C. However, eight of the 17 horseshoe crabs were even more sensitive and responded to a temperature change of only 1 °C. In general, the higher the rate of temperature change, the faster they responded. Out of the 17 horseshoe crabs tested, nine responded to a rate of temperature change <1.5 °C/min, while the remaining eight did not respond unless the rate of change was at least 6 °C/min.

Given the rapid response time to changes in temperature, we conducted an additional experiment to try and determine if their thermoreceptors were located externally. In this experiment, with five horseshoe crabs, in addition to the temperature probe located in the experimental tank (Fig. 1b), we also placed a small thermocouple inside their shells. In each of the five trials the horseshoe crabs exhibited a change in heart rate before their internal temperature had changed, suggesting that the relevant receptors were external.

## 4 Discussion

Whereas many previous studies have focused on characterizing the cues that horseshoe crabs use to "select" a particular beach for spawning, the goal of this study was to determine what cues they might use to guide their large-scale migrations to the vicinity of spawning beaches. Specifically, we aimed to determine if they might use the temperature or salinity gradients that are characteristic of most estuaries to guide their movements further up into the estuaries in the spring, just before spawning commences. The results from both our Y-maze and cardiac assay experiments indicate that horseshoe crabs can sense small changes in water temperature and, when given a choice, they prefer slightly warmer water. Thus, we propose that temperature is one of the most important environmental cues that horseshoe crabs utilize to guide their spring migrations, especially in locations such as estuaries where the thermal gradients can strong and consistent.

Horseshoe crabs were able to perceive small differences in water temperature in the Y-maze and in the cardiac assay experiments – almost all the horseshoe crabs sensed a temperature increase of ~3 °C and some responded to even smaller changes. Interestingly, a few horseshoe crabs responded to decreases in water temperature, but they did not appear to be as sensitive as they were to increases in temperature. Unlike Crustacea horseshoe crabs do not use antennae to detect environmental cues (Barber 1961) but must rely on other appendages or other means for thermoreception and chemoreception. It is likely that at least some of the receptors mediating these responses to changes in water temperature are located on their walking legs as Wyse demonstrated in 1971. However, while the temperature changes in Wyse's experiments were typically on the order of ±10 °C (and in Barber's experiments (1956) with water of 3–5 °C above room temperature (30 °C) elicited a response), in our study, we documented responses in some horseshoe crabs to increases of only 1 °C. While it appears as if they have external thermoreceptors on their legs (and perhaps elsewhere) that are responsible for triggering a change in heart rate when they detect a small increase in water temperature, it is possible they also have some internal thermoreceptors. We attempted to address this question by placing a temperature sensor inside the carapace while conducting some of the cardiac assays. In general, at the time horseshoe crabs first exhibited a cardiac response to an increase in water temperature, their internal temperature had changed very little compared with the external water temperature. Therefore, while it is certainly possible that they have internal temperature receptors, our data suggests that the thermoreceptors that are involved in triggering the startle responses we monitored with our cardiac assay are located externally. Furthermore, it is certainly possible, given the timeframe of the Y-maze experiments, that internal thermoreceptors are involved in guiding their movements in their natural habitat.

A horseshoe crab's ability to detect warmer water enables it to "behaviorally thermoregulate," which was manifested as a preference for warmer water when given a choice in the Y-maze. When given a choice between water that was 5–10% warmer than ambient water and ambient estuarine water, horseshoe crabs spent more time on the warmer side of the Y-maze. When ice packs and heaters were used to create a larger difference in temperature between the channels of the Y-maze (a 10–20% difference), they also preferred the warm water side. This is not the first time the thermal preferences of horseshoe crabs have been investigated. In 1979 Reynolds and Casterlin (1979a, b) used a shuttle box apparatus to examine the thermal preferences of 10 juvenile horseshoe crabs collected in Florida, USA. They expressed a wide range of thermal preferences ranging from ~24 to 36 °C (mean of 29.4 °C) and tended to avoid water temperatures below 10 °C and above 40 °C. In our experiments, we did not provide them with enough of a temperature range to determine their "preferred" water temperature, and it would have been interesting to see if horseshoe crabs from New Hampshire have a cooler preferred water temperature than those from Florida. Rather, our goal was to determine if they could detect the small differences in water temperature that are common in the Great Bay Estuary in the spring and if they would show a preference for warmer water which would lead them to the waters in the vicinity of the spawning beaches where they are found in great abundances in this estuary (Cheng et al. 2016).

This preference for warm water sheds light on previous field studies of seasonal horseshoe crab movements and their choices of their spring-summer locations in the estuary. For example, acoustic telemetry studies done in the Great Bay Estuary showed that when water temperatures exceeded 11 °C during the transition from the winter to spring, horseshoe crabs initiated large-scale movements up into the estuary where the water warms faster than at near coast (Schaller et al. 2010, Watson et al. 2016). This type of spring migration has also been seen within other New England estuaries, such as Taunton Bay and Pleasant Bay (Moore and Perrin 2007; James-Pirri, 2010, respectively). In a study investigating the temporal distribution of spawning activity in the Great Bay Estuary, a year that experienced unusually warm temperatures in early spring also yielded in an early spawning activity (Cheng 2014, Cheng et al. 2016, Watson et al., 2016). Furthermore, peaks of spawning activity throughout the spawning season typically coincided with increases in water temperature, and the largest spawning densities were found in areas of the Great Bay Estuary that were generally the warmest (Cheng et al. 2016). This pattern has also been seen in other areas of the geographic range of *L. polyphemus*, but at different temperature thresholds. For example, Thompson (1998) observed that when water temperatures in South Carolina USA reached ≥20 °C, spawning activity greatly increased, which is a much higher temperature than the 11–15 °C threshold in New Hampshire. Overall, field data throughout the range of *L. polyphemus* and the present laboratory studies support the hypothesis that temperature has a very strong influence on activity, especially spawning activity and seasonal migrations, even though the thresholds might differ from North to South along the US Atlantic coast.

In this study we also sought to determine if horseshoe crabs had a preference for low or high salinity water. We found that over the very small range of salinities we tested, they do not seem to have a salinity preference. In addition, based on some preliminary cardiac assays, they do not appear to be very sensitive to changes in salinity (Watson unpub. data) even though they appear to have receptors on their legs that are sensitive to changes in salt concentrations and osmolarity (Barber 1961, Wyse 1971). Furthermore, in the field, large decreases in salinity associated with heavy precipitation events appear to reduce their tendency to approach beaches to mate during the spawning season (Watson et al. 2009, Cheng 2014, Cheng et al. 2016). However, in both sets of experiments the salinity differences we tested were much smaller than the large gradients that are present in many of the estuaries where horseshoe crabs are common, so it is very possible they can detect and respond to larger salinity differences. That said, given their broad distribution, which includes areas that do not have large salinity gradients, it is not surprising that horseshoe crabs would not rely on salinity gradients as much as thermal gradients to guide them during their annual migrations to optimal spawning areas.

While the data presented indicate that horseshoe crabs are sensitive to thermal gradients and may use them to guide their movements up into the estuary during the onset of the spring spawning season, it remains to be seen why horseshoe crabs seek areas in the Great Bay Estuary that are warmer and tend to have a lower salinity. One hypothesis is these environmental conditions may affect their overall reproductive success, including embryonic development, larval hatching, and post-hatch larval horseshoe crab development, and larval and juvenile survival (Jegla and Costlow 1982, Laughlin 1983, Ehlinger and Tankersley 2003, Vasquez et al. 2015a, b). While salinities lower than 5–10 psu may prolong development and/or create less viable eggs (Jegla and Costlow 1982, Sugita 1988, Vasquez et al. 2015b), and hyposalinity is the most significant stressor to developing embryos (Vasquez et al. 2017), moderately low salinity water can have some positive impacts. When embryos reach their last stages of embryonic molting and are exposed to low salinities, the probability of hatching increases due to the swelling and rupture of the egg membrane (Ehlinger and Tankersley 2003). Additionally, development time and molting of post-hatch larval horseshoe crabs can increase at salinities as low as 10 psu (Jegla and Costlow 1982, Laughlin 1981, 1983 and Sugita 1988). When temperature is added to the process, it becomes even more complicated. Research looking at the combined effects of salinity and temperature on embryo development rates found that there was a significant interaction between increased temperature (35 °C) and moderately decreased salinity (15 psu), causing a reduced rate in reaching an egg size of 5 $mm^2$, and the combined effect of elevated temperatures and low salinities was more pronounced in Delaware USA compared with Florida (Vasquez et al. 2015b, Vasquez et al. 2017). Once larval horseshoe crabs have hatched, Laughlin (1983) suggests that temperature has a more profound effect than salinity. Optimal temperatures for development may vary, but as long as it does not exceed 35 °C, which negatively impacts larval growth and development (Ehlinger and Tankersley 2004, Vasquez et al. 2015a), warmer water should increase the rate of development and growth. In contrast, moderately cool temperatures (for example, ~20 °C for Florida) are tolerable for embryos, but if the conditions are too cold, development rate and ability to molt may be compromised (Laughlin 1983, Vasquez et al. 2015a). Vasquez et al. (2015a) found sediment temperatures reaching a maximum of 42 °C are lethal to Florida horseshoe crab nests.

All of the aforementioned temperature temperature effects and adaptations to those temperatures are dependent on *L. polyphemus'* geographic location. At the northern end of their range, where these studies were conducted, "warm" areas are clearly not considered as warm as in the southern part of their geographic range. For example, when comparing the development success of embryos from Florida with Delaware, Vasquez et al. 2017 found embryos from Delaware performed significantly better at 35 °C than those from Florida. We know there are genetically distinct populations along the geographic range of *L. polyphemus* (King et al. 2005), and thus it is likely that each population has evolved to respond most appropriately to the thermal gradients they experience throughout the year.

A great deal of attention has been focused on the potential impacts of climate change and the associated warming of the oceans and coasts on commercially important species (Le Bris et al. 2018). Estuaries in particular, where many fished species spend their early life history, are highly capricious and complex systems; these ecosystems experience dramatic changes in water quality and physical characteristics depending on tides, weather conditions, and seasonality. While many of these species have adapted to these changes, estuarine ecosystems are experiencing rapid and extreme changes in climate and are highly sensitive to these events (Robinson et al. 2013), with impacts cascading the dynamics of fisheries populations. Computer models that rely on information about the thermal preferences of commercially important species have been developed to predict how warmer water might shift their distribution (Morley et al. 2018). Our data suggest that the same types of models might be useful for determining how the warming of coastal and estuarine waters might influence the movements and activities of horseshoe crabs. For example, it is important to know if shifts in their annual spawning events could lead to a temporal disconnect between their mating activities and the annual migrations of the birds that rely on horseshoe crab eggs for nutrition during their journeys northward (Dyer 1995, Scavia et al. 2002). Or, if the upper reaches of the estuary, where most horseshoe crabs spawn, become so warm that horseshoe crabs avoid them, will they be able to find appropriate alternative spawning locations closer to the ocean? And, if they can find appropriate beaches, will the larvae they produce be able to recruit to suitable habitat for juvenile growth and survival (Cheng et al. 2021). With drastic and rapid changes in weather, water and climate conditions, and additional pressures from coastal development and over-harvesting, these may place further stress upon all the life stages of horseshoe crabs and their life activities. The data presented in this paper, and additional studies of this nature, could help provide the information necessary to run these models and thus predict how climate change might impact horseshoe crab populations and their associated ecosystems.

**Acknowledgements** We thank all the research assistants who helped us with both studies and Allison Nash for producing the illustrations used in Fig. 1. We also thank David Shay for providing space for holding horseshoe crabs and carrying out the experiments at the University of New Hampshire Jackson Estuarine Laboratory. We greatly appreciate all the insightful comments and feedback from colleagues and anonymous reviewers who have reviewed previous iterations of this manuscript. This work was funded by New Hampshire Sea Grant Development Funds, the University of New Hampshire School of Marine Science and Ocean Engineering, and a grant from the National Science Foundation (NSF-IOS 0920342) to CCC and WHW III.

## References

S.B. Barber, Chemoreception and proprioception in *limulus*. J. Exp. Zool. **131**(1), 51–73 (1956)

S.B. Barber, Chemoreception and thermoreception, in *The physiology of the Crustacea. Vol. II. Sense organs, integration and behavior.*, ed. by T. H. Waterman, (Academic Press, New York, 1961), pp. 109–131

M.L. Botton, R.E. Loveland, T.R. Jacobsen, Beach erosion and geochemical factors: Influence on spawning success of horseshoe crabs (*Limulus polyphemus*) in Delaware Bay. Mar. Biol. **99**, 325–332 (1988)

H. Cheng, The environmental influences on American horseshoe crab (*Limulus polyphemus*) behavior and distribution in the Great Bay estuary, in *New Hampshire, U.S.A. M.S. thesis, University of new Hampshire*, (Durham, New Hampshire, 2014)

H. Cheng, W.H. Watson, C.C. Chabot, Influence of environmental factors on spawning of the American horseshoe crab (*Limulus polyphemus*) in the Great Bay estuary, New Hampshire, USA. Estuar. Coasts **39**(4), 1142–1153 (2016)

H. Cheng, C.C. Chabot, W.H. Watson III, Distribution of juvenile American horseshoe crabs Limulus polyphemus in the Great Bay Estuary, New Hampshire, USA. Mar. Ecol. Prog. Ser. **662**, 199–203 (2021)

J.A. Cohen, H.J. Brockmann, Breeding activity and mate selection in the horseshoe crab, *Limulus polyphemus*. Bull. Mar. Sci. **33**, 274–281 (1983)

M.H. Depledge, Photoplethysmography--a non-invasive technique for monitoring heart beat and ventilation rate in decapod crustaceans. Comp. Biochem. Physiol. **77A**, 369–371 (1984)

C.G. Dufort, S.H. Jury, J.M. Newcomb, et al., Detection of salinity by the lobster, *Homarus americanus*. Biol. Bull. **201**, 424–434 (2001)

K.R. Dyer, Chapter 5: Response of estuaries to climate change, in *Climate change: impact on coastal habitation*, ed. by E. Eisma, (CRC Press, Florida, 1995), pp. 85–110

G.S. Ehlinger, R.A. Tankersley, Larval hatching in the horseshoe crab, *Limulus polyphemus*: Facilitation by environmental cues. J. Exp. Mar. Biol. Ecol. **292**(2), 199–212 (2003)

G.S. Ehlinger, R.A. Tankersley, Survival and development of horseshoe crab (*Limulus polyphemus*) embryos and larvae in hypersaline conditions. Biol. Bull. **206**(2), 87–94 (2004)

V.P. Fedotov, S.V. Kholodkivich, A.G. Strochilo, Study of contractile activity of the crayfish heart with the aid of a new non-invasive technique. J. Evol. Biochem. Physiol. **36**(3), 288–293 (2000)

K.M. Ferrari, N.M. Targett, Chemical attractants in horseshoe crab, *Limulus polyphemus*, eggs: The potential for an artificial bait. J. Chem. Ecol. **29**(2), 477–496 (2003)

B.A. Fulton, E.A. Fairchild, R. Warner, The green crab Carcinus maenas in two New Hampshire estuaries. Part 1: spatial and temporal distribution, sex ratio, average size, and mass. Journal of Crustacean Biology, **33**(1), 25–35 (2013)

C. Hassler, H.J. Brockmann, Evidence for use of chemical cues by male horseshoe crabs when locating nesting females (*Limulus polyphemus*). J. Chem. Ecol. **27**, 2319–2335 (2001)

W.H. Howell, W.H. Watson III, S.H. Jury, Skewed sex ratio in an estuarine lobster (*Homarus americanus*) population. J. Shellfish Res. **18**, 193–201 (1999)

M.J. James-Pirri, Seasonal movement of the American horseshoe crab *Limulus polyphemus* in a semi-enclosed bay on Cape Cod, Massachusetts (USA) as determined by acoustic telemetry. Curr. Zool. **56**(5), 575–586 (2010)

T.C. Jegla, J.D. Costlow, Temperature and salinity effects on developmental stages of limulus. Prog. Clin. Biol. Res. **81**, 103–114 (1982)

S.H. Jury, W.H. Watson III, Thermosensitivity of the lobster, *Homarus americanus*, as determined by cardiac assay. Biol. Bull. **199**(3), 257–264 (2000)

T.L. King, M.S. Eackles, A.P. Spidle, et al., Regional differentiation and sex-biased dispersal among populations of the horseshoe crab *Limulus polyphemus*. Trans. Am. Fish. Soc. **134**, 441–465 (2005)

R. Laughlin, Sodium, chloride and water exchange in selected developmental stages of the horseshoe crab *Limulus polyphemus*. J. Exp. Mar. Biol. Ecol. **52**, 135–146 (1981)

R. Laughlin, The effects of temperature and salinity on larval growth of the horseshoe crab *Limulus polyphemus*. Biol. Bull. **164**(1), 93–103 (1983)

A. Le Bris, K.E. Mills, R.A. Wahle, et al., Climate vulnerability and resilience in the most valuable north American fishery. Proc. Natl. Acad. Sci. **115**(8), 1831–1836 (2018)

S. Moore, S. Perrin, Seasonal movement and resource-use patterns of resident horseshoe crab (*Limulus polyphemus*) populations in a Maine, USA estuary. Estuar. Coasts **30**(6), 1016–1026 (2007)

J.W. Morley, R.L. Selden, R.J. Latour, et al., Projecting shifts in thermal habitat for 686 species on the north American continental shelf. PLoS One **13**(5), e0196127 (2018)

W. Patten, On the morphology and physiology of the brain and sense organs of *limulus*. Q. J. Microsc. Sci. **35**, 1–96 (1894)

W.W. Reynolds, M.E. Casterlin, Thermoregulatory behaviour of the primitive arthropod *Limulus polyphemus* in an electronic shuttlebox. J. Thermal Biol. **4**, 165–166 (1979a)

W.W. Reynolds, M.E. Casterlin, Thermoregulatory behavior and diel activity of *Limulus polyphemus*, in *Biomedical Applications of the Horseshoe Crab (Limulidae)*, ed. by E. Cohen, (Alan R. Liss, Inc., New York, 1979b), pp. 47–59

P. Robinson, A.K. Leight, D.D. Trueblood, et al. Climate sensitivity of the National Estuarine Research Reserve System. Report to NOAA's Climate Program Office., pp.79 (2013)

A.E. Rudloe, W.F. Herrnkind, Orientation of *Limulus polyphemus* in the vicinity of breeding beaches. Marine & Freshwater Behaviour & Phy. **4**, 75–89 (1976)

A.E. Rudloe, W.R. Herrnkind, Orientation by horseshoe crabs, *Limulus polyphemus*, in a wave tank. Marine Freshwater Behav. Phy. **7**, 199–211 (1980)

K.M. Saunders, H.J. Brockmann, W.H. Watson III, et al., Male horseshoe crabs *Limulus polyphemus* use multiple sensory cues to locate mates. Curr. Zool. **56**, 611–617 (2010)

D. Scavia, J.C. Field, D.F. Boesch, et al., Climate change impacts on U.S. coastal and marine ecosystems. Estuaries **25**, 149–164 (2002)

S.Y. Schaller, C.C. Chabot, W.H. Watson III, Seasonal movements of American horseshoe crabs *Limulus polyphemus* in the Great Bay estuary, New Hampshire (USA). Curr. Zool. **56**, 587–598 (2010)

R.L. Schwab, H.J. Brockmann, The role of visual and chemical cues in the mating decisions of satellite male horseshoe crabs, *Limulus polyphemus*. Anim. Behav. **74**, 837–846 (2007)

F.L. Short. The ecology of the Great Bay Estuary, New Hampshire and Maine: An estuarine profile and bibliography. Piscatiqua Region Estuaries Partnership (PREP) Reports & Publications (1992)

D.R. Smith, P.S. Pooler, B.L. Swan, et al., Spatial and temporal distribution of horseshoe crab (*Limulus polyphemus*) spawning in Delaware Bay: Implications for monitoring. Estuaries **25**, 115–125 (2002a)

D.R. Smith, P.S. Pooler, R.W. Loveland, et al., Horseshoe crab (*Limulus polyphemus*) reproductive activity on Delaware Bay beaches: Interactions with beach characteristics. J. Coast. Res., 730–740 (2002b)

H. Sugita, Environmental adaptation of embryos, in *Biology of horseshoe crabs*, ed. by K. Sekiguchi, (Science House, Tokyo, Japan, 1988), pp. 195–122

B.L. Swan, Migrations of adult horseshoe crabs, *Limulus polyphemus*, in the middle Atlantic bight: A 17-year tagging study. Estuaries **28**, 28–40 (2005)

M. Thompson, Assessments of the population biology and critical habitat for the horseshoe crab, *Limulus polyphemus*, in the South Atlantic bight, in *M.S. thesis. Medical University of South Carolina, University of Charleston*, (Charleston, SC, 1998)

M.C. Vasquez, S.L. Johnson, et al., Nest site selection minimizes environmental stressor exposure in the American horseshoe crab, *Limulus polyphemus*. J. Exp. Mar. Biol. Ecol. **463**, 105–114 (2015a)

M.C. Vasquez, A. Murillo, et al., Multiple-stressor interactions influence embryo development rate in the American horseshoe crab, *Limulus polyphemus*. J. Exp. Biol. **218**, 2355–2364 (2015b)

M.C. Vasquez, H.J. Brockmann, et al., Between-population differences in multi-stressor tolerance during embryo development in the American horseshoe crab, *Limulus polyphemus*. Estuar. Coasts **40**, 1487–1501 (2017)

W.H. Watson III, A. Vetrovs, W.H. Howell, Lobster movements in an estuary. Mar. Biol. **134**(1), 65–75 (1999)

W.H. Watson III, S.Y. Schaller, C.C. Chabot, The relationship between small- and large-scale movements of horseshoe crabs in the Great Bay estuary and *limulus* behavior in the laboratory, in *Biology and conservation of horseshoe crabs*, ed. by D. Smith, M. L. B. Botton, J. T. Tanacredi, (Springer, New York, 2009), pp. 131–148

W.H. Watson III, S.K. Johnson, C.D. Whitworth, et al., Rhythms of locomotion and seasonal changes in activity expressed by horseshoe crabs in their natural habitat. Mar. Ecol. Prog. Ser. **542**, 109–121 (2016)

G.A. Wyse, Receptor organization and function in *limulus* chelae. Z. Vergl. Physiologie **73**, 249–273 (1971)

# The Effects of Roundup® (a Glyphosate-Based Herbicide) on the Survival and Development of American Horseshoe Crab, *Limulus polyphemus*

**Hope VanDerwater, Karine Khoder, and Mark L. Botton**

## 1 Introduction

Horseshoe crab populations in North America and Asia are in widespread decline (e.g., Smith et al. 2016a; John et al. 2018). Of the four extant species, the American horseshoe crab, *Limulus polyphemus*, is listed as vulnerable on the IUCN Red List (Smith et al. 2016b); the tri-spine horseshoe crab, *Tachypleus tridentatus*, is endangered (Laurie et al. 2019); and the coastal horseshoe crab, *Tapinauchenius gigas*, and mangrove horseshoe crab, *Carcinoscorpius rotundicauda*, have also declined precipitously in at least portions of their range (John et al. 2018). The decline in horseshoe crabs is significant because of their seminal importance in the ecology of marine benthic food webs, especially the importance of American horseshoe crab eggs as food for migratory shorebirds in Delaware Bay, USA (Botton 2009; Mizrahi and Peters 2009).

The erosion and/or armoring of estuarine breeding beaches, overfishing for the eel and whelk bait fisheries, and mortality associated with the bleeding of crab to produce LAL for endotoxin testing are generally considered as the primary factors leading to the decline in American horseshoe crabs (ASMFC 1998; Smith et al.

H. VanDerwater
Environmental Science Program, Fordham University, New York, NY, USA

K. Khoder
Department of Natural Sciences, Fordham University, New York, NY, USA

M. L. Botton (✉)
Environmental Science Program, Fordham University, New York, NY, USA

Department of Natural Sciences, Fordham University, New York, NY, USA
e-mail: botton@fordham.edu

 
J. T. Tanacredi et al. (eds.), *International Horseshoe Crab Conservation and Research Efforts: 2007-2020*, https://doi.org/10.1007/978-3-030-82315-3_34

2016a, 2016b; Botton et al. this volume). Water pollution has also been shown to have detrimental effects on the survival and normal development of horseshoe crab eggs and embryos (e.g., Botton and Itow 2009). Horseshoe crabs spend their spawning and early life stages on beaches near many major cities in North America and Asia, and their populations can be affected by pollutants. Their role as benthic predators means that they are surrounded by sediment and ocean water, and because they ingest small mollusks, polychaete worms, and even plant detritus (Botton 2009), they are susceptible to ingesting or absorbing harmful environmental toxins from their prey items or sediment. Moreover, contaminants found in females may be transferred to their eggs during oogenesis (Bakker et al. 2017) and be passed up the food chain to shorebirds (Burger et al. 2017).

Adult horseshoe crabs are an infamously bad indicator species when it comes to identifying the effects of changes in their ecosystem (Jegla and Costlow Jr 1982; Palumbi and Johnson 1982), but a number of researchers have studied the effects of pollutants on the development of their embryos and larvae. Heavy metals (copper and zinc) cause slower development and higher mortality in the embryo stage than in the trilobite larval stages (Botton et al. 1998a). A similar study focused on the effects of the biocide tributyltin (TBT) on the embryonic and larval stages of horseshoe crabs and found that the embryos exhibited higher levels of mortality at lower doses than the larval stages, and that development was delayed at sublethal concentrations of TBT (Botton et al. 1998b). When embryos were reared in polluted water, Itow et al. (1998) found that there was a higher frequency of embryonic abnormalities relative to embryos developing in clean water. Exposure to TBT and cadmium adversely affected hemolymph quality in juvenile *T. tridentatus*, with respect to plasma protein level and amoebocyte viability and morphology (Kwan et al. 2015).

Herbicides are the most widely used class of pesticide in the United States, far surpassing the use of insecticides, fungicides, fumigants, and other chemicals (Atwood and Paisley-Jones 2017). Glyphosate is the active ingredient in the commercial herbicide Roundup® and the most heavily used pesticide in the United States. Roundup®, as it is marketed commercially, is composed of two main ingredients. Glyphosate, technically N-(phosphonomethyl) glycine, is the active herbicide, typically applied in combination with a surfactant, a chemical designed to increase the effectiveness of the biocide by allowing it to cover more of the plant's surface area. The most popular formulation of glyphosate is Roundup®, first introduced in the 1970s by the Monsanto Corporation (now as part of the crop science division of Bayer AG). The increasing popularity of genetically modified "Roundup®-ready" crops (i.e., plants that are genetically resistant to the effects of glyphosate), including soybeans, corn, alfalfa, and cotton, has greatly expanded the worldwide application of this herbicide (Myers et al. 2016; Richmond 2018). In 2012, an estimated 270–290 million pounds of glyphosate were used in US commercial agriculture, and another 4–six million pounds were used in homes and gardens (Atwood and Paisley-Jones 2017). Glyphosate-based herbicides have also been applied to eradicate the invasive form of the common reed, *Phragmites australis* (Saltonstall 2002), in various salt marsh restoration projects (Teal and Peterson 2005). Because the primary target of glyphosate involves a metabolic pathway unique to plants, bacteria, and fungi (Pérez et al. 2011; Myers et al. 2016), it was initially believed that adverse effects on animals would be minimized. However, numerous studies have now shown that Roundup® can act as a neurotoxin, endocrine disruptor, and potential carcinogen (Kissane and Shephard 2017) and have detrimental effects on a variety of aquatic animals (Pérez et al. 2011).

The objective of this study was to examine the toxicity of Roundup® to horseshoe crab embryos from Jamaica Bay, New York using laboratory bioassays and to document unusual developmental abnormalities encountered at sublethal concentrations. This work expands the scope of a previous report (Khoder 2018).

## 2 Methods

Freshly laid horseshoe crab eggs were collected at the eastern section of Plumb Beach, Jamaica Bay, New York (40°34’53″, 72°54’42″) during May and early June in 2017 and 2019. Although receiving little agricultural runoff, Jamaica Bay is bordered by heavily populated communities in Brooklyn, Queens, and Western Nassau County, and Roundup® could potentially be introduced to the bay from homes and golf courses (Marine Park, Inwood Country Club). Plumb Beach, part of Gateway National Recreation Area, has a substantial population of spawning adult horseshoe crabs in late spring, and the collection site at Plumb Beach-East is approximately 2 km east of a recent beach nourishment project at Plumb Beach-West (Botton et al. 2018).

In the laboratory, we separated the eggs by rinsing them gently through stacked 2 mm and 1 mm sieve to remove the majority of sand, shell fragments, and other debris. Procedures for raising eggs and embryos were described in previous reports (Botton et al. 1998a, 1998b). Eggs were maintained in circular glass dishes (115 mm diameter and 50 mm height) with approximately 200 ml of artificial seawater (ASW) at 20 psu salinity, which was made by adding Instant Ocean™ to distilled water. Each bowl contained approximately 200 eggs, and all were maintained at ambient room temperature (ca. 22–25 °C). ASW was changed 2–3 times per week, but because of the low density of eggs relative to the volume of ASW, no supplemental aeration was required. After fertilization, the eggs are usually an opaque bluish-green color; as they develop, the outer opaque chorion is shed, allowing the larger transparent inner membrane containing the embryo to expand as the embryo continues growing and limbs become visible. The embryo next goes through three embryonic molts, developing further with each one.

Embryos used in all bioassays were “Stage 20,” just prior to hatching (Botton et al. 2010). Stage 20 embryos were sorted from the glass fingerbowls and transferred to labeled Petri dishes, with 40 individuals per dish. We set up static renewal bioassays with ASW controls and treatments for 24 hours, a dish for 48 hours, and a dish for 72 hours for each concentration. Roundup® Weed and Grass Killer III, with a 2% concentration of glyphosate, isopropylamine salt, was purchased from a local home and garden retailer, and dilutions were made using 20 psu ASW. Twenty milliliter of the Roundup® dilution was pipetted into each dish. Concentrations (expressed as % Roundup®) were 0.01%, 0.015%, 0.05%, 0.075%, 0.10%, 0.50%, 1.00%, 2.50%, 5.00%, 7.5%, and 10.00%. After 24 hours, we removed the treatment solutions from the 24 hr. dishes with fresh pipets, rinsed them with ASW, and the dishes were then maintained for the duration of the experiment in ASW. Similarly, Roundup® solutions were removed after 48 and 72 hours for these treatments, and the embryos in these dishes were rinsed with ASW and maintained in ASW until the conclusion of the experiment. All dishes were examined several times per week; dead embryos, dead trilobites, and live trilobites were removed and recorded. Dead trilobites and embryos were identified by their continued lack of gill and limb movement after they were gently nudged with a pipette. An individual dish was completed when all embryos either hatched or died. During the course of the study, any abnormally developed embryos were preserved in 70% isopropanol and photographed under a dissecting microscope.

Data on mortality and date of hatching were compiled on Excel and/or Google Sheets. To determine the median lethal concentration ($LC_{50}$), we used IBM SPSS to perform a probit analysis of the $\log_{10}$ of the concentrations and the mortality rates of the various treatment groups.

## 3 Results

There was an increase in the mortality rate of horseshoe crab embryos as the Roundup® concentration increased, but exposure time had little influence on mortality (Fig. 1). Notably, there was only minimal mortality at concentrations of ≤5% Roundup®. There was 100% mortality at the 10% Roundup® at all three time intervals. LC50s were estimated to be between 8.3% and 8.9% Roundup® (Table 1).

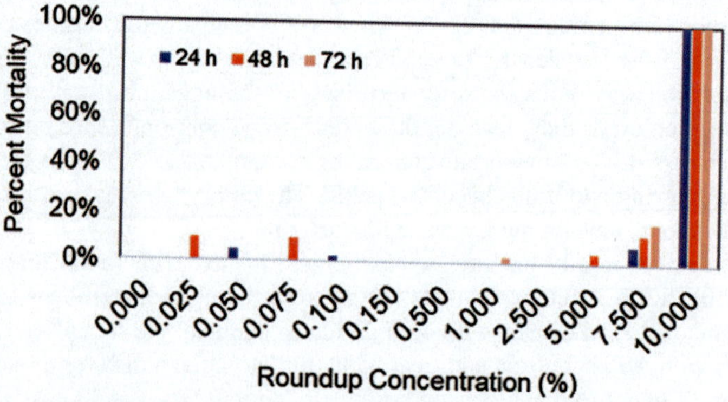


**Fig. 1** Mortality of Stage 20 horseshoe crab embryos following 24 h, 48 h, and 72 h exposure to different concentrations of Roundup®

**Table 1** Median lethal (LC50) concentrations of a glyphosate-based herbicide (as % Roundup®) to stage 20 horseshoe crab embryos exposed for 24, 48, or 72 hr

| 24 hr | 48 hr | 72 hr |
|---|---|---|
| 8.49% | 8.89% | 8.29% |

Embryonic development was prolonged at higher concentrations of Roundup® (Fig. 2), most notably at concentrations of 0.50% or higher. In the ASW control group and in Roundup® concentrations of 0.15% or below, the mean hatching time was significantly shorter than in Roundup® concentrations of 0.50% or above (Table 2). The hatching time for Stage 20 embryos exposed to Roundup® concentrations of 0.50% or higher was delayed by a factor of about 2, regardless of the duration of exposure.

Several developmental anomalies were noted in the course of this study (Fig. 3) in the 5% and 7.5% Roundup® concentrations, but not among ASW controls or low Roundup® concentrations. The inner egg membrane appears to be fully intact and connected to the embryo, almost pinching it, and still completely round and firm when pinched with forceps, and full of normal perivitelline fluid. We found 13 specimens that contained deformed membranes as described above, 2 in the 5% 48 hr. dish, 3 in the 7.5% 48 hr. dish, 2 in the 10% 48 hr. dish, and 6 in the 10% 72 hr. dish. We also found several other dead embryos that were deformed in other ways; however, deformed embryos were not found in the control dishes.

## 4 Discussion

There was little to no mortality of Stage 20 horseshoe crab embryos exposed to Roundup® concentrations ≤5.0% for 24, 48, or 72 h (Fig. 1), and hatching into the trilobite larval stage was significantly delayed at concentrations above 5.0% Roundup (Fig. 2 and Table 2). Abnormal embryonic development was also seen at Roundup® concentrations of 5% and 7.5% (Fig. 3). However, it should be noted that the concentrations that caused these effects are environmentally unrealistic, given the dilution that would occur when runoff containing Roundup® would mix with the waters of Jamaica Bay. Glyphosate, or its degradation product aminomethylphosphonic acid (AMPA) is commonly detectable in samples of treated sewage effluents or in downstream samples, but typically at levels below 2 μg/L (Kolpin et al. 2006). In comparison, the Roundup® product used in our experiments contained 2% glyphosate, meaning that even our lowest concentration (0.01% Roundup®) would have approximately 100 ppm glyphosate, which is well above levels that are typically found in water samples.

Although the effects of herbicides on horseshoe crab development have not been previously studied, Weis and Ma (1987) studied the effects of the insecticide Dimilin® (diflubenzuron) on the survival and developmental rate of *Limulus*

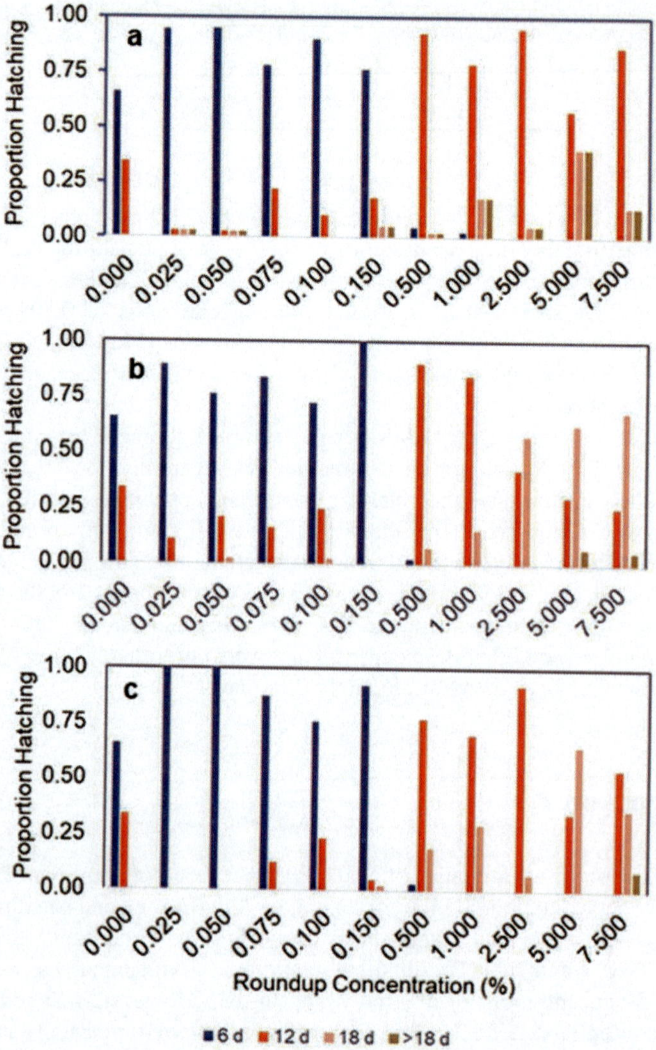


**Fig. 2** Time to hatching as a function of Roundup® concentration at exposure times of (**a**) 24 h, (**b**) 48 h, and (**c**) 72 h. Embryos hatching between 1 and 6 days after treatment were grouped together as 6 d, those hatching between day 7 and day 12 were grouped together as 12 d, those hatching between day 13 and day 18 were grouped together as 18 d, and those that took longer than 18 d to hatch were grouped together as >18 d. Note that there was 100% mortality among embryos exposed to 10% Roundup®

trilobite larvae. Similar to the results of our study, these authors found that lethal effects were found at only the highest pesticide concentration (50 ppm), but metamorphosis was delayed at a lower concentration (5 ppm). This and other studies suggest that the *Limulus* embryos and larvae are highly tolerant of a variety of

**Table 2** Comparison of hatching times for Stage 20 horseshoe crab embryos exposed to low (0.15% or below) and high (0.50% or above) concentrations of Roundup®

| Concentration | Exposure time | Mean (d) | Standard deviation | Median (d) | Range (d) | Significance (t-test) |
|---|---|---|---|---|---|---|
| ≤0.15% | 24 h | 5.13 | 2.23 | 5 | 3–13 | $P < 0.0001$ |
| ≥0.50% | | 9.95 | 3.93 | 10 | 4–55 | |
| ≤0.15% | 48 h | 5.21 | 2.10 | 5 | 3–13 | $P < 0.0001$ |
| ≥0.50% | | 11.75 | 2.81 | 11 | 5–21 | |
| ≤0.15% | 72 h | 4.98 | 1.96 | 5 | 3–13 | $P < 0.0001$ |
| ≥0.50% | | 10.75 | 2.69 | 10 | 5–23 | |

**Fig. 3** Photomicrographs of malformed treatment specimens taken from two dishes: 5% Roundup® at 48 hours (left 3 specimens) and 7.5% Roundup® at 48 hrs (right specimen.) Pictured are three deformed embryos developing outside of their inner membrane

pollutants, including heavy metals, oils, and organic compounds, especially in comparison with similar life-history stages of marine crustaceans (Botton and Itow 2009).

There are several resources examining the effects of this environmental pollutant on the development and survival of other marine arthropods. Researchers in Brazil studied the effects of glyphosate on the cellular responses of the hepatopancreas, an organ in mollusks and arthropods that helps with the detoxification processes, in a freshwater prawn, *Macrobrachium potiuna*. The researchers found that acute exposure to this chemical resulted in impairments to R-cells and induced tissue differentiation in the hepatopancreas, showing that this chemical not only can affect the normal functions of an arthropod but also can damage the organism's ability to detoxify itself, making certain concentrations of the chemical exponentially harmful (de Melo et al. 2019). A 2003 study examined the effects of Roundup® and its formulated products on several species, finding that the marine crustaceans studied, *Ceriodaphnia dubia* and *Acartia tonsa*, were 4–5 times more sensitive to Roundup® toxicity than protozoa and bacteria. Interestingly, they also found that polyoxyethylene amine (POEA), the surfactant in commercial Roundup®, was more toxic to the organisms than glyphosate alone (Tsui and Chu 2003). A similar conclusion was reached by Janssens and Stoks (2017), who found that Roundup® had higher toxicity and caused more sublethal behavioral effects on damselfly larvae compared with glyphosate alone. Likewise, Turhan et al. (2020) showed that pure glyphosate had no lethal or developmental effects on amphibian (*Xenopus laevis*) embryos at concentrations up to 500 mg/L, whereas Roundup® caused mortality or growth inhibition at much lower levels.

Itow et al. (1998) described several kinds of embryonic abnormalities in horseshoe crab embryos caused by exposure to contaminants such as mercury, zinc, copper, lead, cadmium, chromium, and organotin. These abnormalities included segment-defective embryos, double embryos, embryos with abnormal lateral eyes, and embryos without book-gills on the opisthosoma. We did not observe these particular developmental anomalies in embryos exposed to Roundup®, nor did Itow et al. (1998) describe the anomaly noted in our study (Fig. 3). The exact cause of this malformation requires further investigation, and it is possible that the surfactant rather than the glyphosate component of Roundup® is the cause.

Horseshoe crab eggs are a vital source of food for many thousands of shorebirds during Spring migration (Mizrahi and Peters 2009), so the potential transfer of contaminants from eggs to higher trophic levels is of concern. It is possible that chemicals such as Roundup® and other glyphosate-based herbicides could enter horseshoe crabs across their gills or through the ingestion of food and sediment. Toxins in adult females can be transferred to their oocytes and subsequently to their mature eggs, as Bakker et al. (2017) have shown to be the case for trace metals. Levels of heavy metals in the blood and feathers of shorebirds directly reflect the levels of these same metals in horseshoe crab eggs (Burger et al. 2017). Although it may not harm the adult horseshoe crabs or their embryos themselves, metals, pesticides, and other chemicals in the eggs could be potentially dangerous to shorebirds and other species that rely on these as a food source.

**Acknowledgements** HV thanks Dr. Mary Hamilton and the Fordham Natural Sciences Department for so generously funding this research and Dr. Deborah Luckett for her help securing the grant, her humor, and her crab name suggestions throughout the summer. We thank Royall McMahon-Ward, Kylie Rothwell, Christina Colón, and Adam Aly for assistance in the field and laboratory. We are also grateful to the New York City Department of Parks and Dr. George Frame from the National Park Service, Gateway National Recreation Area for issuing the necessary permits to MLB to conduct research at Plumb Beach. We also thank the New York State Department of Environmental Conservation for issuing a Scientific Collecting License to MLB to collect horseshoe crab eggs.

## References

Atlantic States Marine Fisheries Commission (ASMFC), *Interstate Fishery Management Plan for Horseshoe Crab. Atlantic States Marine Fisheries Commission, Fishery Management Report No. 32* (Washington, DC, 1998)

D. Atwood, C. Paisley-Jones, *Pesticides industry sales and usage 2008-2012 market estimates. Biological and economic analysis division, Office of Pesticide Programs, Office of Chemical Safety and Pollution Prevention. U.S. Environmental Protection Agency* (https://www.epa.gov/sites/production/files/2017-01/documents/pesticides-industry-sales-usage-2016_0.pdf, Washington, DC, 2017), p. 24

A.K. Bakker, J. Dutton, M. Sclafani, N. Santangelo, Maternal transfer of trace elements in the Atlantic horseshoe crab (*Limulus polyphemus*). Ecotoxicology **26**, 46–57 (2017)

M.L. Botton, The ecological importance of horseshoe crabs in estuarine and coastal communities: A review and speculative summary, in *Biology and Conservation of Horseshoe Crabs*, ed. by J. T. Tanacredi, M. L. Botton, D. R. Smith, (Springer, New York, 2009), pp. 45–64

M.L. Botton, C.P. Colón, J. Rowden, S. Elbin, D. Kriensky, K. McKown, M. Sclafani, R. Madden, Effects of a beach nourishment project in Jamaica Bay, New York on horseshoe crab (*Limulus polyphemus*) spawning activity and egg deposition. Est. Coasts **41**, 974–987 (2018). https://doi.org/10.1007/s12237-017-0337-8

M.L. Botton, T. Itow, The effects of water quality on horseshoe crab embryos and larvae, in *Biology and Conservation of Horseshoe Crabs*, ed. by J. T. Tanacredi, M. L. Botton, D. R. Smith, (Springer, New York, 2009), pp. 439–454

M.L. Botton, K. Johnson, L. Helleby, Effects of copper and zinc on embryos and larvae of the horseshoe crab, *Limulus polyphemus*. Arch. Environ. Contam. Toxicol. **35**, 25–32 (1998a). https://doi.org/10.1007/s002449900344

M.L. Botton, M. Hodge, T.I. Gonzalez, High tolerance to tributyltin in embryos and larvae of the horseshoe crab, *Limulus polyphemus*. Estuaries **21**, 33–40 (1998b). https://doi.org/10.2307/1352480

M.L. Botton, R.E. Loveland, D. Munroe, D. Bushek, J.F. Cooper, Identifying the major threats to American horseshoe crab populations. In: Tanacredi et al. (this volume)

M.L. Botton, R.A. Tankersley, R.E. Loveland, Developmental ecology of the American horseshoe crab, *Limulus polyphemus*. Curr. Zool. **56**, 550–562 (2010)

J. Burger, N. Tsipoura, M. Gochfeld, Metal levels in blood of three species of shorebirds during stopover on Delaware Bay reflect levels in their food, horseshoe crab eggs. Toxics **2017**(5), 20, 9 (2017). https://doi.org/10.3390/toxics5030020

M.S. de Melo, T.P.G. dos Santos, M. Jaramillo, L. Nezzi, Y.M. Rauh Muller, E.M. Nazari, Histopathological and ultrastructural indices for the assessment of glyphosate-based herbicide cytotoxicity in decapod crustacean hepatopancreas. Aquat. Toxicol. **210**, 207–214 (2019). https://doi.org/10.1016/j.aquatox.2019.03.007

T. Itow, R.E. Loveland, M.L. Botton, Developmental abnormalities in horseshoe crab embryos caused by exposure to heavy metals. Arch. Environ. Contam. Toxicol. **35**, 33–40 (1998). https://doi.org/10.1007/s002449900345

L. Janssens, R. Stoks, Stronger effects of roundup than its active ingredient glyphosate in damselfly larvae. Aquat. Toxicol. **193**, 210–216 (2017). https://doi.org/10.1016/j.aquatox.2017.10.028

T.C. Jegla, J.D. Costlow Jr., Temperature and salinity effects on developmental and early post-hatch stages of *limulus*, in *Physiology and Biology of Horseshoe Crabs: Studies on Normal and Environmentally Stressed Animals*, ed. by J. Bonaventura, C. Bonaventura, S. Tesh, (Liss, New York, NY, 1982), pp. 103–113

B. A. John, B. R. Nelson, H. I. Sheikh, S. G. Cheung, Y. Wardiatno, B. P. Dash et al. A review on fisheries and conservation status of Asian horseshoe crabs. Biodivers Conserv **27**(14), 3573–3598 (2018). https://doi.org/10.1007/s10531-018-1633-8

K. Khoder, Toxicity of glyphosate, decylamine, and copper to horseshoe crab cmbryos. Fordham Undergraduate Res. J. **8**(2), 23–27 (2018)

Z. Kissane, J.M. Shephard, The rise of glyphosate and new opportunities for biosentinel early-warning studies. Conserv. Biol. **31**, 1293–1300 (2017)

D.W. Kolpin, E.M. Thurman, E.A. Lee, M.T. Meyer, E.T. Furlong, S.T. Glassmeyer, Urban contributions of glyphosate and its degradate AMPA to streams in the United States. Sci. Tot. Environ. **354**, 191–197 (2006)

B.K.Y. Kwan, A.K.Y. Chan, S.G. Cheung, P.K.S. Shin, Responses of growth and hemolymph quality in juvenile Chinese horseshoe crab *Tachypleus tridentatus* (Xiphosura) to sublethal tributyltin and cadmium. Ecotoxicology **24**, 1880–1895 (2015)

K. Laurie, C.-P. Chen, S.G. Cheung, V. Do, H. Hsieh, A. John, F. Mohamad, S. Seino, S. Nishida, P. Shin, M. Yang, *Tachypleus tridentatus*. The IUCN Red List of Threatened Species 2019, e.T21309A149768986 (2019). https://doi.org/10.2305/IUCN.UK.2019-1.RLTS.T21309A149768986.en. Accessed 8 August 2019

D.S. Mizrahi, K.A. Peters, Relationships between sandpipers and horseshoe crabs in Delaware Bay: A synthesis, in *Biology and Conservation of Horseshoe Crabs*, ed. by J. T. Tanacredi, M. L. Botton, D. R. Smith, (Springer, New York, 2009), pp. 65–87

J.P. Myers, M.N. Antoniou, B. Blumberg, L. Carroll, T. Colborn, L.G. Everett, M. Hansen, P.J. Landrigan, B.P. Lanphear, R. Mesnage, L.N. Vandenberg, F.S. Vom Saal, W.V. Welshons, C.M. Benbrook, Concerns over use of glyphosate-based herbicides and risks associated with exposures: A consensus statement. Environ. Health **15**, 19 (2016). https://doi.org/10.1186/s12940-016-0117-0

S.R. Palumbi, B.A. Johnson, A note on the influence of life-history stage on metabolic adaptation: The responses of *limulus* eggs and larvae to hypoxia, in *Physiology and Biology of Horseshoe Crabs: Studies on Normal and Environmentally Stressed Animals*, ed. by J. Bonaventura, C. Bonaventura, S. Tesh, (Liss, New York, NY, 1982), pp. 115–124

G.L. Pérez, M.S. Vera, L.A. Miranda, Effects of herbicide glyphosate and glyphosate-based formulations on aquatic ecosystems, in *Herbicides and environment*, ed. by A. Kortekamp, (Croatia, InTech Europe, Rijeka, 2011), pp. 343–368

M.E. Richmond, Glyphosate: A review of its global use, environmental impact, and potential health effects on humans and other species. J. Environ. Stud. Sci. **8**, 416–434 (2018). https://doi.org/10.1007/s13412-018-0517-2

K. Saltonstall, Cryptic invasion by a non-native genotype of the common reed, *Phragmites australis,* into North America. Proc. Natl. Acad. Sci. **99**, 2445–2449 (2002)

D.R. Smith, H.J. Brockmann, M.A. Beekey, T.L. King, M.J. Millard, J. Zaldivar-Rae, Conservation status of the American horseshoe crab, (*Limulus polyphemus*): A regional assessment. Rev. Fish Biol. Fisheries **27**, 135–175 (2016a)

D.R. Smith, H.J. Brockmann, M.A. Beekey, T.L. King, M.J. Millard, J. Zaldivar-Rae, *Limulus polyphemus*. The IUCN red list of threatened species 2016: E.T11987A80159830. (2016b). https://doi.org/10.2305/IUCN.UK.2016-1.RLTS.T11987A80159830.en. Downloaded on 04 January 2019

J.M. Teal, S. Peterson, The interaction between science and policy in the control of *Phragmites* in oligohaline marshes of Delaware Bay. Restor. Ecol. **13**, 223–227 (2005)

D.Ö. Turhan, A. Güngördü, M. Ozmen, Developmental and lethal effects of glyphosate and a glyphosate-based product on *Xenopus laevis* embryos and tadpoles. Bull Environ. Contamin. Toxicol. **104**, 173–179 (2020)

M.T.K. Tsui, L.M. Chu, Aquatic toxicity of glyphosate-based formulations: Comparison between different organisms and the effects of environmental factors. Chemosphere **52**, 1189–1197 (2003). https://doi.org/10.1016/s0045-6535(03)00306-0

J.S. Weis, A. Ma, Effects of the pesticide diflubenzuron on larval horseshoe crabs, *Limulus polyphemus*. Bull. Environ. Contam. Toxicol. **39**, 224–228 (1987)

# Feeding Ecology and Dietary Preferences of *Tachypleus gigas* from East Malay Peninsula

**Akbar John, Bryan Raveen Nelson, Hassan I. Sheikh, S. Hajisamae, and Jalal Khan**

## 1 Introduction

Out of the four extant species for horseshoe crabs, namely *Limulus polyphemus*, *Tachypleus gigas*, *T. tridentatus*, and *Carcinoscorpius rotundicauda*, the latter three species are native to Malaysian coastal waters (John et al. 2018). Moreover, sympatric presence of *T. gigas* and *C. rotundicauda* populations in Balok, Pahang on the coast of East Peninsular Malaysia may indicate their habitat overlaps (Nelson et al. 2015; Nelson et al. 2016a). The populations are currently exploited for exotic protein (Nelson et al. 2016b, 2019, John et al. 2021). On a global scale, biomedical applications have

The original version of this chapter was revised. The correction to this chapter is available at https://doi.org/10.1007/978-3-030-82315-3_41

A. John (✉)
Institute of Oceanography and Maritime Studies, Kulliyyah of Science, International Islamic University Malaysia, Kuantan, Pahang, Malaysia

B. R. Nelson
Institute of Tropical Biodiversity and Sustainable Development, Universiti Malaysia, Kuantan, Pahang, Malaysia
e-mail: bryan.nelson@umt.edu.my

H. I. Sheikh
Faculty of Fisheries and Food Science, Universiti Malaysia Terengganu, Kuala Nerus, Terengganu, Malaysia

S. Hajisamae
Faculty of Science and Technology, Prince of Songkla University, Muang Pattani, Thailand

J. Khan
Department of Marine Science, Kulliyyah of Science, International Islamic University Malaysia, Kuantan, Pahang, Malaysia



J. T. Tanacredi et al. (eds.), *International Horseshoe Crab Conservation and Research Efforts: 2007-2020*, https://doi.org/10.1007/978-3-030-82315-3_35

increased the demand for horseshoe crab blue blood, especially *L. polyphemus* and *T. tridentatus*. Export of *T. gigas* from Malaysia and Thailand is anticipated apart from their local use (Zauki et al. 2019a, 2019b). Other threats to horseshoe crabs in Malaysia include impoverished food sources (Tan et al. 2011; John et al. 2012; John et al. 2013), water quality (John et al. 2011), and habitat degradation (John et al. 2012a; Fairuz-Fozi et al. 2018). Feeding ecology and behavior of wild and captive *L. polyphemus* (Able et al. 2019), *T. gigas* (Razak et al. 2017), *T. tridentatus* (Chiu and Morton 2004; Kwan et al. 2015a,b), and *C. rotundicauda* (John et al. 2012) were also examined. Electivity is used to indicate food availability while gastrosomatic index indicates the *T. gigas's* diets (Razak et al. 2017). However, only 20 samples were used in the study, and the results are preliminary and may be inconclusive. The present study was based on the approach described in Razak et al. (2017) but in a one-year timescale. This study enhances our understanding on *T. gigas* feeding ecology and food preference and also the way the population adapts during transitional weather at their spawning grounds.

# 2 Materials and Methods

## 2.1 Description of Study Area

Balok (3°56.194′ N; 103°22.608′ E) and Tanjung Selangor (Pekan) (3°36.181′ N; 103°23.946′ E) in Pahang state are important *T. gigas* spawning sites along the coast of East Malay Peninsula (John et al. 2018; Zauki et al. 2019a; 2019b) (Fig. 1). Adult horseshoe crabs migrate from the offshore to shallow continental shelf waters and spawn on intertidal sandy or sandy mud flats during every full and new moon period. Pahang has an average monthly rainfall between 250 and 300 mm and reaches

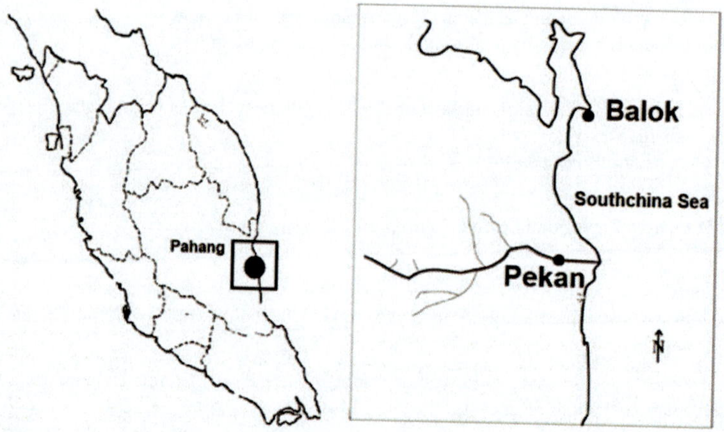


**Fig. 1** Location of the sampling area. Distance between two sampling zones (Balok and Pekan) is 67 km. These two stations were identified to be the major nesting and spawning grounds of horseshoe crabs in Pahang state, Malaysia

>3000 mm annually. The sediments at the sampling location were observed to be soft which may facilitate laying of eggs and their burial by the female horseshoe crab. Meteorological data from November to January reflected the wet monsoon (MMD 2011). Balok and Pekan coastal waters have active vessel movement connecting an upstream jetty and South China Sea to serve fisheries practices.

## 2.2 Sample Collection and Preparation

A total of 93 adult *T. gigas*, including 41 males and 52 females, were collected for gut content analysis by either handpicking or sampled using gill net placed across the nesting ground a day prior to the full and new moon days. The horseshoe crabs were euthanized injecting 50 ml of 70% ethanol at the opisthosoma-carapace hinge using a 16-gauge needle. At the laboratory, sterile scissors and forceps were used to separate gut contents from the gastrointestinal tract before transferring them into petri dishes containing 90% ethanol. After measuring the gastrointestinal length and weight, its gut contents were washed individually with 90% ethanol using dissecting and compound microscopes. The gut content was sorted under the stereoscopic microscope, identified to the lowest possible taxon and counted. Samples that were partially digested were considered as individuals of a representative group of animals and were counted accordingly. To photograph and measure the size of the ingested food stuffs, an epifluorescence microscope was used. In addition to horseshoe crabs, macrobenthos samples were also collected from the two nesting grounds in multiple random quadrats (1 $m^2$) using hand scoop during low tide, and the samples were identified using standard taxonomic keys to the lowest possible taxon.

## 2.3 Data Analysis

The relative abundance of identifiable gut remnants in *T. gigas* was expressed as a percentage of total number of food items in the samples. Unidentified organic matters were considered as macrophytes based on microscopic observation following the method of Ferreira and Vasconcellos (Ferreira Soraia and João 2001). The intensity of feeding was studied by determining the gastrosomatic index (GSI), i.e., gut weight expressed as percentage of body weight. Electivity index (*E*) was calculated to understand *T. gigas* dietary composition and preferences (Chatterji et al. 1992). The *E* value ranged from +1 to −1, with a positive value indicates that a prey type is found in a higher proportion in the diet than in the prey community. A negative value indicates that a prey type is found in lower proportion in the diet than in the prey community. The *E* was calculated using the following formula:

$$E = \frac{r_i - P_i}{r_i + P_i}$$

Where,
$E$ = electivity index
"$r_i$" = the relative abundance (%) of any food item in the gut, expressed as a percentage of total amount of food item.
"$P_i$" = the relative abundance (%) of the same food item in the environment.
GSI = gastrosomatic index was calculated using the following formula:

$$GSI = \frac{Weight\ of\ stomach\ contents}{Weight\ of\ the\ horseshoe\ crab} \times 100$$

# 3 Results and Discussion

## 3.1 Macrobenthos Composition at the Nesting Grounds

Gut content of *T. gigas* was predominated with bivalves, gastropods, polychaetes and crustaceans. Besides these major macrobenthos, there were other miscellaneous groups such as larval insects, amphipods, isopods, unknown larvae, fish larvae, foraminifera, and annelid worms. Among the macrobenthos, bivalves were encountered throughout the sampling periods in both sampling sites. In Balok, bivalve items accounted for 37.2% and 31.5% during the non-monsoon and monsoon seasons, respectively, while they constituted 28% and 21.5% during new moon days and on full moon days, respectively. In Pekan, bivalves accounted for 30.6% and 24.9% during the non-monsoon and monsoon seasons, respectively, while they accounted for 23.9% and 23.3% during new moon and full moon days, respectively. Gastropods were the second most dominant macrobenthic community followed by other organisms (i.e., insects, amphipods, isopods, larval and juvenile fishes, foraminifera, and annelidan worms) and polychaetes (Video 1). Crustaceans were the least dominant in both sampling locations. In general, the macrobenthic organisms were present in higher abundances during full moon compared with the new moon days. It was also apparent that the benthic percentage abundance was higher in Balok than that in Pekan. This might probably be due to the presence of nutrient-rich habitat together with better water quality along the Balok station compared with Pekan (Zauki et al. 2019a; 2019b). Some of the dominant macrobenthos in Balok include bivalves (*Anadara granosa, A. antiquata, Circe* sp., *Tellina virgata, Donax cuneatus, D. variabilis, Myadora striata* and *Codakia orbicularis*), gastropods (*Cerithidea cingulata, C. obtuse, Cerithium litteratum, Neritina* sp., *Littorina scabra, L. fasciata, L. undulata, Cliton oualaniensis*, and *Umbonium vestiarium*), polychaetes (*Ganoderma australe, Lentinus squarrosulus, Microporellus* sp., *Pycnoporus sanguineus, Rigidoporus* sp., *Lenzites elegans, Microporus* sp., and *Trametes* sp.) and crustaceans (juvenile penaeid shrimp, common crabs, and hermit crabs). Previous studies demonstrated that some polychaetes would feed on horseshoe crab eggs (John et al. 2013; Nelson et al. 2016b, Video 2) after habitat disturbance occurred at Balok (Nelson et al. 2019).

## 3.2 Feeding Composition of *T. Gigas*

Feeding composition of *T. gigas* from Balok and Pekan were pooled since similar food items were observed in their guts. Bivalve, gastropod, crustacean, and polychaete were the main food items for male *T. gigas*, while female horseshoe crabs primarily consumed bivalves during the non-monsoon and with the addition of polychaete during the monsoon (Table 1). Despite the fact that male *T. gigas* are smaller (311–383 g body weight, BW) compared with the females (725.7–1025 g BW), the males consumed in greater amount than the females, as observed between March 2010 and January 2011 (Fig. 2). This is probably due to the availability of foregut space in males, and their difference in metabolism and energy demands between sexes (Kyomo 1992). The maximum GSI value in the males was observed during July 2010 (2.56), followed by December 2010 (2.55) and April 2010 (2.53). The minimum GSI value in the males, however, was noted in February 2011 (1.57). The maximum value of GSI in the females was observed in June 2010 (2.07), followed by July 2010 (1.84) and February 2011 (1.79), while the minimum GSI in the

**Table 1** Relative composition of different food items in the gut of male (M) and female (F) *T. gigas* during the monsoon (MS) and non-monsoon (NMS) seasons. Data expressed as the number of individual prey items represented in percentage abundance in the gut

| | | | | Relative composition of food items (%) | | | | |
|---|---|---|---|---|---|---|---|---|
| | Seasons | | GSI | Bivalve | Gastropod | Crustacean | Polychaete | Others |
| M | NMS | Average | 2.24 | 33.66 | 17.08 | 9.81 | 26.11 | 13.34 |
| | | Max | 2.56 | 39.93 | 22.18 | 19.92 | 40.14 | 30.50 |
| | | Min | 1.86 | 22.19 | 8.00 | 0.00 | 16.90 | 7.71 |
| | | SD | 0.40 | 6.47 | 5.88 | 5.48 | 11.61 | 6.73 |
| | MS | Average | 2.04 | 25.80 | 11.91 | 4.73 | 30.45 | 27.11 |
| | | Max | 2.55 | 33.39 | 19.01 | 10.09 | 42.12 | 34.45 |
| | | Min | 1.57 | 19.68 | 6.69 | 0.00 | 18.11 | 20.67 |
| | | SD | 0.40 | 6.47 | 5.88 | 5.48 | 11.61 | 6.73 |
| F | NMS | Average | 1.52 | 35.95 | 17.21 | 10.60 | 27.41 | 8.84 |
| | | Max | 2.07 | 48.19 | 22.11 | 14.32 | 34.21 | 14.14 |
| | | Min | 1.09 | 27.91 | 11.22 | 5.11 | 21.00 | 1.77 |
| | | SD | 0.23 | 4.86 | 2.17 | 9.63 | 2.34 | 12.72 |
| | MS | Average | 1.48 | 26.68 | 11.61 | 11.13 | 31.01 | 19.58 |
| | | Max | 1.79 | 31.20 | 14.12 | 22.21 | 33.18 | 31.92 |
| | | Min | 1.30 | 20.10 | 8.89 | 0.00 | 27.71 | 6.34 |
| | | SD | 0.23 | 4.86 | 2.17 | 9.63 | 2.34 | 12.72 |

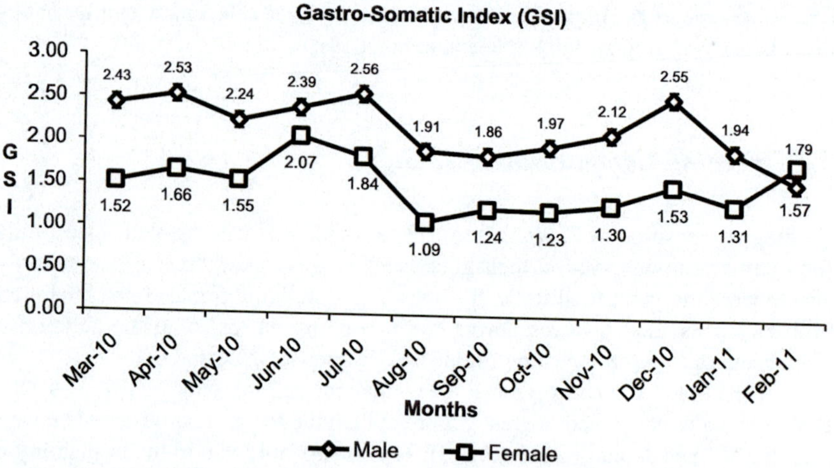


**Fig. 2** Gastrosomatic index shows the monthly variation in the feeding intensity of male and female *Tachypleus gigas*

females was noted in August 2010 (1.09) (Fig. 2). This observation was due to seasonal variation in feeding intensity of the crabs as recorded in the previous studies (Prasad and Neelakantan 1988; Pramanik and Mohanty 2016). Average feeding intensity values of the males during the monsoon and non-monsoon seasons were 2.04 ± 0.4 and 2.24 ± 0.4, respectively (Table 1). Mean feeding intensity values of the females during the monsoon and non-monsoon seasons were 1.48 ± 0.23 and 1.52 ± 0.23, respectively. Both male and female *T. gigas* endure exertive currents when migrating from deep into shallow waters. This also implies that they must feed to satiation before attempting the long-distance travel to the spawning grounds. Horseshoe crabs were previously reported to rely on bottom currents for long-distance locomotion (John et al. 2018). Their body weight will increase along with feeding activity regardless of their sex. For the males, a higher body weight may allow the increase of gravity to maintain an upright position when moving along inshore bottom currents. On the other hand, female *T. gigas* are larger in size to store more eggs in the bodies. Consuming more food increases chances of sinking and energy when traveling. Therefore, feeding less allows the females to remain buoyant when moving along the inshore bottom currents. Earlier research has indicated that intertidal sand flats with low energy currents are preferred by horseshoe crab as foraging grounds (Botton 1984; John et al. 2012; Botton et al. 2003). In addition, seabed perturbations cause flocculation and promote crustacean and polychaete thriving (Pinotti et al. 2014). Consequently, gastrosomatic index rose in February 2011 and reached 1.79, possibly due to additional crustacean and polychaete portions in female *T. gigas* diets. This finding indicated that both male and female *T. gigas* may migrate to and feed intensively in the shallow waters during the wet monsoon season (November to March) when the wave action is turning stronger.

### 3.3 Natural Food Selectivity

The electivity index (*E*) calculated for the frequent food items during monsoon and non-monsoon seasons is presented in Fig. 3. Throughout the sampling period, the *E* was predominantly negative for gastropods and crustaceans, whereas positive values were found for bivalves, polychaetes, and miscellaneous food items (i.e., insects, amphipods, isopods, larval and juvenile stages of fishes, foraminifera, and annelid worms). During the non-monsoon season, *T. gigas* foraged on bivalves instead of polychaetes. This trend was reversed during the wet monsoon season. It was also apparent that during the peak mating season (June to August 2010), *T. gigas* may become less selective and prey on more food items. Seasonal variations in food composition showed that mollusks, especially bivalves, remained as the dominant food item. Unidentified organic matters in the gut of *T. gigas* were mostly macrophytes, apart from the presence of a considerable amount of sand particles in the foregut. Considering the uncertainty about the foraging time for *T. gigas*, we assumed that the negative values in *E* attained in this study would mean *T. gigas* has become selective and concentrates its effort on that resource. Comparisons in Table 1 indicated the gastropods and crustaceans are alternative but encompass major food portions for horseshoe crabs.

Recent studies on horseshoe crabs suggest that it consumes a variety of food items, including particulate organic matter (POM, such as algae), which may play a significant role in their diet (Gaines et al. 2002; Carmichael et al. 2004; Carmichael and Valiela 2005). We observed floc deposition with sedimentation, an attribute for organic matter enrichment in Balok (Nelson et al. 2016a; Zauki et al. 2019b). POM was reported to be major component in the diet of juvenile *L. polyphemus*,

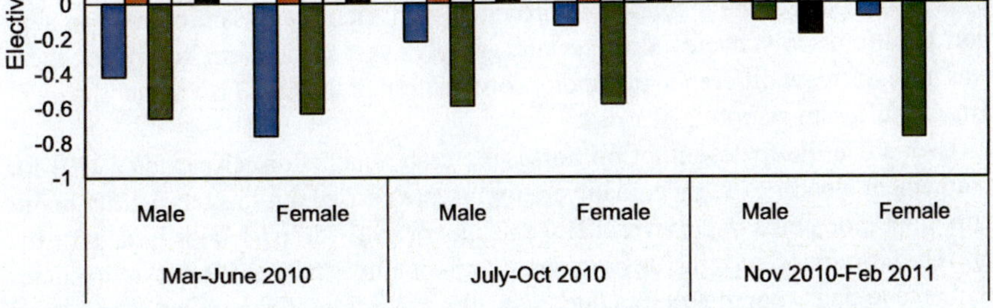


**Fig. 3** The electivity index (*E*) of different major food items consumed by *Tachypleus gigas* during March 2010 to February 2011

especially for the second to seventh instar, while the larger instars (8th–11th) relied less on POM and fed mainly on benthic organisms such as amphipods, isopods, and polychaete (Carmichael et al. 2009; Fan et al. 2017). POM was also found to be essential for *T. tridentatus*, *C. rotundicauda*, and *L. polyphemus*, especially in early stages. Seagrass and seagrass-derived organic matter were also crucial for juvenile *T. tridentatus* and *C. rotundicauda*; however, POM does not contribute as the major food item for the juveniles (Carmichael et al. 2009; Fan et al. 2017; Noor Jawahir et al. 2017). Our findings reported a dietary "calendar" for *T. gigas* where portions of food items changed with season due to the prey availability.

The relative percentage of food items in *T. gigas* guts varied monthly. This was correlated to the abundance of that particular food item in the environment. Such case in *T. gigas* was similar to selective feeding behavior as previously reported in many marine animals, despite their multiple feeding strategies, including filter feeding, deposit feeding, and hunting (Hughes 1980). To the best of our knowledge, studies concerning feeding preference and behavior of *T. gigas* are largely lacking, and therefore, we have to refer the relevant knowledge from their counterparts in Indian waters (Chatterji et al. 1992). The relationship between prey size and mouth size of horseshoe crabs were addressed in literature (Botton 1984; Botton and Haskin 1984; Botton et al. 2003; Botton and Shuster 2004). The highest values for gastrosomatic indices in our study confirmed that 87% of adult *T. gigas* diet comprises bivalves. Feeding behavior and food preference study on mangrove horseshoe crabs, *C. rotundicauda*, demonstrated that they prefer polychaetes more than bivalves during the non-monsoon season, while in monsoon season, their preferred gastropods (John et al. 2012). We hypothesize that the spawning season of gastropods and polychaetes was prior to the monsoon seasons (Hughes 1980; Pinotti et al. 2014), leading to their increase in abundance and subsequently contributing to horseshoe crab diets. Nevertheless, macrobenthos availability and their flourishing (Kwan et al. 2015a; b) may influence the shift in *T. gigas* feeding regimes.

Monthly variation in feeding intensity of *T. gigas* showed that male horseshoe crabs fed more intensely than the females. GSI analysis also showed that the adults fed more intensely during their mating season. No considerable variation was observed in the feeding intensity between the monsoon and non-monsoon seasons. An increased in food preference during mating season might be due to the increased availability of preferred food items in the nesting grounds. Kwan et al. (Kwan et al. 2015a, b) reported that a healthy environment with rich seagrass biomass can support the life of polychaetes, crustaceans, and bivalves. In such environments, there was no significant difference in the diet composition of the 6th–11th instar *T. tridentatus* in different sampling seasons.

Ever since the first attempt on horseshoe crab aquaculture (Kropach 1979) for biomedical bleeding practice, many attempts were carried out to successfully maintain horseshoe crabs in captivity (Tinker-Kulberg et al. 2020). Major issues for the present husbandry system for horseshoe crabs include the nutritional deficiencies that lead to panhypoproteinemia due to an administration of either single or limited number of mixed feed (Nolan and Smith 2009). However, in general, wild horseshoe crabs do not face this problem due to the availability and accessibility to varying food sources. In such condition, gut content analysis data presented in this study could also be used to formulate better feed by understanding the feeding behavior and food preference of horseshoe crabs in the natural environment.

## 4 Conclusion

Adult *T. gigas* had different diets throughout the year as changes in seasonal weather and habitat condition influenced their foraging preferences. Our study showed the annual dietary "calendar" for *T. gigas*, which feed mainly on bivalves during the non-monsoon, while polychaetes were the major diet constituents during the monsoon seasons. Female fed on a mixed diet containing bivalves and polychaetes, while males fed mostly on mollusk (i.e., bivalves and gastropods). During the monsoon season when rough tides and heavy raining occur, *T. gigas* preferred polychaetes than bivalves, which was probably due to the easier access to polychaetes in the season. In general, adult horseshoe crabs were demonstrated to feed on different mollusks across the ranges. More research work can be carried out to determine the role of *T. gigas* in coastal food webs by addressing the dietary differences between adult and juvenile crabs. The obtained information in this study can also be used in long-term culture of *T. gigas*.

**Acknowledgement** This research was financially supported under Fundamental Research Grant Scheme (FRGS 15- 210-0451 and FRGS19-021-0629, Ministry of Higher Education, Malaysia.

**Conflict of Interest** Authors declare no conflict of interest.

## References

K.W. Able, P.C. López-Duarte, T.M. Grothues, et al., Limulus polyphemus (Atlantic horseshoe crab) habitats in a small estuary and the adjacent inner continental shelf: Linkages across mid-Atlantic bight seascapes. Northeast. Nat. **26**(2), 287–304 (2019)

B.K.Y. Kwan, S.G. Cheung, P.K.S. Shin, A dual stable isotope study for diet composition of juvenile Chinese horseshoe crab *Tachypleus tridentatus* (Xiphosura) on a seagrass-covered intertidal mudflat. Mar. Biol. **162**, 1137–1143 (2015a)

M.L. Botton, Diet and food preferences of the adult horseshoe crab *Limulus polyphemus* in Delaware Bay, New Jersey, USA. Mar. Biol. **81**, 199–207 (1984)

M.L. Botton, C.N. Shuster Jr., J.A. Keinath, Horseshoe crabs in a food web: Who eats whom? in *The American Horseshoe Crab*, ed. by C. N. Shuster Jr., R. B. Barlow, H. J. Brockmann, (Harvard University Press, Cambridge, 2003), pp. 135–153

M.L. Botton, H.H. Haskin, Distribution and feeding of the horseshoe crab, *Limulus polyphemus*, on the continental shelf off New Jersey. Fish. Bull. **82**, 383–389 (1984)

R.H. Carmichael, E. Gaines, Z. Sheller, A. Tong, A. Clapp, I. Valiela, Diet composition of juvenile horseshoe crabs: Implications for growth and survival of natural and cultured stocks, in *Biology and Conservation of Horseshoe Crabs*, ed. by J. T. Tanacredi, M. L. Botton, D. Smith, (Springer, Boston, MA, 2009), pp. 521–534

R.H. Carmichael, I. Valiela, Coupling of near-bottom seston and surface sediment composition: Changes with nutrient enrichment and implications for estuarine food supply and biogeochemical processing. Limnol. Oceanogr. **50**(1), 97–105 (2005)

R.H. Carmichael, D. Rutecki, B. Annett, E. Gaines, I. Valiela, Position of horseshoe crabs in estuarine food webs: N and C stable isotopic study of foraging ranges and diet composition. J. Exp. Mar. Biol. Ecol. **299**, 231–253 (2004)

A. Chatterji, J.K. Mishra, A.H. Parulekar, Feeding behaviour and food selection in the horseshoe crab, *Tachypleus gigas* (Müller). Hydrobiologia **246**, 41–48 (1992)

N. Fairuz-Fozi, B. Satyanarayana, N.A.M. Zauki, A.M. Muslim, M.L. Husain, S. Ibrahim, B.R. Nelson, Carcinoscorpius rotundicauda (Latreille, 1802) population status and spawning behaviour at pendas coast, peninsular Malaysia. Glob. Ecol. Conserv. **15**, e00422 (2018)

L. Fan, C. Chen, M.C. Yang, G. Qiu, Y. Liao, H.L. Hsieh, Ontogenetic changes in dietary carbon sources and trophic position of two co-occurring horseshoe crab species in southwestern China. Aquat. Biol. **26**, 15–26 (2017)

A. Ferreira Soraia, V.-N. João, Host plants of the grasshopper *Cornops aquaticum* (Bruner) (Orthoptera: Acrididae) in the wetland of Poconé, MT, Brazil. Neotrop. Entomol. **30**(4), 523–533 (2001)

E.F. Gaines, R.H. Carmichael, S.P. Grady, I. Valiela, Stable isotopic evidence for changing nutritional sources of juvenile horseshoe crabs. Biol. Bull. **203**(2), 228–230 (2002)

R.N. Hughes, Optimal foraging theory in the marine context. Oceanogr. Mar. Biol. Ann. Rev. **18**, 423–481 (1980)

H.M. Chiu, B. Morton, The behaviour of juvenile horseshoe crabs, *Tachypleus tridentatus* (Xiphosura), on a nursery beach at Shui Hau wan, Hong Kong. Hydrobiologia **523**(1–3), 29–35 (2004)

B.A. John, B.Y. Kamaruzzaman, K.C.A. Jalal, K. Zaleha, Hydrology of the horseshoe crab nesting grounds at Pahang coast, Malaysia. Orient. J. Chem. **27**(4), 1475–1483 (2011)

B.A. John, B. Kamaruzzaman, K. Jalal, K. Zaleha, Feeding ecology and food preferences of *Carcinoscorpius rotundicauda* collected from the Pahang nesting grounds. Sains Malaysiana **41**(7), 855–861 (2012)

B.A. John, B.R. Nelson, H.I. Sheikh, S.G. Cheung, Y. Wardiatno, B.P. Dash, K. Tsuchiya, Y. Iwasaki, S. Pati, A review on fisheries and conservation status of Asian horseshoe crabs. Biodivers. Conserv. **27**(14), 3573–3598 (2018)

B.A. John, K.C.A. Jalal, B.Y. Kamaruzzaman, Macrobenthic diversity in horseshoe crab nesting ground–Balok station, Pahang, Malaysia. Orient. J. Chem. **29**(4), 1311–1324 (2013)

John, A., Shin, P.K.S., Botton, M.L. et al. Conservation of Asian horseshoe crabs on spotlight. Biodivers. Conserv. **30**, 253–256 (2021). https://doi.org/10.1007/s10531-020-02078-3

B.A. John, B.Y. Kamaruzzaman, K.C.A. Jalal, K. Zaleha, Sediment profiling of the nesting grounds of horseshoe crabs at east peninsular Malaysia. Internat. J. Biol. **4**(2), 159–165 (2012a)

C. Kropach, Observations on the potential of *limulus* aquaculture in Israel, in *Biomedical Applications of the Horseshoe Crab (Limulidae)*, ed. by E. Cohen, F. B. Bang, J. Levin, J. J. Marchalonis, T. G. Pistole, R. A. Prendergast, C. Shuster, S. W. Watson, (Alan R. Liss, Inc., New York, 1979), pp. 103–106

B.K. Kwan, S.G. Cheung, P.K. Shin, A dual stable isotope study for diet composition of juvenile Chinese horseshoe crab *Tachypleus tridentatus* (Xiphosura) on a seagrass-covered intertidal mudflat. Mar. Biol. **162**(5), 1137–1143 (2015b)

J. Kyomo, Variations in the feeding habits of males and females of the crab *Sesarma intermedia*. Mar. Ecol. Progr. Ser. **83**(2), 151–155 (1992)

MMD (Malaysian Meteorological Department), Monthly rainfall review (2011). http://www.met.gov.my/index.php?option=com_content&task=view&id=31&Itemid=1012 (Accessed on 13 July 2020)

B.R. Nelson, B. Satyanarayana, J.H.Z. Moh, M. Ikhwanuddin, A. Chatterji, F. Shaharom, The final spawning ground of *Tachypleus gigas* (Müller, 1785) on the east peninsular Malaysia is at risk: A call for action. Peer. J. **4**, e2232 (2016a)

B.R. Nelson, B. Satyanarayana, J.H.Z. Moh, F. Shaharom, Does human infringement at the spawning grounds challenge horseshoe crab eggs and their embryogenesis. J. Sustain. Sci. Manage. **1**, 1–10 (2016b)

B.R. Nelson, B. Satyanarayana, J.M.H. Zhong, F. Shaharom, M. Sukumaran, A. Chatterji, Episodic human activities and seasonal impacts on the *Tachypleus gigas* (Müller, 1785) population at Tanjung Selangor in peninsular Malaysia. Estuar. Coast. Shelf Sci. **164**, 313–323 (2015)

B.R. Nelson, J.M.H. Zhong, N.A.M. Zauki, B. Satyanarayana, A.J.K. Chowdhury, Effects of shore sedimentation to *Tachypleus gigas* (Müller, 1785) spawning activity from Malaysian waters. J. Sustain. Sci. Manage. **14**(1), 41–60 (2019)

A.R.M. Noor Jawahir, M.L. Samsur, A.R. Shabdin, A.R. Khairul-Adha, Distribution of two species of Asian horseshoe crabs at west coast of Sarawak's waters, East Malaysia. Egyptian. J. Aquat. Res. **43**, 135–140 (2017)

R.M. Pinotti, D.M. Minasi, L.A. Colling, C.E. Bemvenuti, A review on macrobenthic trophic relationships along subtropical sandy shores in southernmost Brazil. Biota. Neotropica. **14**(3), 1–12 (2014)

D.S. Pramanik, S.S. Mohanty, Length-weight relationship and biology of some common edible fish species at Chandipur, bay of Bengal, Odisha. Internat. J. Fish. Aquat. Studi. **4**, 335–340 (2016)

P.N. Prasad, B. Neelakantan, Food and feeding of the mud crab *Scylla serrata* Forskal (Decapoda: Portunidae) from Karwar waters. Indian J. Fish. **35**, 164–170 (1988)

M.R.M. Razak, Z. Kassim, A.A. Sabuti, A. Ismail, Feeding ecology and food preferences of Cherok Paloh, Pahang horseshoe crab, *Tachypleus gigas*. Malaysian J. Fundament. Appl. Sci. **13**(3), 198–202 (2017)

A.N. Tan, A. Christianus, M.A. Satar, Epibiont infestation on horseshoe crab *Tachypleus gigas* (Müller) at Pantai Balok in peninsular Malaysia. Our Nature **9**(1), 9–15 (2011)

R. Tinker-Kulberg, K. Dellinger, T.E. Brady, L. Robertson, J.H. Levy, S.K. Abood, F.M. LaDuca, C.L. Kepley, A.L. Dellinger, Horseshoe crab aquaculture as a sustainable endotoxin testing source. Front. Mar. Sci. **7**, 153 (2020)

N.A.M. Zauki, B. Satyanarayana, N. Fairuz-Fozi, B.R. Nelson, M.B. Martin, B. Akbar-John, A.J.K. Chowdhury, Citizen science frontiers horseshoe crab population regain at their spawning beach in east peninsular Malaysia. J. Environ. Manag. **232**, 1012–1020 (2019a)

N.A.M. Zauki, B. Satyanarayana, N. Fairuz-Fozi, B.R. Nelson, M.B. Martin, B. Akbar-John, A.J.K. Chowdhury, Horseshoe crab bio-ecological data from Balok, east coast peninsular Malaysia. Data Brief **22**, 458–463 (2019b)

# Patterns of Flatworm (*Bdelloura candida*) Infestation in a Population of Horseshoe Crabs (*Limulus polyphemus*) from Jamaica Bay, New York

**Royall McMahon Ward and Mark L. Botton**

## 1 Introduction

The American horseshoe crab, *Limulus polyphemus*, is a marine arthropod that inhabits the Atlantic coastline of North America, from Mexico to Maine (Botton et al. 2018). *L. polyphemus* spawns on estuarine beaches as described by many authors (e.g., Brockmann 2003). Young-of-the-year and early juveniles are commonly found on intertidal flats and nearshore benthic habitats (Botton et al. 2003a); as they mature, they inhabit coastal bays or the inshore continental shelf (Smith et al. 2016).

Throughout their distribution along the Gulf and Atlantic coasts, the exoskeleton of adult horseshoe crabs serves as a substrate for numerous marine organisms. Epibionts, such as algae, barnacles, bivalves, gastropods, sea anemones, polychaetes, and bryozoans (Botton et al. 2003b), are believed to have minimal effects on their host in the majority of occurrences. More serious conditions result from green algae invasions of the carapace and cyanobacterial (*Oscillatoria* spp.) infections of the book gills (Leibovitz and Lewbart 2003). The digenetic trematode, *Microphallus limuli*, uses *Limulus polyphemus* as its intermediate host, and encysted metacercaria have been found in various tissues including the nervous system, connective tissue, and muscle (Stunkard 1951).

Several species of triclad flatworms (Platyhelminthes, Tricladida, Maricola) are associated with all four living species of horseshoe crabs (Leibovitz and Lewbart

R. M. Ward
Environmental Science Program, Fordham University, New York, NY, USA

M. L. Botton (✉)
Environmental Science Program, Fordham University, New York, NY, USA

Department of Natural Sciences, Fordham University, New York, NY, USA
e-mail: botton@fordham.edu



J. T. Tanacredi et al. (eds.), *International Horseshoe Crab Conservation and Research Efforts: 2007-2020*, https://doi.org/10.1007/978-3-030-82315-3_36

2003). Two of these species (*Bdelloura candida* and *Syncoelidium pellucidum*) reside on *L. polyphemus* (Leibovitz and Lewbart 2003), especially on the book gills and on the legs surrounding the mouth. *S. pellucidum*, the smaller of the two species, is considered to be a commensal that does not cause gill lesions, whereas *B. candida* "…is a true parasite and a very important pathogen of the modern adult Atlantic horseshoe crab" (Leibovitz and Lewbart 2003, p. 264). Damage to the host ensues from the feeding by adult flatworms on hemolymph within the gill lamellae and from their egg cocoons, which are attached to the gills and damage the integrity of the gill surface, potentially interfering with respiration (Leibovitz and Lewbart 2003; Nolan and Smith 2009; Tan et al. 2012). All horseshoe crab populations sampled from the Gulf of Mexico and Atlantic coasts of the United States were found to harbor *B. candida* (Riesgo et al. 2017).

Members of the genus *Bdelloura* are "characterized by the absence of eye lenses, a posterior adhesive caudal disk set off from the rest of the body, and numerous testes distributed throughout the body" (Riesgo et al. 2017). The genus includes three described species, *Bdelloura candida*, *B. propinqua*, and *B. wheeleri*, where *B. candida* is the most widespread and easily identified (Roux and Giribet 2020). *B. candida* generally has a whitish coloration with an oval-shaped body, where they are approximately 15 × 4 mm (Sluys 1989). *B. candida* lives its entire life cycle on the ventral surfaces of its host, *L. polyphemus*. *B. candida* depends on its host for nutrition, where they aggregate on the appendages of the horseshoe crab, feeding on the food scraps that the horseshoe crab produces as it eats. *B. candida* reproduces by depositing stalked cocoons inside the inner wall of the gill lamellae (Groff & Leibovitz 1982; Riesgo et al. 2017). *B. candida* has very limited dispersal capabilities, partly because the egg cocoons are attached to the gill lamellae, and adult *B. candida* cannot survive independently of its host. Thus, they rely on the mating behavior of *L. polyphemus* and chemical signaling to recolonize the same host or switch to an adjacent individual (Roux and Giribet 2020).

*B. candida* is concomitant with *L. polyphemus* along the Atlantic and Gulf of Mexico coasts of the United States (Riesgo et al. 2017; Roux and Giribet 2020). Yet, despite its extensive distribution, there is no previous study of the prevalence of infestation within a population, or whether infestation correlates with the sex, size, or carapace condition of the hosts. In this study, we document the extent of flatworm infestation of adult and juvenile horseshoe crabs from Jamaica Bay, New York. For adult horseshoe crabs, in addition to the quantitative analysis of the extent of infestation, we tested three hypotheses: (1) Are poorer condition horseshoe crabs more susceptible to being infected? (2) Does the size of the host crab affect the prevalence of infestation? and (3) Are adult male and female horseshoe crabs infected at similar frequencies?

# 2 Methods

## *2.1 Data Collection of Adult Horseshoe Crabs*

All research was conducted at Plumb Beach in Jamaica Bay, New York (Fig. 1), which has a large population of spawning horseshoe crabs from mid-May to mid-June (Botton et al. 2018). Adult horseshoe crabs were sampled between 3 June and 1 July 2019. We randomly selected horseshoe crabs as we walked the length of Plumb Beach, either collecting animals during high tide spawning activity, or when they were buried on the tidal flats at Plumb Beach-East during low tide. We determined the sex of the animal based on the presence of claspers, a male appendage used for amplexus. Prosoma width was measured to the nearest millimeter using a measuring tape across the ventral surface.

After an initial scouting trip to gauge the general extent of flatworm infestation, we developed an infestation scale from 0 to 4 for the presence of adult flatworms in the legs or gills and for the presence of egg cocoons in the gill leaflets. The large number of adult worms and egg cocoons made it impractical to count them individually. For adult flatworms, a score of 0 was given if no adults were present (Fig. 2a), and a score of 4 indicates a high number (many hundreds) of adult flatworms in the legs or surrounding the mouth (Fig. 2b). For egg cocoons, a score of 0 was given if none were present, and a score of 4 indicates the highest degree of infestation (every gill leaflet 70% or more covered with egg cocoons). Scores of 2

**Fig. 1** Plumb Beach, Jamaica Bay. (Modified from Botton et al. 2018)

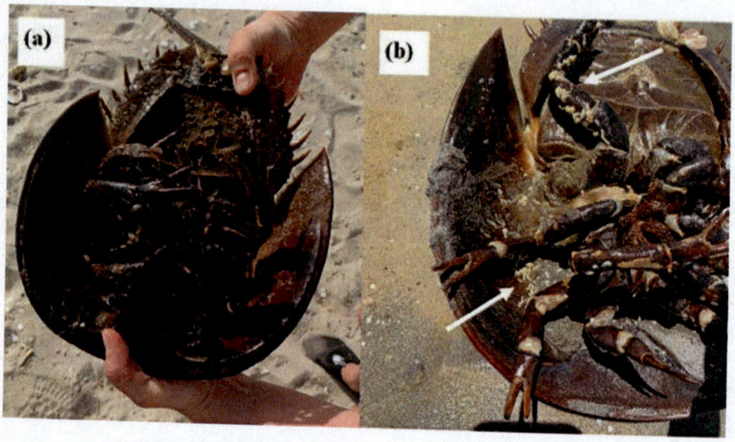


**Fig. 2** Ventral surfaces of adult *L. polyphemus*: (**a**) a *L. polyphemus* with no visible infestation by adult *B. candida* in legs (score of 0) and (**b**) a *L. polyphemus* with a very heavy infestation by adult *B. candida* in legs (score of 4). Arrows show concentrations of adult *B. candida*

or 3 represented intermediate levels of infestation. Each adult was scored separately on the 1–4 scale for adult flatworms in gills, adult flatworms in legs, and egg cocoons in gills. We assessed adult carapace condition scale on a scale of 1–3 based on Duffy et al. (2006). A score of 1 was given for a good condition horseshoe crab, characterized by a shiny and smooth carapace, clear eyes, and only a small number of fouling organisms such as barnacles, a score of 3 was given to crabs that had extensive carapace wear and erosion, often with decay of one or both lateral eyes, and a score of 2 crab was given to crabs with intermediate carapace condition. All statistical analysis was done using Excel.

## 2.2 *Data Collection for Juvenile Horseshoe Crabs*

During low tides from late May to mid-June, 2019, when the tidal flats on Plumb Beach-East were exposed, we collected and examined 40 juvenile horseshoe crabs of various ages. Prosoma width was measured using Vernier calipers. Using a pocket magnifying glass, we inspected the ventral side of the crabs to look for the presence either adult *B. candida* or egg cocoons in the gill lamellae.

## 2.3 *Confirmation of Flatworm Species and Photography*

Samples of adult *B. candida* and the egg cocoons embedded in the gill leaflets were collected throughout June 2019. From heavily infested horseshoe crabs, 5–10 flatworms were taken using a plastic spatula and placed in a capped plastic tube with bay water to ensure that the flatworm remained alive and viable for microscopy. Samples of the egg cocoons were retrieved from freshly deceased horseshoe crabs found on the beach. Using scissors, 2–3 whole book gills were removed and placed in a tube with 70% ethyl alcohol. In the laboratory, adult flatworms were sedated in a 10% ethyl alcohol solution to reduce their movements, then placed in a small glass dish with seawater for examination. Adult flatworms and egg cocoons were photographed using an Olympus dissecting microscope with a digital camera, and their size and morphology were compared with descriptions in Leibovitz and Lewbart (2003) to confirm species identification.

## 3 Results

### 3.1 General Extent

We examined a total of 93 adult and 40 juvenile American horseshoe crabs between June 3 and July 1, 2019. There was a 100% infestation of all 93 adult horseshoe crabs with *B. candida*, either adult flatworms or egg cocoons. There was a 0% infestation of the 40 juvenile horseshoe crabs, although both juvenile and adult horseshoe crabs co-occur on the same tidal flats.

### 3.2 Flatworm Species Identification

We used the number of developing embryos, size, and the short pedicle at the base attaching the cocoon to the gill leaflet to identify the species as *Bdelloura candida* (Fig. 3). The species of flatworm was further confirmed through the dissection of the unhatched flatworm egg cocoon, which revealed eight developing embryos, as is specific to *B. candida* (Leibovitz and Lewbart 2003).

### 3.3 Correlation of Prosoma Width to Infestation

For males, we found no significant correlations between prosoma width size and adult flatworms in gills, legs, or egg cocoons in gills (Fig. 4). We found the same results for just females. When the data was pooled for both sexes, we found a positive correlation between prosoma width and adult flatworms in the legs ($R = 0.331$, 82 df, $P < 0.01$). Similarly, there were positive correlations between prosoma width and adult flatworms in the gills ($R = 0.435$, 90 df, $p < 0.01$) and cocoons on gills ($R = 0.421$, 91 df, $p < 0.01$). It is important to note that female horseshoe crabs are significantly larger than males, where the average prosoma width of the 67 males

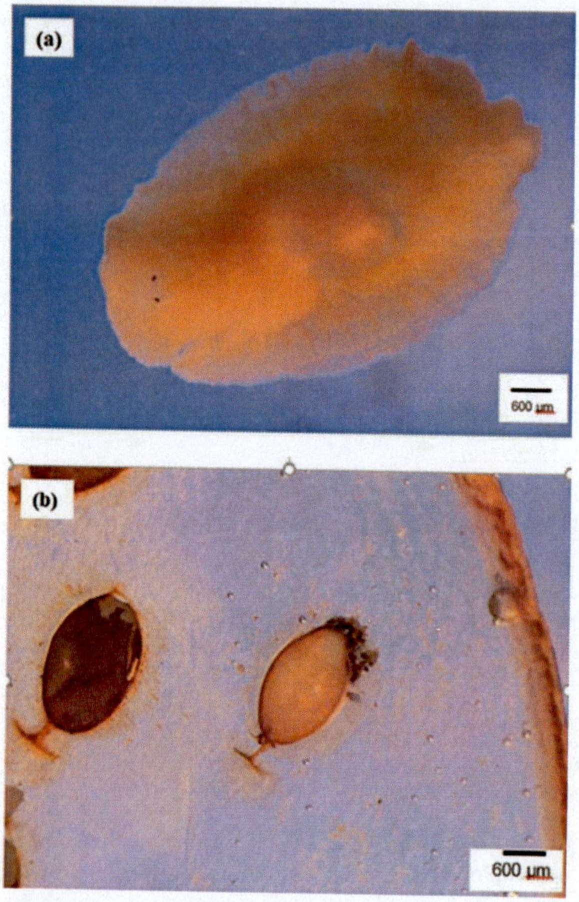


**Fig. 3** (**a**) Adult *Bdelloura candida;* (**b**) *B. candida* egg cocoons attached to the inner wall of the gill lamellae. The hatched egg cocoon is a darker brown coloration (*left*), whereas the lighter tan egg cocoon (*right*) still has *B. candida* embryos

was 214.7 mm and the average width of 26 females was 258.6 mm, 43.9 mm wider than the average male (Student's t-test, 92 df, $P < 0.0001$).

## 3.4 Sex and Carapace Condition and Flatworm Infestation

To measure the relationship between the sex of the horseshoe crab and the presence of flatworms in the legs, gills, and flatworm eggs in the gill leaflets, we completed three tests for each category. We were only able to use condition 2 and condition 3 crabs because of a lack of condition 1 female horseshoe crabs in our data sample. Most horseshoe crabs at Plumb Beach tend to be older crabs and thus have poorer

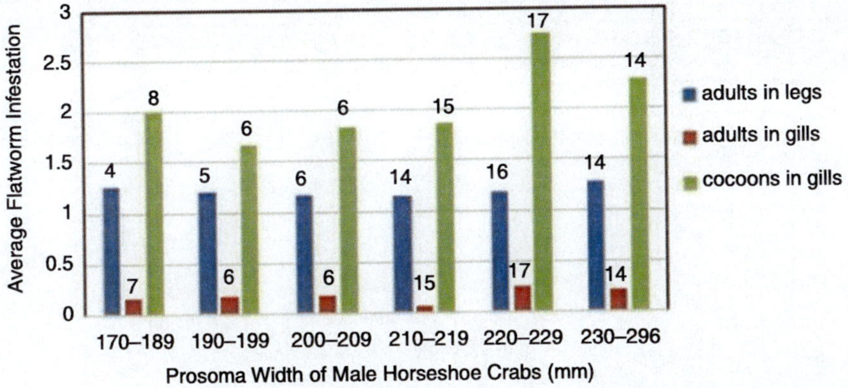


**Fig. 4** Prosoma width of male horseshoe crabs compared to average flatworm infestation (*Numbers above bars = sample size)*

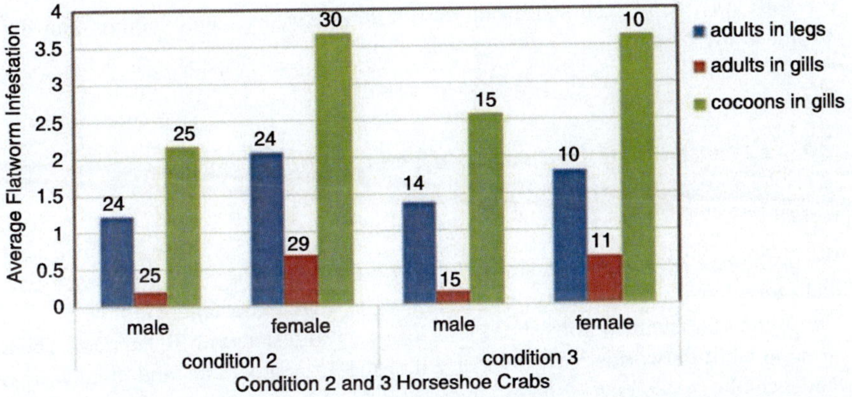


**Fig. 5** Horseshoe crab sex compared to average flatworm infestation (*Numbers above bars = sample size)*

carapace condition. Females had significantly heavier infestation of adult flatworms in gills, legs, and egg cocoons in gills than males (t-test, $P < 0.001$) (Fig. 5). All three measures of infestation were approximately twice as high for condition 2 females than condition 2 males. This same trend is consistent with condition 3 horseshoe crabs, but is less extreme, with females slightly less than double the infestation score of males.

Condition 1 males had the lightest degree of flatworm infestation (Fig. 6) (for this analysis, we only used male horseshoe crabs because there were no condition 1 female horseshoe crabs in our sample). We found a significant correlation between male condition and adult flatworms on legs ($R = 0.324$, 58 df, $p < 0.01$), with condition 3 males having the heaviest infestation. There was a significant correlation between male condition and flatworm egg cocoons in gills ($R = 0.435$, 65 df,

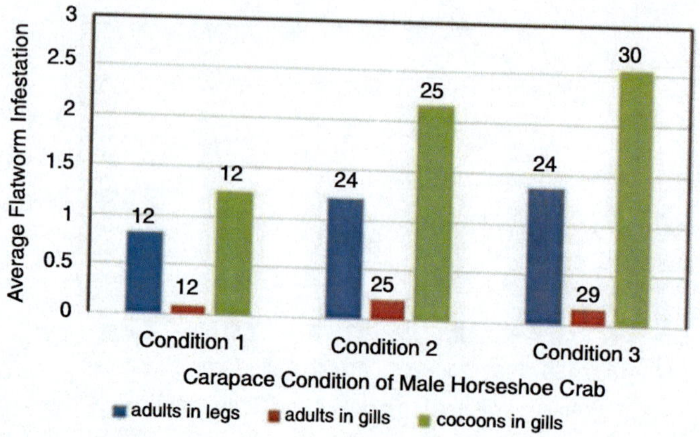


**Fig. 6** Carapace condition of male horseshoe crabs compared to average flatworm infestation (*Numbers above bars = sample size*)

$p < 0.01$). There was no significant correlation between male condition and flatworms in the gills.

### 3.5 *Prevalence of Flatworm Infestation on Juvenile Horseshoe Crabs*

We examined 40 juvenile horseshoe crabs on the tidal flats of Plumb Beach-East that ranged in size from 12 mm to 53 mm wide, with an average width of 24 mm. Out of the 40 examined horseshoe crabs, 0% were infested with *B. candida*. There were no adult flatworms in the legs or gill of the horseshoe crab, and there was no evidence of egg cocoons in the gill leaflets.

## 4 Discussion

In the Plumb Beach population, 100% of the adult horseshoe crabs had a flatworm infestation, regardless of sex or condition. Prosoma width, when using the combined population of males and females, had a significant relationship to the severity of infestation. Based on the $R^2$ value, prosoma width explains 11% of the presence of flatworm infestation in the legs, 19% of flatworm infestation in gills, and 18% of egg cocoons in gills. As the prosoma width increases, the overall presence of flatworm infestation also increases. However, within sexes, there was no significant correlation between prosoma width and infestation of flatworms.

Females had a significantly heavier infestation of flatworms in gills, legs, and egg cocoons in gills (Fig. 5). When completing the comparison between female and male horseshoe crabs, only condition 2 and 3 crabs could be used due to the lack of condition 1 females found at Plumb Beach. Condition 1 horseshoe crabs were rare in our sample, which is typical of horseshoe crabs at this site (Mattei et al. 2015). Females showed a heavier infestation than males for both the condition 2 and condition 3 animals. Males in poorer condition were more heavily infected with adult flatworms in legs and flatworm egg cocoons in the gills. As the carapace condition of the crab gets progressively worse, the infestation of *B. candida* increases for all categories except adult flatworms in gills. As carapace condition is an approximation of the relative age of the crab, or at least the length of time since its last molt (Shuster 2009), this strongly suggests that flatworm infestation of adults accumulates over time.

At Plumb Beach, adult and juvenile horseshoe crabs co-occur on the same tidal flats and share water and sediments. The 100% infection of flatworms on adult horseshoe crabs and their complete absence from juveniles suggests that the transference of *B. candida* among hosts is probably through prolonged physical attachment and interaction along with chemical signaling (Roux and Giribet 2020). One action that could facilitate the spread could be the spawning behavior of the crabs themselves, where male horseshoe crabs attach to the female's opisthosoma using their clasper appendage, and satellite males tend to aggregate around amplexed pairs (Brockmann 2003). Spawning in Jamaica Bay occurs from early May through at least mid-June (Botton et al. 2018); thus, there are ample opportunities for adult flatworms to move from host to host.

We consistently saw a smaller extent of infestation for the adult flatworms in gills compared with legs (Figs. 4 and 5). This could be explained by the nature of the life cycle of *B. candida*. The only time *B. candida* travels to the gill leaflets of the horseshoe crab is when they go to lay their eggs inside the gill lamellae. In contrast, most of the flatworm's life is spent at the base of the mouth and around the leg appendages, where it feeds. Adult *B. candida* have a caudal disk have helps them adhere to the horseshoe crab (Sluys 1989). We also saw that there was heavier average infestation of egg cocoons in gills than any other category. The stalk and the location of the egg cocoons, inside the inner surface of the gill lamellae, keep them in place even during burrowing, spawning, and rough waters.

The degree to which flatworm infestation adversely affects the health of the Jamaica Bay population is uncertain, but based its prevalence and the correlations between condition and extent of infestation, we believe this is worth further study. Heavy infestation by egg cocoons inside the gill lamellae of the host appears to make the gills more fragile and susceptible to tearing. The egg cocoons also take up space on the gill lamellae, which may impair respiratory capacity or interfere with normal gill cleaning behavior (Watson 1980). Furthermore, although *B. candida* is known to be associated with horseshoe crabs throughout their geographic range (Riesgo et al. 2017; Roux and Giribet 2020), the high prevalence and intensity of flatworm infestation in the Jamaica Bay population may not be typical of other horseshoe crab populations.

**Acknowledgements** We thank the New York State Department of Environmental Conservation, New York City Parks Department, and the National Park Service-Gateway National Recreation Area for providing necessary permits and licenses to MLB. RMW was supported by the Department of Natural Sciences Summer Research Award. The authors thank K. Rothwell and H. VanDerwater for their field assistance.

## References

M.L. Botton, C.P. Colón, J. Rowden, S. Elbin, D. Kriensky, K. McKown, M. Sclafani, R. Madden, Effects of a beach nourishment project in Jamaica Bay, New York on horseshoe crab (*Limulus polyphemus*) spawning activity and egg deposition. Estuaries Coasts **41**, 974–987 (2018). https://doi.org/10.1007/s12237-017-0337-8

M.L. Botton, R.E. Loveland, A. Tiwari, Recruitment and survival of young-of-the-year in a commercially exploited population of horseshoe crabs (*Limulus polyphemus*). Mar. Ecol. Prog. Ser. **265**, 175–184 (2003b)

M.L. Botton, C.N. Shuster Jr., J.A. Keinath, Horseshoe crabs in a food web: Who eats whom, in *The American Horseshoe Crab*, ed. by C. N. Shuster Jr., R. B. Barlow, H. J. Brockmann, (Harvard University Press, Cambridge, MA, 2003a), pp. 133–153

H.J. Brockmann, Nesting behavior: A shoreline phenomenon, in *The American Horseshoe Crab*, ed. by C. N. Shuster, R. B. Barlow, H. J. Brockmann, (Harvard University Press, Cambridge, 2003), pp. 33–49

E.E. Duffy, D. Penn, M.L. Botton, H.J. Brockmann, R.E. Loveland, Eye and clasper damage influence male mating tactics in the horseshoe crab, *Limulus polyphemus*. J. Ethol. **24**, 67–74 (2006). https://doi.org/10.1007/s10164-005-0163-5

J.F. Groff, L. Leibovitz, A gill disease of *Limulus polyphemus* associated with triclad turbellarid worm infections. Biological Bulletin **163**, 392 (1982)

L. Leibovitz, G. Lewbart, Diseases and symbionts: Vulnerability despite tough shells, in *The American Horseshoe Crab*, ed. by C. N. Shuster, R. B. Barlow, H. J. Brockmann, (Harvard University Press, Cambridge, 2003), pp. 245–275

J.H. Mattei, M.L. Botton, M.A. Beekey, C.P. Colón, Horseshoe crab research in urban estuaries: Challenges and opportunities, in *Changing Global Perspectives on Horseshoe Crab Biology*, ed. by R. H. Carmichael, M. L. Botton, P. K. S. Shin, S. G. Cheung, (Conservation and Management, Switzerland, Springer International, 2015), pp. 537–555

M.W. Nolan, S.A. Smith, Clinical evaluation, common diseases, and veterinary care of the horseshoe crab, *Limulus polyphemus*, in *Biology and Conservation of Horseshoe Crabs*, ed. by J. T. Tanacredi, M. L. Botton, D. R. Smith, (Springer, New York, 2009), pp. 479–511

A. Riesgo, E.A. Burke, C. Laumer, G. Giribet, Genetic variation and geographic differentiation in the marine triclad *Bdelloura candida* (Platyhelminthes, Tricladida, Maricola), ectocommensal on the American horseshoe crab *Limulus polyphemus*. Mar. Biol. **164**, 111 (2017). https://doi.org/10.1007/s00227-017-3132-y

E.L. Roux, G. Giribet, Cryptic speciation in the ectocommensal *Bdelloura candida* (Platyhelminthes, Tricladida, Maricola) follows habitat specialization of the American horseshoe crab, *Limulus polyphemus*. Invertebrate Biology **00**, e12284 (2020). https://doi.org/10.1111/ivb.12284

C.N. Shuster, Public participation in studies on horseshoe crab populations, in *Biology and Conservation of Horseshoe Crabs*, ed. by J. T. Tanacredi, M. L. Botton, D. R. Smith, (Springer, New York, 2009), pp. 585–594

R. Sluys, *A Monograph of the Marine Triclads* (A.A. Balkema, Rotterdam, The Netherlands, 1989)

D.R. Smith, H.J. Brockmann, M.A. Beekey, T.L. King, M.J. Millard, J. Zaldivar-Rae, Conservation status of the American horseshoe crab, (*Limulus polyphemus*): A regional assessment. Reviews in fish biology and fisheries. **27**, 135–175 (2016). https://doi.org/10.1007/s11160-016-9461-y

H.W. Stunkard, Observations on the morphology and life-history of *microphallus* n. sp. (Trematoda: Microphallidae). Biol. Bull. **101**, 307–318 (1951)

A.N. Tan, A. Christianus, M.K. Satar, Epibiont infestation on horseshoe crab *Tachypleus gigas* (Müller) at Pantai Balok in Peninsular Malaysia. Our Nature **9** (2012). https://doi.org/10.3126/on.v9i1.5725

W.H. Watson III, Limulus gill cleaning behavior. J. Comp. Physiol. **141**, 67–75 (1980)

# Trophic Relationship of Sympatric Juvenile Asian Horseshoe Crabs in Beibu Gulf, Southwestern China

**Menghong Hu, Kit Yue Kwan, Zhe Xie, Siu Gin Cheung, Youji Wang, and Paul K. S. Shin**

## 1 Introduction

Resource partitioning is a key factor in structuring the assemblages of marine mammals (Kiszka et al. 2010), fishes (Young et al. 2010), and invertebrates (Van den Brink et al. 2012) in various habitats. The principle of competitive exclusion postulates that ecologically similar species are expected to segregate their use of available food, space, and time resources, resulting in niche divergence (Gause 1934; Pianka 1974). Along the coast of southern China, northeastern Borneo and northern Java, the tri-spine horseshoe crab *Tachypleus tridentatus* and mangrove horseshoe crab *Carcinoscorpius rotundicauda* coexist in the same intertidal habitat (Kwan et al. 2016; Manca et al. 2017). Although the juveniles of both species can be found on

Menghong Hu and Kit Yue Kwan contributed equally as first authors.

M. Hu · Z. Xie · Y. Wang (✉)
International Research Center for Marine Biosciences at Shanghai Ocean University, Ministry of Science and Technology, Shanghai, People's Republic of China

Key Laboratory of Exploration and Utilization of Aquatic Genetic Resources, Ministry of Education, Shanghai Ocean University, Shanghai, People's Republic of China
e-mail: yj_wang@shou.edu.cn

K. Y. Kwan
Guangxi Key Laboratory of Beibu Gulf Marine Biodiversity Conservation, College of Marine Sciences, Beibu Gulf University, Qinzhou, People's Republic of China

S. G. Cheung · P. K. S. Shin (✉)
Department of Chemistry (Formerly Biology and Chemistry), City University of Hong Kong, Hong Kong SAR, People's Republic of China

State Key Laboratory in Marine Pollution, City University of Hong Kong, Hong Kong SAR, People's Republic of China



J. T. Tanacredi et al. (eds.), *International Horseshoe Crab Conservation and Research Efforts: 2007-2020*, https://doi.org/10.1007/978-3-030-82315-3_37

the same shore, *C. rotundicauda* are observed to gather mainly within the fringes of mangroves, whereas *T. tridentatus* prefer to stay at the open sandy-mudflat (Chen et al. 2015; Kwan et al. 2016). It may be of interest to understand the diet composition and ecological relationships of these horseshoe crabs along with other species that utilize the same natural environment and to demonstrate their trophic segregation.

The present information on the foraging biology of these Asian horseshoe crab species is primarily derived from discrete reports at different sampling locations. For example, gut content analyses demonstrated that insect larvae, polychaetes, oligochaetes, small crabs, and thin-shelled bivalves were noted in the guts of juvenile *T. tridentatus* and *C. rotundicauda* in Hong Kong (Zhou and Morton 2004). Similar gut composition, such as mollusks, insects, crustaceans, and polychaetes, was also found in varying growth stages of another species of horseshoe crab *T. gigas* from India (Chatterji et al. 1992) and *C. rotundicauda* from Malaysia (Akbar John et al. 2012). The use of stable isotope analyses to assess trophic relationships of Asian horseshoe crab species, however, is very limited. Newell et al. (1995) compared the relative importance of benthic microalgae, phytoplankton, and mangrove as nutrition sources for marine invertebrates in intertidal areas of Malaysia and observed that benthic diatoms contributed largely to the diets of *C. rotundicauda*. The food sources of juvenile *T. tridentatus* on a seagrass sandy-mudflat in Hong Kong were found to include crustaceans, polychaetes, and bivalves, in which their abundance was in turn supported by seagrass biomass (Kwan et al. 2015). A recent study on co-occurring juvenile *T. tridentatus* and *C. rotundicauda* in Pearl Bay, southwestern China, also demonstrated that seagrass, but not mangrove, contributed the most to the dietary carbon of both horseshoe crab species (Fan et al. 2017). However, the question whether there is a degree of trophic segregation between these Asian horseshoe crab species remained unsolved.

In this study, we aimed to investigate the diet compositions and trophic positions of various growth stages of juvenile *T. tridentatus* and *C. rotundicauda* on three intertidal beaches in Beibu Gulf, southwestern China. We hypothesize that there is resource partitioning between the two horseshoe crab species foraging on the same intertidal mudflat, which can be reflected in the dual stable isotope $\delta^{13}C$ and $\delta^{15}N$ values in their tissue and fecal samples.

# 2 Materials and Methods

## 2.1 *Study Locations*

Juvenile horseshoe crabs and their potential food source specimens were collected from three nursery beaches, named Xibeiling (XBL), Jinhaiwan (JHW), and Xichang (XC) along the coastline of Beibu Gulf, southwestern China (17°00′ – 21°45′N, 105°40′ – 110°10′E) (Fig. 1). Beibu Gulf, situated in the

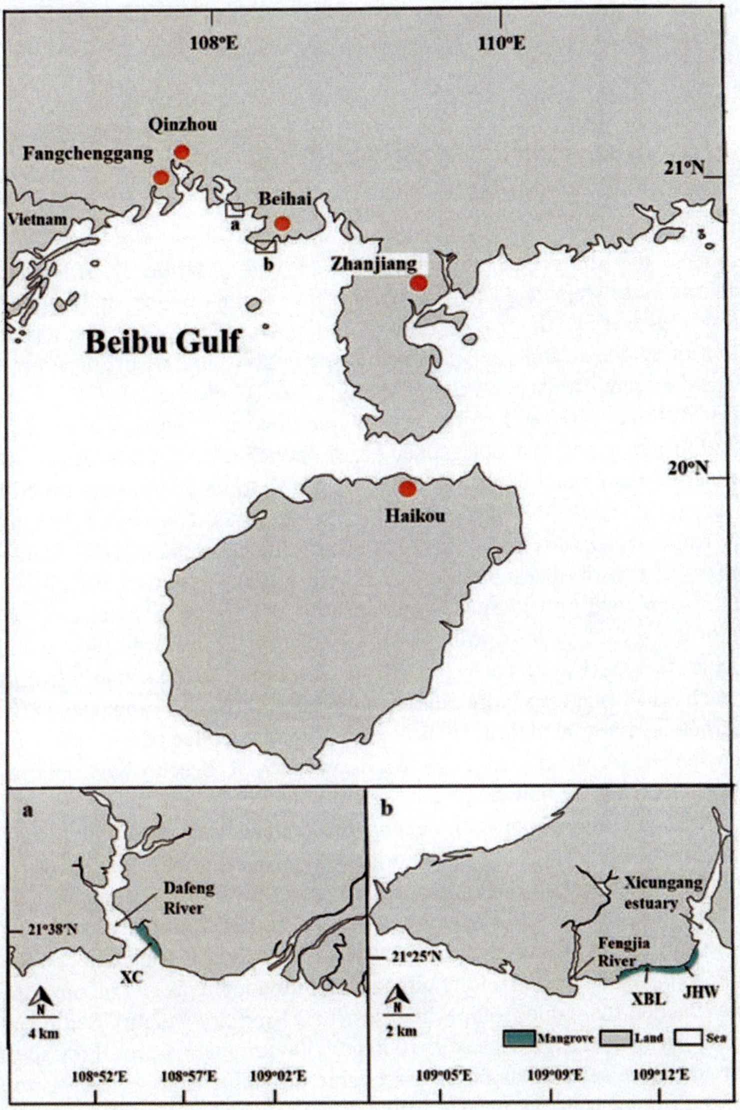


**Fig. 1** Location of the study beaches: Xibeiling (XBL), Jinhaiwan (JHW), and Xichang (XC) in Beibu Gulf, southwestern China

northern South China Sea between China and Vietnam, is a semi-enclosed gulf receiving a considerable load of freshwater and nutrients from seven major rivers (namely, Fangcheng, Qin, Dafeng, Nanliu, Red, Beilun, and Changhua; Chen et al. 1991). JHW and XBL are listed as part of the Beihai National Wetland Park, whereas XC is a fishing village. These study shores are fringed by mangroves *Avicennia marina* and *Aegiceras corniculatum* with an expanse of intertidal sandy mudflats during low tides.

## 2.2 Horseshoe Crab and Potential Prey Tissue Sample Collection

At each of the study beaches, specimens of both juvenile *T. tridentatus* and *C. rotundicauda* were collected in summer for stable isotope analysis. Juvenile *T. tridentatus* were distinguished from *C. rotundicauda* by possessing a triangular cross-sectional telson and three immovable spines above the telson insertion. As the isotopic signature of a species may vary with size (Gaines et al. 2002), horseshoe crabs of different prosomal width were divided into three groups per species and sampled. Instar stages were determined based on the previous study (Hu et al. 2015) and divided into three size groups (Group 1: fourth to seventh instars for both species, Group 2: eighth to tenth instars for both species, and Group 3: 11th–12th instars for *T. tridentatus* and 11th–13th instars for *C. rotundicauda*). Early-stage juveniles (first to third instars) were rare and difficult to collect on all the study shores. A nonlethal procedure was applied in which tissue from the last two segments of the second or third walking leg of horseshoe crabs was clipped and kept cool in an icebox (Carmichael et al. 2004). A single tissue sample was collected from each horseshoe crab individual, and ten horseshoe crab samples were pooled as one replicate. A total of three replicate samples were collected for each size group of both species for subsequent analyses, except at XBL beach where only juvenile *T. tridentatus* could be found.

Macrobenthic animals at each beach were collected using a random sampling method during low tides along three horizontal transects at tidal heights of 1.2, 1.6, and 2.0 m above the chart datum. For each transect per beach, the top 5 cm of sediment from five quadrats (20 × 20 cm) was sampled. In addition, five sediment core (D: 8 cm and L: 20 cm) samples per transect were collected. The number of quadrat and core samples depended on the biomass of individual epi- and infaunal animals that was needed for stable isotope analyses (i.e., 1 mg dry weight). Sediment samples were washed through a stack of 1000 and 500 μm mesh sieves. Prey species of similar size were selected as far as practicable for stable isotope analysis to avoid potential differences in isotopic signatures.

To collect the information on the recent diets of the juveniles, feces from horseshoe crabs were also sampled. At each beach, ten juvenile horseshoe crabs of the same species at a similar instar stage were captured and placed in holding tanks with filtered seawater for 1 hour. Feces were collected using a pipette immediately after they were deposited.

## 2.3 Organic Matter, Microalgal, and Plant Sample Collection

Particulate organic matter (POM), sedimentary organic matter (SOM), and benthic microalgae (BMA) that may be ingested by horseshoe crabs or their prey were sampled. For POM, about 30 l of seawater were sampled from near-bottom water during high tide at each beach and divided into triplicate for subsequent filtration in the laboratory to obtain the particulate matter. For SOM and BMA, triplicate cores (D: 8 cm and L: 20 cm) of sediment were collected at each beach. In addition, leaves of the seagrass *Halophila* sp. on the three nursery beaches were sampled together with the potential prey of the juvenile horseshoe crabs.

## 2.4 Sample Preparation and Stable Isotope Analyses

In the laboratory, macrofaunal samples were sorted and identified under a stereoscopic microscope. For gastropods and bivalves, the shell was removed, and all soft tissues were used for analysis. For polychaetes and crustaceans, the whole animal was used. Specimens were pooled if individual biomass was too small for subsequent isotopic analysis. The animal tissues were soaked in 1.2 N HCl for a few minutes to remove traces of carbonates, dried at 60 °C for 24 hours to a constant weight and kept frozen at −20 °C before isotopic analysis (Yokoyama and Ishihi 2007).

For analysis of POM, the method of Gao et al. (2006) was applied. The water samples were sieved with a 200 μm stainless screen prior to filtration to remove large particles and filtered with pre-combusted GF/C glass fiber filters under a vacuum suction of <1/3 atmospheric pressure. The residue on the filter papers was rinsed with distilled water to remove salt that had adsorbed on the particle surface. Part of the residue was treated with 1.2 N isotonic HCl to remove carbonates and dried at 60 °C, to a constant weight, for stable isotope analysis.

For analysis of SOM, the method from Riera et al. (1996) was followed. Approximately 15 g of sediment were homogenized, freeze-dried, and ground using a mortar and pestle. Then, 200 mg were acidified with 1.2 N HCl to remove inorganic carbon for stable isotope analysis. To prevent any loss of dissolved organics, the sample was not rinsed, but was dried overnight at 60 °C under a fume extractor to evaporate the acid. Once dried, the sediment was mixed with distilled water, freeze-dried, ground again to a fine powder, and kept frozen (−20 °C) until analysis.

Extraction of BMA was carried out with the Ludox technique (Moseman et al. 2004). About 15 $cm^3$ of sediment sample were suspended in colloidal Si (Ludox HS-40), agitated, and stirred. Distilled water (5 ml) was added to the top of the sediment-Ludox mixture, and then the sample was centrifuged; the resulting suspended algal layers were removed with a pipette and filtered through pre-combusted 0.45 μm Whatman GF/C glass fiber filters. The procedure was repeated five times to obtain enough biomass. All benthic microalgal samples were examined through a stereoscopic microscope to remove detrital particles and animals. Only samples with visible algal cells were dried at 60 °C to a constant weight and kept at −20 °C prior to stable isotopic analysis.

Seagrass leaf samples collected from the field were rinsed with deionized water thoroughly before being dried at 80 °C. The dried leaves were ground to a fine powder using a mortar and pestle and stored until further analysis.

Carbon and nitrogen isotopic ratios of animal tissues and other samples were determined at the University of California-Davis Stable Isotope Facility using a PDZ Europa ANCA-GSL elemental analyzer interfaced to a PDZ Europa 20–20 isotope ratio mass spectrometer (Sercon Ltd., Cheshire, UK). A laboratory working standard (glycine) was run. Data were expressed in standard δ-unit notation, where $\delta X = [(R_{sample}/R_{reference}) - 1] \times 10^3$, where X is $^{13}\delta C$ or $^{15}\delta N$ and $R$ is either the $^{13}\delta C{:}^{12}\delta C$ ratio for carbon or the $^{15}\delta N{:}^{14}\delta N$ ratio for nitrogen. These values were reported relative to the international Pee Dee Belemnite standard for carbon and to air $N_2$ for nitrogen. The number of replicates of animal tissues for isotopic determination depended on the total biomass obtained for each species collected, whereas three replicates of each size group of both horseshoe crab species and POM, SOM, BMA and seagrass leaves were analyzed for each study beach.

Based on Schoeninger and Deniro's (1984) study that the enrichment of $\delta^{13}C$ is 1‰ and that of $\delta^{15}N$ is 3‰ per trophic level, the mean values of $\delta^{13}C$ (‰) and $\delta^{15}N$ (‰) of the diets from that of the horseshoe crab tissue were estimated.

## 2.5 *Data Analysis*

Data were first examined for normality and homogeneity of variance using Shapiro–Wilk and Levene's tests, respectively. Differences in stable isotopic signatures of tissues and feces in the three size groups of juvenile *T. tridentatus* and *C. rotundicauda* collected at the study beaches were compared using one-way ANOVA followed by post hoc Tukey test for multiple comparisons. The stable isotopic differences in tissues and feces between the two juvenile horseshoe crab species were also compared using Student's *t*-test. All statistical analyses were undertaken using SPSS version 16.0.

The contribution of potential food sources to juvenile horseshoe crabs was quantitatively determined using the Bayesian isotopic mixing model in the IsotopeR 0.4 package under R 3.10 software (Hopkins III and Ferguson 2012). As an omnivore, the discrimination factors used for computations in the case of horseshoe crabs were based on the published values 0.91 ± 1.04‰ for $\delta^{13}C$ and 3.23 ± 0.41‰ for $\delta^{15}N$ (Vander Zanden and Rasmussen 2001). In the present study, the food sources assigned to horseshoe crabs in the mixing model included eight categories: POM, SOM, BMA, seagrass, polychaetes, bivalves, gastropods, and crustaceans. For isotopic values of different animal groups, data from each individual species were treated as a replicate sample in the mixing model analysis.

# 3 Results

## 3.1 $\delta^{13}C$ and $\delta^{15}N$ Signatures in Juvenile Horseshoe Crabs

The mean values of $\delta^{13}C$ in *T. tridentatus* tissues collected from the three study beaches ranged from −12.76‰ to −11.52‰, while that of $\delta^{15}N$ varied from 10.71‰ to 11.13‰ (Table 1). For *C. rotundicauda* tissues, which were only sampled from two study beaches, the lowest $\delta^{13}C$ (−14.25‰) and highest $\delta^{13}C$ (−13.27‰) values were found in the same size group 1 at JHW and XC, respectively. The mean values of $\delta^{15}N$ in *C. rotundicauda* tissues ranged from 11.23‰ to 12.92‰. There was no significant difference among various size groups in $\delta^{13}C$ of *T. tridentatus* tissues at XBL and *C. rotundicauda* tissues at XC. For $\delta^{15}N$, no significant difference was found in *T. tridentatus* tissues from JHW (Table 1). For *T. tridentatus*, the mean tissue $\delta^{13}C$ values of size group 1 were significantly lower than that of size groups 2 and 3 at both JHW and XC beaches (Table 1). A similar situation was found in tissue samples of *C. rotundicauda*, in which the mean $\delta^{13}C$ values of size group 1 were significantly lower than that of size groups 2 and 3 at JHW. For tissue samples in both species, $\delta^{15}N$ of size group 1 did not show significant differences from size group 2, but it was significantly lower than that of size group 3 (Table 1).

Significant differences in isotopic signatures between the two species were found in tissue samples. At both JHW and XC beaches, tissue $\delta^{13}C$ values of *T. tridentatus* were significantly heavier than that of *C. rotundicauda* (Student's *t*-test, JHW: $t = 44.292$; $p < 0.001$; XC: $t = 5.586$; $p = 0.001$), while tissue $\delta^{15}N$ values in *T. tridentatus* were significantly lighter than that in *C. rotundicauda* (Student's *t*-test, JHW: $t = -5.985$; $p < 0.001$; XC: $t = -9.795$; $p < 0.001$).

The mean values of $\delta^{13}C$ in *T. tridentatus* feces were from −13.77‰ to −12.96‰, while that of *C. rotundicauda* feces ranged from −16.29‰ to −14.41‰. The mean values of $\delta^{15}N$ in *T. tridentatus* feces ranged from 5.57‰ to 7.91‰. For *C. rotundicauda* feces, the highest mean value of $\delta^{15}N$ (7.81‰) was found in size group 1 of JHW, and the lowest mean value of $\delta^{15}N$ (5.01‰) was found in size group 3 of XC (Table 1). For both species, $\delta^{13}C$ and $\delta^{15}N$ of feces did not show significant differences among various size groups at the three study beaches and between the two species.

## 3.2 Potential Food Sources of Juvenile Horseshoe Crabs

Isotopic signatures of food sources available to juvenile horseshoe crabs at each beach are shown in Table 2. A total of nine species of polychaetes ($\delta^{13}C$: −20.76‰ – −11.31‰; $\delta^{15}N$: 8.56‰ – 14.36‰), four gastropods ($\delta^{13}C$: −15.40‰ – −12.16‰; $\delta^{15}N$: 7.05‰ – 14.91‰), 16 bivalves ($\delta^{13}C$: −21.93‰ – −12.55‰; $\delta^{15}N$: 6.04‰ – 11.78‰), and nine crustaceans ($\delta^{13}C$: −16.26‰ – −10.35‰; $\delta^{15}N$: 4.48‰ – 13.34‰) were found at the study sites. One sipunculid species, *Golfingia* sp., was

**Table 1** Mean (± SE) of $\delta^{13}C$ (‰) and $\delta^{15}N$ (‰) signatures in tissues from juvenile horseshoe crabs of the three study beaches. N = 3, ND = no data

| | XBL | | JHW | | XC | |
|---|---|---|---|---|---|---|
| | $\delta^{13}C$ | $\delta^{15}N$ | $\delta^{13}C$ | $\delta^{15}N$ | $\delta^{13}C$ | $\delta^{15}N$ |
| **Tissues** | | | | | | |
| *T. tridentatus* | | | | | | |
| Size group 1 | −12.66 ± 0.02 | $10.71 \pm 0.03^{a}$ | $-12.29 \pm 0.05^{a}$ | 10.87 ± 0.06 | $-12.76 \pm 0.02^{a}$ | $10.83 \pm 0.12^{a}$ |
| Size group 2 | −12.68 ± 0.03 | $10.95 \pm 0.08^{ab}$ | $-11.79 \pm 0.04^{b}$ | 10.81 ± 0.04 | $-11.52 \pm 0.04^{c}$ | $10.72 \pm 0.04^{a}$ |
| Size group 3 | −12.68 ± 0.01 | $11.13 \pm 0.06^{b}$ | $-11.79 \pm 0.04^{b}$ | 11.03 ± 0.06 | $-11.92 \pm 0.05^{b}$ | $11.12 \pm 0.08^{b}$ |
| *C. rotundicauda* | | | | | | |
| Size group 1 | ND | ND | $-14.25 \pm 0.03^{a}$ | $11.44 \pm 0.02^{a}$ | −13.27 ± 0.34 | $11.23 \pm 0.06^{a}$ |
| Size group 2 | ND | ND | $-13.69 \pm 0.07^{b}$ | $11.92 \pm 0.14^{a}$ | −13.34 ± 0.09 | $11.37 \pm 0.10^{a}$ |
| Size group 3 | ND | ND | $-13.84 \pm 0.04^{b}$ | $12.92 \pm 0.14^{b}$ | −13.97 ± 0.04 | $11.73 \pm 0.03^{b}$ |
| **Feces** | | | | | | |
| *T. tridentatus* | | | | | | |
| Size group 1 | −13.55 ± 0.15 | 6.25 ± 0.62 | −13.46 ± 0.21 | 6.96 ± 0.55 | −13.77 ± 0.15 | 5.57 ± 1.25 |
| Size group 2 | −13.01 ± 0.03 | 6.04 ± 0.08 | −12.96 ± 0.04 | 6.03 ± 0.87 | −12.98 ± 0.16 | 6.46 ± 0.30 |
| Size group 3 | −13.01 ± 0.56 | 6.43 ± 0.32 | −12.97 ± 0.50 | 7.91 ± 0.84 | −13.37 ± 0.64 | 7.22 ± 2.34 |
| *C. rotundicauda* | | | | | | |
| Size group 1 | ND | ND | −15.94 ± 0.91 | 7.81 ± 0.40 | −14.62 ± 0.68 | 6.89 ± 0.99 |
| Size group 2 | ND | ND | −16.29 ± 0.70 | 7.49 ± 0.35 | −15.13 ± 0.97 | 6.93 ± 0.07 |
| Size group 3 | ND | ND | −14.96 ± 0.04 | 6.32 ± 0.43 | −14.41 ± 0.88 | 5.01 ± 1.85 |

Note: Juvenile *T. tridentatus* and *C. rotundicauda* were divided into three groups by size (prosomal width): Group 1: 14–30 mm, Group 2: 31–50 mm, and Group 3: 31–50 mm. No *C. rotundicauda* was found at XBL. For each species, the values within the same column with different superscripts (a, b, c) are significantly different at $p < 0.05$ based on Tukey multiple comparison test after one-way ANOVA found at JHW ($\delta^{13}C = -16.23‰$, $\delta^{15}N = 9.23‰$), whereas one brachiopod species, *Lingula anatina*, was recorded at XBL ($\delta^{13}C = -18.38‰$, $\delta^{15}N = 6.58‰$) and JHW ($\delta^{13}C = -18.52‰$, $\delta^{15}N = 6.78‰$). The mean values of $\delta^{13}C$ in POM were around −22‰, while that in SOM and BMA were about −21‰. The mean $\delta^{15}N$ values of POM, SOM, and BMA (5.82‰, 5.74‰ and 5.71‰, respectively) at XC were higher than those at XBL and JHW (XBL: 3.59‰, 4.24‰ and 4.22‰; JHW: 3.23‰, 4.38‰ and 4.22‰). The mean $\delta^{13}C$ value of seagrass at XBL (−9.51‰) was considerably higher than that at JHW and XC (−10.27‰ and −10.89‰, respectively). The mean $\delta^{15}N$ values of seagrass, however, were relatively the same, with data of 6.15‰, 6.03‰ and 6.06‰ for XBL, JHW, and XC, respectively.

Based on the general trophic relationship that the values of $\delta^{13}C$ and $\delta^{15}N$ of prey should be 1‰ and 3‰ less than the predator, respectively, the estimated diet of juvenile horseshoe crabs was identified (Fig. 2a–c) with the isotopic position being closely related to that of invertebrates (polychaetes, gastropods, bivalves, and crustaceans) collected from the field. Results of the isotopic mixing model provided a quantitative estimate of contribution from each potential food source to juvenile horseshoe crabs (Table 3). At XBL, *T. tridentatus* primarily fed on bivalves, especially for size group 1 (over 90% contribution, fourth to seventh instars). For size groups 2 (eighth to tenth instars) and 3 (11th–12th instars), apart from bivalves, they also fed on crustaceans (over 20% contribution) (Table 3). At JHW, *T. tridentatus* also tended to feed on bivalves, particularly for size groups 1 and 2 (over 90% contribution), whereas for size group 3, the consumption of bivalves tended to decrease (some 60% contribution) with the supplement of other food sources such as gastropods. However, at the same beach (JHW), the three size groups of *C. rotundicauda* appeared to be a generalist feeding on a variety of food sources including POM, BMA, polychaetes, gastropods, and crustaceans (Table 3). At XC, the trophic niches of the two species tended to overlap (Fig. 2c), with both species consuming a wide range of food sources (Table 3). For *T. tridentatus*, size group 1 mainly relied on POM (over 80% contribution), whereas size group 2 fed on POM (about 50% contribution) and polychaetes (about 30% contribution) and size group 3 on crustaceans (over 80% contribution). For *C. rotundicauda*, POM and polychaetes constituted some 80% to the diet of size group 1 (fourth to seventh instars), whereas size group 2 (eighth to tenth instars) shifted to feed on crustaceans and polychaetes (about 70% contribution) and size group 3 (11th–13th instars) on POM, BMA, and crustaceans (100% contribution) (Table 3). At the three study beaches, $\delta^{13}C$ signatures of estimated diets (Fig. 2a–c) were similar to horseshoe crab feces, while $\delta^{15}N$ signatures of estimated diets were approximately 2‰ higher than that of feces (Table 1).

## 4 Discussion

Food webs show an enrichment of both $\delta^{13}C$ and $\delta^{15}N$ within different trophic levels (Michener and Kaufman 2007). The differences in $\delta^{15}$ N values of various species link to differences in trophic position of taxa, with predators having heavier

**Table 2** δ$^{13}$C (‰) and δ$^{15}$N (‰) signatures in tissues from potential food items of juvenile horseshoe crabs at the three study beaches

| | XBL | | JHW | | XC | |
|---|---|---|---|---|---|---|
| | δ$^{13}$C | δ$^{15}$N | δ$^{13}$C | δ$^{15}$N | δ$^{13}$C | δ$^{15}$N |
| **Polychaetes** | | | | | | |
| *Ampharete* sp. | −17.78 | 9.93 | −15.97 | 10.06 | −19.26 | 11.13 |
| Capitellidae sp. | −14.97 | 10.39 | −14.42 | 9.69 | – | – |
| Eunicidae sp. | −14.14 | 9.35 | −12.11 | 9.91 | −14.27 | 13.59 |
| *Glycera tridactyla* | – | – | −15.40 | 9.28 | – | – |
| *Nectoneanthes* sp. | −16.76 | 10.37 | −15.95 | 10.60 | −16.76 | 10.36 |
| *Onuphis eremita* | −13.04 | 11.51 | −11.31 | 8.56 | −15.48 | 9.92 |
| *Pectinaria* sp. | −14.85 | 9.48 | −15.68 | 14.36 | −20.76 | 12.01 |
| Spionidae sp. | −13.71 | 8.56 | −17.09 | 9.62 | −13.62 | 9.17 |
| *Tylorrhynchus heterochaetus* | – | – | −17.56 | 11.55 | – | – |
| Mean value for polychaetes | −15.04 | 9.94 | −15.05 | 10.40 | −16.69 | 11.03 |
| **Gastropods** | | | | | | |
| *Batillaria zonalis* | – | – | −12.92 | 7.09 | −15.11 | 14.91 |
| *Cerithidea cingulata* | −12.16 | 7.05 | −12.84 | 7.69 | – | – |
| *Cerithidea djadjariensis* | −12.52 | 7.22 | −12.54 | 7.22 | −12.51 | 12.20 |
| *Nassarius festivus* | – | – | −12.41 | 9.38 | −15.40 | 12.38 |
| Mean value for gastropods | −12.34 | 7.14 | −12.48 | 8.30 | −14.34 | 13.36 |
| **Bivalves** | | | | | | |
| *Anomalocardia squamosa* | – | – | −17.26 | 8.48 | – | – |
| *Corbula crassa* | – | – | −16.86 | 6.23 | – | |
| *Cycladicama cumingi* | −13.49 | 9.06 | −12.55 | 6.04 | −20.05 | 10.98 |
| *Cyclina sinensis* | −18.74 | 10.28 | −17.46 | 8.71 | −18.77 | 9.56 |
| *Dosinia japonica* | −18.07 | 7.03 | −17.34 | 10.99 | – | – |
| *Dosinia orbiculata* | −17.14 | 8.70 | – | – | – | – |
| *Laternula anatina* | −18.81 | 6.34 | −17.27 | 7.10 | −17.69 | 8.07 |
| *Macalia bruguieri* | −19.19 | 7.37 | −18.27 | 9.41 | −19.03 | 9.03 |
| *Merisca capsoides* | −17.08 | 7.19 | −16.85 | 6.72 | −17.26 | 7.70 |
| *Modiolus metcalfei* | – | – | – | – | −17.13 | 7.24 |
| *Modiolus vagina* | – | – | – | – | −21.01 | 11.78 |
| *Moerella iridescens* | −16.70 | 7.01 | −21.08 | 8.28 | −21.93 | 8.99 |
| *Solen gracilis* | −19.57 | 8.18 | – | – | −15.32 | 8.07 |
| *Solen linearis* | −18.45 | 9.39 | −17.52 | 8.30 | −14.00 | 7.35 |
| *Ruditapes philippinarum* | −18.15 | 9.50 | −21.66 | 9.48 | – | – |
| *Tegillarca granosa* | – | – | −20.96 | 8.29 | −17.69 | 8.18 |
| Mean value for bivalves | −17.76 | 8.19 | −17.98 | 8.14 | −18.17 | 8.81 |
| **Crustaceans** | | | | | | |
| *Camptandrium sexdentatum* | – | – | – | – | −12.10 | 4.48 |
| *Cythura* sp. | – | – | −17.35 | 13.34 | −16.26 | 10.57 |
| *Diogenes deflectomanus* | −11.70 | 8.91 | −11.07 | 10.15 | −11.80 | 8.09 |
| *Helice tientsinensis* | −11.29 | 9.26 | – | – | – | – |

(continued)

**Table 2** (continued)

| | XBL | | JHW | | XC | |
|---|---|---|---|---|---|---|
| | $\delta^{13}C$ | $\delta^{15}N$ | $\delta^{13}C$ | $\delta^{15}N$ | $\delta^{13}C$ | $\delta^{15}N$ |
| *Heptacarpus geniculatus* | – | – | – | – | −11.31 | 6.39 |
| *Macrophthalmus abbreviatus* | – | – | – | – | −12.59 | 10.95 |
| *Philyra pisum* | – | – | – | – | −10.35 | 8.63 |
| *Portunus haanii* | −12.03 | 9.03 | −13.07 | 8.79 | −11.24 | 6.97 |
| *Uca vocans* | −11.95 | 9.76 | −11.83 | 9.58 | – | – |
| Mean value for crustaceans | −11.74 | 9.24 | −13.33 | 10.47 | −11.37 | 8.24 |
| **Others** | | | | | | |
| Sipunculid *Golfingia* sp. | – | – | −16.23 | 9.23 | – | – |
| Brachiopod *Lingula anatina* | −18.38 | 6.58 | −18.52 | 6.78 | – | – |
| Particulate organic matter (POM) | −22.04 | 3.59 | −22.58 | 3.23 | −22.88 | 5.82 |
| Sedimentary organic matter (SOM) | −20.53 | 4.24 | −20.43 | 4.38 | −21.51 | 5.74 |
| Benthic microalgae (BMA) | −21.31 | 4.22 | −21.01 | 4.22 | −21.51 | 5.71 |
| Seagrass *Halophila* sp. | −9.51 | 6.15 | −10.27 | 6.03 | −10.89 | 6.06 |

SOM, POM, BMA, and seagrass show mean values of $N = 3$, whereas other species show values of $N = 1$ only due to insufficient individuals collected. A dash symbol (–) means the particular species was not found at the study site

signatures than that of prey (Fry 2006). The average $\delta^{15}N$ enrichment per trophic level is 3‰ (Minagawa and Wada 1984; Schoeninger and Deniro 1984). Although $\delta^{13}C$ enrichment is generally small (around 1‰), $\delta^{13}C$ is more informative, regarding the base of the food chain (Fry 2006), which can reveal feeding ecology information, such as inshore versus offshore feeding preferences (Hobson et al. 1994). Michener and Kaufman (2007) indicated that juvenile animals that migrate offshore from estuaries tend to change their isotopic composition as they feed on a new diet. Fry (1983) demonstrated that juvenile shrimp collected at the beginning of their life had isotope values of $\delta^{13}C$ ranging from −11‰ to −14‰ and $\delta^{15}$ N values of approximately 6‰. As shrimp grew and migrated offshore, $\delta^{13}C$ changed to −16‰ and $\delta^{15}N$ reached to 11.5‰. The stable isotopic data of Darnaude (2005) also revealed that the diet of benthic flatfish switched from riverine POM to offshore marine POM as they grew. Juvenile horseshoe crabs have similar spatial distribution patterns as the abovementioned juvenile shrimp (Fry 1983) and benthic flatfish (Darnaude 2005), in that they prefer to move offshore from their natal area as they grow (Sekiguchi 1988; Shuster Jr et al. 2003; Hu et al. 2009). Previous studies showed that the diets of juvenile horseshoe crabs changed according to such migration (Gaines et al. 2002; O'Connell et al. 2003; Carmichael et al. 2009). Carmichael et al. (2009) revealed that American horseshoe crab, *Limulus polyphemus*, at second to seventh instar stages relied heavily on benthic and suspended POM, while 8th–11th instars switched to animal-based food sources. Similar findings were also observed in the present study that POM constituted a major component of the diet in size groups 1 (fourth to seventh instars) and 2 (eighth to tenth instars) of *T. tridentatus* and size group 1 (fourth to seventh instars) of *C. rotundicauda*, notably at the study beach XC. In contrast, group size 3 juveniles (11th–12th instars for *T.*

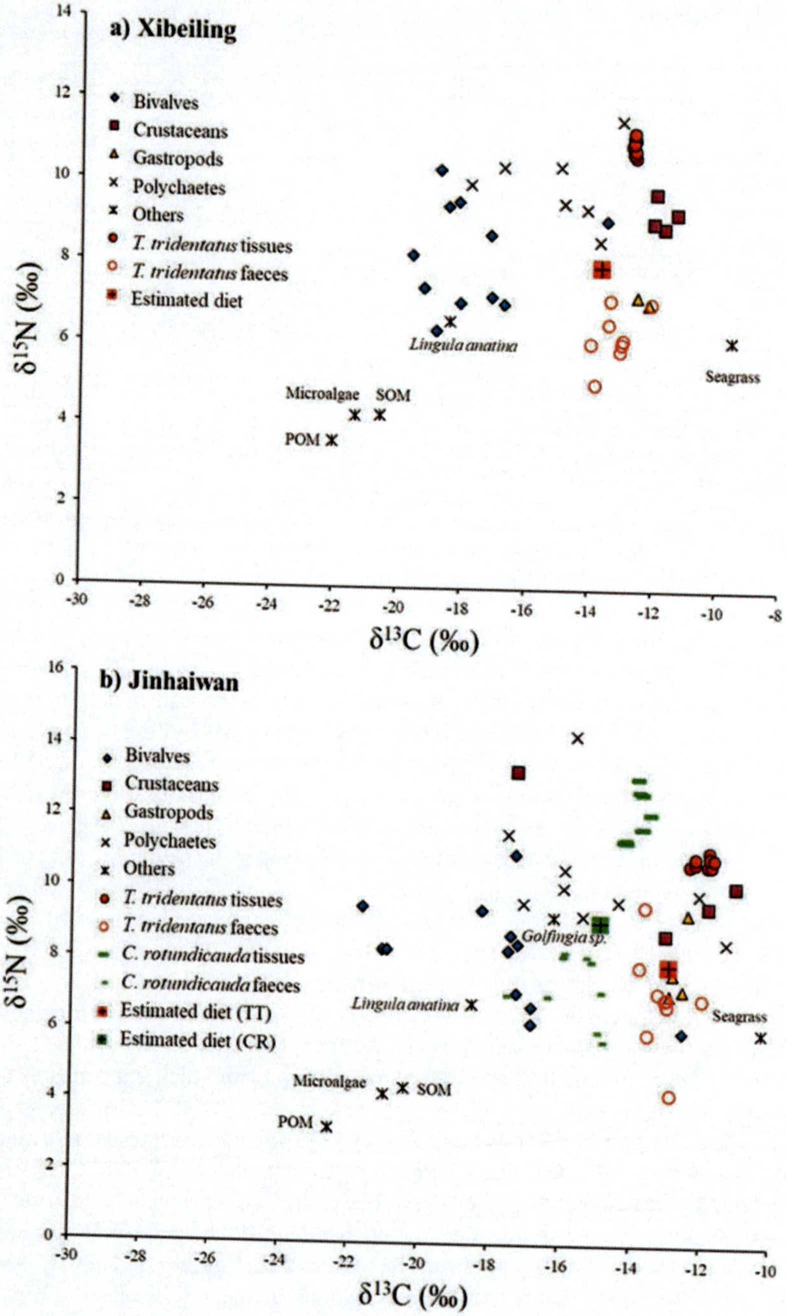


**Fig. 2** The mean $\delta^{13}C$ (‰) and $\delta^{15}N$ (‰) signature of horseshoe crab tissues, feces, and their estimated diet compared with the mean signatures of potential food items at three study beaches: (**a**) Xibeiling (XBL), (**b**) Jinhaiwan (JHW), and (**c**) Xichang (XC). TT *T. tridentatus*, CR *C. rotundicauda*

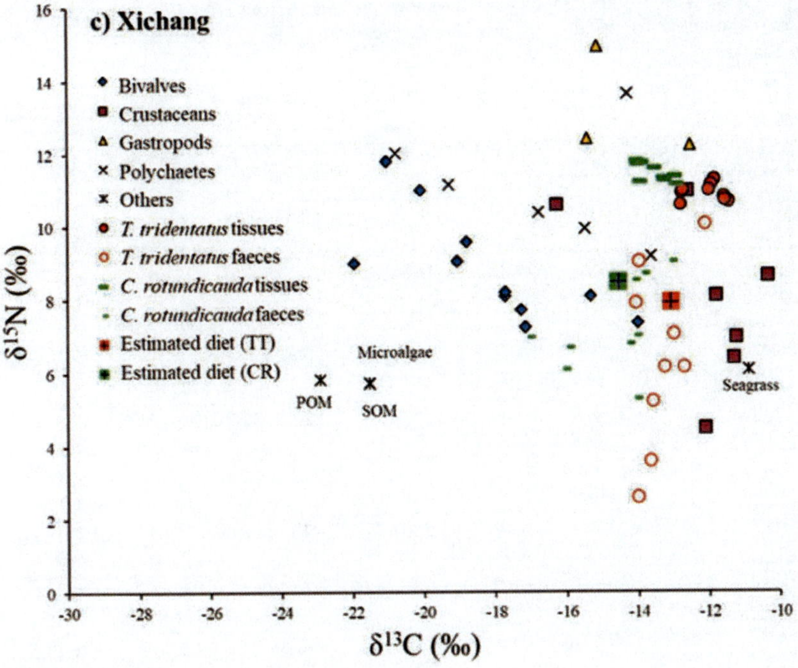


**Fig. 31.2** (continued)

*tridentatus* and 11th–13th instars for *C. rotundicauda*) tended to feed on a range of invertebrates including polychaetes, bivalves, and crustaceans. The present findings also demonstrated that $\delta^{13}C$ values of tissues in size group 1 (fourth to seventh instars) of *T. tridentatus* and *C. rotundicauda* were significantly lighter than other size groups at JHW. For $\delta^{15}$ N values, the two smaller size groups (fourth to seventh and eighth to tenth instars) were significantly lighter than larger size group (11th–12th instars) in *T. tridentatus* from two beaches (XBL and XC) and in *C. rotundicauda* (11th–13th instars) from JHW and XC.

The isotopic signatures of the tissues in juvenile horseshoe crabs were derived from their long-term diet (Fry 2006). In the present study, food sources of different horseshoe crab species varied at different beaches. At XBL, *T. tridentatus* were found to feed on bivalves and, to a lesser extent, crustaceans. According to the biplot (Fig. 2b), isotopic signatures of *T. tridentatus* tissue did not overlap with that of *C. rotundicauda* at JHW, implying possible trophic segregation at this beach. At JHW, the mixing model data revealed that *T. tridentatus* tended to prey on bivalves, whereas *C. rotundicauda* showed a different feeding strategy in which the juveniles encompassed a wide range of diet including both plant- and animal-based food sources. Nevertheless, at XC, both species fed on relatively similar food sources, largely composed of POM, BMA, polychaetes, and crustaceans, and the isotopic signatures of tissues in both species overlapped. In terms of fecal samples, the

**Table 3** Results of isotopic mixing model showing mean percentage (SD) of potential food sources contributed to the diet of juvenile *T. tridentatus* and *C. rotundicauda* at the three study beaches

| | POM | SOM | BMA | Seagrass | Polychaetes | Gastropods | Bivalves | Crustaceans |
|---|---|---|---|---|---|---|---|---|
| **Xibeiling (XBL)** | | | | | | | | |
| Tt 1 | 0.49 (2.45) | 0.19 (2.26) | 5.71 (11.91) | 0 (0.05) | 0.30 (1.36) | 0.88 (5.92) | 92.43 (14.47) | 0 (0.03) |
| Tt 2 | 0.09 (1.25) | 0.03 (0.34) | 0 (0) | 0.02 (0.63) | 0 (0) | 0.44 (3.78) | 66.16 (46.96) | 33.26 (47.00) |
| Tt 3 | 0.18 (2.89) | 0.72 (6.87) | 0.03 (1.42) | 0.23 (4.34) | 0.13 (3.56) | 0.53 (6.90) | 75.66 (41.81) | 22.54 (41.08) |
| **Jinhaiwan (JHW)** | | | | | | | | |
| Cr1 | 14.73 (30.90) | 5.05 (10.35) | 3.73 (16.42) | 0.10 (0.51) | 27.33 (37.52) | 0.06 (2.04) | 43.87 (44.00) | 5.14 (15.40) |
| Cr2 | 0.86 (5.53) | 0.65 (5.57) | 32.08 (45.87) | 0.04 (0.49) | 18.39 (36.73) | 2.48 (6.67) | 30.97 (43.22) | 14.53 (34.63) |
| Cr3 | 0.52 (5.05) | 0 (0) | 65.69 (46.90) | 0 (0) | 11.36 (20.05) | 0.53 (4.23) | 0 (0.01) | 21.91 (33.26) |
| Tt 1 | 0 (0) | 0.01 (0.14) | 1.81 (6.42) | 0.14 (0.83) | 0 (0) | 0.38 (1.61) | 97.43 (7.86) | 0.23 (4.40) |
| Tt 2 | 0.17 (3.88) | 0.81 (7.82) | 0 (0) | 0.10 (0.70) | 0.53 (6.08) | 0.06 (2.17) | 98.34 (11.87) | 0 (0) |
| Tt 3 | 8.02 (26.95) | 6.69 (24.40) | 0 (0.02) | 0 (0) | 4.75 (20.96) | 13.88 (34.12) | 66.40 (47.10) | 0.26 (3.57) |
| **Xichang (XC)** | | | | | | | | |
| Cr1 | 52.12 (28.03) | 0.23 (2.29) | 12.42 (24.52) | 0 (0) | 29.30 (30.01) | 0 (0) | 5.88 (17.19) | 0.04 (0.22) |
| Cr2 | 0.14 (3.16) | 0.05 (1.74) | 3.86 (16.46) | 2.32 (11.84) | 17.67 (37.00) | 1.73 (11.94) | 4.45 (20.47) | 69.78 (43.93) |
| Cr3 | 20.04 (30.10) | 0 (0) | 47.76 (35.82) | 0 (0) | 0 (0) | 0 (0) | 0 (0) | 32.20 (40.83) |
| Tt 1 | 81.03 (17.07) | 4.75 (8.22) | 10.17 (14.19) | 0.47 (1.77) | 0 (0) | 0 (0) | 0.80 (3.02) | 2.78 (7.43) |
| Tt 2 | 54.16 (49.08) | 0.03 (1.82) | 1.94 (11.91) | 0 (0) | 29.46 (45.14) | 0.03 (1.30) | 2.35 (14.62) | 12.03 (31.62) |
| Tt 3 | 0.05 (0.73) | 0 (0) | 0.36 (4.57) | 0.19 (1.47) | 14.15 (28.24) | 1.45 (7.44) | 0.21 (2.58) | 83.60 (28.75) |

Tt1, Tt2, Tt3 = size groups 1, 2, 3 of *T. tridentatus*; Cr1, Cr2, Cr3 = size groups 1, 2, 3 of *C. rotundicauda*

present results demonstrated that isotopic signatures of *T. tridentatus* feces overlapped with that of *C. rotundicauda* at JHW and XC, reflecting that the two species might recently consume similar food sources. This may imply that potential competition between the two sympatric horseshoe crab populations, not only in space but also in food sources. A lack of resource partitioning in juvenile Asian horseshoe crab populations at XC may be due to the great abundance of prey available in the intertidal areas; otherwise, resource partitioning can occur, as observed at JHW.

It has been reported that seagrass could be a food source for juvenile horseshoe crabs especially when animal-based food items are scarce in the field (Kwan et al. 2015, 2020). However, the present findings did not reveal such an observation as the contribution of seagrass to the diet of both juvenile horseshoe crab species was very low (Table 3). This could be attributed to the ample presence of a variety of potential food sources at the three study beaches and/or to the fact that seagrass leaves may be difficult for intake and digestion (Ramos et al. 2018) by the juvenile horseshoe crabs, especially when they are at a young stage.

## 5 Conclusion

The two Asian horseshoe crab species co-occurring at JHW demonstrated possible trophic segregation as reflected by $\delta^{13}C$ and $\delta^{15}N$ values in their tissue samples, in which *T. tridentatus* primarily fed on bivalves, whereas *C. rotundicauda* tended to consume available food sources on the shore, including POM, BMA, polychaetes, gastropods, and crustaceans. However, both horseshoe crab species tended to forage on POM, BMA, polychaetes, and crustaceans at XC. Isotopic signatures of fecal samples indicated that both species recently foraged on similar food sources at JHW and XC. A lack of resource partitioning in juvenile Asian horseshoe crab populations may be due to the high abundance of prey available in the intertidal areas of XC; otherwise, resource partitioning can occur, as observed at JHW. Although the food sources of the juveniles varied among beaches, the present findings suggested that the two juvenile horseshoe crab species are cosmopolitan omnivores with a wide range of diet on the nursery beaches in southwestern China.

**Acknowledgements** This study was funded by the Ocean Park Conservation Foundation, Hong Kong. We also appreciate the assistance from Ms. Q.Z. Li and staff from the Guangxi Institute of Fisheries in the field and Dr. Wenzhe Xu from the Tianjin University of Science and Technology for running the Bayesian mixing model.

## References

R.H. Carmichael, D. Rutecki, B. Annett, E. Gaines, I. Valiela, Position of horseshoe crabs in estuarine food webs: N and C stable isotopic study of foraging ranges and diet composition. J. Exp. Mar. Biol. Ecol. **299**, 231–253 (2004)

R.H. Carmichael, E. Gaines, Z. Sheller, A. Tong, A. Clapp, I. Valiela, Diet composition of juvenile horseshoe crabs: Implications for growth and survival of natural and cultured stocks, in *Biology and Conservation of Horseshoe Crabs*, ed. by J. T. Tanacredi, M. L. Botton, D. R. Smith, (Springer, New York, 2009), pp. 521–534

A. Chatterji, J.K. Mishra, A.H. Parulekar, Feeding behaviour and food selection in the horseshoe crab, *Tachypleus gigas* (Müller). Hydrobiologia **246**, 41–48 (1992)

G. Chen, X. Gu, H. Gao, *Marine Fishery Environment in China*, 1st edn. (Zhejiang Scientific and Technological Press, Hangzhou, China, 1991) (In Chinese with English abstract)

C.P. Chen, M.C. Yang, L.F. Fan, G.L. Qiu, Y.Y. Liao, H.L. Hsieh, Co-occurrence of juvenile horseshoe crabs *Tachypleus tridentatus* and *Carcinoscorpius rotundicauda* in an estuarine bay, southwestern China. Aquat. Biol. **24**, 117–126 (2015)

A.M. Darnaude, Fish ecology and terrestrial carbon use in coastal areas: Implications for marine fish production. J. Anim. Ecol. **74**, 864–876 (2005)

L.F. Fan, C.P. Chen, M.C. Yang, G. Qiu, Y.Y. Liao, H.L. Hsieh, Ontogenetic changes in dietary carbon sources and trophic position of two co-occurring horseshoe crab species in southwestern China. Aquat. Biol. **26**, 15–26 (2017)

B. Fry, Fish and shrimp migrations in the northern Gulf of Mexico analyzed using stable C, N, and S isotope ratios. Fish. Bull. **81**, 789–801 (1983)

B. Fry, *Stable Isotope Ecology*, 1st edn. (Springer, New York, 2006)

E.F. Gaines, R.H. Carmichael, S.P. Grady, I. Valiela, Stable isotopic evidence for changing nutritional sources of juvenile horseshoe crabs. Biol. Bull. **203**, 228–230 (2002)

Q.F. Gao, P.K.S. Shin, G.H. Lin, S.P. Chen, S.G. Cheung, Stable isotope and fatty acid evidence for uptake of organic waste by green-lipped mussels *Perna viridis* in a polyculture fish farm system. Mar. Ecol. Prog. Ser. **317**, 273–283 (2006)

G.F. Gause, *The Struggle for Existence* (Hafner Publishing Company, London, 1934)

K.A. Hobson, J.F. Piatt, J. Pitocchelli, Using stable isotopes to determine seabird trophic relationships. J. Anim. Ecol. **63**, 786–798 (1994)

J.B. Hopkins III, J.M. Ferguson, Estimating the diets of animals using stable isotopes and a comprehensive Bayesian mixing model. PLoS One **7**, e28478 (2012)

M. Hu, Y. Wang, Y. Chen, S.G. Cheung, P.K.S. Shin, Q. Li, Summer distribution and abundance of juvenile Chinese horseshoe crabs *Tachypleus tridentatus* along an intertidal zone in southern China. Aquat. Biol. **7**, 107–112 (2009)

M. Hu, B.K.Y. Kwan, Y. Wang, S.G. Cheung, P.K.S. Shin, Population structure and growth of juvenile horseshoe crabs *Tachypleus tridentatus* and *Carcinoscorpius rotundicauda* (Xiphosura) in southern Chin, in *Changing Global Perspectives on Horseshoe Crab Biology, Conservation and Management*, ed. by R. H. Carmichael, M. L. Botton, P. K. S. Shin, S. G. Cheung, (Springer, Switzerland, 2015), pp. 167–180

B.A. John, B.Y. Kamaruzzaman, K.C.A. Jalal, K. Zaleha, Feeding ecology and food preferences of *Carcinoscorpius rotundicauda* collected from the Pahang nesting grounds. Sains Malays **41**, 855–861 (2012)

J. Kiszka, M. Oremus, P. Richard, M. Poole, V. Ridoux, The use of stable isotope analyses from skin biopsy samples to assess trophic relationships of sympatric delphinids off Moorea (French Polynesia). J. Exp. Mar. Biol. Ecol. **395**, 48–54 (2010)

B.K.Y. Kwan, S.G. Cheung, P.K.S. Shin, A dual stable isotope study for diet composition of juvenile Chinese horseshoe crab *Tachypleus tridentatus* (Xiphosura) on a seagrass-covered intertidal mudflat. Mar. Biol. **162**, 1137–1143 (2015)

B.K.Y. Kwan, H.L. Hsieh, S.G. Cheung, P.K.S. Shin, Present population and habitat status of potentially threatened Asian horseshoe crabs *Tachypleus tridentatus* and *Carcinoscorpius rotundicauda* in Hong Kong: A proposal for marine protected areas. Biodivers. Conserv. **25**, 673–692 (2016)

K.Y. Kwan, J. Bopp, S. Huang, Q. Chen, C.-C. Wang, X. Wang, W. Zhen, J. Zhu, X. Huang, Ontogenetic Resource Use and Trophic Dynamics of Endangered Juvenile *Tachypleus Tridentatus* Among Diversified Nursery Habitats in the Northern Beibu Gulf, China. Integr Zool (2020) In press

A. Manca, F. Mohamad, A. Ahmad, M.F.A.M. Sofa, N. Ismail, Tri-spine horseshoe crab, *Tachypleus tridentatus* (L.) in Sabah, Malaysia: The adult body sizes and population. J. Asia Pac. Biodivers **10**, 355–361 (2017)

R.H. Michener, L. Kaufman, Stable isotope ratios as tracers in marine food webs: An updated, in *Ecological Methods and Concepts Series*, ed. by M. Gessner, (Blackwell Publishing, New York, 2007), pp. 238–282

M. Minagawa, E. Wada, Stepwise enrichment of δ15N along food chains: Further evidence and the relation between δ15N and animal age. Geochim. Cosmochim. Acta **48**, 1135–1140 (1984)

S.M. Moseman, L.A. Levin, C. Currin, C. Forder, Colonization, succession, and nutrition of microbenthic assemblages in a restored wetland at Tijuana Estuary, California. Estuar. Coast. Shelf Sci. **60**, 755–770 (2004)

R.I.E. Newell, N. Marshall, A. Sasekumar, V.C. Chong, Relative importance of benthic microalgae, phytoplankton, and mangroves as sources of nutrition for penaeid prawns and other coastal invertebrates from Malaysia. Mar. Biol. **123**, 595–606 (1995)

C.W. O'Connell, S.P. Grady, A.S. Leschen, R.H. Carmichael, I. Valiela, Stable isotopic assessment of site loyalty and relationships between size and trophic position of the Atlantic horseshoe crab, *Limulus polyphemus*, within Cape Cod estuaries. Biol. Bull. **205**, 254–255 (2003)

E.R. Pianka, Niche overlap and diffuse competition. Proc. Natl. Acad. Sci. U. S. A. **71**, 2141–2145 (1974)

R.J. Ramos, F.G. Brun, L.G. Egea, J.J. Vergara, Food choice effects on herbivory: Intra-specific seagrass palatability and inter-specific macrophyte palatability in seagrass communities. Estuar. Coast. Shelf Sci. **204**, 31–39 (2018)

P. Riera, P. Richard, A. Gremare, G. Blanchard, Food source of intertidal nematodes in the Bay of Marennes Oleron (France), as determined by dual stable isotope analysis. Mar. Ecol. Prog. Ser. **142**, 303–309 (1996)

M.J. Schoeninger, M.J. Deniro, Nitrogen and carbon isotopic composition of bone collagen from marine and terrestrial animals. Geochim. Cosmochim. Acta **48**, 625–639 (1984)

K. Sekiguchi, *Biology of Horseshoe Crabs*, 1st edn. (Science House Co. Ltd., Tokyo, Japan, 1988)

C.N. Shuster Jr., R.B. Barlow, H.J. Brockmann, *The American Horseshoe Crab*, 1st edn. (Harvard University Press, Cambridge, MA/London, 2003)

A.M. Van den Brink, S. Wijnhoven, C.L. McLay, Competition and niche segregation following the arrival of *Hemigrapsus takanoi* in the formerly *Carcinus maenas* dominated Dutch delta. J. Sea Res. **73**, 126–136 (2012)

M.J. Vander Zanden, J.B. Rasmussen, Variation in $\delta^{15}N$ and $\delta^{13}C$ trophic fractionation: Implications for aquatic food web studies. Limnol. Oceanogr. **46**, 2016–2066 (2001)

H. Yokoyama, Y. Ishihi, Variation in food sources of the macrobenthos along a land–sea transect: A stable isotope study. Mar. Ecol. Prog. Ser. **346**, 127–147 (2007)

J.W. Young, M.J. Lansdell, R.A. Campbell, S.P. Cooper, F. Juanes, M.A. Guest, Feeding ecology and niche segregation in oceanic top predators off eastern Australia. Mar. Biol. **157**, 2347–2368 (2010)

H. Zhou, B. Morton, The diets of juvenile horseshoe crabs, *Tachypleus tridentatus* and *Carcinoscorpius rotundicauda* (Xiphosura), from nursery beaches proposed for conservation in Hong Kong. J. Nat. Hist. **38**, 1915–1925 (2004)

# *Tachypleus gigas* and *Carcinoscorpius rotundicauda* Selection and Consumption of Tissue from Four Different Bivalve Species in Captivity

**Anis Syahira Abdul Halim, Farah Najihah Razali, Fisal Ahmad, Noraznawati Ismail, Tuan Chilek Tuan Zainazor, Ahmad Shamsuddin Ahmad, and Faridah Mohamad**

## 1 Introduction

In general, horseshoe crabs are benthic omnivores and feed on variety organisms. Previous studies reported several types of prey that were commonly found in the gut of horseshoe crabs such as bivalves, gastropods, crustaceans, echinoderms, and polychaetes (Botton et al. 1984; Zhou and Morton 2004; John et al. 2012; Razak et al. 2017). Horseshoe crabs were also reported to consume decayed organic matter and plant materials apart from a presence of sand particles in their gut (Chatterji et al. 1992; Carmichael et al. 2004). Among the reported food items, bivalves were favored as they contributed to the largest percentage of the total amount of food in the gut of wild caught horseshoe crabs (Botton and Ropes 1989; Chatterji et al. 1992).

Based on this information, in captivity, horseshoe crabs were commonly fed with bivalve tissue, fish meat, squid, small crabs, and brine shrimp (Smith and Berkson 2005; Kwan et al. 2017). These diets were considered optimum food to enhance

A. S. A. Halim · A. S. Ahmad · F. Mohamad (✉)
Horseshoe Crab Research Group, Universiti Malaysia Terengganu, Terengganu, Malaysia

Faculty of Science and Marine Environmental, Universiti Malaysia Terengganu, Terengganu, Malaysia
e-mail: mfaridah@umt.edu.my

F. N. Razali · N. Ismail
Horseshoe Crab Research Group, Universiti Malaysia Terengganu, Terengganu, Malaysia

Institute of Marine Biotechnology, Universiti Malaysia Terengganu, Terengganu, Malaysia

F. Ahmad · T. C. T. Zainazor
Horseshoe Crab Research Group, Universiti Malaysia Terengganu, Terengganu, Malaysia

Faculty of Fisheries and Food Science, Universiti Malaysia Terengganu, Terengganu, Malaysia



J. T. Tanacredi et al. (eds.), *International Horseshoe Crab Conservation and Research Efforts: 2007-2020*, https://doi.org/10.1007/978-3-030-82315-3_38

growth performance and increases horseshoe crab survival rate especially for juvenile individuals (Kwan et al. 2017). Ample intakes of a variety of food are essential for normal physiological function, growth, and reproductive success of horseshoe crabs (Birnie-Gauvin et al. 2017; Kwan et al. 2017). For example, proteins and amino acids are important for gonad development and egg production in adult horseshoe crabs (Marsh and Watts 2007; Razak and Kassim 2018).

Natural diets consisting of bivalves were commonly used for brood stock in aquaculture production (Ikhwanuddin et al. 2015; Ghazali et al. 2017). Bivalves are nutrient and mineral rich food source that are consumed by human and many animals (Venugopal and Gopakumar 2017; Prato et al. 2019). For captive animals, they can provide a significant level of nutrients and minerals such as protein; vitamins D, A, and B; and polysaturated fatty acid like Omega-3 (Venugopal and Gopakumar 2017; Prato et al. 2019). According to Azra and Ikhwanuddin (2016), who mainly worked on mud crab genus *Scylla*, the natural diets such as bivalve tissue, fish meat, squid, and shrimp can increase fecundity of the brood stock and improved mating success by increasing the quality and quantity of the eggs. Moreover, this diet can reduce production cost since it required less labor and time for feed preparation (Azra and Ikhwanuddin 2016; Ghazali et al. 2017).

The knowledge of adult horseshoe crab diets in captivity is still limited because most of the previous studies mainly focus on captive-rear juvenile (Carmichael et al. 2009; Tzafrir-Prag et al. 2010; Carmichael and Brush 2012; Hu et al. 2014; Kwan et al. 2017). Identifying the diet of adult horseshoe crabs is also important for hatchery management as the adults were kept in captivity for research purposes. Like other captive animals, horseshoe crabs might prefer tissue of certain types of bivalves, which might be related to the nutritional values of the food. The purpose of this study was to demonstrate the behavioral preference and consumption of two horseshoe crab species, *Tachypleus gigas* and *Carcinoscorpius rotundicauda* on four species of bivalve, *Scapharca cornea*, *Meretrix meretrix*, *Polymesoda erosa*, and *Magallana bilineata* in captivity. These four species of bivalves were selected based on their abundance in an area where the horseshoe crabs were normally sighted. The feeding experiment was tested with and without choice and reported as number of bivalves (tissue) consumed per day. Then, the proximate composition of the tissue for each species of bivalve was also determined in order to associate the nutritional values of the bivalves with the horseshoe crab selection.

## 2 Materials and Methods

### 2.1 Sample Collection

Both male and female for adult *T. gigas* and *C. rotundicauda* were collected from local fishermen at Kuala Kemaman and Setiu, Terengganu. The horseshoe crabs were transported to the hatchery of Institute of Tropical Aquaculture, Universiti

**Fig. 1** Four species of bivalves used as horseshoe crab feed. From left is *Magallana bilineata*, *Polymesoda erosa*, *Scapharca cornea*, and *Meretrix meretrix*

Malaysia Terengganu. They were cleaned from epibionts and placed in 152 × 121 × 120 cm (length × width × height) holding tank for two weeks acclimatization period prior to the experiment based on protocols by Botton (1984). During this period, horseshoe crabs were fed with prawn meat purchased from fish market once a day.

There are four bivalve species used in this experiments which are *S. cornea*, *M. meretrix*, *P. erosa*, and *M. bilineata* (Fig. 1). The fresh live bivalves were purchased from local fish sellers in Setiu, Terengganu, and transported immediately to the laboratory. The bivalves were cleaned and rinsed thoroughly with tap water to remove sand and mud on the shell surface. Then, bivalves were carefully opened using knife and cockle opener to separate the tissue from the shell.

## 2.2 *Feeding Experiment*

Experiments were divided into two parts, horseshoe crabs (i) fed with either bivalve tissue separately and (ii) fed with a mix of four types of bivalve tissue simultaneously. At each trial, a single horseshoe crab was fed with either *S. cornea* only, *M. meretrix* only, *P. erosa* only, *M. bilineata* only, or the tissues of all four bivalve

**Table 1** Mean ± Standard Error (SE) of tissue weight and shell length of bivalves used in feeding experiments

| Parameters | *S. cornea* | *M. meretrix* | *P. erosa* | *M. bilineata* |
|---|---|---|---|---|
| Tissue weight (g) | 1.73 ± 0.16 | 1.84 ± 0.06 | 4.02 ± 0.29 | 4.21 ± 0.41 |
| Shell length (cm) | 4.32 ± 0.49 | 4.11 ± 0.21 | 5.71 ± 0.64 | 8.63 ± 0.23 |

Note: Measurements were taken using the electronic balance and measuring tape to the nearest 0.01 g and 0.1 cm, respectively

species involving five males and females of *T. gigas* and *C. rotundicauda*, respectively. This experiment is repeated nine times (replications = 9) for both species of horseshoe crabs using a total of 90 adult individuals ($n = 45$ males, 45 females) of each *T. gigas* and *C. rotundicauda*.

The experiments took place in feeding tanks of two different size based on the species' size, 103 × 69 × 23 cm for *T. gigas* and 60 × 48 × 16 cm for *C. rotundicauda*. All tanks were aerated and filled with treated seawater at 12 cm deep that ensured the crabs remained submerged. The temperature and salinity of the seawater were maintained between 26 to 28 °C and 29 to 31 ppt, respectively. No sand was added to keep all food and residue in view during the observation to avoid miscalculation. Only single horseshoe crab individual were placed in each feeding tank. Then, the horseshoe crab was starved for three days prior to the experiment.

Seven pieces of the same-sized (Table 1), whole individual bivalve tissues were placed separately in each tank to determine the consumption rate when no choice was available for five days. More pieces were added if decreased and removed when it floated to minimize fouling. The number of individual bivalves consumed was recorded twice daily (8 am and 7 pm). To determine their selection, the horseshoe crabs were fed with a mix of four types of bivalves simultaneously and observed for 3 hours. The number of individual pieces of each bivalve species consumed was recorded. We avoided the experiment during full moon to avoid food restriction phase (especially for the females) due to circumlunar spawning behavior. Control experiments ($n =$ four males, four females for each horseshoe crab species) that fed with prawn meat (similar during the acclimatization period) were set up simultaneously to confirm that the horseshoe crab sample used in this experiment were in their feeding phase.

## 2.3 *Proximate Analysis of Bivalve Tissue*

Fifty samples of each type of bivalve tissue were left on absorbent surface for 30 minutes to remove excess water (Mohammad and Yusuf 2016). The samples were oven-dried at 100 °C for 24 hours until constant weights were obtained (Swapna and Ravinder 2015). All samples were analyzed after drying except for moisture analysis (Swapna and Ravinder 2015). The dried samples were ground to fine powder and kept in labeled air-tight container until further analysis. Moisture, ash, protein, fat, and carbohydrate contents were analyzed in triplicates according to the standard methods by Association of Official Chemists (AOAC 1990).

## 2.4 Data Analysis

All data were expressed as mean ± SE. The number of bivalve individuals consumed by a horseshoe crab was calculated using the following formula:

$$\textit{Mean bivalve individual consumed} = \frac{\sum\left(\textit{number of bivalve consumed / day of experiment}\right)}{\textit{Number of replicates}} \quad (1)$$

For the bivalve choice experiment, the "day of experiment" was excluded since the observations were made over a 3-hour time period. Data were first tested for normality and homogeneity of variance using Kolmogorov-Smirnov test and Levene's test. The one-way analysis of variance (ANOVA) was further used to test for significant differences among the analyzed species with 95% significance level followed with Tukey's multiple comparison test. If the data were not normal even after transformation, the nonparametric Kruskal-Wallis analyses were employed to test for significant differences in food consumed, and *p*-values of less than 0.05 were considered as statistically significant. All statistical analyses were performed using SPSS 20.0 statistical software.

# 3 Results

The highest number of bivalve tissue consumed per day by a single male and female *T. gigas* was 1.93 ± 0.31 individuals per day of *M. meretrix* and 1.62 ± 0.37 individuals per day of *M. bilineata,* respectively (Fig. 2a; Kruskal-Wallis, $p > 0.05$). Male *C. rotundicauda* consumed 2.51 ± 0.57/d *M. meretrix* and 2.51 ± 0.48/d of and *P. erosa*, while female *C. rotundicauda* consumed 2.02 ± 0.25 individuals of *M. meretrix* per day (Fig. 2b; Kruskal-Wallis, $p > 0.05$).

Overall, *M. bilineata* and *S. cornea* were the most consumed by male and female *T. gigas* when offered with four types of bivalves simultaneously (Fig. 3a). Male *T. gigas* consumed 1.67 ± 0.29 individuals of *M. bilineata*, whereas female *T. gigas* consumed 1.89 ± 0.39 individuals of *S. cornea* throughout the feeding experiments. However, no distinct selection was shown by *T. gigas* as both sexes consumed approximately similar amount for each types of bivalve tissue (Kruskal-Wallis, $p > 0.05$). A similar trend was shown by female *C. rotundicauda* (Kruskal-Wallis, $p > 0.05$), though female consumed more tissue of *S. cornea*, 0.89 ± 0.26 (Fig. 3b). Male *C. rotundicauda* showed preference in tissue consumption as they consumed

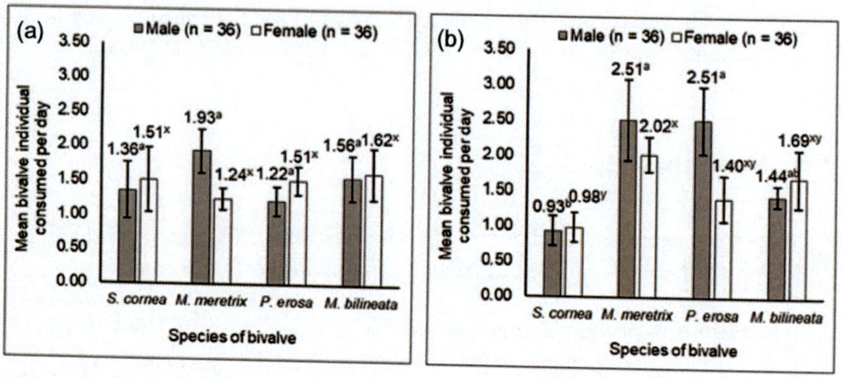


**Fig. 2** Mean consumption of bivalve for (**a**) *T. gigas* and (**b**) *C. rotundicauda* when fed with four species of bivalve tissue separately (no choice). The similar superscript letters indicate that mean values are not significantly different ($p > 0.05$)

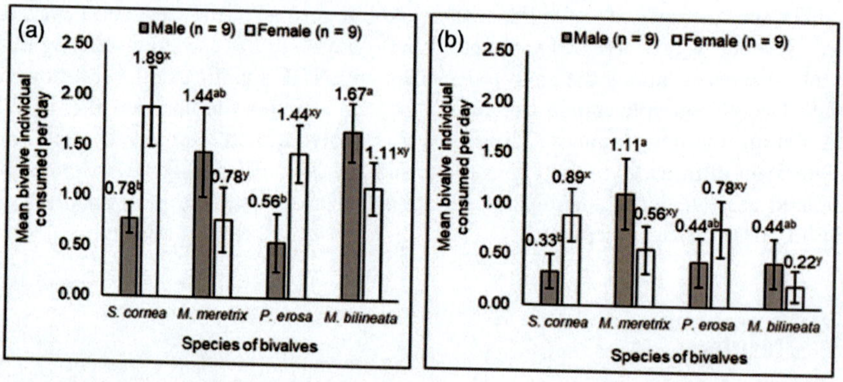


**Fig. 3** Mean of consumption for (**a**) *T. gigas* and (**b**) *C. rotundicauda* when fed with a mix of four species of bivalve tissue simultaneously (choice available). The similar superscript letters indicate that mean values are not significantly different ($p > 0.05$)

1.11 ± 0.35 individuals of *M. meretrix*, higher than other bivalve tissue available to them (Fig. 3b; Kruskal-Wallis, $p < 0.05$).

Proximate composition analysis for each species of bivalve tissue is shown in Table 2. The percentage for moisture, protein, fat, ash, and carbohydrate varied among species. The moisture content ranged between 82.85 and 86.23% for all species of bivalve tissue. The highest protein content was 70.50% recorded by *P. erosa*. *Magallana bilineata* had the highest fat and ash content with 18.45% and 26.10%, respectively. The carbohydrate content had similar values ranging from 7.44% to 7.90% for *S. cornea*, *M. meretrix* and *P. erosa*, while *M. bilineata* recorded the lowest carbohydrate value, 4.39%. There are significant differences recorded in the proximate composition elements between all types of bivalve tissue (one-way ANOVA, $p < 0.05$).

**Table 2** Variation in proximate nutritional composition analysis for dry-basis matter of tissue of four bivalve species. Values in the same column with different superscripts indicate significant difference among the species ($p < 0.05$)

| Species | Composition % | | | | |
|---|---|---|---|---|---|
| | [a]Moisture | Protein | Fat | Ash | Carbohydrate |
| *S. cornea* | $82.85 \pm 0.11^{b}$ | $68.16 \pm 0.31^{ab}$ | $7.55 \pm 0.12^{d}$ | $16.39 \pm 0.22^{b}$ | $7.90 \pm 0.10^{a}$ |
| *M. Meretrix* | $86.23 \pm 0.23^{a}$ | $52.05 \pm 1.97^{b}$ | $8.31 \pm 0.06^{c}$ | $20.81 \pm 2.31^{ab}$ | $7.78 \pm 0.69^{a}$ |
| *P. erosa* | $83.10 \pm 0.13^{ab}$ | $70.50 \pm 1.54^{a}$ | $12.44 \pm 0.03^{b}$ | $9.63 \pm 1.49^{c}$ | $7.44 \pm 0.11^{a}$ |
| *M. bilineata* | $84.39 \pm 1.41^{ab}$ | $50.00 \pm 0.60^{c}$ | $18.45 \pm 0.10^{a}$ | $26.10 \pm 0.63^{a}$ | $4.39 \pm 0.01^{b}$ |

[a]Note: Moisture content was independent from the rest proximate analysis and measured separately

## 4 Discussion

In general, our findings showed no sex-influence in the quantity of bivalve tissue consumed by *T. gigas* and *C. rotundicauda*. Our findings differed from previous studies on field and in-lab experiment by Razak et al. (2017) and Razak and Kassim (2018) who reported female *T. gigas* to have higher feeding intensity compared with males, thus consumed more. In the wild, females might consume more to meet energy requirement to support their unique biological functions especially during the spawning season, carrying the male for several months during amplexus and for egg production. Whereas males may need to consume more before and after the spawning season due to feeding restriction they experience for several months during amplexus (Razak and Kassim 2018). In our case, energy expenditure may be reduced because of constrained movement in captivity and continuous access to food that led to the insignificant results. Based on the gut content from individuals captured from the wild, *C. rotundicauda* males were reported to have higher consumption than females (John et al. 2012), which also contradicts with our findings. In the wild, there is a possibility of males' instinct to gorge on food to the maximum when found in order to prepare enough food storage for the coming spawning season of amplexus and food restriction. However, this instinct might be reduced in captivity as food was made available without competition as the horseshoe crabs were placed in separate tanks.

Horseshoe crab especially *T. gigas* were fed equal numbers of each type of bivalve tissue and were found to be nonselective for bivalves of different species. They are selective feeders (Chatterji et al. 1992) in the wild where a vast number of benthic organisms are available. In line with a previous study, the observed hunger response may have triggered the animals into generalist feeding mode compared with the animals in a satiated condition (Emlen 1966). According to Razak and Kassim (2018), horseshoe crabs started to defecate only after 12–24 hours, emptying their entire gut within a day. Thus, three days of starvation prior to our feeding experiment might provide enough time to trigger the horseshoe crab to switch the feeding mode into generalist feeder in order to satisfy their hunger. This trend was similar to the findings by Botton (1984) where both sexes of *Limulus* had similar preferences toward two species of bivalves, which were *Mulinia lateralis* and *Mya arenaria* both in the wild and in captivity. Therefore, we conclude that there is little to no influence of sex of the horseshoe crab on feeding selectivity for different bivalve species.

When offered simultaneously, the tissue of *S. cornea* was the most consumed feed by females of both species of horseshoe crab, whereas *M. bilineata* and *M. meretrix* were consumed more by the males. We suggest that nutritional values are linked with the selection. Among the three elements, proteins, lipid, and carbohydrates, proteins are the most distinct elements that may play a role. Despite the differences in proteins concentration among different bivalves, it did not explain the statistical insignificance of the bivalve selection. Previously, juveniles were reported to prefer ~40% of proteins in the formulated feed (Hu et al. 2014). These protein values varied upon species-specific for energy sustainability (Lall and Tibbetts 2009). The present study found that bivalves had >50% protein content that exceeded the basic protein requirement by horseshoe crabs. Therefore, any value beyond 40% might have been equally selected.

Another factor that influenced the feeding activity is the presence of amino acids (Lemme 2010). Certain amino acids act as stimulant to trigger the feeding activity and thus provide distinct selection in feed such as glycine was favored by cold water fish (Floreto et al. 2000; Jobling et al. 2012; Kasumyan 2016; Zhang et al. 2018). Bivalves differed in amino acid content depending on the environmental condition they were collected from and species specific differences (Swapna and Ravinder 2015). Since our four bivalves were collected from the same niche area with the same environmental condition, we assumed their amino acid content to be similar and do not influence selection in feed. However, no attempt in profiling the amino acid in the bivalves was carried out in the present study. Previous report of horseshoe crabs' amino acid preference was also lacking.

## 5 Conclusion

Our findings showed that *T. gigas* and *C. rotundicauda* consumed tissues from all bivalve species offered, but they only consume small amount of these feeds. Therefore, in the hatchery, the number of food given to the horseshoe crabs should be reduced to two or three pieces per individual daily to reduce food wastage and fouling and save costs. The nutrients in natural feed might not suit the nutritional requirement of horseshoe crabs; however, this is the most practical and reliable way to temporarily rear horseshoe crabs in captivity for research purposes. To better understand the bivalve selection as food by horseshoe crabs, we suggest future studies to include amino acid profiling and texture analysis as they were previously reported to influence selection in feeding activities for other animals.

**Acknowledgements** We thank other HCRG members for their support in this study. Special thanks to Nor Amyra Fazira, Nur Ema Ashaini, Alisa Shahfiqa, and Nurul Fatin Syafiqah for their assistance in data collection. We also acknowledge the funding by The Ministry of Higher Education of Malaysia under Translational Research Grant (Vot 53387).

## References

AOAC, *Official methods of analysis. Association of Official Analytic Chemists*, 15th edn. (Washington DC, 1990)

M.N. Azra, M. Ikhwanuddin, A review of maturation diets for mud crab genus *Scylla* broodstock: Present research, problems and future perspective. Saudi J. Biol. Sci. **23**, 257–267 (2016)

K. Birnie-gauvin, K.S. Peiman, D. Raubenheimer, et al., Nutritional physiology and ecology of wildlife in a changing world. Conserv. Physiol. **5**(1), 1–18 (2017)

M.L. Botton, Diet and food preferences of the adult horseshoe crab *Limulus polyphemus* in Delaware Bay, New Jersey, USA. Mar. Biol. **81**, 199–207 (1984)

M.L. Botton, J.W. Ropes, Feeding ecology of horseshoe crabs on the continental shelf, New Jersey to North Carolina. Bull. Mar. Sci. **45**(3), 637–647 (1989)

R.H. Carmichael, E. Brush, Three decades of horseshoe crab rearing: A review of conditions for captive growth and survival. Aquaculture **4**, 32–43 (2012)

R.H. Carmichael, E. Gaines, Z. Sheller, et al., Diet composition of juvenile horseshoe crabs: Implications for growth and survival of natural and cultured stocks, in *Biology and Conservation of Horseshoe Crabs*, ed. by J. T. Tanacredi, M. L. Botton, D. Smith, (Springer International Publishing, Switzerland, 2009), pp. 521–534

R.H. Carmichael, D. Rutecki, B. Annett, et al., Position of horseshoe crabs in estuarine food webs: N and C stable isotopic study of foraging ranges and diet composition. J. Exp. Mar. Biol. Ecol. **299**(2), 231–253 (2004)

A. Chatterji, J.K. Mishra, A.H. Parulekar, Feeding behaviour and food selection in the horseshoe crab, *Tachypleus gigas* (Muller). Hydrobiologia **246**, 41–48 (1992)

J.M. Emlen, The role of time and energy in food preference. Am. Nat. **100**(916), 611–617 (1966)

E.A.T. Floreto, R.C. Bayer, P.B. Brown, The effects of soybean-based diets, with and without amino acid supplementation, on growth and biochemical composition of juvenile American lobster, *Homarus americanus*. Aquaculture **189**, 211–235 (2000)

A. Ghazali, N.M. Noordin, A.B. Abol-Munafi, et al., Ovarian maturation stages of wild and captive mud crab, *Scylla olivacea* fed with two diets. Sains Malaysiana **46**(12), 2273–2280 (2017)

M. Hu, Y. Wang, S.G. Cheung, et al., Digestible dietary protein and energy requirements of juvenile Asian horseshoe crabs, *Tachypleus tridentatus* and *Carcinoscorpius rotundicauda*. Aquac. Res. **45**, 1621–1633 (2014)

M. Ikhwanuddin, S. Mohamed, A.I.A. Rahim, et al., Observation on the effect of natural diets on ovarian rematuration in blue swimming crab *Portunus pelagicus* (Linnaeus, 1758). Indian J. Fish **62**, 124–127 (2015)

M. Jobling, A. Alanärä, S. Kadri, et al., Feeding biology and foraging, in *Aquaculture and Behaviour*, ed. by F. Huntingford, M. Jobling, S. Kadri, (Blackwell Publishing Ltd., New Jersey, 2012), pp. 121–149

B.A. John, B.Y. Kamaruzzaman, K.C.A. Jalal, et al., Feeding ecology and food preferences of *Carcinoscorpius rotundicauda* collected from the Pahang nesting grounds. Sains Malaysiana **41**(7), 855–861 (2012)

A.O. Kasumyan, Taste attractiveness of free amino acids and their physicochemical and biological properties (as exemplified by fishes). J. Evol. Biochem. Physiol. **52**(4), 271–281 (2016)

B.K.Y. Kwan, A.K.Y. Chan, S.G. Cheung, et al., Marine microalgae as dietary supplements in the culture of juvenile Chinese horseshoe crabs, *Tachypleus tridentatus* (Xiphosura). Aquac. Res. **48**, 3910–3924 (2017)

S.P. Lall, S.M. Tibbetts, Nutrition, feeding, and behavior of fish. Vet. Clin. Exot. Anim. **12**, 361–372 (2009)

A. Lemme, Availability and effectiveness of free amino acid in aquaculture, in *Avances en Nutriecion Acuicola X-Memorias del Decimo Simposio Internacional de Nutricion Acuicola*, ed. by L. E. Cruz-Suarez, D. Ricque-Marie, M. Tapia-Salazar, et al., (Universidad Autónoma de Nuevo León, Monterrey, 2010), pp. 264–275

A.G. Marsh, S.A. Watts, Biochemical and energy requirements of gonad development, in *Edible Sea Urchins: Biology and Ecology*, ed. by J. M. Lawrence, (Elsevier Science, London, 2007), pp. 35–53

S.H. Mohammad, M.S. Yusuf, Proximate evaluation of some economical seafood as a human diet and as an alternative prospective valuable of fish meal. J. Fish. Aquat. Sci. **11**(1), 12–27 (2016)

E. Prato, F. Biandolino, I. Parlapiano, et al., Proximate, fatty acids and metals in edible marine bivalves from Italian market: Beneficial and risk for consumers health. Sci. Total Environ. **648**, 153–163 (2019)

M.M.R. Razak, Z. Kassim, Food intake, gut transit time and defecation pattern of Asian horseshoe crab, *Tachypleus gigas*. ASM Sci. J. **11**(2), 56–66 (2018)

M.R.M. Razak, Z. Kassim, A.A. Sabuti, et al., Feeding ecology and food preferences of Cherok Paloh, Pahang horseshoe crab, *Tachypleus gigas*. Mal. J. Fund. Appl. Sci. **13**(3), 198–202 (2017)

S.A. Smith, J. Berkson, Laboratory culture and maintenance of the horseshoe crab (*Limulus polyphemus*). Resource **34**(7), 27–34 (2005)

P. Swapna, R.T. Ravinder, Proximate, amino acid, fatty acid and mineral analysis of bivalve *Parreysia cylindrica* from Waddepally and Kaleshwaram Lake. World J. Pharm. Pharm. Sci. **4**(4), 1388–1401 (2015)

T. Tzafrir-Prag, M.P. Schreibman, I. Lupatch, et al., Preliminary studies of energy and protein requirements of Atlantic horseshoe crabs, *Limulus polyphemus*, grown in captivity. J. World Aquac. Soc. **41**(6), 874–883 (2010)

V. Venugopal, K. Gopakumar, Shellfish: Nutritive value, health benefits, and consumer safety. Compr. Rev. Food Sci. Food Saf. **16**, 1219–1242 (2017)

D. Zhang, P. Hu, T. Liu, et al., GC bias lead to increased small amino acids and random coils of proteins in cold-water fishes. BMC Genomics **19**, 1–10 (2018)

H. Zhou, B. Morton, The diets of juvenile horseshoe crabs, *Tachypleus tridentatus* and *Carcinoscorpius rotundicauda* (Xiphosura), from nursery beaches proposed for conservation in Hong Kong. J. Nat. Hist. **38**, 1915–1925 (2004)

# Research Progress and Prospect of Tachypleus Amebocyte Lysate in China

**Chang Liu, Ximei Liu, Youji Wang, Jie Song, Jinfeng He, Zongguang Tai, Quangang Zhu, and Menghong Hu**

Menghong Hu and Youji Wang conceived this project. Jie Song and Jinfeng He provided some first-hand data. Chang Liu and Ximei Liu wrote the paper. Chongguang Tai, Quangang Zhu, and Menghong Hu participated in the revision of this paper by providing comments and editing.

Horseshoe crabs are arthropods of the family Limulidae, suborder Xiphosurida, and order Xiphosura. Only four species belonging to three genera still exist, and they are *Carcinoscorpius rotundicauda*, *Tachypleus tridentatus*, *Tachypleus gigas*, and *Limulus polyphemus*. Among them *T. tridentatus* and *C. rotundicauda* can be found in China. In 1968, Bang and Levin discovered the blood coagulation mechanism of arthropods and developed a method by using limulus amebocyte lysate (LAL) to detect the presence of endotoxins (Levin and Bang 1968). In 1973, the US Food and Drug Administration admitted LAL as a biological agent. It was first written into American Pharmacopoeia in 1980. In 1983, Chinese Ministry of Health authorized the National Institute for the Control of Pharmaceutical and Biological

Chang Liu and Ximei Liu contributed equally with all other contributors.

C. Liu · X. Liu · Y. Wang · M. Hu (✉)
National Demonstration Center for Experimental Fisheries Science Education, Shanghai Ocean University, Shanghai, People's Republic of China

Key Laboratory of Exploration and Utilization of Aquatic Genetic Resources, Ministry of Education, Shanghai Ocean University, Shanghai, People's Republic of China

Shanghai Engineering Research Center of Aquaculture, Shanghai Ocean University, Shanghai, People's Republic of China
e-mail: mhhu@shou.edu.cn

J. Song
Tianjin Era Biology Technology Co., Ltd., Hebei, People's Republic of China

J. He
Beihai Product Quality Testing Institute, Beihai, People's Republic of China

Z. Tai · Q. Zhu
Shanghai Skin Disease Hospital, Tongji University School of Medicine, Shanghai, People's Republic of China



J. T. Tanacredi et al. (eds.), *International Horseshoe Crab Conservation and Research Efforts: 2007-2020*, https://doi.org/10.1007/978-3-030-82315-3_39

Products to conduct large-scale TAL tests throughout China. The endotoxin detection method was revised several times from 1983 to 1995. The endotoxin test, for the first time, was included in the second Chinese Pharmacopoeia in 1993. Revisions were mainly focused on the types of drugs and the requirement of test water. In 2003, the quantitative detection of bacterial endotoxin began to be applied in China. Compared with the rabbit pyrogen test, TAL is more sensitive, highly specific, faster, and more convenient to use (Chinese Pharmacopoeia Commission 2015).

Currently amebocyte lysate has two natural sources. Tachypleus amebocyte lysate (TAL) is extracted from the blood of the *Tachypleus tridentatus species*, while LAL is extracted from the blood of the *Limulus polyphemus species*. It should be noted that TAL properties and testing methods are the same as LAL. More than 95% TAL come from China (Hong 2011). Currently, Chinese TAL suppliers mainly distributed in Xiamen, Fujian, and Zhanjiang and Guangdong provinces. Based on our previous investigations, there are six largest manufacturers in China, including Beihai Xinglong Biological Products Co Ltd., Xiamen Bioendo Technology Co Ltd., Fuzhou Xinbei Biochemical Industry Co Ltd., Zhanjiang A & C Biological Ltd., Zhanjiang Bokang Marine Biological Co Ltd., and Dynamiker Biotechnology (Tianjin) Co Ltd.

## 1 Principles and Mechanisms of TAL

Biochemical basis of TAL is built upon the blood coagulation mechanism in arthropods. A large number of high-density particles in the amoebocytes of horseshoe crab blood contain clotting and anti-lipopolysaccharide factors. Once an endotoxin comes in contact with these particles, it activates Factor C and Factor B, causing proclotting enzymes to turn into clotting enzymes. Finally, coagulogen could be digested into coagulin, and coagulin could produce an agglutination reaction by cross-linking enzymes. The aforementioned pathway is for the traditional TAL. Some manufacturers add a synthetic chromogenic matrix into the TAL lysate. They use clotting enzyme to decompose the matrix into a polypeptide and yellow p-nitroaniline, which allows analyzation to be made base upon color. However, TAL has an alternate coagulation pathway to react with 1,3-β-D-glucan. The 1,3-β-D-glucan reacts with the Factor G, and it will cause the proclotting enzyme to turn into clotting enzyme, and the following process is similar with endotoxin (Wang et al. 2007; Morita et al. 1981). The specific molecular weight, function (Zhou 2005), and principle (Liu 2008) of each component were summarized in Table 1 and Fig. 1.

## 2 TAL Applications

The manufacturing process of TAL begins with extracting the cell lysates from the horseshoe crab's blood cell via a centrifugalizing and washing process. Extracting components by chloroform, adjusting pH values, and finally vacuum drying are conducted step by step by (Wang et al. 2007). Currently, TAL has two main applications. TAL is applied to test dialysis water, dialysate, and dialysis-related products in hospitals and pharmaceutical applications. Moreover, it is used for testing biological products like gene drugs, protein drugs, monoclonal drugs, therapeutic vaccines, and small molecule chemicals, which need to be free of endotoxin residue. In addition, TAL may be used to detect "environmental pollution," for instance, detection for endotoxin presence in fresh water, air, and outer space (Novitsky 2009).

## 3 Endotoxin Test Classification

Endotoxin test has two methods: gel clot method and photometric method. The photometric method could be divided into the turbidimetric method and chromogenic method. When different testing methods conflict, it is the gel clot method that resolves the conflict. The gel clot method is a qualitative endotoxin test method based on the blood coagulation mechanism. When using this method, it is the gel formation that is observed. The photometric method is a quantitative method for endotoxin testing that includes four specific tests, turbidimetric kinetic test, turbidimetric end-point test, chromogenic kinetic test, and chromogenic end-point test. All these four tests require corresponding instruments.

Turbidimetric test is based on the development of an endogenous substrate, and the chromogenic test is based on the development of the color after cleavage of a synthetic peptide-chromogen complex. Both tests can be widely used in endotoxin determination, but they have different scopes of application. The gel clot method applies to tests, which have no need for high accuracy, such as the examination of medicine and medical instruments, while photometric method applies to the test of the patient.

## 4 TAL Disadvantages

Although the TAL has a lot of advantages in endotoxin testing, the test is not perfect and may result in false results. First, due to the principle activity of TAL, 1,3-β-D-glucan can also react with TAL, specifically with factor G, producing false positive result, while the physical properties and bioactivity of glucan are very

**Table 1** The molecular weight and function of TAL (Zhou 2005)

| Aggregation factor | Molecular weight/kDa | Function/character |
|---|---|---|
| Factor C | 123 | Serine protease |
| Factor B | 64 | Serine protease |
| Factor G | 110 | Serine protease |
| Proclotting enzyme | 54 | Serine protease |
| Coagulogen | 20 | Gel formation |

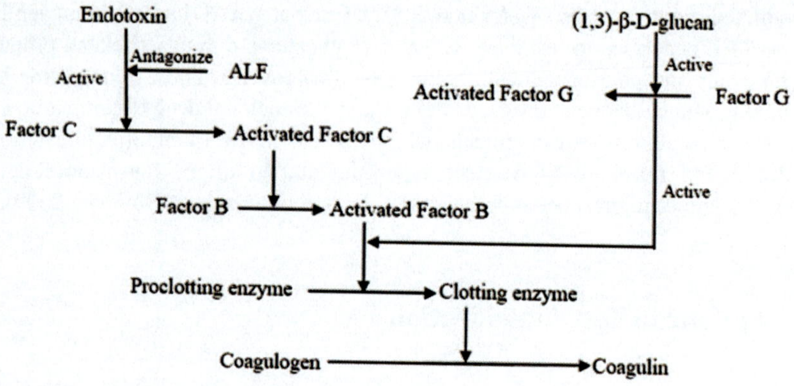


**Fig. 1** The principle of bacterial endotoxin test (Liu 2008)

heterogeneous different. In some cases, glucan can be detected in some gram-positive patients, but not in gram-negative patients (Shen 2018). Therefore, it is hard to distinguish whether endotoxin or 1,3-β-D-glucan is the cause of the positive result. Positive rate of endotoxin test is high in patients with long-term hemodialysis treatment, which may include false positive results due to the influence of (1,3)-β-D-glucan (Wong et al. 2016). The pH value also plays a vital role in test. Compared with LAL, TAL has different suitable pH value for reaction. Gao and Yan (1998) reported that the optimal pH value of TAL experiment is 6.68–7.12, and that of LAL experiment is 6.68–7.83. Peracid or peralkaline may cause false negative results (Ma et al. 2013). Gnauck et al. (2015) showed that the examination of using TAL on healthy women can sometimes be inaccurate. Inconsistent product quality during production also may affect TAL results, but the manufacturing process has improved over time.

# 5 Current Difficulties of China's TAL Industry

## 5.1 *Declining Horseshoe Crab Resources*

Information about the current population size of *Tachypleus tridentatus* is lacking. The largest horseshoe crab habitat in China is found along Guangxi Beibu Gulf. However, the number of horseshoe crabs in the Beibu Gulf has decreased rapidly from 600,000 to 700,000 pairs in 1990s to about 300,000 pairs in 2010 (Chen et al. 2015). The tri-spine horseshoe crab is a key protected species in some provinces, such as Guangdong, Guangxi, and Fujian. No trade is allowed in these provinces (Li et al. 2018). In 2019, *Tachypleus tridentatus* was listed on the IUCN Red List as endangered species and at risk of extinction throughout its spawning range.

The decline of the population is mainly caused by two factors. First, the coastal habitat has been largely destroyed. The sea south of the Yangtze river once was the important habitats for horseshoe crabs such as Hangzhou Bay, Dongshan Bay in Fujian, Zhoushan Archipelago and so on, which have been destroyed. Today, it is hard to find horseshoe crab in the sea north of the Pearl River estuary. The beaches are widely used for residential and commercial developments (housing/tourism and recreation areas). And large amounts of industrial and domestic waste water are discharged into the sea, polluting the water and influencing reproduction (Xu 2016; Liang et al. 2001). Second, overexploitation and overfishing diminish the present number of the horseshoe crab. According to recent information, thousands of horseshoe crabs are killed annually for their meat, blood, or carapace (Liao and Li 2001; Wong et al. 2012), which has been made easier due to improvements in fishing gear and changes in harvesting from manual beach capture to trawling with large gill nets and other advanced methods. Large numbers of horseshoe crabs regardless of sizes are also caught on the shore. Some are sold to fish markets, and others are killed for their meat and carapace (Liao and Li 2001), resulting in the decline in the horseshoe crab population.

### 5.2 *Uneven Quality of TAL Products*

Although China has set a standard for TAL and revised it several times, the quality of TAL is not consistent. Individual difference between manufactures is the primary factor for this inconsistency. Most horseshoe crabs are caught in the wild and manufacturers release them after bleeding. So, individual difference influences the quality greatly. Sometimes, horseshoe crabs are recaptured and bleed again within a short period of time. In addition, during the production process, the lack of a sanitized environment may also result in uneven product quality (Yan et al. 2018), and the TAL quality standards are not strictly implemented.

## 6 Future Prospects of TAL in China

### 6.1 *Substitutes Development for TAL*

As the natural horseshoe crab resources are plummeting and the demand for TAL in the global market is increasing, it is necessary to find an alternative, such as developing a synthetic TAL. Factor C is the first component reacting with endotoxin in TAL. Ding and Ho (2001) proposed replacing traditional TAL with recombinant C factors (rFC). rFC is a kind of endotoxin-inducible coagulation enzyme, which functions as the combination of the Factor C, B, proclotting enzyme, and the coagulogen. Recombinant factor C constructed by Tan et al. (2000) has a high affinity with endotoxin and multiple endotoxin binding sites. Alwis and Milton (2009) have realized the detection of endotoxin in room dust with recombinant factor C. And a zymogen-based sensor containing recombinant factor C has been developed to achieve electrochemical detection of endotoxin (Inoue et al. 2010). The accuracy of the reagent they developed has reached 0.005EU/ml (Ding and Ho 2001). Currently, the highest accuracy of natural TAL we can get in the market is 0.003EU/ml (http://www.houshiji.com/Product/detail/classid/159/id/148.html). Compared between two reagents, the rFC has similar accuracy and does not require the harvest of horseshoe crabs. rFC is still at the experimental stage (Gao et al. 2020). It is quite new, and it is not used in China until recently. However, the Chinese pharmacopeia has not listed it as a common drug. Nevertheless, it may be a new and practical way to replace traditional TAL.

### 6.2 *Horseshoe Crab Population Restoration by Artificial Breeding*

With more attention being placed on the protection of horseshoe crabs through restrictions on overfishing and abuse, it may be possible to restore the population by artificial intervention. Adult horseshoe crabs have few predators in nature except humans, but eggs and juveniles have many; therefore, artificial breeding and release programs may be an effective way to restore the population. Cheng et al. (2006) found that releasing the larvae of horseshoe crab along sandy beaches can effectively help them avoid being killed by natural aquatic predators. Xie et al. (2017) also believe that releasing third instars and above could significantly increase their survival rate in the wild. According to previous studies (Liang 1987; Xie 2012), funding and space are available for horseshoe crabs to hatch, and mature indoors survival may be improved when released into the wild. Hence, hatching and rearing horseshoe crabs in the lab and releasing juvenile horseshoe crabs in protected areas may prove to be an effective way to help restore the population.

**Acknowledgements** This research was supported by a grant from 2017 Beihai City 13th Five-Year Plan Marine Economic Innovation and Development Demonstration Project--Tachypleus Amebocyte Lysate and Chinese Horseshoe Crab Ecological Utilization Industry Chain Collaborative Innovation Project (Grant No.: Bhsfs006).

## References

K.U. Alwis, D.K. Milton, Recombinant factor C assay for measuring endotoxin in house dust: Comparison with LAL, and (1-3)-β-D-glucans. Am. J. Ind. Med. **49**(4), 296–300 (2009). https://doi.org/10.1002/ajim.20264

Z.B. Chen, H.Q. Fan, Y.Y. Liao, et al., Limulus, a living animal fossil facing the dilemma of survival. Shanghai Sci. **67**(3), 60–62 (2015). https://doi.org/10.3969/j.issn.0368-6396.2015.03.015. (in Chinese)

P. Cheng, A.N. Zhou, S.F. Huo, et al., Study on the adaptability of the larva of Limulus amebocyte in China to different environments. J. Xiamen Univ. (Nat. Sci.). **45**(3), 404–407 (2006). https://doi.org/10.3321/j.issn:0438-0479.2006.03.025. (in Chinese)

Chinese Pharmacopoeia Commission, *Pharmacopoeia of the People's Republic of China* (Chinese Medical Science and Technology Press, Beijing, 2015), pp. 95–98. (in Chinese)

J.L. Ding, B. Ho, A new era in pyrogen testing. Trends Biotechnol. **19**(8), 277–281 (2001). https://doi.org/10.1016/s0167-7799(01)01694-8

H. Gao, Y.S. Pei, T. Cai, C. Chen, H. Gao, X. Wei, Y. Liu, T.T. Wang, Q.F. Zhu, Y.H. Lu, J.H. Piao, Q. Liu, H. Zhang, S.M. Li, Validation of a determination method for recombinant factor C of bacterial endotoxin. Chinese J. Biol. **33**(01), 76–79 (2020). https://doi.org/10.13200/j.cnki.cjb.002967

G.Z. Gao, J. Yan, Effects of various pH in testing bacterial endotoxins. Chin. Pharm. J. **33**(3), 162–165 (1998) http://www.cnki.com.cn/Article/CJFDTOTAL-ZGYX803.012.htm

A. Gnauck, R.G. Lentle, M.C. Kruger, The limulus amebocyte lysate assay may be unsuitable for detecting endotoxin in blood of healthy female subjects. J. Immunol. Methods **416**, 146–156 (2015). https://doi.org/10.1016/j.jim.2014.11.010

K.Y. Inoue, K. Ino, H. Shiku, et al., Electrochemical detection of endotoxin using recombinant factor C zymogen. Electrochem. Commun. **12**(8), 1066–1069 (2010). https://doi.org/10.1016/j.elecom.2010.05.028

S.G. Hong, *Biology of horseshoe crabs. Tachypleus tridentatus* (Xiamen University Press, Xiamen, China, 2011) (in Chinese)

Y.H. Li, X.Y. Xie, K.Y. Kwan, Endangered status and declaration on conservation of the "Living Fossil" *Tachypleus tridentatus*. Wetland Sci. **16**, 690–692 (2018)

J. Levin, F.B. Bang, Clottable protein in Limulus; its localization and kinetics of its coagulation by endotoxin. Thromb. Diath. Haemorrh. **19**(1), 186–197 (1968) https://www.ncbi.nlm.nih.gov/pubmed/5690028

G.Y. Liang, The research of artificial incubation of Chinese horseshoe crab. Mar. Sci. **1**, 40–47 (1987) http://www.cnki.com.cn/Article/CJFDTotal-HYKX198701008.htm

J.R. Liang, J. Wang, Y.Q. Su, D.X. Wang, Y.G. Yao, Effects of metals on embryodevelopment of *Tachypleus tridentatus*. Acta Ecologica Sinica **21**(6), 1009–1012 (2001) CNKI:SUN:STXB.0.2001-06-025

Y.Y. Liao, X.M. Li, Present situation of horseshoe crab resources in the sea area of China and tactics of preservation. Resources Science **23**(2), 53–57 (2001) CNKI:SUN:ZRZY.0.2001-02-011 (in Chinese)

C.T. Liu, Limulus amebocyte lysate and its application. Chin. J. Mod. Drug Appl. **2**(9), 110–111 (2008). https://doi.org/10.3969/j.issn.1673-9523.2008.09.097. (in Chinese)

L. Ma, S. Zhang, A.H. Wang, M.X. Gong, et al., Analysis of the common problems in the quantitative detection of bacterial endotoxin. Chin. J. Hosp. Pharm. **33**(9), 733–734 (2013). https://doi.org/10.13286/j.cnki.chinhosppharmacyj.2013.09.022. (in Chinese)

T. Morita, S. Tanaka, T. Nakamura, et al., A new 1,3-β-D-glucan-mediated coagulation pathway found in *Limulus* amebocytes. FEBS Letter **129**(2), 0–321 (1981). https://doi.org/10.1016/0014-5793(81)80192-5

T.J. Novitsky, *Biomedical Applications of Limulus Amebocyte Lysate. Biology and Conservation of Horseshoe Crabs* (Springer, Boston, MA, 2009)

Product introduction data source from Xiamen Bioendo Technology Co., LTD: http://www.houshiji.com/Product/detail/classid/159/id/148.html

Y.Z. Shen, Bacterial endotoxin test and its clinical application. Shanghai Med. Pharm. J. **39**(21), 3–6 (2018). CNKI:SUN:SYIY.0.2018-21-002

N.S. Tan, M.L.P. Ng, Y.H. Yau, et al., Definition of endotoxin binding sites in horseshoe crab factor c recombinant sushi proteins and neutralization of endotoxin by sushi peptides. FASEB J. **14**(12), 1801 (2000). https://doi.org/10.1096/fj.99-0866com

L. Wang, H. Gao, T. Cai, G.L. Zhang, et al., Research and development of tachypleus amebocyte lysate. Chin J. Pharm. Anal. **27**(6), 938–942 (2007). https://doi.org/10.16155/j.0254-1793.2007.06.045

J. Wong, Y. Zhang, A. Patidar, et al., Is endotoxemia in stable hemodialysis patients an artefact? Limitations of the limulus amebocyte lysate assay and role of (1→3)-β-D glucan. PLoS One **11**(10) (2016). https://doi.org/10.1371/journal.pone.0164978

C.H. Wong, Y.J. Xie, Z.Q. Xiao, L.M. Huang, J. Li, S.H. Wang, Y.Z. Zhang, Distribution and Resource of Chinese Horseshoe Crab (*Tachypleus tridentatus*) in Fujian and Other Coast Water of China. Chinese Journal of Zoology, **47**(3), 40–48 (2012) https://doi.org/10.13859/j.cjz.2012.03.004

Y. Xie, *The Industrial Seed Production of Tachypleus Tridentatus and Effect of Key Ecological Factors on it, and Research of Tetraodaotoxin (TTX) on Horseshoe Crabs in Water of China* (Guangdong Ocean University, 2012) http://cdmd.cnki.com.cn/Article/CDMD-10566-1012522661.htm (in Chinese)

H.L. Xie, H.Q. Fan, J.Y. Guan, et al., Horseshoe crabs conservation: A triple winning strategy. Guangxi Sci. **24**(05), 509–515 (2017) (in Chinese). http://www.en.cnki.com.cn/Article_en/CJFDTotal-GXKK201705013.htm

D.Q. Xu, Tissue Distributions of Heavy Metals in Horseshoe Crabs and their Relationships with Major Environmental Factors from the Intertidal Zone along Beibu Gulf. Master theses. Guangxi University (2016)

M.Y. Yan, Y. Wang, H.X. Gu, et al., Current situation of tachypleus amebocyte lysate industry in China and relevant suggestions. J. Biol. **35**(2), 88–91 (2018). https://doi.org/10.3969/j.issn.2095-1736.2018.02.088. (in Chinese)

H.J. Zhou, *Biological Examination of Pharmaceutica* (People's Medical Publishing House, Beijing, China, 2005), p. 448. (in Chinese)

# Ecology of Juvenile American Horseshoe Crabs (*Limulus polyphemus*) at Plumb Beach, Jamaica Bay, New York

**Christina P. Colon, Mark L. Botton, Peter Funch, Emil Hoffgaard, Kaur Mandeep, and Kera Mansfield**

## 1 Introduction

Horseshoe crabs worldwide are facing precipitous population declines due to overharvesting for food, bait and biomedical industries, increased pollution levels, habitat loss, beach erosion, shoreline hardening, and sea level rise (Smith et al. 2016a; Akbar John et al. 2018). This lineage of marine chelicerates has survived over 480 million years of environmental turbulence, but today, their future is uncertain. In 2016, the American horseshoe crab (*Limulus polyphemus*) was listed as vulnerable by the IUCN Red List of Threatened Species (Smith et al. 2016b), and the tri-spine horseshoe crab (*Tachypleus tridentatus*) was listed as endangered in 2019 (Laurie et al. 2019). The remaining two species (*T. gigas* and *Carcinoscorpius rotundicauda*) are generally perceived as declining (Akbar et al. 2018; Vestbo et al. 2018). This rapid decline and lack of information threatens the complex ecological food web of which these species are an integral part (Botton 2009).

Adult horseshoe crabs are hardy and relatively impervious to predation, but juveniles are vulnerable and experience high mortality (Botton 2009). While *Limulus* spawning behavior and mate selection and the factors influencing nesting locations have been studied extensively (e.g., Botton et al. 1988; Brockmann and Smith 2009; Brockmann et al. 2015), far fewer studies have been carried out on juveniles.

C. P. Colon (✉) · K. Mansfield
Department of Biological Sciences Kingsborough Community College CUNY, Brooklyn, NY, USA
e-mail: christina.colon@kbcc.cuny.edu

M. L. Botton · K. Mandeep
Department of Natural Sciences Fordham College at Lincoln Center, New York, NY, USA

P. Funch · E. Hoffgaard
Department of Bioscience, Aarhus University, Aarhus, Denmark



J. T. Tanacredi et al. (eds.), *International Horseshoe Crab Conservation and Research Efforts: 2007-2020*, https://doi.org/10.1007/978-3-030-82315-3_40

Females bury thousands of eggs in the sand but few survive the two- to four-week incubation period due to predation from migratory shorebirds who rapidly increase body mass on protein rich eggs prior to their spring migrations (e.g., Mizrahi and Peters 2009). Only three larvae per 100,000 survive their first summer under natural conditions (Botton et al. 2003). Once hatched, juveniles are extremely vulnerable as they feed in the sediment of shallow intertidal flats over the next few years. Young-of-the-year appear to subsist on particulate organic matter and meiofauna before transitioning to a diet composed mainly of benthic invertebrates such as polychaetes, crustaceans, and mollusks as they grow (Carmichael et al. 2004; Gaines et al. 2002). During and immediately after they molt, juveniles are particularly vulnerable, burrowing into the sediment until their new shell hardens. After several years of seasonal growth, juveniles move into deeper waters. Aside from occasional carapaces that wash ashore, almost nothing is known about the larger juveniles (Estes Jr et al. 2015).

Spawning locations selected by females include sandy estuarine beaches (Brockmann 2003) that contain well oxygenated sheltered and shallow sand of medium to coarse grains, away from the influence of hydrogen sulfide (Vasquez et al. 2015). Optimal beaches have at least 15–20 cm of sand with specific temperature, oxygen, and moisture regime for egg development (Penn and Brockmann 1994). Few tidal flats support healthy populations of juvenile crabs, and these are not well understood in terms of the conditions necessary for juvenile populations to flourish. Because newly hatched larvae spend their first week in a free-swimming phase before molting and settling, there is a tenuous connection between hatching location and where juveniles ultimately settle.

Juvenile horseshoe crabs tend to be cryptic and difficult to find, often aggregating in shallow pools, while adjacent pools remain empty. Foraging juveniles burrow just below the soft sediment only visible by the signature spiraling trails they leave in their wake. The exact physical and biotic characteristics of the sediment they occupy or their dietary preferences are not well understood. Laboratory studies indicate that larvae and early juveniles can withstanding wide fluctuations in temperature, salinity, and dissolved oxygen (Funch et al. 2016; Jegla and Costlow 1982; Palumbi and Johnson 1982) and are also capable of surviving relatively high levels of pollutants including heavy metals (Botton and Itow 2009; Botton et al. 2010). The paucity of easy to locate juvenile populations combined with their patchy distribution and high vulnerability make increased knowledge of this life stage a critical component of a comprehensive management and conservation plan for horseshoe crabs worldwide.

This study seeks to expand our knowledge of the ecology of juvenile horseshoe crabs in Jamaica Bay, an urban estuary in New York City. We address three principal research questions. First, what is the seasonal abundance and size distribution of juvenile horseshoe crabs in three potential juvenile habitats: a broad undisturbed intertidal sand flat, a disturbed sand flat adjacent to a recently nourished beach, and a salt marsh tidal creek? Second, what is the diet and food selectivity of juvenile horseshoe crabs in Jamaica Bay, and what is their position within the broader food web of the estuary? Lastly, do burrowing and righting behaviors on sediments of differing textures help to explain the spatial distribution of juvenile horseshoe crabs within the mosaic of available intertidal habitats in this region?

## 2 Research Site

A summer monitoring program of juvenile horseshoe crabs begun in 2011 at Plumb Beach at the mouth of Jamaica Bay. Located on federally protected land and managed in collaboration with New York City and the Department of Transportation, Plumb Beach was made part of Gateway National Recreation Area in 1972 (see Fig. 1).

Plumb Beach is a favored spawning site for adult crabs who come ashore on the eastern end of the beach in large numbers to deposit eggs in the sand (Botton et al. 2018). Researchers from New York City Audubon have been monitoring spawning adults in Jamaica Bay since 2009. The eastern span (approximately 1 km) is favored by females due to the relatively calm waters and coarse, well-aerated sand (Botton et al. 2018). Expansive tidal flats are exposed at low tide, and the foraging trails of juvenile horseshoe crabs are especially common in the shallow pools. These shallow pools cover an area approximately 2623m$^2$ and were populated by a few hermit crabs (*Pagurus sp.*) and polychaetes. The eastern beach terminates at the entrance to

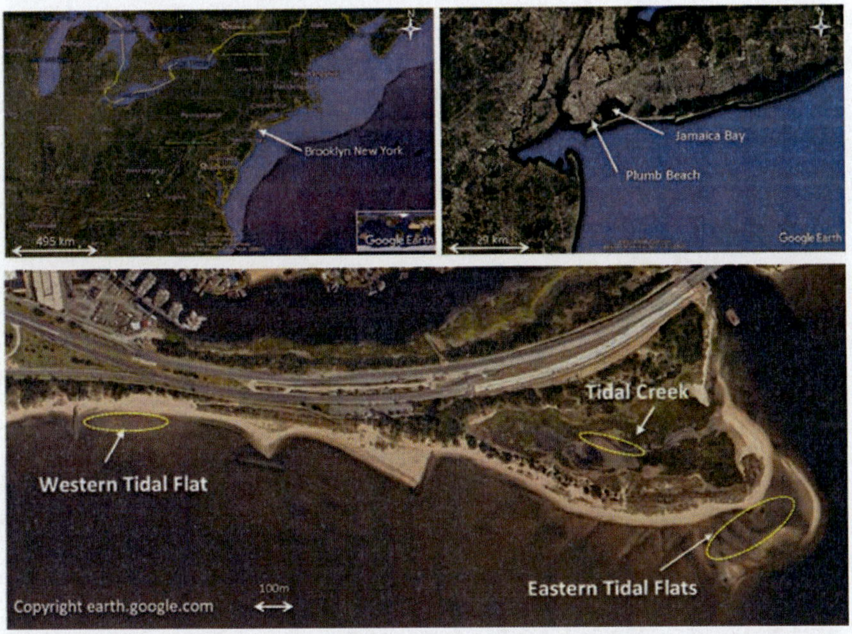


**Fig. 1** Google Earth image of Plumb Beach and indication of the three main research locations: Western Tidal Flats, Eastern Tidal Flats (2623m$^2$) and Tidal Creek (347m$^2$)

a salt marsh tidal lagoon, from which flows a tidal creek. The central portion occupied by horseshoe crabs was approximately 63 m long and 5.74 m wide creating a foraging area approximately 347m$^2$.

According to former director of New York City Audubon, Ron Bourque (2000) at one time this salt marsh was a spawning hotspot, with amplexed pairs jockeying for position throughout the summer. The longshore drift in this portion of Jamaica Bay is typically from west to east (Botton et al. 2018); thus, as sediment has washed eastward, the entrance to this area has silted up and restricted access to spawning crabs. *Spartina alterniflora* beds encrusted with ribbed mussels (*Geukensia demissa*) and barnacles (*Balanus* sp.) rim the basin of the tidal creek. This area is home to a large population of fiddler crabs (*Leptuca pugilator*) and killifish (*Fundulus* sp.) and frequented by a variety of wading birds.

The western portion of Plumb Beach (also approximately 1 km) adjacent to the parking lot and Belt Parkway is in stark contrast to the eastern portion. This stretch experiences stronger wave action and more direct pollution and is subject to heavy erosion. Several beach restoration projects have been carried out over the years by the US Army Corps of Engineers (Botton et al. 2018). Beach nourishment took place in 1992 and again in Fall 2012, only weeks prior to when SuperStorm Sandy made landfall carrying away a portion of the 97,098 cubic meters of sand that was deposited. The sand within the nourished section of Plumb Beach tends to be highly compacted and poorly oxygenated, with a black layer bearing an odor of hydrogen sulfide. It is usually avoided by all but a handful of solo males, although in 2019 mated pairs were spotted at the eastern margin (China Moore, pers. com). Despite these nourishment projects, sand continues to flow eastward, as the western beach grows thinner each year. The most abundant organisms visible among the ridges on the narrow tidal flats are a large population of mud snails (*Tritia (Ilyanassa) obsoleta*) and a late summer layer of macroalgae, primarily *Ulva* sp.

## 3 Population Dynamics, Growth, and Size Distribution of Juvenile Horseshoe Crabs

### *3.1 Methods*

Juvenile population surveys were conducted opportunistically in May through August approximately every 2 weeks around the new and full moon daytime low tides ±2 hr. These were initially carried out only on the eastern and western tidal flats of Plumb Beach where spawning adults, eggs, and juveniles were previously encountered. In 2014 a population of juveniles was located in a tidal creek, so this third area was subsequently added to the survey effort. Timed visual surveys entailed a team of trained student observers simultaneously searching the shallows for juvenile crabs for 10–15 minutes depending on the number of researchers. Search areas were not restricted to specific quadrats but rather entailed searching all shallow pools present. All crabs were measured with dial calipers to the nearest 0.1 mm at the widest part of the prosoma. After measurement, crabs were returned to the habitat. From 2011 to 2018, a total of 48 surveys were conducted, and each site surveyed within 2 hours of the first site. Raw counts were converted into number of crabs per man-hour search effort, or catch per unit effort (CPUE) in order to normalize the data for comparison between surveys, years, and locations.

Shed carapaces were collected opportunistically along the entire beach, at the wrack line, usually visible in August and September, when juveniles are generally known to molt. In 2018, an unusually large number of shed carapaces were found from June to September. All carapaces collected were measured at the widest part of the prosoma. Partial carapaces found were measured from the mid line then doubled. Visual surveys of both live crabs and juveniles are inherently biased toward larger specimens and individuals at the surface, so buried and/or smaller individuals are likely less well represented in this study.

Prosoma width of live crabs and shed carapaces was converted to instar based on the growth data in Sekiguchi (1988b). Mean prosoma width for each instar was used to create size bins equidistant from the mean size of the next instar. Values that fell between these parameters were attributed to the next lower size class. Density was calculated as number of crabs per 100m$^2$ based on only areas still submerged with several centimeters of water at low tide where crabs were located. Sampled areas were estimated using linear and polygon measurement tools in Google Earth based on satellite images of the sand flat tidal pools and tidal creek taken at low tide on November 3, 2017. While size of individual pools varied considerably within and between years, size of the flats and the tidal creek remained consistent from year to year. Because the size of individual pools was not measured, density values here should be considered an estimate and less precise than CPUE. A t-test was used to compare CPUE, density, and average prosoma width in the tidal creek and eastern tidal flats for all years of the study and 2014–2018 when both areas were sampled.

## 3.2 *Results*

### Live Juveniles

Overall catch per unit effort (CPUE) of juvenile horseshoe crabs on the eastern tidal flats ranged up to 77 crabs per hour, averaging 41.78 (SD ± 23.87) and was not significantly higher than in the tidal creek (t-test, $p = 0.229$), where peak CPUE was 65.9 crabs per hour, averaging 20.44 (±27.75) (Table 1). Peak counts were similarly higher on the eastern beach ($n$ = 160 on 6/27/2018) compared with the tidal creek ($n$ = 123 on7/29/2015). Density was significantly higher in the tidal creek (t-test, $p = 0.029$) with an average density of 18.62 crabs per 100 m2 (±11.54) compared with the 1.75 crabs per 100 m$^2$ at the more expansive eastern flats (±2.69). Total counts of foraging juvenile crabs appear to peak in July and August although these were not necessarily synchronous between locations. There appears to be high

**Table 1** Average catch per unit effort (CPUE) seasonal peak and density (per 100 m²) of juvenile horseshoe crabs on the eastern tidal flats (BE) and tidal creek (TC) of Plumb Beach

| YEAR | Avg CPUE | Avg CPUE | Peak day | Peak day | Peak # | Peak # | Density 100 m² | Density 100 m² |
|---|---|---|---|---|---|---|---|---|
| | BE | TC | BE | TC | BE | TC | BE | TC |
| 2011 | 9.986 | | 7/13/2011 | | 35 | | 1.33 | |
| 2012 | 17.5 | | 6/18/2012 | | 52 | | 1.98 | |
| 2013 | 10 | | 7/23/2013 | | 48 | | 1.83 | |
| 2014 | 0.4 | 43.7 | 7/22/2014 | | 1 | 89 | 0.04 | 25.50 |
| 2015 | 0 | 77.0 | N/A | 7/29/2015 | 0 | 123 | 0.00 | 35.24 |
| 2016 | 8.4 | 37.7 | 8/17/2016 | 8/1/2016 | 0 | 52 | 0.00 | 14.90 |
| 2017 | 27.5 | 40.6 | 8/21/2017 | 8/21/2017 | 71 | 30 | 2.71 | 8.60 |
| 2018 | 65.9 | 9.9 | 8/9/2018 | 6/27/2018 | 160 | 31 | 6.10 | 8.88 |
| **OVERAL** | **17.46** | **41.78** | | | **45.88** | **65.00** | **1.75** | **18.62** |

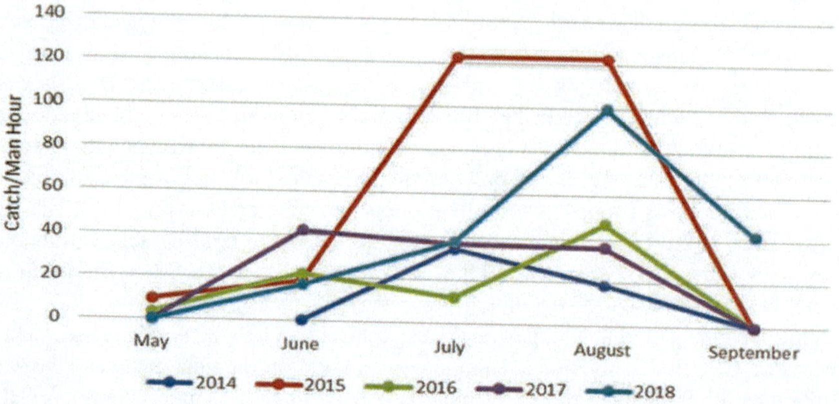


**Fig. 2** Seasonal peaks in catch per unit effort (CPUE) of juvenile horseshoe crabs at Plumb Beach from 2014 to 2018

interannual variation in juvenile crab populations overall with both viable habitats exhibiting different peaks and declines.

The seasonal timing of juvenile crab abundance at both the eastern tidal flats and adjoining tidal creek within Plumb Beach has been fairly consistent over the years, with low CPUE's in May, gradually increasing during June and July (Fig. 2). The peak in August coincides with appearance of young of the year in their second or third instar as they settle onto the tidal flats. Sightings typically trail off by September as young of the year either decline due to predation, burrow below the surface, or emigrate to deeper waters as temperatures cool. Lee and Morton (2009) found that under laboratory conditions, fewer individuals emerge to feed when sediment temperature decline.

Initially the juvenile abundance on the intertidal sand flats of the eastern beach was relatively robust (Fig. 3). In 2011, a total of 109 juvenile crabs ranging from 4

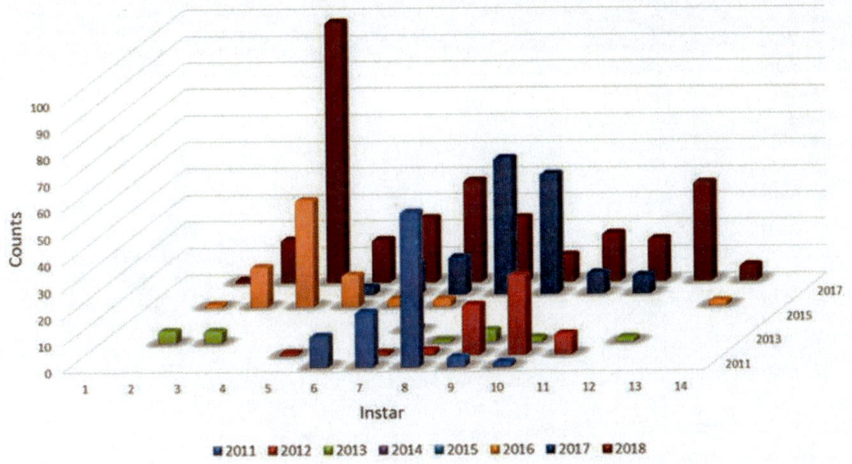


**Fig. 3** Juvenile horseshoe crab counts sorted by instar at Eastern side of Plumb Beach from 2011 to 2018

to 40 mm were found on the eastern tidal flats during four surveys from 7/13/2011 to 8/25/2011. CPUE declined linearly from 21.33 crab per hour to 1 by the end of that summer. Most juveniles fell into the seventh to ninth instar and were likely in their second year. Total counts in 2012 and 2013 were 61 (69.3 CPUE) and 20 (29 CPUE), which declined to zero by 2015. Smaller juveniles were only seen in late summer and presumed to be young of the year. The disappearance of juveniles was unexpected given the thousands of spawning adults and tens of thousands of eggs observed at Plumb Beach every year since 2009. Continued surveys on the eastern flats in late summer 2016, revealed a small pocket of 71 one year old juveniles (58.9 CPUE) in a single tide pool close to the shore. By 2018, we observed 323 juvenile crabs over eight surveys from 6/11/2018 to 9/22/2018 with an average CPUE of 527.5. Preliminary data from 2019 indicates a continued population use on the eastern flats.

In mid-summer 2014 after failing to locate more than two juveniles on the eastern tidal flats, an expanded search within a salt marsh creek behind the eastern tidal flats revealed a population of 158 juveniles (31.4 CPUE) over three surveys (Fig. 4). These juveniles ranged in prosoma width from 4.0 to 65.0 mm with the smaller instars being young of the year that likely hatched in mid-July. It is notable that no further recruitment was observed in subsequent years. The following summer in 2015, a total of 499 juveniles (76.7 CPUE) were found during eight surveys from 5/20/2015 to 9/12/2015. These ranged in size from 10 to 78.9 mm. The smaller individuals were likely fifth instar individuals in their second year, and the largest were likely 14th instar in their sixth year. In 2016, eight surveys revealed 294 juveniles (44.0 CPUE) ranging from 0.5 to 157.6 mm and in 2017, 112 juveniles were found during seven surveys (56.8 CPUE) ranging from 11.7 to 75.5 mm. By the summer of 2018, this population too was on a precipitous decline. Only 54 juveniles

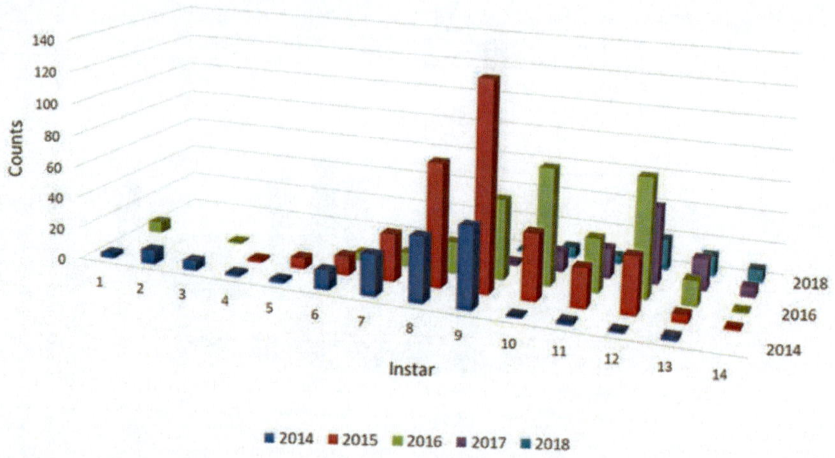


**Fig. 4** Juvenile horseshoe crab counts sorted by instar in a tidal creek at Plumb Beach from 2014 to 2018

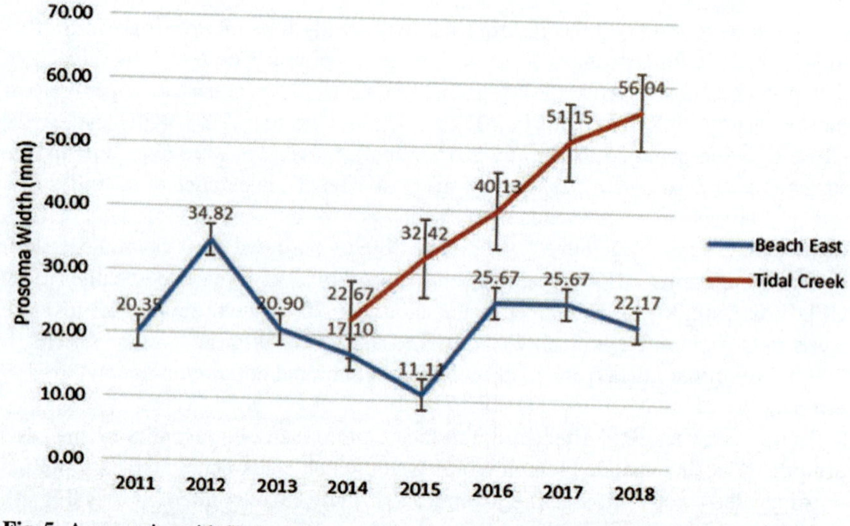


**Fig. 5** Average size with SD bars of juvenile horseshoe crabs on a tidal flat and in a tidal creek at Plumb Beach from 2011 to 2018

(9.9 CPUE) were found over eight surveys ranging from 29.1 to 81.1 mm. In 2016 and 2017, coring efforts revealed no eggs despite seeing the occasional adult crab. Despite declines, population density was overall highest in the tidal creek, which had a peak of 35.24 crabs per 100m$^2$ in 2015. Preliminary analysis of the data from 2019 indicates a continuation of the population decline and lack of recruitment.

No eggs have been found in the tidal creek and little or no recruitment has been observed. Size changes in the tidal creek were therefore used to approximate growth rate of crabs. From 2015 to 2018, average prosoma width increased linearly from 22.67 to 56.04 mm (see Fig. 5). Annual growth ranged from 4.89 to 11.02 mm and averaged 8.34 mm per year. Not surprisingly, overall size in the tidal creek averaged 35.83 (± 14.53) differed significantly (t-test, $p = 0.026$) from the average sizes of juveniles on the eastern tidal flats of 22.43 (± 22.43) where recruitment accounted for the nonlinear variability.

Few spawning females have been observed on the western beach, which may explain why low egg counts and virtually no juveniles have been observed on the western tidal flats despite thorough surveys since 2011. On 7/9/2014, a single juvenile was found measuring 17.1 mm and another measuring 17.8 mm on 8/6/2014. The same year, thousands of second and third instar young of the year were observed blanketing the western tidal flats on 7/22/2014. This observation led to initial optimism, but virtually none of these appear to have survived the winter. Another survey conducted on 7/26/2019 revealed first and second instar young of the year (ranging in size from 3 to 10.4 mm) foraging in very high densities ($n = 141$). Clearly the western intertidal habitat can support a population of small juvenile crabs, but whether these can persist beyond the first year is unknown.

## Shed Carapaces

Shed carapaces were normally found in late summer and predominantly along the wrack line of the eastern half of Plumb Beach. From June through August of 2018, a total 615 carapaces were found on the eastern side of Plumb Beach, while only nine were found on the western portion (Fig. 6). This by far surpasses numbers observed in previous years (84 in 2014, 10 in 2015, 112 in 2016, and 24 in 2017) indicating an unusually large population of juvenile crabs this season. Over the

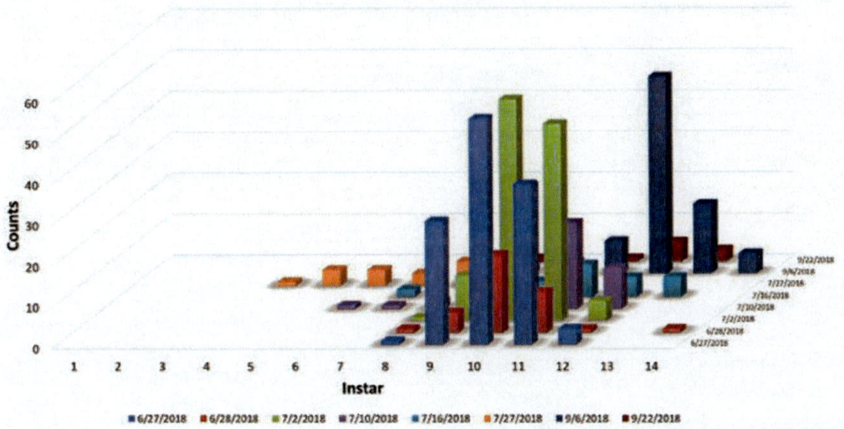


**Fig. 6** Shed carapaces of juvenile horseshoe crabs from Plumb Beach 2018 sorted by instar and date

summer, the average size of carapaces tripled, indicating continuous growth of individuals in the population. In late July, a minor downturn in mean size indicated molting by young of the year. Size range of carapaces throughout the summer spanned from 10 to 180 mm, with most individuals in the 20–50 mm range. A trend toward both smaller (young of the year) and larger individuals as the summer progressed indicates recruitment and growth.

### 3.3 Discussion

Although current literature makes scant reference to the potential importance of salt marsh habitats or restored intertidal flats to support juvenile horseshoe cabs, our investigation indicates that these areas provide adequate habitat for crabs. Small habitat pockets and asynchronous fluctuations among the number of crabs in these locations and between years make it difficult to determine predictable patterns or trends. It remains to be determined what events or conditions cause or contribute to these fluctuations. These may be normal and may be explained by interannual variation in spawning densities, egg deposition, predation intensity, or spatial shift driven by a variety of abiotic factors such as temperature or pollution.

Despite highly uneven distribution and highly variable densities, our observed densities of juvenile crabs compare favorably to reported densities of two Asian species, but are considerably lower than the densities of *L. polyphemus* reported in both Pleasant Bay and Delaware Bay (see Table 2). These differences may be due to

**Table 2** Reported densities of juvenile horseshoe crabs in other studies and other locations

| Species | Density ($100\ m^2$) | Location | Authors | Year |
|---|---|---|---|---|
| *T. tridentatus* | 1.47 | Palawan, Philippines | Almendral and Schoppe | 2005 |
| *T. tridentatus* | 2.0 | Palawan, Philippines | Kaiser | 2002 |
| *T. tridentatus* | 0.10 to 1.97 | Hong Kong | Morton and Lee | 2003 |
| *T. tridentatus* | 1.1 to 0.3 | Hong Kong | Li | 2008 |
| *T. tridentatus* | 0.00006 to 0.00169 | Taiwan | Hsieh and Chen | 2009 |
| *T. tridentatus* | 0.09 to 2.43 | Beibu Gulf, Southern China | Hu et al. | 2009 |
| *C. rotundicauda* | 2.05 | Singapore | Burton et al (per. Hong, 2004) | 2009 |
| *L. polyphemus* | 158 | Pleasant Bay, Cape Cod Massachusetts | Carmichael et al. | 2003 |
| *L. polyphemus* | 538.1 | Delaware Bay, New Jersey | Burton et al. | 2009 |
| *L. polyphemus* | 1.75 | Plumb Beach East, Jamaica Bay, New York | this study | 2019 |
| *L. polyphemus* | 18.62 | Plumb Beach Tidal Creek, Jamaica Bay, NY | this study | 2019 |

methodological differences as Carmichael et al. 2003 excavated quadrats to 5 cm and Burton et al. 2009 used a suction dredge.

The density of foraging juvenile crabs in our study is highly patchy and ephemeral. While CPUE was not significantly different between sites, it is likely that the tidal creek density appears higher because the habitat consists of a long narrow stream bed, whereas the intertidal flats on the eastern beach consists of multiple pools, which vary in size and in which juveniles tend to aggregate.

Growth rates of juvenile horseshoe crabs in our study are based on average size of the tidal creek cohort and not on repeat measurements of individuals over time. However, the annual increases observed in this study of 8.34 mm are in keeping with repeated cohort sampling from July to October by Burton et al. (2009) in Delaware Bay, which increased between 6.5 and 9 mm. Juvenile crabs become more visible during late summer, particularly after hatching events, which occur in late July or early August. Older juveniles (> 60 mm prosoma width) are rarely seen alive on the intertidal flats, but their presence as molts indicates that they are present but may be distributed further offshore in subtidal waters. Use of a suction dredge or underwater ROV might shed additional light into the location and abundance of these larger juveniles.

The observed recovery of juvenile counts on the eastern beach coincided with a deepening and expansion of the tidal pools, which had been largely reduced by sand deposition after Superstorm Sandy. However, the initial decline began long before Superstorm Sandy, so the cause of this decline remains unknown.

It is also not known whether the juveniles observed in the tidal creek were always present, or if the storm surge or interannual advection pushed crabs from the eastern flats into this refuge. We have seen no evidence of horseshoe crab nesting within the tidal creek, and no freshly laid eggs have been found here. The lack of recruitment, spawning adults, or eggs and the steady decline of juvenile crabs in the tidal creek indicate that advection may be the case. Alternately, the entrance to the tidal creek could have been partially blocked by sand accretion from the 2012 western beach restoration, as occurred after the previous restoration (Bourque 2000). This may limit spawning adults and or newly hatched juveniles from entering and utilizing this habitat. As the sand continues to migrate eastward and as the tidal creek continues to fill in and close off due to natural and human causes, this refuge habitat may be lost (Botton et al. 2018). Efforts to conduct a mark recapture study could help determine the extent to which the tidal creek population mixes with that of the eastern tidal flats.

# 4 Diet and Food Preferences of Juvenile *Limulus polyphemus*

## 4.1 Methods

Juvenile horseshoe crabs and sediment cores were collected from a tidal creek at Plumb Beach on June 20th, July fifth, and July sixth 2017 during the 11 AM–1 PM low tides. Twenty-one juvenile crabs were located by following their signature spiraling trails, collected, placed in a cooling box for transportation, and stored in a freezer until dissection and identification of the contents in their digestive tracts. Each juvenile crab was weighed, and the length of the prosoma was measured and used to group the juveniles into instars according to Sekiguchi (1988) (Table 3). The chelicerae, pedipalps, walking legs, and book gills were then removed, and a cut was made through the integument of the dorsal cephalothorax and along the flanges to either side making it possible to remove part of the carapace and expose the digestive tract. The crop and gizzard were removed intact by inserting forceps into the mouth, holding the esophagus and carefully pulling it out. The midgut was removed by carefully lifting the nervous system and digestive tract and pulling the midgut away from the remaining intestine. Crop, gizzard, and midgut with content were weighed after the dissection (Table 3).

Thirteen sediment cores (8.5 cm diameter, 1.5 cm depth, and 85.1 $cm^3$ volume) were collected, placed in plastic bags, and brought back to the laboratory where they were washed through a 500 μm sieve. The material that passed the sieve was preserved in 70% ethanol and stained with Rose Bengal (200 mg/l) for 48 hours.

The meiofauna, macroalgae, and other organic content were extracted from the preserved sediment samples using an isopycnic separation technique with LUDOX® HS-40 colloidal silica (Burgess 2001). Each sample was rinsed with demineralized water on a 63 μm sieve to prevent gelling with silica sols and divalent cations, transferred to a 50 ml polypropylene centrifuge tubes, where 22 ml colloidal silica was added giving a total volume of 45 ml per sample. The samples were mixed using a vortex mixer at 3200 rpm for 30 s, followed by a slow decrease in speed, which allowed heavier items and shell hash to fall out of suspension. The samples were then mixed at a slow speed for 4 min to allow lighter materials to float up through the fluidized sediment and come to rest at the surface followed by centrifugation at

**Table 3** Data on the 21 collected *Limulus polyphemus* juveniles

| Prosomal width (cm) | Average weight (g) | Digestive tract weight (g) | Instar | *n* |
|---|---|---|---|---|
| 3.9 | 7.74 | 0.39 | 10 | 1 |
| 4.1–4.7 | 12.05 | 0.43–0.88 | 11 | 5 |
| 4.8–5.4 | 15.72 | 0.46–1.45 | 12 | 6 |
| 6–7.5 | 39.33 | 1.04–2.23 | 13 | 7 |
| 8–9.15 | 74.54 | 2.06–2.86 | 14 | 2 |

900 x g (6000 rpm) for 5 min to allow the sedimentation of the fine silt and clay. The suspended material including the meiofauna and colloidal silica was transferred onto a 63 μm sieve and rinsed with demineralized water, transferred into another 50 ml tube, and preserved in 70% ethanol for later identification.

Food item selectivity was estimated using Strauss' linear index (Strauss 1979) based on food types found in the digestive tracts in relation to their availability and abundance in the sediment samples as follows:

$$\mathbf{L} = \mathbf{r}_i - \mathbf{p}_i$$

Where ($r_i$) is the relative abundance of food items (i) in the digestive tract and ($p_i$) is the relative abundance of this item in the habitat. The values of Strauss's food selectivity index (L) ranges from – 100 (total avoidance) to +100 (strongly preferred) with (L) having an expected value of 0 (no selection) for random feeding.

**Table 4** *Limulus polyphemus,* juveniles. Food items from the digestive tracts of 21 crabs and the meiofauna and other content recovered from 13 sediment samples from the tidal creek, summer 2017

| Food item | *L. polyphemus* | | Sediment samples | |
|---|---|---|---|---|
| | 21 collected | | 13 collected | |
| | 21 with items | | 13 with items | |
| | Freq | *n* | Freq | *n* |
| **Bivalvia** | | | | |
| *Mulinia* | 4.8% | 1 | 38% | 6 |
| **Gastropoda** | | | | |
| *Tritia obsoleta* | 0% | 0 | 77% | 42 |
| **Polychaeta** | | | | |
| *Nereis* | 71% | 84 | 77% | 34 |
| **Heteromastus** | 100% | 933 | 77% | 143 |
| Setae | 100% | 21 | 0% | 0 |
| **Nematoda** | 95% | 182 | 100% | 2384 |
| **Malacostraca** | | | | |
| Ampeliscidae | 19% | 10 | 0% | 0 |
| *Idotea* | 0% | 0 | 8% | 1 |
| **Chelicerata** | | | | |
| *Limulus* trilobite | 0% | 0 | 8% | 1 |
| **Macroalgae** | | | | |
| *Ulva* | 52% | 11 | 100% | 13 |

Freq = percentage of samples containing one or more of a given item. n = total number of items found

## 4.2 Results

### Gut and Sediment Content

The polychaete *Nereis* sp. was found in 15 out of 21 digestive tracts of juveniles, while the polychaete *Heteromastus* sp. was found in all digestive tracts (Table 4). Amphipods (*Ampeliscidae spp.*) were found in 4 out of 21 digestive tracts of juveniles, while none were found in the sediment samples. A single *Limulus* trilobite and one isopod (*Idotea* sp.) were found in the sediment samples, but none were found in the digestive tracts of juvenile horseshoe crabs. Approximately 200 nematodes were found in 20 out of 21 digestive tracts of juveniles, and approximately 2400 nematodes were found in all 13 sediment samples. The macroalgae *Ulva* spp., sediment, and polychaete setae were found in all digestive tracts and, apart from setae, in all sediment core samples.

### Electivity

Electivity was only positive for the polychaete *Heteromastus* sp. (20.6) and the amphipods Ampeliscidae spp. (4.3), while the other taxa had negative values, *Nereis* sp. (−6.8), *Mulinia lateralis* (−2.4), and *Tritia (Ilyanassa) obsoleta* (−18.6), indicating avoidance (Fig. 7).

## 4.3 Discussion

In this study, the digestive tracts of 21 juvenile *L. polyphemus* (instars 10–14) from the tidal creek in Plumb Beach, NY, mainly contained polychaetes (Table 2). The electivity was only positive for the polychaete *Heteromastus* sp. and for the amphipods Ampeliscidae. However, since many species of amphipods are mobile and

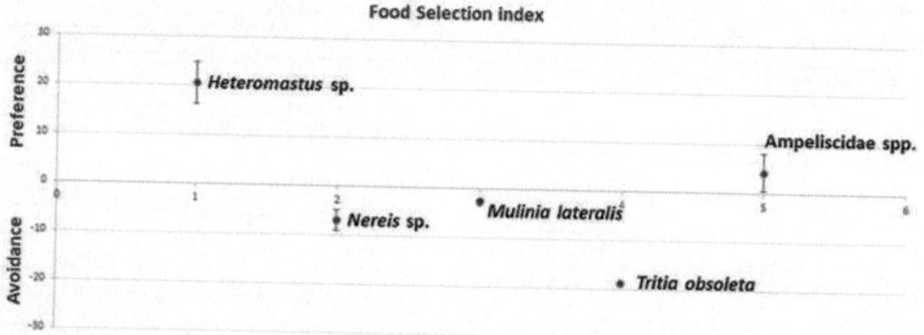


**Fig. 7** Values of Strauss' Linear Food Selectivity Index (L) for five of the most frequently occurring prey items, with SD bars

could escape predation by the juvenile horseshoe crabs, it is likely that the amphipods found in the digestive tracts are the result of scavenging on dead amphipods or exuviae, rather than active predation. Also, the density of amphipods in the sediment samples could be underestimated, because amphipods could escape sampling during sediment coring. Gaines et al. (2002) studied the diet composition of juvenile *L. polyphemus* (instars 2–11), but based on stable isotopes. They inferred that instars 2–3, feed on sedimentary organic material and meiofauna, while larger juveniles, instars 5–11, had a mixed diet mainly consisting of macrofaunal crustaceans and polychaetes, which is in congruence with our study and a study on instars 6–11 of the tri-spine horseshoe crab, *Tachypleus tridentatus* based on stable isotopes (Kwan et al. 2015) suggesting that juveniles of these two species exhibit the same shift in diet composition and preference with size and age.

The diet of adult *L. polyphemus* is dominated by mollusks (Botton 1984), which shows that as horseshoe crabs grow in size their diet change from meiofauna to macrofauna. While larger juveniles (instars 5–14) mostly feed on soft-bodied polychaetes and smaller crustaceans, adults are capable of handling and feeding on shelled mollusks, especially bivalves. Data on the diet of instars 15–17 are unavailable, so it is still unclear when *L. polyphemus* becomes capable of feeding on the hard-bodied shelled macrofauna. Here, we studied the diet of juvenile *L. polyphemus* in one location in July, but food availability for juvenile horseshoe crabs vary over seasons, yearly and with geographic location. To obtain a better understanding of the diet of juvenile horseshoe crabs, future studies should include more localities and longer time-series.

## 5 Righting and Burrowing Behavior of Juvenile *Limulus* in Different Sand Substrates

### *5.1 Methods*

We conducted a laboratory study to understand if burrowing and righting behavior of crabs can help explain the distribution of juvenile horseshoe crabs. Righting and burrowing behavior of 18 juvenile horseshoe crabs collected at Plumb Beach (average prosoma width = 43.2 mm, range 28.9–55.5 mm) was observed in aquaria (36 x 22 cm) with 5 L of artificial sea water (20 psu) and 2 cm of either coarse (> 2 mm) or fine (250 μm) sand. During the week prior to experimentation, juveniles were maintained in holding aquaria and fed pieces of chopped clams or mussels three times per week. For the righting experiments, each animal was placed in an aquarium with either no sand, coarse sand, or fine sand. To begin, the crab was placed upside down in the aquarium and timed until it righted. A trial was terminated if the crab failed to right within 5 min. For the burrowing experiments, the crab was placed right side up on either coarse or fine sand and timed until it burrowed to a depth that covered its lateral eyes. The treatment order in both the righting and burrowing experiments was randomly selected using Randbetween in Excel and each treatment was run in triplicate. The Kaplan-Meier method (SPSS) was used to plot the time-course of righting and burrowing success over time for the proportion of animals that burrowed or righted themselves over time. Initially, the proportion of animals that were upside-down in the righting experiments or level with the sediment in the burrowing experiments was 1.0, and as an individual turned upright or burrowed, the proportion remaining in the initial condition decreased until the termination of the trials at 5 min.

## 5.2 Results

Juvenile horseshoe crabs burrowed much more readily in fine sand (92.6% success rate) than coarse sand (24.1% success rate) (Table 5). Most of the juveniles on fine sand were able to burrow in <1 min, whereas those crabs that burrowed in coarse sand generally took between 2 and 4 min (Fig. 8). For the burrowing experiments, there was a significant difference between coarse and find sand (Chi-square = 21.73, $P < 0.001$).

In contrast, righting success was the same (33.3%) in both coarse and fine sand (Table 4) but significantly different than no sand (0%) (Chi-square = 18.00, $P < 0.001$). Some crabs tried righting themselves almost immediately, while others would make little effort other than flexing their telsons. When the telson made contact with the substrate, they first used the telson as an anchor point as they attempted to right themselves using rapid movements of their legs and book gills (Vosatka 1970). The rate at which horseshoe crabs righted themselves on coarse and fine sand was very similar (Fig. 9). In aquaria with no sand, all attempts at righting themselves within 5 min failed (Table 4). Some individuals made multiple attempts; some spun around in circles or occasionally pushed themselves along until they struck the wall of the aquarium.

**Table 5** Results of the juvenile horseshoe crab burrowing and righting experiments on coarse (>2 mm) vs. fine (250 μm) sand

| Substrate | N Trials | N successfully burrowing (%) | N successfully righting (%) |
|---|---|---|---|
| Coarse sand | 54 | 13 (24.1%) | 18 (33.3%) |
| Fine sand | 54 | 50 (92.6%) | 18 (33.3%) |
| No sand | 54 | – | 0 (0.0%) |

A crab that burrowed to the depth of its lateral eyes within 5 minutes was considered successful. The "no sand" treatment was only used in the righting experiments

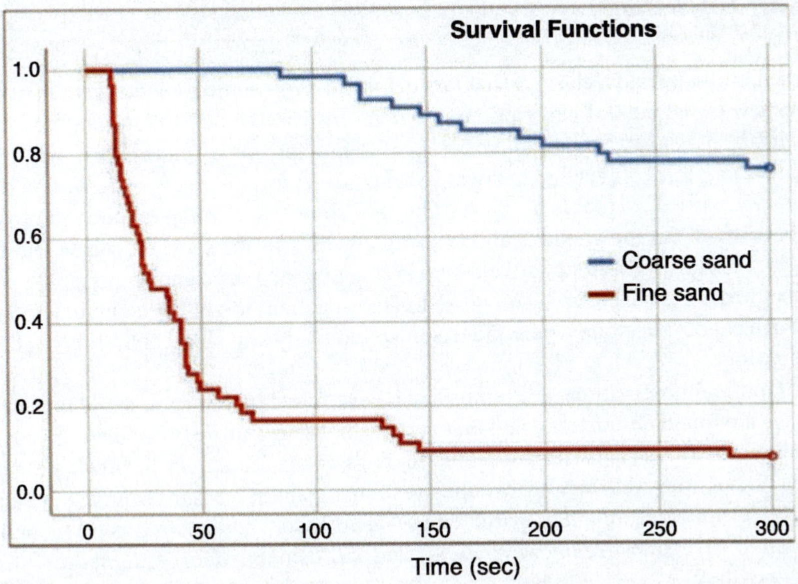


**Fig. 8** Kaplan-Meier plots of burrowing success vs. time in coarse (blue line) and fine (red line) substrates. Survival refers to the proportion of crabs remaining on the sediment surface

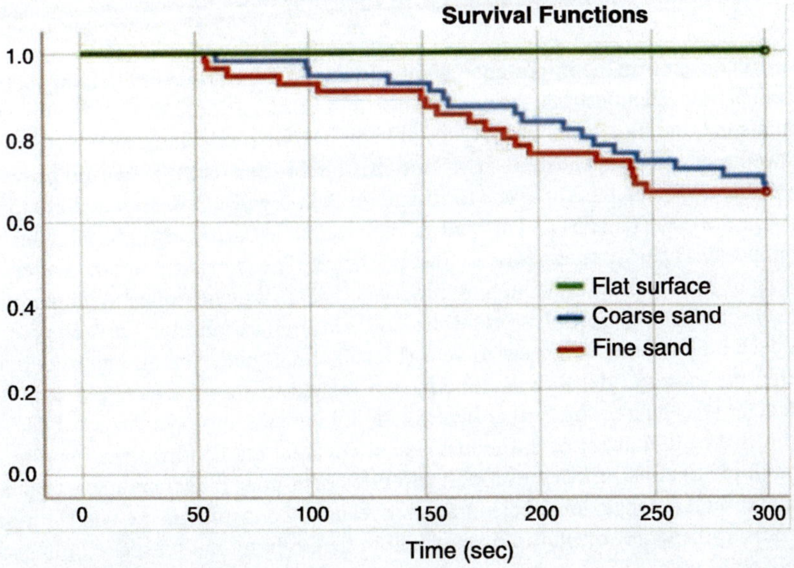


**Fig. 9** Kaplan-Meier plots of righting success vs. time in coarse (blue line), fine (red line), and no sand (green line) substrates. Survival refers to the proportion of crabs remaining upside-down

## 5.3 Discussion

The burrowing and righting behaviors exhibited by juvenile *Limulus* were in agreement with the careful descriptions provided by Vosatka (1970), although he used animals of somewhat larger size (50–100 mm prosoma width) than our trials (28.9–55.5 mm). Juvenile burrowing was considerably more effective in fine sand than in coarse sand (Table 5, Fig. 8). Our data show that righting requires some type of substrate, but the animal's ability to right itself was the same on coarse and fine sand. Presumably, as long as the telson is able to establish an anchor point, the sediment texture is secondary. The no-sediment treatment might be analogous to very coarse cobble substrate where the telson would likely slip along, inhibiting righting behavior.

Our field observations at Plumb Beach have found that most juvenile horseshoe crabs are found on fine sand and that they are not found in areas of pebble or coarse shell hash. This is consistent with the findings of our laboratory burrowing experiments. Possible predators of juvenile horseshoe crabs at our site include gulls, blue crabs (*Callinectes sapidus*), hermit crabs (*Pagurus* sp.), and various fishes including Atlantic silversides (*Menidia menidia*) and killifish (*Fundulus heteroclitus*) (e.g., Botton 2009).

# 6 General Discussion and Conclusions

Our results confirm the importance of intertidal sand flats as nursery habitat and, for the first time, demonstrate the importance of salt marsh tidal creeks for juvenile American horseshoe crabs. Both types of habitats have moderately well sorted fine-skewed medium sand substrates (M. Botton, unpublished data). This sand provides favorable opportunities for crabs to forage on the polychaete worms that comprise most of the diet (Table 5) and also allow more efficient burrowing (Fig. 8), which is an important behavioral defense against predation. The consistent annual pattern of timing of peak activity and hatching observed (Fig. 2) may be linked to temperature and/or seasonal availability of preferred food. The high variability of juvenile counts for each tidal habitat from year to year (Figs. 3 and 4) and even among tidal pools within the eastern tidal flats may be in part related to storm surges and sediment characteristics but require further investigation. Likewise, the asynchrony of decline and recovery in number of individuals is not apparent but confirms the importance of multiple tidal flats that can support juvenile horseshoe crabs on nursery beaches. An important management consideration revealed by our study is the need to monitor nursery beaches for multiple years before determining the presence or absence of juvenile crabs. Shed carapaces can be a useful supplement to surveys of live juveniles (Fig. 6), but at Plumb Beach, they appear to reflect older age-classes of juveniles that are primarily subtidal, in contrast to the younger age-classes that are found on the tidal flats and marsh creek. That said, at Plumb Beach, larger juvenile crabs appear to be exclusively subtidal and absent from the intertidal flats where smaller juveniles forage. Their presence was detected solely based on the collection of shed carapaces in August and September (Fig. 6) filling in an important cohort that live juvenile surveys cannot. Therefore, if resources and time are limited and permits not easily obtained, carapace data can provide a rapid, low cost and non-intrusive snapshot of juvenile presence and size categories. They require no special handling and do not evade observers, and shed carapaces can be collected quickly and measured offsite at a later time. This removal allows subsequent surveys with no risk of accidental duplication of observations.

Juvenile horseshoe crabs represent an understudied life stage of this long-lived species. Data collection on the juveniles at Plumb Beach is ongoing and will continue into the foreseeable future. Comparisons with other similar populations of *L. polyphemus* would be helpful and the authors encourage such research. We can conclude that while sandy intertidal flats are an important habitat, adjacent tidal creek and wetland habitat are a vital part of a management program. Shallow subtidal areas are more difficult to sample, but may have substantial numbers of juvenile horseshoe crabs (Burton et al. 2009).

We encourage additional studies to look at the rate and nature of predation on juveniles and further study their feeding ecology, particularly at different instars and under different conditions. Observation of physical characteristics such as sediment type, pool depth, or co-occurrence of other species could reveal correlating variables useful in predicting suitable juvenile habitat. Exclusion studies may lend insight into the effect of organisms such as *Ulva* or mud snails on the survival or foraging site selection of juvenile crabs.

Observed declines in counts as the summer progresses could be attributed to mortality through predation, emigration to deeper water, or burrowing into deeper sediment. We hope to research these possibilities in future years with an underwater ROV. Mark recapture efforts are planned to better assess the number of crabs present and to determine movement patterns between locations.

Juvenile crab surveys and behavioral studies can be easily done with small groups of volunteers and provide excellent citizen science and civic engagement opportunities. Another learning opportunity exists in maintaining captive populations of hatchlings and juveniles. These can also be used for future behavioral studies in captive conditions. We therefore hope similar work to ours will be implemented elsewhere and help fill the gaps in our knowledge of the life cycle of horseshoe crabs.

**Acknowledgements** Gratitude is extended to the many colleagues who assisted with this research, including J. Petersen from Queensborough Community College who assisted with field work and dissections, D. Lo and E. Perri from the New York Aquarium for collecting and sharing their water quality data, Y. Iwasaki, L.J. Moore and C. Moore, for volunteering their time and expertise in the field. Dr. R. Madden provided statistical advice. This work would not be possible without the hard work of many students from Kingsborough Community College (R Roman, J Martinez, A Waul, K. Harry, K. Grosvenor, N. Campos, R. Rivera, M. Ortikova, K. Gutierez, and A. Khoja) and Fordham University (S. Gopal, K. Lee, R. Ramsundar, C. Muñoz, C. Ferullo, E. Hall, R. Lamy, A. Aly, K. Khoder, and S. Zarakrzewski). Thanks to New York City Department of Parks, New York State Department of Environmental Conservation, Dr. George Frame from the

National Parks Service Gateway National Recreation Area for issuing permits to conduct this research, and the Army Corps of Engineers New York District for its cooperation. This work was supported by grant 2R25GM06003 of the Bridge Program of NIGMS under the supervision of Dr. Ed Catapane, grant 0537121091 of the CSTEP Program of NYSED, and grant R/CMB-39 from the New York Sea Grant Program.

## References

B. Akbar John, B.R. Nelson, H.I. Sheikh, S.G. Cheung, Y. Wardiatno, B.P. Dash, K. Tsuchiya, Y. Iwasaki, S. Pati, A review on fisheries and conservation of Asian horseshoe crabs. Biodiver. Conserv. **27**, 3573–3598 (2018). https://doi.org/10.1007/s105331-018-1633-8

M.L. Botton, Diet and food preferences of the adult horseshoe crab *Limulus polyphemus* in Delaware Bay, New Jersey, USA. Mar. Biol. **33**, 29–33 (1984)

M.L. Botton, C.N. Shuster Jr., J.A. Keinath, Horseshoe crabs in a food web: Who eats whom, in *The American Horseshoe Crab*, ed. by C. N. Shuster Jr., R. B. Barlow, H. J. Brockmann, (Harvard University Press, Cambridge, 2003), pp. 133–153

M.L. Botton, The ecological importance of horseshoe crabs in estuarine and coastal communities: A review and speculative summary, in *Biology and Conservation of Horseshoe Crabs*, ed. by J. T. Tanacredi, M. L. Botton, D. R. Smith, (Springer, New York, 2009), pp. 45–64

M.L. Botton, T. Itow, The effects of water quality on horseshoe crab embryos and larvae, in *Biology and Conservation of Horseshoe Crabs*, ed. by J. T. Tanacredi, M. L. Botton, D. R. Smith, (Springer, New York, 2009), pp. 439–454

M.L. Botton, R.A. Tankersley, R.E. Loveland, Developmental ecology of the American horseshoe crab, *Limulus polyphemus*. Curr. Zool. **56**, 550–562 (2010)

M.L. Botton, R.E. Loveland, T.R. Jacobsen, Beach erosion and geochemical factors: Influence on spawning success of horseshoe crabs (*Limulus polyphemus*) in Delaware Bay. Mar. Biol. **99**, 325–332 (1988)

M.L. Botton, C.P. Colón, J. Rowden, S. Elbin, D. Kriensky, K. McKown, M. Sclafani, R. Madden, Effects of a beach nourishment project in Jamaica Bay, New York on horseshoe crab (*Limulus polyphemus*) spawning activity and egg deposition. Estuar. Coasts **41**, 974–987 (2018)

R. Bourque, *Plumb Beach on the Move; Ecological Changes at Plumb Beach*, vol 4 (The Urban Audubon, 2000)

H.J. Brockmann, Nesting behavior: A shoreline phenomenon, in *The American Horseshoe Crab*, ed. by C. N. Shuster, R. B. Barlow, H. J. Brockmann, (Harvard University Press, Cambridge, 2003), pp. 33–49

H.J. Brockmann, M.D. Smith, Reproductive competition and sexual selection in horseshoe crabs, in *Biology and Conservation of Horseshoe Crabs*, ed. by J. T. Tanacredi, M. L. Botton, D. R. Smith, (Springer, New York, 2009), pp. 199–221

H.J. Brockmann, S.L. Johnson, M.D. Smith, D. Sasson, Mating tactics of the American horseshoe crab (*Limulus polyphemus*), in *Changing Global Perspectives on Horseshoe Crab Biology, Conservation and Management*, ed. by R. H. Carmichael, M. L. Botton, P. K. S. Shin, S. G. Cheung, (Springer, Switzerland, 2015), pp. 321–351

R. Burgess, An improved protocol for separating meiofauna from sediments using colloidal silica sols. Mar. Ecol. Prog. Ser. **214**, 161–165 (2001)

W.H. Burton, F.S. Kelley, E.A. Franks, Distribution of juvenile horseshoe crabs in subtidal habitats of Delaware Bay using a suction-dredge sampling device, in *Biology and Conservation of Horseshoe Crabs*, ed. by J. T. Tanacredi, M. L. Botton, D. R. Smith, (Springer, New York, 2009), pp. 285–293

R.H. Carmichael, D. Rutecki, I. Valiela, Abundance and population structure of the Atlantic horseshoe crab *Limulus polyphemus* in Pleasant Bay, Cape Cod. Mar. Ecol. Prog. Ser. **246**, 225–239 (2003)

R.H. Carmichael, D. Rutecki, B. Annett, E. Gaines, I. Valiela, Position of horseshoe crabs in estuarine food webs: N and C stable isotopic study of foraging ranges and diet composition. J Exp Marine Biol Ecol **299**, 231–253 (2004)

M.G. Estes Jr., R.H. Carmichael, P.D.M. Macdonald, A.J. Brady, J. McFadyen, Molts reveal life-history patterns for juvenile American horseshoe crabs in fringe habitats, in *Changing Global Perspectives on Horseshoe Crab Biology, Conservation and Management*, ed. by R. H. Carmichael, M. L. Botton, P. K. S. Shin, S. G. Cheung, (Springer, Switzerland, 2015), pp. 255–278

P. Funch, T. Wang, C. Pertoldi, A.B. Middelbo, Low oxygen levels slow embryonic development of *Limulus polyphemus*. Biol. Bull. **231**, 113–119 (2016)

E.F. Gaines, R.H. Carmichael, S.P. Grady, I. Valiela, Stable isotopic evidence for changing nutritional sources of juvenile horseshoe crabs. Biol. Bull. **203**(2), 228–230 (2002)

R.F. Hong, Population and Distribution of Horseshoe Crab *Carcinoscopius Rotundicauda* at the Kranji Nature Trail Estuaries, Western Johor Straits. Undergraduate Research Opportunities in Science Project Report Submitted to the Department of Biological Sciences The National University of Singapore (2004)

M. Hu, Y. Wang, Y. Chen, S.G. Cheung, P.K.S. Shin, Q. Li, Summer distribution and abundance of juvenile Chinese horseshoe crabs along an intertidal zone in southern China. Aquat. Biol. **117**, 107–112 (2009)

T.C. Jegla, J.D. Costlow, Temperature and salinity effects on developmental and early posthatch stages of *Limulus*, in *Biomedical Applications of the Horseshoe Crab (Limulidae)*, ed. by E. Cohen, (Alan R. Liss, Inc., New York, 1982)

C.N. Lee, B. Morton, Emergence behavior of juvenile *Tachypleus tridentatus* under simulated tidal conditions in the laboratory and at two different sediment temperatures, in *Biology and Conservation of Horseshoe Crabs*, ed. by J. T. Tanacredi, M. L. Botton, D. R. Smith, (Springer, New York, 2009), pp. 275–283

K. Laurie, C.P. Chen, S.G. Cheung, V. Do, H. Hsieh, A. John, F. Mohamad, S. Seino, S. Nishida, P. Shin, M. Yang, *Tachypleus tridentatus*. The IUCN Red List of Threatened Species 2019: e.T21309A149768986. https://doi.org/10.2305/IUCN.UK.2019-1.RLTS.T21309A149768986.en. Accessed 18 Aug 2019 (2019)

B.K.Y. Kwan, S.G. Cheung, P.K.S. Shin, A dual stable isotope study for diet composition of juvenile Chinese horseshoe crab *Tachypleus tridentatus* (Xiphosura) on a seagrass-covered intertidal mudflat. Mar. Biol. **162**, 1137–1143 (2015)

D.S. Mizrahi, K.A. Peters, Relationships between sandpipers and horseshoe crabs in Delaware Bay: A synthesis, in *Biology and Conservation of Horseshoe Crabs*, ed. by J. T. Tanacredi, M. L. Botton, D. R. Smith, (Springer, New York, 2009), pp. 65–87

S.R. Palumbi, B.A. Johnson, A note on the influence of life-history stage on metabolic adaptation: The responses of *Limulus* eggs and larvae to hypoxia, in *Biomedical Applications of the Horseshoe Crab (Limulidae)*, ed. by E. Cohen, (Alan R. Liss, Inc., New York, 1982)

D. Penn, H.J. Brockmann, Nest-site selection in the horseshoe crab. Biol. Bull. **187**, 373–384 (1994)

K. Sekiguchi, *Biology of Horseshoe Crabs* (Tokyo, Science House, 1988b), p. 188

D.R. Smith, H.J. Brockmann, M.A. Beekey, T.L. King, M.J. Millard, J. Zaldivar-Rae, Conservation status of the American horseshoe crab, (*Limulus polyphemus*): A regional assessment. Rev. Fish Biol. Fish. **27**, 135–175 (2016a)

D.R. Smith, H.J. Brockmann, M.A. Beekey, T.L. King, M.J. Millard, J. Zaldivar-Rae, *Limulus polyphemus*. The IUCN Red List of Threatened Species 2016: e.T11987A80159830. https://doi.org/10.2305/IUCN.UK.2016-1.RLTS.T11987A80159830.en. Downloaded on 04 January 2019 (2016b)

R.E. Strauss, Reliability estimates for Ivlev's electivity index, the forage ratio, and a proposed linear index of food selection. Trans. Am. Fish. Soc. **108**, 344–352 (1979)

C. Vasquez, A. Murillo, H.J. Brockmann, D. Julian, Multiple-stressor interactions influence embryo development rate in the American horseshoe crab, *Limulus polyphemus*. J. Exp. Biol. **218**, 2355–2364 (2015)

S. Vestbo, M. Obst, F.J. Quevedo Fernandez, I. Intanai, P. Funch, Present and potential future distributions of Asian horseshoe crabs determine areas for conservation. Front. Mar. Sci. **5**, 164 (2018)

E.D. Vosatka, Observations on the swimming, righting, and burrowing movements of young horseshoe crabs, *Limulus polyphemus*. Ohio J. Sci. **70**, 276–283 (1970)

# Correction to: Feeding Ecology and Dietary Preferences of *Tachypleus gigas* from East Malay Peninsula

Akbar John, Bryan Raveen Nelson, Hassan I. Sheikh, S. Hajisamae, and Jalal Khan

**Correction to:**
**Chapter 35 in: J. T. Tanacredi et al. (eds.), *International Horseshoe Crab Conservation and Research Efforts: 2007-2020*, https://doi.org/10.1007/978-3-030-82315-3_35**

On page 609

In chapter 35, the author's name has been misspelt as "Br. Nelson".

It has been updated as "Bryan Raveen Nelson".

The updated version of this chapter can be found at
https://doi.org/10.1007/978-3-030-82315-3_35



J. T. Tanacredi et al. (eds.), *International Horseshoe Crab Conservation and Research Efforts: 2007-2020*, https://doi.org/10.1007/978-3-030-82315-3_41